# Strategies to Improve Water-Use Efficiency in Plant Production—Volume I

# Strategies to Improve Water-Use Efficiency in Plant Production—Volume I

Guest Editors

**Yang Gao**
**Aiwang Duan**

MDPI

Basel • Beijing • Wuhan • Barcelona • Belgrade • Novi Sad • Cluj • Manchester

*Guest Editors*

Yang Gao
Institute of Farmland Irrigation
Chinese Academy of Agricultural Sciences
Xinxiang
China

Aiwang Duan
Institute of Farmland Irrigation
Chinese Academy of Agricultural Sciences
Xinxiang
China

*Editorial Office*
MDPI AG
Grosspeteranlage 5
4052 Basel, Switzerland

This is a reprint of the Special Issue, published open access by the journal *Plants* (ISSN 2223-7747), freely accessible at: https://www.mdpi.com/journal/plants/special_issues/PZB50B5M96.

For citation purposes, cite each article independently as indicated on the article page online and as indicated below:

Lastname, A.A.; Lastname, B.B. Article Title. *Journal Name* **Year**, *Volume Number*, Page Range.

**Volume I**
**ISBN 978-3-7258-6928-2 (Hbk)**
**ISBN 978-3-7258-6929-9 (PDF)**
**https://doi.org/10.3390/books978-3-7258-6929-9**

**Set**
**ISBN 978-3-7258-6870-4 (Hbk)**
**ISBN 978-3-7258-6871-1 (PDF)**

# Contents

# About the Editors

**Yang Gao**

Yang Gao, Ph.D., is a research professor and Chief Scientist of the Deficit Irrigation Theory and Technology Research Team, and serves as Deputy Director of the Crop Water-Use Efficiency Research Center. His research interests focus on crop–water relations, deficit irrigation, and strategies for improving crop water-use efficiency under water-limited conditions. He has led and participated in multiple national and regional research projects, including a National Key Research and Development Program of China during the 14th Five-Year Plan on improving crop water-use efficiency in groundwater-overexploited areas of North China, a National Natural Science Foundation of China project on water–heat–carbon–nitrogen processes and coupled modeling in walnut–wheat intercropping systems in arid oasis regions, and several provincial and ministerial projects on water-saving and yield-enhancing technologies for major cropping systems. He has published 162 scientific papers, including 52 indexed in SCI or EI, serves as an Editorial Board Member of *Agronomy* and *Journal of Integrative Agriculture*, has published four academic books, holds seven patents and eight software copyrights, has received multiple national and regional science and technology awards, and has supervised 29 graduate students.

**Aiwang Duan**

Aiwang Duan, Ph.D., is a research professor at the Farmland Irrigation Research Institute of the Chinese Academy of Agricultural Sciences (CAAS), former Director of the Institute, Honorary Chief Scientist of the CAAS Agricultural Science and Technology Innovation Program "Deficit Irrigation Theory and New Technologies", and Director of the Irrigation Experimental Station network of the Ministry of Water Resources of China. His research interests focus on crop water requirements and irrigation scheduling, theories and technologies for improving crop water-use efficiency, soil moisture monitoring, and irrigation forecasting. He has led numerous major national research projects, including a National Science and Technology Basic Resources Survey on updating and supplementing water requirement data for major crops in China, national public welfare research projects on crop water requirements and high-efficiency water-use technologies in high-yield farmlands of the Huang–Huai–Hai Plain, a National "863" Program project on crop water information acquisition and intelligent irrigation control technologies, National Natural Science Foundation of China projects on irrigation effects on $N_2O$ emissions in winter wheat fields and crop water-use modeling under intercropping systems, and a key expert project of the National Agriculture Research System. He has also undertaken more than 20 provincial and ministerial research projects, received 12 science and technology awards at or above the provincial level, published over 280 academic papers, and edited or authored more than 10 academic books.

*plants*

MDPI

*Article*

# The Effect of Deficit Irrigation on the Quality Characteristics and Physiological Disorders of Pomegranate Fruits

**Rossana Porras-Jorge [1], José Mariano Aguilar [2], Carlos Baixauli [2], Julián Bartual [3], Bernardo Pascual [4] and Nuria Pascual-Seva [4,*]**

[1] Departamento de Producción Vegetal, Universitat Politècnica de València, Camí de Vera, s/n, 46022 València, València, Spain; zrporras@doctor.upv.es

[2] Centro Experiencias Cajamar Paiporta, C. Cementerio Nuevo s/n, 46200 Paiporta, València, Spain; josemarianoaguilar@fundacioncajamar.com (J.M.A.); carlosbaixauli@fundacioncajamar.com (C.B.)

[3] Estación Experimental Agraria de Elche (SST), CV-855, Km. 1, 03290 Elche, Alicante, Spain; bartual_jul@gva.es

[4] Centro Valenciano de Estudios Sobre el Riego, Universitat Politècnica de València, Camí de Vera, s/n, 46022 València, València, Spain; bpascual@upv.es

[*] Correspondence: nupasse@upv.es

**Abstract:** This study assesses the impact of two regulated deficit irrigation (RDI) and one sustained deficit irrigation (SDI) strategies on the fruit quality characteristics of pomegranate (*Punica granatum* L.) compared to a fully irrigated control in a Mediterranean climate. Field trials were conducted over two growing seasons at the Cajamar Experimental Center in Paiporta, Valencia, Spain. The SDI strategy, which achieved considerable water savings of approximately 50%, led to a reduction in yield (both total and marketable), as well as a decrease in the size and unit weight of the fruits. However, it also produced arils with higher dry matter content and aril juice with higher soluble solids content, all without altering the maturity index. Notably, the SDI approach resulted in increased non-marketable production due to a higher incidence of cracking, particularly during the exceptionally hot and dry summer of 2023. Although the maturity index remained unchanged across the irrigation strategies, the SDI yielded a greater percentage of pink-red rind on marketable fruits compared to the other strategies. This is important because 'Mollar de Elche' pomegranates are typically harvested based on their external colour. Thus, the SDI strategy could allow for earlier harvesting, potentially enhancing the commercial value, as earlier harvests often command higher prices, which may partially offset some of the reduction in marketable yield. Conversely, both RDI strategies achieved a slight water saving without compromising marketable yield or the quality characteristics of the fruit.

**Keywords:** 'Mollar de Elche'; sustained deficit irrigation; regulated deficit irrigation; sensory characteristics of fruits; cracking

## 1. Introduction

'Mollar de Elche' (referred to as 'ME') is the predominant varietal group of pomegranates (*Punica granatum* L.) cultivated in Spain, accounting for over 60% of the land dedicated to pomegranate cultivation. This popularity stems from its excellent adaptation to the local edaphoclimatic conditions, high productivity, and favourable organoleptic and nutraceutical attributes. Globally, 'ME' is highly regarded for its exceptional flavour and antioxidant properties [1].

This study complements a companion article by the same authors published in the same Special Issue [2], which examines the physiological and agronomic responses of pomegranates to deficit irrigation. The current study focuses on how these irrigation strategies affect fruit quality and physiological disorders. Together, both studies aim to provide growers with effective irrigation management practices. In the companion paper, the growing interest in pomegranate cultivation, both in Spain and worldwide, is highlighted, particularly emphasising techniques such as deficit irrigation (DI). DI comprises irrigation practices where crops receive less water than optimal for their growth. This approach includes continuous or sustained deficit irrigation (SDI) and controlled or regulated deficit irrigation (RDI) [3,4]. SDI maintains a consistent water deficit throughout the entire crop cycle, thereby preventing substantial water stress during key phenological phases that could adversely impact marketable yield (MY). Conversely, RDI involves managing water deficits during specific stages when crops are less susceptible to water stress ([2] and references therein).

Fruit cracking and sunburn are important physiological disorders that can considerably reduce the marketable yield of pomegranate crops. Cracking, also referred as fruit split (hereinafter referred as cracking), which is the rupture of the rind, typically occurs during the final stages of fruit development [5] and is primarily caused by the pressure from rapidly expanding arils, which grow at a faster rate than the rind (also referred as skin or peel; hereinafter referred as rind) itself [6]. Other researchers [7] discovered that strategies applying the least irrigation water inputs led to the highest incidence of fruit cracking across all experimental seasons. Similarly, others [8] found that implementing RDI during the linear fruit growth phase increased cracking incidence in only one of the three tested growing seasons. This season was characterised by a sudden recovery in the plant's water status, highlighting that fruit cracking is typically linked to abrupt changes in plant–water relationships. Furthermore, ref. [9] observed that when rainfall impacted previously water-stressed pomegranate plants during the critical final stages of fruit growth and ripening shortly before harvest, the increased pressure from the arils on the rind promoted cracking. This was due to the aril turgor rising significantly more than the rind turgor. Iscimen et al. [10] established that potassium levels and irrigation rates had a considerable effect on yield but also influenced the incidence of cracking and specific fruit quality parameters, such as shell thickness. Other contributing factors may include genetic predispositions, inadequate nutrient levels, sunburn on the rind, surface lesions from pathogens, and insect damage [6,11,12]. Sunburn, another physiological disorder, manifests as burns on the fruit's rind when it is exposed to intense sunlight during extremely hot days, leading to a transpiration rate that exceeds the water uptake. Initially, this condition results in discolouration of small areas of the rind, but as it progresses, it can cause severe necrosis and cracking of the rind, ultimately rendering the underlying arils discoloured and unpalatable, thereby classifying them as non-marketable [11,12]. In the study by [8], sunburn on fruit increased with the SDI strategy and decreased with RDI during flowering, fruit set, and early fruit growth, compared to the control treatment in only one of three tested growing seasons. Additionally, windy conditions can exacerbate fruit damage. The friction between fruits and thorns may create wounds that can become suberified, making the fruit unsuitable for the fresh market. In severe cases, thorns can inflict deep wounds that serve as entry points for pathogens, leading to fruit decay. The primary diseases impacting pomegranate are mainly caused by several pathogens, including *Botrytis cinerea* Pers.:Fr., *Alternaria alternata* (Fr.) Keissl., *Coniella granati* (Sacc.) Petr. & Syd., and *Aspergillus niger* (Tiegh.) Speg. These pathogens typically infect the fruit during its developmental stages [13]. Other factors that may lead to non-marketable fruit include malformations and small size, but these are usually of minor significance due to the common practice of fruit thinning.

Irrigation is a critical production factor that influences the quality of pomegranate fruits, impacting key attributes such as soluble solids content, titratable acidity, dry matter, and firmness, all of which affect sensory qualities. The gross income generated from pomegranate crops is closely linked to both the marketable yield and the quality characteristics of the fruit. In a review [14] on the effect of water deficit on the physiology, phenology, vegetative growth, fruit growth, yield, and fruit quality of pomegranate, the author concluded that the results of one study could not simply be transferred to another area where conditions may be different given the variable results obtained with different systems and cultivars in different countries, which require conducting research under local conditions. The purpose of this study was to evaluate the quality characteristics of the fruit, alongside the incidence of physiological disorders, from seven-year-old pomegranate trees subjected to an SDI strategy and two RDI strategies in the early stages, ultimately aiming to establish an effective irrigation schedule for growers in the Valencian Region.

## 2. Results and Discussion

### 2.1. Fruit Characteristics

Yield, both total and marketable, was significantly influenced by the growing season ($p \leq 0.01$), irrigation treatment ($p \leq 0.01$), and their interaction ($p \leq 0.05$; $p \leq 0.01$; Table 1). Notably, marketable yields in 2023 (32.5 t ha$^{-1}$) were higher than those in 2022 (23.3 t ha$^{-1}$). When examining total yield, SDI recorded the lowest yield (35.5 t ha$^{-1}$), with significant differences compared to the other irrigation strategies in both growing seasons ($p \leq 0.05$). Similarly, for marketable yield, SDI presented the lowest values (on average 18.1 t ha$^{-1}$); however, when compared to the control strategy, the difference was only statistically significant in 2023.

Both the growing season and irrigation strategy significantly influenced ($p \leq 0.01$) the unit weight of the fruits. Notably, the lightest fruits were recorded in 2023 under the SDI strategy (456 g). This trend extended to fruit size, with the smallest fruits (as indicated by lower diameter measurements) also produced in 2023 and in SDI (96.9 cm). The year 2022 led to a lower overall fruit yield compared to 2023, averaging 153.6 fruits per tree in 2022 versus 255.5 in 2023. This reduced competition between fruits for photoassimilates in 2022, as suggested by [11], resulting in heavier fruits. When irrigation was restored in RDI, fruit growth accelerated, as fruits function as a strong sink for photosynthates [15], reaching sizes comparable to those found in the control strategy. The shape of the fruits was solely influenced by the growing season; on average, the fruits from 2022 exhibited a flatter shape (indicated by a higher shape index; 1.19) compared to those harvested in 2023 (1.15).

The weight of the rind and septal membranes (referred to here as the rind) was also affected by both the growing season and irrigation strategy ($p \leq 0.01$). The lowest rind weights were associated with the fruits harvested in 2023 (213 g) and those grown using SDI (198 g), aligning with the trend in smaller fruit size and reduced unit weight. The weight of the arils was influenced by both the growing season ($p \leq 0.01$) and the irrigation strategy ($p \leq 0.05$), as well as the interaction between these factors ($p \leq 0.05$). The fruits harvested in 2022 under the RDI1 strategy had the highest aril weight (360 g), while there were no significant weight differences among the irrigation strategies in 2023. The rind fraction (representing the percentage of the fruit accounted for by the rind) was only influenced by the irrigation strategy ($p \leq 0.05$; Table 2). Fruits produced under the SDI strategy had, on average, an 11% lower rind fraction compared to those from the control strategy, indicating a 9% higher aril fraction and, consequently, a greater proportion of edible fruit. Fruits harvested in 2022 exhibited a thicker rind (4.5 mm) compared to those from 2023 (3.6 mm), which was anticipated given that 2023's fruits were both smaller and lighter (Table 1). This observation aligns with the findings of [16], who also reported variations in rind thickness across different growing seasons. Furthermore, among the fruits collected in 2023, those produced under SDI exhibited a thinner rind (3.25 mm) compared to those from the RDI1 (3.95 mm) strategy. The dry matter content of the fruit rind remained constant, unaffected by the growing season or irrigation strategy employed.

**Table 1.** Effect of the growing season and the irrigation strategy on the total (TY; t ha$^{-1}$) and marketable yield (MY; t ha$^{-1}$), unit weight (UW; g) with its partitioning into rind (RW) and arils (AW) weight, and size [maximum diameter of the equatorial section (D; mm)] and shape index (D/L, being L the length, mm) of the fruits.

| | TY | MY | UW | D | D/L | RW | AW |
|---|---|---|---|---|---|---|---|
| Growing season (GS) | | | | | | | |
| 2022 | 36.74 b | 23.33 b | 572.78 a | 105.27 a | 1.19 b | 241.17 a | 331.61 a |
| 2023 | 52.54 a | 32.54 a | 500.86 b | 100.00 b | 1.15 a | 213.02 b | 287.83 b |
| LSD | 2.89 | 2.12 | 22.37 | 1.44 | 0.02 | 19.38 | 12.86 |
| Irrigation strategy (IS) | | | | | | | |
| Control | 46.35 a | 29.40 b | 539.61 a | 103.72 a | 1.17 | 239.77 a | 299.84 b |
| RDI1 | 48.10 a | 32.53 a | 566.66 a | 104.05 a | 1.16 | 239.09 a | 327.56 a |
| RDI2 | 48.56 a | 31.73 ab | 540.04 a | 103.20 a | 1.18 | 231.13 a | 308.91 b |
| SDI | 35.54 b | 18.09 c | 500.98 b | 99.56 b | 1.17 | 198.39 b | 302.58 b |
| LSD | 4.08 | 2.99 | 31.64 | 2.04 | 0.02 | 27.41 | 18.19 |
| GS × IS | | | | | | | |
| 2022-Control | 38.37 b | 22.91 bc | 550.30 | 105.68 | 1.20 | 245.53 | 304.77 cd |
| 2022-RDI1 | 39.36 b | 25.57 b | 613.44 | 106.27 | 1.16 | 253.89 | 359.55 a |
| 2022-RDI2 | 38.20 b | 26.09 b | 581.46 | 106.87 | 1.20 | 253.48 | 327.98 bc |
| 2022-SDI | 31.02 c | 18.75 cd | 545.94 | 102.25 | 1.18 | 211.78 | 334.15 ab |
| 2023-Control | 54.34 a | 35.88 a | 528.92 | 101.75 | 1.15 | 234.01 | 294.91 de |
| 2023-RDI1 | 56.83 a | 39.48 a | 519.87 | 101.83 | 1.16 | 224.30 | 295.58 de |
| 2023-RDI2 | 58.92 a | 37.37 a | 498.62 | 99.54 | 1.15 | 208.78 | 289.84 de |
| 2023-SDI | 40.05 b | 17.43 d | 456.01 | 96.88 | 1.15 | 185.00 | 271.01 e |
| LSD | 5.77 | 4.23 | 44.75 | 2.89 | 0.03 | 38.77 | 25.72 |
| Source (df) | | | | % sum of squares | | | |
| S (1) | 44.04 ** | 23.75 ** | 16.22 ** | 19.66 ** | 9.19 ** | 3.99 ** | 17.92 ** |
| IS (3) | 19.96 ** | 37.70 ** | 6.88 ** | 9.18 ** | 1.42 ns | 5.77 ** | 4.37 * |
| GS × IS (3) | 3.22 * | 10.60 ** | 2.72 ns | 1.19 ns | 2.68 ns | 0.70 ns | 4.59 * |
| Residual (88 $^{Z}$/184 $^{Y}$) | 32.78 | 27.95 | 74.19 | 69.97 | 86.71 | 89.54 | 73.11 |
| Standard deviation | 7.12 | 5.22 | 78.57 | 5.07 | 0.06 | 68.07 | 45.16 |

Mean values followed by different lowercase letters in each column indicate significant differences at $p \leq 0.05$ using the LSD test. Degrees of freedom (df): $^{Z}$ for TY and MY and $^{Y}$ for UW, D, D/L, RW, and AW. ns: no significant difference; ** (*): significant differences at $p \leq 0.01$ ($p \leq 0.05$).

In 2023, the percentage of pink-red rind surface on marketable pomegranates was significantly higher than in 2022 (29.8% in 2023 and 24.6% in 2022; $p \leq 0.01$) and greater in the SDI strategy (35.3%) compared to the other irrigation strategies (22.0%, 25.6%, and 25.8 in control, RDI1, and RDI2, respectively; $p \leq 0.01$). The colour of pomegranate rind is primarily attributed to anthocyanins, which are a class of flavonoid pigments [17]. Generally, a deeper red tone corresponds to a higher content of these pigments. As noted by [15], the colouration of pomegranate rind—a key commercial quality indicator—depends on both the concentration and type of anthocyanins present. Several factors can influence the levels of secondary metabolites, including anthocyanins, which may fluctuate seasonally. These factors include temperature, water availability, ultraviolet radiation, nutrient levels, atmospheric pollution, mechanical stress, and pathogen attack [18]. They can vary from season to season, consequently causing a variation in the colour of the pomegranate rind, both on the surface of the coloured area and in the colour of the rind itself.

**Table 2.** Effect of the growing season and irrigation strategy on the rind characteristics [rind fraction (RF, %), rind thickness (RT, mm), rind dry matter content (RDM, %), the pink-red zone of the rind (% with respect to the entire surface), and the CIE L*C*h* colour space parameters of the pink-red zone of the rind (L* indicates lightness, C* represents chroma, and h* is the hue angle)].

| | RF | | RT | | RDM | | Pink-Red Zone of the Rind | | | | | | | |
|---|---|---|---|---|---|---|---|---|---|---|---|---|---|---|
| | | | | | | | % | | L* | | C* | | h* | |
| Growing season (GS) | | | | | | | | | | | | | | |
| 2022 | 41.26 | | 4.51 | a | 26.25 | | 24.58 | b | 60.04 | b | 45.90 | | 51.62 | |
| 2023 | 42.28 | | 3.61 | b | 24.37 | | 29.79 | a | 62.07 | a | 45.65 | | 49.05 | |
| LSD | 2.26 | | 0.22 | | 1.93 | | 3.47 | | 1.46 | | 0.88 | | 2.66 | |
| Irrigation strategy (IS) | | | | | | | | | | | | | | |
| Control | 43.93 | a | 3.96 | | 26.46 | | 21.98 | b | 61.20 | | 45.48 | | 49.89 | |
| RDI1 | 41.92 | ab | 4.19 | | 23.72 | | 25.63 | b | 60.68 | | 46.01 | | 50.12 | |
| RDI2 | 42.08 | ab | 4.13 | | 26.27 | | 25.83 | b | 62.59 | | 45.18 | | 52.15 | |
| SDI | 39.16 | b | 3.96 | | 24.81 | | 35.31 | a | 59.76 | | 46.41 | | 49.19 | |
| LSD | 3.19 | | 0.32 | | 2.73 | | 4.91 | | 2.06 | | 1.24 | | 3.76 | |
| GS × IS | | | | | | | | | | | | | | |
| 2022-Control | 43.83 | | 4.33 | ab | 27.88 | | 20.00 | | 58.97 | | 46.32 | | 48.46 | |
| 2022-RDI1 | 40.69 | | 4.43 | a | 24.07 | | 22.71 | | 60.21 | | 45.91 | | 51.97 | |
| 2022-RDI2 | 42.62 | | 4.63 | a | 27.56 | | 23.54 | | 62.41 | | 45.14 | | 51.60 | |
| 2022-SDI | 37.90 | | 4.67 | a | 25.50 | | 32.08 | | 58.58 | | 46.21 | | 51.45 | |
| 2023-Control | 44.03 | | 3.60 | cd | 25.03 | | 23.96 | | 63.43 | | 44.65 | | 51.32 | |
| 2023-RDI1 | 43.15 | | 3.95 | bc | 23.36 | | 28.54 | | 61.14 | | 46.11 | | 48.27 | |
| 2023-RDI2 | 41.53 | | 3.64 | cd | 24.99 | | 28.13 | | 62.78 | | 45.23 | | 49.69 | |
| 2023-SDI | 40.41 | | 3.25 | d | 24.12 | | 38.54 | | 60.95 | | 46.62 | | 46.93 | |
| LSD | 4.51 | | 0.45 | | 3.86 | | 6.94 | | 2.91 | | 1.76 | | 5.31 | |
| Source (df) | % sum of squares | | | | | | | | | | | | | |
| GS (1) | 0.39 | ns | 41.27 | ** | 3.83 | ns | 2.10 | ** | 1.90 | * | 0.08 | ns | 0.93 | ns |
| IS (3) | 4.37 | * | 2.14 | ns | 5.47 | ns | 7.57 | ** | 1.94 | ns | 1.18 | ns | 0.69 | ns |
| GS × IS (3) | 0.88 | ns | 5.98 | * | 0.81 | ns | 0.08 | ns | 1.15 | ns | 0.90 | ns | 1.42 | ns |
| Residual (184) | 94.35 | | 50.61 | | 89.90 | | 90.21 | | 95.02 | | 97.85 | | 96.97 | |
| Standard deviation | 8.08 | | 0.52 | | 4.75 | | 17.28 | | 7.25 | | 4.38 | | 13.23 | |

Mean values followed by different lowercase letters in each column indicate significant differences at $p \leq 0.05$ using the LSD test. df: degrees of freedom; ns: no significant difference; ** (*): significant differences at $p \leq 0.01$ ($p \leq 0.05$).

The percentage of the pink-red rind area observed in pomegranates harvested in 2023 exceeded that of the previous year ($p \leq 0.01$; Table 2). This year's fruits exhibited a higher L* value (62.1) compared to those from 2022 (60.0), indicating a darker pigmentation, as lower luminosity values signify greater light absorption. However, neither the growing season nor the irrigation treatment appeared to significantly influence C* or h* parameters. The increased proportion of pink-red rind in SDI fruits can likely be attributed to moderate water stress experienced by the plants under this irrigation strategy. This stress could directly, as suggested by [18], or indirectly alter secondary metabolism by reducing vegetative growth, leading to the enhanced exposure of the fruits to sunlight. According to [19], this increased exposure can result in a redder appearance. These findings align with research from [20], who noted that moderate water deficits during the growth of 'ME' fruit only significantly affected ($p \leq 0.05$) the h* value in the first harvest, without impacting rind colour in subsequent harvests. Additionally, [15] discovered that both SDI and, to a lesser extent, RDI fruits, exhibited lower L* values than control fruits, resulting in a darker and redder appearance. Similarly, [19] found statistical differences in L* values between growing seasons, but no significant variation between irrigation treatments in 'ME'. Rind colouration is a critical commercial quality attribute since pomegranates of the 'ME' varietal group are typically harvested based on their external appearance. Thus, the increased percentage of pink-red surface on SDI fruits could allow for earlier harvests, thereby enhancing their commercial value, as earlier harvesting tends to command higher prices.

The aril fraction, which represents the weight ratio of arils to fruit, was significantly influenced by the irrigation strategy ($p \leq 0.01$; Table 3). As anticipated, the fruits from the SDI treatment exhibited a higher aril fraction (60.8%) compared to those from the control strategy (56.1%), aligning with the findings regarding the rind-to-fruit ratio (Table 2).

**Table 3.** Effect of the growing season and irrigation strategy on the characteristics of the arils [aril fraction (AF, %), dry matter content (ADM, %), juice/fruit ratio (J/F, % based on weight), and on the CIE L*C*h* colour space parameters (L* indicates lightness. C* represents chroma and h* is the hue angle)].

| | AF | | ADM | | J/F | | L* | | C* | | h* | |
|---|---|---|---|---|---|---|---|---|---|---|---|---|
| Growing season (GS) | | | | | | | | | | | | |
| 2022 | 58.74 | | 17.08 | a | 29.03 | a | 33.95 | a | 33.37 | b | 11.43 | a |
| 2023 | 57.72 | | 16.13 | b | 22.65 | b | 33.31 | b | 36.81 | a | 5.38 | b |
| LSD | 2.30 | | 0.41 | | 1.37 | | 0.63 | | 1.54 | | 2.23 | |
| Irrigation strategy (IS) | | | | | | | | | | | | |
| Control | 56.07 | b | 16.57 | b | 25.37 | | 33.98 | | 35.10 | | 10.79 | |
| RDI1 | 58.08 | ab | 16.51 | b | 27.19 | | 33.74 | | 36.56 | | 7.45 | |
| RDI2 | 57.92 | ab | 15.96 | c | 24.89 | | 33.30 | | 33.46 | | 6.92 | |
| SDI | 60.84 | a | 17.37 | a | 25.92 | | 33.51 | | 35.24 | | 8.45 | |
| LSD | 3.26 | | 0.58 | | 1.93 | | 0.88 | | 2.17 | | 3.15 | |
| GS × IS | | | | | | | | | | | | |
| 2022-Control | 56.17 | | 17.09 | | 26.65 | bc | 34.36 | | 34.60 | | 13.39 | |
| 2022-RDI1 | 59.31 | | 17.16 | | 30.90 | a | 34.01 | | 34.03 | | 10.91 | |
| 2022-RDI2 | 57.38 | | 16.21 | | 27.08 | b | 33.23 | | 32.60 | | 9.17 | |
| 2022-SDI | 62.1 | | 17.85 | | 31.50 | a | 34.22 | | 32.25 | | 12.25 | |
| 2023-Control | 55.97 | | 16.05 | | 24.08 | cd | 33.61 | | 35.60 | | 8.19 | |
| 2023-RDI1 | 56.85 | | 15.86 | | 23.47 | d | 33.46 | | 39.08 | | 4.00 | |
| 2023-RDI2 | 58.47 | | 15.71 | | 22.70 | de | 33.38 | | 34.32 | | 4.68 | |
| 2023-SDI | 59.59 | | 16.89 | | 20.35 | e | 32.80 | | 38.24 | | 4.64 | |
| LSD | 4.60 | | 0.82 | | 2.73 | | 1.25 | | 3.07 | | 4.46 | |
| Source (df) | | | | | % sum of squares | | | | | | | |
| GS (1) | 0.39 | ns | 15.88 | ** | 28.59 | ** | 2.80 | * | 11.45 | ** | 16.68 | ** |
| IS (3) | 4.37 | ns | 17.65 | ** | 2.07 | ns | 1.76 | ns | 4.68 | ns | 4.01 | ns |
| GS × IS (3) | 0.88 | ns | 1.45 | ns | 7.42 | ** | 2.17 | ns | 4.36 | ns | 0.72 | ns |
| Residual (184) | 94.35 | | 65.02 | | 61.92 | | 93.26 | | 79.51 | | 78.59 | |
| Standard deviation | 8.08 | | 1.00 | | 4.80 | | 1.90 | | 4.66 | | 6.76 | |

Mean values followed by different lowercase letters in each column indicate significant differences at $p \leq 0.05$ using the LSD test. df: degrees of freedom; ns: no significant difference; ** (*): significant differences at $p \leq 0.01$ ($p \leq 0.05$).

Both the growing season and the irrigation strategy had a significant impact ($p \leq 0.01$) on aril dry matter content (ADM); the fruit harvested in 2022 showed higher ADM (17.1%) compared to that harvested in 2023 (16.1%). Furthermore, the fruits from the SDI treatment presented greater ADM (17.4%) than those from the control (16.6%) and RDI1 strategies (16.5%), which were, in turn, superior to the RDI2 (16.0%) fruits.

The juice-to-fruit ratio (expressed as a percentage based on weight) was also significantly influenced by the growing season ($p \leq 0.01$), with higher values recorded in

2022 (29.0%) compared to 2023 (22.7%). In contrast, the irrigation strategy did not exert a significant effect on this ratio, likely due to the varying responses of different strategies (particularly control and SDI) across the two growing seasons. This finding is consistent with the results reported by [15], who also found no significant effect of irrigation treatments on juice yield percentage.

The colour of the arils in the fruits was found to be influenced solely by the growing season, which affected all measured parameters. Specifically, the arils from 2023 exhibited a lighter colour (lower L*; 33.3; $p \leq 0.01$), were more vibrant (higher C*; 36.8; $p \leq 0.01$), and appeared redder (lower h*; 5.4; $p \leq 0.05$) compared to those from 2022 (34.0, 33.4, and 11.4 for L*, C*, and h*, respectively). These findings align with the research by [19], which similarly attributed the colour variation in the arils of 'ME' fruits to the growing season. Ref. [20] reported that a moderate water deficit during fruit development influenced the hue (h*), but primarily in the first harvest. They suggested that the shift towards a more intense red colour in the arils correlates with an increase in total anthocyanin content. Additionally, [21] identified six anthocyanins contributing to the red colouration of the aril juice, specifically delphinidin 3-glucoside and 3,5-diglucoside, cyanidin 3-glucoside and 3,5-diglucoside, and pelargonidin 3-glucoside and 3,5-diglucoside. When discussing the relationship between water stress and secondary metabolite production, [20] and the references therein noted that establishing a linear correlation was not possible. Mild osmotic stress from minor water limitations typically leads to reduced plant growth (as observed in the current study) while increasing the concentration of non-nitrogenous secondary metabolites. Conversely, heightened water stress triggers stomatal regulation, reducing $CO_2$ assimilation due to diminished stomatal conductance, which was also evidenced in the present research. In these conditions, carbon allocation is primarily directed towards synthesising primary metabolites, thereby limiting the production of carbon-based secondary metabolites. Furthermore, the irrigation strategy played an important role in determining the rind colouration (specifically the extent of the pink-red area) of the fruits and the colour of the arils, as reported by [15]. In their studies, the SDI enhanced the coloured surface of the rind but did not significantly alter the colour of the arils. Abiotic stress from moderate water deficiency may indeed boost anthocyanin synthesis, a phenomenon previously observed in grapevines. This direct effect of water stress on the anthocyanin biosynthesis pathway has been documented by [22] and might also influence specific timing aspects of the ripening process [23]. According to [15], the timing of water stress during the growing season is crucial for the accumulation of anthocyanins in both rind and aril tissues.

Both the growing season and irrigation strategy significantly influenced the soluble solid content (SSC) of pomegranate aril juice ($p \leq 0.01$, Table 4). Fruits harvested in 2022 exhibited a higher SSC (15.9° Brix, on average) compared to those harvested in 2023 (15.3° Brix, on average), with the SDI fruits achieving the highest SSC values (16.5° Brix, on average). This phenomenon can be attributed to a decrease in water accumulation in the SDI arils, which showed greater dry matter content (Table 3) without a corresponding increase in accumulated sugars. This observation aligns with findings by [24] in processing tomatoes and [25] in processing cherry tomatoes. The observed difference in SSC, averaging 1.25° Brix above the control arils, is remarkable, as a 1° Brix increase is considered a meaningful enhancement in fruit flavour perception [26]. According to various standards for pomegranate fruit quality [19], SSC values exceeding 14° Brix are classified as very good for sweetness. Thus, all irrigation strategies employed in this research yielded fruits with SSC values surpassing this threshold, indicating high quality. The predominant sugars in 'ME' pomegranate juice are fructose and glucose (reducing sugars), while the concentration of sucrose (a non-reducing sugar) generally remains low [19,27].

**Table 4.** Effect of growing season and irrigation strategies on pomegranate aril juice soluble solids content (SSC, °Brix), SSC yield (kg $ha^{-1}$), titratable acidity (TA, g citric acid $L^{-1}$), and maturity index (MI = SSC/TA).

| | | SSC | | SSC Yield | | TA | | MI | |
|---|---|---|---|---|---|---|---|---|---|
| Growing season (GS) | | | | | | | | | |
| | 2022 | 15.90 | a | 1544 | b | 0.35 | | 47.21 a | |
| | 2023 | 15.28 | b | 1779 | a | 0.35 | | 43.50 b | |
| | LSD | 0.17 | | 124 | | 0.02 | | 3.15 | |
| Irrigation strategy (IS) | | | | | | | | | |
| | Control | 15.23 | b | 1657 | a | 0.33 | | 46.59 | |
| | RDI1 | 15.29 | b | 1827 | a | 0.35 | | 45.83 | |
| | RDI2 | 15.36 | b | 1792 | a | 0.35 | | 44.40 | |
| | SDI | 16.48 | a | 1371 | b | 0.37 | | 44.61 | |
| | LSD | 0.23 | | 175 | | 0.03 | | 4.46 | |
| GS × IS | | | | | | | | | |
| | 2022-Control | 15.65 | | 1481 | cd | 0.31 | | 50.82 | |
| | 2022-RDI1 | 15.61 | | 1717 | bc | 0.36 | | 48.35 | |
| | 2022-RDI2 | 15.76 | | 1579 | cd | 0.37 | | 43.90 | |
| | 2022-SDI | 16.60 | | 1398 | d | 0.37 | | 45.76 | |
| | 2023-Control | 14.82 | | 1833 | ab | 0.35 | | 42.36 | |
| | 2023-RDI1 | 14.97 | | 1936 | ab | 0.35 | | 43.31 | |
| | 2023-RDI2 | 14.96 | | 2005 | a | 0.34 | | 44.90 | |
| | 2023-SDI | 16.36 | | 1343 | d | 0.38 | | 43.45 | |
| | LSD | 0.33 | | 248 | | 0.04 | | 6.31 | |
| Source (df) | | % sum of squares | | | | | | | |
| S (1) | | 18.47 | ** | 9.87 | ** | 0.04 | ns | 5.47 | * |
| IS (3) | | 50.02 | ** | 22.98 | ** | 7.76 | ns | 1.29 | ns |
| GS × IS (3) | | 2.65 | ns | 6.00 | * | 6.90 | ns | 4.83 | ns |
| Residual (184) | | 28.86 | | 61.15 | | 85.30 | ns | 88.4 | |
| Standard deviation | | 0.41 | | 306 | | 0.05 | | 7.78 | |

Mean values followed by different lowercase letters in each column indicate significant differences at $p \leq 0.05$ using the LSD test. df: degrees of freedom; ns: no significant difference; ** (*): significant differences at $p \leq 0.01$ ($p \leq 0.05$).

SSC yield was influenced by both the growing season ($p \leq 0.01$) and irrigation strategy ($p \leq 0.01$), as well as their interaction ($p \leq 0.05$). On average, SSC yield in 2023 (1779 kg $ha^{-1}$) was higher than in 2022 (1544 kg $ha^{-1}$), while the SDI strategy yielded the lowest SSC values (1371 kg $ha^{-1}$), consistent with expectations. The interaction analysis revealed that the lower SSC yield associated with SDI in 2023 (1343 kg $ha^{-1}$) can be attributed to diminished overall yield, which was 26% lower than that of the control strategy. The SDI strategy not only resulted in the lowest SSC yield but also contributed to reduced plant size, consequently leading to lower soluble solid accumulation, which is related to decreased stomatal conductance [2].

Titratable acidity remained unaffected by the growing season or irrigation strategy, likely due to a high residual percentage of the sum of squares (85.3%), indicating substantial variability. As noted by [28], SSC is a poor predictor of the sweetness intensity perceived by consumers. Conversely, measuring acidity provides a more reliable estimate of acidity intensity. The acidity present in fruits and juices can obscure the relationship between SSC and sweetness perception. Consequently, the perception of sweetness is contingent on the acidity of the aril juice, making it essential to interpret SSC in conjunction with acidity, specifically through the maturity index (MI). When SSC and titratable acidity values are considered, the MI ratio compensates, revealing similar values across strategies. The MI is influenced solely by the growing season ($p \leq 0.05$), with the highest MI corresponding to the fruits from 2022 (47.2), although in both seasons the same criterion was followed (SSC $\geq$ 14° Brix) and the harvest dates were practically identical.

Mellisho et al. [20] found that in the 'ME' cultivar, a moderate water deficit during fruit growth increased the SSC, but this effect was noted only in the first harvest and did not significantly impact titratable acidity. They also observed that the harvest date significantly influenced TA, with lower values recorded for the first harvest. Similarly, [15] reported higher SSC in 'ME' fruit arils using the SDI strategy; however, variations in titratable acidity resulted in notable differences in maturity index, with the highest values associated with fruits from the SDI treatment. Other authors [29] found that the effects of irrigation treatment on fruit weight, total soluble solids, total acidity, and pH of the aril juice were not significant. More recently, [30] assayed two sustained deficit irrigation treatments applying 70% and 50% ETc and found different results depending on the genotype. In the case of 'Sefi', differences were found with respect to the control treatment in SSC and pH (with the lowest values corresponding to the lowest water supply), while in 'Wonderful' no significant differences were found in these parameters; water restriction did not affect the titratable acidity in either cultivar tested. In our current study, the titratable acidity of 'ME49' fruit arils was comparable to the values reported for 'ME' by [19] and was higher than those reported by [15]. The SSC and titratable acidity values obtained led to maturity index values ranging from 42.4 to 50.8, which are consistent with the findings of [19] and surpass those of [15]. These values fall within the normal range considered for sweet varieties (31–98; [19]).

Table S1 presents the matrix of Pearson correlation coefficients for the main parameters analysed. A very strong correlation (r = 0.93; [31]) is observed between fruit size (diameter) and unit weight. According to [32], unit weight can be used as an indicator of size. The shape index (D/L; [11]) is more closely related to the maximum diameter (D) with a correlation of r = 0.72, compared to the length (L) with a correlation of r = 0.17, which justifies that the size is determined by the maximum diameter and not by the length [32].

Pomegranate fruit size, volume, and weight correlate with the number of arils per fruit, but not with the size of the arils [33]. Each aril corresponds to a fertilised ovule. Since pollination and fertilisation occur early in fruit development and each aril requires a fertilisation event, effective crop management, particularly irrigation strategies during this period, is crucial for achieving large fruits and high marketable yields [33]. Future studies will investigate the relationship between the number of arils obtained per fruit and different irrigation strategies.

The SSC shows a strong negative correlation with both total yield (r = −0.76) and marketable yield (r = −0.83). This means that as yield increases, the juice's SSC in the arils decreases. Additionally, the dry matter content of the arils has a moderately poor negative correlation with both total yield (r = −0.66) and marketable yield (r = −0.65), indicating that the highest yields are produced by fruits with arils that have a higher water content.

The percentage of pink-red rind surface on commercial fruits exhibits a moderately poor negative correlation with both unit weight (r = −0.65) and diameter (r = −0.66). This suggests that larger and heavier fruits tend to have less pink/red on their rind surface. No correlation has been found between the colour parameters of the rind and the arils of the fruits, reinforcing the observations mentioned earlier.

### 2.2. Fruit Physiological Disorders

In terms of non-marketable yield, expressed as a percentage of the total yield by weight (Table 5), no significant differences ($p \leq 0.05$) were found between growing seasons. However, both the irrigation strategy and the interaction between the two factors were significant ($p \leq 0.01$). Notably, the SDI strategy resulted in a high percentage of non-marketable yield (48.4%) due to physiological disorders and decay. In 2023, the non-marketable yield for this strategy reached a peak of 57%, while the value for 2022 was also considerably high at 40%.

**Table 5.** Effect of growing season and irrigation strategy on non-marketable yield (in terms of % yield by weight) and its partitioning into sunburned, cracked, scratched, deformed, and small fruits.

| | Total | | Sunburned | | Cracked | | Scratched | | Deformed | | Small Fruits | |
|---|---|---|---|---|---|---|---|---|---|---|---|---|
| Growing season (GS) | | | | | | | | | | | | |
| 2022 | 36.71 | | 10.27 | b | 5.70 | b | 9.37 | a | 2.48 | b | 8.90 | |
| 2023 | 39.25 | | 15.41 | a | 7.99 | a | 3.42 | b | 4.40 | a | 8.04 | |
| LSD | 3.01 | | 1.54 | | 2.05 | | 1.19 | | 0.84 | | 1.39 | |
| Irrigation strategy (IS) | | | | | | | | | | | | |
| Control | 36.85 | b | 12.93 | | 5.09 | b | 5.81 | | 4.37 | a | 8.65 | |
| RDI 1 | 32.54 | c | 11.96 | | 3.39 | b | 5.92 | | 3.08 | b | 8.19 | |
| SDI | 34.14 | bc | 12.38 | | 3.24 | b | 6.58 | | 3.68 | ab | 8.27 | |
| RDS | 48.40 | a | 14.08 | | 15.67 | a | 7.25 | | 2.63 | b | 8.78 | |
| LSD | 4.26 | | 2.17 | | 2.90 | | 1.69 | | 1.18 | | 1.97 | |
| GS × IS | | | | | | | | | | | | |
| 2022-Control | 40.35 | b | 10.99 | | 8.53 | b | 8.45 | | 3.39 | | 9.00 | ab |
| 2022-RDI 1 | 34.81 | bcd | 10.17 | | 4.62 | bcd | 8.77 | | 2.50 | | 8.75 | ab |
| 2022-RDI 2 | 31.57 | cd | 8.72 | | 3.41 | cd | 10.03 | | 2.26 | | 7.15 | b |
| 2022-SDI | 40.12 | b | 11.20 | | 6.23 | bc | 10.23 | | 1.77 | | 10.70 | a |
| 2023-Control | 33.35 | cd | 14.87 | | 1.65 | d | 3.18 | | 5.36 | | 8.30 | ab |
| 2023-RDI 1 | 30.27 | d | 13.76 | | 2.15 | cd | 3.07 | | 3.66 | | 7.64 | b |
| 2023-RDI 2 | 36.71 | bc | 16.04 | | 3.06 | cd | 3.14 | | 5.09 | | 9.38 | ab |
| 2023-SDI | 56.69 | a | 16.97 | | 25.11 | a | 4.28 | | 3.49 | | 6.85 | b |
| LSD | 6.02 | | 3.08 | | 4.10 | | 2.39 | | 1.67 | | 2.78 | |
| Source (df) | | | | | % sum of squares | | | | | | | |
| GS (1) | 1.44 | ns | 31.48 | ** | 1.75 | * | 51.46 | ** | 17.24 | ** | 1.52 | ns |
| IS (3) | 34.36 | ** | 3.02 | ns | 35.08 | ** | 1.94 | ns | 8.05 | * | 0.50 | ns |
| GS × IS (3) | 19.18 | ** | 2.72 | ns | 32.21 | ** | 0.51 | ns | 1.68 | ns | 9.52 | * |
| Residual (88) | 45.02 | | 62.78 | | 30.96 | | 46.10 | | 73.03 | | 88.46 | |
| Standard deviation | 7.42 | | 3.79 | | 5.05 | | 2.94 | | 2.06 | | 3.43 | |

Mean values followed by different lowercase letters in each column indicate significant differences at $p \leq 0.05$ using the LSD test. df: degrees of freedom; ns: no significant difference; ** (*): significant differences at $p \leq 0.01$ ($p \leq 0.05$).

When analysing the incidence of various types of blemished fruits (Table 5), it is evident that sunburn and scratches were predominantly influenced by the growing season ($p \leq 0.01$); deformed fruits were affected by the irrigation strategy ($p \leq 0.05$), while small fruits were influenced by the interaction of both factors ($p \leq 0.05$). Notably, the incidence of diseases, attributed to different pathogens, was negligible.

In the 2023 season, which experienced exceptionally high summer temperatures (with average and maximum temperatures of 26.1 °C and 46.8 °C, respectively), the incidence of sunburned fruits increased to 15%, compared to 10% in 2022, when average temperatures reached 26.5 °C and maximum temperatures were 39.6 °C. Conversely, 2022 saw a higher incidence of decay due to scratches (9.4%) compared to 2023 (3.4%), likely influenced by observed wind speeds. During August 2022, there were four significant gusts classified as category five on the Beaufort scale [34], reaching speeds of up to 35 km $h^{-1}$. According to [11], windy conditions can cause fruits to rub against thorns, resulting in rind injuries that, although may suberify, render the fruits unsuitable for fresh consumption. These injuries can also lead to deeper wounds, making the fruits vulnerable to pathogenic agents that contribute to decay.

The highest incidence of deformed fruits ($p \leq 0.05$) was noted in 2023 and in the control strategy (4.4% in both cases). The percentage of small fruits was influenced by the interaction between the growing season and the irrigation strategy, showing only significant differences between seasons under the SDI strategy.

Cracking incidence was affected by both the growing season ($p \leq 0.05$) and the irrigation strategy ($p \leq 0.01$), along with their interaction ($p \leq 0.01$). Notably, the SDI strategy in 2023 exhibited a cracking incidence of 25.1%, exceeding rates from the other strategies during the same year by over eight times. In contrast, the control strategy recorded the highest cracking incidence in 2022 (8.5%) but the lowest in 2023 (1.7%).

Figure 1 illustrates the evolution of cracking incidence across the four irrigation strategies evaluated over the two experimental years.

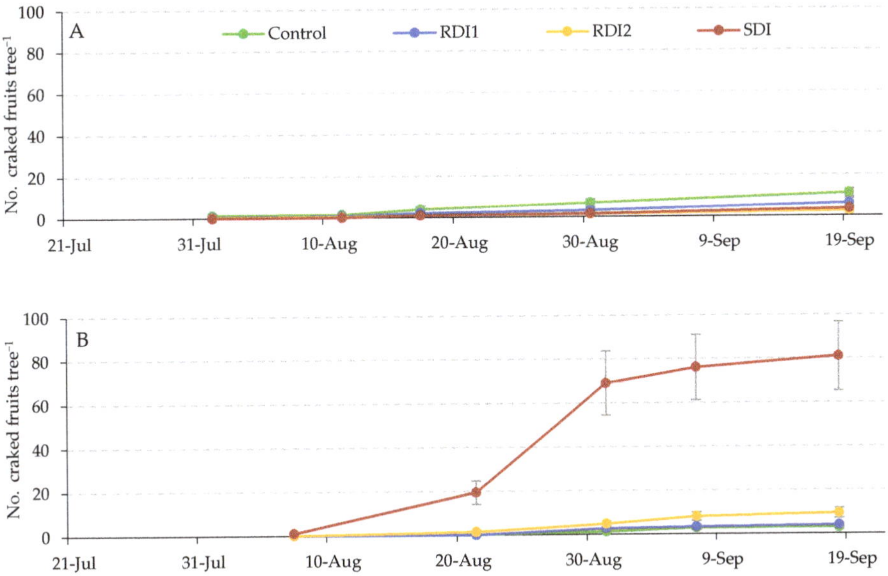


**Figure 1.** Evolution of the incidence of cracking in the four irrigation strategies tested in the two years of experimentation, 2022 (**A**) and 2023 (**B**). Vertical bars represent the standard error; their absence indicates that the bar size was less than that of the symbol used. SDI: sustained deficit irrigation; RDI1: regulated deficit irrigation (appearance of the lower buds—young fruits); RDI2: regulated deficit irrigation (open flower—fruit growth).

Figure 2 presents the evolution of relative soil water content (volumetric soil water content/volumetric soil water content at field capacity at a depth of 0.25 m; %) for the four tested irrigation strategies, showcasing daily rainfall from 21 July to 21 September in 2023. The most significant incidences of cracking were observed with the SDI strategy in 2023, particularly during sampling on 21 and 31 August. During this period, temperatures peaked at 41.4 °C on 24 July and reached 46.8 °C on 10 August, with an average temperature of 26.9 °C from 10 August to 31 August. An increase in relative volumetric soil water content in early and late August (Figure 2) likely stemmed from irrigation management strategies defined in relation to the control strategy, as described in the Material and Methods section.

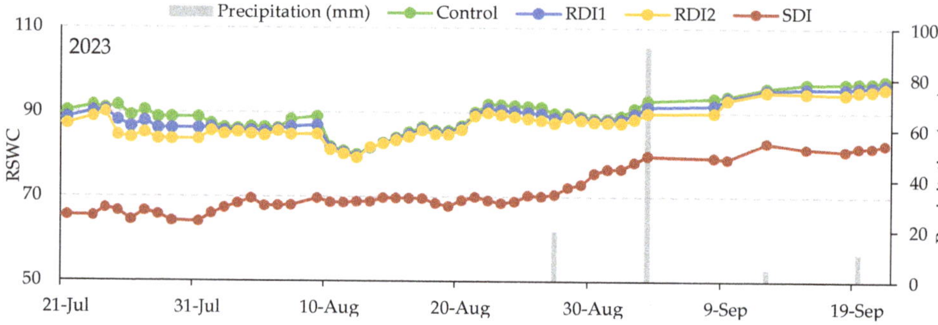


**Figure 2.** Relative soil water content [RVSWC; volumetric soil water content/volumetric soil water content at field capacity (%) at a 0.25 m depth for the four irrigation strategies tested before each irrigation event], and daily rainfall (vertical bars) from 21 July to 21 September in 2023. SDI: sustained deficit irrigation; RDI1: regulated deficit irrigation (appearance of the lower buds—young fruits); RDI2: regulated deficit irrigation (open flower—fruit growth).

The primary cause of pomegranate cracking is fluctuations in soil moisture. This phenomenon typically arises from discrepancies in growth rates between the fruit's rind and its arils, which is influenced by the pressure from the rapid expansion of the arils against the stretched rind [6]. The imbalance in pressure can result from a high value in aril pressure or a low value in rind resistance. Galindo et al. [9] noted that as fruit development progresses toward maturation, pomegranates become particularly vulnerable to water stress. During significant water shortages, while leaf turgor may be maintained, the fruit's turgor can decline, leading to inhibited expansion. When rainfall or irrigation occurs after a period of water stress, the arils experience a disproportionate increase in turgor pressure compared to the rind. This disparity can cause an increase in pressure on the rind, making it more prone to cracking. These authors found strong first-order correlations between fruit water potential (Ψfruit) and stem water potential (Ψstem), indicating that Ψstem is a reliable predictor for Ψfruit. They determined that rind turgor loss and aril turgor loss occurred at approximately 1.95 and 2.46 MPa, respectively. In our 2023 experiment [2], the Ψstem did not fall below −1.125 MPa in the control strategy and −1.67 MPa in the SDI strategy. Although we did not observe a total loss of turgor in the fruit, it appears that the intensity of turgor loss may have been greater in the SDI fruits compared to those in other irrigation strategies. Similarly, in the 2022 experiment, Ψstem remained above −1.63 MPa across all strategies, indicating that fruit turgor loss was not achieved during that season either. Other authors [10] found that the number of uncracked fruits per tree increased with increasing amounts of irrigation water applied. Fruit obtained with SDI strategies that involved insufficient irrigation input showed a significant thinning of the rind, which caused the fruits to crack. They recommended that available soil water in the root zone should not fall below 50% to avoid poor fruit size and shape and increase the yield of 'Hicaznar' pomegranate.

The mechanical properties of fruit rind play a crucial role in sustaining internal pressure and resisting fruit cracking. According to [9], water stress can significantly alter these properties, leading to decreased extensibility and increased stiffness. Environmental factors, including temperature and relative humidity, also impact these mechanical properties. Matas et al. [35] discovered that the mechanical traits of the tomato fruit's cuticular membrane are influenced by temperature and humidity; as temperatures rise, both stiffness and strength tend to decrease. In the summer of 2023, notably high temperatures were recorded, with an average of 26.9 °C in August and a peak of 46.8 °C on 10 August. In comparison, the maximum temperature in 2022 was considerably lower at 39.6 °C (on

15 August). The elevated temperatures seen in 2023 likely affected the cuticle, diminishing its stiffness and strength, thus increasing its vulnerability to cracking. Other environmental conditions, such as high evapotranspiration, low humidity, and drastic temperature fluctuations during the day, may also contribute to higher cracking rates [36]. The summer of 2023 experienced high evapotranspiration levels (190 mm), low average relative humidity (66%), and sharp daily temperature variations (up to 22.2 °C), which are likely linked to the increased incidence of cracking observed during fruit growth in August. As already mentioned, [35] explored how environmental conditions affect the mechanical properties of the tomato's cuticular membrane. They found that immersing the cuticular membrane in water reduced its stiffness, relating their findings to key environmental factors that contribute to cracking: moisture on the fruit's rind from rain or condensation coupled with high temperatures. In our 2023 study, we noted not only the high temperatures but also four rainfall events during fruit growth and ripening: 19.8 mm on 27 August, 92.5 mm on 3 September, followed by 4.7 mm on 12 September, and 11 mm on 19 September.

Genetic factors can also significantly influence fruit cracking. Different pomegranate cultivars exhibit varying levels of sensitivity to cracking [11,14,36–38]. Crack-resistant varieties tend to show greater water use efficiency compared to those that are more susceptible. In the current study, apart from the SDI strategy in 2023, the observed cracking levels (1.7% to 8.5% of total yield) fell below the threshold generally considered normal in Spain for the 'ME' varietal group of pomegranates, which is prone to cracking (5–20%; [38]).

Moreover, both fruit volume and shape—flat versus spheroidal—have been linked to susceptibility to cracking. Larger fruits, which contain more arils and exhibit a shape index greater than 1, demonstrate increased sensitivity to cracking [39]. The pressure exerted by the arils on the rind escalates with the fruit's radius, exposing the rind of larger fruits to greater stress, thus heightening their risk of cracking. However, our study did not find a significant correlation between the incidence of cracking and parameters such as fruit weight, size (diameter and length), or shape index. According to [39], the reduction in rind elasticity seems more critical to cracking incidence than fruit volume.

In the literature, relationships have also been noted between pomegranate rind traits—such as rind thickness and the proportion of rind to aril (on a weight basis)—and cracking incidence, but these findings are not consistently observed. Mellisho et al. [20] suggested that a decreased rind fraction may correspond with higher cracking rates, while [39] found no correlation between rind thickness and cracking susceptibility. In the present study, no significant relationship was established between the incidence of cracking and either rind thickness or rind fraction; this is despite the fact that, in 2023, the SDI fruits exhibited the lowest rind thickness ($p \leq 0.05$), a reduced rind fraction (which lacked statistical significance due to high variability between samples), and they also showed the highest incidence of cracking ($p \leq 0.01$). It is noteworthy that the variables measured (size, unit weight, shape index, rind thickness, and rind fraction) did not demonstrate a significant correlation with cracking.

Another factor linked to fruit cracking is nutrient deficiency [5,14,36], particularly involving calcium (Ca), potassium (K), and magnesium (Mg). These nutrients are crucial as they influence the biomechanical behaviour of the fruit rind. For effective cell expansion, the accumulation of K, Mg, and Ca in the vacuole is essential for maintaining osmotic potential [40]. In litchi [41] and citrus fruits [42], low concentrations of Ca in the pericarp have been shown to contribute to fruit cracking. However, [43] reported that excessive Ca levels in pomegranate fruits could actually increase the rate of cracking. Yılmaz and Özgüven [44] discovered that cracked fruit rinds exhibited lower levels of K, as well as reduced K/Ca and K/(Ca + Mg) ratios. Cybulska et al. [45] indicated that high concentrations of Ca (greater than 2%) alter the biomechanical properties of plant tissues, making them harder and more brittle. The effect of Ca can be attributed to its ability to form bridges between pectin chains [39], whereas K imparts elasticity to the cell wall [5,11], because part of the Mg is bound to pectate in the middle lamella of cell walls, contributing to their overall structure [40].

Recently, [10] found that K application improved rind thickness of uncracked fruit and decreased fruit cracking. Chater and Garner [5] found that single foliar applications of zinc (Zn), Mg, and K in pomegranate trees reduced the incidence of fruit cracking. However, their results suggest that the physiological or biomechanical effects of these nutrient applications on the rind's ability to withstand internal pressure were limited. They also observed a negative correlation between fruit cracking incidence and leaf concentrations of copper (Cu) and boron (B) but did not identify any other statistically significant relationships between fruit cracking and leaf nutrient concentrations. In the current study (Table 6), foliar analyses showed no significant differences in nutrient content among the various irrigation strategies, with all levels falling within the normal ranges established for 'ME' pomegranate [46]. Moreover, no statistically significant relationship was found between the incidence of fruit cracking and the concentrations of the analysed leaf nutrients. Given that fruit cracking is primarily caused by pressure on the rind due to the expansion of the arils within each fruit, it is important to note the considerable variability in cracking incidence both among different trees and among individual fruits on the same tree. Future research could benefit from a deeper analysis of nutrient content in fruits, particularly considering the reduced translocation of Ca within the plant.

**Table 6.** Effect of growing season and irrigation strategy on leaf macronutrient (nitrogen, phosphorus, potassium, calcium, and magnesium; %) and micronutrient (iron, copper, manganese, and zinc; mg $L^{-1}$) concentrations in adult leaves (collected on 15 July of each year) of pomegranate trees of clone 49 'Mollar de Elche'. Average values of four trees per replication block.

| | Macronutrients | | | | | Micronutrients | | | |
|---|---|---|---|---|---|---|---|---|---|
| | N | P | K | Ca | Mg | Fe | Cu | Mn | Zn |
| Growing season (GS) | | | | | | | | | |
| 2022 | 1.82 b | 0.15 | 0.85 | 1.53 b | 0.35 | 145.3 a | 6.33 | 26.7 b | 20.2 |
| 2023 | 2.01 a | 0.16 | 0.89 | 1.84 a | 0.41 | 114.2 b | 8.42 | 40.0 a | 16.2 |
| LSD | 0.18 | 0.02 | 0.11 | 0.14 | 0.07 | 24.54 | 2.34 | 4.47 | 4.40 |
| Irrigation strategy (IS) | | | | | | | | | |
| Control | 2.04 | 0.16 | 0.86 | 1.71 | 0.38 | 124.3 | 8.50 | 34.0 | 19.3 |
| RDI1 | 1.88 | 0.15 | 0.88 | 1.73 | 0.35 | 131.3 | 6.50 | 33.3 | 17.7 |
| RDI2 | 1.92 | 0.15 | 0.91 | 1.62 | 0.38 | 144.8 | 7.50 | 33.3 | 18.5 |
| SDI | 1.82 | 0.14 | 0.83 | 1.68 | 0.41 | 118.5 | 7.00 | 32.7 | 17.2 |
| LSD | 0.25 | 0.02 | 0.16 | 0.20 | 0.10 | 34.71 | 3.31 | 6.32 | 6.22 |
| GS × IS | | | | | | | | | |
| 2022-Control | 1.99 | 0.16 | 0.79 | 1.56 | 0.33 | 140.0 | 7.33 | 27.3 | 19.7 |
| 2022-RDI1 | 1.75 | 0.14 | 0.91 | 1.66 | 0.36 | 149.7 | 5.00 | 27.0 | 21.7 |
| 2022-RDI2 | 1.75 | 0.13 | 0.89 | 1.37 | 0.37 | 146.0 | 7.00 | 24.3 | 21.0 |
| 2022-SDI | 1.79 | 0.15 | 0.81 | 1.51 | 0.34 | 145.7 | 6.00 | 28.0 | 18.3 |
| 2023-Control | 2.10 | 0.17 | 0.92 | 1.86 | 0.43 | 108.7 | 9.67 | 40.7 | 19.0 |
| 2023-RDI1 | 2.00 | 0.16 | 0.85 | 1.81 | 0.35 | 113.0 | 8.00 | 39.7 | 13.7 |
| 2023-RDI2 | 2.09 | 0.18 | 0.93 | 1.86 | 0.38 | 143.7 | 8.00 | 42.3 | 16.0 |
| 2023-SDI | 1.86 | 0.13 | 0.85 | 1.84 | 0.48 | 91.3 | 8.00 | 37.3 | 16.0 |
| LSD | 0.36 | 0.03 | 0.22 | 0.28 | 0.14 | 49.09 | 4.69 | 8.93 | 8.79 |
| Source (df) | % sum of squares | | | | | | | | |
| GS (1) | 19.45 * | 10.38 ns | 2.93 ns | 52.29 ** | 15.19 ns | 25.22 * | 16.31 ns | 68.58 * | 16.80 ns |
| IS (3) | 14.04 ns | 11.32 ns | 6.56 ns | 3.81 ns | 6.21 ns | 10.02 ns | 8.22 ns | 0.34 ns | 2.86 ns |
| GS × IS (3) | 5.80 ns | 28.85 ns | 8.05 ns | 7.66 ns | 14.33 ns | 9.08 ns | 1.96 ns | 3.69 ns | 8.11 ns |
| Residual (88) | 60.70 | 49.44 | 82.47 | 36.23 | 64.27 | 55.68 | 73.51 | 27.39 | 72.23 |
| Standard deviation | 0.21 | 0.02 | 0.13 | 0.16 | 0.08 | 28.36 | 2.71 | 5.16 | 5.08 |

Mean values followed by different lowercase letters in each column indicate significant differences at $p \leq 0.05$ using the LSD test. df: degrees of freedom; ns: no significant difference; ** (*): significant differences at $p \leq 0.01$ ($p \leq 0.05$).

Fruit cracking has been associated with the expression of the expansin gene [36]. During plant growth, cells produce proteins known as expansins, which play a critical role in relaxing cell wall tension and enabling turgor-driven cell enlargement. Expansins operate through a non-enzymatic mechanism by disrupting the hydrogen bonds that link cellulose fibrils and other polysaccharides [47], ultimately resulting in fruit cracking [36]. Investigating the relationship between expansin gene expression and deficit irrigation in pomegranate trees would be particularly insightful.

In summary, environmental factors recorded in 2023, such as high temperatures, elevated evapotranspiration, low humidity, significant day–night temperature fluctuations, and rainfall during the growth and ripening phases, uniformly affected all four irrigation strategies. These conditions likely impacted the cuticle, decreasing its stiffness and strength and thereby increasing susceptibility to cracking. Although the total loss of fruit turgor did not occur, there was a more pronounced decrease in turgor among fruits subjected to SDI compared to those under the other strategies. This likely resulted in an uneven increase in fruit turgor pressure; upon turgor recovery, aril turgor would have risen more significantly than that of the rind. The higher aril pressure might have rendered the rind more vulnerable to cracking, a process further exacerbated by increased soil water content, culminating in a higher incidence of fruit cracking.

## 3. Materials and Methods

### 3.1. Experimental Site and Plant Material

The field studies were conducted at the Cajamar Experimental Centre in Paiporta, Valencia, Spain (39.4175 N, 0.4184 W) over two consecutive growing seasons in 2022 and 2023. The soil at the site was uniform in the root zone depth (50 cm), with a silt loam texture, and was very slightly alkaline, slightly saline, and highly fertile (organic matter = 1.4%). Irrigation water was pumped from a well, which averaged EC 1.65 dS $m^{-1}$. The climate is subtropical Mediterranean with hot and dry summers. Detailed information regarding soil, irrigation water, climate characteristics, and agricultural practices is available in the companion paper.

The pomegranate trees used in the experiment belonged to the 'ME49' clone, a selected clone of the 'Mollar de Elche' ('ME') varietal group (*Punica granatum* L.). This cultivar, part of the pomegranate gene bank at the Agricultural Experiment Station of Elche, Spain [48], is recognised for its mid-season harvest, high vigour, robust productivity, and minimal thorns. The fruit produced is large and sweet, noted for its pleasant flavour and soft seeds. Additionally, it features a pink-red rind covering a significant portion of its surface, placing it among the highly recommended varieties [11].

### 3.2. Irrigation Strategies

In the control strategy, irrigation was conducted based on practices commonly adopted by local growers, focusing on meeting crop water needs while ensuring that soil moisture reached a depth of 50 cm, without recorded water loss at that depth. Trees subjected to SDI received 50% of the water allocated to control trees. For the RDI strategies, trees were irrigated with 33.3% of their crop water needs during specific restriction periods (from BBCH 51 to BBCH 71 in RDI1 and from BBCH 61 to BBCH 73 in RDI2; Tables S2 and S3).

Volumetric soil water content (VSWC; $m^3$ $m^{-3}$) was monitored continuously using $ECH_2O$ EC-5 capacitance sensors linked to an Em50 data logger and managed through the $ECH_2O$ Utility software version 1.8.5 (Decagon Devices, Inc., Pullman, WA, USA). One sensor per replicate was placed 0.25 m beneath an emitter, as the highest density of pomegranate tree roots is found at this depth. Additional sensors placed at a depth of 0.50 m enabled the verification that moisture penetrated this deep, and that deep percolation was controlled. VSWC readings were recorded every 15 min, and variations in VSWC were utilised to assess in situ field capacity. VSWC data were standardised as a percentage of field capacity to minimise the effects of sensor calibration discrepancies. Irrigation events commenced when the VSWC in the control fell to 90% of field capacity, triggering the appropriate irrigation dose for each treatment.

### 3.3. Tree and Fruit Assessments

Leaf nutrient concentration analyses were conducted following expert recommendations [38,46]. Eighty fully expanded leaves were collected on 15 July of each year per treatment on branches free of developing fruit, with samples taken from four trees per replication block. Leaf analysis was performed at the "Las Palmerillas" laboratory in Cajamar, adhering to official methodologies [49].

Pomegranate fruits were harvested manually during two commercial pickings once they reached maturity. The fruits were deemed ready for harvest when they exhibited the characteristic colour of their variety, without any green parts, and had a SSC exceeding 14° Brix; additionally, the fruits were considered marketable only when their unit weight exceeded 220 g. The harvest dates were 13 October and 3 November 2022, as well as 16 and 31 October 2023. In this moment, 48 large, representative marketable fruits were selected for each treatment (4 fruits per tree, four trees per block, encompassing three replication blocks) and transported to the Agronomy Laboratory at the Universitat Politècnica de València for further analysis. Fruit metrics, including equatorial diameter (D) and length from calyx to base (L), were measured using a digital calliper (Powerfix Profi+, Ovibell GmbH & Co. KG, Mülheim an der Ruhr, Germany). Fruit weight (FW) was determined using a precision balance (Mettler Toledo AB204–S, Mettler-Toledo S.A.E., Cornellà del Llobregat, Spain). According to the Economic Commission for Europe [32], fruit size is assessed based on either the maximum diameter of the equatorial section or the individual unit weight, both of which were evaluated in this study.

The fruits were then peeled, and the arils were manually separated from the rind; rind thickness was measured with the digital calliper, while the arils were weighed and subsequently homogenised. Half of the arils were blended (Taurus Liquafruit Pro compact, 600 W; Taurus, Oliana, Spain) to extract the juice, following the methods described by [50]. The other half of the arils and the complete rind of each fruit were weighed separately using an analytical balance and dried at 65 °C in a forced air oven (Selecta 297; Barcelona, Spain) until reaching a constant weight, allowing for the calculation of dry weights and dry matter content.

The visual estimation of the percentage of pink-red rind surface on the selected fruits was conducted by a panel of three experts. To quantify the colour attributes of the pink-red zones, colour coordinates (L*, a*, and b*) were measured using a Minolta CR-300 chroma meter (Konica Minolta Sensing Inc., Tokyo, Japan). Measurements were taken in triplicate at three equidistant points from the equatorial region of each fruit. Similarly, the colours of the arils were assessed at four locations along the longitudinal section of the fruit. The L* value indicates lightness, with a scale ranging from 0 to 100. Chroma (C*) and hue angle (h*) were calculated using the equations $C = \sqrt{(a^{*2} + b^{*2})}$ [51] and $h^* = \arctan(b^*/a^*)$ [52], respectively.

The soluble solids content (SSC, °Brix) of the juice was measured with an Atago digital refractometer (model N-20; Atago, Bellevue, WA, USA) at a laboratory temperature of 20 °C. The SSC yield (kg $ha^{-1}$) was derived by multiplying the SSC by the total yield. Titratable acidity (TA) was determined using an acid-base potentiometer (877 Titrino plus; Metrohm Ion Analysis, Herisau, Switzerland) with 0.1 N NaOH to titrate the sample up to pH 8.1, reporting the results as grammes of citric acid per litre (g of citric acid $L^{-1}$). The maturity index (MI) was calculated from these values as the SSC/TA ratio.

Cracked fruits were monitored periodically on each tree leading up to harvest, with any cracked fruits removed following each assessment. Upon harvest, a thorough visual inspection of all fruits was performed to identify physiological disorders such as sunburn, cracking, and other anomalies (including scratches, small size, and deformities). Fruits exhibiting blemishes (including previously harvested cracked fruits) were classified as non-marketable, and this was expressed as a percentage of non-marketable fruit compared to the total of marketable and non-marketable fruit.

### 3.4. Experimental Design and Statistical Analysis

The experimental layout utilised a randomised complete block design with three replicates. Each experimental plot consisted of three rows of trees, each containing six trees. The analysis was conducted on the four central trees of the middle row, selected for their consistent characteristics (height, ground shaded area, trunk cross-section, etc.) in accordance with the recommendations of [53].

The results for the various parameters were analysed using analysis of variance (ANOVA) via Statgraphics Centurion XVII [54]. Prior to analysis, percentage data were arcsin $\sqrt{}$ transformed. The least significant difference (LSD) at a 0.05 probability level was used as the criterion for mean separation. Additionally, the Pearson correlation coefficient was computed to evaluate potential interdependencies among the quantitative parameters.

## 4. Conclusions

The characteristics of fruits are influenced by both the growing season and the irrigation strategy employed. The largest fruits with the highest unit weight are produced during the season with the lowest yield. In comparison to the control strategy, SDI tends to reduce fruit size and unit weight. However, this strategy also results in arils with higher dry matter content and aril juice that possesses greater soluble solids concentration, all without impacting the maturity index. One downside of this SDI is the increase in non-marketable yields, particularly due to cracking incidents, which are more likely during extremely hot and dry summers. The RDI strategies that were tested did not compromise either commercial yield or fruit quality, and they represented a slight water saving, making them recommendable irrigation strategies for pomegranate cultivation. Given the lack of significant differences between both RDI strategies, it would be interesting to continue with this study, lengthening the restriction period between the flowering and early fruit growth phase with colour change (BBCH51-BBCH73).

**Supplementary Materials:** The following supporting information can be downloaded at https://www.mdpi.com/article/10.3390/plants14050720/s1: Table S1. Matrix of Pearson correlation coefficients of the parameter values corresponding to the yield [total (TY) and marketable (MY] and characteristics of the fruits [unit weight (UW) and its partitioning in arils (AW) and rind (RW), size (equatorial diameter (D) and length (L)) and shape index (SI)], rind [R; thick (RT), dry matter content (RDM), pink-red zone of the rind (% with respect to the entire surface), and CIE L*C*h* colour space parameters of the pink-red zone of the rind)], aril [A; dry matter content (ADM), CIE L*C*h* colour space parameters] and aril juice soluble solids content (SSC), titratable acidity (TA) and maturity index (MI). Table S2. Irrigation restriction periods in the regulated deficit irrigation strategies for 2022 and 2023 [2]. Table S3. Irrigation water applied corresponding to irrigation strategies (IS) assayed, expressed as a percentage of the irrigation water requirements applied [2].

**Author Contributions:** Conceptualisation, C.B., J.B., B.P. and N.P.-S.; methodology, R.P.-J., J.M.A., C.B., J.B., B.P. and N.P.-S.; formal analysis, R.P.-J.; investigation, R.P.-J. and J.M.A.; resources, J.M.A., C.B., B.P. and N.P.-S.; data curation, R.P.-J. and J.M.A.; writing—original draft preparation, R.P.-J.,

B.P. and N.P.-S.; writing—review and editing, R.P.-J., J.M.A., C.B., J.B., B.P. and N.P.-S.; visualisation, R.P.-J., J.M.A., C.B., J.B., B.P. and N.P.-S.; supervision, C.B., J.B., B.P. and N.P.-S. All authors have read and agreed to the published version of the manuscript.

**Funding:** Funding for open access charge: Universitat Politècnica de València.

**Data Availability Statement:** Dataset available on request from the authors.

**Conflicts of Interest:** The authors declare no conflicts of interest.

## References

1. Manera, F.J.; Legua, P.; Melgarejo, P.; Brotons, J.M.; Hernández, F.C.A.; Martínez, J.J. Determination of a colour index for fruit of pomegranate varietal group "Mollar de Elche". *Sci. Hortic.* **2013**, *150*, 360–364. [CrossRef]
2. Porras-Jorge, R.; Aguilar, J.M.; Baixauli, C.; Bartual, J.; Pascual, B.; Pascual-Seva, N. Effect of deficit irrigation on agronomic and physiological performance of pomegranate (*Punica granatum* L.). *Plants* **2025**, *14*, 164. [CrossRef] [PubMed]
3. Fereres, E.; Soriano, M.A. Deficit irrigation for reducing agricultural water use. *J. Exp. Bot.* **2007**, *58*, 147–159. [CrossRef] [PubMed]
4. Abdelkhalik, A.; Pascual, B.; Nájera, I.; Domene, M.A.; Baixauli, C.; Pascual-Seva, N. Effects of deficit irrigation on the yield and irrigation water use efficiency of drip-irrigated sweet pepper (*Capsicum annuum* L.) under Mediterranean conditions. *Irrig. Sci.* **2020**, *38*, 89–104. [CrossRef]
5. Chater, J.M.; Garner, L.C. Foliar nutrient applications to 'Wonderful' pomegranate (*Punica granatum* L.). II. Effects on leaf nutrient status and fruit split, yield and size. *Sci. Hortic.* **2018**, *242*, 207–213. [CrossRef]
6. Yilmaz, C.; Özgüven, A.I. The effects of some plant nutrients, gibberellic acid and pinolene treatments on the yield, fruit quality and cracking in pomegranate. *Acta Hortic.* **2009**, *818*, 205–212. [CrossRef]
7. Khattab, M.M.; Shaban, A.E.; El-Shrief, A.H.; Mohamed, A.S.E.-D. Growth and Productivity of Pomegranate Trees Under Different Irrigation Levels. II: Fruit Quality. *J. Hortic. Sci. Ornament. Plants* **2011**, *3*, 259–264.
8. Intrigliolo, D.S.; Bonet, L.; Nortes, P.A.; Puerto, H.; Nicolas, E.; Bartual, J. Pomegranate Trees Performance Under Sustained and Regulated Deficit Irrigation. *Irrig. Sci.* **2013**, *31*, 959–970. [CrossRef]
9. Galindo, A.; Rodríguez, P.; Collado-González, J.; Cruz, Z.N.; Torrecillas, E.; Ondoño, S.; Corell, M.; Moriana, A.; Torrecillas, A. Rainfall Intensifies Fruit Peel Cracking in Water Stressed Pomegranate Trees. *Agric. For. Meteorol.* **2014**, *194*, 29–35. [CrossRef]
10. Iscimen, B.; Sezen, S.M.; Yılmaz, C.; Unlu, M. Combined Impacts of Different Irrigation Levels and Potassium Doses on Drip-Irrigated Pomegranate Yield, Quality, Water Productivity, Cracking Rate, and the Economic Net Return. *Irrig. Sci.* **2023**, *41*, 629–648. [CrossRef]
11. Melgarejo Moreno, P.; Salazar Hernández, D.M. *Tratado de Fruticultura Para Zonas Áridas y Semiáridas (Vol. II). Algarrobo, Granado y Jinjolero*; AMV-Mundi-Prensa: Madrid, Spain, 2003; ISBN 8484761118.
12. Bartual, J.; Bonet, L.; Intrigliolo, D.S.; Palou, L.; Pomares, F. Técnicas de cultivo del granado. *Agrícola Vergel* **2014**, *33*, 359–363.
13. Munhuweyi, K.; Lennox, C.L.; Meitz-Hopkins, J.C.; Caleb, O.J.; Opara, U.L. Major diseases of pomegranate (*Punica granatum* L.), their causes and management—A Review. *Sci. Hortic.* **2016**, *211*, 126–139. [CrossRef]
14. Volschenk, T. Effect of Water Deficits on Pomegranate Tree Performance and Fruit Quality—A Review. *Agric. Water Manag.* **2021**, *246*, 106499. [CrossRef]
15. Laribi, A.I.; Palou, L.; Intrigliolo, D.S.; Nortes, P.A.; Rojas-Argudo, C.; Taberner, V.; Bartual, J.; Pérez-Gago, M.B. Effect of sustained and regulated deficit irrigation on fruit quality of pomegranate cv. 'Mollar de Elche' at harvest and during cold storage. *Agric. Water. Manag.* **2013**, *125*, 61–70. [CrossRef]
16. Bartual, J.; Navarro, M.J.; Pérez-Gago, M.B.; Ortiz, M.; Palou, L. Irrigation strategies affect quality, mineral composition and internal rind browning of 'Mollar de Elche' pomegranate fruits. *Acta Hortic.* **2022**, *1349*, 47–55. [CrossRef]
17. Kayesh, E.; Shangguan, L.; Korir, N.K.; Sun, X.; Bilkish, N.; Zhang, Y.; Han, J.; Song, C.; Cheng, Z.M.; Fang, J. Fruit skin color and the role of anthocyanin. *Acta Physiol. Plant.* **2013**, *35*, 2879–2890. [CrossRef]
18. Gobbo-Neto, L.; Lopes, N.P. Plantas medicinais: Fatores de influência no conteúdo de metabólitos secundários. *Quim. Nova* **2007**, *30*, 374–381. [CrossRef]
19. Martínez-Nicolás, J.J.; Galindo, A.; Griñán, I.; Rodríguez, P.; Cruz, Z.N.; Martínez-Font, R.; Carbonell-Barrachina, A.A.; Nouri, H.; Melgarejo, P. Irrigation water saving during pomegranate flowering and fruit set period do not affect *Wonderful* and *Mollar de Elche* cultivars yield and fruit composition. *Agric. Water. Manag.* **2019**, *226*, 105781. [CrossRef]
20. Mellisho, C.D.; Egea, I.; Galindo, A.; Rodríguez, P.; Rodríguez, J.; Conejero, W.; Romojaro, F.; Torrecillas, A. Pomegranate (*Punica granatum* L.) fruit response to different deficit irrigation conditions. *Agric. Water. Manag.* **2012**, *114*, 30–36. [CrossRef]
21. Hernández, F.; Melgarejo, P.; Tomás-Barberán, F.A.; Artés, F. Evolution of juice anthocyanins during ripening of new selected pomegranate (*Punica granatum*) clones. *Eur. Food Res. Technol.* **1999**, *210*, 39–42. [CrossRef]

22. Castellarin, S.D.; Matthews, M.A.; Di Gaspero, G.; Gambetta, G.A. Water deficits accelerate ripening and induce changes in gene expression regulating flavonoid biosynthesis in grape berries. *Planta* **2007**, *227*, 101–112. [CrossRef] [PubMed]
23. Roby, G.; Harbertson, J.F.; Adams, D.A.; Matthews, M.A. Berry size and vine water deficits as factors in winegrape composition: Anthocyanins and tannins. *Aust. J. Grape Wine Res.* **2004**, *10*, 100–107. [CrossRef]
24. Patanè, C.; Tringali, S.; Sortino, O. Effects of deficit irrigation on biomass, yield, water productivity and fruit quality of processing tomato under semi-arid Mediterranean climate conditions. *Sci. Hortic.* **2011**, *129*, 590–596. [CrossRef]
25. Pék, Z.; Daood, H.G.; Neményi, A.; Helyes, L.; Szuvandzsiev, P. Seasonal and irrigation effect on yield parameters and soluble solids content of processing cherry tomato. *Acta Hortic.* **2017**, *1159*, 45–49. [CrossRef]
26. Abdelfatah, M.A.M. Response of Pomegranate Trees (*Punica granatum* L., cv. Mollar de Elche) to Deficit Irrigation at Different Phenological Stages. Master's Thesis, Universitat Politècnica de València, València, València, Spain, 9 May 2014.
27. Melgarejo, P.; Salazar, D.M.; Artés, F. Organic acids and sugars composition of harvested pomegranate fruits. *Eur. Food Res. Technol.* **2000**, *211*, 185–190. [CrossRef]
28. Scandella, D.; Kraeutler, E.; Venien, S. Anticiper la qualité gustative des pêches et des nectarines. *Infos CTIFL* **1997**, *129*, 16–19.
29. Dinc, N.; Aydinsakir, K.; Isik, M.; Bastug, R.; Ari, N.; Sahin, A.; Buyuktas, D. Assessment of Different Irrigation Strategies on Yield and Quality Characteristics of Drip Irrigated Pomegranate Under Mediterranean Conditions. *Irrig. Sci.* **2018**, *36*, 87–96. [CrossRef]
30. Adiba, A.; Khachtib, Y.; Boutagayout, A.; Hamdani, A.; Kouighat, M.; Haddioui, A.; Hssaini, L.; Razouk, R. Bioactive Compounds and Quality Attributes of Pomegranate Fruit as Affected by Continuous Deficit Irrigation. *Vegetos* **2024**, 1–14. [CrossRef]
31. Raes, D.; Steduto, P.; Hsiao, T.C.; Fereres, E. *Reference Manual, Chapter 2–AquaCrop*; Food and Agriculture Organization of the United Nations: Rome, Italy, 2023. Available online: https://openknowledge.fao.org/handle/20.500.14283/br267e (accessed on 15 January 2025).
32. United Nations Economic Commission for Europe (UNECE). *Fresh Fruit and Vegetables–Standards. Pomegranates*; UNECE: Geneva, Switzerland, 2021. Available online: https://unece.org/trade/documents/2021/02/standards/pomegranates (accessed on 14 January 2025).
33. Wetzstein, H.Y.; Zhang, Z.; Ravid, N.; Wetzstein, M.E. Characterization of attributes related to fruit size in pomegranate. *Hortscience* **2011**, *60*, 908–912. [CrossRef]
34. Fernández García, F. *Manual de Climatología Aplicada: Clima, Medio Ambiente y Planificación*; Editorial Síntesis: Madrid, Spain, 1996; ISBN 8477382751.
35. Matas, A.J.; López-Casado, G.; Cuartero, J.; Heredia, A. Relative humidity and temperature modify the mechanical properties of isolated tomato fruit cuticles. *Am. J. Bot.* **2005**, *92*, 462–468. [CrossRef]
36. Singh, A.; Shukla, A.K.; Meghwal, P.R. Fruit Cracking in Pomegranate: Extent, Cause, and Management—A Review. *Int. J. Fruit Sci.* **2020**, *20* (Suppl. S3), S1234–S1253. [CrossRef]
37. Holland, D.; Hatib, K.; Bar-Yàakov, I. Pomegrante: Botany, horticulture and breeding. *Hortic. Rev.* **2009**, *35*, 127–191. [CrossRef]
38. Bartual, J.; Intrigliolo, D.S.; Pomares, F.; Bonet, L.; Parra, J.; García-González, J.; Pérez-Gago, M.B.; Palou, L. Manejo del riego y la fertilización del granado. *Rev. Frutic.* **2016**, *51*, 40–51.
39. Saei, H.; Sharifani, M.M.; Dehghani, A.; Seifi, E.; Akbarpour, V. Description of biomechanical forces and physiological parameters of fruit cracking in pomegranate. *Sci. Hortic.* **2014**, *178*, 224–230. [CrossRef]
40. Marschner, H. *Marschner's Mineral Nutrition of Higher Plants*; Academic Press: London, UK, 2011; ISBN 0123849063.
41. Huang, X.M.; Wang, H.C.; Zhong, W.L.; Yuan, W.Q.; Lu, J.M.; Li, J.G. Spraying calcium is not an effective way to increase structural calcium in litchi pericarp. *Sci. Hortic.* **2008**, *117*, 39–44. [CrossRef]
42. Mesejo, C.; Reig, C.; Martínez-Fuentes, A.; Gambetta, G.; Gravina, A.; Agustí, M. Tree water status influences fruit splitting in Citrus. *Sci. Hortic.* **2016**, *209*, 96–104. [CrossRef]
43. Volschenck, T. Water use and irrigation management of pomegranate trees. A review. *Agric. Water Manag.* **2020**, *241*, 106375. [CrossRef]
44. Yılmaz, C.; Özgüven, A.I. Physiology of pre-harvest fruit cracking in pomegranate: Mineral contents. *Acta Hortic.* **2019**, *1254*, 205–212. [CrossRef]
45. Cybulska, J.; Zdunek, A.; Konstankiewicz, K. Calcium effect on mechanical properties of model cell walls and apple tissue. *J. Food. Eng.* **2011**, *102*, 217–223. [CrossRef]
46. Melgarejo-Sánchez, P.; Martínez, J.J.; Hernández, F.; Legua, P.; Martínez, R.; Melgarejo, P. The pomegranate tree in the world: New cultivars and uses. *Acta Hortic.* **2015**, *1089*, 327–332. [CrossRef]
47. Quiroz-Castañeda, R.E.; Folch-Mallol, J.L. Proteínas que remodelan y degradan la pared celular vegetal: Perspectivas actuales. *Biotechnol. Appl.* **2002**, *4*, 205–215.
48. Mena, P.; García-Viguera, C.; Navarro-Rico, J.; Moreno, D.A.; Bartual, J.; Saura, D.; Martí, N. Phytochemical characterisation for industrial use of pomegranate (*Punica granatum* L.) cultivars grown in Spain. *J. Sci. Food. Agric.* **2011**, *91*, 1893–1906. [CrossRef] [PubMed]

49. Ministerio de Agricultura Pesca y Alimentación (MAPA). *Métodos Oficiales de Análisis;* MAPA: Madrid, Spain, 1986; ISBN 84-7479-532.X.
50. Galindo, A.; Calín-Sánchez, A.; Collado-González, J.; Ondoño, S.; Hernández, F.; Torrecillas, A.; Carbonell-Barrachina, A.A. Phytochemical and Quality Attributes of Pomegranate Fruits for Juice Consumption as Affected by Ripening Stage and Deficit Irrigation. *J. Sci. Food. Agric.* **2014**, *94*, 2259–2265. [CrossRef]
51. Pathare, P.B.; Opara, U.L.; Al-Said, F.A.J. Colour measurement and analysis in fresh and processed foods: A review. *Food Bioproc. Tech.* **2013**, *6*, 36–60. [CrossRef]
52. McGuire, R.G. Reporting of objective color measurements. *HortScience* **1992**, *27*, 1254–1255. [CrossRef]
53. Galindo, A.; Calín-Sánchez; Griñán, I.; Rodríguez, P.; Cruz, Z.N.; Girón, I.F.; Corell, M.; Martínez-Font, R.; Moriana, A.; Carbonell-Barrachina, A.A.; et al. Water stress at the end of the pomegranate fruit ripening stage produces earlier harvest and improves fruit quality. *Sci. Hortic.* **2017**, *226*, 68–74. [CrossRef]
54. Statistical Graphics Corporation. *Statgraphics Centurion XIX;* Statistical Graphics: Rockville, MD, USA, 2023.

plants

MDPI

Article

# Effect of Deficit Irrigation on Agronomic and Physiological Performance of Pomegranate (*Punica granatum* L.)

**Rossana Porras-Jorge [1], José Mariano Aguilar [2], Carlos Baixauli [2], Julián Bartual [3], Bernardo Pascual [4] and Nuria Pascual-Seva [4,*]**

1 Departamento de Producción Vegetal, Universitat Politècnica de València, Camí de Vera, s/n, 46022 Valencia, Spain; zrporras@doctor.upv.es

2 Centro Experiencias Cajamar Paiporta, C. Cementerio Nuevo s/n, 46200 Paiporta, Spain; josemarianoaguilar@fundacioncajamar.com (J.M.A.); carlosbaixauli@fundacioncajamar.com (C.B.)

3 Estación Experimental Agraria de Elche (SST), CV-855, Km. 1, 03290 Elche, Spain; bartual_jul@gva.es

4 Centro Valenciano de Estudios sobre el Riego, Universitat Politècnica de València, Camí de Vera, s/n, 46022 Valencia, Spain; bpascual@upv.es

* Correspondence: nupasse@upv.es

**Abstract:** Agriculture accounts for over 70% of global freshwater consumption, with increasing competition for water resources due to climate change and rising urban and industrial demands. This study analyzes the effect of deficit irrigation (DI) on the agronomic and physiological performance of pomegranate (*Punica granatum* L.) in a Mediterranean climate. Deficit irrigation strategies, including sustained deficit irrigation (SDI) and regulated deficit irrigation (RDI), were evaluated against a control with full irrigation. The research was conducted over two growing seasons (2022–2023) at the Cajamar Experimental Centre in Paiporta, Valencia, Spain. RDI strategies achieved approximately 10% water savings without compromising marketable yield or fruit weight, while SDI resulted in significant water savings (~50%) but with a notable reduction in marketable yield, particularly in hot and dry conditions. SDI also reduced tree growth in height and trunk diameter compared to RDI and control strategies. The study concludes that RDI is a viable irrigation strategy for pomegranate cultivation under water-limited conditions, whereas SDI should be reserved for situations of severe water scarcity.

**Keywords:** 'Mollar de Elche'; sustained deficit irrigation; regulated deficit irrigation; water productivity

## 1. Introduction

Agriculture consumes over 70% of the planet's freshwater [1]. In recent years, freshwater scarcity has become a significant problem, especially in arid areas, increasing the competition for water among agricultural, industrial, and urban consumers. Rapid population growth, different human activities, and the greater incidence of droughts caused by climate change are factors that increase this problem. As urban and industrial demands for water grow, agriculture will be pressured to make more efficient use of its shrinking allocation. The Mediterranean climate is characterized by mild winter temperatures and long, hot, and dry summers, with rainfall subject to high interannual and seasonal variability, making irrigation essential for crop production. The widespread water limitations for agriculture require developing strategies to make the most of irrigation water.

Deficit irrigation (DI) is an irrigation practice through which crops are irrigated with lower amounts of water than those that plants need for optimal growth and development [2–4]. It includes continuous or sustained deficit irrigation (SDI) and controlled or regulated deficit irrigation (RDI) [5]. SDI imposes water deficit uniformly throughout the crop cycle, thus avoiding significant water stress in any phenological phase that could significantly affect marketable yield. RDI is deficit irrigation based on phases or stages, which consists of imposing water deficits in specific phenological phases in which crops are less sensitive to water stress [5].

The pomegranate tree (*Punica granatum* L.) is one of the oldest known fruit trees; its origin extends from the Balkans to the Himalayas [6]. Pomegranate cultivation began in prehistoric times, and its domestication probably began in the Neolithic era in the Transcaucasian–Caspian region and northern Turkey [7]. The Phoenicians, Greeks, and Egyptians cultivated it. The Phoenicians spread its cultivation along the shores of the Mediterranean [8]. They cultivated it around Carthage, which is why the Romans named this fruit *Malum punicatum* (the Carthaginian apple). Later, due to the attractive appearance of the arils (seeds) packed in the carpels [9], Columella gave it the name *Malum granatum*, or grain apple, from which comes the names pomegranate in English, grenade in French, granada in Spanish, and melagrana in Italian. Due to the many seeds it contains, the pomegranate was considered as an emblem of fertility in ancient times, being mentioned in the Bible (Numbers 13:23; Deuteronomy 8:8; Exodus 28:33–35; 1 Kings 7:18–20; Songs 4:3, 4:13, 8:2) [10].

The pomegranate is a fruit-bearing deciduous tree that can adapt to different climatic conditions, as demonstrated by its wide distribution in the wild form throughout Eurasia to the Himalayas. It is perfectly acclimated to the Mediterranean edaphoclimatic conditions, being considered a crop tolerant of soil water deficit [7,11]. Nevertheless, to reach optimal yield and fruit quality and reduce the incidence of fruit physiological disorders, traditionally, pomegranate trees have been regularly irrigated throughout the season. Although it is considered a minor fruit tree (also known as a neglected and underutilized species) [12], it is acquiring great interest both in Spain and worldwide, being considered an emerging fruit crop [13] for its organoleptic characteristics and its beneficial effects on human health due to its high content of antioxidants [6,11,14]. Thus, it is an interesting option for diversifying fruit production in the Mediterranean basin.

Rodriguez et al. [15] stated that 'Mollar de Elche' ('ME') pomegranate trees face water stress by developing stress avoidance and tolerance mechanisms. From the beginning of the stress period, leaf conductance decreases to control water loss and avoid stress. Near the end of the stress period, maximum stress levels are developed, and an active osmotic adjustment is triggered, favoring the maintenance of leaf turgor (stress tolerance mechanism).

The interest that deficit irrigation has raised in pomegranate cultivation in semi-arid climates is demonstrated in the research works that have been carried out on this subject in the province of Alicante (southeast Spain) on 'ME'. In two companion papers [11,16], authors tested both SDI and RDI in different periods, concluding that irrigation restriction during the flowering and fruit set could be used in water scarcity situations; however, it led to small water savings (9–14%). The flowering period was identified as a non-critical period for RDI application [14], as it generated a 5.25% lower commercial yield, with a water saving of 24%. Pomegranate fruits are sensitive to water restrictions applied during the final phases of fruit growth and ripening [17,18]; the authors stated that fruit ripening is a critical period, and thus, irrigation is essential in this phenological period to achieve maximum yield. When SDI was tested [11,16], it produced more but lighter fruits. Galindo et al. [13] reviewed the state of the art regarding DI and the response of four emerging fruit crops, including pomegranate. In the review, the authors collected the studies by the leading research teams in DI of the pomegranate tree. Volschenk [19] reviewed in detail research conducted on pomegranate trees regarding irrigation methods, crop water requirements, water use efficiency and productivity, and water management strategies under conditions of limited water supply. She subsequently carried out another review [20] focused mainly on the effects of water deficits on physiology, vegetative and fruit growth, and yield and fruit quality of the pomegranate. She concluded that the results of one study could not simply be transferred to another area where conditions may be different given the variable results obtained with different systems and cultivars in different countries, which require conducting research under local conditions.

Given that the response of pomegranate trees to water restrictions differs depending on the different phenological phases in which they are applied and that the published results on water savings and the yields obtained with DI are not always coincident, particularly in the initial stages of the growing season, the objective of this research was to evaluate the agronomic behaviour of seven-year-old pomegranate trees under an SDI and two RDI strategies in the initial stages [one during the flowering–setting phase (RDI1) and another during the open flower–early fruit growth phase with color change (RDI2), which, to the authors' knowledge, has not been studied] and to establish a schedule of irrigation for growers in the Valencia Region. The yield (weight and number of fruits), the fruit setting, the water status of the soil and the trees, and the water productivity were analyzed in this paper.

## 2. Results and Discussion

Soil analyses show that the soil on the experimental site is fertile and has no health problems. Irrigation water also does not present any health problems in general or in the pomegranate crop. Regarding salinity, a 'moderately tolerant' crop such as pomegranate [21] should not have problems with irrigation water with an EC = 1.65 dS $m^{-1}$. Indeed, the value of the EC of the saturated-soil extract (Ke = 2.48 dS $m^{-1}$) is lower than the value of the salinity threshold (5 dS $m^{-1}$ for pomegranate [22]), which is the parameter A (salinity from which the decrease in yield begins) of the Maas and Hoffman equation $Yr = 100 - B\,(Ke - A)$, where Yr is the relative yield for any given soil salinity exceeding the threshold [23].

### 2.1. Growth Stages and Climatology

The growing season duration (considered from the first leaves sprouting to harvest) was 252 d in both 2022 and 2023; the duration, reference evapotranspiration (ETo), and crop coefficient (Kc) of each stage are presented in Table 1.

Figure 1 shows the most significant climatological data for both experimental seasons, maximum and minimum temperature, ETo, and precipitation. Although the difference between the total ETo values throughout the season was not very large, 5.4% (1110 mm in 2022 and 1169 mm in 2023), this difference increases up to 75% when considering the ETo recorded during April and May (147 mm in 2022 and 258 mm in 2023). It should be noted that while 2022 behaved like an average year in terms of rainfall (505 mm during the growing period), with significant rainfall recorded in spring, 2023 was very hot and dry (352 mm), with almost half of this precipitation recorded in September. The Pe recorded in the two seasons was different considered both throughout the season (364 mm in 2022 and 301 mm in 2023) and in seasonality. While the 2022 spring was very rainy (227 mm during March and April), the 2023 one was very dry (less than 2 mm). On the other hand, while in September 2022, only 24 mm was recorded, 107 mm was registered in 2023.

**Table 1.** Duration, reference evapotranspiration (ETo), crop coefficient (Kc) values and irrigation water applied (IWA) in the different irrigation stages during 2022 and 2023.

| Strategy | Period | 2022 | | | | 2023 | | | |
|---|---|---|---|---|---|---|---|---|---|
| | | Duration (d) | ETo (mm) | Kc | IWA (mm) | Duration (d) | ETo (mm) | Kc | IWA (mm) |
| Control | Entire cycle | 252 | 1110.2 | 0.30–0.68 | 316.0 | 252 | 1169.3 | 0.30–0.72 | 463.4 |
| SDI | Entire cycle | 252 | 1110.2 | 0.30–0.68 | 163.4 | 252 | 1169.3 | 0.30–0.72 | 236.6 |
| RDI1 | Full irrigation | 60 | 140.2 | 0.30–0.39 | 0.0 | 56 | 223.1 | 0.30–0.41 | 22.2 |
| | Restriction | 55 | 291.6 | 0.48–0.58 | 16.1 | 47 | 207.3 | 0.51–0.62 | 20.9 |
| | Full irrigation | 137 | 678.3 | 0.39–0.68 | 268.1 | 149 | 738.9 | 0.41–0.72 | 383.4 |
| RDI2 | Full irrigation | 80 | 212.5 | 0.30–0.39 | 3.3 | 66 | 264.8 | 0.30–0.41 | 32.9 |
| | Restriction | 42 | 258.5 | 0.48–0.58 | 19.9 | 44 | 203.1 | 0.51–0.62 | 21.8 |
| | Full irrigation | 130 | 639.2 | 0.39–0.68 | 252.9 | 142 | 701.4 | 0.41–0.72 | 375.7 |

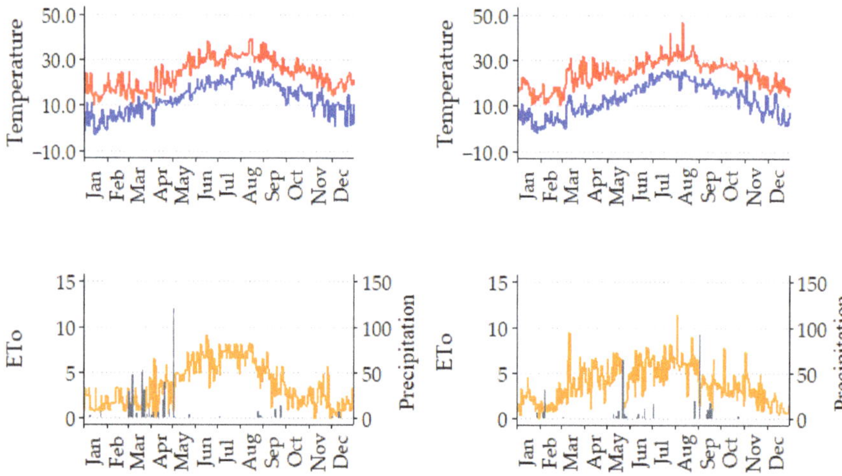


**Figure 1.** Seasonal variation of daily maximum (Tmax; red) and minimum (Tmin; blue) temperatures (°C), reference evapotranspiration (ETo; mm; yellow), and precipitation (mm; gray vertical bars) in 2022 (**left**) and 2023 (**right**).

### 2.2. Soil and Plant Water Status

The differences in the climatology recorded in the two years and the growth of the trees throughout the experiment, with the consequent increase in Kc (Table 1), led to an increase in the IWA over the years of study. In 2022, the IWA was 316, 284, 276, and 163 mm in Control, RDI1, RDI2, and SDI, respectively, while in 2023, it was 463, 426, 430, and 237 mm. SDI meant a water saving of approximately 50% compared to the Control, as expected, indicating that the objective of providing 50% of the Control was met. In RDI1, water savings of 11% in 2022 and 8% in 2023 were achieved compared to Control, while in RDI2, the savings were 13% and 7% in 2022 and 2023, respectively.

Figure 2 shows the seasonal variation of midday volumetric soil water content (VSWC), air vapor pressure deficit (VPD), stem water potential (Ψstem), and stomatal conductance ($g_s$). The values presented correspond to the midday, when stem water potential and stomatal conductance were measured before the corresponding irrigation event.

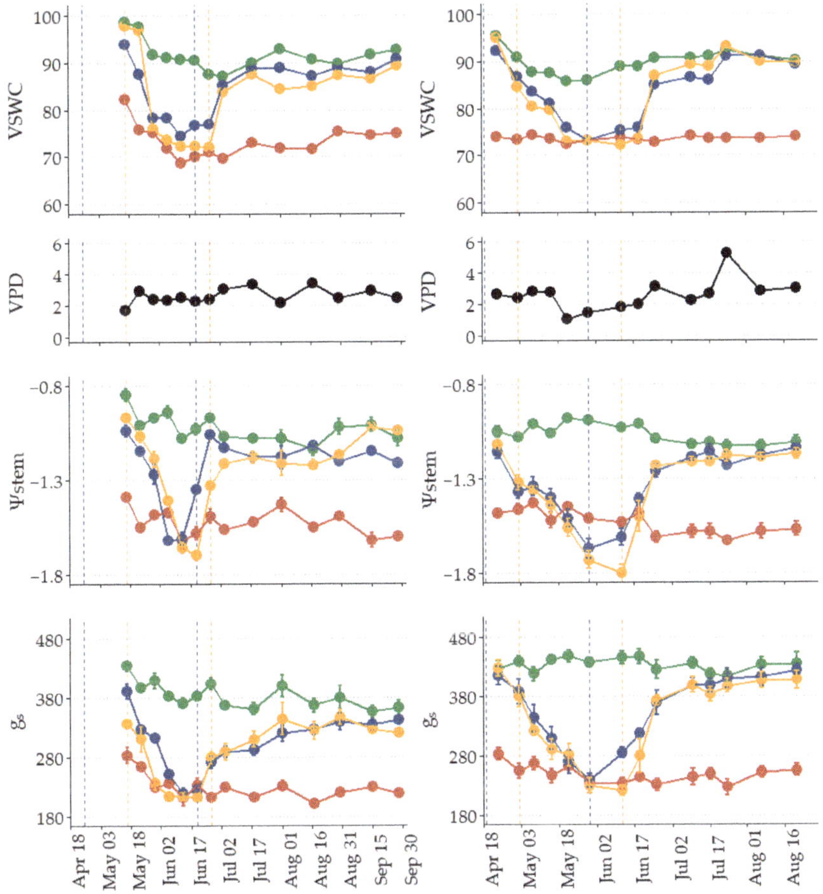


**Figure 2.** Seasonal variation of volumetric soil water content (VSWC), air vapor pressure deficit (VPD in black), stem water potential (Ψstem), and stomatal conductance ($g_s$) registered in the different irrigation treatments (Control in green; RDI1 in blue; RDI2 in yellow; SDI in red) in 2022 (**left**) and 2023 (**right**). All measures correspond to midday. Discontinuous vertical lines represent the start and end of the water restriction periods in RDI1 (blue) and RDI2 (yellow). Vertical bars represent the standard error.

In general terms, it can be stated that the objective of starting irrigation when the VSWC value reached 90% of the field capacity was met. The values of VSWC, Ψstem, and $g_s$ recorded in SDI (73.7%, −1.56 MPa, 226 mmol m$^{-2}$ s$^{-1}$, on average before the evaluated irrigation events, respectively) were markedly lower than those corresponding to the Control strategy (91.9%, −1.04 MPa, 387 mmol m$^{-2}$ s$^{-1}$, respectively). The values corresponding to RDI were close to those of SDI during the water restriction phases and to those of the Control during the rest of the crop cycle. The VPD values ranged between 1.30 and 5.19 kPa, obtaining in 2022 a maximum value of 3.4 kPa (on 17 August) and in 2023 of 5.19 kPa (on 24 July).

Ψstem values at midday decreased as the season progressed (Figure 2), in agreement with what has been reported in the literature for 'ME' pomegranate [11,14,24]. In RDI strategies, a rapid recovery of Ψstem values was recorded when full irrigation was resumed, as also observed by Mellisho et al. (2012) and Martínez-Nicolás et al. (2019). The midday Ψstem values recorded in similar research in 'ME' for Control plants ranged between

−0.6 and −1.02 MPa, while those in RDI plants in the restriction's periods ranged between −1.61 and −2.18 MPa [11,14,24,25]. Parvizi et al. [26] obtained lower values (−2.6 and −3.3 MPa for Control and SDI at 50% ETc, respectively) with 'Rabat' pomegranate in Iran. The minimum value of midday Ψstem recorded in SDI is like the minimum obtained by [11] in the Control treatment; therefore, it does not seem that there was significant water stress in any of the strategies tested.

Regarding $g_s$, there was also a general trend to decrease as the season progressed (Figure 2), oscillating from 435 to 356 mmol $m^{-2}$ $s^{-1}$ for the Control plants and from 276 to 202 mmol $m^{-2}$ $s^{-1}$ for the SDI plants. RDI plants behaved similarly to SDI during the respective water restriction periods and like Control plants during the initial and final phases. These values agree with those obtained by [14,24], but they are higher than those presented by [17] (from 220 to 50 mmol $m^{-2}$ $s^{-1}$ for Control and from 230 to 30 mmol $m^{-2}$ $s^{-1}$ for SDI) and are much higher than those presented by [26] (120 to 20 mmol $m^{-2}$ $s^{-1}$).

Significant relationships ($p \leq 0.05$) have been found between Ψstem and $g_s$ with VSWC (positive) and VPD (negative) for Control and SDI strategies. Figure 3 presents these relationships adjusted with the 2022–2023 data, whose correlation coefficients (≥0.69) are similar to those obtained by [11] with 'ME'. These relationships for RDI presented lower correlation coefficients since they considered both measurements carried out with water restrictions and full irrigation. A positive linear relationship is observed between $g_s$ and Ψstem, with a high correlation coefficient (0.88; Figure 4).

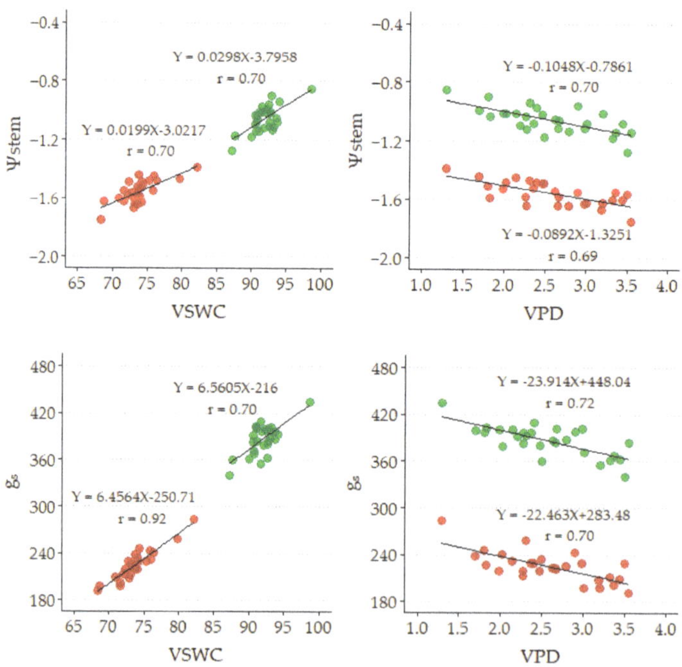


**Figure 3.** Linear correlation between midday stem water potential (Ψstem, MPa) and stomatal conductance ($g_s$, mmol $m^{-2}$ $s^{-1}$) with volumetric soil water content [VSWC, (%); **left**] and air vapor pressure deficit (VPD, kPa; **right**) for Control (green) and sustained deficit irrigation (SDI; red) strategies. Obtained from values corresponding to 2022 and 2023.

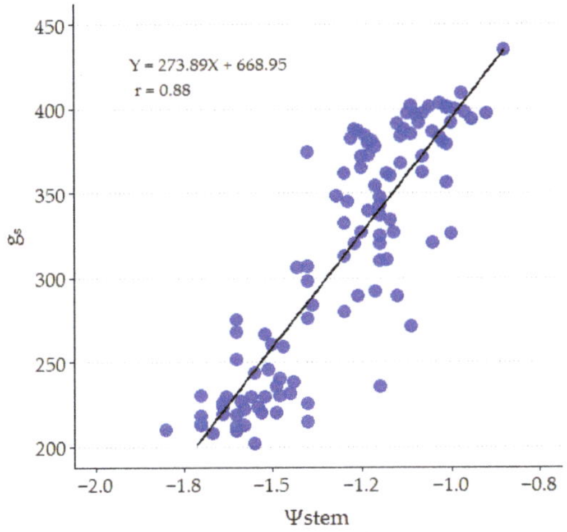


**Figure 4.** Linear correlation between midday stomatal conductance ($g_s$, mmol $m^{-2}$ $s^{-1}$) and stem water potential (Ψstem, MPa), obtained from values corresponding to 2022 and 2023.

With the strategies tested, decreases in Ψstem and $g_s$ are observed when VSWC decreases (Figure 3), so it seems that 'ME49' trees have effective control of the plant water status by reducing transpiration due to stomatal closure, as reported by [27], while maintaining Ψstem in narrow limits during the hydric restrictions in RDI strategies and throughout the crop cycle in the SDI strategy. That is, pomegranate trees 'ME49' presented a typical anisohydric behaviour as reported by [28]. Rodríguez et al. [15] reported that pomegranate plants confront water stress by developing stress avoidance and tolerance mechanisms, which are complementary and take place gradually. When water is restricted, leaf conductance decreases to control water loss via transpiration and avoid leaf turgor loss (stress avoidance mechanism). When the deficit irrigation is more severe and maximum stress levels are developed, active osmotic adjustment is triggered, contributing to the maintenance of leaf turgor (stress tolerance mechanism). According to these authors, only stress avoidance mechanisms would have occurred in the present study since only a mild deficit was obtained.

### 2.3. Tree Performance

Table 2 presents the size of the trees at the end of the 2023 season, expressed in terms of canopy height and diameter and trunk perimeter, as well as growth, expressed by the increase in the values of these parameters throughout the experiment (end 2021–end 2023). The final size of the trees was larger than that of 2021 (height: 286 cm; diameter: 258 cm; trunk perimeter: 33.6 cm; mean values corresponding to the Control trees). RDI1 led to a lower height and trunk perimeter than RDI2, which would reduce the canopy size in the long term, which could result in lower yields. SDI reduced the canopy's height and diameter and the trunk perimeter growth compared to Control and RDI2 (and the canopy diameter compared to RDI1). These results are similar to those obtained by [11] in 'ME' and to those obtained in 'Wonderful' by [29] in Chile and [30] in Italy. Since these values also depend on the values corresponding to the initial situation, the relative growth was analyzed, observing that the trees corresponding to the SDI strategy presented a lower relative growth than the trees corresponding to the other irrigation strategies (with statistical significance in the relative growth of the trunk perimeter). The high values of the percentage of the residual sum of squares (up to 88.4% for canopy diameter relative growth; Table 2) indicate an important variability between replicates, as reported by [31] in

'Wonderful' pomegranate in California. These authors reported that water deficit reduced the tree canopy size, although due to the notable variability between replications, the effect of irrigation strategies on canopy growth was not statistically significant.

**Table 2.** Effect of irrigation strategy on both size [canopy height (H) and diameter (D) and trunk perimeter (TP)] and the corresponding relative growth for the entire experiment.

| | Size (cm) | | | | | | Relative Growth (%) | | | | | |
|---|---|---|---|---|---|---|---|---|---|---|---|---|
| | H | | D | | TP | | H | | D | | TP | |
| Irrigation strategy (IS) | | | | | | | | | | | | |
| Control | 344 | ab | 319.4 | a | 37.13 | a | 20.65 | | 24.17 | | 10.58 | a |
| RDI1 | 324 | bc | 318.0 | a | 35.04 | b | 18.45 | | 23.01 | | 8.22 | a |
| RDI2 | 352 | a | 333.1 | a | 37.06 | a | 21.56 | | 27.57 | | 8.46 | a |
| SDI | 319 | c | 287.0 | b | 34.33 | b | 12.60 | | 15.68 | | 5.04 | b |
| Sources (df) | % Sum of squares | | | | | | | | | | | |
| IS (3) | 19.79 | * | 23.86 | ** | 25.90 | ** | 15.19 | ns | 11.57 | ns | 22.40 | * |
| Residual (44) | 80.21 | | 76.14 | | 74.10 | | 84.81 | | 88.43 | | 77.60 | |
| Standard deviation | 28.86 | | 31.51 | | 2.17 | | 8.61 | | 12.52 | | 3.85 | |

Mean values followed by different lowercase letters in each column indicate significant differences at $p \leq 0.05$ using the LSD test. df: degrees of freedom; ns: no significant difference; ** (*): significant differences at $p \leq 0.01$ ($p \leq 0.05$).

The total number of flowers was affected ($p \leq 0.01$; Table 3) by the growing season and the irrigation strategy but not by their interaction, so that in 2023, many more flowers were produced than in 2022 (more than tripled), highlighting the high number of flowers obtained in the Control strategy.

**Table 3.** Effect of growing season and irrigation strategy on the number of hermaphrodite flowers per tree and percentage of set, thinned, and harvested fruits in relation to the hermaphrodite flowers.

| | Hermaphrodite Flowers (No Tree$^{-1}$) | | Set Fruits (%) | | Thinned Fruits (%) | | Harvested Fruits (%) | |
|---|---|---|---|---|---|---|---|---|
| Growing season (GS) | | | | | | | | |
| 2022 | 219.15 | b | 78.12 | b | 8.03 | b | 70.10 | a |
| 2023 | 723.23 | a | 45.90 | a | 9.92 | a | 36.01 | b |
| Irrigation strategy (IS) | | | | | | | | |
| Control | 537.46 | a | 59.60 | b | 9.82 | | 49.78 | b |
| RDI1 | 436.63 | b | 65.71 | a | 8.75 | | 56.96 | a |
| RDI2 | 476.33 | ab | 63.49 | a | 8.88 | | 54.61 | a |
| SDI | 434.33 | b | 59.25 | b | 8.44 | | 50.87 | b |
| GS × IS | | | | | | | | |
| 2022—Control | 239.50 | | 79.39 | a | 10.00 | ab | 69.39 | bc |
| 2022—RDI1 | 223.50 | | 80.03 | a | 6.16 | c | 73.87 | a |
| 2022—RDI2 | 211.17 | | 79.54 | a | 8.51 | bc | 71.03 | ab |
| 2022—SDI | 202.42 | | 73.52 | b | 7.43 | bc | 66.09 | c |
| 2023—Control | 835.42 | | 39.81 | e | 9.64 | ab | 30.17 | f |
| 2023—RDI1 | 649.75 | | 51.38 | c | 11.33 | a | 40.05 | d |
| 2023—RDI2 | 741.50 | | 47.43 | cd | 9.24 | ab | 38.19 | de |
| 2023—SDI | 666.25 | | 44.99 | d | 9.45 | ab | 35.65 | e |
| Sources (df) | % Sum of squares | | | | | | | |
| GS (1) | 79.09 | ** | 85.73 | ** | 7.42 | ** | 90.33 | ** |
| IS (3) | 2.17 | * | 2.39 | ** | 2.21 | ns | 2.58 | ** |
| GS × IS (3) | 1.31 | ns | 1.69 | ** | 8.90 | * | 0.80 | * |
| Residual (88) | 17.44 | | 10.19 | | 81.47 | | 6.29 | |
| Standard deviation | 123.61 | | 5.78 | | 3.28 | | 4.7 | |

Mean values followed by different lowercase letters in each column indicate significant differences at $p \leq 0.05$ using the LSD test df: degrees of freedom; ns: no significant difference; ** (*): significant differences at $p \leq 0.01$ ($p \leq 0.05$).

Flowering is induced by endogenous and exogenous signals that initiate parallel pathways [32]. Exogenous signals include temperature, photoperiod, and stress, while endogenous signals include plant age, fruit load, and nutritional and hormonal status. In the present experiment, regarding external signals, the photoperiod was similar in the two seasons. Table 4 shows the chill accumulation calculated for each season using the three main and most used models [33]—hours under 7.2 °C [34], Utah [35], and Dynamic [36] —and heat accumulation calculated using the Anderson model [37].

**Table 4.** Chill accumulation calculated for each season using three models: hours under 7.2 °C, Utah, and Dynamic. Heat accumulation calculated using the Anderson model.

| Growing Season | Chill Accumulation | | | Heat Accumulation |
|---|---|---|---|---|
| | Hours < 7.2 °C (h) | Utah (Chill Units) | Dynamic (Chill Portions) | Anderson (GDH) |
| 2021–2022 | 395 | 488.5 | 53.92 | 23,052.3 |
| 2022–2023 | 331 | 333.5 | 49.58 | 24,047.4 |

The most used terminology for the dormancy of fruit trees in temperate zones [38] is that proposed by Lang et al. in 1987 [39], who defined dormancy as the temporary suspension of visible growth of any plant structure containing a meristem. They distinguished between para-, endo-, and eco-dormancy phases. Para-dormancy refers to growth suppression imposed on organs by other tree structures (e.g., apical dominance) due to inhibitory molecules' production and/or action. During the endo-dormancy phase, growth is impossible even under suitable temperature conditions, as buds during winter rest must be exposed to a low-temperature phase (chill needs). Once endo-dormancy has passed, they must be exposed to warm temperatures during the eco-dormancy phase (heat needs) to flower. In 2023, there was less chill accumulation and greater heat accumulation than in 2022. The differences in chill accumulation recorded in this experiment (with higher values in 2022) do not justify the differences in flowering intensity. However, the approximately 4% higher heat accumulation recorded in 2023 compared to 2022 is probably the cause of the earlier sprouting of both vegetative and reproductive buds. The recorded chill and heat accumulation values are in line with those obtained by [40] in 'ME' and confirm that 'ME49' is much less demanding in chill accumulation than other cultivars/accessions, such as those grown in Iran [41,42]. Regarding heat accumulation, the GDH registered was higher than reported in Iran [41] and Himachal Pradesh (India) [43].

The last exogenous signal reported by [32] is the fact that plants are under some kind of stress; no stress was recorded during bud dormancy nor after it, with the only difference being in the irrigation treatment; in this sense, the Control strategy led to a greater number of hermaphrodite flowers than the RDI1 and SDI strategies. Although despite water restriction, the trees did not suffer water stress (as mentioned above), the results obtained are consistent with those obtained by [42] in the sense that with the same chill accumulation, the greatest flowering was obtained with the Control strategy, that is, without any water restriction.

Regarding endogenous signals, during the experiment, the trees grew in height (from 13 to 21%) and diameter (from 16 to 24%), which would have increased their flowering capacity. Our result agrees with that reported by [44], who stated the possibility that SDI and RDI decreased growth sprouting and, consequently, the number of flowers formed on these shoots. On the other hand, the yield obtained by the Control trees in 2021 (27 kg tree$^{-1}$) was much lower than that obtained in 2022 (58 kg tree$^{-1}$), which in turn was much lower than in 2023 (83 kg tree$^{-1}$). It seems that the fruit load of the trees in each season should have been independent of the flowering of the next, not showing biennial (or alternate) bearing. Regarding the nutritional status of the trees, all of them received identical fertilization, and the nutrient content recorded in the foliar analyses had value ranges at levels considered normal [45,46], with no significant differences detected between the values corresponding to the different irrigation strategies. Table 5 shows the average foliar content (for the two seasons) of the nutrients analyzed.

**Table 5.** Leaf macronutrient (nitrogen, phosphorus, potassium, calcium, and magnesium; %) and micronutrient (iron, copper, manganese, and zinc; mg $L^{-1}$) concentrations corresponding to adult leaves (15 July) from 'Mollar de Elche' clone 49 pomegranate trees. Average values of four trees per block of repetition and the two growing seasons.

| | Macronutrients | | | | | Micronutrients | | | |
|---|---|---|---|---|---|---|---|---|---|
| | N | P | K | Ca | Mg | Fe | Cu | Mn | Zn |
| Control | 2.04 | 0.16 | 0.86 | 1.71 | 0.38 | 124.33 | 8.50 | 34.00 | 19.33 |
| RDI1 | 1.88 | 0.15 | 0.88 | 1.73 | 0.35 | 131.33 | 6.50 | 33.33 | 17.67 |
| RDI2 | 1.92 | 0.15 | 0.91 | 1.62 | 0.38 | 144.83 | 7.50 | 33.33 | 18.50 |
| SDI | 1.82 | 0.14 | 0.83 | 1.68 | 0.41 | 118.50 | 7.00 | 32.67 | 17.17 |

About fruit setting (Table 3), it is observed that in 2023 many more fruits set than in 2022 and that RDI1 and RDI2 led to a higher percentage of fruit set ($p \leq 0.01$) than Control and SDI. The interaction analysis ($p \leq 0.01$) shows that the lowest value in 2022 corresponded to SDI, while in 2023, it corresponded to Control. According to [47], with the production of a high number of flowers, there is high competition between flowers for synthesized carbohydrates, which would justify the lowest percentage of fruit set in 2023 in the Control strategy. As expected, due to the high number of hermaphrodite flowers produced, and despite the high percentage of fallen reproductive organs (100—set fruits), the percentage of thinned fruits about the total flowers in 2023 was higher ($p \leq 0.01$) than in 2022, highlighting RDI1, which recorded the lowest value in 2022 ($p \leq 0.01$) and the highest in 2023.

The percentage of fruits harvested compared to the total number of hermaphrodite flowers was greater in 2022 than in 2023 ($p \leq 0.01$; Table 3), and the two RDI strategies led to higher percentages ($p \leq 0.01$) than the Control and SDI; from the analysis of the interaction ($p \leq 0.05$), it is observed that the lowest value in 2022 corresponded to SDI, while in 2023 it corresponded to the Control. In young-bearing pomegranate trees, there appears to be a balance between the water available to the plant and the competition for assimilates between yield and vegetative growth [20] as well as between different fruits [47]. In this sense, the Control strategy led to the highest tree growth and the highest number of flowers, and therefore to the greatest competition, both between shoots and fruits and between fruits, and consequently to the highest percentage of fallen flowers and fruitlets in 2023. The balance between the water available and the competition for assimilation between flowers and fruitlets led to the lowest percentage of fruits harvested compared to the total number of hermaphrodite flowers in SDI in 2022. This result is consistent with [14], who stated that water stress decreased shoot growth but did not affect the number of viable flowers.

### 2.4. Total and Marketable Yield, Water Productivity, and Yield Response Factor

Both irrigation strategies and growing season affected total and marketable yields (Table 6), resulting in significant interactions between both factors for yield in terms of weight ($p \leq 0.05$ for total and $p \leq 0.01$ for marketable yield). The total yield (kg $tree^{-1}$) obtained in 2023 was higher than that obtained in 2022 (on average 43%), and in both seasons, SDI led to the lowest value. As for the number of fruits, the production in 2023 was, on average, 66% greater than in 2022, with the lowest value corresponding to SDI.

**Table 6.** Effect of growing season and irrigation strategy on the total yield, marketable yield (including unit weight: UW), and water productivity (WP, in terms of kg $m^{-3}$).

| | Total Yield | | | | Marketable Yield | | | | | | WP ($kg\ m^{-3}$) | |
|---|---|---|---|---|---|---|---|---|---|---|---|---|
| | (kg $Tree^{-1}$) | | (No $Tree^{-1}$) | | (kg $Tree^{-1}$) | | (No $Tree^{-1}$) | | UW (g) | | | |
| Growing season (GS) | | | | | | | | | | | | |
| 2022 | 57.86 | b | 153.58 | b | 36.75 | b | 81.88 | b | 448.89 | a | 9.29 | a |
| 2023 | 82.74 | a | 255.46 | a | 51.25 | a | 137.00 | a | 372.59 | b | 8.26 | b |
| Irrigation strategy (IS) | | | | | | | | | | | | |
| Control | 73.01 | a | 206.96 | ab | 46.30 | b | 115.75 | a | 409.36 | | 7.50 | b |
| RDI1 | 75.75 | a | 211.42 | a | 51.23 | a | 125.75 | a | 417.32 | | 9.13 | a |
| RDI2 | 76.48 | a | 215.63 | a | 49.98 | ab | 124.58 | a | 416.67 | | 9.06 | a |
| RDS | 55.97 | b | 184.08 | b | 28.49 | c | 71.67 | b | 399.61 | | 9.42 | a |
| GS × IS | | | | | | | | | | | | |
| 2022—Control | 60.43 | b | 165.83 | | 36.09 | bc | 83.33 | b | 431.51 | b | 7.25 | e |
| 2022—RDI1 | 62.00 | b | 164.92 | | 40.28 | b | 89.50 | b | 448.90 | ab | 9.00 | bc |
| 2022—RDI2 | 60.16 | b | 150.42 | | 41.09 | b | 89.25 | b | 465.56 | a | 9.45 | b |
| 2022—RDS | 48.87 | c | 133.17 | | 29.52 | cd | 65.42 | c | 449.58 | ab | 11.48 | a |
| 2023—Control | 85.58 | a | 248.08 | | 56.52 | a | 148.17 | a | 387.21 | c | 7.74 | cde |
| 2023—RDI1 | 89.51 | a | 257.92 | | 62.18 | a | 162.00 | a | 385.73 | c | 9.26 | b |
| 2023—RDI2 | 92.81 | a | 280.83 | | 58.87 | a | 159.92 | a | 367.79 | cd | 8.68 | bcd |
| 2023—RDS | 63.08 | b | 235.00 | | 27.45 | d | 77.92 | bc | 349.64 | d | 7.37 | de |
| Source (df)) | % Sum of squares | | | | | | | | | | | |
| GS (1) | 44.03 | ** | 59.53 | ** | 23.76 | ** | 41.07 | ** | 58.29 | ** | 6.33 | * |
| IS (3) | 19.96 | ** | 3.41 | * | 37.70 | ** | 26.52 | ** | 2.04 | ns | 13.46 | ** |
| GS × IS (3) | 3.22 | * | 1.83 | ns | 10.59 | ** | 8.29 | ** | 5.55 | ** | 20.14 | ** |
| Residual (88) | 32.78 | | 35.23 | | 27.95 | | 24.11 | | 34.12 | | 60.07 | |
| Standard deviation | 11.21 | | 40.93 | | 8.22 | | 22.06 | | 30.48 | | 1.66 | |

Mean values followed by different lowercase letters in each column indicate significant differences at $p \leq 0.05$ using the LSD test. df: degrees of freedom; ns: no significant difference; ** (*): significant differences at $p \leq 0.01$ ($p \leq 0.05$).

In terms of marketable yield, the average value obtained in 2023 exceeded that obtained in 2022 (Table 6), both in weight (by 40%) and in number (by 68%). The high average marketable yield obtained in 2023 (51.3 kg $tree^{-1}$) exceeds that obtained by [14] with 'ME' in 2016 (40.6 kg $tree^{-1}$), while that obtained in 2022 (36.8 kg $tree^{-1}$) slightly exceeds that obtained by these authors in 2015 (35.7 kg $tree^{-1}$). The number of marketable fruits obtained per tree in 2023 (on average 137) was much higher than in 2022 (on average 82), being higher and lower, respectively, than those obtained by [14] (90–99 in the different growing seasons). RDI1 led to the highest marketable yield (not significantly differing from the Control or RDI2), coinciding with the results obtained by Intrigliolo et al. (2013), as well as those obtained by [14], which found no negative effect of the deficit irrigation. The SDI strategy led to the lowest marketable yield ($p \leq 0.01$) both in weight and number. It should be noted that due to the severe incidence of cracking recorded in SDI in 2023, the number of marketable fruits obtained in SDI was only 8% higher than in 2022, while in the rest of the irrigation strategies, the increase was, on average, 80%. Regarding the unit weight, the analysis of the growing season–irrigation strategy interaction ($p \leq 0.01$) shows that the fruits obtained in 2022 were larger (higher unit weight) than those of 2023, which would be related to the higher number of marketable fruits obtained that year. On the other hand, in 2023, SDI led to the lowest value, probably due to the high incidence of cracking, which would affect the largest fruits, leaving the smaller fruits, and therefore those with lower unit weight as commercial fruits.

Regarding the WP, from the analysis of the season–irrigation strategy interaction ($p \leq 0.01$), it is remarkable that SDI led to the highest absolute value (11.5 kg $m^{-3}$) in 2022 and the lowest (7.4 kg $m^{-3}$) in 2023. These different values are mainly due to the higher

IWA in 2023 compared to in 2022 (163 mm in 2022 and 237 mm in 2023) and to a lesser extent to the lower marketable yield obtained in 2023 (18.75 t $ha^{-1}$ in 2022 and 17.43 t $ha^{-1}$ in 2023), a consequence in turn of the different flowering and fruit setting and, above all, the high incidence of cracking in 2023. In the two growing seasons, a lower value was obtained with the control strategy (7.3 and 7.7 kg $m^{-3}$ in 2022 and 2023, respectively) than with the RDI strategies.

Regarding Ky, values greater than 1 indicate that the crop's response is very sensitive to water deficit. In contrast, a Ky value lower than 1 means the crop is more tolerant to water deficit. When Ky is equal to 1, the reduction in yield is directly proportional to the reduction in water use. RDI strategies led to higher marketable yields than Control, so Ky would be negative; thus, it was only determined in SDI. In 2022, with a climate that can be considered normal, Ky was 0.38, meaning that 'ME 49' behaved as tolerant, but instead, in 2023, with a very hot summer, in which a high incidence of cracking was added to the lower performance usually obtained with SDI, Ky was 1.05, meaning that it behaved as sensitive, with a reduction in yield almost directly proportional to the reduction in water use.

### *2.5. Considerations for the Future*

Despite the interesting works published on the pomegranate tree, there are still aspects that should be studied in further depth in the future, such as the use of multispectral cameras installed or not in drones, which would allow researchers to accurately analyze the growth of the trees as a response to the water deficit and also estimate parameters such as leafiness and/or leaf area index.

Given that the percentage of hermaphrodite flowers is proportional to the plant's bearing ability and determines the degree of fruit set [48] and that not all bibliographic references agree on the influence of RDI on the hermaphrodite/male ratio [44], the aim is to continue this study by analyzing the influence of deficit irrigation (both RDI and SDI) on the number and proportion of hermaphrodite flowers and their setting, as well as its influence on yield. Future studies will analyze the exogenous (cold and water stress) and endogenous (particularly hormonal) pathways that control flowering and their effect on changes in gene expression and protein and metabolite levels in pomegranate leaves and shoots. Regulated deficit irrigation with more severe water restrictions would allow us to study, on the one hand, the possible active osmotic adjustment, which would favor the maintenance of leaf turgor (stress tolerance mechanism) and, on the other hand, the corresponding productive response and the water productivity. Sustained deficit irrigation with fewer restrictions is probably also an appropriate option in situations of significant water scarcity that should be addressed.

## 3. Materials and Methods

### *3.1. Experimental Site Description*

The field studies were conducted at the Cajamar Experimental Centre in Paiporta, Valencia, Spain (39.4175 N, 0.4184 W), over two consecutive growing seasons (2022 and 2023). The soil at the site is deep with a silt loam texture; according to the USDA Soil Taxonomy, it is classified as Petrocalcic Calcixerepts [49]. At the beginning of the experiment, the soil was very slightly alkaline (pH = 8.17), slightly saline [EC (1:2.5) = 3.03 dS $m^{-1}$], and highly fertile [on average: organic matter = 1.4%; phosphorous = 171.4 mg $kg^{-1}$ (Olsen); potassium = 581 mg $kg^{-1}$ (ammonium acetate extract)]. The soil was uniform in the root zone depth (50 cm), and volumetric soil water contents (VSWC) at the field capacity and permanent wilting point were (on average) 28.7% and 16.1%, respectively, for 2022 and 28.8% and 16.1% for 2023. Irrigation water was pumped from a well, which averaged EC 1.65 dS $m^{-1}$, pH 7.58, and 23.7 mg $kg^{-1}$ N-$NO_3$- content. According to Papadakis's agroclimatic classification [50], the climate is subtropical Mediterranean (Su, Me) with hot and dry summers. The annual average rainfall is approximately 450 mm, irregularly distributed throughout the year, mostly in autumn and the beginning of spring.

### 3.2. Plant Material and Agronomic Details

The experimental orchard was established in 2015 from vegetatively propagated plants of ME49, a selected clone of 'Mollar de Elche'. This clone is one of the most appreciated and used in the area due to its adequate adaptation to edaphoclimatic conditions, high productivity, and great acceptance. Pomegranate tree spacing followed a 3.5 m × 4.5 m pattern, and trees were trained to the vase-shaped system.

The agricultural practices, including covering the soil with a perennial grass (Arbovert Perenne; 100 kg ha$^{-1}$; Intersemillas, València, Spain), fertilization, pruning, and fruit thinning, were those commonly used by farmers in the area. Specifically, on 17 and 30 June 2022 and 6 and 26 June 2023, a light manual thinning was carried out to eliminate multiple fruits, leaving one–two fruits per node, and to promote an increase in fruit size at harvest by reducing competition between fruits by eliminating late-flowering fruits. The incorporation of nutrients (150–75–180 kg ha$^{-1}$ year$^{-1}$ N–$P_2O_5$–$K_2O$) was carried out through fertigation, integrating the criteria indicated by [14,51].

### 3.3. Irrigation Strategies

In all four strategies, irrigation started with the sprouting of the first leaves (BBCH 10; 25 February 2022 and 22 February 2023) and finished with the second harvest (BBCH 81; 3 November 2022 and 31 October 2023).

The irrigation in the Control strategy was performed according to the criteria commonly used by the growers in the area [15], consisting of replacing the crop water needs [11,14] and at the same time allowing soil moisture to reach a depth of 50 cm, without any water loss being recorded at depth. Trees subjected to SDI received 50% of the water applied to Control trees in each irrigation event. RDI trees were irrigated with 33.3% of the crop water needs during the corresponding restriction period (Table 7) and with 100% outside this period. Differential irrigation started at the beginning of the 2021 season to not interfere with the establishment of trees. Table 8 presents the percentage of irrigation water requirements applied during the irrigation season in each irrigation strategy.

**Table 7.** Irrigation restriction periods in the regulated deficit irrigation strategies for 2022 and 2023.

| Strategy | Beginning | End | 2022 | | 2023 | |
|---|---|---|---|---|---|---|
| | | | From | To | From | To |
| RDI1 | The end of stage 1 (BBCH 51) | Young fruit of second flowering (BBCH 71) | 26 April | 19 June | 19 April | 4 June |
| RDI2 | First open flower (BBCH 61) | Fruit growth of second flowering (BBCH 73) | 16 May | 26 June | 29 April | 11 June |

**Table 8.** Irrigation water applied corresponding to irrigation strategies (IS) assayed, expressed as a percentage of the irrigation water requirements applied.

<table>
<tr><th>IS</th><th>February</th><th>March</th><th>April</th><th>May</th><th>June</th><th>July</th><th>August</th><th>September</th><th>October</th></tr>
<tr><td>Control</td><td colspan="9">100%</td></tr>
<tr><td>RDI1</td><td colspan="2">100%</td><td colspan="2">33.3%</td><td colspan="5">100%</td></tr>
<tr><td>RDI2</td><td colspan="3">100%</td><td colspan="2">33.3%</td><td colspan="4">100%</td></tr>
<tr><td>SDI</td><td colspan="9">50%</td></tr>
</table>

The irrigation water requirement (IWR) for each irrigation event was determined with the following equation: IWR = (ETc-Pe)/Ef, where ETc (mm) is the crop evapotranspiration, Pe is the effective precipitation (mm), determined from rainfall data using the US Bureau of Reclamation method [52], and Ef is the irrigation efficiency. This Ef was 0.97, considering the distribution uniformity (0.97; in situ determined) and negligible leaching requirement [EC of the irrigation water 1.65 dS $m^{-1}$; EC of saturated-soil extract = 2.48 dS $m^{-1}$]. The irrigation frequency ranged between daily when the water requirements were maximum (summer) and every three days when the water requirements were minimum (spring). Irrigation dose was determined retrospectively to replace the water requirements of the previous period.

Following the criteria described by [53], ETc was determined by equation ETc = ETo × Kc, using the reference evapotranspiration (ETo) and the single crop coefficient (Kc), with values ranging from 0 to 0.72, which were proposed for local conditions by the *Instituto Valenciano de Investigaciones Agrarias* (IVIA) [54], adapting the duration of each stage to the growing cycle. Daily ETo (mm $day^{-1}$) was determined using the FAO-56 Penman–Monteith equation (Equation (1); [53]):

$$ET_0 = \frac{0.408\Delta(R_n - G) + \gamma\frac{900}{T+273}u_2(e_s - e_a)}{\Delta + \gamma(1 + 0.34u_2)} \quad (1)$$

where Δ is the slope of the saturation vapor pressure–temperature relationship at mean air temperature (kPa °$C^{-1}$), $R_n$ is the net radiation at the crop surface (MJ $m^{-2}$ $d^{-1}$), G is the soil heat flux (MJ $m^{-2}$ $d^{-1}$), T is the average air temperature (°C), $u_2$ is the wind speed at 2 m height (m $s^{-1}$), (es-ea) is the vapor pressure deficit (kPa), and γ is the psychrometric constant (kPa °$C^{-1}$).

Plants were irrigated by a drip irrigation system with two lateral lines (Netafim Standard Irrigation PE Pipes 16/4, Netafim, Valencia, Spain) with six self-compensating emitters per tree (Netafim PCJ, 4.0 L $h^{-1}$) spaced 0.50 m apart. In each strategy, the irrigation water applied was recorded by a water flow meter (MJ-LFC, Dn 22, NWM, Ningbo Water Meter Co., Ltd., Ningbo, China).

### 3.4. Volumetric Soil Water Content

Volumetric soil water content (VSWC; $m^3$ $m^{-3}$) was continuously monitored by $ECH_2O$ EC-5 capacitance sensors connected to an Em50 data logger using the $ECH_2O$ Utility software (Decagon Devices, Inc., Pullman, WA, USA). Factory calibration provides ± 3% accuracy for mineral soils and was, therefore, used directly. Two sensors per replicate were placed below an emitter at a 0.25 and 0.5 m depth, respectively. In previous experiments performed at the same experimental plot, it was stated that the maximum root density of pomegranate trees occurred at a depth of 0.25 m. The sensor located at 0.50 m depth made it possible to verify that the humidity reached this depth and that deep percolation was controlled. VSWC was measured and stored every 15 min, and its variation was used to determine the in situ field capacity. The irrigation event for each IS began when the VSWC in the Control dropped to 90% of the field capacity, applying the corresponding irrigation dose to each IS. This criterion was already satisfactorily used in preliminary studies. VSWC data were managed and presented as a percentage of field capacity, reducing the importance of sensor calibration.

### 3.5. Plant Water Status

Midday (12 h solar time) stem water potential (Ψstem) was assessed with a pressure chamber (Soil Moisture Equip. Corp. mod. 5100A, Santa Barbara, CA, USA), following the procedure described by Intrigliolo et al. (2013). Mature (fully expanded) leaves from the north-facing side and middle third of the tree were covered with silver foil and enclosed in plastic bags two hours before measurements. Under the same conditions, the midday leaf conductance ($g_s$) was measured with a porometer (Delta T AP4, Delta-T Devices, Cambridge, UK) on the abaxial surface of the leaves. Both determinations were performed simultaneously in twelve leaves per treatment (two leaves per tree, two trees per block, and three replication blocks). In each growing season, eight measurements were taken before the corresponding irrigation event.

*3.6. Tree Performance Determinations*

Tree size was estimated by measuring the canopy height and diameter at three locations in four trees per replicate at the end of each growing season (after leaf fall). Similarly, the trunk perimeter was measured at the beginning of the experiment and the end of each growing season at 0.2 m above the ground. Relative growth in canopy height and in diameter, as well as trunk perimeter, was calculated as the ratio between the increment of each parameter during the experiment and their initial values.

The fallen reproductive organs (flowers and fruitlets) per tree were collected and recorded weekly and considered together with the number of set fruits to determine the number of hermaphrodite flowers. The number of set fruits was, in turn, defined as the number of fruits thinned and harvested.

Sampling to analyze the leaf nutrient concentrations was carried out on July 15 of each year, following the recommendations of experts [45,46]. Eighty fully expanded leaves per treatment were collected 20 leaves per tree (five leaves from each orientation: north, south, east, and west) at 1.5 m above ground level) from branches without developing fruits [55] and from 4 trees per replication block. The leaf analyses were carried out in the 'Las Palmerillas' laboratory in Cajamar, following the official analysis methods [56].

*3.7. Chill and Heat Accumulation*

Chill accumulation was calculated for each season from 1 November until 28 February, as reported by [40], using the three most used models [33] (hours under 7.2 °C [34], Utah [35], and Dynamic [36]). Heat accumulation was calculated as growing degree hours (GDH) for each season, from 1 January to 8 April [40], considering a base temperature of 4 °C, an optimum temperature of 25 °C, and a critical temperature of 36 °C (the temperature above which no appreciable growth occurs [37]).

*3.8. Yield, Water Productivity, and Yield Response Factor*

When fruit commercial maturity was reached (unit weight > 220 g [57] and total soluble solids > 14° Brix [14]), pomegranate fruits were manually harvested in two commercial pickings carried out on 13 October and 3 November 2022 and 16 and 31 October 2023. The total and marketable yield of fruits destined for the fresh market was determined, both in weight and in number, visually evaluating rejection due to alterations in the fruits (cracking, sunburn, small size, and other alterations, which mainly included scratches and deformed fruits).

Water productivity was calculated in physical terms as the relationship between marketable yield and the IWA, according to [5]. The yield response to water deficits during each growing season was determined according to [58] $(1\text{-}Ya/Ym) = Ky\ (1 - ETa/ETm)$, where Ky is the yield response factor; Ya and Ym are the actual and maximum marketable yield (kg $m^{-2}$), respectively; and ETa and ETm are the actual and maximum ETc (mm), respectively, being in turn calculated from soil water balance, considering negligible both the drainage and the variation in the volumetric soil water content [4].

### 3.9. Experimental Design and Statistical Analysis

The experiment was laid out in a randomized complete block design with three replicates. Each experimental plot was composed of three rows of trees with six trees each, of which the four central trees of the central row were analyzed; these trees were very similar in appearance (height, ground shaded area, and trunk cross-sectional), as [18] reported.

The results for the different parameters were evaluated by analysis of variance (ANOVA) using Statgraphics Centurion 19 [59]. Percentage data were arcsin transformed before analysis. Least significant difference (LSD) at a 0.05 probability level was used as the mean separation test. Initially, the data from each growing season were analyzed separately, with the result that in no case was the repetition block significant, so they were analyzed together without considering the repetition block to avoid third-order interactions.

## 4. Conclusions

With the strategies tested, a decrease in stem water potential (Ψstem) and leaf conductance ($g_s$) is observed when soil moisture decreases, so 'ME49' pomegranate trees have effective control of the plant water status by reducing transpiration due to stomatal closure, maintaining the stem water potential in narrow limits during the hydric restrictions. With the regulated deficit irrigation strategies tested, a small water saving is achieved, of the order of 10% of the irrigation water requirement, without reducing the marketable yield or the unit weight of the fruits. Specifically, when the reduction is applied in the flowering–fruit set stage, a slight increase in marketable yield is obtained; this irrigation strategy leads to a slight reduction in the growth of both the height and the perimeter of the tree trunk. With the tested sustained deficit irrigation, savings equivalent to 50% of the irrigation water requirement are obtained but at the cost of a slight reduction in the growth of height and perimeter of the trunk and a significant reduction in commercial yield. The magnitude of this reduction depends on the climate of the growing season, so with very hot and dry summers, the reduction in commercial yield can reach 50%. On average, regulated and sustained deficit irrigation increases water productivity. Regulated deficit irrigation is an advisable irrigation strategy for pomegranate cultivation regardless of whether or not there is water restriction, while the tested sustained deficit irrigation would only be advisable in situations of significant water scarcity.

**Author Contributions:** Conceptualization, C.B., J.B., B.P. and N.P.-S.; methodology, R.P.-J., J.M.A., C.B., J.B., B.P. and N.P.-S.; formal analysis, R.P.-J.; investigation, R.P.-J. and J.M.A.; resources, J.M.A., C.B., B.P. and N.P.-S.; data curation, R.P.-J. and J.M.A.; writing—original draft preparation, R.P.-J., B.P. and N.P.-S.; writing—review and editing, R.P.-J., J.M.A., C.B., J.B., B.P. and N.P.-S.; visualization, R.P.-J., J.M.A., C.B., J.B., B.P. and N.P.-S.; supervision, C.B., J.B., B.P. and N.P.-S. All authors have read and agreed to the published version of the manuscript.

**Funding:** This research received no external funding.

**Data Availability Statement:** Dataset available on request from the authors.

**Conflicts of Interest:** The authors declare no conflicts of interest.

## References

1. Food and Agriculture Organization of the United Nations (FAO). *The State of Food and Agriculture. Overcoming Water Challenges in Agriculture*; Food and Agriculture Organization of the United Nations (FAO): Rome, Italy, 2020. [CrossRef]
2. Costa, J.M.; Ortuño, M.F.; Chaves, M.M. Deficit irrigation as a strategy to save water: Physiology and potential application to horticulture. *J. Integr. Plant Biol.* **2007**, *49*, 1421–1434. [CrossRef]
3. Chai, Q.; Gan, Y.; Zhao, C.; Xu, H.L.; Waskom, R.M.; Niu, Y.; Siddique, K.H.M. Regulated deficit irrigation for crop production under drought stress. A review. *Agron. Sustain. Dev.* **2016**, *36*, 3. [CrossRef]

4. Abdelkhalik, A.; Pascual, B.; Nájera, I.; Domene, M.A.; Baixauli, C.; Pascual-Seva, N. Effects of deficit irrigation on the yield and irrigation water use efficiency of drip-irrigated sweet pepper (*Capsicum annuum* L.) under Mediterranean conditions. *Irrig. Sci.* **2020**, *38*, 89–104. [CrossRef]
5. Fereres, E.; Soriano, M.A. Deficit irrigation for reducing agricultural water use. *J. Exp. Bot.* **2007**, *58*, 147–159. [CrossRef] [PubMed]
6. Melgarejo, P.; Núñez-Gómez, D.; Legua, P.; Martínez-Nicolás, J.J.; Almansa, M.S. Pomegranate (*Punica granatum* L.) a dry pericarp fruit with fleshy seeds. *Trends Food Sci. Technol.* **2020**, *102*, 232–236. [CrossRef]
7. Holland, D.; Hatib, K.; Bar-Yàakov, I. Pomegrante: Botany, horticulture and breeding. *Hortic. Rev.* **2009**, *35*, 127–191. [CrossRef]
8. Zurriaga, F. El Magraner; la tradició perduda. *Mètode* **2008**, *59*, 27–33.
9. Melgarejo, P.; Núñez-Gómez, D.; Melgarejo-Sánchez, P.; Martínez-Nicolás, J.J.; Legua, P. Pomegranate definitions and research methodologies: Importance of using correct and unified terminology. *Acta Hortic.* **2022**, *1349*, 83–84. [CrossRef]
10. Worcester, J. *Correspondence of the Bible*; Swedenborg Foundation Press: West Chester, PA, USA, 1888.
11. Intrigliolo, D.S.; Bonet, L.; Nortes, P.A.; Puerto, H.; Nicolás, E.; Bartual, J. Pomegranate trees performance under sustained and regulated deficit irrigation. *Irrig. Sci.* **2013**, *31*, 959–970. [CrossRef]
12. Llácer, G. Importancia de las especies minoritarias en la alimentación y en la biodiversidad. XI Congreso nacional de Ciencias Hortícolas "Agriculturas minoritarias". *Vida Rural* **2007**, *260*, 48–50.
13. Galindo, A.; Collado-González, J.; Griñán, I.; Corell, M.; Centeno, A.; Martín-Palomo, M.J.; Girón, I.F.; Rodríguez, P.; Cruz, Z.N.; Memmi, H.; et al. Deficit irrigation and emerging fruit crops as a strategy to save water in Mediterranean semiarid agrosystems. Agric. *Water Manag.* **2018**, *202*, 311–324. [CrossRef]
14. Martínez-Nicolás, J.J.; Galindo, A.; Griñán, I.; Rodríguez, P.; Cruz, Z.N.; Martínez-Font, R.; Carbonell-Barrachina, A.A.; Nouri, H.; Melgarejo, P. Irrigation water saving during pomegranate flowering and fruit set period do not affect Wonderful and Mollar de Elche cultivars yield and fruit composition. *Agric. Water Manag.* **2019**, *226*, 105871. [CrossRef]
15. Rodríguez, P.; Mellisho, C.D.; Conejero, W.; Cruz, Z.N.; Ortuño, M.F.; Galindo, A.; Torrecillas, A. Plant water relations of leaves of pomegranate trees under different irrigation conditions. *Environ. Exp. Bot.* **2012**, *77*, 19–24. [CrossRef]
16. Laribi, A.I.; Palou, L.; Intrigliolo, D.S.; Nortes, P.A.; Rojas-Argudo, C.; Taberner, V.; Bartual, J.; Pérez-Gago, M.B. Effect of sustained and regulated deficit irrigation on fruit quality of pomegranate cv. 'Mollar de Elche' at harvest and during cold storage. *Agric. Water Manag.* **2013**, *125*, 61–70. [CrossRef]
17. Galindo, A.; Rodríguez, P.; Collado-González, J.; Cruz, Z.N.; Torrecillas, E.; Ondoño, S.; Corell, M.; Torrecillas, A. Rainfall intensifies fruit peel cracking in water stressed pomegranate trees. *Agric. For. Meteorol.* **2014**, *194*, 29–35. [CrossRef]
18. Galindo, A.; Calín-Sánchez, A.; Griñán, I.; Rodríguez, P.; Cruz, Z.N.; Girón, I.F.; Corell, M.; Martínez-Font, R.; Moriana, A.; Carbonell-Barrachina, A.A.; et al. Water stress at the end of the pomegranate fruit ripening stage produces earlier harvest and improves fruit quality. *Sci. Hortic.* **2017**, *226*, 68–74. [CrossRef]
19. Volschenck, T. Water use and irrigation management of pomegranate trees. A review. *Agric. Water Manag.* **2020**, *241*, 106375. [CrossRef]
20. Volschenck, T. Effect of water deficits on pomegranate tree performance and fruit quality. A review. *Agric. Water Manag.* **2021**, *246*, 106499. [CrossRef]
21. Maas, E.V. Salt tolerance of plants. In *CRC Handbook of Plant Science in Agriculture*; Christie, R.B., Ed.; CRC Press: Boca Ratón, FL, USA, 1987; Volume 2, pp. 57–76. ISBN 9780367251796.
22. Instituto Valenciano de Investigaciones Agrarias (IVIA). Tolerancia de los Cultivos a la Salinidad. Available online: http://www.agrosal.ivia.es/tolerancia.html (accessed on 14 October 2024).
23. Maas, E.V.; Hoffman, G.J. Crop salt tolerance: Current assessment. *J. Irrig. Drain.* **1977**, *103*, 115–134. [CrossRef]
24. Mellisho, C.D.; Egea, I.; Galindo, A.; Conejero, W.; Rodríguez, P.; Rodríguez, J.; Romojaro, F.; Torrecillas, A. Pomegranate (*Punica granatum* L.) fruit response to different deficit irrigation conditions. *Agric. Water Manag.* **2012**, *114*, 30–36. [CrossRef]
25. Galindo, A.; Rodríguez, P.; Mellisho, C.D.; Torrecillas, E.; Moriana, A.; Cruz, Z.N.; Conejero, W.; Moreno, F.; Torrecillas, A. Assessment of discretely measured indicators and maximum daily trunk shrinkage for detecting water stress in pomegranate trees. *Agric. For. Meteorol.* **2013**, *180*, 58–65. [CrossRef]
26. Parvizi, H.; Sepaskhah, A.R.; Ahmadi, S.H. Physiological and growth responses of pomegranate tree (*Punica granatum* L. cv. Rabab) under partial root zone drying and deficit irrigation regimes. *Agric. Water Manag.* **2016**, *163*, 146–158. [CrossRef]
27. Intrigliolo, D.S.; Nicolás, E.; Bonet, L.; Ferrer, P.; Alarcón, J.J.; Bartual, J. Water relations of field grown Pomegranate trees (*Punica granatum*) under different drip irrigation regimes. *Agric. Water Manag.* **2011**, *98*, 1462–1468. [CrossRef]
28. Tardieu, F.; Simonneau, T. Variability among species of stomatal control under fluctuating soil water status and evaporative demand: Modelling isohydric and anisohydric behaviours. *J. Exp. Bot.* **1998**, *49*, 419–432. [CrossRef]
29. Bugueño, F.; Livellara, N.; Varas, F.; Undurraga, P.; Castro, M.; Salgado, E. Responses of young *Punica granatum* plants under four different water regimes. *Cienc. Investig. Agrar.* **2016**, *43*, 49–56. [CrossRef]
30. Tarantino, A.; Frabboni, L.; Disciglio, G. Water-yield relationship and vegetative growth of Wonderful young pomegranate trees under deficit irrigation conditions in Southeastern Italy. *Horticulturae* **2021**, *7*, 79. [CrossRef]

31. Zhang, H.; Wang, D.; Ayars, J.E.; Phene, C.J. Biophysical response of young pomegranate trees to surface and sub-surface drip irrigation and deficit irrigation. *Irrig. Sci.* **2017**, *35*, 425–435. [CrossRef]
32. Goldschmidt, E.E.; Sadka, A. Yield alternation: Horticulture, physiology, molecular biology, and evolution. *Hortic. Rev.* **2021**, *48*, 363–418. [CrossRef]
33. López Alcolea, J. *Evaluación de Necesidades de Frío Para la Salida del Letargo y Necesidades de Calor Para Florecer en Nuevas Variedades y Selecciones de Albaricoquero del CEBAS-CSIC y Estudio del Efecto de Tratamientos Químicos Para Acelerar la Ruptura del Letargo Invernal*; Trabajo Fin de Grado E.P.S.O. Universidad Miguel Hernández: Orihuela, Spain, 2018.
34. Weinberger, J.H. Chilling requirements of peach varieties. *Proc. Am. Soc. Hortic. Sci.* **1950**, *56*, 122–128.
35. Richardson, E.A.; Seeley, S.D.; Walker, D.R. A model for estimating the completion of rest for 'Redhaven' and 'Elberta' peach trees. *HortScience* **1974**, *9*, 331–332. [CrossRef]
36. Fishman, S.; Erez, A.; Couvillon, G.A. The temperature dependence of dormancy breaking in plants: Mathematical analysis of a two-step model involving a cooperative transition. *J. Theor. Biol.* **1987**, *124*, 473–483. [CrossRef]
37. Anderson, J.L.; Richardson, E.A.; Kesner, C.D. Validation of chill unit and flower bud phenology models for "Montmorency" sour cherry. *Acta Hortic.* **1986**, *184*, 71–78. [CrossRef]
38. Fadón, E.; Fernandez, E.; Behn, H.; Luedeling, E. A conceptual framework for winter dormancy in deciduous trees. *Agronomy* **2020**, *10*, 241. [CrossRef]
39. Lang, G.A.; Early, J.D.; Martin, G.C.; Darnell, R.L. Endodormancy, paradormancy, and ecodormancy. Physiological terminology and classification for dormancy research. *HortScience* **1987**, *22*, 371–377. [CrossRef]
40. Egea, J.A.; Caro, M.; García-Brunton, J.; Gambín, J.; Egea, J.; Ruiz, D. Agroclimatic metrics for the main stone fruit producing areas in Spain in current and future climate change scenarios: Implications from an adaptive point of view. *Front. Plant Sci.* **2022**, *13*, 842628. [CrossRef]
41. Soloklui, A.A.G.; Gharaghani, A.; Oraguzie, N.; Eshghi, S.; Vazifeshenas, M. Chilling and heat requirements of 20 Iranian pomegranate cultivars and their correlations with geographical and climatic parameters, as well as tree and fruit characteristics. *HortScience* **2017**, *52*, 560–565. [CrossRef]
42. Nasrabadi, M.; Ramezaniana, A.; Eshghia, S.; Sarkhoshb, A. Chilling and heat requirement of pomegranate (*Punica granatum* L.) trees grown under sustained deficit irrigation. *Sci. Hortic.* **2020**, *263*, 109117. [CrossRef]
43. Mehta, D.; Gupta, T.; Gupta, R.K.; Thakur, S.; Sharma, J.P.; Baweja, P.K.; Sharma, P. Chilling hours and growing degree days of Himalayan wild pomegranate. *World J. Agric. Sci. Techn.* **2023**, *1*, 106–113. [CrossRef]
44. Selahvarzi, Y.; Zamani, Z.; Fatahi, R.; Talaei, A.R. Effect of deficit irrigation on flowering and fruit properties of pomegranate (*Punica granatum* cv. Shahvar). *Agric. Water Manag.* **2017**, *192*, 189–197. [CrossRef]
45. Melgarejo, P.; Gimeno, H.; Hernández, F.; Martínez, J.J.; Martínez-Font, R.; Melgarejo-Sánchez, P.; Giménez, M. Determinación de los rangos de normalidad para la interpretación foliar en el granado en el Sureste español. *Actas Hortic.* **2015**, *71*, 75–78.
46. Bartual, J.; Intrigliolo, D.S.; Pomares, F.; Bonet, L.; Parra, J.; García-González, J.; Pérez-Gago, M.B.; Palou, L. Manejo del riego y la fertilización del granado. *Frutic* **2016**, *51*, 40–51.
47. Melgarejo, P.; Salazar, D.M. *Tratado de Fruticultura Para Zonas Áridas y Semiáridas*; AMV-Mundi-Prensa: Madrid, Spain, 2003; pp. 167–365. ISBN 84-8476-111-8.
48. Hussain, F.; Murthy, B.N.S.; Reddy, M.L.N.; Satisha, J.; Upreti, K.K.; Laxman, R.H.; Srinivasulu, B. Induction of flowering and improvement of fruit yield pomegranate (*Punica granatum*) cv. Bhagwa by various regulators. *Environ. Conserv. J.* **2021**, *22*, 49–58. [CrossRef]
49. Soil Survey Staff. *Keys to Soil Taxonomy*, 13th ed.; USDA-NRCS: Washington, DC, USA, 2022.
50. Verheye, W. Agro-climate-based land evaluation systems. In *Land Use, Land Cover and Soil Sciences*; Verheye, W., Ed.; Encyclopedia of Life Support Systems (EOLSS) Publishers Co.: Paris, France, 2009; Volume 2, pp. 130–159.
51. Bartual, J.; Navarro, M.J.; Pérez-Gago, M.B.; Ortiz, M.; Palou, L. Irrigation strategies affect quality, mineral composition and internal rind browning of 'Mollar de Elche' pomegranate fruits. *Acta Hortic.* **2022**, *1349*, 47–55. [CrossRef]
52. Stamm, G.G. Problems and procedures in determining water supply requirements for irrigation projects. In *Irrigation of Agricultural Lands*; Hagan, R.M., Haise, H.R., Edminster, T.W., Eds.; Agronomy monograph 11; American Society of Agronomy: Madison, WI, USA, 1968; pp. 771–785.
53. Allen, R.G.; Pereira, L.S.; Raes, D.; Smith, M. *Crop Evapotranspiration-Guidelines for Computing Crop Requirements*; FAO Irrigation and drainage paper 56; FAO: Rome, Italy, 1998; ISBN 92-5-104219-5.
54. Instituto Valenciano de Investigaciones Agrarias (IVIA). Cálculo de Necesidades de Riego. Available online: http://riegos.ivia.es/calculo-de-necesidades-de-riego (accessed on 29 September 2024).
55. Chater, J.M.; Garner, L.C. Foliar nutrient applications to 'Wonderful' pomegranate (*Punica granatum* L.) II. Effects on leaf nutrient status and fruit split, yield and size. *Sci. Hortic.* **2018**, *242*, 207–213. [CrossRef]
56. Ministerio de Agricultura Pesca y Alimentación (MAPA). *Métodos Oficiales de Análisis*; MAPA: Madrid, Spain, 1986; ISBN 84-7479-532.X.

57. United Nations Economic Commission for Europe (UNECE). *Fresh Fruit and Vegetables–Standards. Pomegranates;* UNECE: Geneva, Switzerland, 2021. Available online: https://unece.org/trade/documents/2021/02/standards/pomegranates (accessed on 29 September 2024).
58. Doorenbos, J.; Kassam, A.H. *Efectos del Agua Sobre el Rendimiento de los Cultivos;* Estudio FAO: Riego y drenaje 33; FAO: Rome, Italy, 1979; ISBN 9253007443, 9789253007448.
59. Statistical Graphics Corporation. *Statgraphics Centurion XIX;* Statistical Graphics: Rockville, MD, USA, 2023.

plants

MDPI

*Article*

# Calibration and Validation of the CSM-CROPGRO-Peanut Model Under Mulched Drip Irrigation Conditions in Xinjiang

**Junwei Chen [1,2], Qiang Li [3], Xiaopei Zhang [1], Jianshu Dong [4], Xianfei Hou [3], Haocui Miao [3], Haiming Li [1], Yuchao Zhang [1], Xiaojun Shen [1,2], Zhuanyun Si [2,5,*] and Zhijie Shan [6,*]**

1 College of Water Conservancy Engineering, Tianjin Agricultural University-China Agricultural University Joint Smart Water Conservancy Research Center, Tianjin Agricultural University, Tianjin 300392, China; chenjunwei202204@163.com (J.C.); xiaopeihb@163.com (X.Z.); 15228177307@163.com (H.L.); 13820627850@163.com (Y.Z.); shenxiaojun@tjau.edu.cn (X.S.)
2 Institute of Western Agriculture, The Chinese Academy of Agricultural Sciences, Changji 831100, China
3 Institute of Economic Crops, Xinjiang Academy of Agricultural Sciences, Urumqi 830091, China; lq19820302@126.com (Q.L.); hou544805196@163.com (X.H.); mc09876@163.com (H.M.)
4 College of Water Conservancy & Architectural Engineering, Shihezi University, Shihezi 832000, China; dongjianshu_1010@163.com
5 Institute of Farmland Irrigation, Chinese Academy of Agricultural Sciences, Xinxiang 453002, China
6 State Key Laboratory of Simulation and Regulation of Water Cycle in River Basin, China Institute of Water Resources and Hydropower Research, Beijing 100048, China
* Correspondence: sizhuanyun@caas.cn (Z.S.); shanzhj@iwhr.com (Z.S.)

**Abstract:** In order to explore the applicability of the peanut growth simulation model CSM-CROPGRO-Peanut under conditions of mulched drip irrigation in Xinjiang, and to determine the optimal scenario for parameter estimation and model validation, field experiments were conducted in 2022 and 2023 on the water and nitrogen regulation of peanut. Based on the water requirements during the stages of peanut growth, three irrigation levels (low, medium, and high) and two nitrogen application levels (100% N and 50% N) were set, resulting in six treatments. An additional control treatment (CK) with a medium irrigation level and no nitrogen application was also included. In this study, four different parameter estimation and validation protocols were designed, and different parameter estimation results were obtained using the DSSAT-GLUE parameter estimation module. The results showed that the FL-SH (time between first flower and first pod), FL-SD (time between first flower and first seed), SIZLF (time between first flower and first seed), XFRT (maximum size of full leaf), and WTPSD (maximum weight per seed) parameters exhibited strong variability, with coefficients of variation of 24.33%, 22.9%, 19.78%, 14.47%, and 23.82%, respectively, and were significantly affected by environment–management interactions. Other parameters showed weaker variability, with coefficients of variation that were all less than 10%. The model outputs varied significantly among different parameter estimation protocols. Scenario 3, which used data from the adequate irrigation and adequate fertilization treatment (W3N2) environment across both years for parameter estimation and data from other treatments for validation, showed the highest model calibration and validation accuracy. The average absolute relative error (ARE) and normalized root mean square error (nRMSE) for model calibration and validation were the lowest at 9.1% and 10.1%, respectively. The CSM-CROPGRO-Peanut model effectively simulated peanut growth and development as well as soil moisture dynamics under mulched drip irrigation conditions in Xinjiang, with the highest simulation accuracy observed under full irrigation conditions. The findings provide a basis for using the CSM-CROPGRO-Peanut model to develop suitable irrigation and nitrogen application regimes for peanuts under mulched drip irrigation in Xinjiang.

 https://doi.org/10.3390/plants14040614

**Keywords:** DSSAT model; Arachis hypogaea; peanut; water and nitrogen regulation; parameter estimation; model validation

## 1. Introduction

Peanut (*Arachis hypogaea* L.), as a crucial economic and oilseed crop in China, holds a significant position in the national economy and is of great importance for its supply of edible oils [1]. The Xinjiang region, characterized by its unique climatic conditions—dry with minimal rainfall, abundant sunshine, and rich light and heat resources—coupled with gray desert soil as the primary soil type, provides a superior natural environment for peanut cultivation. The limited rainfall in this area effectively inhibits the growth of *Aspergillus flavus*, thereby ensuring the high quality of peanut oil, and the peanut crop exhibits strong drought tolerance, which is conducive to achieving a high quality and yield [2]. The peanut growing season in the Xinjiang region spans from April to September, with an annual average precipitation ranging between 100 and 200 mm. Precipitation is primarily concentrated between June and August, with monthly rainfall typically ranging from 10 to 30 mm. During July and August, daytime temperatures generally range from 30 °C to 35 °C, occasionally exceeding 35 °C, while nighttime temperatures typically range between 15 °C and 20 °C. In May and September, temperatures may drop to 5 °C to 10 °C. Due to the limited precipitation, peanut growth is heavily dependent on irrigation.

In recent years, numerous scholars have conducted extensive research on modes of irrigation and nitrogen application for peanuts under film drip irrigation in Xinjiang [3–5]. However, traditional field experiment methods are time-consuming, labor-intensive, and costly, and their results have limited general applicability. With the development of information technology and mathematical models, crop growth models are now widely used in agricultural research. Among them, the Decision Support System for Agrotechnology Transfer (DSSAT) is one of the most widely used crop models globally, capable of accurately simulating the daily growth and development processes of crops and responding to various factors such as crop genetic characteristics, field management practices, and environmental conditions [6–10].

Numerous studies have successfully been conducted based on the DSSAT model to optimize the water and nitrogen regulation schedules for different crops. For instance, Si et al. [11] used the DSSAT-CERES-Wheat model to develop optimized irrigation and nitrogen application strategies for drip-irrigated winter wheat in North China. Du et al. [12] determined the optimal irrigation quota for mulched-drip-irrigated cotton in Southern Xinjiang using the DSSAT model. Additionally, Sezen et al. [13] effectively simulated the impact of irrigation levels and frequencies on peanut growth, development, and yield in Turkey in the Eastern Mediterranean region by adjusting and evaluating the CSM-CROPGRO-Peanut model, achieving efficient utilization of irrigation water.

However, model calibration and validation are crucial steps to ensuring the simulation accuracy and reliability of crop models. Different model parameter estimation scenarios yield different parameter values and simulation results [14–16]. In terms of model parameter calibration, the traditional trial-and-error method, while simple and straightforward, is subjective and significantly less efficient when dealing with a large number of parameters [17]. Therefore, researchers have developed various systematic algorithms, such as the simplex method [18], the simulated annealing method [19], genetic algorithms [20,21], Bayesian methods [22], and the particle swarm optimization method [23], to optimize model parameter estimation. Among them, the generalized likelihood uncertainty estimation (GLUE) method, a type of Bayesian method, can effectively estimate model parameters by combining prior information with experimental data [24] and has been integrated into DSSAT v4.8 software as a universal method for estimating the cultivation parameters of different crops [25,26].

This study intended to utilize experimental data on the water and nitrogen regulation of peanuts in Xinjiang over two consecutive years (2022 and 2023) to run the CSM-CROPGRO-Peanut model and estimate model genotype parameters using the GLUE method. By comparing and analyzing different model parameter estimation and validation scenarios, the aim was to assess the reliability of the CSM-CROPGRO-Peanut model in simulating peanut growth, development, and yield formation under mulched drip irrigation in Xinjiang and to select the best simulation scenario. The results of this study will provide a scientific basis for using the model to develop suitable irrigation and nitrogen application regimes for peanuts under mulched drip irrigation in Xinjiang and support the widespread application of this model in simulating agricultural ecosystems in arid and semi-arid regions of China.

## 2. Materials and Methods

### 2.1. Experimental Site

Field experiments were conducted from 2022 to 2023 at the Anningqu Comprehensive Experimental Station of Xinjiang Academy of Agricultural Sciences, located in Urumqi, Xinjiang Uyghur Autonomous Region (43°58′ N, 87°30′ E, altitude 590 m). The study area is situated in the middle of the northern foot of the Tianshan Mountains, in an alluvial plain in the northern outskirts of Urumqi, and is characterized by a gentle terrain. It is a typical region of the Tianshan Northern Slope Economic Belt. During the growing season in 2022, the maximum and minimum temperatures ranged from 3.7 to 38.5 °C and from −1.4 to 26.6 °C, respectively, with a total rainfall of 63.1 mm. In 2023, the maximum and minimum temperatures during the growing season were 7.1 to 38.7 °C and 0.9 to 28.0 °C, respectively, with a total rainfall of 95.8 mm, as depicted in Figure 1. The region follows a one-crop-per-year system, exhibiting a state of "insufficient for two crops, surplus for one". The soil at the experimental site is classified as grey desert soil, with a pH value ranging from 7.8 to 8.0 in the 0–60 cm layer before sowing, an average bulk density of 1.41 g/cm$^3$, and a groundwater table depth of 7.5 m. The basic physicochemical parameters of the soil are presented in Table 1.

**Table 1.** Soil physical and chemical properties.

| Depth (cm) | Clay/% | Silt/% | Sand/% | Field Capacity (cm$^3$·cm$^{-3}$) | Saturated Water Content (cm$^3$·cm$^{-3}$) | Permanent Wilting Point (cm$^3$·cm$^{-3}$) | Available P (mg·kg$^{-1}$) | Available K (mg·kg$^{-1}$) | Alkaline Hydrolysis N (mg·kg$^{-1}$) | Organic Matter (g·kg$^{-1}$) | Bulk Density (g·cm$^{-3}$) |
|---|---|---|---|---|---|---|---|---|---|---|---|
| 0–20 | 11.85 | 43.74 | 44.42 | 0.270 | 0.405 | 0.11 | 23.77 | 199.06 | 42.98 | 14.86 | 1.35 |
| 20–40 | 13.05 | 45.71 | 41.25 | 0.286 | 0.429 | 0.10 | 22.17 | 169.33 | 32.04 | 14.28 | 1.43 |
| 40–60 | 12.38 | 42.98 | 44.64 | 0.288 | 0.432 | 0.10 | 22.86 | 114.46 | 43.41 | 9.73 | 1.44 |

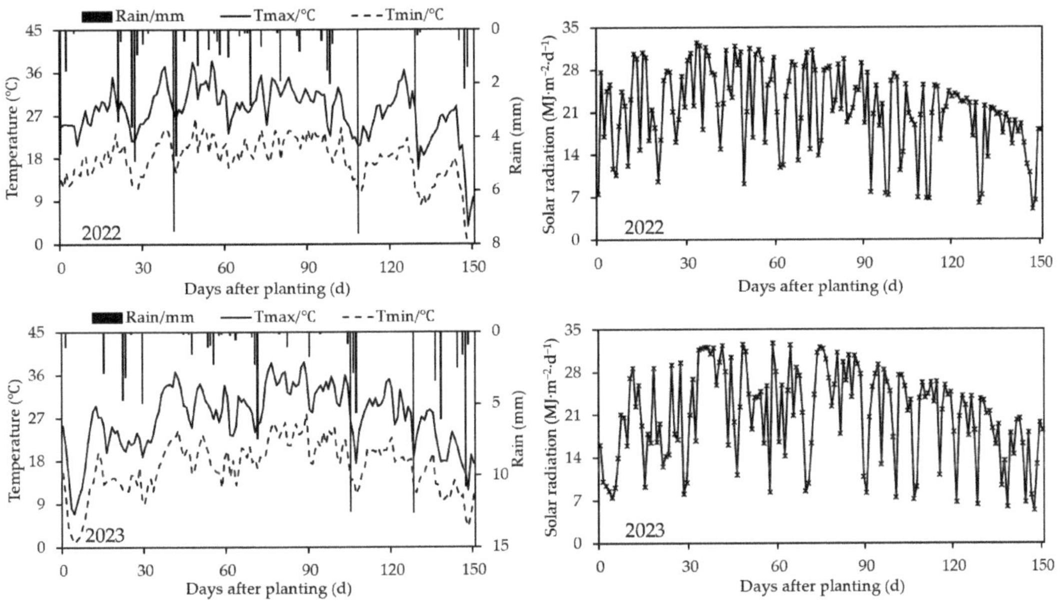


**Figure 1.** Meteorological data over whole peanut growth period in 2022 and 2023.

*2.2. Experimental Design*

The test crop was peanut (variety "Huayu 9610"), planted in a wide–narrow row configuration with 1 film covering 2 strips and 4 rows (Figure 2). The film width was 1.25 m, with 4 rows of peanuts planted within each film. The row spacings for the peanuts were 30 + 40 + 30 cm, and the hole spacing was 15 cm. The spacing between peanut rows across two films was 60 cm. The total width for 4 rows of peanuts under a single film was 1.5 m, with a planting density of approximately $1.67 \times 10^5$ holes/ha. The sowing dates for peanuts in 2022 and 2023 were May 7 and May 1, respectively. Before sowing, a base compound fertilizer ($N$-$P_2O_5$-$K_2O$ = 15-15-15) was applied at a rate of 300 kg/ha. During the growth period, nitrogen fertilizer (urea ($CO(NH_2)_2$, with nitrogen content $\geq$ 46%)) was applied using a Venturi fertilizer applicator. No additional micronutrients were applied to peanuts during the growth period. Irrigation was carried out using drip irrigation under the film, with water volumes for each treatment measured using flow meters. The irrigation water source was groundwater.

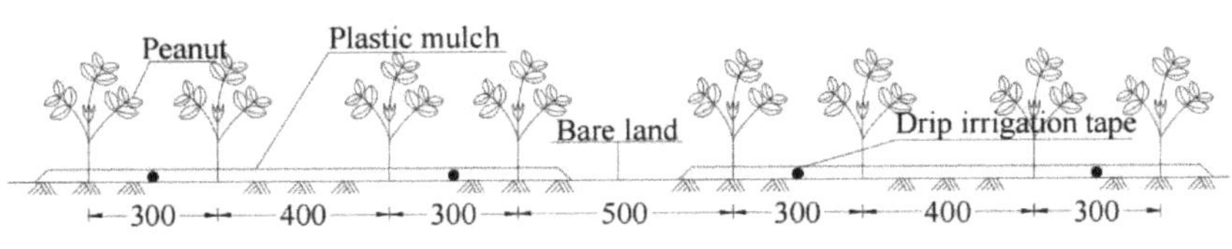


**Figure 2.** Layout of drip irrigation in peanut field under film mulching (mm).

Based on the growth habits of peanuts, the growth period was divided into four stages: seedling, flowering and pegging, pod-setting, and pod-filling. The field experiment was a split-plot design with water and nitrogen as the two factors. The main plot factor was nitrogen application, with two levels: conventional nitrogen application during the growth period (1) and half the conventional nitrogen application during the growth period (0.5) (denoted as N2 and N1, respectively). The subplot factor was irrigation, with three levels set as low, medium, and high water (denoted as W1, W2, and W3, respectively). A complete combination of the two factors resulted in 6 treatments. An additional control treatment (CK) with a medium irrigation level and no nitrogen application was also included, resulting in a total of 7 treatments (Table 2). Each treatment was replicated 3 times. According to local peanut production practices, no irrigation was applied during the seedling stage. To ensure smooth emergence, the peanuts were irrigated with 45 mm of water after sowing to facilitate seedling emergence.

**Table 2.** Irrigation and fertilization regimes for peanuts.

| Treatment | Irrigation Quota (mm) | | | | | Irrigation Cycle (d) | | | | Nitrogen Fertilizer Rate |
|---|---|---|---|---|---|---|---|---|---|---|
| | Sowing–Emergence Stage | Seedling Stage | Flowering–Pegging Stage | Pod-Setting Stage | Pod-Filling Stage | Seedling Stage | Flowering–Pegging Stage | Pod-Setting Stage | Pod-Filling Stage | |
| W1N2 | 45 | - | low | low | low | - | 10 | 10 | 15 | 1 |
| W2N2 | 45 | - | medium | medium | medium | - | 10 | 10 | 15 | 1 |
| W3N2 | 45 | - | high | high | high | - | 10 | 10 | 15 | 1 |
| W1N1 | 45 | - | low | low | low | - | 10 | 10 | 15 | 0.5 |
| W2N1 | 45 | - | medium | medium | medium | - | 10 | 10 | 15 | 0.5 |
| W3N1 | 45 | - | high | high | high | - | 10 | 10 | 15 | 0.5 |
| CK | 45 | - | medium | medium | medium | - | 10 | 10 | 15 | 0 |

Note: The irrigation quotas for 2022 and 2023 are 22.5, 30, 37.5 mm and 30, 37.5, 45 mm for low, medium, and high water, respectively. A nitrogen fertilizer rate of 1 indicates conventional nitrogen application, 0.5 indicates halved nitrogen application, and 0 indicates no nitrogen application.

## 2.3. *Measurement Items and Methods*

### 2.3.1. Determination of Soil Water Content

The soil water content was determined at a depth of 0–60 cm, in layers of 10 cm each, before sowing, before and after irrigation, and after harvest using the soil drying method. The sampling point was selected directly below the drip irrigation line for each treatment. The method described by Shen et al. [27] was used to calculate the average soil water content of the profile, which represented the average soil water content in the peanut field.

### 2.3.2. Measurement of Growth Indicators

From the time of peanut sowing, three representative peanut plants from each treatment were selected for measurement at each growth stage, with each treatment replicated three times. The leaf area index (LAI) was determined and calculated using the specific leaf weight method [28]. The aboveground biomass of peanut plants was measured separately for stems and leaves. They were first killed at 105 °C for 30 min and then dried at 75 °C until a constant weight was reached, and the dry weight was measured using an electronic balance with an uncertainty of $\pm$ 0.01 g.

### 2.3.3. Measurement of Pod Yield

After peanut harvest, three representative 6.67 $m^2$ plots were selected from each treatment. The pods were picked, placed into mesh bags, air-dried naturally, and weighed. The yield was then converted to pod yield per hectare.

## 2.4. *Model Input Data*

### 2.4.1. Meteorological Data

The meteorological data required by the CSM-CROPGRO-Peanut model were obtained from the automatic meteorological monitoring station (HOBO U30 Weather Station, HOBOware 3.7.25 Software, Onset Computer Corporation, Bourne, MA, USA) at the experimental site. The data mainly included daily maximum and minimum temperatures, solar radiation, and rainfall. The Weatherman program built into the model was used to import the data and create WTH meteorological files. The meteorological data for 2022 and 2023 are shown in Figure 1.

#### 2.4.2. Soil Data

Before the start of the field experiment, soil samples were taken at different depths using a soil auger to measure the physical properties and initial conditions of the soil at a depth of 0–60 cm. The SBuild module in the model was used to create the SOL soil file required by the CSM-CROPGRO-Peanut model. The physicochemical properties of the soil in the experimental area are shown in Table 1.

#### 2.4.3. Field Management Data

The field management data mainly included sowing date, planting density and depth, irrigation date and amount, fertilization date and amount, etc. The field management data used in this study were derived from the 2022 and 2023 field experiments on peanut cultivation under film-covered drip irrigation in Xinjiang.

### *2.5. Description, Calibration, and Validation of the CSM-CROPGRO-Peanut Model*

The CSM-CROPGRO-Peanut model is a component of the DSSAT software application. It is embedded in the DSSAT-CSM (crop system model) platform and can invoke the generic modules of weather, soil, soil–plant atmosphere (SPAM), and management to simulate crop growth, yield and carbohydrate balance [29]. The main function of the weather module is to read or generate daily weather data. The soil module integrates information from four sub-modules—soil water, soil temperature, soil carbon and nitrogen, and soil dynamics—and calculates and updates data for each soil layer on a daily basis [29]. The SPAM module brings together soil, plant, and atmospheric inputs and calculates root water uptake, potential evapotranspiration (ET), actual soil evapotranspiration, and plant transpiration. ET can be calculated using the Priestley Taylor method [30] or the Penman-FAO method. The default Priestley and Taylor methods were used in our study, requiring only daily solar radiation and temperature. Actual soil evapotranspiration and plant transpiration were calculated by ET, and detailed information on these aspects can be found in Jones et al. [29]. The SPAM module determines when to perform field operations (planting, harvesting, applying inorganic fertilizers, irrigating, and applying crop residues and organic materials) by invoking the sub-modules. The data required to run the model typically consist of four types: weather data, soil data, crop genetic coefficients, and management information.

The DSSAT-GLUE parameter calibration tool was used in this study to estimate the genetic parameters of the peanut variety "Huayu 9610". The parameters included CSDL, PPSEN, EM-FL, FL-SH, FL-SD, SD-PM, FL-LF, LFMAX, SLAVR, SIZLF, XFRT, WTPSD, SFDUR, SDPDV, PODUR, THRSH, SDPRO, and SDLIP (Table 3). The DSSAT-GLUE tool is based on the generalized likelihood uncertainty estimation (GLUE) method [6,31]. The GLUE method evaluates the credibility of parameters by comparing simulated values with observed data, using likelihood values rather than simply accepting or rejecting potential parameters. This approach is particularly effective for parameter estimation under conditions of significant model and observational uncertainty, thus enhancing the scientific rigor of the estimation [14].

**Table 3.** Genetic parameters of the CSM-CROPGRO-Peanut model for the variety "Huayu 9610".

| Parameter | Description | Unit |
|---|---|---|
| CSDL | Critical short day length below which reproductive development progresses with no day length effect (for short-day plants) | hour |
| PPSEN | Slope of the relative response of development to photoperiod with time (positive for short-day plants) | 1/hour |
| EM-FL | Time between plant emergence and flower appearance (R1) | photothermal days |
| FL-SH | Time between first flower and first pod (R3) | photothermal days |
| FL-SD | Time between first flower and first seed (R5) | photothermal days |
| SD-PM | Time between first seed (R5) and physiological maturity (R7) | photothermal days |
| FL-LF | Time between first flower (R1) and end of leaf expansion | photothermal days |
| LFMAX | Maximum leaf photosynthesis rate at 30 C, 350 vpm $CO_2$, and high light | mg $CO_2 \cdot m^2 \cdot s^{-1}$ |
| SLAVR | Specific leaf area of cultivar under standard growth conditions | $cm^2 \cdot g^{-1}$ |
| SIZLF | Maximum size of full leaf (three leaflets) | $cm^2$ |
| XFRT | Maximum fraction of daily growth that is partitioned into seed + shell | g |
| WTPSD | Maximum weight per seed | g |
| SFDUR | Seed filling duration for pod cohort at standard growth conditions | photothermal days |
| SDPDV | Average seed per pod under standard growing conditions | grain/pod |
| PODUR | Time required for cultivar to reach final pod load under optimal conditions | photothermal days |
| THRSH | The maximum ratio of [seed/(seed + shell)] at maturity | % |
| SDPRO | Fraction protein in seeds | g(protein)/g(seed) |
| SDLIP | Fraction oil in seeds | g(oil)/g(seed) |

In the GLUE framework, the likelihood function serves as the criterion for evaluating genetic parameters. The program compares the simulated values of each parameter set with the corresponding observed values and calculates their likelihood values. The higher the likelihood value, the better the match between the simulated and observed values. Subsequently, the total likelihood of all related simulations and observations is computed, and Bayesian inference is used to calculate the posterior distribution of the parameter sets [32]. To ensure the accuracy of the parameter estimation and the reliability of the posterior distribution, each GLUE run must involve at least 3000 iterations [33].

The phenological period (including flowering and maturation), final aboveground biomass, pod yield, and maximum peanut leaf area index were taken as the output variables of the model, and the parameters were estimated and verified by comparing the simulated values of the model with the observed values in the field. During the calibration process, initial default values required for the program were set, and 10,000 random searches were conducted using the GLUE tool to obtain a more reliable set of parameter combinations. Considering the influence of temporal and environment–management interactions on field experiments, this study employed different parameter calibration–validation scenarios (Table 4), and through the comparison of simulation results from these scenarios, the optimal scenario was determined.

**Table 4.** Different calibration–validation scenarios for the CSM-CROPGRO-Peanut model.

| Scenario | Data for Model Calibration | Data for Model Validation |
|---|---|---|
| 1 | All treatments in 2022 | All treatments in 2023 |
| 2 | All treatments in 2023 | All treatments in 2022 |
| 3 | Two years of full water and fertilizer treatment (W3N2) | All other processing |
| 4 | Two years of low-water and low-fertilizer treatment (W1N1) | All other processing |

## 2.6. Model Error Detection

In this study, the model calibration and validation processes were evaluated based on the absolute relative error (ARE), root mean square error (RMSE), normalized root mean square error (nRMSE), and correlation coefficient ($R^2$) between the simulated and observed values. These metrics quantify the relative discrepancies between simulated and observed values and are dimensionless statistics, allowing for comparisons across different variables [34]. Lower values of ARE and nRMSE indicate higher model simulation accuracy.

$$ARE = \frac{|S_i - O_i|}{O_i} \times 100\% \quad (1)$$

$$RMSE = \sqrt{\sum_{i=1}^{n} \frac{(S_i - O_i)^2}{n}} \quad (2)$$

$$nRMSE = \frac{\sqrt{\sum_{i=1}^{n} \frac{(S_i - O_i)^2}{n}}}{\bar{O}} \times 100 \quad (3)$$

$$R^2 = \frac{\left[\sum_{i=1}^{n} \left(O_i - \bar{O}\right) \times \left(S_i - \bar{S}\right)\right]^2}{\sum_{i=1}^{n} \left(O_i - \bar{O}\right)^2 \times \sum_{i=1}^{n} \left(S_i - \bar{S}\right)^2} \quad (4)$$

where $S_i$ and $O_i$ represent the i-th simulated and observed values, respectively; $\bar{S}$ and $\bar{O}$ are the simulated and observed mean values, respectively; and n is the number of data points. The ARE value reveals the relative deviation between the simulated and observed data. A lower ARE value indicates higher model accuracy and precision. When nRMSE < 10%, the simulation accuracy is excellent; when 10% ≤ nRMSE ≤ 20%, the simulation accuracy is good; when 20% ≤ nRMSE ≤ 30%, the simulation accuracy is moderate; and when nRMSE > 30%, the simulation accuracy is poor [8].

## 3. Results

### 3.1. Comparison of Model Parameters Derived from Different Calibration–Validation Scenarios

The parameter estimates obtained from different model parameter estimation scenarios are presented in Table 5.

**Table 5.** Genetic parameters of peanuts under different model parameter estimation scenarios.

| Parameter | Calibration Result | | | | Mean | Standard Deviation | Coefficient of Variation/% |
|---|---|---|---|---|---|---|---|
| | 1 | 2 | 3 | 4 | | | |
| CSDL | 11.84 | 11.84 | 11.84 | 11.84 | 11.84 | 0 | 0 |
| PPSEN | 0 | 0 | 0 | 0 | 0 | 0 | — |
| EM-FL | 15.52 | 15.04 | 15.19 | 15.17 | 15.23 | 0.20 | 1.3 |
| FL-SH | 6.499 | 5.549 | 6.198 | 9.344 | 6.90 | 1.68 | 24.33 |
| FL-SD | 22.99 | 23.66 | 13.87 | 23.86 | 21.10 | 4.83 | 22.9 |
| SD-PM | 63.79 | 59.1 | 69.84 | 60.25 | 63.25 | 4.83 | 7.63 |
| FL-LF | 73.0 | 73.0 | 73.0 | 73.0 | 73.0 | 0 | 0 |
| LFMAX | 1.496 | 1.488 | 1.438 | 1.492 | 1.479 | 0.027 | 1.84 |
| SLAVR | 278.2 | 284.4 | 286.2 | 274 | 280.7 | 5.63 | 2.01 |
| SIZLF | 19.46 | 13.2 | 18.41 | 13.7 | 16.2 | 3.20 | 19.78 |
| XFRT | 0.719 | 0.645 | 0.632 | 0.857 | 0.713 | 0.10 | 14.47 |
| WTPSD | 1.162 | 0.736 | 0.886 | 0.708 | 0.873 | 0.21 | 23.82 |
| SFDUR | 38.28 | 37.88 | 41.15 | 45.39 | 40.68 | 3.46 | 8.5 |
| SDPDV | 1.824 | 1.628 | 1.949 | 1.655 | 1.764 | 0.15 | 8.55 |
| PODUR | 15.0 | 15.0 | 15.0 | 15.0 | 15.0 | 0 | 0 |
| THRSH | 74.5 | 74.5 | 74.5 | 74.5 | 74.5 | 0 | 0 |
| SDPRO | 0.27 | 0.27 | 0.27 | 0.27 | 0.27 | 0 | 0 |
| SDLIP | 0.51 | 0.51 | 0.51 | 0.51 | 0.51 | 0 | 0 |

Note: Specific meaning of the parameters are shown in Table 3.

A comparison of the results across scenarios reveals significant variability in the following parameters: FL-SH (ranging from 5.549 to 9.344), FL-SD (ranging from 13.87 to 23.86), SIZLF (ranging from 13.2 to 19.46), XFRT (ranging from 0.632 to 0.857), and WTPSD (ranging from

0.708 to 1.162). Their coefficients of variation are 24.33%, 22.9%, 19.78%, 14.47%, and 23.82%, respectively, all exceeding 10%. This indicates substantial changes in the values of FL-SH, FL-SD, SIZLF, XFRT, and WTPSD with different model parameter estimation scenarios. Different parameter estimation scenarios essentially represent different peanut growth contexts (such as different irrigation and nitrogen application treatments and different years). Therefore, the estimates of FL-SH, FL-SD, SIZLF, XFRT, and WTPSD are highly dependent on the specific crop growth context, suggesting that environment–management interactions significantly influence the estimated values of these five parameters. The coefficients of variation for other parameters are all below 10%. Specifically, the standard deviations of CSDL, PPSEN, FL-LF, PODUR, THRSH, SDPRO, and SDLIP are all zero, indicating that these parameters are rarely affected by environment–management interactions. The values of these other parameters are relatively consistent across different crop growth contexts, suggesting that the model's estimates of various parameters for peanuts with subsurface drip irrigation are reliable.

### 3.2. Model Calibration and Validation Results Across Different Scenarios

In this study, we compared and analyzed the differences between simulated and measured values for the flowering stage, maturity stage, pod yield, aboveground biomass, and maximum leaf area index corresponding to four model calibration and validation scenarios. Based on this, we judged the simulation errors of each scenario and selected the optimal model calibration and validation scenario. The results are shown in Table 6. As can be seen from the table, the ARE and nRMSE of the model calibration results for Scenario 3 are 5.8% and 6.2%, respectively, while the ARE and nRMSE of the model validation results are 12.3% and 13.9%, respectively. The average ARE and nRMSE for the calibration and validation of Scenario 3 are 9.1% and 10.1%, respectively, which are the lowest and most stable compared to other scenarios. Therefore, Scenario 3 has the highest accuracy in simulating the growth, development, and yield of peanuts with subsurface drip irrigation in Xinjiang. The specific comparison results for Scenario 3 are shown in Table 7. The results of the calibration and validation of the CSM-CROPGRO-Peanut model under different scenario conditions (Tables S1–S4) can be found in the Supplementary Materials.

**Table 6.** Simulation accuracies of different scenarios of model calibration and validation.

| Scenario | Model Calibration | | Model Validation | | Average ARE/% | Average nRMSE/% |
|---|---|---|---|---|---|---|
| | ARE/% | nRMSE/% | ARE/% | nRMSE/% | | |
| 1 | 6.8 | 8.2 | 13.3 | 14.8 | 10.0 | 11.5 |
| 2 | 8.2 | 9.4 | 10.0 | 11.2 | 9.1 | 10.3 |
| 3 | 5.8 | 6.2 | 12.3 | 13.9 | 9.1 | 10.1 |
| 4 | 14.3 | 16.8 | 12.8 | 15.4 | 13.5 | 16.1 |

Under the model calibration conditions of Scenario 3, the average ARE values for the peanut flowering stage, maturity stage, pod yield, and aboveground biomass are 1.2%, 3.0%, 4.4%, and 6.6%, respectively, all less than 10%, indicating the high accuracy of the simulation results. However, the average ARE value for the maximum leaf area index is 14.0%, greater than 10%, indicating the lower accuracy of these simulation results. Under model validation conditions, the average ARE values for the peanut flowering stage, maturity stage, pod yield, aboveground biomass, and maximum leaf area index are 1.7%, 3.8%, 18.0%, 18.5%, and 19.7%, respectively, all higher than the average ARE values under model calibration conditions. This suggests that the simulation accuracy under model validation conditions is slightly lower than that under calibration conditions. Among them, Scenario 4 has the largest error in the model calibration results, which is because the treatments used for model calibration were all low-water treatments. This further demonstrates that the CSM-CROPGRO-Peanut model is less accurate in simulating peanut growth under water stress conditions. Overall, Scenario 3 has a relatively high simulation accuracy, and the CSM-CROPGRO-Peanut model performs well under subsurface drip irrigation conditions in Xinjiang, making it the most ideal model calibration and validation scenario.

**Table 7.** Results of calibration and validation of the CSM-CROPGRO-Peanut model under Scenario 3 conditions.

| Process | Year | Treatment | Anthesis Date (d) | | | Maturity Date (d) | | | Pod Yield (kg·ha$^{-1}$) | | | Aboveground Biomass (kg·ha$^{-1}$) | | | Maximum Leaf Area Index (cm$^2$·cm$^{-2}$) | | |
|---|---|---|---|---|---|---|---|---|---|---|---|---|---|---|---|---|---|
| | | | Sim. | Obs. | ARE/% | Sim. | Obs. | ARE/% | Sim. | Obs. | ARE/% | Sim. | Obs. | ARE/% | Sim. | Obs. | ARE/% |
| Model calibration | 2022 | W3N2 | 32 | 32 | 0 | 126 | 132 | 4.5 | 5056 | 5297 | 4.5 | 11,208 | 11,869 | 5.6 | 6.47 | 7.34 | 11.9 |
| | 2023 | W3N2 | 43 | 42 | 2.4 | 146 | 144 | 1.4 | 5784 | 5544 | 4.3 | 15,156 | 14,078 | 7.7 | 7.44 | 8.88 | 16.2 |
| | | Average | | | 1.2 | | | 3.0 | | | 4.4 | | | 6.6 | | | 14.0 |
| Model validation | 2022 | W1N2 | 32 | 31 | 3.2 | 122 | 132 | 7.6 | 1689 | 2749 | 38.6 | 4894 | 8712 | 43.8 | 3.18 | 4.47 | 28.9 |
| | | W2N2 | 32 | 31 | 3.2 | 124 | 132 | 6.1 | 3275 | 4523 | 27.6 | 8287 | 10,409 | 20.4 | 5.34 | 6.56 | 18.6 |
| | | W1N1 | 32 | 31 | 3.2 | 122 | 132 | 7.6 | 1679 | 2249 | 25.3 | 4837 | 6709 | 27.9 | 3.1 | 3.88 | 20.1 |
| | | W2N1 | 32 | 32 | 0 | 124 | 132 | 6.1 | 3186 | 3898 | 18.3 | 8129 | 9257 | 12.2 | 5.32 | 5.78 | 8.0 |
| | | W3N1 | 32 | 31 | 3.2 | 126 | 132 | 4.5 | 4890 | 4698 | 4.1 | 11,167 | 10,354 | 7.9 | 6.62 | 6.86 | 3.5 |
| | | CK | 32 | 33 | 3.0 | 124 | 132 | 6.1 | 2801 | 3398 | 17.6 | 7649 | 8255 | 7.3 | 4.83 | 5.63 | 14.2 |
| | 2023 | W1N2 | 43 | 43 | 0 | 141 | 144 | 2.1 | 2974 | 4495 | 33.8 | 8630 | 12,442 | 30.6 | 5.26 | 7.38 | 28.7 |
| | | W2N2 | 43 | 42 | 2.4 | 143 | 144 | 0.7 | 5056 | 5242 | 3.5 | 11,860 | 13,630 | 13.0 | 6.71 | 8.82 | 23.9 |
| | | W1N1 | 43 | 43 | 0 | 141 | 144 | 2.1 | 2850 | 4347 | 34.4 | 8439 | 12,352 | 31.7 | 5.14 | 7.33 | 29.9 |
| | | W2N1 | 43 | 43 | 0 | 143 | 144 | 0.7 | 4879 | 5128 | 4.9 | 11,826 | 14,289 | 17.2 | 6.66 | 8.58 | 22.4 |
| | | W3N1 | 43 | 42 | 2.4 | 146 | 144 | 1.4 | 5752 | 5414 | 6.2 | 15,154 | 14,343 | 5.7 | 7.49 | 9.03 | 17.1 |
| | | CK | 43 | 43 | 0 | 143 | 144 | 0.7 | 4717 | 4618 | 2.1 | 11,692 | 12,183 | 4.0 | 6.49 | 8.19 | 20.8 |
| | | Average | | | 1.7 | | | 3.8 | | | 18.0 | | | 18.5 | | | 19.7 |

Note: Sim. and Obs. represent the simulated and observed values.

### 3.3. Simulation Results of Dynamic Variables Under Scenario 3

To further explore the differences in simulation accuracy across different model calibration–validation scenarios, this study compared the simulation results of some important dynamic variables across scenarios. Due to the high simulation accuracy of Scenario 3, it was taken as an example. The genetic parameters obtained from model calibration under this scenario and the relevant experimental data from 2023 were used to run the CSM-CROPGRO-Peanut model. A comparative analysis was conducted on the dynamic changes in peanut aboveground biomass and soil moisture (Figures 3 and 4). Since peanut roots are mostly distributed in the surface soil layer (0–30 cm) [35], for brevity, this study selected the soil moisture in the middle 10–20 cm soil layer for dynamic analysis.

Under the same nitrogen application rate, the normalized root mean square error (nRMSE) for simulating peanut aboveground biomass across treatments was as follows: W3N2 < W2N2 < W1N2 < 20%, W3N1 < W2N1 < W1N1 < 25%, and CK < 10%, with all $R^2$ values exceeding 0.9. For soil moisture, the nRMSE followed a similar pattern: W3N2 < W2N2 < W1N2 < 20%, W3N1 < W2N1 < W1N1 < 20%, and CK < 20%, with all $R^2$ values exceeding 0.8. Among these treatments, the W3 treatment exhibited the lowest nRMSE and the highest $R^2$ values for both aboveground biomass and soil moisture compared to W1 and W2. Specifically, in the W3N2 and W3N1 treatments, the nRMSE for aboveground biomass was 4.1% and 5.9%, respectively, with corresponding $R^2$ values of 0.998 and 0.993. For soil moisture, the nRMSE was 9.8% and 8.0%, respectively, with $R^2$ values of 0.950 and 0.963. These results demonstrate that under adequate irrigation conditions, the dynamic simulation of peanut aboveground biomass and soil moisture was highly accurate, with the simulation accuracy for aboveground biomass exceeding that for soil moisture. The overall simulation trend in soil moisture was consistent with the observations, and most measured values closely matched the simulated values (Figure 4). Therefore, the CSM-CROPGRO-Peanut model is effective in simulating dynamic changes in peanut aboveground biomass and soil moisture under adequate water supply. At the same irrigation level, the nRMSE for simulating aboveground biomass across treatments was observed to follow this pattern: W1N2 < W1N1, CK < W2N1 < W2N2, and W3N2 < W3N1. For soil moisture, the nRMSE followed a different pattern: W1N1 < W1N2, W2N1 < CK < W2N2, and W3N1 < W3N2.

This indicates that increasing the nitrogen application rate reduces the model's accuracy in simulating soil moisture.

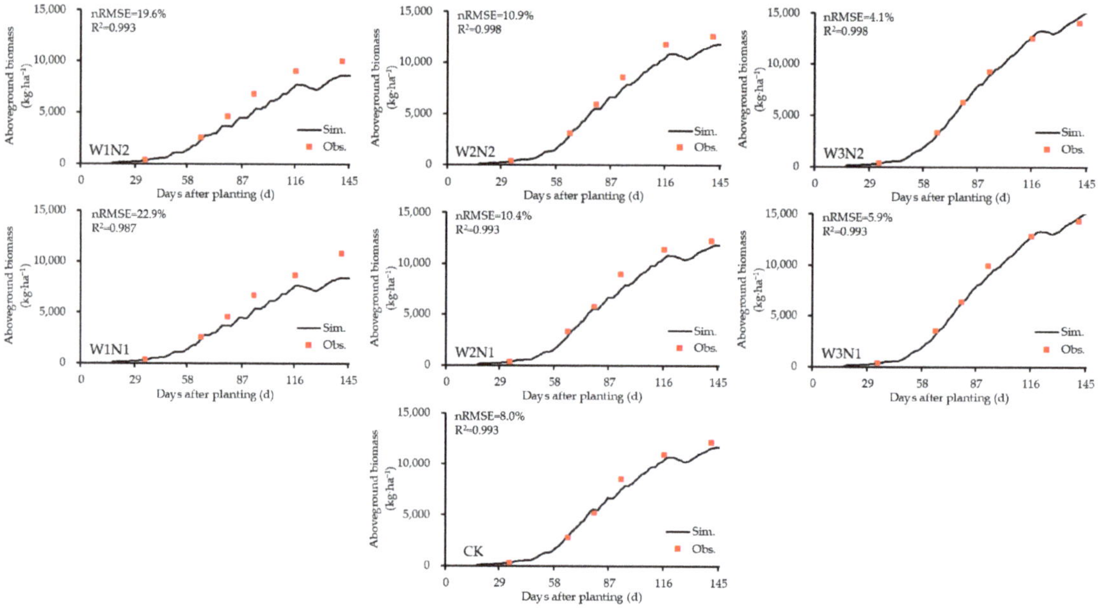


**Figure 3.** Dynamic simulation of peanut aboveground biomass under mulched drip irrigation in Xinjiang in 2023. Note: Sim. and Obs. represent the simulated and observed values, the same below.

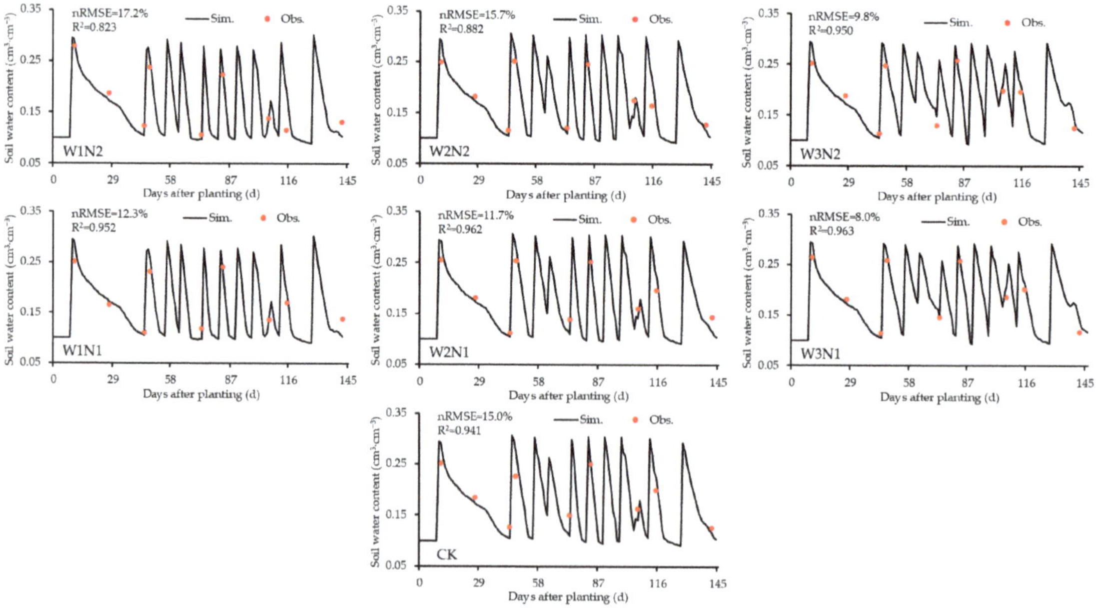


**Figure 4.** Simulation of peanut soil moisture dynamics under mulched drip irrigation in Xinjiang in 2023.

In summary, except for the aboveground biomass in the low-water, low-nitrogen treatment (W1N1), the nRMSE for simulating aboveground biomass and soil moisture across all other treatments using the CSM-CROPGRO-Peanut model was less than 20%, with all $R^2$ values exceeding 0.8. These findings confirm that the model meets the requirements for accurately simulating peanut growth and development under mulched drip irrigation conditions in Xinjiang.

## 4. Discussion

### *4.1. Impact of Environment-Management Interactions on Genetic Parameter Estimation in Peanut and Enhancement of Model Accuracy*

During model operation, the selection of crop variety parameters directly influences the effectiveness of model simulations. Therefore, the determination of crop variety parameters plays a crucial role in the accuracy of crop model simulations [36]. Crop variety parameters are generally not directly accessible and are often obtained through various parameter estimation methods. Among these methods, the GLUE method has been integrated into the DSSAT model as a built-in parameter estimation tool, allowing its direct use [37]. In this study, DSSAT-GLUE was employed to estimate the genetic parameters of peanuts under film-covered drip irrigation conditions in Xinjiang. A total of four different model calibration–validation scenarios (Table 4) were adopted. The results from the different calibration–validation scenarios underscore the importance of environment–management interactions in model parameterization, especially for parameters such as FL-SH, FL-SD, SIZLF, XFRT, and WTPSD, which exhibited substantial variability across different scenarios. These parameters demonstrated coefficients of variation (CV) exceeding 10% [14], which is indicative of their sensitivity to varying environmental conditions, such as irrigation and nitrogen application regimes. This finding is consistent with previous studies emphasizing the dynamic nature of crop growth under diverse agronomic management practices [14,37]. The significant variation in these parameters suggests that their accurate estimation requires a robust calibration process tailored to specific environmental contexts. It is evident that under varying field management practices, particularly irrigation and nitrogen treatments, the model's predictive accuracy is enhanced by re-estimating these parameters to capture the complexity of environment–management interactions. The DSSA-GLUE parameter estimation module has good convergence and high reliability when estimating parameters with a coefficient of variation of less than 10%, indicating that some parameters (such as CSDL, PPSEN, FL-LF, PODUR, THRSH, SDPRO, and SDLIP) are relatively unaffected by environmental variability. This stability highlights the robustness of the model in predicting these parameters across different growth conditions. This is consistent with the findings of Fan et al. [38], who observed that certain genetic and physiological parameters remain relatively constant under varying environmental conditions. Thus, these parameters can be reliably used in the model without the need for frequent re-estimation, ensuring stable predictions for long-term simulations or extrapolations across similar environments.

### *4.2. Optimizing Peanut Growth Simulation Under Mulched Drip Irrigation in Xinjiang*

Further analysis of the model calibration and validation scenarios revealed that Scenario 3 exhibited the highest accuracy in simulating peanut growth dynamics. The fact that it has the lowest ARE and nRMSE in both calibration and validation (9.1% and 10.1%, respectively) supports the conclusion that this scenario is the most suitable for simulating peanut growth under subsurface drip irrigation in Xinjiang. It is worth noting that the model performed best under fully irrigated conditions, which is reflected in the lower simulation error in the calibration phase. However, a marked decrease in accuracy was observed when the model was validated using water-stressed treatments, particularly in the simulation of aboveground biomass and maximum leaf area index. This indicates that using treatments under fully irrigated conditions for model calibration can achieve a relatively high simulation accuracy, but some errors will be encountered when simulating peanut growth under water stress conditions. This aligns with the findings of Yao et al. [14] and Boote et al. [39], who observed that crop simulation models often struggle with water stress conditions due to their complex effects on plant growth dynamics. The model's performance, as indicated by the nRMSE values for aboveground biomass and soil moisture, was generally excellent for most treatments under Scenario 3. The fact that the lowest nRMSE and highest $R^2$ values for aboveground biomass and soil moisture were observed under W3 treatments confirms the model's robustness under optimal irrigation conditions. This high accuracy can be attributed to the model's ability to simulate the dynamic changes in peanut growth under subsurface drip irrigation, a crucial method for improving water use efficiency in semi-arid regions such as Xinjiang [12,40]. The results show that the CSM-CROPGRO-Peanut model can effectively simulate crop growth and water dynamics under controlled irrigation systems, which is of great significance for optimizing water resource utilization and improving agricultural sustainability in water-scarce areas. However, while the model performs well under normal irrigation conditions, its low accuracy under water stress scenarios, especially for aboveground biomass and soil moisture, suggests limitations in its current form [41,42]. This highlights the need to further improve the model's representation of the effects of water stress on crop growth, especially under variable nitrogen supply conditions. Under the same irrigation level, the accuracy of soil moisture simulations decreased with the increase in nitrogen application. This effect may be due to the complex interaction between nitrogen application and the soil water holding capacity, which makes it difficult for the model to accurately capture the water dynamics after nitrogen application [43]. Therefore, future research could focus on improving the model's water stress simulation capabilities by incorporating more-dynamic soil–plant–atmosphere interactions and better accounting for nitrogen management strategies.

In conclusion, the CSM-CROPGRO-Peanut model is an effective tool for simulating peanut growth under drip irrigation in Xinjiang under Scenario 3, which is helpful for understanding the dynamic interaction between irrigation, nitrogen application, and crop performance. The model's accuracy in simulating aboveground biomass and soil water dynamics further supports its potential for application in precision agriculture, where accurate predictions are critical for optimizing water and nutrient management strategies. However, improvements are needed in simulating growth under water constraints, especially in predicting biomass and soil moisture under water stress.

## 5. Conclusions

This study calibrated and validated the CSM-CROPGRO-Peanut model through four distinct scenarios utilizing field experiment data on peanut cultivation under film-covered drip irrigation in Xinjiang from 2022 and 2023. Based on the research findings, the following key conclusions can be drawn:

(1) The genetic parameters FL-SH, FL-SD, SIZLF, XFRT, and WTPSD exhibited significant variability and were significantly affected by environment–management interactions. Therefore, when applying the model to simulation scenarios with substantial variations in field irrigation and nitrogen management conditions, these five parameters should be re-estimated. Otherwise, they may introduce considerable errors in model simulations.

(2) When employing parameter estimation Scenario 3 (using data from the well-watered and well-fertilized treatment W3N2 samples in both years for parameter estimation and data from other treatments for validation), the average ARE and nRMSE for model calibration and validation were 9.1% and 10.1%, respectively, representing the

lowest average values for both ARE and nRMSE for both calibration and validation. Scenario 3 was simulated with high accuracy, while the other scenarios were simulated with a relatively low accuracy.

(3) The CSM-CROPGRO-Peanut model effectively simulates the growth and development processes of peanuts and the dynamic changes in soil moisture under film-covered drip irrigation conditions in Xinjiang. The model achieves the highest simulation accuracy under full irrigation conditions, providing a theoretical foundation for predicting the growth, development, and yield of peanuts under varying levels of film-covered drip irrigation and nitrogen application in Xinjiang.

**Supplementary Materials:** The following supporting information can be downloaded at: https://www.mdpi.com/article/10.3390/plants14040614/s1%E2%80%934, Tables S1–S4: Results of calibration and validation of the CSM-CROPGRO-Peanut model under different scenario conditions.

**Author Contributions:** Conceptualization, J.C.; methodology, J.C. and Z.S. (Zhuanyun Si); investigation, J.C., J.D. and H.L.; data curation, J.C., J.D., H.L. and Y.Z.; formal analysis, J.C. and X.Z.; writing—original draft preparation, J.C. and X.Z.; writing—editing, J.C., Q.L. and Z.S. (Zhijie Shan); supervision, Q.L., X.H., H.M. and X.S.; project administration, Q.L. and H.M.; funding acquisition, X.S. and Q.L. All authors have read and agreed to the published version of the manuscript.

**Funding:** This work was supported by the National Natural Science Foundation of China (52179052), the China Agriculture Research System of MOF and MARA (CARS-13), the Autonomous Region "Tianshan Talents" Cultivation Program—Modern Agriculture and Animal Husbandry Development High-level Talents (2023SNGGGCC018), and the Henan Province high-end foreign experts introduction plan project (HNGD2023054).

**Data Availability Statement:** The data that support the findings of this study are available from the corresponding author, [Z.S. (Zhuanyun Si)].

**Conflicts of Interest:** The authors declare no conflicts of interest.

## References

1. Deng, L.; Miao, J.L.; Guo, M.J.; Yin, J.H.; Li, Y.; Lu, Z.H.; Li, S.W.; Hu, J.P.; Ren, L. Correlation and path analysis of yield composition of large fruit-type peanut varieties with high oleic acid. *Fenzi Zhiwu Yuzhong (Mol. Plant Breed.)* **2024**, *22*, 1978–1985. [CrossRef]
2. Wan, S.B.; Zhang, J.L. Development strategy and countermeasure of peanut industry in Xinjiang. *J. Peanut Sci.* **2019**, *48*, 66–68. [CrossRef]
3. Xia, G.M.; Chu, F.Y.; Chen, J.X.; Chen, G.M.; Yao, Z.Z. Effects of different irrigation quota on yield and water use efficiency of black peanut with drip irrigation under film mulch. *J. Shenyang Agric. Univ.* **2015**, *46*, 119–123.
4. Qin, W.J.; Guo, R.Z.; Zou, X.X.; Zhang, X.J.; Yu, X.N.; Wang, Y.F.; Si, T. Effects of fertilizer topdressing on growth and yield of peanut with drip irrigation under film. *Chin. J. Oil Crop Sci.* **2021**, *43*, 891–897. [CrossRef]
5. Dong, J.S.; Xue, Z.; Shen, X.J.; Yi, R.C.; Chen, J.W.; Li, Q.; Hou, X.F.; Miao, H.C. Effects of different water and nitrogen supply modes on peanut growth and water and nitrogen use efficiency under mulched drip irrigation in Xinjiang. *Plants* **2023**, *12*, 3368. [CrossRef] [PubMed]
6. Tan, H.; Lu, X.; Zhang, Z.; Yin, C.X.; Ma, L.L.; Su, W. Influence of climate change on cotton production potential based on DSSAT model simulation. *Cotton Sci.* **2020**, *32*, 113–120.
7. Liu, T.; Wan, N.H.; Zhang, Z.T.; WAng, X.Y.; Sun, S.; Liu, Z.J.; Zhao, J.; Yang, X.G. Adaptability of DSSAT crop model to main crops in Lishu County, Jilin Province. *Chin. J. Ecol.* **2023**, *42*, 1264–1272. [CrossRef]
8. Wang, B.; Yang, H.X.; Feng, J.; Guo, Z.Y.; Guo, Y.W. Rice cultivar coefficient optimization of DSSAT based on PSO. *Trans. Chin. Soc. Agric. Mach.* **2023**, *54*, 369–375.
9. Zhang, F.L.; Liu, W.Y.; Tian, J.; Wang, J.J.; Liu, D.; Yang, J.; Li, Y.C.; Zhang, Y.Z. Simulating effects of climate change on growth season and yield of double cropping rice in Jiangxi province based on DSSAT model. *Acta Agron. Sin.* **2024**, *50*, 2614–2624.
10. Li, M.; Du, Y.J.; Zhang, F.C.; Bai, Y.G.; Fan, J.L.; Zhang, J.H.; Chen, S.M. Simulation of cotton growth and soil water content under film-mulched drip irrigation using modified CSM-CROPGRO-cotton model. *Agric. Water Manag.* **2019**, *218*, 124–138. [CrossRef]

11. Si, Z.Y.; Zain, M.; Li, S.; Liu, J.M.; Liang, Y.P.; Gao, Y.; Duan, A.W. Optimizing nitrogen application for drip-irrigated winter wheat using the DSSAT-CERES-Wheat model. *Agric. Water Manag.* **2021**, *244*, 106592. [CrossRef]
12. Du, J.T.; Zhang, N.; Gong, K.N.; Du, M.Z.; Yang, Y.P.; Wang, X.P. Optimization of cotton irrigation schedule under mulch drip irrigation in southern Xinjiang based on DSSAT model. *Chin. J. Ecol.* **2021**, *40*, 3760–3768. [CrossRef]
13. Sezen, S.M.; Ahmad, I.; Habib-Ur-Rahman, M.; Amiri, E.; Tekin, S.; Oz, K.C.; Maambo, C.M. Growth and productivity assessments of peanut under different irrigation water management practices using CSM-CROPGRO-Peanut model in Eastern Mediterranean of Turkey. *Environ. Sci. Pollut. Res. Int.* **2022**, *29*, 26936–26949. [CrossRef]
14. Yao, N.; Zhou, Y.G.; Song, L.B.; Liu, J.; Li, Y.; Wu, S.F.; Feng, H.; He, J.Q. Parameter estimation and verification of DSSAT-CERES-Wheat model for simulation of growth and development of winter wheat under water stresses at different growth stages. *Trans. Chin. Soc. Agric. Eng.* **2015**, *31*, 138–150.
15. Yang, Q.; Zhu, X.H.; Ouyang, G.Z.; Ma, L.L.; Liu, W.; Wang, K. Optimization of the simulated parameters of grassland productivity using CLM5.0 in Hulunbuir of Inner Mongolia. *Trans. Chin. Soc. Agric. Eng.* **2023**, *39*, 139–148.
16. Wei, X.H.; Nie, Z.G. Sensitivity analysis and optimization of leaf area index related parameters of dryland wheat based on APSIM model. *Chin. J. Eco-Agric.* **2024**, *32*, 119–129. [CrossRef]
17. Li, Z.H.; He, J.Q.; Xu, X.G.; Jin, X.l.; Huang, W.J.; Clark, B.; Yang, G.J.; Li, Z.H. Estimating genetic parameters of DSSAT-CERES model with the GLUE method for winter wheat (*Triticum aestivum* L.) production. *Comput. Electron. Agric.* **2018**, *154*, 213–221. [CrossRef]
18. Inoue, Y.; Moran, M.S.; Horie, T. Analysis of spectral measurements in paddy field for predicting rice growth and yield based on a simple crop simulation model. *Plant Prod. Sci.* **1998**, *1*, 269–279. [CrossRef]
19. Soldevilla-Martinez, M.; Quemada, M.; López-Urrea, R.; Muñoz-Carpena, R.; Lizaso, J.I. Soil water balance: Comparing two simulation models of different levels of complexity with lysimeter observations. *Agric. Water Manag.* **2014**, *139*, 53–63. [CrossRef]
20. He, J.Q.; Le Gouis, J.; Stratonovitch, P.; Allard, V.; Gaju, O.; Heumez, E.; Orford, S.; Griffiths, S.; Snape, J.W.; Foulkes, M.J.; et al. Simulation of environmental and genotypic variations of final leaf number and anthesis date for wheat. *Eur. J. Agron.* **2012**, *42*, 22–33. [CrossRef]
21. He, J.Q.; Dukes, M.D.; Hochmuth, G.J.; Jones, J.W.; Graham, W.D. Identifying irrigation and nitrogen best management practices for sweet corn production on sandy soils using CERES-Maize model. *Agric. Water Manag.* **2012**, *109*, 61–70. [CrossRef]
22. Ceglar, A.; Črepinšek, Z.; Kajfež-Bogataj, L.; Pogačar, T. The simulation of phenological development in dynamic crop model: The Bayesian comparison of different methods. *Agric. For. Meteorol.* **2011**, *151*, 101–115. [CrossRef]
23. Li, Z.H.; Wang, J.H.; Xu, X.G.; Zhao, C.J.; Jin, X.L.; Yang, G.J.; Feng, H.K. Assimilation of two variables derived from hyperspectral data into the DSSAT-CERES model for grain yield and quality estimation. *Remote Sens.* **2015**, *7*, 12400–12418. [CrossRef]
24. Wallach, D.; Makowski, D.; Jones, J.W.; Brun, F. *Working with Dynamic Crop Models: Evaluation, Analysis, Parameterization, and Applications*; Elsevier: Amsterdam, The Netherlands, 2006.
25. He, J.Q.; Jones, J.W.; Graham, W.D.; Dukes, M.D. Influence of likelihood function choice for estimating crop model parameters using the generalized likelihood uncertainty estimation method. *Agric. Syst.* **2010**, *103*, 256–264. [CrossRef]
26. Hoogenboom, G.; Jones, J.W.; Traore, P.C.S.; Boote, K.J. Experiments and data for model evaluation and application. In *Improving Soil Fertility Recommendations in Africa Using the Decision Support System for Agrotechnology Transfer (DSSAT)*; Springer: Dordrecht, The Netherlands, 2012; pp. 9–18.
27. Shen, X.J.; Sun, J.S.; Zhang, J.Y.; Song, N.; Liu, X.F.; Yang, L. Study on calculation method of soil moisture content under drip irrigation. *J. Soil Water Conserv.* **2011**, *25*, 241–244+253. [CrossRef]
28. Dong, J.S.; Shen, X.J.; Yi, R.C.; Li, Q.; Miao, H.C.; Hou, X.F.; Chen, J.W.; Xue, Z. Study on accurate and rapid monitoring method of peanut leaf area index. *Water Sav. Irrig.* **2024**, *02*, 88–94+104.
29. Jones, J.W.; Hoogenboom, G.; Porter, C.H.; Boote, K.J.; Batchelor, W.D.; Hunt, L.A.; Wilkens, P.W.; Singh, U.; Gijsman, A.J.; Ritchie, J.T. The DSSAT cropping system model. *Eur. J. Agron.* **2003**, *18*, 235–265. [CrossRef]
30. Ritchie, J.T. Model for predicting evaporation from a row crop with incomplete cover. *Water Resour. Res.* **1972**, *8*, 1204–1213. [CrossRef]
31. Wang, X.P.; Xin, L.; Du, J.T.; Li, M.F. Simulation of cotton growth and yield under film drip irrigation condition based on DSSAT model in Southern Xinjiang. *Trans. Chin. Soc. Agric. Mach.* **2022**, *53*, 314–321.
32. Song, L.B.; Chen, S.; Yao, N.; Feng, H.; Zhang, T.B.; He, J.Q. Parameter estimation and verification of CERES-Maize model with GLUE and PEST methods. *Trans. Chin. Soc. Agric. Mach.* **2015**, *46*, 95–111.
33. Li, B.; Zhao, Z.L.; Wang, T.L.; Feng, X.; Yao, M.Z.; Xing, J.W. Simulation of maize growth under water stress during jointing and tasseling in Shenyang natural climate condition. *J. Shenyang Agric. Univ.* **2017**, *48*, 567–575.
34. Sun, Y.B.; Sun, S.; Li, M.Y.; Chang, J.F.; Wang, L.W.; Gao, X.N. Study on high and stable yield areas of soybean in northeast china based on dssat model. *Chin. J. Agric. Resour. Reg. Plan.* **2024**, *45*, 106–120.

35. Wang, J.G.; Zhang, J.L.; Guo, F.; Yang, S.; Gao, H.X.; Zhang, C.Y.; Zhao, H.J.; Li, X.G.; Wan, S.B. Band application of peanut-specific slow-release compound fertilizer at different soil depths promotes the root growth, yield performance and nitrogen utilization of peanut. *J. Plant Nutr. Fertil.* **2022**, *28*, 2274–2286.
36. Meng, J.H.; Wang, Y.N.; Lin, Z.X.; Fang, H.T. Progress and perspective of crop growth models. *Trans. Chin. Soc. Agric. Mach.* **2024**, *55*, 1–15+27.
37. Zhao, Z.L.; Li, B.; Feng, X.; Yao, M.Z.; Xie, Y.; Xing, J.W.; Li, C.X. Parameter estimation and verification of DSSAT-CROPGRO-Tomato model under different irrigation levels in greenhouse. *Chin. J. Appl. Ecol.* **2018**, *29*, 2017–2027. [CrossRef]
38. Fan, Y.C.; Li, Y.; Wei, L.J.; Zhao, X.; Zhou, H. Comparison of responsiveness of three oil flax growth models to drought stress at budding stage and green fruit stage. *Jiangsu J. Agric. Sci.* **2023**, *39*, 423–433.
39. Boote, K.J. Concepts for calibrating crop growth models. In *DSSAT Version 3*; Hoogenboom, G., Wilkens, P.W., Tsuji, G.Y., Eds.; University of Hawaii: Honolulu, HI, USA, 1999; Volume 4–6, pp. 179–200.
40. Shen, Y.J.; Qi, Y.Q.; Luo, J.M.; Zhang, Y.C.; Liu, C.M. The combined pathway to sustainable agricultural water saving and water resources management: Anintegrated geographical perspective. *Acta Geogr. Sin.* **2023**, *78*, 1718–1730.
41. Wu, L.F.; Zhang, F.C.; Fan, J.L.; Zhou, H.M.; Xing, Y.Y.; Qiang, S.C. Sensitivity and uncertainty analysis for CROPGRO-cotton model at different irrigation levels. *Trans. Chin. Soc. Agric. Eng.* **2015**, *31*, 55–64.
42. Li, B.; Li, C.X.; Yao, M.Z.; Wei, X.G.; Bao, Z.R.; Sun, X.L. Sensitivity and uncertainty analysis for CROPGRO-Tomato model at different irrigation levels. *J. Shenyang Agric. Univ.* **2020**, *51*, 153–161.
43. Zhang, J.; Gou, L.N.; He, Z.J.; Xiao, X.Y.; Liang, F.; Ren, H.X.; Fu, Y.L. Effect of fertilizer concentration on soil infiltration characteristics and water transport under double point source in bubbled-root irrigation. *Water Sav. Irrig.* **2023**, *08*, 10–16.

plants

MDPI

Article

# Synergistic Effects of Biochar and Irrigation on Sugar Beet Growth, Yield, Quality, and Economic Benefit in Arid Regions

**Fuchang Jiang [1], Yi Li [1,2,*], Liwei Li [1], Xiangwen Xie [2], Wanli Xu [2], Yang Gao [3] and Asim Biswas [4]**

1 Key Laboratory of Agricultural Soil and Water Engineering in Arid and Semiarid Areas at Ministry of Education, College of Water Resources and Architectural Engineering, Northwest A&F University, Yangling 712100, China; jiangfuchang@nwafu.edu.cn (F.J.); 18119906527@163.com (L.L.)

2 Institute of Soil Fertilizer and Agricultural Water Saving, Xinjiang Academy of Agricultural Sciences, Urumqi 830091, China; xiexw@sina.cn (X.X.); wlxu2006@163.com (W.X.)

3 Institute of Farmland Irrigation, Chinese Academy of Agricultural Sciences, Xinxiang 453002, China; gaoyang@caas.cn

4 School of Environmental Sciences, University of Guelph, Guelph, ON N1G 2W1, Canada; biswas@uoguelph.ca

* Correspondence: liyi@nwafu.edu.cn

**Abstract:** Water scarcity hinders sustainable agriculture in arid and semi-arid regions. This study investigated the combined effects of trickle irrigation and biochar application on sugar beet cultivation in northwest China's arid and semi-arid regions, addressing challenges of water scarcity. Three-year field experiments were conducted using plastic film mulch, four irrigation levels based on crop evapotranspiration (0.6–1.2 *ETc*), and four biochar application rates (0–30 t ha$^{-1}$). Results showed that biochar application increased sugar beet germination rates by 7.2–24.5% and enhanced relative chlorophyll content by 3.1–22.1%. Optimal combinations of irrigation and biochar significantly improved growth indices and yield, with maximum values observed under the highest irrigation (1.2 *ETc*) and 10 t ha$^{-1}$ biochar treatment. However, the 1.0 *ETc* irrigation treatment with 10 t ha$^{-1}$ biochar demonstrated superior water use efficiency (14.8% higher), sustainable yield index (1.1% higher), and economic benefits (1.4% higher) compared to the highest irrigation treatment. Considering growth, yield, quality, water use efficiency, sustainability, and economic factors, an irrigation level of 1.0 *ETc* combined with a biochar application rate of 10 t ha$^{-1}$ is recommended for sugar beet cultivation in Xinjiang. This study provides valuable insights and practical strategies for water conservation, high yield, and quality improvement in sugar beet cultivation under arid and semi-arid conditions, contributing to sustainable agricultural practices in water-scarce regions.

**Keywords:** sugar beet; *SPAD*; germination rate; sustainable yield index

## 1. Introduction

Sugar beet holds a crucial position in agricultural production as a primary sugar and cash crop, accounting for approximately 30% of the world's sugar production [1]. Beyond its role in sugar production, molasses derived from sugar beet is extensively utilized across various industries, including chemicals (methanol, ethanol, butanol, and glycerol), pharmaceuticals (vitamins and minerals), and food production (MSG and citric acid) [2]. The cultivation of sugar beet is particularly significant in arid and semi-arid regions where water resources are limited and soil salinity is prevalent. The shortage of water and the issue of soil salinization pose significant challenges to sustainable agricultural development in these areas. Thus, enhancing soil quality and developing effective irrigation strategies for sugar beet cultivation are essential for improving both yield and quality. By addressing these challenges, sustainable sugar beet production can be ensured, contributing to the broader agricultural sustainability in water-scarce regions.

Biochar is a solid material rich in carbon that is produced from organic matter at temperatures ranging from 300 to 1000 °C in the absence of oxygen [3]. As a novel soil amendment, biochar has garnered significant attention for its potential to enhance the soil environment, primarily due to the following reasons: (1) With high specific surface area and porosity, the application of biochar can reduce soil bulk density and increased porosity [4]. (2) Abundant carbon could increase soil activity and soil organic matter content [5]. (3) With good adsorption capacity, it can improve soil cation exchange rate and the utilization efficiency of nutrients and reduce nutrient loss [6]. (4) It can improve the structure of soil aggregates and increase the types and activities of enzymes in soil [7]. (5) It can repair polluted soil while regulating soil acidity and alkalinity [8]. Additionally, biochar also has strong water retention properties, which can inhibit soil water evaporation [9], thus achieving the effect of enhancing soil water content [10].

Biochar not only enhances soil properties but also holds significant potential for increasing crop productivity. Numerous studies have demonstrated that appropriate biochar application rates can enhance the productivity of various crops, including cotton, wheat, maize, rice, and tomato [11–15]. Wang et al. [16] conducted a three-year study on the application of biochar, which showed that biochar had a significant impact on the growth, development, yield, and quality of cotton. However, the application of biochar exceeding 30 t ha$^{-1}$ was not conducive to cotton growth; Liu et al. [17] confirmed in their experiment of intercropping fava bean and ryegrass that the application of biochar can increase leaf moisture content and total aboveground biomass in the intercropping system. Secondly, studies have shown that different types of biochar can significantly affect the application efficiency. Ren et al. [18] found that both the type and the application level of biochar can affect the level of soil improvement and nitrogen uptake by plants; the study by Tiong et al. [19] showed that biochar made from coconut coir significantly enhanced the biological nitrification process and significantly increased the chlorophyll content of tomato plants; Shang et al. [20] found that pyrolysis temperature significantly affects the properties of biochar, especially its surface characteristics.

For sugar beet, there are significant differences in the amount of biochar applied under different planting conditions. Wang et al. [21] found that an application rate of biochar at 20 t ha$^{-1}$ had positive effects on the growth and sugar production of sugar beet in Heilongjiang Province. Li et al. [2] (2022) observed that an application rate of 10 t ha$^{-1}$ resulted in the most significant increase in sugar beet yield, while rates between 10 and 25 t ha$^{-1}$ also showed a positive impact. However, application rates exceeding 30–100 t ha$^{-1}$ led to a decrease in sugar beet yield. Lebrun et al. [22] believed that adding 2% biochar to soil was the most beneficial to the growth of sugar beet, doubling the yield compared to no biochar application. Alves et al. [23] found that when 5% biochar was used instead of conventional fertilizer, sugar beet yield decreased by 45% in sandy loam soils of the Brazilian savanna. Therefore, it is necessary to study the effects of biochar on the growth characteristics of sugar beet in arid and semi-arid salinized soil, especially in the Xinjiang region of China, a region of strong agricultural, economic, and cultural significance.

In addition, as a high-water-demand crop, a reasonable irrigation strategy for sugar beets can not only improve yield and quality but also enhance water use efficiency and reduce water resource waste [24]. However, there are significant differences in the water requirements of sugar beets in different regions. For example, in Central Europe, the annual water demand for sugar beet is about 200 mm. In some Mediterranean countries, the annual water demand for sugar beet is 300–500 mm, while in Asia, North Africa, and

South America, the annual water demand for sugar beet is about 600 mm [25]. Specifically, sugar beet yield can only be guaranteed under fully irrigated conditions (approximately 700 mm) in the western United States [26]. In the southern Xinjiang region of China, Yan et al. [27] believed that the highest sugar beet yield was achieved under 0.6 ETc (about 200 mm) irrigation. These studies show that the amount of water required by sugar beets varies greatly from region to region. However, as an important region for sugar beet production in China, there are relatively few studies on the water demand of sugar beet. This study hypothesized that the synergistic effect of biochar and irrigation can effectively promote the growth of sugar beet and enhance yield and quality, resulting in a significant economic effect of sugar beet cultivation. Therefore, the objectives of this study were (1) to explore the effects of different amounts of biochar application and irrigation on the growth, physiological indexes, yield, and quality of sugar beet; (2) to determine the effects of irrigation and biochar application on the stability of sugar beet harvest; and (3) based on economic benefits, to determine the amount of biochar application and irrigation suitable for sugar beet cultivation in Xinjiang.

## 2. Materials and Methods

### 2.1. Experimental Site

The study area was situated in 8th Division, 146th Regiment, 13th company, Shihezi, Xinjiang, China (85°97′05″ E; 44°47′83″ N). The region experiences an arid climate year-round, typical of the temperate continental climate prevalent in this area. The long-term average temperature is 6.2 °C, with a frost-free period lasting between 168 and 171 days, and the precipitation ranges from 180 to 270 mm. The soil tape is sandy loam (clay, silt, and sand accounted for 3.1%, 69.1%, and 27.8%, respectively), with an average density of 1.53 g $cm^{-3}$ at 0–100 cm soil depth. The soil's pH is 8.3; the soil's organic carbon is 8.5 g $kg^{-1}$, and the total nitrogen is 0.4 g $kg^{-1}$. Groundwater is located at >3 m below the surface. Field meteorological data were collected using a portable automatic meteorological station (HOBO U30, MA, USA) which recorded parameters such as rainfall, mean wind speed, average temperature, solar radiation intensity, sunshine duration, and relative humidity levels. During the sugar beet growth periods in 2021, 2022, and 2023, respectively, total precipitations were 91.7 mm, 81.8 mm, and 69.8 mm with corresponding mean temperatures of 20.13 °C, 21.05 °C, and 19.63 °C.

### 2.2. Experimental Design

#### 2.2.1. Biochar Preparation and Experimental Plot Arrangement

The raw material for biochar was derived from the empty shells of palm fruit branches. Biochar was made as follows: to the palm fruit shells in the collection, a blower was used for debris separation measures, followed by the use of dryers to dry the water in the shells that were crushed, and then the raw materials were processed into the carbonization furnace in a closed oxygen-deficient environment for heating to 600 °C for the pyrolysis reaction; for the biocarbon in the carbonization furnace to complete the pyrolysis reaction, it needs to be cooled for more than 8 h so that the temperature is reduced to room temperature. This process was carried out by Zhengzhou Yongbang New Energy Equipment Technology Co. Ltd. (Zhengzhou, China). The diameter of the biochar particles measured approximately 2 mm, effectively enhancing the contact area between the biochar and soil. To mitigate any potential impact on salinized soil pH levels, ferrous sulfate was used to acidify the biochar to a pH value of 6.7. The bulk density was 0.5 g $cm^{-3}$; the specific surface area was 217 $m^2$ $g^{-1}$; the water drop penetration time was 8 s, and the cation exchange capacity was 12.2 cmol $ka^{-1}$; the initial electrical conductivity ($EC_{1:5}$) was 11.02 mS $cm^{-1}$; the dissolved organic carbon was 143.5 mg $kg^{-1}$; nitrate nitrogen was 2.54 mg $kg^{-1}$; ammonia nitrogen was 1.15 mg $kg^{-1}$; the available potassium was 2575 mg $kg^{-1}$; the total nitrogen was 2.30 g $kg^{-1}$, and the total organic carbon was 472.2 g $kg^{-1}$.

The experiment was conducted in a completely randomized two-factor arrangement consisting of four irrigation rates and four biochar application rates as treatments, with each repeated three times. There were a total of 16 treatments and 48 plots. The dimensions of each test plot were 8 m× 6 m, with a 0.5 m aisle between plots to minimize interferences between the treatments. Biochar application rates of 0, 10, 20, and 30 t $ha^{-1}$ were applied once before sowing in the year 2021, and samples were monitored for three consecutive years.

The sugar beet variety was "Detian 2". The planting and harvest times were 19 April and 28 September, respectively, in 2021; 15 April and 25 September, respectively, in 2022; and 18 April and 23 September, respectively, in 2023. The planting pattern was "one layer of plastic film, two droppers, four rows", which is a common practice of local farmers. The row spacing of the sugar beet and the spacing between individual plants were 40 cm (Figure 1).

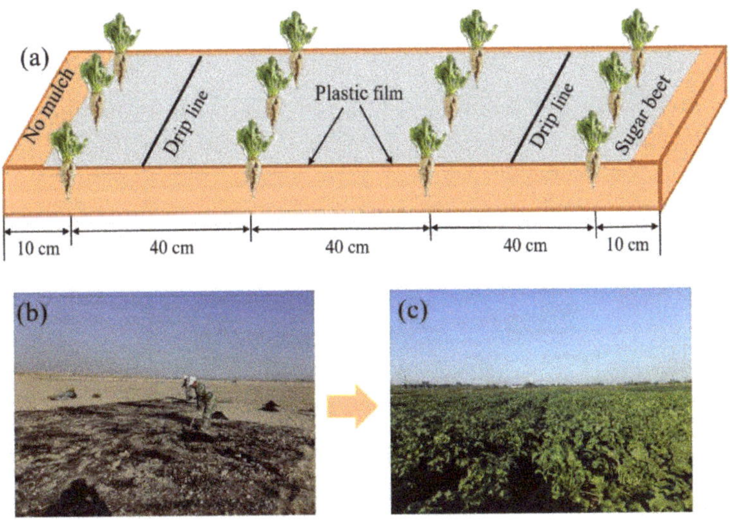


**Figure 1.** Schematic diagram (**a**) and experimental planting (**b**,**c**) of sugar beet planting under plastic mulched-drip irrigation.

2.2.2. Irrigation Schedule

The time and amount of irrigation varied annually based on local climatic conditions. In 2021, 2022, and 2023, there were 10, 8, and 9 irrigation events, respectively. The first irrigation was applied five days after sowing with an initial amount of 50 mm per event—a practice referred to as "dry sowing wet seedling." Detailed descriptions of specific irrigation schemes are provided in Table 1.

The irrigation quota of this experiment was determined based on the actual crop evapotranspiration (*ETc*). Irrigation gradients were set as fractions I1 (0.6 *ETc*), I2 (0.8 *ETc*), I3 (1.0 *ETc*), and I4 (1.2 *ETc*) and calculated using the formula developed by Allen et al. [27]: $ETc = Kc \times ET_0$, where the crop coefficient (*Kc*) of sugar beet varied across different growth stages, while the daily reference crop evapotranspiration ($ET_0$) was computed using the standard formula [28]:

$$ET_0 = \frac{0.408\Delta(R_n - G) + \gamma\frac{900}{T+273}\mu_2(e_s - e_a)}{\Delta + \gamma(1 + 0.34\mu_2)} \quad (1)$$

where $R_n$ is the net radiation amount ($MJ \cdot m^{-2} \cdot d^{-1}$); $G$ is soil heat flux density, and it is set to 0 at the daily scale ($MJ \cdot m^{-2} \cdot d^{-1}$); $\gamma$ is the humidity meter constant ($kPa \cdot C^{-1}$); $T$ is the daily average temperature (°C); $\mu_2$ is the wind speed at 2 m above the ground ($m \cdot s^{-1}$); $e_s$ and $e_a$ are the saturated and actual water vapor pressures (kPa); and $\Delta$ is the slope of the saturation vapor pressure curve at air temperature ($kPa\ C^{-1}$).

**Table 1.** Irrigation amounts over the three growth seasons of sugar beet in 2021, 2022, and 2023.

| Irrigation Times | Days After Sowing (d) | $ET_0$ (mm) | $K_C$ | Irrigation Amount (mm) | | | |
|---|---|---|---|---|---|---|---|
| | | | | I1 (0.6 *ETc*) | I2 (0.8 *ETc*) | I3 (1.0 *ETc*) | I4 (1.2 *ETc*) |
| 2021 | | | | | | | |
| 1 | 5 | - | - | 50 | 50 | 50 | 50 |
| 2 | 20 | - | - | 25 | 25 | 25 | 25 |
| 3 | 55 | 36 | 0.7 | 16 | 20 | 25 | 30 |
| 4 | 66 | 37 | 0.7 | 16 | 21 | 26 | 31 |
| 5 | 73 | 51 | 1.05 | 32 | 43 | 54 | 64 |
| 6 | 86 | 41 | 1.05 | 26 | 35 | 43 | 52 |
| 7 | 98 | 41 | 1.05 | 26 | 35 | 43 | 52 |
| 8 | 107 | 41 | 0.95 | 23 | 31 | 39 | 47 |
| 9 | 113 | 23 | 0.95 | 13 | 17 | 22 | 26 |
| Total | | | | 227 | 277 | 327 | 377 |
| 2022 | | | | | | | |
| 1 | 5 | - | - | 50 | 50 | 50 | 50 |
| 2 | 55 | 50 | 0.7 | 21 | 28 | 35 | 42 |
| 3 | 66 | 55 | 0.7 | 23 | 31 | 39 | 46 |
| 4 | 73 | 35 | 1.05 | 22 | 29 | 38 | 44 |
| 5 | 86 | 55 | 1.05 | 35 | 46 | 58 | 69 |
| 6 | 98 | 56 | 1.05 | 35 | 47 | 59 | 70 |
| 7 | 107 | 52 | 0.95 | 30 | 40 | 49 | 59 |
| 8 | 113 | 32 | 0.95 | 18 | 24 | 30 | 36 |
| Total | | | | 234 | 295 | 358 | 416 |
| 2023 | | | | | | | |
| 1 | 5 | - | - | 50 | 50 | 50 | 50 |
| 2 | 44 | 50 | 0.7 | 21 | 28 | 35 | 42 |
| 3 | 54 | 38 | 0.7 | 16 | 21 | 27 | 32 |
| 4 | 66 | 46 | 1.05 | 29 | 39 | 48 | 58 |
| 5 | 79 | 33 | 1.05 | 21 | 28 | 35 | 42 |
| 6 | 91 | 40 | 1.05 | 25 | 34 | 42 | 50 |
| 7 | 103 | 31 | 0.95 | 18 | 24 | 30 | 35 |
| 8 | 110 | 23 | 0.95 | 13 | 18 | 22 | 26 |
| 9 | 117 | 24 | 0.95 | 14 | 18 | 23 | 27 |
| Total | | | | 207 | 260 | 312 | 362 |

Fertilization was integrated into the irrigation processes representing the types and quantities of fertilizers commonly used locally. Specifically, nitrogen (450 kg $ha^{-1}$), phosphorus (265 kg $ha^{-1}$), and potassium (100 kg $ha^{-1}$) were applied using urea (with a minimum nitrogen content of 46%), diammonium phosphate (with at least 46% $P_2O_5$), and potassium sulfate (containing no less than 52% $K_2O$). Weeding and pest control of sugar beets during the reproductive period were consistent with the management of local farmers. Due to the mulch cover in the early stage, no weeding operation was required during the seedling stage, the rapid growth stage of sugar beets, and in the late stage of sugar beet growth; manual weeding was required to clear the weeds in the sugar beets as the coverage of sugar beets reached more than 90%. Diseases and pests, such as flea beetles and their disease, of sugar beet mainly appeared in July and August when the weather was hot, and the method of control was to use a sprayer to spray 1200 times diluted bifenthiacloprid and 24% jinggangmycin every 15 days, with a continuously spray pattern for 2–3 times to control the diseases and pests.

### 2.3. Sampling and Measurement of Sugar Beet

#### 2.3.1. Growth Index of Sugar Beet

On the 25th day after seeding, the germination rate (*GR*) of sugar beet was calculated using the following formula [29]:

$$GR = \frac{T_{SE}}{T_{SN}} \times 100 \tag{2}$$

where *GR* is germination rate, $T_{SE}$ is total number of seedlings emerging, and $T_{SN}$ is total seeding number.

Five sugar beet plants were randomly selected from each plot, and their height and root lengths were measured using a steel ruler with an accuracy of 1 mm. The tuber thickness of sugar beet root blocks was measured using a vernier caliper with an accuracy of 0.01 mm. The measurement dates were the 42nd, 71st, 96th, 118th, and 142nd days after sowing (*DAS*) in 2021; the 41st, 60th, 90th, 120th, and 154th DAS in 2022; and the 41st, 58th, 86th, 111th, and 131st DAS in 2023. All the leaves of the selected plant were measured to calculate the leaf area index (*LAI*). *LAI* was calculated using the following formula [30]:

$$LAI = \frac{N_P \times \sum_{i=1}^{m}\left(\sum_{j=1}^{n} L_{ij} \times W_{ij} \times 0.84\right)}{m \times 10^4} \tag{3}$$

where $N_P$ is the number of sugar beet planting density (plant $m^{-2}$), m is total number of measured plants, n is total leaf number for single plant, $L_{ij}$ and $W_{ij}$ are the leaf length and width (cm) on the *ith* plant, *i* and *j* are the *jth* leaf on the *ith* plant, and 0.84 is the conversion coefficient.

The relative chlorophyll content (presented as *SPAD* value) of sugar beet leaves was measured with a handheld chlorophyll instrument (model SPAD-502, Konica Minolta, Tokyo, Japan), and the measurement time was consistent with *LAI*. A portable photosynthetic instrument li-6400xt (Li COR, Lincoln, NE, USA) was used to monitor the photosynthetic capacity of sugar beet leaves. The measurement position was the middle part of sugar beet leaves, and 6 leaves were measured in each treatment.

#### 2.3.2. Yield, Biomass, and Sugar Content

During the sugar beet harvest period, three 6 $m^2$ (2 m × 3 m) sample areas were selected in each plot. All sugar beets in the sample area were excavated and weighed to measure the yields. The aboveground and underground parts of sugar beet were dried to a constant weight at 75 °C, and the biomass was weighed and recorded. During the sugar beet harvest period, a Japan Atago PAL-1 (Tokyo, Japan) portable digital display sugar meter was used to measure the sugar content. Sugar beet juice was extracted and placed into the instrument for sugar content analysis. The sugar yield was further calculated as [27]

$$\text{Sugar yield}\left(\text{t ha}^{-1}\right) = \text{taproot yield}\left(\text{t ha}^{-1}\right) \times \text{sugar content}(\%) \tag{4}$$

2.3.3. Irrigation Water Use Efficiency Indices

Taproot yield irrigation water use efficiency ($IWUE_T$) and sugar yield irrigation water uses efficiency ($IWUE_S$) were calculated as follows [31]:

$$IWUE_T = \frac{\text{Taproot yield}\left(\text{kg ha}^{-1}\right)}{\text{Irrigation amount}\left(\text{kg ha}^{-1}\right)} \quad (5)$$

$$IWUE_S = \frac{\text{Sugar yield}\left(\text{kg ha}^{-1}\right)}{\text{Irrigation amount}\left(\text{kg ha}^{-1}\right)} \quad (6)$$

2.3.4. Yield Sustainability

The sustainable yield index (*SYI*), the sustainable biomass index (*SBI*) and the coefficient of variance (*CV*) of the sugar beet yields were used to assess the stability and sustainability of grain yields and biomass. The *SYI* and *SBI* values were calculated as [32]

$$\text{Sustainable yieldt index } (SYI) = \frac{\overline{Y} - \sigma}{Y_{max}} \quad (7)$$

$$\text{Sustainable biomass index } (SBI) = \frac{B - \sigma}{B_{max}} \quad (8)$$

$$\text{Coefficient of variance}(CV) = \left(\sigma/\overline{Y}\right) \times 100 \quad (9)$$

where $\overline{Y}$ is the average sugar beet yield (t ha$^{-1}$), σ represents the standard deviations, B is the average biomass (t ha$^{-1}$), and $Y_{max}$ and $B_{max}$ are the maximum sugar beet yield and biomass, respectively.

2.3.5. Economic Benefit Analysis

Net return was calculated as follows [33]:

$$Net\ return = G_b - I_w - B_w - F_w - O \quad (10)$$

where $G_b$ is the sugar beet yield benefit (CNY ha$^{-1}$), $I_w$ is the irrigation water cost (CNY ha$^{-1}$), $B_w$ is the cost of biochar (CNY ha$^{-1}$), $F_w$ is the fertilizer input (CNY ha$^{-1}$), and O represents the other inputs (CNY ha$^{-1}$). The price of sugar beet was 505, 520, and 530 (CNY t$^{-1}$) in 2021, 2022, and 2023, respectively. The prices of irrigation water, biochar, and fertilizer were 0.3 CNY m$^{-3}$, 2100 CNY ha$^{-1}$, and 2140 CNY ha$^{-1}$, respectively. The cost of the other inputs including machinery, labor, seeds, drip tapes, plastic mulch, and material was about 11,100 CNY ha$^{-1}$.

2.4. *Statistical Analysis*

Microsoft Excel 2019 (Microsoft Corporation, NM, USA) was used to summarize and analyze the test data. The SPSS 26.0 statistical software (IBM, SPSS, Statistics 26.0) was used to analyze a two-factor (Irrigation amount and biochar application) analysis of variance (*ANOVA*) and Duncan multiple range test ($p < 0.05$ or $p < 0.01$). Utilizing the R package (corrplot) for obtaining correlation coefficients and analyzing the interrelationships among indicators, the entropy–TOPSIS multi-objective decision analysis method was employed to ascertain the weightings of indices and comprehensive evaluation levels for each treatment. Figures were drawn using the Origin 9.0 software.

## 3. Results

### *3.1. Effects of Biochar and Irrigation on the Germination Rate and Growth of Sugar Beet*

#### 3.1.1. Influence of Biochar Application Amount on Germination Rate

Under consistent irrigation amounts, the one-time biochar application in 2021 led to a gradual increase in the sugar beet germination rate from 2021 to 2023 (Figure 2), ranging 60.3–75.0%, 68.3–84.7%, and 79.6–95.4%, respectively. The highest germination rate was consistently observed under the B10 treatment. Compared to the control (B0), the germination rates for the B10, B20, and B30 treatment groups increased by 19.9–24.5%, 14.1–21.4%, and 7.2–18.9% over the three years, respectively.

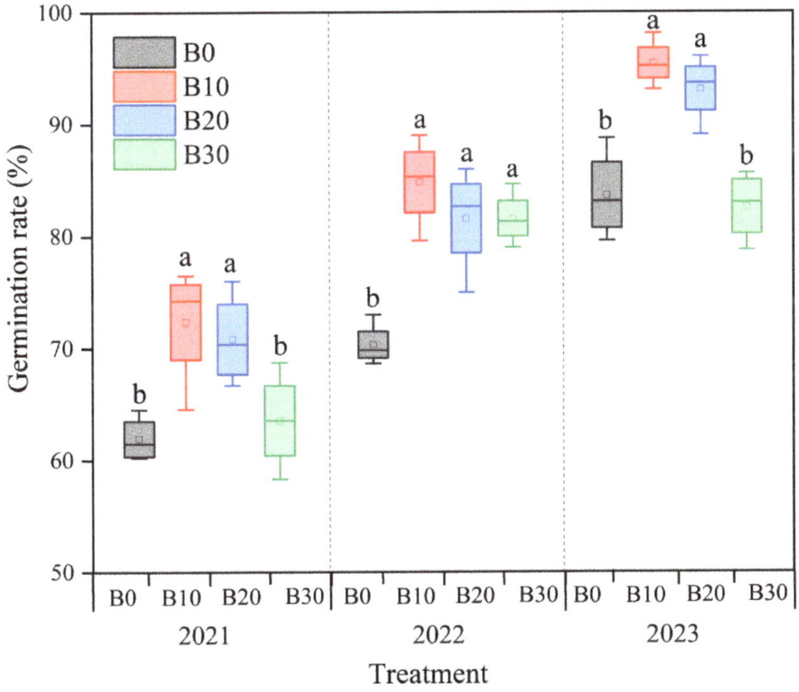


**Figure 2.** Effect of biochar application on the germination rate of sugar beet. Different letters above the bars indicate statistical differences among treatments at the significance level $p < 0.05$ with an LSD test.

#### 3.1.2. Synergistic Effects of Biochar and Irrigation on Growth of Sugar Beet

The changes in sugar beet plant height under varying irrigation levels and biochar application rates are presented in Figure 3. Throughout the three-year experiment, sugar beet plant height initially increased with the number of sowing days, peaking before eventually declining. The highest plant height was consistently observed under the I4B10 treatment. Biochar application significantly contributed to increased plant height under consistent irrigation conditions. In 2021, the plant heights for the B10, B20, and B30 treatments increased by 20.6%, 13.8%, and 6.9%, respectively, compared to the B0 treatment. Similarly, under the same biochar application rates, plant height increased with higher irrigation levels. In 2021, plant heights under the I2, I3, and I4 irrigation treatments increased by 13.1%, 17.3%, and 26.7%, respectively, compared to the I1 treatment. This trend was consistently observed in 2022 and 2023, demonstrating the beneficial effects of both biochar application and increased irrigation on sugar beet plant height.

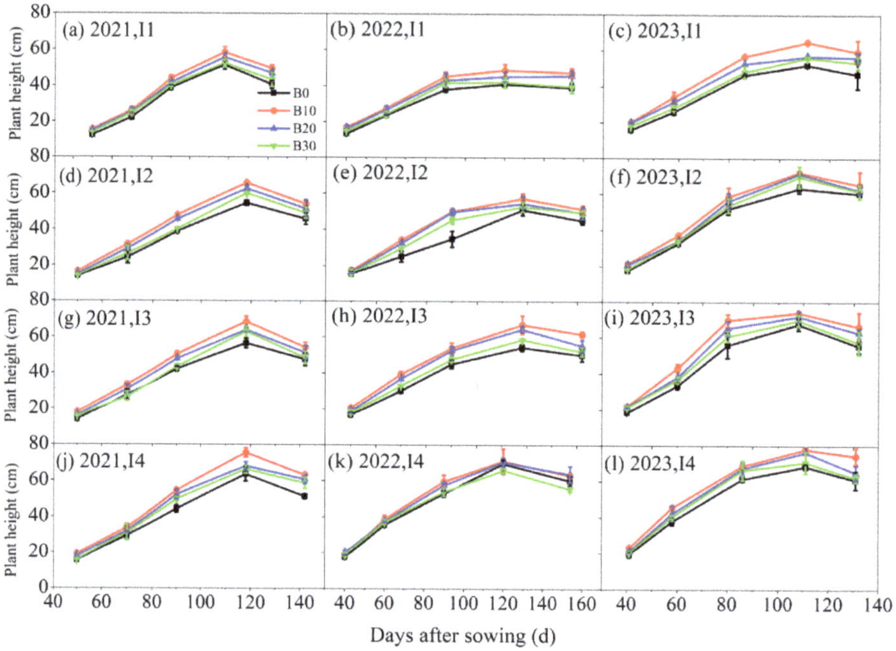


**Figure 3.** Plant height of sugar beet under different irrigation amounts and biochar application rates in 2021, 2022, and 2023.

The *LAI* of sugar beet was influenced by both irrigation amounts and biochar application rates (Figure 4). Over the three-year experiment, *LAI* increased with higher irrigation levels and exhibited an initial rise followed by a decline with increasing biochar application rates, peaking under the I4B10 treatment. Under consistent irrigation conditions, *LAI* in the B10, B20, and B30 biochar treatments increased by 32.6–42.1%, 20.2–35.1%, and 2.8–8.7%, respectively, compared to the B0 treatment without biochar. Similarly, at a constant biochar application rate, *LAI* under the I2, I3, and I4 irrigation treatments increased by −6.8–29.5%, 0.7–15.3%, and 3.2–16.8%, respectively, compared to the I1 treatment. These data highlight the synergistic effects of irrigation and biochar application on improving the *LAI* of sugar beet.

Under varying irrigation levels and biochar application rates, the tuber diameter and tuber length of sugar beet reached their maximum under the I4B10 treatment, both increasing with the number of sowing days (Figures 5 and 6). Biochar application led to a 5.6%, 19.2%, and 6.3% increase in tuber diameter under the I2, I3, and I4 irrigation treatments, respectively, compared to the I1 treatment. Similarly, with the same irrigation amount, the B10, B20, and B30 biochar treatments resulted in a 31.2%, 22.4%, and 12.3% increase in tuber diameter, respectively, compared to the B0 treatment. Additionally, under the same irrigation amount, root length followed the trend B10 > B20 > B30 > B0. Conversely, with the same biochar application rate, tuber length increased with irrigation amounts in the order I4 > I3 > I2 > I1.

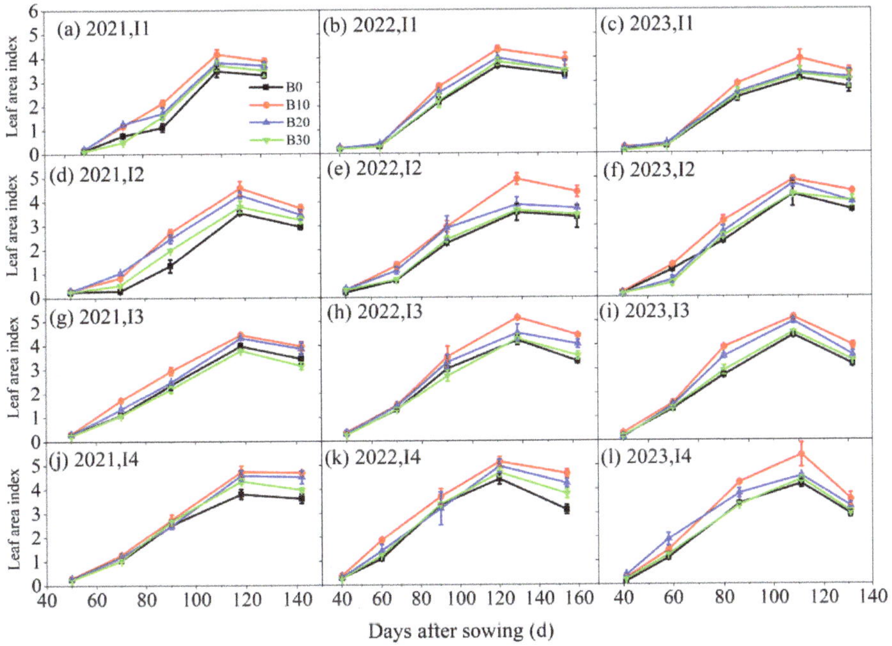


**Figure 4.** Leaf area index of sugar beet under different irrigation amounts and biochar application rates in 2021, 2022, and 2023.

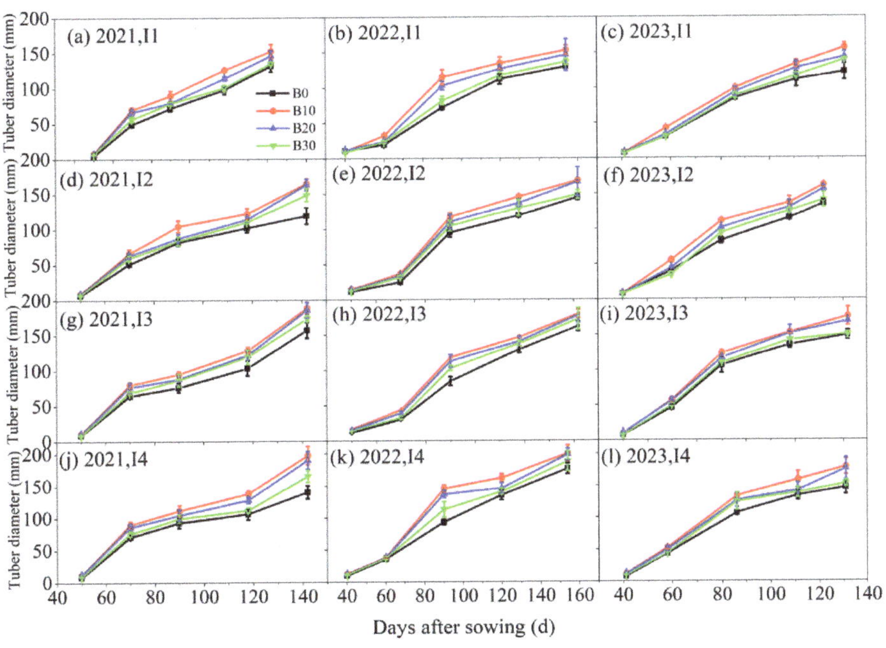


**Figure 5.** Tuber diameter under different irrigation amounts and biochar application rates in 2021, 2022, and 2023.

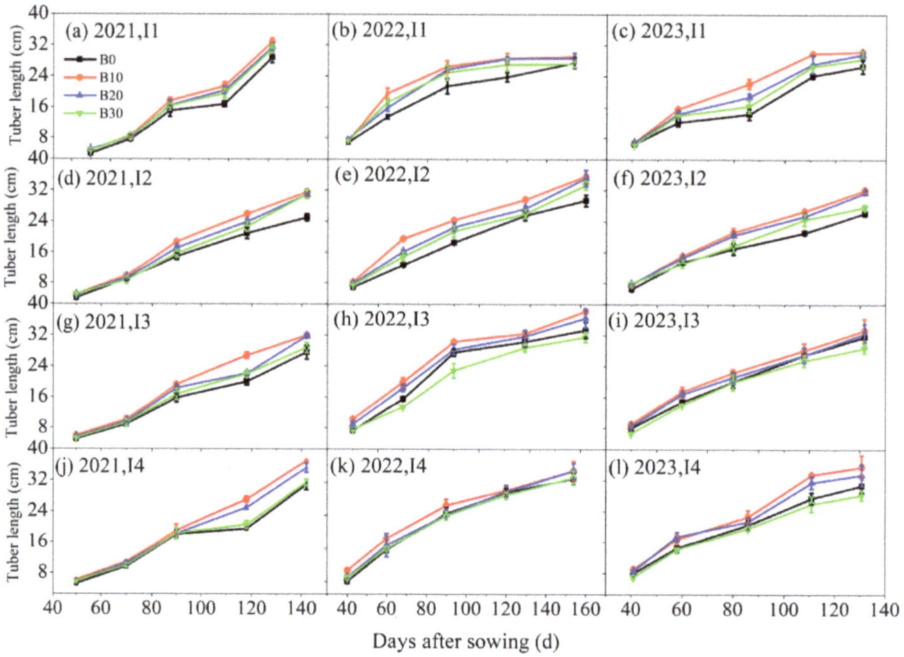


**Figure 6.** Tuber length of sugar beet under different irrigation amounts and biochar application rates in 2021, 2022, and 2023.

### 3.2. *Effects of Biochar and Irrigation on Physiological Indexes*

Effects of Irrigation and Biochar Application on *SPAD*

Table 2 presents the effects of irrigation and biochar application on the *SPAD* values of sugar beet during the seedling stage, the rapid leaf growth stage, the sugar accumulation stage, and the harvest stage in 2021, 2022, and 2023. Throughout the growth stages, *SPAD* values generally increased initially and then decreased, peaking during the sugar accumulation period. Under consistent irrigation conditions, increasing biochar application significantly enhanced *SPAD* values. In 2021, *SPAD* values for the B10, B20, and B30 treatments increased by 8.1–13.5%, 11.5–14.8%, and 13.1–22.1%, respectively, compared to the B0 treatment. In 2022 and 2023, SPAD values increased by 6.2–9.3%, 6.8–13.4%, and 11.9–14.1%, and by 3.1–7.7%, 10.0–13.6%, and 11.3–17.8%, respectively. With the same biochar application rates, *SPAD* values under the I2, I3, and I4 irrigation treatments increased by 1.6% and 0.3% and decreased by 0.3% compared to the I1 treatment, respectively. Statistical analysis over the three-year period indicated that the irrigation amount significantly affected *SPAD* values only during the rapid leaf growth and sugar accumulation stages, while biochar application had a significant impact on *SPAD* values throughout all growth stages. The interaction between irrigation and biochar (I × B) showed significant differences in *SPAD* values only during the rapid leaf growth period.

Figure 7 shows the effects of different irrigation amounts and biochar application rates on net photosynthetic rate (*Pn*), stomatal conductance (*Gs*), intercellular carbon dioxide concentration (*Ci*), and transpiration rate (*Tr*) of sugar beet. With the same amount of irrigation, *Pn* first increased and then decreased with increasing biochar application rates, reaching its maximum value under the B10 treatment (Figure 7a). In contrast, *Ci* exhibited an opposite trend, increasing with higher biochar application rates. Under the B10, B20, and B30 treatments, *Ci* increased by an average of −5.6%, 6.5%, and 10.1% compared to the

B0 treatment (Figure 7b). Under consistent irrigation conditions, both *Tr* and *Gs* achieved their highest values under the B10 treatment, increasing by 18.7–42.9% and 16.4–44.8%, respectively, compared to the B0 treatment. *Tr* decreased by 3.7–36.8% and 14.6–33.1% in the B20 and B30 treatments, respectively, compared to the B0 treatment, while *Gs* decreased by 0.7–13.1% and 10.4–18.2%, respectively (Figure 7c,d). With the same biochar application rate, Tr and *Gs* initially increased and then decreased with higher irrigation amounts, reaching their maximum values under the I3 irrigation level.

**Table 2.** *SPAD* values under different irrigation and biochar combinations in 2021, 2022, and 2023.

| Treatment | 2021 | | | | 2022 | | | | 2023 | | | |
|---|---|---|---|---|---|---|---|---|---|---|---|---|
| | Seedling | Leaf Fast Growth | Root Sugar Increment | Harvesting | Seedling | Leaf Fast Growth | Root Sugar Increment | Harvesting | Seedling | Leaf Fast Growth | Root Sugar Increment | Harvesting |
| I1B0 | 34.7 c | 38.4 c | 44.3 b | 41.0 a | 37.9 c | 53.3 b | 53.4 b | 46.1 c | 37.3 c | 54.7 c | 50.1 c | 45.2 c |
| I1B10 | 36.6 b | 40.8 b | 51.4 a | 45.8 c | 47.2 a | 55.2 a | 54.5 b | 51.6 b | 43.3 a | 59.9 b | 52.3 b | 47.2 b |
| I1B20 | 36.8 b | 40.6 b | 52.4 a | 50.9 a | 44.7 b | 55.2 a | 56.4 ab | 54.1 a | 42.9 a | 59.8 b | 56.9 a | 49.6 ab |
| I1B30 | 38.8 a | 43.2 a | 52.3 a | 48.3 b | 45.9 ab | 56.5 a | 57.6 a | 53.3 a | 41.2 b | 60.0 a | 57.2 a | 51.2 a |
| I2B0 | 37.9 c | 40.5 c | 42.3 b | 41.3 c | 39.2 b | 51.5 c | 56.2 b | 47.3 b | 38.5 c | 51.8 d | 52.8 c | 44.8 c |
| I2B10 | 39.1 b | 42.1 b | 48.8 a | 45.3 b | 44.4 a | 55.5 b | 60.8 a | 51.4 a | 41.8 b | 55.6 c | 55.3 b | 47.1 b |
| I2B20 | 40.5 ab | 43.2 ab | 49.3 a | 47.6 a | 43.3 a | 59.6 a | 60.6 a | 52.3 a | 44.5 a | 63.1 b | 57.2 ab | 48.7 b |
| I2B30 | 41.8 a | 44.6 a | 49.6 a | 47.2 a | 42.7 ab | 60.8 a | 61.3 a | 52.9 a | 44.2 a | 68.6 a | 58.0 a | 50.8 a |
| I3B0 | 40.0 b | 40.0 c | 44.3 c | 40.4 c | 37.9 b | 53.2 c | 52.5 c | 44.0 c | 40.3 c | 49.6 d | 49.2 b | 45.0 c |
| I3B10 | 42.1 a | 42.9 b | 48.2 b | 45.0 b | 41.2 a | 55.8 b | 57.6 b | 49.6 b | 42.4 b | 53.5 c | 50.6 b | 46.5 c |
| I3B20 | 42.5 a | 44.7 a | 48.6 b | 48.1 a | 42.2 a | 58.7 a | 58.8 ab | 53.0 a | 42.3 b | 55.6 b | 55.9 a | 48.8 b |
| I3B30 | 43.2 a | 45.3 a | 52.2 s | 49.8 a | 42.6 a | 59.1 a | 59.1 a | 53.3 a | 46.6 a | 59.5 a | 56.1 a | 53.9 a |
| I4B0 | 36.1 c | 38.9 c | 42.4 c | 40.5 c | 39.2 b | 56.2 b | 55.9 c | 43.6 c | 37.7 c | 50.1 b | 50.3 b | 46.7 c |
| I4B10 | 42.8 a | 44.4 b | 46.6 b | 46.3 b | 42.1 ab | 59.3 ab | 58.7 b | 47.4 b | 40.3 b | 52.1 b | 50.6 b | 47.5 bc |
| I4B20 | 41.4 b | 45.1 ab | 47.3 b | 46.9 b | 42.2 ab | 61.4 a | 58.8 b | 45.7 b | 44.2 a | 56.1 a | 54.2 a | 48.8 b |
| I4B30 | 43.1 a | 46.8 a | 51.2 a | 51.6 a | 43.4 a | 61.8 a | 62.6 a | 50.7 a | 44.8 a | 54.8 a | 55.4 a | 53.0 a |
| ANOVA | | | | | | | | | | | | |
| I | * | ** | ** | NS | * | ** | ** | ** | NS | ** | ** | NS |
| B | ** | ** | * | ** | ** | ** | ** | ** | ** | ** | ** | ** |
| I × B | ** | ** | ** | ** | NS | * | NS | NS | * | ** | NS | NS |

Note: Different letters represent statistical differences between treatments with a significance level of $p < 0.05$ in the LSD test; * indicates that it is significant at the level of 0.05; ** indicates that it is significant at the level of 0.01; I is the irrigation amount; B represents the biochar application rates.

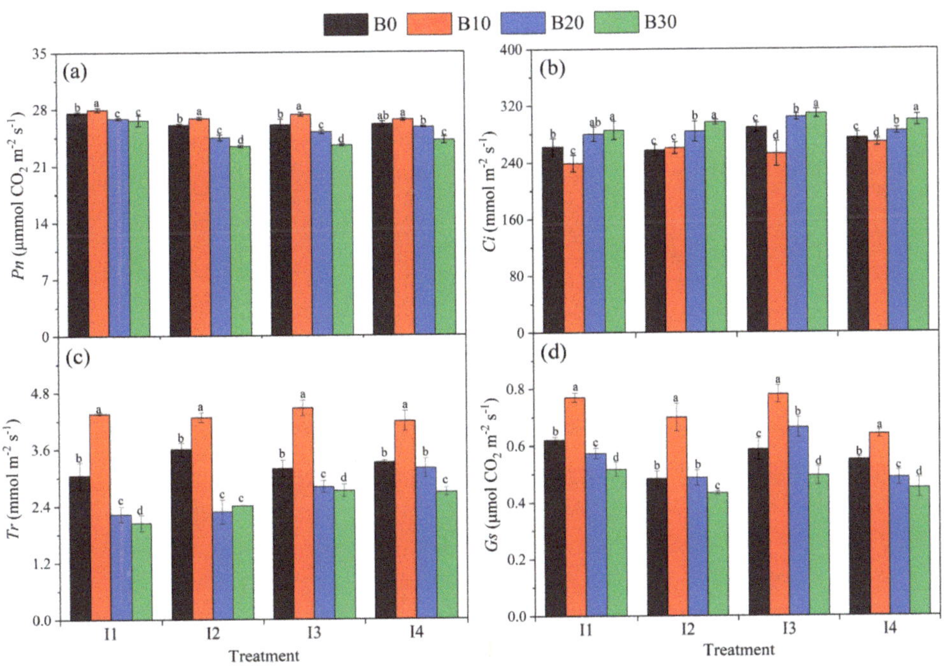


**Figure 7.** (**a–d**) Net photosynthetic rate (*Pn*), stomatal conductance (*Gs*), intercellular carbon dioxide (*Ci*), and transpiration rate (*Tr*) of sugar beet under different irrigation amounts and biochar application rates. Error bars represent standard errors. Different letters above the bars indicate statistical differences among treatments at the significance level $p < 0.05$ with an LSD test.

### 3.3. Effects of Irrigation and Biochar Application Yield, Biomass, Quality, and IWUE

The irrigation amount and biochar application significantly impacted the yield and biomass of sugar beet during the harvesting period, although the interaction between irrigation amount (I) and biochar application amount (B) did not significantly affect the dry matter content (Table 3). Across all treatments, the highest sugar beet production was observed at I4B10, with yields of 130.7, 147.6, and 141.4 t ha$^{-1}$ in 2021, 2022, and 2023, respectively. Under consistent biochar application rates, increased irrigation water enhanced sugar beet yield. Compared to the I1, I2, and I3 treatments, the yield under the I4 treatment increased by an average of 68.4%, 39.9%, and 12.8%. Interestingly, both $IWUE_T$ and $IWUE_S$ decreased with higher irrigation amounts. Compared to I1, I2, and I3, $IWUE_T$ under the I4 treatment decreased by 2.8%, 0.6%, and 2.7%, while $IWUE_S$ decreased by 19.4%, 10.5%, and 10.7%. The use of biochar also increased sugar beet yield; however, when the biochar application rate exceeded 20 t ha$^{-1}$, there was a decrease in yield with further increases in biochar application. Conversely, this trend was observed for $IWUE_T$ and $IWUE_S$ (Table 4).

**Table 3.** Sugar beet yield, biomass, and sugar yield under different irrigation and biochar combinations in 2021, 2022, and 2023.

| Treatment | 2021 | | | 2022 | | | 2023 | | |
|---|---|---|---|---|---|---|---|---|---|
| | Taproot Yield (t ha$^{-1}$) | Biomass (t ha$^{-1}$) | Sugar Yield (t ha$^{-1}$) | Taproot Yield (t ha$^{-1}$) | Biomass (t ha$^{-1}$) | Sugar Yield (t ha$^{-1}$) | Taproot Yield (t ha$^{-1}$) | Biomass (t ha$^{-1}$) | Sugar Yield (t ha$^{-1}$) |
| I1B0 | 58.5 d | 15.7 c | 7.5 c | 53.7 d | 16.2 ab | 8.3 d | 54.8 b | 13.6 a | 8.9 c |
| I1B10 | 89.5 b | 16.9 b | 11.8 b | 66.2 c | 17.6 a | 10.9 c | 87.4 a | 14.2 a | 14.8 a |
| I1B20 | 108.6 a | 20.2 a | 14.6 a | 90.8 a | 17.2 a | 15.4 a | 79.8 a | 13.1 a | 13.5 ab |
| I1B30 | 79.4 c | 16.7 b | 11.2 b | 80.1 b | 16.7 b | 13.7 b | 76.4 ab | 11.3 b | 13.1 b |
| I2B0 | 61.5 b | 13.9 d | 7.4 c | 85.5 c | 17.2 b | 11.8 c | 76.6 c | 14.2 c | 11.0 c |
| I2B10 | 93.2 a | 16.4 c | 11.1 b | 120.3 a | 19.5 a | 19.4 a | 98.3 a | 18.6 a | 13.5 b |
| I2B20 | 99.9 a | 20.2 b | 13.4 a | 117.2 b | 19.9 a | 19.6 a | 90.8 b | 15.7 b | 13.7 b |
| I2B30 | 86.2 ab | 21.1 a | 11.8 ab | 85.2 c | 18.0 b | 14.1 b | 90.6 b | 18.8 a | 14.8 a |
| I3B0 | 74.5 c | 17.1 d | 7.7 c | 106.4 c | 18.1 b | 14.7 c | 87.4 c | 13.4 c | 12.8 c |
| I3B10 | 128.5 a | 26.5 b | 14.9 a | 147.3 a | 19.9 a | 23.2 a | 140.5 a | 24.6 a | 20.5 a |
| I3B20 | 117.8 a | 28.3 a | 14.1 a | 128.0 b | 19.8 a | 19.6 b | 132.7 b | 23.8 a | 20.3 a |
| I3B30 | 95.4 b | 19.7 c | 12.3 b | 115.5 b | 18.9 b | 17.9 b | 106.7 b | 17.7 b | 17.1 b |
| I4B0 | 112.8 b | 18.7 c | 11.0 b | 115.8 c | 18.5 ab | 15.2 b | 106.3 b | 22.7 a | 14.1 c |
| I4B10 | 130.7 ab | 26.6 a | 13.2 a | 147.6 a | 21.5 ab | 22.6 a | 141.3 a | 22.8 a | 19.4 a |
| I4B20 | 128.5 a | 27.6 a | 13.5 a | 143.6 b | 21.8 a | 20.9 a | 138.7 a | 20.5 b | 19.7 a |
| I4B30 | 119.1 c | 22.1 b | 13.1 a | 140.1 b | 19.5 b | 21.9 a | 117.2 b | 19.3 b | 17.5 b |
| ANOVA | | | | | | | | | |
| I | ** | * | ** | ** | * | ** | ** | ** | ** |
| B | ** | ** | ** | * | * | ** | ** | ** | ** |
| I×B | * | NS | ** | NS | NS | ** | ** | NS | ** |

Note: Different letters represent statistical differences between treatments with a significance level of $p < 0.05$ in the LSD test; * indicates that it is significant at the level of 0.05; ** indicates that it is significant at the level of 0.01; I is the irrigation amount; B represents the biochar application rates.

**Table 4.** Irrigation and biochar application treatments on $IWUE_T$ and $IWUE_S$ in 2021, 2022, and 2023.

| Treatment | 2021 | | 2022 | | 2023 | |
|---|---|---|---|---|---|---|
| | $IWUE_T$ | $IWUE_S$ | $IWUE_T$ | $IWUE_S$ | $IWUE_T$ | $IWUE_S$ |
| I1B0 | 25.79 d | 3.32 c | 22.98 d | 3.56 d | 21.09 c | 3.44 c |
| I1B10 | 39.43 b | 5.22 b | 28.29 c | 4.70 c | 33.63 a | 5.72 a |
| I1B20 | 47.86 a | 6.46 a | 38.84 a | 6.62 a | 30.72 b | 5.19 b |
| I1B30 | 34.98 c | 4.96 b | 34.25 b | 5.87 b | 29.39 b | 5.05 b |

**Table 4.** *Cont.*

| Treatment | 2021 | | 2022 | | 2023 | |
|---|---|---|---|---|---|---|
| | $IWUE_T$ | $IWUE_S$ | $IWUE_T$ | $IWUE_S$ | $IWUE_T$ | $IWUE_S$ |
| I2B0 | 22.26 c | 2.71 v | 28.99 b | 4.02 c | 23.57 b | 3.39 b |
| I2B10 | 33.71 b | 4.00 b | 43.49 a | 6.60 a | 27.94 a | 4.17 a |
| I2B20 | 36.15 a | 4.85 a | 41.77 a | 6.65 a | 27.81 a | 4.21 a |
| I2B30 | 31.18 b | 4.29 ab | 28.91 b | 4.8 b | 27.89 a | 4.57 a |
| I3B0 | 22.8 b | 2.36 c | 29.75 c | 4.13 d | 22.41 c | 3.3 c |
| I3B10 | 39.31 a | 4.58 a | 43.39 a | 6.49 a | 34.51 a | 5.26 a |
| I3B20 | 36.02 a | 4.31 a | 35.75 b | 5.49 b | 32.96 a | 5.2 a |
| I3B30 | 29.19 b | 3.77 b | 32.27 bc | 5.00 c | 26.56 b | 4.38 b |
| I4B0 | 29.92 b | 2.92 b | 27.85 c | 3.66 b | 23.45 c | 3.1 b |
| I4B10 | 34.68 a | 3.52 a | 37.16 a | 5.44 a | 31.07 a | 4.27 a |
| I4B20 | 34.10 a | 3.59 a | 34.52 b | 5.04 a | 30.50 a | 4.34 a |
| I4B30 | 31.6 b | 3.48 a | 33.67 b | 5.29 a | 25.76 b | 3.85 ab |
| ANOVA | | | | | | |
| I | * | * | ** | * | ** | * |
| B | ** | ** | * | * | * | ** |
| I × B | * | NS | * | * | ** | ** |

Note: Different letters represent statistical differences between treatments with a significance level of $p < 0.05$ in the LSD test; * indicates that it is significant at the level of 0.05; ** indicates that it is significant at the level of 0.01; I is the irrigation amount; B represents the biochar application rates.

### 3.4. Variation in Sugar Content

Both irrigation levels and biochar application significantly influenced the sugar content in sugar beet (Figure 8). Under the same irrigation level, sugar content increased with higher biochar application rates, with the B30 treatment yielding significantly higher sugar content than the B0 (no biochar) treatment. In 2021, the sugar content of sugar beet under the B10, B20, and B30 treatments increased by 4.3%, 9.7%, and 15.3%, respectively, compared to the B0 treatment. In 2022, these treatments resulted in increases of 9.0%, 11.7%, and 15.4%, and in 2023, by 3.7%, 5.7%, and 10.9%, respectively. Conversely, under the same biochar application rate, sugar content decreased with increasing irrigation amounts over the three years. In 2021, sugar content under the I1 irrigation level increased by an average of 4.9%, 12.9%, and 22.8% compared to the I2, I3, and I4 treatments, respectively. This trend continued in 2022 and 2023, with increases of 7.1%, 9.8%, and 12.4%, and 9.8%, 7.6%, and 16.7%, respectively.

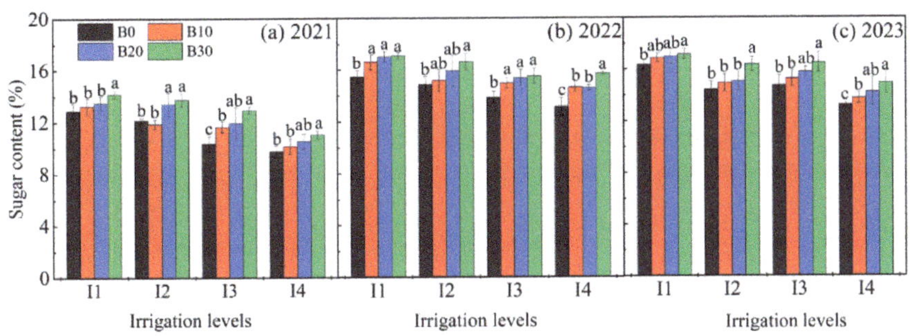


**Figure 8.** Sugar content under different irrigation amounts and biochar application rates in 2021, 2022, and 2023. Error bars represent standard errors. Different letters above the bars indicate statistical differences among treatments at the significance level $p < 0.05$ with an LSD test.

### 3.5. Effects of Irrigation and Biochar Application on SYI

The variations in the *SYI*, *SBI*, and the *CV* of sugar beet yield and biomass under different irrigation amounts and biochar application rates are presented in Table 5. Under consistent irrigation conditions, both SYI and SBI initially increased and then decreased with increasing biochar application rates, reaching their highest values under the B10 treatment. Conversely, *CV* for yield (*CVy*) and biomass (*CVb*) first decreased and then increased, hitting their lowest points under the B10 treatment. With a fixed biochar application rate, *SYI* and *SBI* achieved their maximum values under the I3 irrigation treatment, while *CV* reached its minimum under the same treatment. Compared to the I1, I2, and I4 treatments, the I3 irrigation treatment increased SYI and SBI by 1.1–11.6% and decreased *CV* by 0.3–8.2%.

**Table 5.** Stability and sustainability index of sugar beet yield and biomass.

| Treatment | Sugar Beet Yield | | Sugar Beet Biomass | |
|---|---|---|---|---|
| | *SYI* | *CV* | *SBI* | *CV* |
| I1B0 | 0.68 | 13.77 | 0.56 | 26.15 |
| I1B10 | 0.82 | 8.98 | 0.75 | 14.32 |
| I1B20 | 0.78 | 9.68 | 0.66 | 20.75 |
| I1B30 | 0.73 | 11.48 | 0.63 | 24.11 |
| I2B0 | 0.66 | 16.75 | 0.63 | 19.76 |
| I2B10 | 0.83 | 6.75 | 0.76 | 9.99 |
| I2B20 | 0.79 | 9.12 | 0.75 | 11.99 |
| I2B30 | 0.76 | 12.39 | 0.57 | 17.98 |
| I3B0 | 0.64 | 17.17 | 0.4 | 32.02 |
| I3B10 | 0.89 | 4.77 | 0.81 | 11.44 |
| I3B20 | 0.85 | 7.55 | 0.76 | 11.88 |
| I3B30 | 0.77 | 11.05 | 0.61 | 20.64 |
| I4B0 | 0.74 | 9.75 | 0.52 | 26.41 |
| I4B10 | 0.84 | 7.35 | 0.72 | 12.21 |
| I4B20 | 0.82 | 8.09 | 0.64 | 19.72 |
| I4B30 | 0.78 | 10.16 | 0.58 | 18.6 |

Note: *SYI* is the sustainable yield index; *SBI* is the sustainable biomass index; *CV* is the coefficient of variance.

### 3.6. Economic Benefit (EB) Analysis

Figure 9 shows the cost, revenue, and net profit of sugar beet production from 2021 to 2023 under different treatments. Under the same irrigation levels, *EB* initially increased and then decreased with higher biochar application. The highest values under both the I1 and I2 irrigation amounts were reached under the B20 treatment; however, the maximum EB was reached with the B10 treatment under the I3 and I4 irrigation levels. Over the three years, the *EB* for the B10, B20, and B30 biochar treatments increased by 38.8%, 13.2%, and −47.4%, respectively, compared to the B0 treatment. Under the same biochar application rate, increasing irrigation amounts enhanced the *EB* of sugar beet and reached its maximum at the I4 irrigation level.

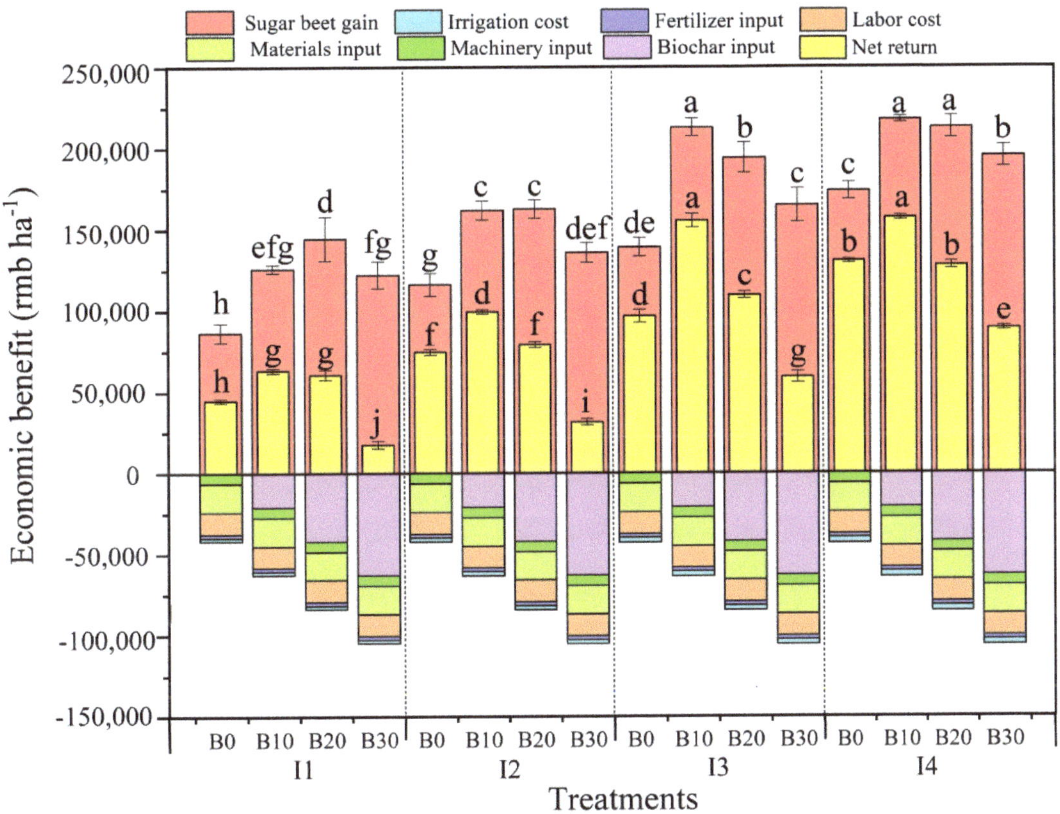


**Figure 9.** Economic benefit of sugar beet cotton under different irrigation amounts and biochar application rates in 2021, 2022, and 2023. Error bars represent standard errors. Different letters above the bars indicate statistical differences among treatments at the significance level $p < 0.05$ with an LSD test.

### 3.7. Comprehensive Evaluation

Figure 10 shows the relationships among various indices, such as beet emergence rate, yield, *WUE*, *EB*, and *SYI*, under different combinations of irrigation and biochar treatments. The data indicate that the I3B10 treatment consistently resulted in higher levels of yield, biomass, and sugar content. Treatments I3B10, I3B20, I4B10, and I4B20 clustered together, with I3B10 exhibiting superior *WUE*, *EB*, and *SYI*. Conversely, treatments I1B0, I2B0, and I3B0 formed a distinct group, underscoring the catalytic effects of biochar on sugar beet growth (Figure 10a). Principal component analysis (*PCA*) revealed that yield, biomass, *WUE*, *SPAD*, sugar content, and *SYI* were closely interrelated. Although *GR* can enhance sugar beet yield, its effect was not significant (Figure 10b). A comprehensive analysis of various indices identified the B10 biochar application as the most favorable treatment (Figure 10c).

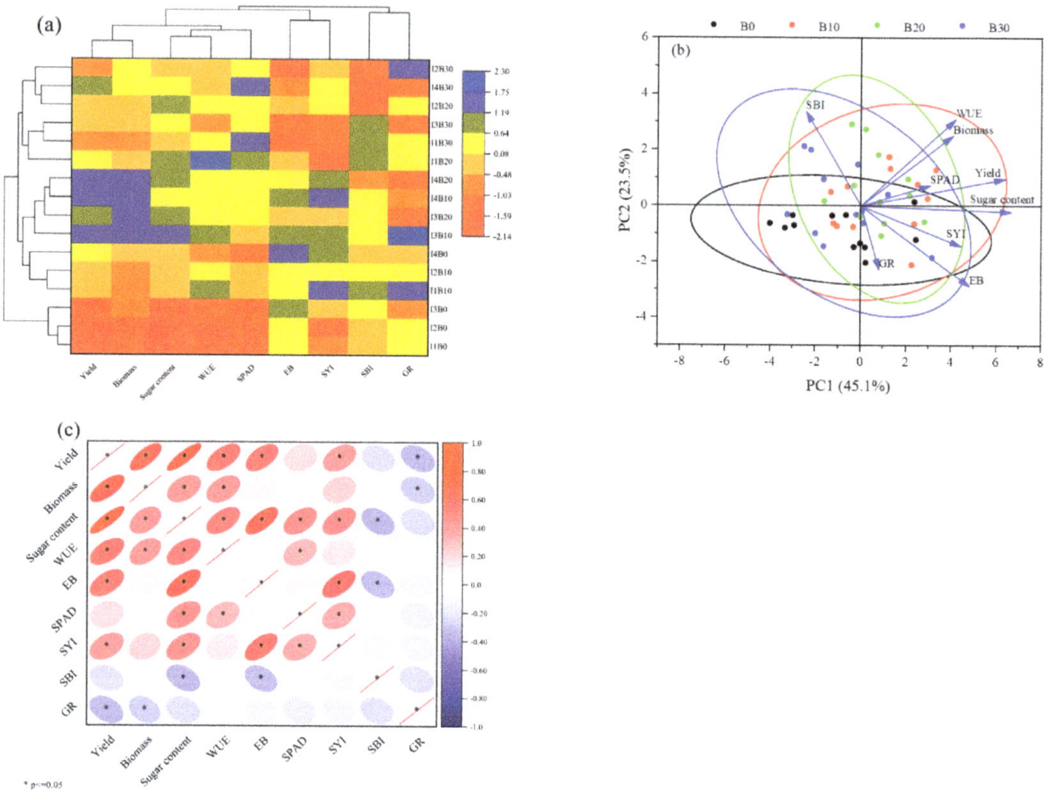


**Figure 10.** Heat maps (**a**), principal component analysis (**b**), and correlation analysis (**c**) for different relevant indicators under different irrigation and biochar treatments in 2021, 2022, and 2023. *WUE*, irrigation water use efficiency; *EB*, economic benefit; *SYI*, sustainable yield index; *SBI*, sustainable biomass index; *GR*, germination rate; *, $p < 0.05$.

## 4. Discussion

### *4.1. Effects of Biochar Application on Germination Rate*

Germination rate is a critical factor influencing crop yield, as it is affected by various factors [34,35]. Understanding the impact of biochar application on crop germination rate is essential for promoting crop growth, increasing yield, and achieving higher economic benefits. Bai et al. [36] demonstrated that biochar produced through pyrolysis at 200 °C had a more pronounced stimulatory effect on seed germination compared to biochar produced at 500 °C and 800 °C. Additionally, they found that higher application rates of biochar significantly decreased crop germination rates. Conversely, Wang et al. [37] discovered that biochar produced from almond shells at 550 °C was more conducive to cucumber germination, with a significantly higher germination rate at 60 t ha$^{-1}$ compared to 30 t ha$^{-1}$.

Das et al. [38] studied the impact of biochar on maize and rye germination rates, revealing significant increases at an application rate of 10 t ha$^{-1}$ compared to 5 t ha$^{-1}$ and no biochar treatment. Similarly, Bouqbis et al. [39] reported that biochar application significantly enhanced barley germination rates and increased the fresh weight of barley seedlings. Interestingly, Quilliam et al. [40] found that a single application of biochar did not significantly affect weed germination rates, but continuous biochar application significantly reduced weed germination rates.

Li et al. [2] observed that after three years of continuous biochar application, the germination rate of sugar beet significantly improved at an application rate of 10 t $ha^{-1}$ but decreased with higher application rates. However, continuous biochar application did not lead to a significant improvement in the germination rate of sugar beet. Our study corroborates these findings, as shown in Figure 2. The results indicate that biochar application can enhance the germination rate of sugar beet, with the optimal application amount being 10 t $ha^{-1}$. Additionally, we found that the germination rate of sugar beet increased with prolonged biochar application duration, underscoring the potential benefits of biochar in sustainable agriculture practices.

### 4.2. Synergistic Effects of Biochar and Irrigation on Chlorophyll

Chlorophyll serves as a crucial indicator of nutrient absorption capacity and crop growth status [41]. Its production is influenced by various factors, including irrigation, fertilizers, and environmental conditions [42]. The role of irrigation in plant growth has been extensively studied for its impact on chlorophyll levels [43,44]. Bekhradi et al. [45] found no significant difference in basic chlorophyll content with varying irrigation amounts, but under deficit irrigation conditions (50% of field capacity), chlorophyll content showed a decreasing trend. Wei et al. [46] reported that nitrogen levels and crop growth periods were determinants of chlorophyll content, whereas irrigation levels did not significantly affect it. Conversely, Dos et al. [47] discovered that chlorophyll content significantly increased with higher irrigation levels under conditions ranging from 25% to 100% *ETc*. Similarly, Tourajzadeh et al. [48] demonstrated that chlorophyll content in quinoa was significantly higher under 100% crop water irrigation compared to 60% irrigation, with no significant difference observed compared to 80% irrigation.

Dos et al. [49] also found that increasing both irrigation and nitrogen application elevated chlorophyll content and photosynthetic rates in *Tithonia diversfolia*. This trend was similarly observed by Li et al. [50] throughout the growth period of winter wheat. Our study corroborates these findings, showing that chlorophyll content increases with higher irrigation levels (60–100% *ETc*) but reaches its lowest point at 120% irrigation level (Table 3). Excessive irrigation led to overgrowth of branches and leaves, likely due to insufficient nutrient supply, consistent with findings by Norouzi et al. [51].

Moreover, our study revealed that chlorophyll content increased with higher biochar application, consistent with the results reported by Li et al. [52]. Biochar application has been shown to significantly increase leaf chlorophyll content. Previous research indicated that biochar is rich in various organic compounds, enhancing soil organic matter and facilitating nutrient uptake in crops [53]. Feng et al. [54] demonstrated a nearly 30% increase in maize chlorophyll content with a biochar application rate of 50 t $ha^{-1}$. Tiong et al. [19] reported a significant increase in tomato chlorophyll content under biochar-treated conditions compared to those without biochar treatment. However, Akhtar et al. [55] observed a significant decrease in tomato chlorophyll content with a biochar application rate of 5 t $ha^{-1}$, consistent with Kammann et al. [56] in their study of wild quinoa. The increase in the soil C/N ratio following biochar application can result in nitrogen fixation, reducing nitrogen absorption by plants, which may be the primary reason for the observed decrease in chlorophyll content [57].

### 4.3. Synergistic Effects of Biochar and Irrigation on Growth and Yield of Sugar Beet

The amount of irrigation and biochar application significantly affected the yield and biomass of sugar beet (Table 3). Appropriate irrigation promotes the development of crop roots, accelerates plant growth, and improves yield and *WUE* [45]. Li et al. [58] demonstrated that increasing irrigation can enhance crop yield, but excessive irrigation significantly reduces WUE. Liu et al. [59] also confirmed that increased irrigation improves water and fertilizer utilization efficiency in wheat, thereby increasing its production. Conversely, water deficits hinder growth and crop development, shorten the growth cycle, and reduce yield and biomass [60].

This experiment's findings indicated that higher irrigation levels contribute to enhanced sugar beet yield, with yields under 1.2 *ETc* irrigation significantly surpassing those under the 0.6 *ETc* and 0.8 *ETc* irrigation levels, but the difference is not significant compared to the 1.0 *ETc* irrigation level. The results are consistent with Yan et al. [26]. And differences in soil types, climatic conditions, and management practices can influence sugar beet water demand [61].

Biochar application has been shown to enhance plant growth and increase yield [55,62]. The promotional effect of biochar on plant growth is largely contingent on its physical structure and properties. Increased porosity and reduced bulk density effectively enhance gas exchange and nutrient absorption by plant roots [4]. Additionally, biochar augments soil water holding capacity and improves plant water status [63]. It also enhances soil organic matter and other nutrients, facilitating plant absorption and leading to increased yield [13,64]. However, excessive biochar application can be detrimental. Li et al. [2] observed a decline in sugar beet yield when biochar application exceeded 10 t $ha^{-1}$. Similarly, Wang et al. [36] found that cotton yield initially increased with biochar application rates up to 25 t $ha^{-1}$ but decreased beyond this threshold. Feng et al. [65] noted a similar trend in maize, with yields declining when biochar application exceeded 30 t $ha^{-1}$. These studies confirm that while biochar can benefit soil and crop growth, it is essential to apply the appropriate amount. This study also found that the application of biochar significantly increased sugar beet yield and biomass (Table 3), and the application of biochar treatment resulted in an average increase of 15.73% in yield compared to the treatment without biochar addition.

Considering factors such as crop yield, economic benefits, *WUE*, and yield stability index, we recommend an irrigation level of 1.0 *ETc* and a biochar application rate of 10 t $ha^{-1}$ as the optimal combination for arid and semi-arid regions. However, to fully understand the long-term effects of biochar application in Xinjiang and similar ecosystems, factors such as biochar type, raw material availability, and preparation process must be evaluated. Additionally, subsequent long-term experiments should address soil properties, nutrient availability, water management, and fertilizer use efficiency.

The application of biochar increased the seedling rate of sugar beet by 7.2–24.5%. The growth indexes of beet (stem diameter, plant height, leaf area index, and root length) increased with the increase in irrigation amount; it first increased and then decreased with the increase in biochar application amount, and all reached the maximum value under the I4B10 treatment. With the increase in biochar application amount, *SPAD* increased by 3.1–22.1%, but irrigation amount had no significant effect on *SPAD*. Under the combination of biochar and irrigation, the sugar beet yield reached the maximum under the I4B10 treatment, and there was no significant difference in sugar beet yield between I4B10 and I3B10. However, under the I3 irrigation treatment, *WUE*, *SYI*, and *EB* increased by 14.8%, 1.1%, and 1.14% compared with the I4 treatment, respectively. Considering the factors of sugar beet growth, yield, quality, *WUE*, and *EB*, the irrigation level set to 1.0 *ETc* and the application amount of biochar set to 10 t $ha^{-1}$ could be used as the recommended scheme for sugar beet planting in Xinjiang region. In order to comprehensively assess the specific impacts of irrigation and biochar application on crops, it is essential to conduct longer-term observations and consider additional factors such as nutrient availability and fertilizer utilization.

**Author Contributions:** F.J.: Conceptualization, methodology, software, formal analysis, investigation, and Writing—original draft. Y.L.: Conceptualization, methodology, writing—review and editing, and supervision. A.B.: Conceptualization, supervision, and writing—review and editing. L.L.: Data curation. X.X.: Supervision. W.X.: Methodology. Y.G.: Methodology. All authors have read and agreed to the published version of the manuscript.

**Funding:** This research was jointly supported by the Key Research and Development Program of Xinjiang (No. 2022B02020-2), the High-end Experts Introduction Projects (Nos. 20240082, H20240401 and S20240161), the National Key Research and Development Program of China (No. 2022YFD1900401), NYHXGG project (No. 2023AA303), and the High-end Foreign Experts Introduction Project (No. G2022172025L).

**Data Availability Statement:** Data will be made available upon request.

**Conflicts of Interest:** The authors declare that they have no known competing financial interests or personal relationships that could have appeared to influence the work reported in this paper.

## Abbreviations

$ET_0$, daily reference crop evapotranspiration; ETc, actual crop evapotranspiration; DAS, days after sowing; LAI, leaf area index; IWUE, irrigation water use efficiency; SYI, sustainable yield index; SBI, sustainable biomass index; CV, coefficient of variance; EB, economic benefit; GR, germination rate.

## References

1. Ahmad, M.; Akbar, H.; Jan, M.T.; Khattak, M.J.K. Effect of Seeding Depth, Nitrogen Placement Method and Biochar on the Growth, Yield and Its Related Parameters of Sugar Beet. *Sarhad J. Agric.* **2015**, *31*, 224–231. [CrossRef]
2. Baptista, S.L.; Soares, P.O.; Romaní, A.; Alonso, J.L.; Domingues, L. Engineering Arabinose-to-Arabitol Conversion in Industrial Saccharomyces Cerevisiae for Sugar Beet Pulp Valorization. *Ind. Crops Prod.* **2024**, *222*, 119718. [CrossRef]
3. Luo, Q.; Zhao, X.; Li, Y.; Deng, Y.; He, Q.; Dai, W. Aging Alters the Physicochemical Properties of Biochar, Enhances Its Adsorption Performance for Tris-(1-Chloro-2-Propyl) Phosphate, and Changes the Adsorption Mechanism. *Environ. Technol. Innov.* **2025**, *37*, 104053. [CrossRef]
4. Liu, Y.; Yang, S.; Lu, H.; Wang, Y. Effects of Biochar on Spatial and Temporal Changes in Soil Temperature in Cold Waterlogged Rice Paddies. *Soil Tillage Res.* **2018**, *181*, 102–109. [CrossRef]
5. Xiao, L.; Yuan, G.; Feng, L.; Bi, D.; Wei, J.; Shen, G.; Liu, Z. Coupled effects of biochar use and farming practice on physical properties of a salt-affected soil with wheat–maize rotation. *J. Soils Sediment.* **2020**, *20*, 3053–3061. [CrossRef]
6. Wang, X.; Riaz, M.; Xia, X.; Babar, S.; El-Desouki, Z.; Li, Y.; Wang, J.; Jiang, C. Alleviation of Cotton Growth Suppression Caused by Salinity through Biochar Is Strongly Linked to the Microbial Metabolic Potential in Saline-Alkali Soil. *Sci. Total Environ.* **2024**, *922*, 171407. [CrossRef]
7. Zhang, Q.; Song, Y.; Wu, Z.; Yan, X.; Gunina, A.; Kuzyakov, Y.; Xiong, Z. Effects of six-year biochar amendment on soil aggregation, crop growth, and nitrogen and phosphorus use efficiencies in a rice-wheat rotation. *J. Clean. Prod.* **2020**, *242*, 118435. [CrossRef]
8. Zhang, T.; Zhan, X.; Kang, Y.; Wan, S.; Feng, H. Improvements of Soil Salt Characteristics and Nutrient Status in an Impermeable Saline–Sodic Soil Reclaimed with an Improved Drip Irrigation While Ridge Planting *Lycium barbarum* L. *J. Soils Sediments* **2017**, *17*, 1126–1139. [CrossRef]
9. Wu, X.; Wang, D.; Riaz, M.; Zhang, L.; Jiang, C. Investigating the Effect of Biochar on the Potential of Increasing Cotton Yield, Potassium Efficiency and Soil Environment. *Ecotoxicol. Environ. Saf.* **2019**, *182*, 109451. [CrossRef]
10. Liu, C.; Tian, J.; Chen, L.; He, Q.; Liu, X.; Bian, R.; Zheng, J.; Cheng, K.; Xia, S.; Zhang, X.; et al. Biochar Boosted High Oleic Peanut Production with Enhanced Root Development and Biological N Fixation by Diazotrophs in a Sand-Loamy Primisol. *Sci. Total Environ.* **2024**, *932*, 173061. [CrossRef]
11. Han, Y.; Zhang, J.; Chen, P.; Li, H.; Li, W.; Liu, J.; Zong, R.; Wang, D.; Liang, Y.; Wang, Z. Biochar Improves Water and Nitrogen Use Efficiency of Cotton under Mulched Drip Irrigation in Arid Regions. *Ind. Crops Prod.* **2024**, *222*, 119830. [CrossRef]

12. Liao, X.; Niu, Y.; Liu, D.; Chen, Z.; He, T.; Luo, J.; Lindsey, S.; Ding, W. Four-year continuous residual effects of biochar application to a sandy loam soil on crop yield and $N_2O$ and NO emissions under maize-wheat rotation. *Agr. Ecosyst Environ.* **2020**, *302*, 107109. [CrossRef]
13. Huang, M.; Wang, C.; Qi, W.; Zhang, Z.; Xu, H. Modelling the integrated strategies of deficit irrigation, nitrogen fertilization, and biochar addition for winter wheat by AquaCrop based on a two-year field study. *Field Crops Res.* **2022**, *282*, 108510. [CrossRef]
14. Hossain, M.; Jotisankasa, A.; Aramrak, S.; Kamchoom, V.; Nishimura, S.; Nontananandh, S.; Muanlhao, T.; Semmad, S. Increased Erosion in Biochar-Amended Soil: Importance of Integrating Erosion Control Blankets and Vegetation. *Biogeotechnics* **2025**, 100161. [CrossRef]
15. Almaroai, Y.A.; Eissa, M.A. Effect of biochar on yield and quality of tomato grown on a metal-contaminated soil. *Sci. Hortic.* **2020**, *265*, 109210. [CrossRef]
16. Wang, S.; Wang, C.; Xie, L.; Li, Y.; Siddique, K.H.; Qi, X.; Luo, H.; Yang, G.; Hou, Z.; Wang, X.; et al. Optimizing Biochar Application for Enhanced Cotton and Sugar Beet Production in Xinjiang: A Comprehensive Study. *J. Sci. Food Agric.* **2024**, *104*, 6626–6639. [CrossRef]
17. Liu, X.; Manevski, K.; Liu, F.; Andersen, M.N. Biomass Accumulation and Water Use Efficiency of Faba Bean-Ryegrass Intercropping System on Sandy Soil Amended with Biochar under Reduced Irrigation Regimes. *Agric. Water Manag.* **2022**, *273*, 107905. [CrossRef]
18. Ren, H.; Warnock, D.D.; Tiemann, L.K.; Quigley, K.; Miesel, J.R. Evaluating Foliar Characteristics as Early Indicators of Plant Response to Biochar Amendments. *For. Ecol. Manag.* **2021**, *489*, 119047. [CrossRef]
19. Tiong, Y.W.; Sharma, P.; Xu, S.; Bu, J.; An, S.; Foo, J.B.L.; Wee, B.K.; Wang, Y.; Lee, J.T.E.; Zhang, J.; et al. Enhancing Sustainable Crop Cultivation: The Impact of Renewable Soil Amendments and Digestate Fertilizer on Crop Growth and Nutrient Composition. *Environ. Pollut.* **2024**, *342*, 123132. [CrossRef]
20. Shang, Q.; Chi, J. Mechanistic Insight into the Effects of Interaction between Biochar and Soil with Different Properties on Phenanthrene Sorption. *J. Environ. Manag.* **2024**, *367*, 121961. [CrossRef]
21. Wang, Q.; Hu, X.; Ma, Y.; Li, Y. Enhancing sugar beet yield and quality in Northeast China: Investigating the synergistic impact of sugar mill filter mud and biochar on black soil. *Sci. Hortic.* **2024**, *326*, 112680. [CrossRef]
22. Lebrun, M.; Bouček, J.; Bímová, K.B.; Kraus, K.; Haisel, D.; Kulhánek, M.; Omara-Ojungu, C.; Seyedsadr, S.; Beesley, L.; Soudek, P.; et al. Biochar in manure can suppress water stress of sugar beet (*Beta vulgaris*) and increase sucrose content in tubers. *Sci. Total Environ.* **2022**, *814*, 152772. [CrossRef] [PubMed]
23. Alves, B.S.Q.; Zelaya, K.P.S.; Colen, F.; Frazão, L.A.; Napoli, A.; Parikh, S.J.; Fernandes, L.A. Effect of sewage sludge and sugarcane bagasse biochar on soil properties and sugar beet production. *Pedosphere* **2021**, *31*, 572–582. [CrossRef]
24. Ebmeyer, H.; Fiedler-Wiechers, K.; Hoffmann, C.M. Drought tolerance of sugar beet—Evaluation of genotypic differences in yield potential and yield stability under varying environmental conditions. *Eur. J. Agron.* **2021**, *125*, 126262. [CrossRef]
25. Morillo-Velarde, R. Water Management In Sugar Beet. *Sugar Tech.* **2011**, *12*, 299–304. [CrossRef]
26. Haghverdi, A.; Yonts, C.D.; Reichert, D.L.; Irmak, S. Impact of irrigation, surface residue cover and plant population on sugar beet growth and yield, irrigation water use efficiency and soil water dynamics. *Agric. Water Manag.* **2017**, *180*, 1–12. [CrossRef]
27. Yan, F.; Liu, X.; Bai, W.; Fan, J.; Zhang, F.; Xiang, Y.; Hou, X.; Pei, S.; Dai, Y.; Zeng, H.; et al. Multi-objective optimization of water and nitrogen regimes for drip-fertigated sugar beet in a desert climate. *Field Crops Res.* **2022**, *288*, 108703. [CrossRef]
28. Allen, R.G.; Pereira, L.S.; Raes, D.; Smith, M. Crop evapotranspiration: Guidelines for computing crop water requirements. In *Irrigation and Drainage Paper 56*; United Nations FAO: Rome, Italy, 1998.
29. Saxena, J.; Rana, G.; Pandey, M. Impact of addition of biochar along with *Bacillus* sp. on growth and yield of French beans. *Sci. Hortic.* **2013**, *162*, 351–356. [CrossRef]
30. García-Vila, M.; Fereres, E.; Mateos, L.; Orgaz, F.; Steduto, P. Deficit irrigation optimization of cotton with AquaCrop. *Agron. J.* **2009**, *101*, 477–487. [CrossRef]
31. Mahmoud, E.A.; Hassanin, M.A.; Borham, T.I.; Emara, E.I.R. Tolerance of some sugar beet varieties to water stress. *Agric. Water Manag.* **2018**, *201*, 144–151. [CrossRef]
32. Han, X.; Hu, C.; Chen, Y.; Qiao, Y.; Liu, D.; Fan, J.; Li, S.; Zhang, Z. Crop yield stability and sustainability in a rice-wheat cropping stuber based on 34-year field experiment. *Eur. J. Agron.* **2020**, *113*, 125965. [CrossRef]
33. Guo, Q.; Huang, G.; Guo, Y.; Zhang, M.; Zhou, Y.; Duan, L. Optimizing irrigation and planting density of spring maize under mulch drip irrigation sytuber in the arid region of Northwest China. *Field Crops Res.* **2021**, *266*, 108141. [CrossRef]
34. Guan, X.; Ramaswamy, H.; Zhang, B.; Lin, B.; Hou, L.; Wang, S. Influence of moisture content, temperature and heating rate on germination rate of watermelon seeds. *Sci. Hortic.* **2020**, *272*, 109528. [CrossRef]
35. Jiao, S.; Yu, Z.; Deng, Y. Hot air-assisted radio frequency heating effects on wheat and corn seeds: Quality change and fungi inhibition. *J. Stored Prod. Res.* **2016**, *69*, 265–271. [CrossRef]
36. Bai, X.; Zhang, S.; Shao, J.; Chen, A.; Jiang, J.; Chen, A.; Luo, S. Exploring the negative effects of biochars on the germination, growth, and antioxidant sytuber of rice and corn. *J. Environ. Chem. Eng.* **2022**, *10*, 107398. [CrossRef]

37. Wang, Q.; Gao, S.; Wang, D.; Cao, A. Biochar significantly reduced fumigant emissions and benefited germination and plant growth under field conditions. *Environ. Pollut.* **2022**, *303*, 119113. [CrossRef]
38. Das, S.K.; Ghosh, G.K.; Avasthe, R. Ecotoxicological responses of weed biochar on seed germination and seedling growth in acidic soil. *Environ. Technol. Innov.* **2020**, *20*, 101074. [CrossRef]
39. Bouqbis, L.; Daoud, S.; Koyro, H.W.; Kammann, C.I.; Ainlhout, F.Z.; Harrouni, M.C. Phytotoxic effects of argan shell biochar on salad and barley germination. *Agric. Nat. Resour.* **2017**, *51*, 247–252. [CrossRef]
40. Quilliam, R.S.; Marsden, K.A.; Gertler, C.; Rousk, J.; DeLuca, T.H.; Jones, D.L. Nutrient dynamics, microbial growth and weed germination in biochar amended soil are influenced by time since application and reapplication rate. *Agric. Ecosyst. Environ.* **2012**, *158*, 192–199. [CrossRef]
41. Zhou, B.; Liang, C.; Chen, X.; Ye, S.; Peng, Y.; Yang, L.; Duan, M.; Wang, X. Magnetically-treated brackish water affects soil water-salt distribution and the growth of cotton with film mulch drip irrigation in Xinjiang, China. *Agric. Water Manag.* **2022**, *263*, 107487. [CrossRef]
42. Tan, S.; Wang, Q.; Zhang, J.; Chen, Y.; Shan, Y.; Xu, D. Performance of AquaCrop model for cotton growth simulation under film-mulched drip irrigation in southern Xinjiang, China. *Agric. Water Manag.* **2018**, *196*, 99–113. [CrossRef]
43. Zhou, B.; Yang, L.; Chen, X.; Ye, S.; Peng, Y.; Liang, C. Effect of magnetic water irrigation on the improvement of salinized soil and cotton growth in Xinjiang. *Agric. Water Manag.* **2021**, *248*, 106784. [CrossRef]
44. Liao, Z.; Zeng, H.; Fan, J.; Lai, Z.; Zhang, C.; Zhang, F.; Wang, H.; Cheng, M.; Guo, J.; Li, Z.; et al. Effects of plant density, nitrogen rate and supplemental irrigation on photosynthesis, root growth, seed yield and water-nitrogen use efficiency of soybean under ridge-furrow plastic mulching. *Agric. Water Manag.* **2022**, *268*, 107688. [CrossRef]
45. Bekhradi, F.; Luna, M.C.; Delshad, M.; Jordan, M.J.; Sotomayor, J.A.; Martínez-Conesa, C.; Gil, M.I. Effect of Deficit Irrigation on the Postharvest Quality of Different Genotypes of Basil Including Purple and Green Iranian Cultivars and a Genovese Variety. *Postharvest Biol Technol.* **2015**, *100*, 127–135. [CrossRef]
46. Wei, K.; Zhang, J.; Wang, Q.; Guo, Y.; Mu, W. Irrigation with ionized brackish water affects cotton yield and water use efficiency. *Ind. Crops Prod.* **2022**, *175*, 114244. [CrossRef]
47. Dos Santos Farias, D.B.; Da Silva, P.S.O.; Tadeu Lucas, A.A.; De Freitas, M.I.; De Jesus Santos, T.; Nascimento Fontes, P.T.; De Oliveira Júnior, L.F.G. Physiological and productive parameters of the okra under irrigation levels. *Sci. Hortic.* **2019**, *252*, 1–6. [CrossRef]
48. Tourajzadeh, O.; Piri, H.; Naserin, A.; Cahri, M.M. Effect of Nano Biochar addition and deficit irrigation on growth, physiology and water productivity of quinoa plants under salinity conditions. *Environ. Exp. Bot.* **2024**, *217*, 105564. [CrossRef]
49. Dos Santos Silva, A.M.; Santos, M.V.; Da Silva, L.D.; Dos Santos, J.B.; Ferreira, E.A.; Santos, L.D.T. Effects of irrigation and nitrogen fertilization rates on yield, agronomic efficiency and morpho physiology in Tithonia divers folia. *Agric. Water Manag.* **2021**, *248*, 106782. [CrossRef]
50. Li, J.; Zhang, Z.; Yao, C.; Liu, Y.; Wang, Z.; Fang, B.; Zhang, Y. Improving winter wheat grain yield and water-/nitrogen-use efficiency by optimizing the micro-sprinkling irrigation amount and nitrogen application rate. *J. Integr. Agric.* **2021**, *20*, 606–621. [CrossRef]
51. Norouzi, M.A.; Benakashani, F.; Soltani, E.; Karimi, S.; Akbari, G.A. Comparative study of influence of water stress on effectiveness of sulfosulfuron to two species of weeds through biomass and photo-physiological parameters. *Crop Prot.* **2024**, *178*, 106589. [CrossRef]
52. Li, Y.; Yao, N.; Liang, J.; Wang, X.; Niu, B.; Jia, Y.; Jiang, F.; Yu, Q.; Liu, D.L.; Feng, H.; et al. Rational biochar application rate for cotton nutrient content, growth, yields, productivity, and economic benefits under film-mulched trickle irrigation. *Agric. Water Manag.* **2023**, *276*, 108079. [CrossRef]
53. Xiao, Q.; Zhu, L.; Shen, Y.; Li, S. Sensitivity of soil water retention and availability to biochar addition in rainfed semi-arid farmland during a three-year field experiment. *Field Crops Res.* **2016**, *196*, 284–293. [CrossRef]
54. Feng, L.; Dai, J.; Tian, L.; Zhang, H.; Li, W.; Dong, H. Review of the technology for high-yielding and efficient cotton cultivation in the northwest inland cotton-growing region of China. *Field Crops Res.* **2017**, *208*, 18–26. [CrossRef]
55. Akhtar, S.S.; Li, G.; Andersen, M.N.; Liu, F. Biochar Enhances Yield and Quality of Tomato under Reduced Irrigation. *Agric. Water Manag.* **2014**, *138*, 37–44. [CrossRef]
56. Kammann, C.I.; Sebastian, L.; Johannes, W.G.; Hans-Werner, K. Influence of biochar on drought tolerance of Chenopodium quinoa Willd and on soil–plant relations. *Plant Soil* **2011**, *345*, 195–210. [CrossRef]
57. Lehmann, J.; da Silva, J.P.; Rondon, M.; Cravo, M.S.; Greenwood, J.; Nehls, T.; Steiner, C.; Glaser, B. Slash-and-char: A feasible alternative for soil fertility management in the central Amazon? In Proceedings of the 17th World Congress of Soil Science, Bangkok, Thailand, 14–21 August 2002; pp. 14–21.
58. Li, Z.; Wan, S.; Chen, G.; Han, Y.; Lei, Y.; Ma, Y.; Xiong, S.; Mao, T.; Feng, L.; Wang, G.; et al. Effects of irrigation regime on soil hydrothermal microenvironment, cotton biomass, and yield under non-film drip irrigation sytuber in cotton fields in southern Xinjiang, China. *Ind. Crops Prod.* **2023**, *198*, 116738. [CrossRef]

59. Liu, J.; Si, Z.; Li, S.; Kader Mounkaila Hamani, A.; Zhang, Y.; Wu, L.; Gao, Y.; Duan, A. Variations in water sources used by winter wheat across distinct rainfall years in the North China Plain. *J. Hydrol.* **2023**, *618*, 129186. [CrossRef]
60. Chen, Y.Z.; Dong, H.Z. Mechanisms and regulation of senescence and maturity performance in cotton. *Field Crops Res.* **2016**, *189*, 1–9. [CrossRef]
61. Hou, J.; Zhang, J.; Liu, X.; Ma, Y.; Wei, Z.; Wan, H.; Liu, F. Effect of biochar addition and reduced irrigation regimes on growth, physiology and water use efficiency of cotton plants under salt stress. *Ind. Crops Prod.* **2023**, *198*, 116702. [CrossRef]
62. Faloye, O.T.; Alatise, M.O.; Ajayi, A.E.; Ewulo, B.S. Effects of biochar and inorganic fertiliser applications on growth, yield and water use efficiency of maize under deficit irrigation. *Agric. Water Manag.* **2019**, *217*, 165–178. [CrossRef]
63. Hou, J.; Liu, X.; Zhang, J.; Wei, Z.; Ma, Y.; Wan, H.; Liu, J.; Cui, B.; Zong, Y.; Chen, Y.; et al. Combined Application of Biochar and Partial Root-Zone Drying Irrigation Improves Water Relations and Water Use Efficiency of Cotton Plants under Salt Stress. *Agric. Water Manag.* **2023**, *290*, 108584. [CrossRef]
64. Zhang, P.; Li, M.; Fu, Q.; Singh, V.P.; Du, C.; Liu, D.; Li, T.; Yang, A. Dynamic regulation of the irrigation-nitrogen-biochar nexus for the synergy of yield, quality, carbon emission and resource use efficiency in tomato. *J. Integr. Agric.* **2024**, *23*, 680–697. [CrossRef]
65. Feng, W.; Yang, F.; Cen, R.; Liu, J.; Qu, Z.; Miao, Q.; Chen, H. Effects of straw biochar application on soil temperature, available nitrogen and growth of corn. *J. Environ. Manag.* **2021**, *277*, 111331. [CrossRef]

*plants*

MDPI

*Article*

# Optimizing Irrigation and Nitrogen Application to Enhance Millet Yield, Improve Water and Nitrogen Use Efficiency and Reduce Inorganic Nitrogen Accumulation in Northeast China

**Tangzhe Nie [1], Jianfeng Li [1], Lili Jiang [1,*], Zhongxue Zhang [2,3], Peng Chen [4], Tiecheng Li [5], Changlei Dai [1], Zhongyi Sun [6], Shuai Yin [7] and Mengxue Wang [8]**

1 School of Water Conservancy and Electric Power, Heilongjiang University, Harbin 150080, China; 2019036@hlju.edu.cn (T.N.); 2222014@s.hlju.edu.cn (J.L.)
2 Key Laboratory of Effective Utilization of Agricultural Water Resources, Ministry of Agriculture and Rural Affairs, Northeast Agricultural University, Harbin 150030, China
3 School of Water Conservancy and Civil Engineering, Northeast Agricultural University, Harbin 150030, China
4 College of Agricultural Science and Engineering, Hohai University, Nanjing 211100, China
5 Institute of Geographic Sciences and Natural Resources Research, Chinese Academy of Sciences, Beijing 100101, China
6 College of Ecology and Environment, Hainan University, Haikou 570208, China
7 State Key Laboratory of Remote Sensing Science, Aerospace Information Research Institute, Chinese Academy of Sciences, Beijing 100101, China
8 College of Agriculture, Heilongjiang Bayi Agricultural University, Daqing 163319, China
* Correspondence: 2023086@hlju.edu.cn

**Abstract:** Enhancing irrigation and nitrogen fertilizer application has become a vital strategy for ensuring food security in the face of population growth and resource scarcity. A 2-year experiment was conducted to determine to investigate the effects of different irrigation lower limits and nitrogen fertilizer application amounts on millet growth, yield, water use efficiency (WUE), N utilization, and inorganic nitrogen accumulation in the soil in 2021 and 2022. The experiment was designed with four irrigation lower limits, corresponding to 50%, 60%, 70%, and 80% of the field capacity (FC), referred to as $I_{50}$, $I_{60}$, $I_{70}$, and $I_{80}$. Four nitrogen fertilizer application were also included: 0, 50, 100, and 150 kg·hm$^{-2}$ (designated as $F_{00}$, $F_{50}$, $F_{100}$, and $F_{150}$), resulting in a total of 16 treatments. Binary quadratic regression equations were established to optimize the irrigation and nitrogen application. The results demonstrated that the plant height, stem diameter, leaf area index, aboveground biomass, yield, spike diameter, spike length, spike weight, WUE, and nitrogen agronomic efficiency for millet initially increased before subsequently decreasing as the irrigation lower limit and nitrogen fertilizer application increased. Their maximum values were observed in the $I_{70}F_{100}$. However, the nitrogen partial factor productivity (PFPN) exhibited a gradual decline with increasing nitrogen application, reaching its peak at $F_{50}$. Additionally, PFPN displayed a pattern of initial increase followed by a decrease with rising irrigation lower limits. The accumulation of $NO_3^-$-N and $NH_4^+$-N in the 0~60 cm soil layer increased with the increase of nitrogen fertilizer application in both years, while they tended to decrease as the irrigation lower limit increased. An optimal irrigation lower limit of 64% FC to 74% FC and nitrogen fertilizer application of 80 to 100 kg ha$^{-1}$ was recommended for millet based on the regression equation. The findings of this study offer a theoretical foundation and technical guidance for developing a drip irrigation and fertilizer application for millet cultivation in Northeast China.

**Keywords:** millet; yield; water use efficiency; nitrogen agronomic efficiency; nitrogen accumulation; water and nitrogen optimization

## 1. Introduction

Food security has emerged as one of the major challenges confronting the world today, driven by a dramatic increase in food demand due to the continuous expansion of the population and sustained increase in consumption levels [1,2]. Meanwhile, global agriculture is facing the dual challenges of declining production growth and rising environmental costs [3]. As the fourth most widely cultivated food crop globally, millet's high-yielding nature play an important role in addressing the substantial gap between the production growth of staple crops like rice and maize and the increasing global population [4–8]. The implementation of scientific management and the efficient use of water and nitrogen are becoming pivotal factors in achieving high millet yield targets [9]. Improving water use efficiency has emerged as the key objective for raising crop yields in light of the severe worldwide scarcity of water resources [10–12]. Additionally, agricultural nitrogen inputs also have to cope with the issue of high costs and relatively low yields [13,14]. Excessive application of nitrogen fertilizer influences crop absorption as well as nitrogen loss, causing instability in yield [15–17]. Moreover, nitrogen overapplication also exacerbates environmental pollution [18,19].

Currently, a great proportion of millet is cultivated using rainfall as the primary water source; however, the implementation of effective irrigation strategies remains essential for maintaining millet yield [20,21]. Research by Matsuura et al. [22] and Jahansuz et al. [23] indicated that millet yields might decline due to insufficient water supply. Additionally, changes in irrigation quota have also impacted crop water use efficiency (WUE). For instance, Zhang et al. [24] and Nagaz et al. [25] demonstrated that reduced irrigation quota significantly increased WUE of millet. Seghatoleslami et al. [26] indicated that while WUE improved under diminished irrigation, millet yields significantly declined. Existing drip irrigation technologies have significantly improved the soil moisture environment by regulating moisture evaporation and infiltration [27,28]. This enhancement facilitated better water uptake and use by crops [29,30]. Ensuring high grain yields while enhancing WUE through drip irrigation technology has become crucial. Additionally, identifying an optimal irrigation interval within range of 50% to 80% of saturated soil moisture content represented an effective strategy for minimizing agricultural water use and promoting millet production [31].

Increased nitrogen application significantly enhanced millet yield [32,33]. Wenqian et al. [34] found that applying nitrogen fertilizer during the jointing stage led to significant improvement in nitrogen partial factor productivity (PFPN), nitrogen agronomic efficiency (AEN), and millet yield. However, it is noteworthy that the effective use of applied nitrogen by crops in agricultural production often exceeds less than half of the nitrogen input [35]. Rosolem [36] noted that increasing nitrogen fertilizer inputs led to elevated levels of $NO_3^-$-N and $NH_4^+$-N in the soil profile. The excessive nitrogen in the soil would be lost through processes such as nitrification and volatilization, which not only reduced the crop's absorption and transformation efficiency of nitrogen fertilizer, but also deteriorated the soil structure [37–39]. The integration of drip irrigation with precise nitrogen application has been shown to substantially improve crop yields while effectively minimizing the depth and volume of inorganic nitrogen (including $NO_3^-$-N and $NH_4^+$-N) leaching [40]. Consequently, the management of soil $NO_3^-$-N and $NH_4^+$-N, alongside the optimization of crop yield, PFPN, and AEN, has emerged as a critical component for developing effective nitrogen application strategies in agriculture [41,42].

The response surface model (RSM) is a statistical method for solving multivariate problems [43]. This method allows for the optimization of multiple impact factors on the response variable by establishing relationships among these factors with a minimal number of experiments. By optimizing the strategies for irrigation and nitrogen in agriculture, RSM is able to achieve a joint maximum value for grain yield, as well as for water and nitrogen use efficiency, with a 95% confidence interval. For example, Zou et al. [44] optimized the water and fertilizer management system by fitting a quadratic equation, achieving 95% of the maximum values for maize yield, WUE, and net return simultaneously. Similarly,

Yan et al. [45] employed RSM to identify the optimal range of irrigation and fertilizer application amounts, aiming to maintain high maize yields while minimizing $NO_3^-$-N in the soil profile. Furthermore, RSM has been extensively utilized to improve agricultural conditions for a diverse range of crops [46].

China ranks as the sixth largest producer of millet globally, with significant potential for further production expansion [47]. However, accelerated industrialization and urbanization in China have led to a continuous reduction in available land for grain cultivation. Consequently, increasing millet yields has become a crucial strategy to address the challenges of food security [48]. Heilongjiang Province plays a vital role in millet production in China [49]. Heilongjiang Province experiences significant seasonal and regional variations in rainfall patterns, highlighting the critical importance of irrigation in satisfying the water requirements of crops and achieving high yields [50]. Additionally, Heilongjiang Province is characterized by its rich black soil resources, renowned for their high productivity [51]. Maintaining soil health and ensuring the sustainable utilization of these resources are crucial for the advancement of regional agricultural development.

The objectives of this study were: (1) to determine the effects of different irrigation lower limits and nitrogen fertilizer application amounts on millet growth, yield, and its components, water and nitrogen use efficiency, and the accumulation of $NO_3^-$-N and $NH_4^+$-N in the soil profile through a 2-year experiment, and (2) to optimize irrigation and nitrogen application strategies by considering soil inorganic nitrogen accumulation, millet yield, and water and nitrogen use efficiency, aim at enhancing millet yield and improving the use of water and nitrogen resources. The findings of the study will provide a strong theoretical foundation and technical support for the implementation and widespread adoption of drip irrigation and fertilizer application systems for millet in Northeast China.

## 2. Results

### 2.1. Growth Indicators

As shown in Table 1, the lower limits of irrigation and nitrogen application amount significantly affected ($p < 0.01$) plant height, stem diameter, leaf area index (LAI), and the aboveground biomass of millet in both 2021 and 2022. Additionally, the interaction between these two factors had highly significant effects ($p < 0.01$) on plant height, stem diameter, and aboveground biomass. The plant height, stem diameter, LAI, and aboveground biomass initially increased and then decreased, with the increase of irrigation lower limits and nitrogen application amount. Their maximum values were observed under the $I_{70}F_{100}$ treatment (Figure 1). In 2021, the aboveground biomass was significantly higher in the $I_{70}F_{100}$ compared to other treatments. In 2022, both the $I_{70}F_{100}$ and $I_{70}F_{50}$ showed significantly greater aboveground biomass than other treatments. Notably, the aboveground biomass in all $I_{70}$ treatments exceeded those of other irrigation treatments by 13.24% to 21.75%. The aboveground biomass in the $F_{100}$ treatments was significantly greater than those of other nitrogen application treatments, showing an increase of 10.29% to 26.21%. Additionally, plant height and stem diameter under the $I_{70}F_{100}$ were increased by 10.33% to 10.52% and 22.26% to 22.54% compared to the $I_{50}F_{00}$. The LAI also exhibited a notable increase of 23.48% to 31.05% under $I_{70}F_{100}$ compared to $I_{50}F_{00}$.

**Table 1.** ANOVA of irrigation lower limit and nitrogen fertilizer amount on plant height, stem diameter, LAI, and aboveground biomass of millet.

| Factors | Plant Height | | Stem Diameter | | LAI | | Aboveground Biomass | |
|---|---|---|---|---|---|---|---|---|
| | 2021 | 2022 | 2021 | 2022 | 2021 | 2022 | 2021 | 2022 |
| Irrigation lower limit | ** | ** | ** | ** | ** | ** | ** | ** |
| Nitrogen fertilizer amount | ** | ** | ** | ** | ** | ** | ** | ** |
| Irrigation lower limit × nitrogen fertilizer amount | ** | ** | ** | ** | ns | ns | ** | ** |

Note: ** means a highly significant difference ($p < 0.01$); and ns means no significant difference ($p > 0.05$).

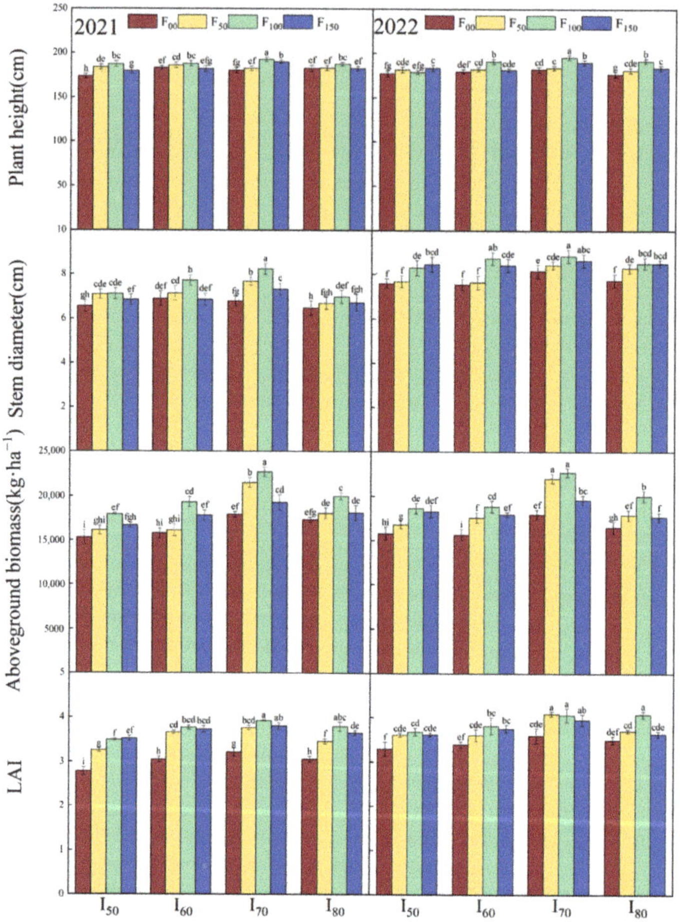

**Figure 1.** Plant height, stem diameter, leaf area index (LAI), and aboveground biomass of millet under different treatments in 2021 and 2022. $I_{50}$, $I_{60}$, $I_{70}$, and $I_{80}$ represent the irrigation lower limit being 50%, 60%, 70%, and 80% of the field capacity, respectively. $F_{00}$, $F_{50}$, $F_{100}$, and $F_{150}$ represent the nitrogen application amount of 0, 50, 100, and 150 kg·hm$^{-2}$, respectively. Error bars represent one standard deviation about the mean. The letters above the bars are the mean separation indicators (LSD0.05), where similar letters indicate no significant difference.

### 2.2. Yield and Its Components

As shown in Table 2, both the irrigation lower limit and nitrogen application amount had a highly significant effect ($p < 0.01$) on the spike length, spike stem, spike weight, and yield. The interaction between these two variables also had a significant effect ($p < 0.01$) on both the spike weight and yield. While the effect of irrigation lower limit and nitrogen application amount on the spike stem were significant ($p < 0.05$), no significant effect was found on the spike length. Overall, the spike length, spike diameter, spike weight, and yield increased with the increase of irrigation lower limit and nitrogen application amount, reaching maximum values under the $I_{70}$ and $F_{100}$ (Figure 2). The yields in $F_{100}$ treatments were significantly higher than those in $F_{00}$ and $F_{50}$ treatments under $I_{50}$ level in both years. In 2021, the yields of $F_{100}$ were also significantly ($p < 0.05$) greater than those of other treatments under the $I_{60}$ level. Over the two years, the $F_{100}$ treatments consistently yielded significantly more than other treatments under the $I_{70}$ and $I_{80}$ levels. Additionally, in 2021, the yield of $I_{70}$ was significantly ($p < 0.05$) greater than those of the other treatments when combined with $F_{100}$ level. Furthermore, the yield of $I_{70}$ was significantly greater than those of the $I_{60}$ and $I_{50}$ treatments when combined with the $F_{100}$ level in 2022. The highest yield was observed in the $I_{70}F_{150}$ in both years. Additionally, the spike length, spike stem, spike weight, and yield were all significantly ($p < 0.05$) greater under $I_{70}F_{100}$ compared to the other treatments.

**Table 2.** ANOVA of irrigation lower limit and nitrogen fertilizer amount on spike diameter, spike length, spike weight, and yield of millet.

| Factors | Spike Diameter | | Spike Length | | Spike Weight | | Yield | |
|---|---|---|---|---|---|---|---|---|
| | 2021 | 2022 | 2021 | 2022 | 2021 | 2022 | 2021 | 2022 |
| Irrigation lower limit | ** | ** | ** | ns | ** | ** | ** | ** |
| Nitrogen fertilizer amount | ** | ** | ** | ** | ** | ** | ** | ** |
| Irrigation lower limit × nitrogen fertilizer amount | * | ** | ns | ns | ** | ** | ** | ** |

Note: * means a significant difference ($p < 0.05$); ** means a highly significant difference ($p < 0.01$); and ns means no significant difference ($p > 0.05$).

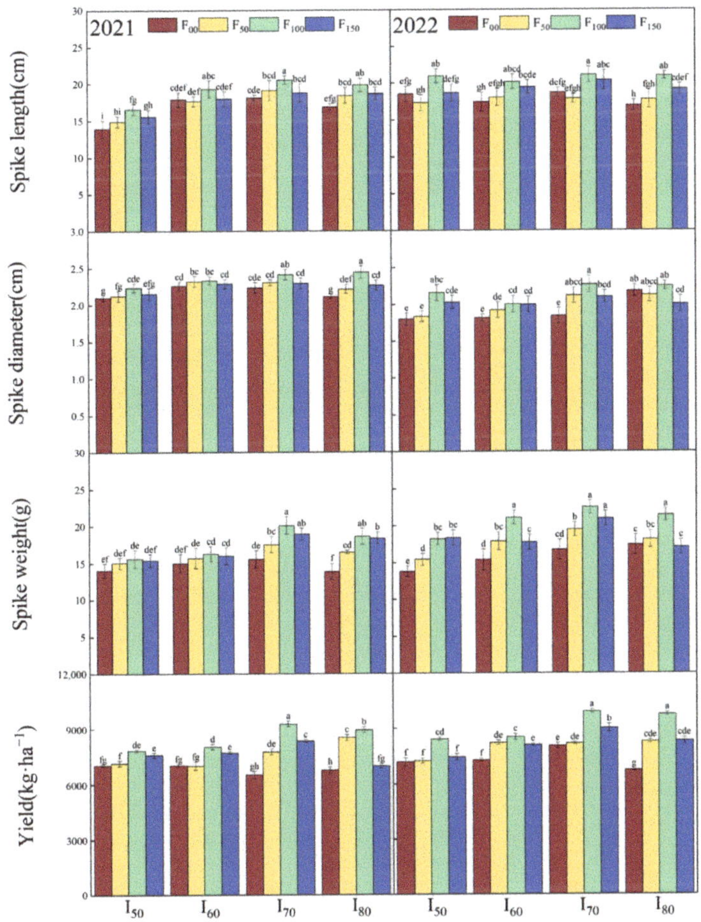


**Figure 2.** Spike diameter, spike length, spike weight, and yield of millet under different treatments in 2021 and 2022. $I_{50}$, $I_{60}$, $I_{70}$, and $I_{80}$ represent the irrigation lower limit being 50%, 60%, 70%, and 80%

of the field capacity, respectively. $F_{00}$, $F_{50}$, $F_{100}$, and $F_{150}$ represent the nitrogen application amount of 0, 50, 100, and 150 kg·hm$^{-2}$, respectively. Error bars represent one standard deviation about the mean. The letters above the bars are the mean separation indicators ($LSD_{0.05}$), where similar letters indicate no significant difference.

### 2.3. WUE, AEN and PFPN

As shown in Table 3, irrigation lower limit and nitrogen application amount significantly affected WUE, AEN, and PFPN ($p < 0.01$). The interaction between these two variables had a highly significant effect on both AEN and PFPN ($p < 0.01$). While this interaction did not significantly affect WUE in 2021, it did have a significant impact in 2022 ($p < 0.05$). Figure 3 illustrated that WUE initially increased then decreased with the increase of the irrigation lower limit and nitrogen application amount. In the $F_{100}$ treatments, WUE was significantly higher under the $I_{70}$ treatment in both 2021 and 2022. The WUE of $I_{70}$ treatments increased by 7.51% to 15.53% compared to other irrigation treatments. Additionally, the WUE of the $F_{100}$ treatment was significantly higher than those of the other nitrogen application treatments, with an increase of 10.12% to 27.33%. Both PFPN and AEN exhibited a trend of initially increasing and then decreasing in response to the irrigation lower limit. Specifically, PFPN displayed a downward trend with higher nitrogen application amount, peaking at the $F_{50}$ level. In contrast, AEN showed an initial increase followed by a decline as nitrogen application increased, reaching its maximum value at the $F_{100}$ level.

**Table 3.** ANOVA of irrigation lower limit and nitrogen fertilizer amount on WUE, AEN, PFPN, $NO_3^-$-N accumulation, and $NH_4^+$-N accumulation.

| Factors | WUE | | AEN | | PFPN | | $NO_3^-$-N Accumulation | | $NH_4^+$-N Accumulation | |
|---|---|---|---|---|---|---|---|---|---|---|
| | 2021 | 2022 | 2021 | 2022 | 2021 | 2022 | 2021 | 2022 | 2021 | 2022 |
| Irrigation lower limit | ** | ** | ** | ** | ** | ** | ** | ** | ** | ** |
| Nitrogen fertilizer amount | ** | ** | ** | ** | ** | ** | ** | ** | ** | ** |
| Irrigation lower limit × nitrogen fertilizer amount | ns | * | ** | ** | ** | ** | ns | ** | ** | * |

Note: * means a significant difference ($p < 0.05$); ** means a highly significant difference ($p < 0.01$); and ns means no significant difference ($p > 0.05$).

### 2.4. $NO_3^-$-N and $NH_4^+$-N Accumulation

As shown in Table 3, both the irrigation lower limit and nitrogen application amount significantly affected ($p < 0.01$) the accumulation of $NO_3^-$-N and $NH_4^+$-N in the soil. In 2021, the interaction between the irrigation lower limit and nitrogen application amount did not have a significant effect on $NO_3^-$-N accumulation. However, in 2022, this interaction was found to be highly significant ($p < 0.01$). Conversely, the interaction between the irrigation lower limit and nitrogen application amount had a highly significant ($p < 0.01$) or significant ($p < 0.05$) effect on $NH_4^+$-N accumulation in 2021 and 2022, respectively. The accumulation of $NO_3^-$-N and $NH_4^+$-N within the 0~60 cm soil layer showed a declining trend with increasing irrigation lower limit and an increasing trend with the increase of nitrogen application amount in both 2021 and 2022 (Figure 4). The percentages of $NO_3^-$-N and $NH_4^+$-N accumulation in the 0~10 cm, 10~20 cm, 20~30 cm, 30~40 cm, 40~50 cm, and 50~60 cm soil layers showed that the percentage of $NH_4^+$-N accumulation increased with soil depth, peaking at 17.44% in the 40~50 cm soil layer. However, $NO_3^-$-N accumulation was highest in the 50~60 cm soil layer, reaching 19.94%.

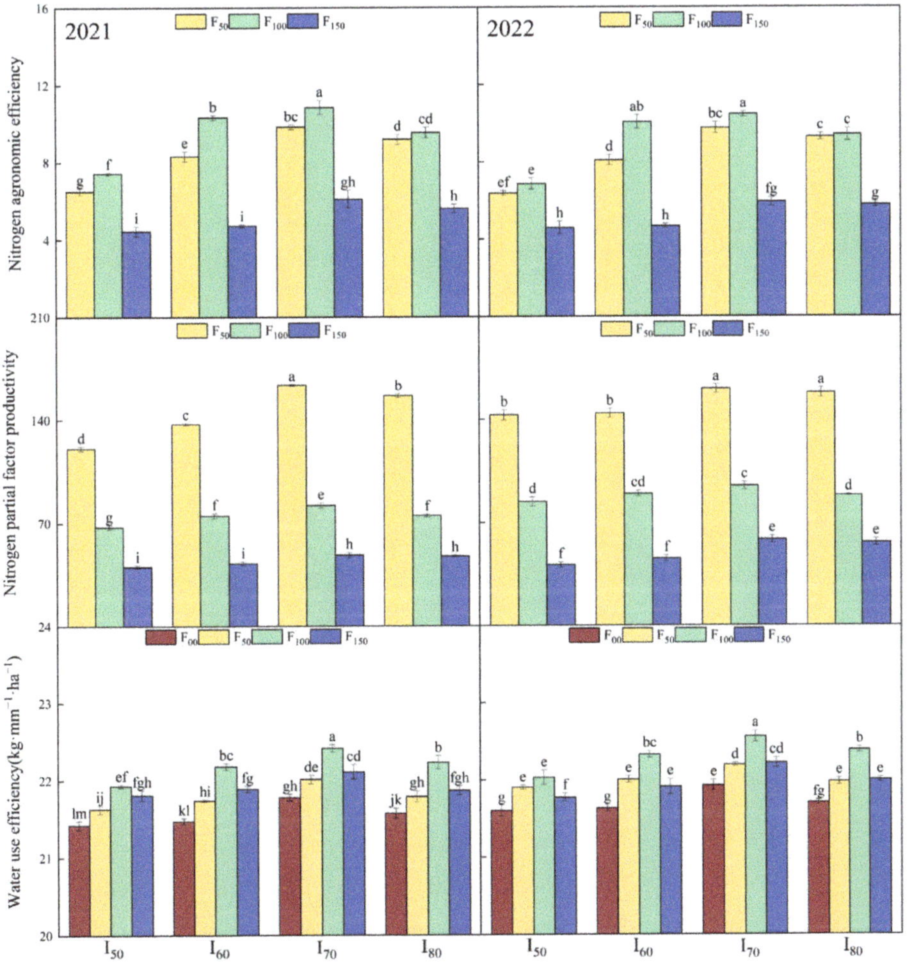


**Figure 3.** WUE, AEN, and PFPN of millet under different treatments in 2021 and 2022. $I_{50}$, $I_{60}$, $I_{70}$, and $I_{80}$ represent the irrigation lower limit being 50%, 60%, 70%, and 80% of the field capacity, respectively. $F_{00}$, $F_{50}$, $F_{100}$, and $F_{150}$ represent the nitrogen application amount of 0, 50, 100, and 150 kg·hm$^{-2}$, respectively. Error bars represent one standard deviation about the mean. The letters above the bars are the mean separation indicators ($LSD_{0.05}$), where similar letters indicate no significant difference.

## 2.5. Optimization of Irrigation and Nitrogen Fertilizer

Binary quadratic regression equations were developed to optimize the irrigation and nitrogen fertilizer, using irrigation lower limits and nitrogen application as independent variables, while millet yield, WUE, AEN, and PFPN were designated as dependent variables (Table 4). The fitting results (Figure 5) indicated that each indicator achieved its optimal value under specified irrigation and nitrogen fertilizer conditions. Millet yield peaked at 9207.3 kg ha$^{-1}$ at an irrigation lower limit of 73% field capacity (FC) and 108 kg ha$^{-1}$ nitrogen fertilizer amount. The WUE peaked at 22.5 at 71% FC with 94 kg ha$^{-1}$ of N. AEN peaked at 10.2 kg·ha$^{-1}$·mm$^{-1}$ at an irrigation lower limit of 74% FC and 79 kg ha$^{-1}$ nitrogen fertilizer amount. Additionally, the PFPN reached a peak value of 158.2 at an irrigation lower limit of 79% FC and 50 kg ha$^{-1}$ nitrogen fertilizer amount.

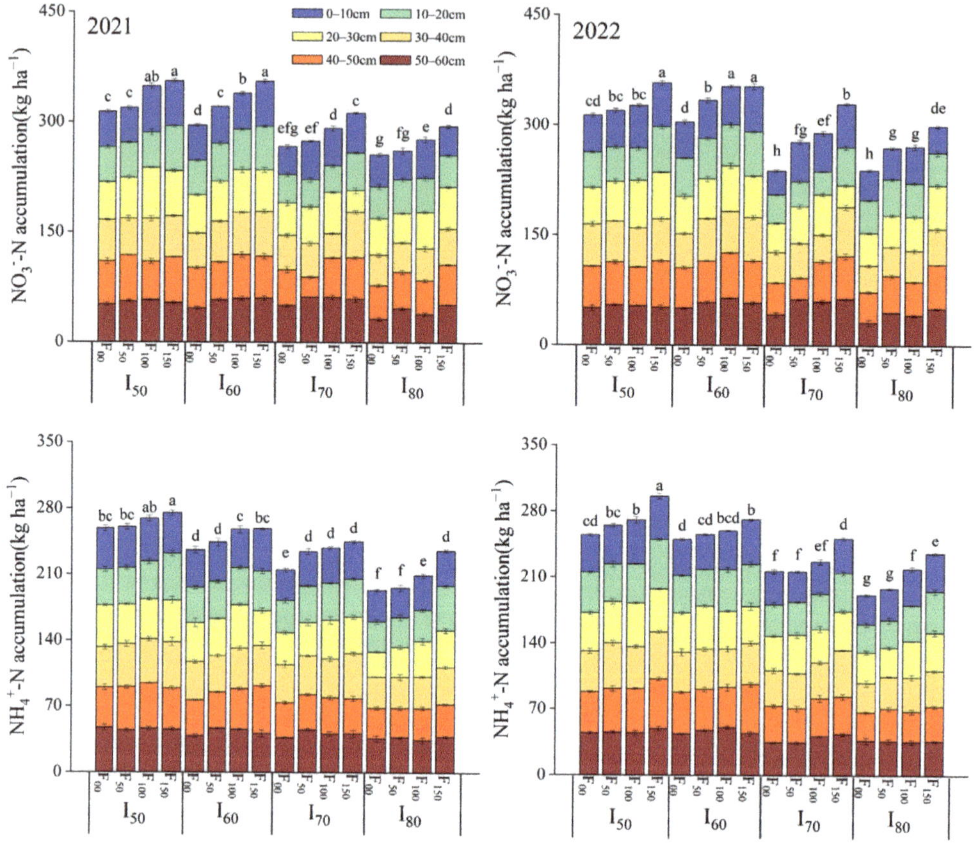


**Figure 4.** $NO_3^-$-N and $NH_4^+$-N accumulation in the 0~60 cm soil profile at harvest in the millet root zone in 2021 and 2022. $I_{50}$, $I_{60}$, $I_{70}$, and $I_{80}$ represent the irrigation lower limit being 50%, 60%, 70%, and 80% of the field capacity, respectively. $F_{00}$, $F_{50}$, $F_{100}$, and $F_{150}$ represent the nitrogen application amount of 0, 50, 100, and 150 kg·hm$^{-2}$, respectively. Error bars represent one standard deviation about the mean. The letters above the bars are the mean separation indicators ($LSD_{0.05}$), where similar letters indicate no significant difference.

**Table 4.** Regression equations for the response variables of yield, PFPN, AEN, WUE under different irrigation and nitrogen fertilizer conditions.

| Response Variable | Regression Equations | $R^2$ |
|---|---|---|
| Yield | $z = -3170.44 + 13.23 \times x + 320.89 \times y + 0.29 \times x \times y - 0.16 \times x^2 - 2.42 \times y^2$ | 0.86 |
| PFPN | $z = 130.01 - 1.91 \times x + 2.86 \times y - 0.36 \times 10^{-3} \times x \times y + 0.0046\, x^2 - 0.18 \times 10^{-1} \times y^2$ | 0.99 |
| AEN | $z = -28.44 + 0.19 \times x + 0.85 \times y - 0.55 \times 10^{-3} \times x \times y - 0.95 \times 10^{-3} \times x^2 - 0.55 \times 10^{-2} \times y^2$ | 0.97 |
| WUE | $z = 17.77 + 9.29 \times 10^{-3} \times x + 0.12 \times y + 0.42 \times 10^{-4} \times x \times y - 0.65 \times 10^{-4} \times x^2 - 0.87 \times 10^{-3} \times y^2$ | 0.92 |

Note: x is the irrigation lower limit, and y is the nitrogen fertilizer amount.

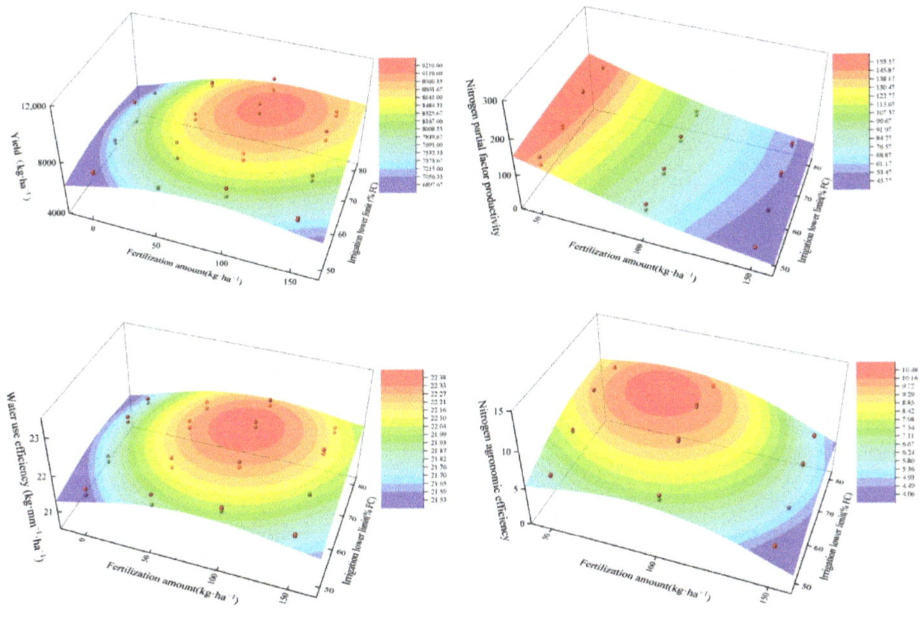


**Figure 5.** Relationship between yield, WUE, AEN, and PFPN under different irrigation lower limits and nitrogen fertilizer amounts. The red dots represent the measured values in 2021 and 2022. The green to red area represents the ≥95% confidence interval. FC: field capacity.

## 3. Discussion

### *3.1. Effects of Irrigation and Nitrogen Application on Millet Growth and Yield*

The results of this experiment demonstrated a significant effect of nitrogen application amount on the aboveground biomass of millet (Table 1). The aboveground biomass continued to increase as the nitrogen application amount rose to $F_{100}$ (Figure 1). This might be because the optimal nitrogen application amount stimulated photosynthesis, which subsequently has a beneficial impact on photosynthetic products and dry matter accumulation [32,52]. When the nitrogen application amount continued to rise until $F_{100}$, the photosynthetic products contributed to higher yields (Figure 1). Gong et al. [53] noted that enhanced nitrogen supply promoted nutrient synthesis in cereal grains, leading to increased yields. In the current study, the observed increasing trend in spike length, spike diameter, and spike weight with the increasing nitrogen application amount was consistent with the increasing trend of yield, indicating a positive correlation between millet growth with yield (Figure 4) [54,55].

The present study demonstrated that irrigation had a highly significant impact ($p < 0.01$) on the yield and aboveground biomass (Table 1). Specifically, the $I_{70}$ treatments resulted in a significantly higher yield compared to other irrigation treatments (Figure 1). This indicating that an irrigation lower limit of 70% FC created a more favorable soil water environment for millet growth, which facilitated better nitrogen absorption from the soil, ultimately leading to increased yield [25,56]. An increase in irrigation lower limit to $I_{80}$ led to a reduction in yield, which likely due to excessive soil moisture content inhibiting root respiration of millet. Additionally, significantly lower yields were observed under $I_{50}$ and $I_{60}$ treatments, highlighting that insufficient irrigation also adversely affected millet yields [57]. The $I_{70}$ treatments demonstrated a significantly higher aboveground biomass compared to other treatments (Figure 1), suggesting that optimal soil moisture conditions positively affect dry matter accumulation [58]. Moreover, the interaction between the irrigation lower limit and nitrogen application amount had a highly significant effect on grain yield (Table 2). By rationally regulating the ratio of these two factors, the efficiency of the nutrient uptake and translocation in grains can be significantly enhanced, leading to higher yields [5,59].

### 3.2. Effects of Irrigation and Nitrogen Application on Millet WUE, AEN and PFPN

The C4 crops exhibited greater resilience to water deficit conditions compared to other crop varieties. For instance, Seghatoleslami et al. [26] and Nagaz et al. [25] found that the WUE of millet improved when reducing irrigation. However, the results of this study showed that the $I_{70}$ treatments demonstrated superior WUE (Figure 3). Zou et al. [44] indicated that increasing irrigation might be beneficial, as it allowed the crop to fully utilize available light and heat resources, ultimately making yield a primary determinant when calculating WUE. In this study, the WUE decreased as irrigation lower limit reduced, hitting a minimum in the $I_{50}$ treatments. This decline could be attributed to the impact of decreased soil moisture on the crop's ability to uptake and utilize soil nitrogen. Additionally, water deficit severely impaired grain growth and reduced yield, which in turn directly suppressed WUE [60–63]. The results of this study revealed that the $I_{80}$ treatments had a significant inhibitory effect on the WUE of millet, suggesting that excessive water supply can be detrimental to grain WUE [64]. Additionally, the amount of nitrogen applied was found to significantly influence WUE in this study. The $F_{100}$ treatments exhibited a higher level of WUE. This indicated that a moderate nitrogen fertilizer supply effectively mitigated the negative effects of water deficit on nitrogen uptake [65].

The results of this study demonstrated a decline in PFPN as the nitrogen application amount increased (Figure 3). Previous research by Lu et al. [16] indicated that higher levels of nitrogen application led to a sustained reduction in PFPN. Additionally, the irrigation lower limit significantly impacted PFPN, highlighting a notable interaction between irrigation and nitrogen application [66]. It was noteworthy that the PFPN of the $I_{70}$ treatments were relatively higher than other treatments. This could be attributed to the favorable soil moisture conditions that improved the effectiveness of the applied nitrogen fertilizer [34,67]. In contrast, the AEN displayed a different pattern, initially increasing then declining as the nitrogen fertilizer application increased (Figure 3) Although the $F_{50}$ treatment produced yields nearly close to those of the $F_{150}$ treatments, their AEN were significantly higher, suggesting that high yields do not necessarily correlate with high AEN.

### 3.3. Effects of Irrigation and Nitrogen Application on $NO_3^-$-N and $NH_4^+$-N Accumulation

The dynamics of inorganic nitrogen in soil were affected by both irrigation and nitrogen management [68,69]. The application of nitrogen fertilizer significantly affected the accumulation of $NO_3^-$-N and $NH_4^+$-N within the 0–60 cm soil layer (Table 3). Furthermore, nitrogen application directly increased the accumulation of $NO_3^-$-N and $NH_4^+$-N in the soil (Figure 4). These findings were consistent with those of Giletto and Echeverría [70], who demonstrated that a higher nitrogen fertilizer amount led to a significant increase in the total accumulation of $NO_3^-$-N and $NH_4^+$-N in soil. Conversely, as the irrigation lower limit increased, the accumulation of $NO_3^-$-N and $NH_4^+$-N decreased. This trend could be attributed to the fact that increased soil moisture enhanced the vertical transport of inorganic nitrogen. Additionally, over-irrigation leads to the gravitational infiltration of $NO_3^-$-N and $NH_4^+$-N into deeper soil layers [45]. The current study revealed that the accumulation of $NO_3^-$-N and $NH_4^+$-N in the soil increased with soil depth. The highest accumulation of $NO_3^-$-N occurred at 50–60 cm, while $NH_4^+$-N peaked at 40–50 cm (Figure 4). This contrasted with the findings of Wang et al. [71], who reported a gradual decline in $NO_3^-$-N and $NH_4^+$-N accumulation with the increasing soil depth. This discrepancy might be due to the highly mobility of $NO_3^-$-N, which was more likely to infiltrate deeper soil layers, thereby leading to its accumulation in 50~60 cm soil layer [72,73]. In addition, $NH_4^+$-N was readily adsorbed by anionic colloids in the soil due to its positive charge; however, its mobility was relatively lower. Nonetheless, there was a tendency for $NH_4^+$-N to migrate to deeper soil layers, with peak accumulation often occurring at depths of 40–50 cm (Figure 4). In contrast, the accumulation of $NO_3^-$-N and $NH_4^+$-N in the shallow soil layer was relatively limited. This could be attributed to the combination of drip irrigation and fertilizer application methods, which enhanced inorganic nitrogen uptake by the grains and minimized nutrient loss. Although there were variations in the soil depths at which $NO_3^-$-N and $NH_4^+$-N were concentrated, both $NO_3^-$-N and $NH_4^+$-N were at risk of migrating to deeper soil layers. This further underscored the importance of nitrogen application and irrigation management. Despite this, our experimental data were analyzed to determine the optimal irrigation and nitrogen application corresponding to each response variable (Table 4). However, the combinations of irrigation lower limits and nitrogen applications associated with these optimal values were not identical. Consequently, this study aimed to identify the optimal irrigation lower limit and nitrogen application range within a 95% confidence interval (Figure 5). Additionally, the alterations in $NO_3^-$-N and $NH_4^+$-N in response to the irrigation lower limit and nitrogen application were evaluated (Figure 4). Based on the comprehensive analysis, this study recommended an optimal irrigation lower limit of 64% FC to 74% FC and a nitrogen fertilizer application of 80 to 100 kg $ha^{-1}$.

This study presented the optimal ranges for irrigation and nitrogen application to support the sustainable production of millet in Northeast China, based on a two-year field experiment. To enhance the economic and ecological efficiency of agricultural production, future research should incorporate cost considerations related to irrigation and fertilization, providing a more comprehensive assessment from multiple perspectives, such as economic outcomes and yield. Furthermore, the use of isotope labeling techniques will enable more precise quantification of the specific impacts of nitrogen application on both yield and soil nitrogen accumulation, thereby establishing a solid foundation for informed scientific decision making [74].

## 4. Materials and Methods

### *4.1. Study Area*

The study was conducted in the Xiangfang District (45°45′16.2″ N, 126°54′39.6″ E), Harbin, Heilongjiang Province, China (Figure 6). The study area experiences a cold–temperate monsoon climate, with precipitation unevenly distributed throughout the year, primarily occurring between July and September. During this period, cumulative precipitation accounts for approximately 70% of the total annual precipitation. The mean air temperature and precipitation during the millet growth period in 2021 and 2022 are presented in Figure 7. The basic experimental soil properties are shown in Table 5.

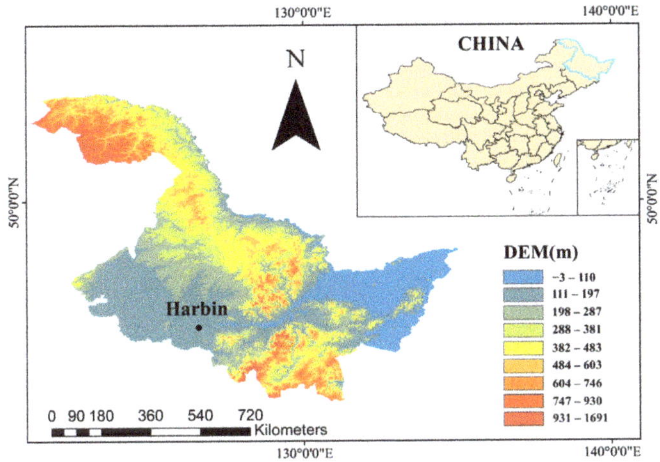


**Figure 6.** The geographical location of the study area.

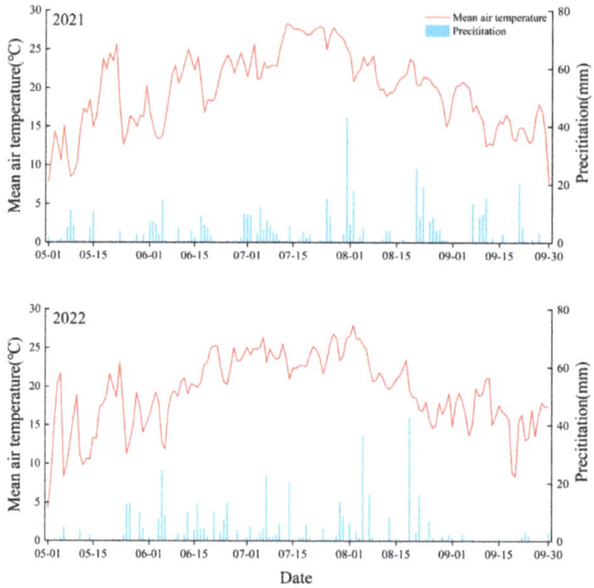


**Figure 7.** The mean air temperature and daily rainfall recorded at the study area during the millet growth periods in 2021 and 2022.

**Table 5.** Soil properties.

| Items | Value |
|---|---|
| Field capacity (%) | 30.93 |
| Total N content ($g \cdot kg^{-1}$) | 0.27 |
| Soil organic matter content ($g \cdot kg^{-1}$) | 32.2 |
| Total P content ($g \cdot kg^{-1}$) | 1.07 |
| Available phosphorus content ($mg \cdot kg^{-1}$) | 24.81 |
| Available potassium content ($mg \cdot kg^{-1}$) | 262.22 |
| pH | 7.12 |
| Soil texture | Loamy soil |

### *4.2. Experimental Design*

The millet variety selected for this experiment was 'Longgu 25', which was sown on 23 May 2021 and 21 May 2022 and harvested on 5 September 2021 and 4 September 2022, respectively. Four lower limits of irrigation were established based on the FC: 50% FC ($I_{50}$), 60% FC ($I_{60}$), 70% FC ($I_{70}$), and 80% FC ($I_{80}$). The upper limit of irrigation was set at 100% FC for all treatments. Nitrogen application amount was divided into four levels: 0 $kg \cdot hm^{-2}$ ($F_{00}$), 50 $kg \cdot hm^{-2}$ ($F_{50}$), 100 $kg \cdot hm^{-2}$ ($F_{100}$), and 150 $kg \cdot hm^{-2}$ ($F_{150}$). The experiment comprised 16 treatments arranged in a completely randomized sequence with three replications, resulting in a total of 48 plots. Each experimental plot measured 60 $m^2$ (10 m × 6 m). Fertilizer was applied in two stages: initially as a base fertilizer, followed by a single application of additional fertilizer. The initial application included 95 $kg \cdot ha^{-1}$ of $P_2O_5$, 55 $kg \cdot ha^{-1}$ of $K_2O$ and one-third of the nitrogen fertilizer, while the remaining two-thirds of the nitrogen fertilizer were applied at the end of the jointing stage of millet.

The experimental field was subdivided into 10 rows of ridges, with a distance of 60 cm between the ridges and 3.5 cm between the plants within each ridge. Drip irrigation tapes were installed in the middle of each ridge, by the roots of the millets, and both irrigation and nitrogen application were regulated using water meters and fertilizer tanks. Soil samples were collected from the center of the experimental field at 7-day intervals throughout the duration of the experiment using a soil auger and the five-point sampling method. Irrigation was initiated when the soil water content reached the lower limit of irrigation. To prevent interaction effects between different treatments, a 1.0 m buffer zone was established between each plot.

### *4.3. Sample Collection and Determination*

#### 4.3.1. Growth Indicators and Yield Components of Millet

Plant samples were collected from five randomly selected plants for each plot. These plants were evenly distributed and well established at the maturity stage. The height of the samples was measured using a tape measure, while the stem diameter was measured using a vernier caliper. The total leaf area of the samples was measured using a YT-YMJ-02 Leaf Area Meter (Shandong Yuntang Intelligent Science and Technology Co., Shandong, China). After measuring the above parameters, the samples were placed in an oven at 105 °C for an initial drying period of 0.5 h. The oven temperature was then adjusted to 75 °C, and drying continued until a constant weight was achieved. Subsequently, the dry weight of each sample was measured using an electronic balance to calculate the aboveground biomass ($kg{\cdot}ha^{-1}$).

For each treatment, one square meter of grain ears was randomly selected for collection, and the following variables were measured: spike length, spike diameter, and spike weight. After threshing, the grains were air-dried and weighed, and the final yield was converted to grain yield at 14% moisture content. The yield was then calculated in $kg{\cdot}ha^{-1}$.

Leaf area index (*LAI*) was calculated as:

$$LAI = \frac{Leaf\ area\ per\ plant}{Land\ area\ per\ plant} \tag{1}$$

#### 4.3.2. WUE

Irrigation (I) was calculated as [71]:

$$I = 10 \times D \times H \times (W_i - W_j) \tag{2}$$

where $I$ is the irrigation volume of each treatment (kg $m^{-2}$), $D$ is the soil bulk density (g $cm^{-3}$), $H$ is the thickness of the soil layer (cm), $W_i$ is the target moisture content (%) of each soil layer from 0 to 60 cm, and $W_j$ is the current moisture content (%) of each soil layer from 0 to 60 cm.

Evapotranspiration (*ET*) was calculated as:

$$ET = P + I + C_r - R_o - D_w - \Delta W \tag{3}$$

where $ET$ is the crop evapotranspiration (mm), $P$ is rainfall during the millet growth period (mm), $\Delta W$ is the change in soil moisture from the beginning to the end of the experiment (mm), $C_r$ is groundwater recharge (mm), $R_o$ is surface runoff (mm), and $D_w$ is deep drainage (mm). In this experiment, the effects of $C_r$, $R_o$, and $D_w$ were not considered [45].

WUE was calculated as [75]:

$$\text{WUE} = \frac{Y}{ET} \tag{4}$$

where WUE is defined as water use efficiency ($kg{\cdot}ha^{-1}{\cdot}mm^{-1}$); $Y$ represents the yield ($kg{\cdot}ha^{-1}$).

#### 4.3.3. AEN and PFPN

AEN was calculated as [76]:

$$\text{AEN} = \frac{Y_N - Y_C}{N_{app}} \tag{5}$$

where $Y_N$ and $Y_C$ refer to the grain yields (kg·ha$^{-1}$) of nitrogen and no nitrogen treatments, respectively. Finally, $N_{app}$ is the total nitrogen applied to each treatment (kg·ha$^{-1}$).

PFPN was calculated as [77]:

$$\text{PFPN} = \frac{Y_N}{N_{app}} \tag{6}$$

#### 4.3.4. $NH_4{}^+$-N and $NO_3{}^-$-N Accumulation in Soil

Soil samples were obtained from depths of 0~10 cm, 10~20 cm, 20~30 cm, 30~40 cm, 40~50 cm, and 50~60 cm, employing the identical sampling methodology utilized for the determination of soil water content. The soil samples were stored at −20 °C and subsequently analyzed for nitrogen content using an UV spectrophotometer (Hach DR5000). The accumulation of $NO_3{}^-$-N ($N_n$) and $NH_4{}^+$-N ($N_a$) was calculated as [78]:

$$N_n \text{or} N_a = 0.1 \times C \times H \times D \tag{7}$$

where $N_n$ and $N_a$ represent soil $NO_3{}^-$-N and $NH_4{}^+$-N accumulations (kg·ha$^{-1}$), respectively. $C$ and $H$ denote the soil $NO_3{}^-$-N or $NH_4{}^+$-N content (mg·kg$^{-1}$) and the thickness of the soil layer (cm), respectively. $D$ is the soil bulk density (g·cm$^{-3}$).

### *4.4. Date Analysis*

SPSS 18.0 (IBM Corp., New York, NY, USA) was employed for analysis of variance (ANOVA) and testing the significance between treatments. Figures were produced using Origin 2023b (Origin Lab Corporation, Northampton, MA, USA). A RSM analysis was conducted to visualize the effects of irrigation lower limit and nitrogen application amount on yield, WUE, AEN, and PFPN. Binary quadratic regression equations were developed by Origin 2023b (Origin Lab Corporation, Northampton, MA, USA). The objective was to identify the optimal regions for each response variable. This involved determining the combinations of irrigation lower limit and nitrogen application amount that would achieve the best response for each variable, as well as identifying ranges that would simultaneously satisfy the optimization requirements for all response variables. The optimization results were compared and overlapped to achieve this goal.

## 5. Conclusions

Millet yield, WUE, and AEN reached their maximum at the irrigation lower limit of 70% and 100 kg ha$^{-1}$ nitrogen fertilizer amount, while the highest PFPN were found at the irrigation lower limit of 70% and 50 kg ha$^{-1}$ nitrogen fertilizer amount. Within the 0~60 cm soil layer, as the nitrogen application rate increased, the accumulation of $NH_4{}^+$-N and $NO_3{}^-$-N showed an increasing trend. However, raising the irrigation lower limit helped reduce the $NH_4{}^+$-N and $NO_3{}^-$-N accumulation in soil. An optimal irrigation lower limit of 64% FC to 74% FC and a nitrogen fertilizer application of 80 to 100 kg ha$^{-1}$ was recommended for enhancing millet yield, improving water and nitrogen use efficiency and reducing inorganic nitrogen accumulation in Northeast China. The agricultural economic and ecological effect worth further consideration when optimizing the irrigation and fertilizer management in future study.

**Author Contributions:** T.N. and J.L.: Investigation, Methodology, Data curation, Writing—original draft. T.N., L.J. and Z.Z.: Conceptualization, Supervision, Writing—review and editing. P.C., T.L., C.D., Z.S., S.Y. and M.W.: Writing—review and editing. All authors have read and agreed to the published version of the manuscript.

**Funding:** This research was funded by the Natural Science Foundation Project of Heilongjiang Province (LH2023E109), the Basic Scientific Research Fund of Heilongjiang Provincial Universities (2023-KYYWF-1498 & 2021-KYYWF-0019), the Opening Project of Key Laboratory of Efficient Use of Agricultural Water Resources, Ministry of Agriculture and Rural Affairs of the People's Republic of China (AWR2021002) and Postdoctoral Fellowship Program of CPSF (GZC20230668).

**Data Availability Statement:** The raw data supporting the conclusions of this article will be made available by the authors on request.

**Acknowledgments:** We thank the anonymous reviewers and the editor for their suggestions, which substantially improved the manuscript.

**Conflicts of Interest:** The authors declare no conflicts of interest.

## References

1. Nelson, W.J.; Lee, B.C.; Gasperini, F.A.; Hair, D.M. Meeting the Challenge of Feeding 9 Billion People Safely and Securely. *J. Agromed.* **2012**, *17*, 347–350. [CrossRef] [PubMed]
2. Barrett, C.B. Measuring Food Insecurity. *Science* **2010**, *327*, 825–828. [CrossRef]
3. Chen, X.; Cui, Z.; Fan, M.; Vitousek, P.; Zhao, M.; Ma, W.; Wang, Z.; Zhang, W.; Yan, X.; Yang, J.; et al. Producing more grain with lower environmental costs. *Nature* **2014**, *514*, 486–489. [CrossRef] [PubMed]
4. Jindal, P.; Nikhanj, P. A review on processing technologies for value added millet products. *J. Food Process Eng.* **2023**, *46*, 14419. [CrossRef]
5. Pavithra, K.S.; Senthil, A.; Babu Rajendra Prasad, V.; Ravikesavan, R.; Djanaguiraman, M. Variations in photosynthesis associated traits and grain yield of minor millets. *Plant Physiol. Rep.* **2020**, *25*, 418–425. [CrossRef]
6. Zhang, Y.; Han, H.; Zhang, D.; Li, J.; Gong, X.; Feng, B.; Xue, Z.; Yang, P. Effects of ridging and mulching combined practices on proso millet growth and yield in semi-arid regions of China. *Field Crops Res.* **2017**, *213*, 65–74. [CrossRef]
7. Muthamilarasan, M.; Prasad, M. Small Millets for Enduring Food Security Amidst Pandemics. *Trends Plant Sci.* **2021**, *26*, 33–40. [CrossRef]
8. Panchal, A.; Singh, R.K.; Prasad, M. Recent advancements and future perspectives of foxtail millet genomics. *Plant Growth Regul.* **2022**, *99*, 11–23. [CrossRef]
9. Sinclair, T.R.; Rufty, T.W. Nitrogen and water resources commonly limit crop yield increases, not necessarily plant genetics. *Glob. Food Secur.* **2012**, *1*, 94–98. [CrossRef]
10. Ma, T.; Sun, S.; Fu, G.; Hall, J.W.; Ni, Y.; He, L.; Yi, J.; Zhao, N.; Du, Y.; Pei, T.; et al. Pollution exacerbates China's water scarcity and its regional inequality. *Nat. Commun.* **2020**, *11*, 650. [CrossRef]
11. Peng, S. Water resources strategy and agricultural development in China. *J. Exp. Bot.* **2011**, *62*, 1709–1713. [CrossRef] [PubMed]
12. An, P.; Ren, W.; Liu, X.; Song, M.; Li, X. Adjustment and Optimization of the Cropping Systems under Water Constraint. *Sustainability* **2016**, *8*, 1207. [CrossRef]
13. Wu, D.; Xu, X.; Chen, Y.; Shao, H.; Sokolowski, E.; Mi, G. Effect of different drip fertigation methods on maize yield, nutrient and water productivity in two-soils in Northeast China. *Agric. Water Manag.* **2019**, *213*, 200–211. [CrossRef]
14. Goyal, M.; Kaur, R. Interactive effect of nitrogen nutrition, nitrate reduction and seasonal variation on oxalate synthesis in leaves of Napier-bajra hybrid (*Pennisetum purpureum* × *P. glaucum*). *Crop Pasture Sci.* **2019**, *70*, 18282. [CrossRef]
15. Cui, Z.; Wang, G.; Yue, S.; Wu, L.; Zhang, W.; Zhang, F.; Chen, X. Closing the N-Use Efficiency Gap to Achieve Food and Environmental Security. *Environ. Sci. Technol.* **2014**, *48*, 5780–5787. [CrossRef]
16. Lu, L.; Wang, Q.; Zhang, W.; Gao, M.; Xv, Y.; Li, S.; Dong, H.; Chen, D.; Yan, P.; Dong, Z. Urea Coated with Polyaspartic Acid-Chitosan Increases Foxtail Millet (*Setaria italica* L. Beauv.) Grain Yield by Improving Nitrogen Metabolism. *Plants* **2024**, *13*, 415. [CrossRef]
17. Bandyopadhyay, T.; Prasad, M. A Precise Method for Analyzing Nitrogen Use in Foxtail Millet. *Methods Mol. Biol.* **2020**, *2057*, 113–118. [CrossRef]
18. Galloway, J.N.; Cowling, E.B. Reactive Nitrogen and The World: 200 Years of Change. *Ambio J. Hum. Environ.* **2002**, *31*, 2. [CrossRef]
19. Shaaban, M.; Peng, Q.-a.; Bashir, S.; Wu, Y.; Younas, A.; Xu, X.; Rashti, M.R.; Abid, M.; Zafar-ul-Hye, M.; Núñez-Delgado, A.; et al. Restoring effect of soil acidity and Cu on $N_2O$ emissions from an acidic soil. *J. Environ. Manag.* **2019**, *250*, 109535. [CrossRef]
20. Crăciun, I.; Crăciun, M. Water and nitrogen use efficiency under limited water supply for maize to increase land productivity. In *Crop Yield Response to Deficit Irrigation*; Moutonnet, P., Hera, C., Nielsen, D.R., Eds.; Kluwer Academic Publishers: Alphen aan den Rijn, The Netherlands, 1996; pp. 87–94.
21. Zhou, S.; Ye, F.; Xia, D.; Liu, Z.; Wu, Y.; Chen, F. Climate change impacts assessment and developing adaptation strategies for rainfed foxtail millet in northern Shanxi, China. *Agric. Water Manag.* **2023**, *290*, 108575. [CrossRef]
22. Matsuura, A.; Tsuji, W.; An, P.; Inanaga, S.; Murata, K. Effect of Pre- and Post-heading Water Deficit on Growth and Grain Yield of Four Millets. *Plant Prod. Sci.* **2015**, *4*, 323–331. [CrossRef]
23. Jahansuz, M.R.; Afshar, R.K.; Heidari, H.; Hashemi, M. Evaluation of Yield and Quality of Sorghum and Millet as Alternative Forage Crops to Corn under Normal and Deficit Irrigation Regimes. *Jordan J. Agric. Sci.* **2014**, *10*, 4.
24. Zhang, J.; Liu, T.; Fu, J.; Zhu, Y.; Jia, J.; Zheng, J.; Zhao, Y.; Zhang, Y.; Wang, G. Construction and application of EST library from Setaria italica in response to dehydration stress. *Genomics* **2007**, *90*, 121–131. [CrossRef]
25. Nagaz, K.; Masmoudi, M.M.; Mechila, N.B. Yield and Water-Use Efficiency of Pearl Millet (*Pennisetum glaucum* (L.) R. Br.) under Deficit Irrigation with Saline Water in Arid Conditions of Southern Tunisia. *Res. J. Agron.* **2009**, *3*, 9–17.

26. Seghatoleslami, M.J.; Kafi, M.; Majidi, E. Effect of Deficit Irrigation on Yield, Wue and Some Morphological and Phenological Traits of Three Millet Species. *Pak. J. Bot.* **2008**, *40*, 1555–1560.
27. Li, Z.; Li, W.; Wang, J.; Zhang, J.; Wang, Z. Drip irrigation shapes the soil bacterial communities and enhances jujube yield by regulating the soil moisture content and nutrient levels. *Agric. Water Manag.* **2023**, *289*, 108563. [CrossRef]
28. Grafton, R.Q.; Williams, J.; Perry, C.J.; Molle, F.; Ringler, C.; Steduto, P.; Udall, B.; Wheeler, S.A.; Wang, Y.; Garrick, D.; et al. The paradox of irrigation efficiency. *Science* **2018**, *361*, 748–750. [CrossRef]
29. Pablo, R.G.; O'Neill, M.K.; McCaslin, B.D.; Remmenga, M.D.; Keenan, J.G.; Onken, B.M. Evaluation of Corn Grain Yield and Water Use Efficiency Using Subsurface Drip Irrigation. *J. Sustain. Agric.* **2007**, *30*, 153–172. [CrossRef]
30. Zhou, L.; Feng, H.; Zhao, Y.; Qi, Z.; Zhang, T.; He, J.; Dyck, M. Drip irrigation lateral spacing and mulching affects the wetting pattern, shoot-root regulation, and yield of maize in a sand-layered soil. *Agric. Water Manag.* **2017**, *184*, 114–123. [CrossRef]
31. Linn, D.M.; Doran, J.W. Effect of water-filled pore space on carbon dioxide and nitrous oxide production in tilled and nontilled soils. *Soil Sci. Soc. Am. J.* **1984**, *48*, 1267–1272. [CrossRef]
32. Jahanzad, E.; Sadeghpour, A.; Hoseini, M.B.; Barker, A.V.; Hashemi, M.; Keshavarz Afshar, R. Competition, Nitrogen Use Efficiency, and Productivity of Millet–Soybean Intercropping in Semiarid Conditions. *Crop Sci.* **2015**, *55*, 2842–2851. [CrossRef]
33. Leblanc, V.; Vanasse, A.; Bélanger, G.; Seguin, P. Sweet Pearl Millet Yields and Nutritive Value as Influenced by Fertilization and Harvest Dates. *Agron. J.* **2012**, *104*, 542–549. [CrossRef]
34. Wenqian, C.; Peng, L.; Shiwei, Z. Effects of Water and Nitrogen Coupling on Growth, Yield and Water and Nitrogen Utilization after Millet Jointing. *Acta Agric. Boreali-Occident. Sin.* **2018**, *27*, 1313–1321. (In Chinese) [CrossRef]
35. Cheng, J.; Tan, H.; Shan, M.; Duan, M.; Ye, L.; Yang, Y.; He, L.; Shen, H.; Yang, Z.; Wang, X. Genome-wide identification and characterization of the NPF genes provide new insight into low nitrogen tolerance in Setaria. *Front. Plant Sci.* **2022**, *13*, 1043832. [CrossRef]
36. Rosolem, C.A. Exchangeable basic cations and nitrogen distribution in soil as affected by crop residues and nitrogen. *Braz. Arch. Biol. Technol.* **2011**, *54*, 441–450. [CrossRef]
37. Bitew, Y.; Alemayehu, G.; Adego, E.; Assefa, A. Impact of precursor crop and nitrogen fertilizer on the productivity of finger millet based cropping system and soil fertility in Lake Tana basin of Ethiopia. *J. Plant Nutr.* **2020**, *43*, 1824–1839. [CrossRef]
38. Nadeem, F.; Ahmad, Z.; Ul Hassan, M.; Wang, R.; Diao, X.; Li, X. Adaptation of Foxtail Millet (*Setaria italica* L.) to Abiotic Stresses: A Special Perspective of Responses to Nitrogen and Phosphate Limitations. *Front. Plant Sci.* **2020**, *11*, 187. [CrossRef]
39. Nascente, A.S.; Crusciol, C.A.C. Cover crops affecting levels of ammonium and nitrate in the soil and upland rice development. *Semin. Ciências Agrárias* **2013**, *34*, 2189–2202. [CrossRef]
40. Gheysari, M.; Mirlatifi, S.M.; Homaee, M.; Asadi, M.E.; Hoogenboom, G. Nitrate leaching in a silage maize field under different irrigation and nitrogen fertilizer rates. *Agric. Water Manag.* **2009**, *96*, 946–954. [CrossRef]
41. Nadeem, F.; Ahmad, Z.; Wang, R.; Han, J.; Shen, Q.; Chang, F.; Diao, X.; Zhang, F.; Li, X. Foxtail Millet [*Setaria italica* (L.) Beauv.] Grown under Low Nitrogen Shows a Smaller Root System, Enhanced Biomass Accumulation, and Nitrate Transporter Expression. *Front. Plant Sci.* **2018**, *9*, 205. [CrossRef]
42. Huang, S.; Zhang, W.; Yu, X.; Huang, Q. Effects of long-term fertilization on corn productivity and its sustainability in an Ultisol of southern China. *Agric. Ecosyst. Environ.* **2010**, *138*, 44–50. [CrossRef]
43. Mao, K.J.; Huang, P.; Zhang, L.; Yu, H.S.; Zhou, X.Z.; Ye, X.Y. Optimisation of extraction conditions for total saponins from *Cynanchum wallichii* using response surface methodology and its anti-tumour effects. *Nat. Prod. Lett.* **2018**, *32*, 2233–2237. [CrossRef]
44. Zou, H.; Fan, J.; Zhang, F.; Xiang, Y.; Wu, L.; Yan, S. Optimization of drip irrigation and fertilization regimes for high grain yield, crop water productivity and economic benefits of spring maize in Northwest China. *Agric. Water Manag.* **2020**, *230*, 105986. [CrossRef]
45. Yan, F.; Zhang, F.; Fan, X.; Fan, J.; Wang, Y.; Zou, H.; Wang, H.; Li, G. Determining irrigation amount and fertilization rate to simultaneously optimize grain yield, grain nitrogen accumulation and economic benefit of drip-fertigated spring maize in northwest China. *Agric. Water Manag.* **2021**, *243*, 106440. [CrossRef]
46. Yang, Y.J.; Guo, M.J.; Pang, Y.F.; Shao, Q.L.; Wu, Y.Z.; Zhao, H.M.; Ji, A.Q.; Ma, J.H.; Song, X.E.; Sun, C.Q.; et al. Foxtail Millet (*Setaria italica* (L.) P. Beauvois) Quality Response to Fertilizer Levels, Herbicide, and Selenium. *Appl. Ecol. Environ. Res.* **2021**, *19*, 4231–4249. [CrossRef]
47. Food and Agriculture Organization. Statistics of Production: Crops. 2022. Available online: https://www.fao.org/faostat/en/#home (accessed on 24 January 2024).
48. Zhu, Z.L.; Chen, D.L. Nitrogen fertilizer use in China—Contributions to food production, impacts on the environment and best management strategies. *Nutr. Cycl. Agroecosyst.* **2002**, *63*, 117–127. [CrossRef]
49. Dong, X.J.; Li, Z.J.; Ma, J.F.; Yu, J.H.; Li, X.Y.; Sun, G.Q.; Zheng, Y.L. Current Situation Analysis of Millet Breeding Varieties from the Spring Sowing Region in Northeast China in 2011–2021. *J. Sichuan Agric. Univ.* **2024**, *42*, 57–65+125. (In Chinese)
50. Lu, X.; Li, Z.; Bu, Q.; Cheng, D.; Sun, Z. Effects of Rainfall Harvesting and Mulching on Corn Yield and Water Use in the Corn Belt of Northeast China. *Agron. J.* **2014**, *106*, 2175. [CrossRef]

51. Pei, J.; Li, H.; Li, S.; An, T.; Farmer, J.; Fu, S.; Wang, J. Dynamics of Maize Carbon Contribution to Soil Organic Carbon in Association with Soil Type and Fertility Level. *PLoS ONE* **2015**, *10*, e0120825. [CrossRef]
52. Araújo, É.d.O.; Tadeu Vitorino, A.C.; Martins Mercante, F. Rates of nitrogen and inoculation of *Herbaspirillum seropedicae* in corn culture in conditions of fertile soil. *Acta Agron.* **2016**, *65*, 16–23. [CrossRef]
53. Gong, X.; Li, J.; Ma, H.; Chen, G.; Dang, K.; Yang, P.; Wang, M.; Feng, B. Nitrogen deficiency induced a decrease in grain yield related to photosynthetic characteristics, carbon–nitrogen balance and nitrogen use efficiency in proso millet (*Panicum miliaceum* L.). *Arch. Agron. Soil. Sci.* **2019**, *66*, 398–413. [CrossRef]
54. Jamali, K.D.; Ali, S.A. Yield and Yield Components with Relation to Plant Height in Semi-Dwarf Wheat. *Pak. J. Bot.* **2008**, *40*, 1805–1808.
55. Kara, B.; Kara, B. Effects of Different Tillage Practices on the Trait Associations in Barley. *Turk. J. Crops* **2008**, *13*, 32–43.
56. Golombek, S. Drought tolerance mechanisms of pearl millet. In Proceedings of the Deutscher Tropentag, Witzenhausen, Germany, 9–11 October 2002; Volume 19, p. 37213.
57. Habiyaremye, C.; Barth, V.; Highet, K.; Coffey, T.; Murphy, K. Phenotypic Responses of Twenty Diverse Proso Millet (*Panicum miliaceum* L.) Accessions to Irrigation. *Sustainability* **2017**, *9*, 389. [CrossRef]
58. Nematpour, A.; Eshghizadeh, H.R.; Zahedi, M. Comparing the Corn, Millet and Sorghum as Silage Crops Under Different Irrigation Regime and Nitrogen Fertilizer Levels. *Int. J. Plant Prod.* **2021**, *15*, 351–361. [CrossRef]
59. Comas, L.H.T.; Thomas, J.; DeJonge, K.C.; Zhang, H.; Gleason, S.M. Water productivity under strategic growth stage-based deficit irrigation in maize. *Agric. Water Manag.* **2019**, *212*, 433–440. [CrossRef]
60. Sadras, V.O. Yield and water-use efficiency of water- and nitrogen-stressed wheat crops increase with degree of co-limitation. *Eur. J. Agron.* **2004**, *21*, 455–464. [CrossRef]
61. Sarr, P.S.; Khouma, M.; Sene, M.; Guisse, A.; Badiane, A.N.; Yamakawa, T. Effect of pearl millet–cowpea cropping systems on nitrogen recovery, nitrogen use efficiency and biological fixation using the15N tracer technique. *Soil Sci. Plant Nutr.* **2008**, *54*, 142–147. [CrossRef]
62. Ahmad, P. (Ed.) Plant growth under drought stress. In *Water Stress and Crop Plants (A Sustainable Approach)*; John Wiley and Sons Inc.: Hoboken, NJ, USA, 2016; Volume 2, pp. 649–668.
63. Shao, H.B.; Chu, L.Y.; Jaleel, C.A.; Manivannan, P.; Panneerselvam, R.; Shao, M.A. Understanding water deficit stress-induced changes in the basic metabolism of higher plants-biotechnologically and sustainably improving agriculture and the ecoenvironment in arid regions of the globe. *Crit. Rev. Biotechnol.* **2009**, *29*, 131. [CrossRef]
64. Payne, W.A.; Hossner, L.R.; Onken, A.B.; Wendt, C.W. Nitrogen and Phosphorus Uptake in Pearl Millet and Its Relation to Nutrient and Transpiration Efficiency. *Agron. J.* **1995**, *87*, 425–431. [CrossRef]
65. Crookston, B.; Blaser, B.; Darapuneni, M.; Rhoades, M. Pearl Millet Forage Water Use Efficiency. *Agronomy* **2020**, *10*, 1672. [CrossRef]
66. Caviglia, O.P.; Melchiori, R.J.M.; Sadras, V.O. Nitrogen utilization efficiency in maize as affected by hybrid and N rate in late-sown crops. *Field Crop. Res.* **2014**, *168*, 27–37. [CrossRef]
67. Liu, X.; Yan, X.; Zhao, B.; Han, X. Study on Relation between Fertilizer Dosage and Its Recovery Efficiency and Their Application. *Chin. J. Soil Sci.* **2012**, *43*, 131–135. (In Chinese)
68. Sun, Y.; Hu, K.; Zhang, K.; Jiang, L.; Xu, Y. Simulation of nitrogen fate for greenhouse cucumber grown under different water and fertilizer management using the EU-Rotate_N model. *Agric. Water Manag.* **2012**, *112*, 21–32. [CrossRef]
69. Ten Berge, H.F.M.; Hijbeek, R.; Van Loon, M.P.; Rurinda, J.; Tesfaye, K.; Zingore, S.; Craufurd, P.; Van Heerwaarden, J.; Brentrup, F.; Schröder, J.J. Maize crop nutrient input requirements for food security in sub-Saharan Africa. *Glob. Food Secur.* **2019**, *23*, 9–21. [CrossRef]
70. Giletto, C.M.; Echeverría, H.E. Nitrogen balance for potato crops in the southeast pampas region, Argentina. *Nutr. Cycl. Agroecosyst.* **2013**, *95*, 73–86. [CrossRef]
71. Wang, Z.; Nie, T.; Lu, D.; Zhang, P.; Li, J.; Li, F.; Zhang, Z.; Chen, P.; Jiang, L.; Dai, C.; et al. Effects of Different Irrigation Management and Nitrogen Rate on Sorghum (*Sorghum bicolor* L.) Growth, Yield and Soil Nitrogen Accumulation with Drip Irrigation. *Agronomy* **2024**, *14*, 215. [CrossRef]
72. Badr, M.A.; Abou Hussein, S.D.; El-Tohamy, W.A.; Gruda, N. Nutrient uptake and yield of tomato under various methods of fertilizer application and levels of fertigation in arid lands. *Gesunde Pflanz.* **2010**, *62*, 11–19. [CrossRef]
73. Dong, Q.; Dang, T.; Guo, S.; Hao, M. Effect of different mulching measures on nitrate nitrogen leaching in spring maize planting system in south of Loess Plateau. *Agric. Water Manag.* **2019**, *213*, 654–658. [CrossRef]
74. Chen, P.; Nie, T.; Chen, S.; Zhang, Z.; Qi, Z.; Liu, W. Recovery efficiency and loss of 15N-labelled urea in a rice-soil system under water saving irrigation in the Songnen Plain of Northeast China. *Agric. Water Manag.* **2019**, *222*, 139–153. [CrossRef]
75. Yan, J.; Zhang, N.; Kang, F.; Wang, J.; Wang, X. Cultivar replacement increases water use efficiency in foxtail millet in Shaanxi Province, China. *Plant Physiol. Biochem.* **2021**, *164*, 73–81. [CrossRef] [PubMed]
76. Rathore, S.S.; Babu, S.; Singh, V.K.; Shekhawat, K.; Singh, R.K.; Upadhyay, P.K.; Hashim, M.; Sharma, K.C.; Jangir, R.; Singh, R. Sulfur Sources Mediated the Growth, Productivity, and Nutrient Acquisition Ability of Pearlmillet–Mustard Cropping Systems. *Sustainability* **2022**, *14*, 14857. [CrossRef]

77. Liu, Y.; Bai, M.; Shen, F.; Wu, Z.; Yang, J.; Li, N.; Liu, H.; Dai, J.; Han, X. Enhancing Soybean and Maize Yields through Improved Nitrogen and Soil Water Use Efficiencies: A 40-Year Study on the Impact of Farmyard Manure Amendment in Northeast China. *Plants* **2024**, *13*, 500. [CrossRef] [PubMed]
78. Lu, J.; Bai, Z.; Velthof, G.L.; Wu, Z.; Chadwick, D.; Ma, L. Accumulation and leaching of nitrate in soils in wheat-maize production in China. *Agric. Water Manag.* **2019**, *212*, 407–415. [CrossRef]

plants

MDPI

*Article*

# Characterizing Droughts During the Rice Growth Period in Northeast China Based on Daily SPEI Under Climate Change

**Tangzhe Nie [1], Xiu Liu [1], Peng Chen [2], Lili Jiang [1], Zhongyi Sun [3], Shuai Yin [4], Tianyi Wang [5], Tiecheng Li [6] and Chong Du [1,*]**

1 School of Water Conservancy and Electric Power, Heilongjiang University, Harbin 150080, China; 2019036@hlju.edu.cn (T.N.); 2222019@s.hlju.edu.cn (X.L.); jianglili@hlju.edu.cn (L.J.)
2 College of Agricultural Science and Engineering, Hohai University, Nanjing 211100, China
3 College of Ecology and Environment, Hainan University, Haikou 570208, China
4 State Key Laboratory of Remote Sensing Science, Aerospace Information Research Institute, Chinese Academy of Sciences, Beijing 100101, China
5 School of Water Conservancy and Civil Engineering, Northeast Agricultural University, Harbin 150030, China
6 Institute of Geographic Sciences and Natural Resources Research, Chinese Academy of Sciences, Beijing 100101, China
* Correspondence: duchong@hlju.edu.cn

**Abstract:** In agricultural production, droughts occurring during the crucial growth periods of crops hinder crop development, while the daily-scale standardized precipitation evapotranspiration index (*SPEI*) can be applied to accurately identify the drought characteristics. In this study, we used the statistical downscaling method to obtain the daily precipitation ($P_r$), maximum air temperature ($T_{max}$) and minimum air temperature ($T_{min}$) during the rice growing season in Heilongjiang Province from 2015 to 2100 under the SSP1-2.6, SSP2-4.5 and SSP5-8.5 in CMIP6, to study the spatial and temporal characteristics of drought during the rice growing season in cold region and the effect of climate change on drought characteristics. The potential evapotranspiration ($PET_0$) was calculated using the regression correction method of the Hargreaves formula recommended by the FAO, and the daily *SPEI* was calculated to quantitatively identify the drought classification. The Pearson correlation coefficient was used to analyze the correlation between the meteorological factors ($P_r$, $T_{max}$, $T_{min}$), $PET_0$ and *SPEI*. The results showed that: (1) Under 3 SSP scenarios, $P_r$ showed an increasing trend from the northwest to the southeast, $T_{max}$ showed an increasing trend from the northeast to the southwest, and higher $T_{min}$ was mainly distributed in the east and west regions. (2) $PET_0$ indicated an overall interannual rise in the three future SSP scenarios, with higher values mainly distributed in the central and western regions. The mean daily $PET_0$ values ranged from 4.8 to 6.0 mm/d. (3) Under SSP1-2.6, rice mainly experienced mild drought and moderate drought ($-0.5 \geq SPEI > -1.5$). The predominant drought classifications experienced were mild, moderate, and severe drought under SSP2-4.5 and SSP8.5 ($-0.5 \geq SPEI > -2.0$). (4) The tillering stage experienced the highest drought frequency and drought intensity, with the longest drought lasting 24 days. However, the heading flower stage had the lowest drought frequency and drought intensity. The drought barycenter was mainly in Tieli and Suihua. (5) The $PET_0$ was most affected by the $T_{max}$, while the *SPEI* was most affected by the $P_r$. This study offers a scientific and rational foundation for understanding the drought sensitivity of rice in Northeast China, as well as a rationale for the optimal scheduling of water resources in agriculture in the future.

**Keywords:** rice; potential evapotranspiration ($PET_0$); standardized precipitation evapotranspiration index (*SPEI*); climate change; drought characteristics

## 1. Introduction

The Intergovernmental Panel on Climate Change's Sixth Assessment Report (IPCC AR6) forecasts a hastened pace of global warming, indicating that the rise in global temperature is anticipated to attain or surpass 1.5 °C [1]. Moreover, precipitation will experience a reduction of around 2% for each 1 °C increment in temperature [2,3]. Under this scenario, further alterations in the global water cycle will result in more intense precipitation and more severe droughts [4]. Droughts are more likely to deteriorate as a result of ongoing climate change and increased human activity, which presents serious dangers to the environment, agriculture and human livelihoods [5]. Agriculture is crucial for socio-economic development. Severe agricultural drought may imperil food security and social stability [6].

In the context of global warming, the climate has been affecting the global water cycle by altering the patterns of precipitation, temperature, surface runoff, potential evapotranspiration ($PET_0$) and increasing atmospheric water vapor content [7]. These alterations generally result in a further intensification of various climate extremes, such as droughts, floods, heat waves, cold waves, and extreme precipitation events, with respect to their duration, frequency, and intensity [8]. Among these alterations, precipitation, temperature, and evapotranspiration are the primary factors contributing to the occurrence of drought events. For instance, Wang et al. conducted an assessment of the global drought characteristics by employing multiple indicators from the CMIP6 model. Their findings revealed that drought was expected to intensify in the majority of regions across the globe in the future, with the alterations in $PET_0$ being among the key contributing factors [9]. Ren et al. and Zeng et al. demonstrated that an increase in $PET_0$ would increase the frequency, duration and severity of droughts [10,11]. Gond et al. assessed the changes in drought in Northern India and indicated that drought prediction was highly reliant on the prediction of precipitation and temperature [12]. Ren et al. showed that the difference between precipitation and $PET_0$ would increase since $PET_0$ was significantly affected by the temperature increase [10].

Drought is a type of meteorological disaster that is marked by its high occurrence frequency, long duration, and extensive impact range. Notably, it exerts a particularly serious impact on agricultural production [13]. Generally, droughts can be categorized into four distinct types specifically meteorological drought, agricultural drought, hydrological drought, and socio-economic drought [14]. A considerable number of drought indices have been devised on the basis of diverse drought analysis parameters. Examples of these indices include the Palmer Drought Severity Index (*PDSI*), the Standardized Precipitation Evapotranspiration Index (*SPEI*), the Standardized Precipitation Index (*SPI*), and the Socio-Economic Drought Index (*SEDI*) [15–17]. Regarding the application of these drought indices, the *PDSI*, which is founded on soil water balance, demands relatively high-quality data. Moreover, it lacks the adaptability to accommodate the inherent multi-scale characteristics of drought. The *SPEI* and *SPI* are important indicators for the assessment of regional droughts in a warming climate and can be used to identify meteorological drought characteristics with multi-scale features [18]. Nonetheless, the *SPI* solely identifies drought based on precipitation. In contrast, the *SPEI* represents an improvement over *SPI*. The principal advantage of *SPEI* compared to *SPI* lies in its ability to identify alterations in evapotranspiration and temperature during drought assessment [19,20]. The *SPEI* has been extensively employed in the fields of drought monitoring and assessment [21,22]. Ling et al. employed *SPEI* to investigate the drought characteristics of maize within the growing period in the Huang-Huai-Hai Plain spanning from 1960 to 2020 and indicated that droughts in the maize seeding period were the most frequent and lasted the longest events. The lowest frequency and shortest duration of drought occurred during the jointing-flowering period, and the barycenter of drought shifted eastward from 1970 to 2000 [1]. Qin et al. used *SPEI* to study the effect of drought on sugarcane yield, and the conclusion showed that the drought stages of sugarcane were the seeding and tillering stages most heavily impacted [23]. Currently, numerous scholars have opted to employ *SPEIs* on larger time scales to characterize drought. However, the monthly-scale *SPEIs* exhibited relatively lower sensitivity towards short-term droughts that take place during the crop growth period [24,25]. When the drought is sporadic and transiently severe, diagnosing the drought on a monthly scale becomes challenging. Consequently, in order to more effectively disclose the drought characteristics during the crop growth period and accurately identify the short-term droughts that occur throughout this period, drought indices on short-term time scales ought to be utilized for monitoring and analysis [1].

A noticeable trend of drought has happened in Northeast China in recent years, with a notable increase in the frequency and intensity of droughts. This has negatively impacted both national food security and agricultural productivity due to the uneven spatial and temporal distribution of precipitation and the ongoing rise in temperature [26]. Heilongjiang Province is a major grain-producing region in Northeastern China's Songnen Plain [27]. By 2020, 289.615 million tons of rice were produced, corresponding to 13.67% of the country's total [28]. Changes in precipitation and temperature have been found to affect rice yields in a number of studies [29,30]. A decrease in rice yields, for instance, has been attributed to warmer temperatures along with greater precipitation, suggesting that rice production is extremely susceptible to floods and drought events brought on by climate change [31]. Currently, the controlled irrigation mode is being promoted in Heilongjiang Province. Under this mode, except for the returning green stage, no water layers are set in other growth stages of rice. Therefore, in this context, rice is more likely to be directly affected by drought [32]. The regular occurrence of droughts during the rice-growing season is endangering food security.

Therefore, this study employs daily scale *SPEI* to quantitatively identify the short-term drought characteristics during the rice growth period in Heilongjiang Province from 2015 to 2100 under three scenarios: SSP1-2.6, SSP2-4.5 and SSP5-8.5. We aim to clarify the changing trends of $PET_0$ and *SPEI* during the rice growth period and reveal the meteorological drought characteristics during the rice growth period. This study will offer a scientific and reasonable basis for assessing the sensitivity of rice in Northeast China to future drought and provide a reasonable basis for the formulation of water resource management measures and regional water resource security.

## 2. Results

### 2.1. Temporal and Spatial Trends in Climatic Factors

From the 2030s to the 2090s, $P_r$ exhibited an increasing trend from northwest to southeast (Figure S1(a1–a3,b1–b3,c1–c3)). The range of mean $P_r$ under SSP1-2.6, SSP2-4.5 and SSP5-8.5 were 3.61–4.12 mm/d, 3.66–4.05 mm/d, and 3.81–4.52 mm/d, respectively (Figure 1(a1–a3)). In both the B and D growth stages, $P_r$ showed an increasing trend from the 2030s to the 2090s under SSP2-4.5 and SSP5-8.5. Under the SSP2-4.5, $P_r$ in the A growth stage demonstrated a distinct trend of initially increasing and then decreasing from the 2030s to the 2090s (Figure 1(a2)). Under SSP5-8.5, $P_r$ in the C growth stage exhibited an increasing trend from the 2060s to the 2090s, with an increase of 0.7 mm/d (Figure 1(a3)).

$T_{max}$ and $T_{min}$ showed an increasing trend of from the northeast to the southwest from the 2030s to 2090s (Figures S2 and S3), the ranges of mean $T_{max}$ under SSP1-2.6, SSP2-4.5 and SSP5-8.5 was 25.82–26.87 °C, 25.58–27.76 °C and 25.9–30.04 °C, respectively (Figure 1(b1–b3)). Similarly, the mean values of $T_{min}$ under SSP1-2.6, SSP2-4.5 and SSP5-8.5 were 15.44–16.24 °C, 15.23–17.23 °C, and 15.68–19.48 °C, respectively (Figure 1(c1–c3)). The $T_{max}$ and $T_{min}$ in each growth stage exhibited an increased trend, with greater $T_{max}$ and Tmin in the C and D stages. In the 2090s, the $T_{max}$ and $T_{min}$ in stage C under the SSP5-8.5

reached their maximum temperature and minimum temperature of 32.5 °C and 23 °C, respectively (Figure 1(b1–b3,c1–c3)).

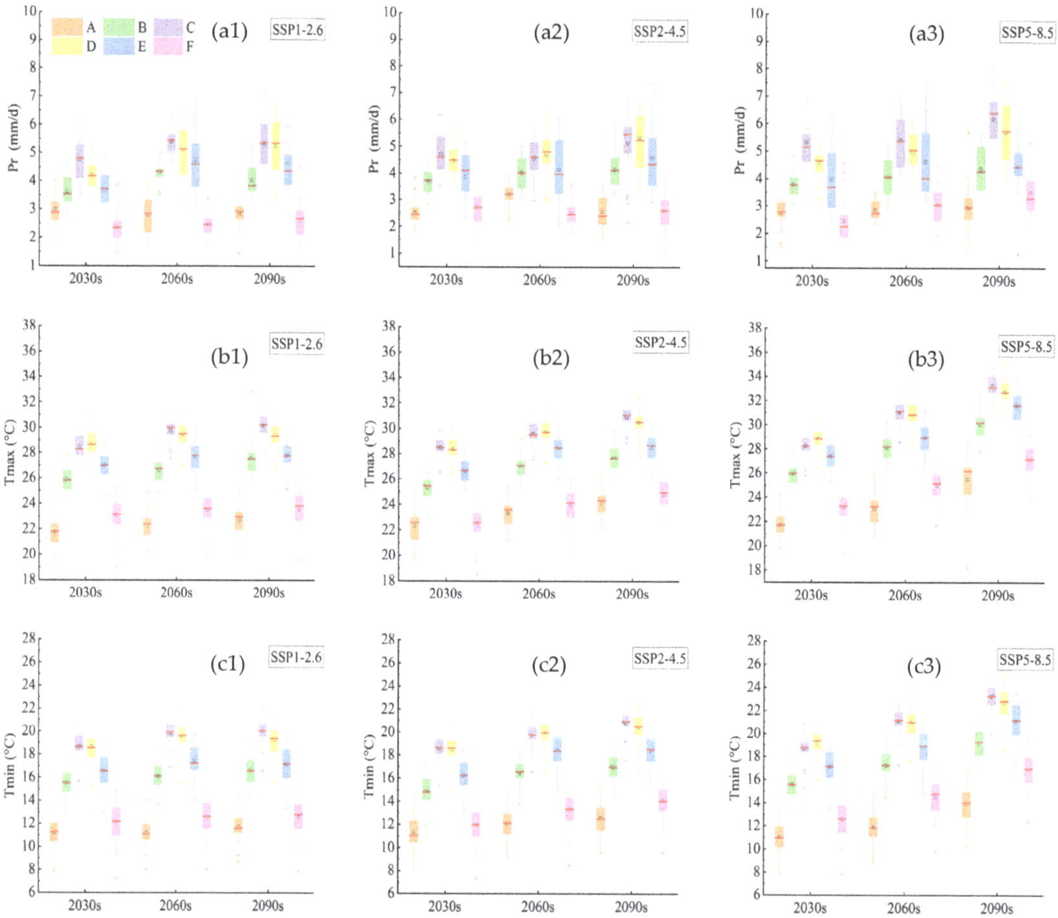


**Figure 1.** Trends of $P_r$ (**a1–a3**), $T_{max}$ (**b1–b3**) and $T_{min}$ (**c1–c3**) under SSP1-2.6, SSP2-4.5 and SSP5-8.5 during each growth period of rice from 2015 to 2100. $P_r$, $T_{max}$ and $T_{min}$ represent precipitation, maximum air temperature and minimum air temperature, respectively. The 2030s, 2060s and 2090s represent the period of 2015–2040, 2041–2070 and 2071–2100, respectively. A, B, C, D, E and F represent the returning green stage, tillering stage, jointing booting stage, heading flower stage, milk stage and yellow ripening stage of rice, respectively. In the box-plot, the red line, the black square and the diamond represent the median, the mean and the outliers, respectively.

### 2.2. Spatial and Temporal Distribution and Trends of $PET_0$

From the 2030s to 2090s, the high $PET_0$ was mainly concentrated in the southwest and central regions. Notably, the $PET_0$ under SSP5-8.5 was greater than those under SSP1-2.6 and SSP2-4.5 (Figure 2). The mean $PET_0$ under SSP1-2.6, SSP2-4.5 and SSP5-8.5 were 5.21–5.38 mm/d, 5.18–5.47 mm/d and 5.18–5.74 mm/d, respectively. Additionally, the $PET_0$ at each rice growth stage showed an increasing trend from the 2030s to the 2090s. The higher $PET_0$ were observed in the B and C stages, and in the B stage, the $PET_0$ reached 6.09 mm/d, 6.1 mm/d, and 6.4 mm/d under SSP1-2.6, SSP2-4.5, and SSP5-8.5 at 2090s, respectively. (Figure 2(a4–c4)).

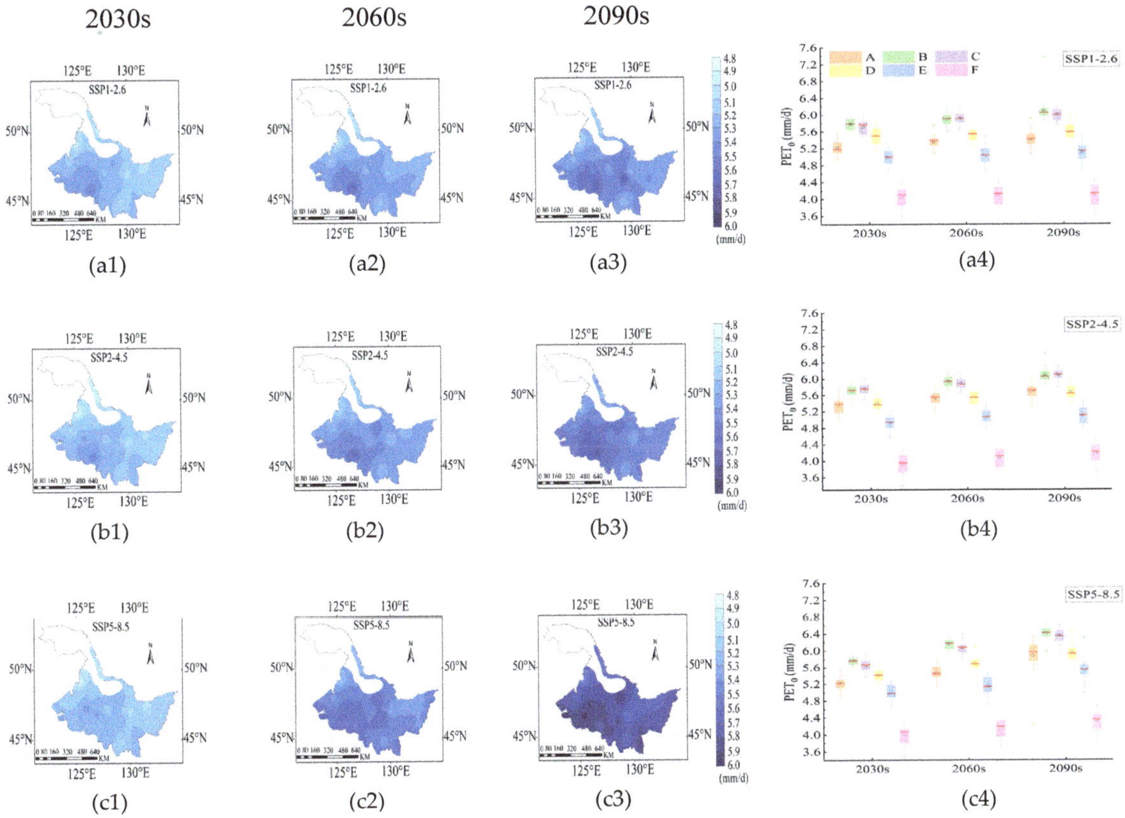


**Figure 2.** Spatiotemporal distribution of $PET_0$ under SSP1-2.6 (**a1–a4**), SSP2-4.5 (**b1–b4**) and SSP5-8.5 (**c1–c4**) during the rice growth period from 2015 to 2100. The 2030s, 2060s and 2090s represent the period of 2015–2040, 2041–2070 and 2071–2100, respectively. A, B, C, D, E and F represent the returning green stage, tillering stage, jointing booting stage, heading flower stage, milk stage and yellow ripening stage of rice, respectively. In the box-plot, the red line, the black square and the diamond represent the median, the mean and the outliers, respectively.

## 2.3. Trend of SPEI

From 2030s to 2090s, alternating wet and dry conditions were manifested. Evidently, the number of days with drought exceeded that of days without drought (Figure 3), and the daily mean *SPEI* ranged from −1.1 to 0.1, with the *SPEI* lower in the south region (Figure S4). Under SSP1-2.6, *SPEI* was mainly concentrated between no drought, mild drought and moderate drought (Figure 3(a1)), while under SSP2-4.5 and SSP5-8.5, *SPEI* was mainly concentrated between no drought, mild drought, moderate drought and severe drought (Figure 3(b1,c1)). Under both SSP1-2.6 and SSP5-8.5, *SPEI* showed a significant upward trend ($Z > 1.96$), with the *SPEI* mutation point interval ranging from 2017 to 2093 and 2017 to 2099, respectively (Figure 3(a2,c2)).

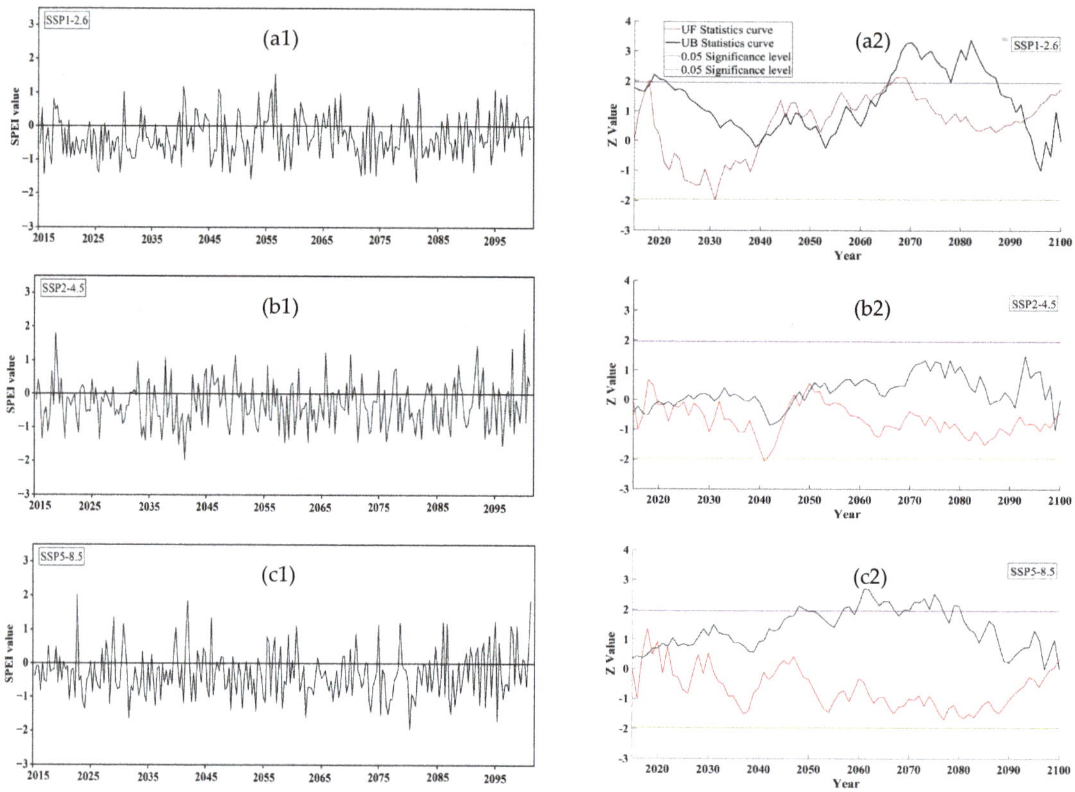


**Figure 3.** Changing trend in *SPEI* and Mann-Kendall mutation analysis under SSP1-2.6 (**a1**,**a2**), SSP2-4.5 (**b1**,**b2**) and SSP5-8.5 (**c1**,**c2**) during the rice growth period from 2015 to 2100.

### *2.4. Drought Frequency ($P_{ij}$)*

From the 2030s to 2090s, the mean values of $P_{ij}$ during the rice growth period were 14.18–15.23%, 14.59–17.15% and 14.82–16.36%, respectively. The higher $P_{ij}$ was mainly concentrated in the western and southern regions (Figure 4). $P_{ij}$ showed a decreasing trend under SSP2-4.5 and SSP5-8.5 from 2030s to 2090s (Figure 4(b1–b3,c1–c3)). From the 2030s to the 2090s, $P_{ij}$ was highest in the B stage and lowest in the D stage (Figure 4(a4,b4,c4)). The drought frequency in the B and F stages fluctuated greatly, ranging from 63% to 72%, indicating a significant difference in the drought frequency between different stations (Figure 4(a4,b4,c4)).

### *2.5. Drought Duration ($D_u$)*

Under SSP1-2.6, SSP2-4.5 and SSP5-8.5 scenarios, the $D_u$ ranges 10–24, 10–22 and 10–21 days from 2030s to 2090s, respectively (Figure 5(a4,b4,c4)). The longest $D_u$ was mainly concentrated in the western and central parts of the study area (Figure 5(a1–a3,b1–b3,c1–c3). Rice experienced persistent drought conditions in stage B, and $D_u$ values in Fujin reached an extremely high value of 24 days in the 2060s under the SSP1-2.6 scenario (Figure 5(a4); Figure S5(a1)). Anda, Nenjiang, Qiqihar, Tieli, Tonghe, and Mudanjiang, located in the west, south and center regions, suffered from persistent droughts under the SSP2-4.5 and SSP5-8.5, and droughts occurred with the highest frequency at B, C, E and F stages. While $D_u$ at the D stage in most regions was 0 (Figure S5).

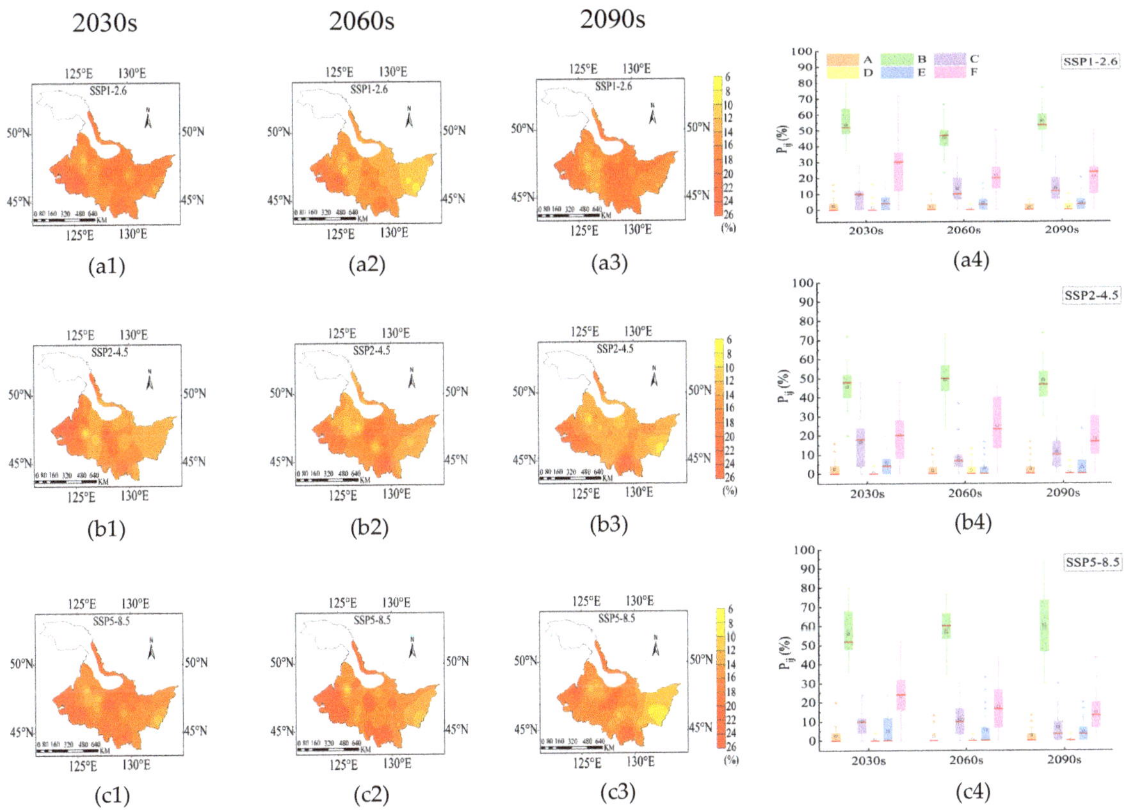


**Figure 4.** Distribution of $P_{ij}$ under SSP1-2.6 (**a1–a4**), SSP2-4.5 (**b1–b4**) and SSP5-8.5 (**c1–c4**) during the rice growth period from 2015 to 2100. The 2030s, 2060s and 2090s represent the period of 2015–2040, 2041–2070 and 2071–2100, respectively. A, B, C, D, E and F represent the returning green stage, tillering stage, jointing booting stage, heading flower stage, milk stage and yellow ripening stage of rice, respectively. In the box-plot, the red line, the black square and the diamond represent the median, the mean and the outliers, respectively.

### 2.6. Drought Intensity (IN) and Drought Barycenter

From 2030s to 2090s, the mean values of *IN* during the rice growth period were 0.87–0.97, 0.89–0.92, and 0.85–0.91 under SSP1-2.6, SSP2-4.5, and SSP5-8.5 (Figure 6), with the high *IN* areas concentrated in the northwest and central regions (Figure 6(a1–a3,b1–b3,c1–c3)). Stage B and D had the highest and lowest *IN*, respectively (Figure 6(a4,b4,c4)). The drought barycenter was mainly distributed around Tieli and Suihua (Figure 7), and the trajectory of the drought barycenter in the D stage migrated from Tieli in the central part to Suihua and Mingshui areas in the west under the SSP1-2.6 and SSP2-4.5 (Figure 7(d1,d2)). Notably, the drought barycenter of the D stage migrated from west to south from the 2060s to 2090s under the SSP5-8.5 scenario (Figure 7(d3)).

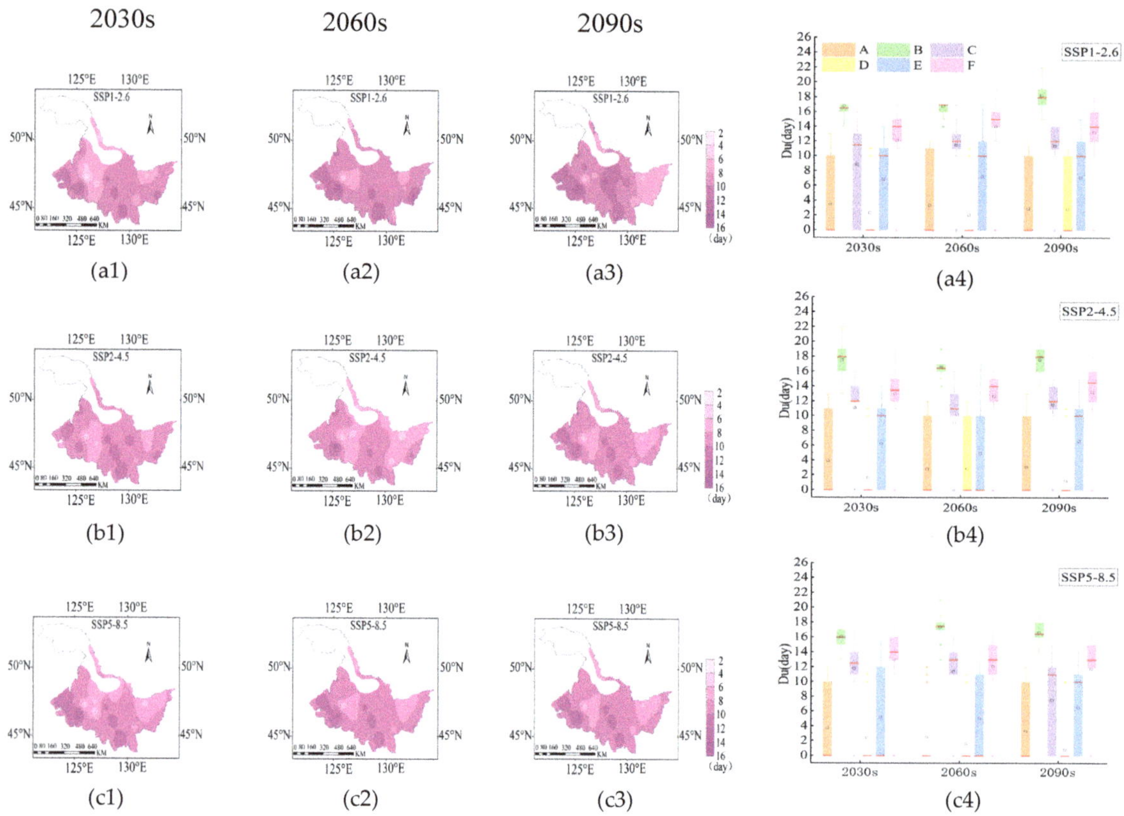


**Figure 5.** Distribution of $D_u$ under SSP1-2.6 (**a1–a4**), SSP2-4.5 (**b1–b4**) and SSP5-8.5 (**c1–c4**) during the rice growth period from 2015 to 2100. The 2030s, 2060s and 2090s represent the period of 2015–2040, 2041–2070 and 2071–2100, respectively. A, B, C, D, E and F represent the returning green stage, tillering stage, jointing booting stage, heading flower stage, milk stage and yellow ripening stage of rice, respectively. In the box-plot, the red line, the black square and the diamond represent the median, the mean and the outliers, respectively.

### 2.7. Correlation of Meteorological Factors with $PET_0$ and SPEI

From the 2030s to 2090s, significant positive correlations between $PET_0$ and $T_{max}$, $T_{min}$ were observed under SSP1-2.6, SSP2-4.5 and SSP5-8.5 (Table S2). Additionally, at B, C, D and F stages, $PET_0$ demonstrated a significant positive correlation with $P_r$. The correlation coefficients of *SPEI* with $T_{max}$ and $T_{min}$ were greater than $P_r$, indicating that air temperature had a stronger effect on $PET_0$ than $P_r$.

Under SSP2-4.5 and SSP5-8.5, *SPEI* was significantly positively correlated with $P_r$ and negatively correlated with $T_{max}$ and $T_{min}$ at the A stage. Under scenarios SSP1-2.6 and SSP5-8.5, *SPEI* was significantly positively correlated with $P_r$ at the B and F stages. Under scenario SSP5-8.5, *SPEI* was significantly positively correlated with $P_r$, $T_{max}$, and $T_{min}$ at the C stage. The correlation coefficients between *SPEI* and $P_r$ were greater than $T_{max}$ and $T_{min}$, indicating that $P_r$ had a greater effect on *SPEI* than air temperature.

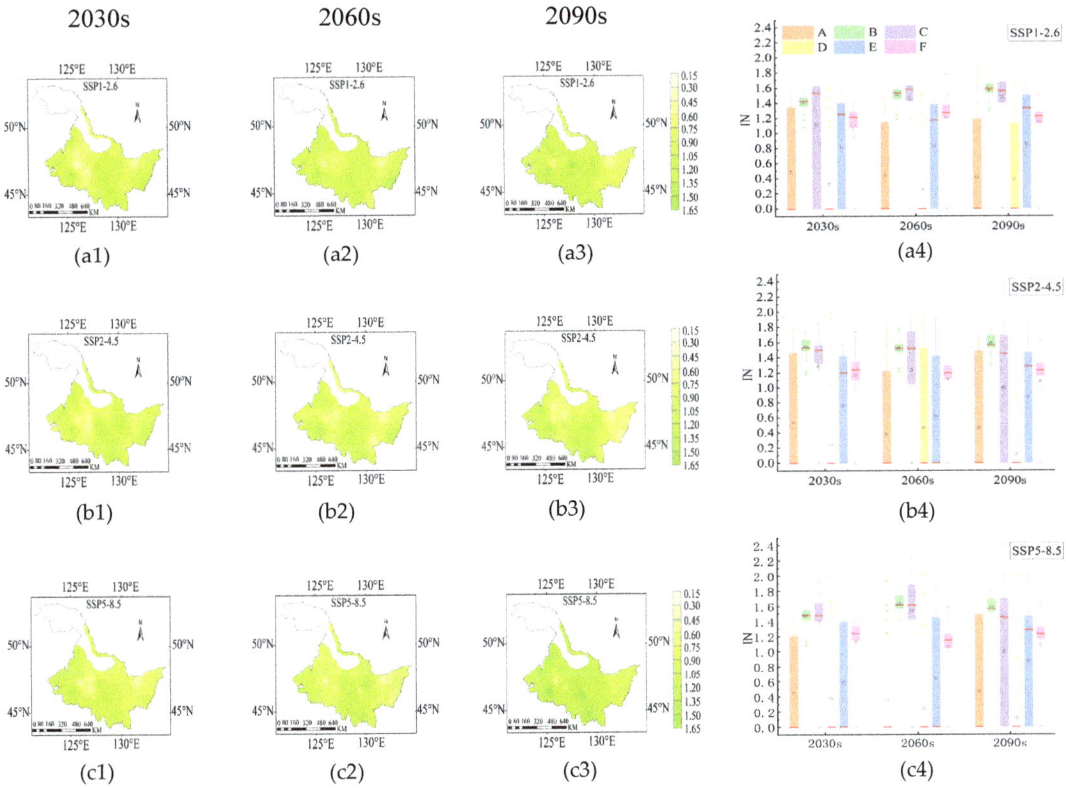

**Figure 6.** Distribution of *IN* under SSP1-2.6 (**a1–a4**), SSP2-4.5 (**b1–b4**) and SSP5-8.5 (**c1–c4**) during the rice growth period from 2015 to 2100. The 2030s, 2060s and 2090s represent the period of 2015–2040, 2041–2070 and 2071–2100, respectively. A, B, C, D, E and F represent the returning green stage, tillering stage, jointing booting stage, heading flower stage, milk stage and yellow ripening stage of rice, respectively. In the box-plot, the red line, the black square and the diamond represent the median, the mean and the outliers, respectively.

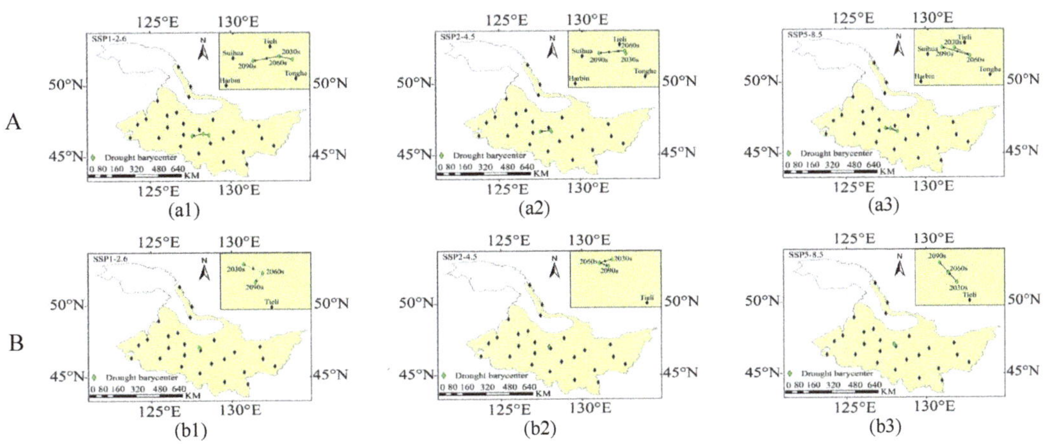

**Figure 7.** *Cont.*

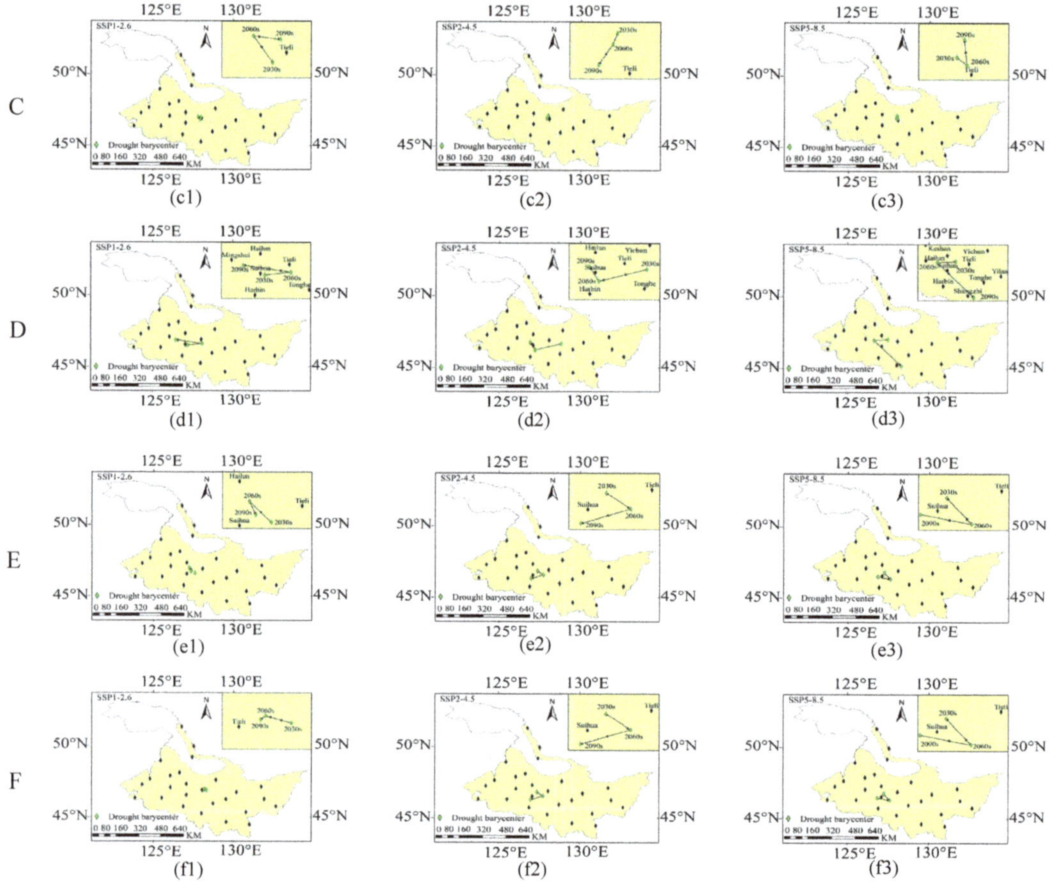

**Figure 7.** The trajectory of the drought barycenter under SSP1-2.6 (**a1–f1**), SSP2-4.5 (**a2–f2**) and SSP5-8.5 (**a3–f3**) during each growth period of rice from 2015 to 2100. A, B, C, D, E and F represent the returning green stage, tillering stage, jointing booting stage, heading flower stage, milk stage and yellow ripening stage of rice, respectively.

## 3. Discussion

Temperature is recognized as a key factor influencing $PET_0$ increase in Bangladesh from 2061 to 2099 [33]. Similarly, Guo et al. revealed that temperature was the primary factor affecting $PET_0$ increase in Australia from 1995 to 2004 [34]. It is crucial to investigate how temperature increases brought on by climate change affect the emergence of droughts, according to Gumus [35]. Research on the Sanjiang Plain and this study both showed that rising temperatures had a greater impact on $PET_0$ than increases in precipitation, with the greatest temperature serving as the primary driver of $PET_0$ growth during the rice growth period [10]. The importance of temperature in escalating drought conditions, particularly in arid and semi-arid regions, was highlighted by Ahmed et al. [36]. This implies that global warming may worsen drought and make water security increasingly challenging. Temperature plays an essential part in the development of drought during the rice-growing season. Figures 2 and 3 also indicate that irrespective of trends in precipitation, rising temperatures increase evapotranspiration and transpiration, which raises the need for water for activities including agriculture and ultimately exacerbates the risk of drought [37].

The causes of spatial and temporal evolution of drought are closely related to land-air interactions [38]. Meteorological factors, as the most fundamental factors, control the formation of climate [39]. A positive feedback process between the land and atmosphere intensifies meteorological drought when it is accompanied by high temperatures and heat waves. Short-term droughts were more frequent in the southern region of Northeast China, and high temperatures may be the primary cause of their occurrence, according to the findings of Xue et al.'s study on the spatial and temporal characteristics and drivers of sudden droughts in northern China from 1978 to 2020 [40]. In our study, although Mudanjiang and other southern regions have high drought frequency, high drought intensity, and long drought duration, they have sufficient precipitation, low temperatures, and lower $PET_0$. This condition could arise from the extreme variations in temperature and precipitation in these locations, which might also be impacted by air circulation and monsoons, with a clear regional pattern of drought [41,42].

Short-term droughts are characterized by strong flashbacks, rapid development, high intensity, and difficulty forecasting compared with traditional droughts [43]. An annual scale *SPEI* analysis by Mei et al. revealed that Heilongjiang Province's drought characteristics from 2000 to 2018 were mostly defined by an east-west pattern of wet and dry in space, with no drought in the summer and moderate drought in autumn [44]. In contrast, the results of our study based on daily-scale *SPEI* showed that moderate and severe drought will occur during the rice growing period; this suggests that the daily-scale *SPEI* may be able to detect more severe potential drought. Especially in the southern regions such as Mudanjiang and Suifenhe, the *SPEI* values during the rice growing period ranged from −1.5 to −0.5 (Figure S4(a1–a3,b1–b3,c1–c3)), with moderate drought. Li et al. indicated that southern regions in Heilongjiang Province, such as Mudanjiang, had *SPEI* values ranging from −2.0–−1.7 with severe drought over the 2011–2020 time period [45]. As a result, in the future, Heilongjiang Province's southern region will experience less drought.

This study quantitatively identified the droughts at each growth stage of rice under future scenarios. According to our study, the tillering stage of rice growth has the highest frequency, duration, and intensity of droughts. The main cause of drought occurrence was found to be $PET_0$, which is higher than precipitation. (Figures 1(a1–a3) and 2(a4,b4,c4)). Prolonged drought during the tillering stage might hinder rice development at the following four growth stages, lowering rice yield [45]. The ability of early warning services should be strengthened to prevent such drought occurrences, along with strategies like rational irrigation [46], modifying the planting and harvesting dates [47], and other efficient drought resistance and disaster reduction methods [42]. The heading stage is characterized by the lowest drought intensity and the most volatility, particularly in the 2090s under SSP1-2.6 and 2060s under SSP2-4.5 (Figure 6(a4,b4)). This variation could be caused by the extreme precipitation of an abrupt rise in precipitation from the 2060s to the 2090s (Figure 1(a1,a2)). The excessive precipitation conditions might also delay rice growth and lower yields [29]. Furthermore, the migration of the drought barycenter may result from variations in precipitation in different regions. The trajectories of the drought barycenter under the SSP1-2.6, SSP2-4.5, and SSP5-8.5 scenarios varied greatly between the 2030s and 2090s at the heading flower stage (Figure 7(d1–d3), indicating variability and fluctuation in drought trajectory characteristics across different SSP scenarios. The severe lack of precipitation and the greater $PET_0$ in the western section of Heilongjiang Province may have caused the drought barycenter to migrate from the central to the western part of the province in the 2060s and 2090s (Figures S1 and S2 and 2). Ng et al. revealed that the drought barycenter in Peninsular Malaysia shifted from the central to the northern and southern regions, and this shift was also attributed to the lack of precipitation in the northern and southern parts [48]. The sudden migration of drought barycenter from Hailun to around Shangzhi from the 2060s to 2090s under the SSP5-8.5 scenario (Figure 7(d3)) might be attributed to the increase in drought intensity and drought frequency around the region (Figures 5 and 6). To alleviate localized drought situations, water supplies in the southeast can be rationally allocated to water-scarce areas in the northwest [49].

In this study, considerable alterations in drought types were caused by short-term drought. For instance, in 2040, the *SPEI* mutation point under the SSP1-2.6 scenario, the drought type changed from mild drought to no drought, and the *SPEI* trend changed from declining to increasing (Figure 3(a1,a2)). Under the SSP2-4.5 scenario from 2015 to 2064, *SPEI* first exhibited a major fall, followed by a slight increase, and then another significant decrease (Figure 3(b1,b2)), which caused the drought to intensify from mild drought to moderate drought, thus increasing the severity of drought. *SPEI* is not sensitive enough to capture extreme drought events in China, although it can generally record large drought events [50]. Research on extreme drought in agriculture should be investigated in the future; this might require a comprehensive assessment of multiple drought indicators.

## 4. Materials and Methods

### 4.1. Study Area, Data Sources and Rice Growth Period Division

The study area is located in Heilongjiang Province in Northeastern China, lies between latitudes 43°26′–53°33′ N and longitudes 121°11′–135°05′ E, with a total jurisdictional area of 473 million km$^2$, the 6th largest in China. Heilongjiang Province belongs to the continental monsoon climate of cold temperate and temperate zones. It can be divided into humid, semi-humid, and semi-arid zones based on dryness indicators from east to west. The average annual temperature is −5–5 °C, and the annual precipitation is between 400 and 800 mm. Figure 8 provides an overview of the study area, while the sixth accumulated temperature zone, which is currently unsuitable for rice production, is represented by the blank area.

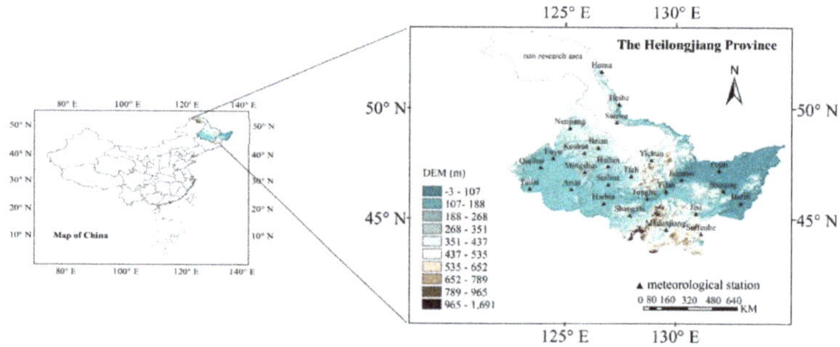


**Figure 8.** Overview of the study area.

Meteorological data were obtained from the NorESM2-MM model in CMIP6 with a resolution of 2.5° × 1.875°. We used the bilinear interpolation method on the statistical downscaling scale to obtain the meteorological datasets of daily precipitation ($P_r$), maximum air temperature ($T_{max}$) and minimum air temperature ($T_{min}$) for three future time periods (2015–2040, 2041–2070 and 2071–2100, defined as 2030s, 2060s and 2090s) under SSP1-2.6, SSP2-4.5 and SSP5-8.5 at 26 meteorological stations in Heilongjiang Province.

The data for the rice growth period emerged from the "Experimental Study on Irrigation of Rice and Evaluation of Irrigation in Zones in Heilongjiang Province" (Table S1). The rice growth periods were classified as follows: returning green stage (A), tillering stage (B), jointing booting stage (C), heading flower stage (D), milk stage (E) and yellow ripening stage (F) [51]. We assumed that the length of each rice growth period would not change in 3 future time periods [52].

*4.2. Standardized Precipitation Evapotranspiration Index (SPEI)*

The daily $PET_0$ is calculated using the FAO-recommended Hargreaves method [53],

$$PET_0{}' = 0.408 \times \alpha \times Ra(T_{mean} + 17.8)(T_{\max} - T_{\min})^{0.5} \quad (1)$$

where $PET_{0'}$ is the potential equivalent evapotranspiration calculated by empirical formula (mm/d), $\alpha$ is a constant, 0.0023; $T_{max}$ and $T_{min}$ are the daily maximum and minimum air temperatures (°C), respectively, $T_{mean}$ is the daily average temperature (°C), $R_a$ is the zenith radiation (MJm$^{-2}$ day$^{-1}$), calculated using reference [54].

Since Equation (1) is obtained from an empirical formula, ref. [55] adjusted $PET_{0'}$ by regression analysis. The fitted $PET_0$ is as follows:

$$PET_0 = a + bPET_0{}' \quad (2)$$

where $PET_0$ is the adjusted $PET_{0'}$ (mm/d), $a$ and $b$ are correction coefficients, which are constants at a particular spatiotemporal scale.

Cumulative moisture deficit ($D_i$):

$$D_i = P_i - (PET_0)_i \quad (3)$$

where $D_i$ is the cumulative moisture deficit (mm/d), $P_i$ is the daily precipitation (mm/d); $(PET_0)_i$ is the daily $PET_0$ (mm/d).

The three-parameter Log-Logistic distribution of the probability density function was used to describe the water deficit time series, and its probability density function $f(x)$ and probability distribution function $F(x)$ are calculated as follows:

$$f(x) = \frac{\beta}{\alpha}(\frac{x-\gamma}{\alpha})[1 + (\frac{x-\gamma}{\alpha})^{\beta}]^{-2} \quad (4)$$

where $\alpha$ is the scale parameter; $\beta$ is the shape parameter; $\gamma$ is the position parameter; the range of $D_i$ value is $(\gamma,\infty)$. Where $\alpha$, $\beta$ and $\gamma$ can be fitted by the linear moment method, the calculation formula is as follows:

$$\beta = \frac{2w_1 - w_0}{6w_1 - w_0 - 6w_2} \quad (5)$$

$$\alpha = \frac{\beta(w_0 - 2w_1)}{\Gamma(1+1/\beta)\Gamma(1-1/\beta)} \quad (6)$$

$$\gamma = w_0 - \alpha\Gamma(1+1/\beta)\Gamma(1-1/\beta) \quad (7)$$

where $\Gamma(\beta)$ is the Gamma distribution function of ($\beta$); $\omega_s$ is the probability-weighted moment of $D_i$ ($s$ = 0, 1, 2), which is calculated as follows:

$$w_s = \frac{1}{n}\sum_{i}^{n}(\frac{j-0.35}{n})^{s} D_i \quad (8)$$

$P(D)$ is the probability of being greater than $D$ for a given daily scale, which is given by:

$$P(D) = 1 - F(x) \quad (9)$$

When $P(D) \leq 0.5$, the $SPEI$ is calculated using Equation (10), while when $P(D) > 0.5$, the $SPEI$ is calculated using Equation (11).

$$SPEI = w - \frac{(c_0 + c_1 w + c_2 w^2)}{1 + d_1 w + d_2 w^2 + d_3 w^3}, w = \sqrt{-2 \ln P(D)} \tag{10}$$

$$SPEI = -(w - \frac{c_0 + c_1 w + c_2 w^2}{1 + d_1 w + d_2 w^2 + d_3 w^3}), w = \sqrt{-2 \ln[1 - P(D)]} \tag{11}$$

where $c_0$ = 2.515517, $c_1$ = 0.802853, $c_2$ = 0.010328, $d_1$ = 1.432788, $d_2$ = 0.189269, $d_3$ = 0.001308 [56].

Table 1 presents the drought classification based on $SPEI$ according to the meteorological drought classification issued by the National Climate Center [57].

**Table 1.** Drought classification based on *SPEI*.

| Drought Classification | *SPEI* |
|---|---|
| No drought | $SPEI > -0.5$ |
| Mild drought | $-0.5 \geq SPEI > -1.0$ |
| Moderate drought | $-1.0 \geq SPEI > -1.5$ |
| Severe drought | $-1.5 \geq SPEI > -2.0$ |
| Extreme drought | $SPEI \leq -2.0$ |

### 4.3. Drought Characteristics

Drought characteristics include drought frequency, drought duration, drought intensity and drought barycenter.

Drought frequency ($P_{ij}$) describes the proportion of years with drought compared to the total number of years. This study calculates the drought frequency based on the $SPEI$ of each meteorological station.

$$P_{ij} = \frac{n_{ij}}{N} \times 100\% \tag{12}$$

where $n_{ij}$ is the number of years in which drought of level $j$ occurred during the i growth period, and N is the total number of years. The larger the value of $P_{ij}$, the higher the frequency of drought occurrence.

A drought process was deemed to have taken place when the $SPEI$ remained less than or equal to −0.5 for 10 successive days. The initial day within these 10 consecutive days marks the commencement of the drought process. Once the $SPEI$ exceeds −0.5 for another 10 consecutive days, the drought process is then considered to have terminated. The day preceding the first day of those 10 days is regarded as the end date of this particular drought event. The time interval stretching from the onset to the conclusion of the drought process forms the drought duration ($D_u$) [1].

Drought intensity ($IN$) is calculated as the ratio of the cumulative $SPEI$ for all days of the drought process above mild drought to the duration of the drought event, with larger values indicating a more intense drought process.

$$IN = \frac{1}{n} |\sum_{i=1}^{n} (\mathrm{SPEI})_i| \tag{13}$$

where n represents the number of days the drought event lasted.

The drought barycenter is proposed based on the barycenter theory, which is expressed as the location of drought events in longitude, latitude, and time. The formula for calculating the drought barycenter in this study is as follows:

$$X_i = \frac{\sum_{s=1}^{n} W_{i,\,s} X_s}{\sum_{s=1}^{n} W_{i,\,s}},\; Y_i = \frac{\sum_{s=1}^{n} W_{i,\,s} Y_s}{\sum_{s=1}^{n} W_{i,\,s}} \tag{14}$$

where $X_i$ and $Y_i$ represent the latitude and longitude of the regional barycenter of drought during the *i*-growth period (°), respectively; $W_{i,s}$ represent the drought intensity of the sth meteorological station in the *i*-growth period; $X_s$ and $Y_s$ represent the latitude and longitude of the sth meteorological station (°).

### 4.4. Pearson Correlation Coefficient

The Pearson correlation coefficient quantitatively describes the degree of correlation between climatic factors and drought characteristics [58]. If there are n data points ($x_i$,$y_i$) ($i$ = 1, 2,. . ., n), the formula is as follows:

$$r = \frac{\sum_{i=0}^{n} (x_i - \overline{x})(y - \overline{y})}{\sqrt{\sum_{i=0}^{n} (x_i - \overline{x})^2 \sum_{i=0}^{n} (y - \overline{y})^2}} \tag{15}$$

In the formula, the value of $r$ ranges from −1 to 1, where −1 and 1 represent positive and negative correlation, respectively, while 0 represents no correlation, the closer $|r|$ is to 0, the weaker the correlation is, and the closer $|r|$ is to 1, the stronger the correlation is [1].

### 4.5. Mann-Kendall Trend Test

The Mann-Kendall trend test, which was introduced by the World Meteorological Organization, is a non-parametric statistical approach utilized to uncover the manner in which a variable changes over time. The positive and negative values of the statistical variable $Z$ signify the trend of the data.

In general, when the significance level $\alpha$ is set to 0.05, the critical value of $Z$ is ±1.96. The curves of the two statistical series, namely UF and UB, along with the two straight lines corresponding to ±1.96, are plotted on a single graph. For detailed analysis, references [59,60] can be consulted.

### 4.6. Data Processing

We extracted the $P_r$, $T_{max}$ and $T_{min}$ datasets from the CMIP6 by employing the bilinear interpolation statistical downscaling method (https://aims2.llnl.gov/search/cmip6/ (accessed on 1 February 2023)). The $PET_0$, drought frequency, drought duration, and drought intensity were computed by VBA macros 7.1(Microsoft Corporation, Redmond, WA, USA), *SPEI* and its Mann -Kendall trend test were calculated and performed by MATLAB R2020b (MathWorks, Natick, MA, USA). Pearson correlation analysis of $P_r$, $T_{max}$, $T_{min}$, $PET_0$ and *SPEI* was conducted using SPSS27.0 (IBM Corporation, Armonk, NY, USA). Drought ephemeral distribution heat maps and box plots were plotted using Origin 2021 (OriginLab Inc., Northampton, MA, USA). The spatiotemporal distribution map of drought features was generated by the inverse distance weight method (IDW) in ArcMap 10.7 (Environmental Systems Institute, Redlands, CA, USA)

## 5. Conclusions

This study explored the temporal and spatial patterns of drought characteristics and their driving factors under different SSP1-2.6, SSP2-4.5, and SSP5-8.5 from 2015 to 2100. *SPEI* showed a decreasing trend sequentially from the northeast to the southwest, and

$P_r$ had the most impact on *SPEI*. $PET_0$ at each rice growth stage showed an increasing trend from the 2030s to 2090s, with the largest $PET_0$ at the tillering stage and the lowest $PET_0$ at the yellow ripening stage. Additionally, $PET_0$ was most affected by $T_{max}$. The tillering stage experienced the highest drought frequency and drought intensity covering the whole study area, with the longest drought duration being 24 days. While the heading flower stage experienced the highest precipitation, lowest drought frequency, and lowest drought intensity. Developing more accurate and efficient drought monitoring and early warning systems, as well as promoting the implementation of sustainable water resource management strategies, will help better address the challenges of short-term drought and ensure the sustainable development of agriculture and water resources.

**Supplementary Materials:** The following supporting information can be downloaded at: https://www.mdpi.com/article/10.3390/plants14010030/s1, Figure S1. Spatial and temporal distribution of $P_r$ under SSP1-2.6 (a1–a3), SSP2-4.5 (b1–b3) and SSP5-8.5 (c1–c3) during rice growth period from 2015 to 2100; Figure S2. Spatial and temporal distribution of $T_{max}$ under SSP1-2.6 (a1–a3), SSP2-4.5 (b1–b3) and SSP5-8.5 (c1–c3) during rice growth period from 2015 to 2100; Figure S3. Spatial and temporal distribution of $T_{min}$ under SSP1-2.6 (a1–a3), SSP2-4.5 (b1–b3) and SSP5-8.5 (c1–c3) during rice growth period from 2015 to 2100; Figure S4. Spatial and temporal distribution of *SPEI* under SSP1-2.6 (a1–a3), SSP2-4.5 (b1–b3) and SSP5-8.5 (c1–c3) during rice growth period from 2015 to 2100; Figure S5. Distribution of $D_u$ under SSP1-2.6 (a1), SSP2-4.5 (a2) and SSP5-8.5 (a3) during the rice growth period from 2015 to 2100. The 2030s, 2060s and 2090s represent the period of 2015–2040, 2041–2070 and 2071–2100, respectively. A, B, C, D, E and F represent the returning green stage, tillering stage, jointing booting stage, heading flower stage, milk stage and yellow ripening stage of rice, respectively. Table S1. Duration of the growth period of rice in 26 selected areas in Heilongjiang Province; Table S2. Correlation of $PET_0$ and *SPEI* with $T_{max}$, $T_{min}$ and $P_r$ during the rice growth period.

**Author Contributions:** Conceptualization, T.N. and X.L.; methodology, X.L.; data curation, T.N. and X.L.; writing—original draft preparation, T.N. and X.L.; writing—review and editing, P.C., L.J., Z.S., S.Y., T.W. and T.L.; supervision, C.D.; project administration, T.N.; funding acquisition, T.N. All authors have read and agreed to the published version of the manuscript.

**Funding:** This research was funded by the Natural Science Foundation Project of Heilongjiang Province (LH2023E109), the National Natural Science Foundation Project of China (52079028), the Basic Scientific Research Fund of Heilongjiang Provincial Universities (2021-KYYWF-0019), the Opening Project of Key Laboratory of Efficient Use of Agricultural Water Resources, Ministry of Agriculture and Rural Affairs of the People's Republic of China (AWR2021002) and Postdoctoral Fellowship Program of CPSF (GZC20230668).

**Data Availability Statement:** Data that support the findings of this study are available from the corresponding author upon reasonable request.

**Acknowledgments:** We thank the anonymous reviewers and the editor for their suggestions, which substantially improved the manuscript.

**Conflicts of Interest:** The authors declare no conflicts of interest.

## References

1. Ling, M.H.; Han, H.B.; Hu, X.Y.; Xia, Q.Y.; Guo, X.M. Drought characteristics and causes during summer maize growth period on Huang-Huai-Hai Plain based on daily scale SPEI. *Agric. Water Manag.* **2023**, *280*, 108198. [CrossRef]
2. Liu, S.Y.; Huang, S.Z.; Xie, Y.Y.; Huang, Q.; Leng, G.Y.; Hou, B.B.; Zhang, Y.; Wei, X. Spatial-temporal changes of maximum and minimum temperatures in the Wei River Basin, China: Changing patterns, causes and implications. *Atmos. Res.* **2018**, *204*, 1–11. [CrossRef]
3. Held, I.M.; Soden, B.J. Robust responses of the hydrological cycle to global warming. *J. Clim.* **2006**, *19*, 5686–5699. [CrossRef]
4. Zhou, T.-J.; Chen, Z.-M.; Chen, X.-L.; Zuo, M.; Jiang, J.; Hu, S. Interpreting IPCC AR6: Future global climate based on projection under scenarios and on near-term information. *Adv. Clim. Chang. Res.* **2021**, *17*, 652.

5. Patel, R.; Patel, A. Evaluating the impact of climate change on drought risk in semi-arid region using GIS technique. *Results Eng.* **2024**, *21*, 101957. [CrossRef]
6. Hoque, M.A.A.; Pradhan, B.; Ahmed, N.; Sohel, M.S.I. Agricultural drought risk assessment of Northern New South Wales, Australia using geospatial techniques. *Sci. Total Environ.* **2021**, *756*, 143600. [CrossRef]
7. Huntington, T.G. Climate Warming-Induced Intensification of the Hydrologic Cycle: An Assessment of the Published Record and Potential Impacts on Agriculture. *Adv. Agron.* **2010**, *109*, 1–53.
8. Mondal, S.K.; Tao, H.; Huang, J.; Wang, Y.; Su, B.; Zhai, J.; Jing, C.; Wen, S.; Jiang, S.; Chen, Z. Projected changes in temperature, precipitation and potential evapotranspiration across Indus River Basin at 1.5–3.0 C warming levels using CMIP6-GCMs. *Sci. Total Environ.* **2021**, *789*, 147867. [CrossRef] [PubMed]
9. Wang, T.; Tu, X.J.; Singh, V.P.; Chen, X.H.; Lin, K.R. Global data assessment and analysis of drought characteristics based on CMIP6. *J. Hydrol.* **2021**, *596*, 126091. [CrossRef]
10. Ren, L.H.; Zhang, P.X.; Zhang, L.L. Water requirement characteristics and influencing factors of main crops in the Sanjiang Plain from 2000 to 2015. *Geogr. Arid. Areas* **2019**, *42*, 854–866.
11. Zeng, P.; Sun, F.Y.; Liu, Y.Y.; Wang, Y.K.; Li, G.; Che, Y. Mapping future droughts under global warming across China: A combined multi-timescale meteorological drought index and SOM-means approach. *Weather. Clim. Extrem.* **2021**, *31*, 100304. [CrossRef]
12. Gond, S.; Gupta, N.; Dikshit, P.K.S.; Patel, J. Assessment of drought variability using SPEI under observed and projected climate scenarios over Uttar Pradesh, India. *Phys. Chem. Earth* **2023**, *131*, 103440. [CrossRef]
13. Guan, X.; Shi, R.; Kong, X.; Liu, J.; Gan, Z.; Ma, J.; Luo, W.; Cao, C. An overview of researches on land-atmosphere interaction over semi-arid region under global changes. *Adv. Earth Sci.* **2018**, *33*, 995.
14. Aggarwal, S.K.; Saini, L.M. Solar energy prediction using linear and non-linear regularization models: A study on AMS (American Meteorological Society) 2013-14 Solar Energy Prediction Contest. *Energy* **2014**, *78*, 247–256. [CrossRef]
15. Shi, H.Y.; Chen, J.; Wang, K.Y.; Niu, J. A new method and a new index for identifying socioeconomic drought events under climate change: A case study of the East River basin in China. *Sci. Total Environ.* **2018**, *616*, 363–375. [CrossRef] [PubMed]
16. Mahmoudi, P.; Rigi, A.; Kamak, M.M. A comparative study of precipitation-based drought indices with the aim of selecting the best index for drought monitoring in Iran. *Theor. Appl. Climatol.* **2019**, *137*, 3123–3138. [CrossRef]
17. Shiru, M.S.; Shahid, S.; Chung, E.S.; Alias, N. Changing characteristics of meteorological droughts in Nigeria during 1901–2010. *Atmos. Res.* **2019**, *223*, 60–73. [CrossRef]
18. Zhou, Z.Q.; Shi, H.Y.; Fu, Q.; Li, T.X.; Gan, T.Y.; Liu, S.N. Assessing spatiotemporal characteristics of drought and its effects on climate-induced yield of maize in Northeast China. *J. Hydrol.* **2020**, *588*, 125097. [CrossRef]
19. Yao, N.; Li, Y.; Lei, T.J.; Peng, L.L. Drought evolution, severity and trends in mainland China over 1961–2013. *Sci. Total Environ.* **2018**, *616*, 73–89. [CrossRef] [PubMed]
20. Su, B.; Huang, J.; Fischer, T.; Wang, Y.; Kundzewicz, Z.W.; Zhai, J.; Sun, H.; Wang, A.; Zeng, X.; Wang, G. Drought losses in China might double between the 1.5 C and 2.0 C warming. *Proc. Natl. Acad. Sci. USA* **2018**, *115*, 10600–10605. [CrossRef]
21. Jiang, Y.L.; Wang, R.H.; Peng, Q.; Wu, X.Q.; Ning, H.S.; Li, C. The relationship between drought activity and vegetation cover in Northwest China from 1982 to 2013. *Nat. Hazards* **2018**, *92*, 145–163. [CrossRef]
22. Deng, S.L.; Chen, T.; Yang, N.; Qu, L.; Li, M.C.; Chen, D. Spatial and temporal distribution of rainfall and drought characteristics across the Pearl River basin. *Sci. Total Environ.* **2018**, *619*, 28–41. [CrossRef] [PubMed]
23. Qin, N.X.; Lu, Q.Q.; Fu, G.B.; Wang, J.N.; Fei, K.; Gao, L. Assessing the drought impact on sugarcane yield based on crop water requirements and standardized precipitation evapotranspiration index. *Agric. Water Manag.* **2023**, *275*, 108037. [CrossRef]
24. Li, X.; Ju, H.; Liu, Q.; Li, Y.; Qin, X. Analysis of drought characters based on the SPEI-PM index in Huang-Huai-Hai Plain. *Acta Ecol. Sin.* **2017**, *37*, 2054–2066.
25. Ming, B.; Guo, Y.Q.; Tao, H.B.; Liu, G.Z.; Li, S.K.; Wang, P. SPEI-based research on drought impact on maize yield in North China Plain. *J. Integr. Agric.* **2015**, *14*, 660–669. [CrossRef]
26. Sun, F.; Yang, S.; Chen, P. Climatic warming-drying trend in Northeastern China during the last 44 years and its effects. *Chin. J. Ecol.* **2005**, *7*, 751.
27. Chen, H.; Meng, F.; Yu, Z.N.; Tan, Y.Z. Spatial-temporal characteristics and influencing factors of farmland expansion in different agricultural regions of Heilongjiang Province, China. *Land Use Policy* **2022**, *115*, 106007. [CrossRef]
28. Yang, S.P.; Yan, W.Y.; Zhang, R.; Liu, W.L.; Tang, J.Q.; Zhang, H.J.; Jang, Y.X.; Wang, X.Y. Analysis of grain output change in Heilongjiang Province since the founding of the People's Republic of China. *J. Heilongjiang Grain* **2022**, *7*, 7–10.
29. Joseph, M.; Moonsammy, S.; Davis, H.; Warner, D.; Adams, A.; Oyedotun, T.D.T. Modelling climate variabilities and global rice production: A panel regression and time series analysis. *Heliyon* **2023**, *9*, e15480. [CrossRef]
30. Korres, N.; Norsworthy, J.; Burgos, N.; Oosterhuis, D. Temperature and drought impacts on rice production: An agronomic perspective regarding short-and long-term adaptation measures. *Water Resour. Rural. Dev.* **2017**, *9*, 12–27. [CrossRef]
31. Mahdu, O. The Impacts of Climate Change on Rice Production and Small Farmers' Adaptation: A Case of Guyana. Ph.D. Thesis, Virginia Tech, Blacksburg, VA, USA, 2019.

32. Zhang, P.; Chen, P.; Nie, T.Z.; Zhang, Z.X.; Li, T.C.; Dai, C.L.; Jiang, L.L.; Wu, Y.; Sun, Z.Y.; Yin, S. Long-Term Straw Incorporation under Controlled Irrigation Improves Soil Quality of Paddy Field and Rice Yield in Northeast China. *Plants* **2024**, *13*, 1357. [CrossRef]
33. Rahman, M.A.; Yunsheng, L.; Sultana, N.; Ongoma, V. Analysis of reference evapotranspiration (ET 0) trends under climate change in Bangladesh using observed and CMIP5 data sets. *Meteorol. Atmos. Phys.* **2019**, *131*, 639–655. [CrossRef]
34. Guo, D.; Westra, S.; Maier, H.R. Sensitivity of potential evapotranspiration to changes in climate variables for different Australian climatic zones. *Hydrol. Earth Syst. Sci.* **2017**, *21*, 2107–2126. [CrossRef]
35. Gumus, V. Evaluating the effect of the SPI and SPEI methods on drought monitoring over Turkey. *J. Hydrol.* **2023**, *626*, 130386. [CrossRef]
36. Ahmed, K.; Shahid, S.; Nawaz, N. Impacts of climate variability and change on seasonal drought characteristics of Pakistan. *Atmos. Res.* **2018**, *214*, 364–374. [CrossRef]
37. Mann, M.E.; Gleick, P.H. Climate change and California drought in the 21st century. *Proc. Natl. Acad. Sci. USA* **2015**, *112*, 3858–3859. [CrossRef]
38. Li, Y. *Observational Study on Global Semi-Arid Climate Change*; Lanzhou University: Lanzhou, China, 2015.
39. Yao, R.; Xia, M.; Sun, P.; Wen, Q.; Liu, G.; Liang, Y. Spatio-temporal distribution characteristics of meteorological drought and climate influence factors. *Acta Ecol. Sin.* **2021**, *41*, 333–347.
40. Xue, Z.Y.; Chen, Y.; Yin, Y.X.; Chen, W.L.; Jiao, Y.; Deng, P.; Dai, S.B. Spatio-temporal characteristics and driving factors of flash drought in northern China from 1978 to 2020. *Glob. Planet. Change* **2024**, *232*, 104326. [CrossRef]
41. Liu, X.; Cao, K.H.; Li, M. Assessing the impact of meteorological and agricultural drought on maize yields to optimize irrigation in Heilongjiang Province, China. *J. Clean. Prod.* **2024**, *434*, 139897. [CrossRef]
42. Guo, H.; Wang, R.; Garfin, G.M.; Zhang, A.Y.; Lin, D.G.; Liang, Q.O.; Wang, J.A. Rice drought risk assessment under climate change: Based on physical vulnerability a quantitative assessment method. *Sci. Total Environ.* **2021**, *751*, 141481. [CrossRef] [PubMed]
43. Chen, C.C.; Zhang, Y.Q.; Wang, J.Y.; Song, J.; Luo, L.F.; Yue, Y. Flash drought: Progress and prospect. *South-North Water Transf. Water Sci. Technol.* **2020**, *18*, 26–33.
44. Mei, X.D.; Wang, Q.; Tian, J.; Tian, Z.Y.; Li, D.; Liu, D.D. Analysis of the relationship between drought and topographic factors in Heilongjiang Province based on SPEI. *J. Heilongjiang Inst. Tech.* **2022**, *36*, 1–8.
45. Li, Y.; Cheng, W.; Wenjie, Z.; Guoming, D. Agricultural vulnerability to drought in the black soil region of northeast China and its adaptive strategy: Heilongjiang Province case study. *J. Agric. Resour. Environ.* **2023**, *40*, 953.
46. Bouman, B.; Tuong, T.P. Field water management to save water and increase its productivity in irrigated lowland rice. *Agric. Water Manag.* **2001**, *49*, 11–30. [CrossRef]
47. Bodner, G.; Nakhforoosh, A.; Kaul, H.P. Management of crop water under drought: A review. *Agron. Sustain. Dev.* **2015**, *35*, 401–442. [CrossRef]
48. Ng, C.Y.; Jaafar, W.Z.W.; Othman, F.; Lai, S.H.; Mei, Y.W.; Juneng, L. Assessment of Evaporative Demand Drought Index for drought analysis in Peninsular Malaysia. *Sci. Total Environ.* **2024**, *917*, 170249. [CrossRef] [PubMed]
49. Zhang, Q.; Miao, C.Y.; Gou, J.J.; Zheng, H.Y. Spatiotemporal characteristics and forecasting of short-term meteorological drought in China. *J. Hydrol.* **2023**, *624*, 129924. [CrossRef]
50. Pei, Z.F.; Fang, S.B.; Wang, L.; Yang, W.N. Comparative Analysis of Drought Indicated by the SPI and SPEI at Various Timescales in Inner Mongolia, China. *Water* **2020**, *12*, 1925. [CrossRef]
51. Nie, T.Z.; Wang, Z.L.; Zhang, Z.X.; Chen, P.; Li, T.C.; Wang, T.Y.; Sun, Z.Y.; Yin, S. Spatio-temporal variability in rice water supply and crop coefficients based on lysimeter measurements at 21 stations in Heilongjiang Province. *Irrig. Sci.* **2024**. [CrossRef]
52. Mainuddin, M.; Kirby, M.; Chowdhury, R.A.R.; Shah-Newaz, S.M. Spatial and temporal variations of, and the impact of climate change on, the dry season crop irrigation requirements in Bangladesh. *Irrig. Sci.* **2015**, *33*, 107–120. [CrossRef]
53. Zhao, J.; Yan, D.H.; Yang, Z.Y.; Hu, Y.; Weng, B.S.; Gong, B.Y. Improvement and adaptability evaluation of standardized precipitation evapotranspiration index. *Acta Phys. Sin.* **2015**, *64*, 382–390.
54. Allen, R.G.; Pereira, L.S.; Raes, D.; Smith, M. Crop evapotranspiration-Guidelines for computing crop water requirements-FAO Irrigation and drainage paper 56. *FAO* **1998**, *300*, D05109.
55. Xia, X.-s.; Zhao, X.; Tang, T.; Pan, Y.-z. Regression correction of Hargreaves equation within a year in China's agricultural regions. *J. Drain. Irrig. Mach. Eng.* **2021**, *39*, 81–88.
56. Chen, G.X. *Effects of Meteorological and Agricultural Drought on Winter Wheat Growth and Yield*; Northwest Agriculture & Forestry University: Xianyang, China, 2022.
57. *GB/T 20481-2017*; Meteorological Drought Levels. National Climate Center: Beijing, China, 2017.
58. Wu, K. Research on the Effects of ENSO on Drought in Four Regions of Liaoning Province Based on SPEI. Master's Thesis, Shenyang Agricultural University, Shenyang, China, 2018.

59. Xing, L.T.; Huang, L.X.; Chi, G.Y.; Yang, L.Z.; Li, C.S.; Hou, X.Y. A Dynamic Study of a Karst Spring Based on Wavelet Analysis and the Mann-Kendall Trend Test. *Water* **2018**, *10*, 698. [CrossRef]
60. Nie, T.Z.; Yuan, R.; Liao, S.H.; Zhang, Z.X.; Gong, Z.P.; Zhao, X.; Chen, P.; Li, T.C.; Lin, Y.Y.; Du, C.; et al. Characteristics of Potential Evapotranspiration Changes and Its Climatic Causes in Heilongjiang Province from 1960 to 2019. *Agriculture* **2022**, *12*, 2017. [CrossRef]

*plants*

MDPI

*Article*

# Simulation and Evaluation of Spring Maize Growth Under Drip Irrigation with Fully Biodegradable Film Mulching Based on the DSSAT Model

**Yanhui Jia [1], Haibin Shi [2,3,*], Qingfeng Miao [2,3], Xiulu Sun [4] and Yayang Feng [4,*]**

1 Shandong Facility Horticulture Bioengineering Research Center, Weifang University of Science and Technology, Weifang 262700, China; jyh_5151@126.com
2 College of Water Conservancy and Civil Engineering, Inner Mongolia Agricultural University, Hohhot 010018, China; imaumqf@imau.edu.cn
3 Autonomous Region Collaborative Innovation Center for Integrated Management of Water Resources and Water Environment in the Inner Mongolia Reaches of the Yellow River, Hohhot 010018, China
4 Farmland Irrigation Research Institute, Chinese Academy of Agricultural Sciences (CAAS), Xinxiang 453002, China; sunxiulu@caas.cn
* Correspondence: shi_haibin@sohu.com (H.S.); fengyayang@caas.cn (Y.F.)

**Abstract:** Fully biodegradable mulch film enhances temperature and moisture retention during the early stages of maize growth while naturally degrading in the later stages, providing an environmentally friendly alternative to conventional plastic mulch films. However, there is no consensus on its impact on maize growth and yield. The present study utilized field test data from spring maize covered with fully biodegradable mulch film in the Xiliaohe Plain, aiming to improve the Decision Support System for Agrotechnology Transfer (DSSAT) model while focusing on soil temperature, irrigation, rainfall, and evapotranspiration. The parameters of the DSSAT model were calibrated and validated using field test data from 2016 to 2018. The improved DSSAT model accurately simulated the maize growth process under various induction periods of fully biodegradable mulch film. The simulation accuracy of this model was as follows: MRE < 10%, nRMSE < 12%, and $R^2 \geq 0.80$. Moreover, the yield of spring corn covered with fully biodegradable mulch film was predicted using meteorological data from 2019 to 2023. This study suggests that regions such as the Xiliaohe Plain, which share climatic conditions, should opt for fully biodegradable mulch film with an induction period of approximately 80 days to ensure high yields across different hydrological years.

**Keywords:** fully biodegradable mulch film; subsurface drip irrigation; DSSAT model; yield; water use efficiency

## 1. Introduction

The Xiliao River Plain is located in the Golden Belt of maize production and is an important grain-producing area in China [1]. However, spring maize in this region is highly susceptible to cold damage and suffers from inadequate water resources [2]. The Xiliao River has experienced years of reduced flow, making groundwater extraction the primary irrigation method. This has resulted in the formation of significant groundwater cones in the depression, severely constraining the regional production capacity [3]. Subsurface drip irrigation technology effectively alleviates issues related to drought and cold conditions affecting spring maize in the Xiliao River Plain [4]. However, the long-term use of conventional plastic mulch film leads to "white pollution" [5]. Additionally, rainfall in the Xiliao River Plain is unevenly distributed throughout the year, with over 60% of the annual rainfall occurring from July to September. The use of mulch can hinder rainwater utilization [6,7]. In this context, promoting fully biodegradable mulch film that provides the same warming and moisture retention effects as those of conventional plastic mulches before degradation [8] while also increasing rainwater utilization during the maize growth phase is crucial. These biodegradable films naturally decompose into $CO_2$, $H_2O$, and organic matter in the later growth stages [9], representing an important measure for ensuring the green and sustainable development of subsurface drip irrigation technology in the Xiliao River Plain and safeguarding national food security [10].

Recent research on fully biodegradable mulch film has primarily focused on the induction phase, warming, and moisture retention effects and their impacts on maize yield and water use efficiency (WUE). Chen et al. [11] showed that in the Hetao Plain, after the degradation of fully biodegradable mulch film, soil surface evaporation increases, leading to a reduced water retention capacity for maize in the mid-to-late growth stages. Transpiration under fully biodegradable mulch was 15.6% higher than that under conventional polyethylene (PM) treatment, whereas the yield decreased by 9.9% relative to that for PM. A study of potatoes has shown that evapotranspiration (ET) and yield do not differ significantly between fully biodegradable and conventional mulch, and the limited impact of fully biodegradable mulch on ET could be explained by the high canopy coverage of potatoes [12]. In a study of maize, Saglam et al. [7] found no significant difference in soil moisture between fully biodegradable mulch and PM treatment; after rainfall, soil moisture in areas covered by fully biodegradable mulch was even higher than that in PM-covered areas. Three-year maize trials in the Weinan Plain have demonstrated that the average yield and WUE are significantly higher for moderately degraded fully biodegradable mulch than for conventional mulch and both slowly and rapidly degrading biodegradable film [13]. Feng et al. [10] studied spring maize in eastern Inner Mongolia, revealing that in wet and average water years, fully biodegradable mulch could achieve higher yields; however, yields did not differ significantly from those for the PM treatment. However, during dry years, the yield and WUE were significantly lower for the fully biodegradable mulch than for the PM treatment. These findings highlight the potential for variation in climatic conditions, hydrological year types, and maize types across regions to lead to differences in water consumption, maize yield, and WUE under fully biodegradable mulch coverage. However, field experiments have not thoroughly investigated the effects of these parameters under fully biodegradable mulch use, and such experiments are time-consuming, labor-intensive, and costly. Combining field experiments with modeling approaches can improve the estimation of maize yields.

The DSSAT (decision support system for agrotechnology transfer) model is widely used to predict maize yields owing to its user-friendly interface and high simulation accuracy. In the DSSAT model, ET and maize yield are primarily estimated based on soil temperature, rainfall, and solar radiation. However, this model does not include a mulch cover module and is mainly used to simulate uncovered field plots. Shen et al. [14] utilized an improved MDSSAT model to simulate maize yields under PM mulch, considering the impact of damaged areas caused by corn stover growth holes on soil evaporation. However, the effect of these damaged areas on soil temperature was not considered. Chen et al. [15] proposed a BDSSAT model to simulate the effects of a fully biodegradable mulch cover by considering the mulch degradation rate and its impact on soil temperature and maize yield. This model aimed to predict maize yield changes in Inner Mongolia under future climate change scenarios. Overall, these studies indicate that the DSSAT model can effectively simulate maize growth under film mulching conditions and accurately predict maize yield and water productivity when using fully biodegradable mulch film. However, the reliability of these simulations is highly contingent upon the careful selection of maize genetic parameters, field management practices, and soil characteristics. Therefore, it is crucial to enhance the DSSAT model using field test data, optimize both maize genetic and input parameters, and validate the model to accurately predict maize yield and water productivity under fully biodegradable mulch film. This approach will ultimately facilitate the optimization of the induction period for this film.

The objectives of this study were as follows: (1) improving the DSSAT model to accommodate fully biodegradable mulch conditions, considering the impact of different stages of mulch coverage on rainfall utilization; (2) considering the one-dimensional point-based nature of the DSSAT model and localized drip irrigation system, necessitating adjustments to the irrigation amounts to enhance the accuracy of soil moisture simulations near the drip tape; (3) optimizing genetic parameters for spring maize using field observation data from 2016 to 2018 to calibrate and validate the DSSAT model under fully biodegradable mulch coverage; and (4) utilizing meteorological data from 2019 to 2023 to simulate and predict spring maize yield under different induction periods with fully biodegradable mulch coverage. Based on these simulations, an optimal induction period for biodegradable mulch in the Xiliaohe Plain and similar climatic regions was proposed. This study offers a scientific basis and technical support for predicting maize productivity under fully biodegradable mulch in the semi-arid regions of Northeast China, particularly in the context of climate change and the efficient management of agricultural water and heat resources.

## 2. Results

### 2.1. Fully Biodegradable Mulch Film Damage Ratio

In this study, a seed drill was used to create corn-planting holes with a radius of approximately 4 cm. The fully biodegradable film gradually degraded until the induction period. The coverage based on actual measurements in field trials is illustrated in Figure 1. Prior to the induction period, all three types of fully biodegradable film exhibited high degrees of integrity. However, 30 d after the induction period, the damage ratio for all three types exceeded 40%. Subsequently, the moisture retention effect provided by the biodegradable film was largely lost, and the obstruction effects of the film on rainfall were negligible.

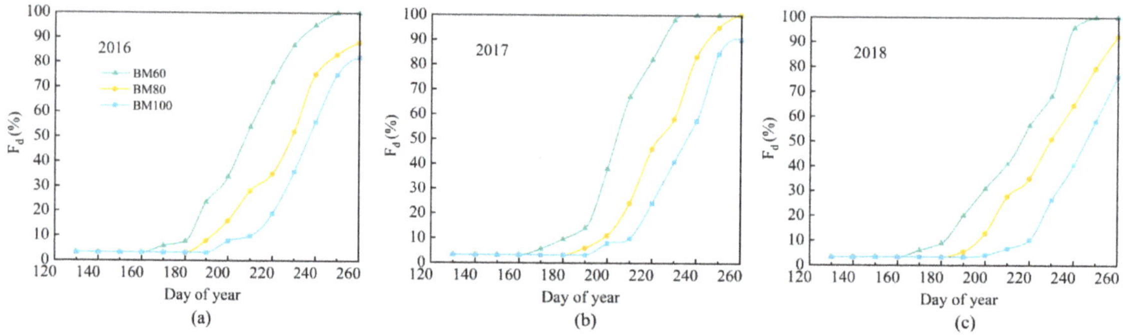


**Figure 1.** Damage ratios of fully biodegradable film at different induction periods in 2016 (**a**), 2017 (**b**), and 2018 (**c**).

### 2.2. Temperature Increase Compensation Coefficient

The dynamic changes in the daily average soil temperature at a depth of 5 cm for the fully biodegradable film-covered treatments and non-covered controls during 2016, 2017, and 2018 are illustrated in Figure 2. From the time of sowing until the corn reached the tasseling stage (approximately 70 days after sowing), the fully biodegradable film coverage exhibited a significant warming effect, with an average soil temperature increase of 2.65 °C compared with that for the non-covered treatment. As spring corn entered the silking stage (about 80 days after sowing for the fully biodegradable film-covered treatment and about 90 days for the non-covered treatment), the canopy coverage of the corn reached its maximum value. Subsequently, the warming effects for both covered and uncovered treatments decreased gradually. Additionally, the warming effect of the biodegradable film in the BM60 and BM80 treatments decreased over time owing to damage and degradation, leading to a gradual loss of the warming effect. The warming effect of the BM100 treatment also began to diminish at 110 d post-sowing.

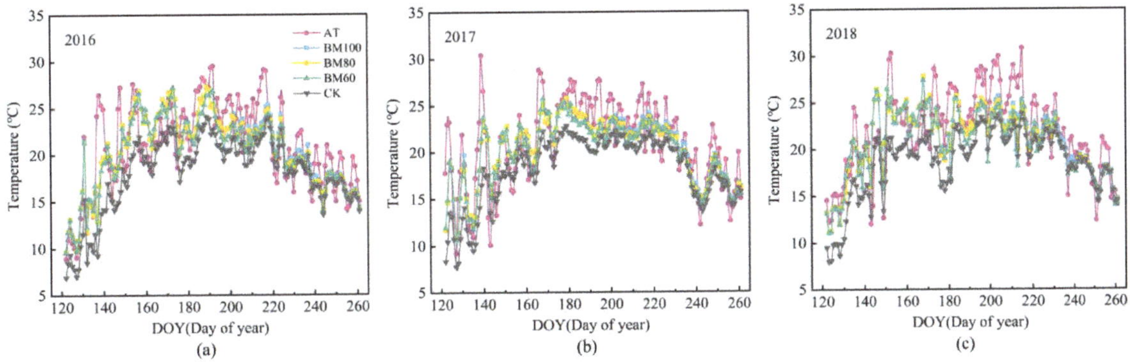


**Figure 2.** Daily changes in daily average temperature and soil temperature at a depth of 5 cm under different mulching treatments in 2016 (**a**), 2017 (**b**), and 2018 (**c**).

Soil temperature is primarily influenced by the air temperature. We set the daily average air temperature as the independent variable and analyzed its linear correlation with the daily average soil temperature at a depth of 5 cm using a 70-day post-sowing period as the threshold (Figure 3). The regression slopes for BM60, BM80, and BM100 from 0 to 70 d post-sowing were 0.71, 0.72, and 0.74 (Figure 3b–d), which were all higher than that of the non-mulched control treatment (CK) at 0.66 (Figure 3a). These results indicate that the use of fully biodegradable mulch significantly enhances warming compared with that for the non-mulched treatment. From 70 to 140 d after sowing, the regression slopes for BM60, BM80, BM100, and CK were 0.61, 0.63, 0.64, and 0.58 (Figure 3e–h). The gap between the CK and mulched treatments decreased, suggesting that the warming effect of the mulch treatments gradually weakened over time. The longer the duration of fully biodegradable mulch coverage, the greater the slope of the regression equation, further confirming the warming effect of mulch. As shown in Figure 3, the correlation coefficients ($R^2$) of the linear equations for the daily average air temperature and daily average soil temperature at a depth of 5 cm for all treatments were greater than 0.8, indicating a strong correlation and high precision. Therefore, using the linear equations shown in Figure 3 and the daily average air temperatures from 2019 to 2023, we calculated the soil temperature at a depth of 5 cm for different fully biodegradable mulch treatments.

Based on the linear relationship between the soil temperature and air temperature in the 5 cm soil layer for fully biodegradable mulch treatments and non-mulched treatments, we utilized air temperature data for 2019 to 2023 to estimate the corresponding soil temperatures. The Temperature Increase Compensation Coefficient was calculated using Equations (8) and (9) (Table 1). The compensation coefficients for the fully biodegradable mulch treatments BM60, BM80, and BM100 during the sowing to emergence stage were 1.50. During the emergence to elongation stage, differences in compensation coefficients began to emerge, with average values of 0.36, 0.40, and 0.42 for BM60, BM80, and BM100, respectively. In the elongation to tasseling stage, the warming effect of the mulch decreased gradually, with average values of 0.11, 0.14, and 0.18 for BM60, BM80, and BM100, respectively. As the growing season progressed, the compensatory effect of air temperature on soil temperature gradually weakened, and the coefficient of variation (CV) increased. This indicates that the coverage ratio is an important factor influencing the soil temperature.

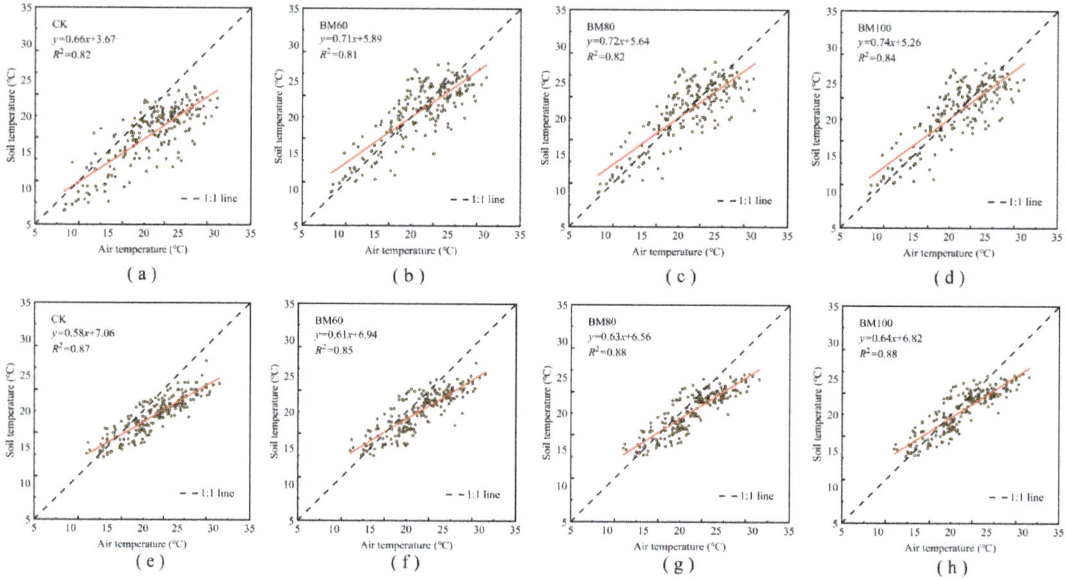


**Figure 3.** Relationship between the daily average air temperature and soil temperature at a depth of 5 cm under different mulch treatments at 0–70 d (**a**–**d**) and 71–140 d (**e**–**h**).

**Table 1.** Compensatory coefficient for the cumulative air temperature under fully biodegradable film mulching.

| Year | Treatments | Temperature Increase Compensation Coefficient | | |
|---|---|---|---|---|
| | | Sowing to Emergence | Emergence to Elongation | Elongation to Tasseling |
| BM60 | 2019 | 1.47 | 0.35 | 0.10 |
| | 2020 | 1.38 | 0.32 | 0.12 |
| | 2021 | 1.59 | 0.43 | 0.13 |
| | 2022 | 1.61 | 0.37 | 0.09 |
| | 2023 | 1.42 | 0.33 | 0.13 |
| | Mean | 1.49 | 0.36 | 0.11 |
| | CV | 5.58% | 9.89% | 13.01% |
| BM80 | 2019 | 1.48 | 0.41 | 0.15 |
| | 2020 | 1.38 | 0.38 | 0.13 |
| | 2021 | 1.59 | 0.45 | 0.16 |
| | 2022 | 1.61 | 0.39 | 0.11 |
| | 2023 | 1.42 | 0.36 | 0.15 |
| | Mean | 1.50 | 0.40 | 0.14 |
| | CV | 5.55% | 7.02% | 11.66% |
| BM100 | 2019 | 1.49 | 0.44 | 0.21 |
| | 2020 | 1.39 | 0.40 | 0.16 |
| | 2021 | 1.59 | 0.47 | 0.20 |
| | 2022 | 1.62 | 0.42 | 0.16 |
| | 2023 | 1.42 | 0.39 | 0.18 |
| | Mean | 1.50 | 0.42 | 0.18 |
| | CV | 5.51% | 6.18% | 10.23% |

### *2.3. Calibration of the DSSAT Model Under Fully Biodegradable Mulch*

Using the improved model, we calibrated various parameters for maize varieties based on field trial data from 2016, specifically for the fully biodegradable mulch treatment, BM60. These parameters included the phenological stage, LAI, yield, and soil moisture content.

First, the Generalized Likelihood Uncertainty Estimation (GLUE) parameter adjustment method for DSSAT was used. Parameters were further refined by manually comparing the simulated and observed values. The genetic coefficients used in this study are listed in Table 2. The CERES-maize model incorporates six genetic parameters for maize. Four of these parameters were related to maize growth and development (P1, P2, P5, and PHINT), and two parameters influenced maize yield (G2 and G3).

**Table 2.** Genetic parameters for Nonghua 106 summer maize.

| Variety | P1 | P2 | P5 | PHINT | G2 | G3 |
|---|---|---|---|---|---|---|
| Nonghua 106 | 196 | 0.55 | 858 | 45 | 980 | 6.2 |

The improved model accurately simulated the LAI and soil moisture content for the fully biodegradable mulch treatment (BM60), as shown in Figures 4 and 5. Additionally, it effectively modeled the phenological stage, yield, and ET, as presented in Table 3. The calibrated model demonstrated a high degree of fit for the number of growing days, yield, and ET, with a mean relative error (MRE) for the simulated and observed values remaining within 5%. The model also showed good performance for the soil moisture content and LAI, with the MRE and normalized root mean square error (nRMSE) under 10% and $R^2$ values exceeding 80%. This indicates that the improved model can accurately simulate soil water movement and spring maize growth under biodegradable mulch. As shown in Figure 5, the soil water content at depths of 0–40 cm varied significantly. Under drip irrigation, maize roots were primarily distributed within this depth range, which is crucial for water uptake. Additionally, irrigation and rainfall primarily affected the moisture content in the top 0–20 cm of the soil layer, resulting in minimal impact on the moisture levels in the soil below 40 cm.

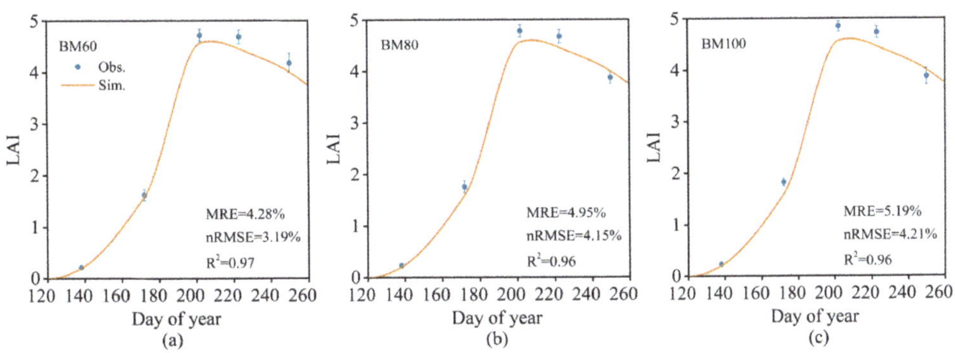


**Figure 4.** Measured values of LAI in 2016BM60 (**a**), 2016BM80 (**b**), and 2016BM100 (**c**) and simulated values based on the improved model.

**Table 3.** Simulation results for the phenological period, yield, and $ET_C$ based on the modified DSSAT model for 2016, 2017, and 2018.

| Treatment | | Emergence | | | Tasseling | | | Filling | | | Harvest | | | Yield (kg/ha) | | | ETc (mm) | | |
|---|---|---|---|---|---|---|---|---|---|---|---|---|---|---|---|---|---|---|---|
| | | Sim. | Obs. | MRE (%) | Sim. | Obs. | MRE (%) | Sim. | Obs. | MRE (%) | Sim. | Obs. | MRE (%) | Sim. | Obs. | MRE (%) | Sim. | Obs. | MRE (%) |
| 2016 | BM60 | 135 | 136 | 0.74% | 203 | 205 | 0.99% | 228 | 229 | 0.44% | 262 | 264 | 0.76% | 13,301 | 13,484 | 1.37% | 443.25 | 450.69 | 1.68% |
| | BM80 | 135 | 136 | 0.74% | 202 | 205 | 1.49% | 227 | 229 | 0.88% | 261 | 264 | 1.15% | 13,015 | 12,995 | 0.15% | 438.22 | 445.52 | 1.67% |
| | BM100 | 135 | 136 | 0.74% | 201 | 205 | 1.99% | 226 | 228 | 0.88% | 260 | 264 | 1.54% | 12,409 | 12,471 | 0.50% | 427.07 | 435.39 | 1.95% |
| 2017 | BM60 | 136 | 138 | 1.47% | 206 | 206 | 0.00% | 231 | 233 | 0.87% | 265 | 265 | 0.00% | 12,890 | 13,379 | 3.79% | 491.74 | 471.35 | 4.15% |
| | BM80 | 136 | 138 | 1.47% | 205 | 206 | 0.49% | 230 | 233 | 1.30% | 264 | 265 | 0.38% | 12,361 | 12,848 | 3.94% | 490.99 | 470.43 | 4.19% |
| | BM100 | 136 | 138 | 1.47% | 204 | 206 | 0.98% | 229 | 233 | 1.75% | 263 | 265 | 0.76% | 12,351 | 12,727 | 3.05% | 484.87 | 464.97 | 4.10% |
| 2018 | BM60 | 135 | 134 | 0.74% | 206 | 204 | 0.97% | 231 | 230 | 0.43% | 265 | 263 | 0.75% | 11,305 | 11,193 | 0.99% | 415.44 | 435.95 | 4.94% |
| | BM80 | 135 | 134 | 0.74% | 206 | 204 | 0.97% | 231 | 230 | 0.43% | 265 | 263 | 0.75% | 12,125 | 12,461 | 2.77% | 411.36 | 431.62 | 4.93% |
| | BM100 | 135 | 134 | 0.74% | 205 | 204 | 0.49% | 230 | 228 | 0.87% | 264 | 263 | 0.38% | 12,330 | 12,732 | 3.26% | 402.26 | 423.19 | 5.20% |

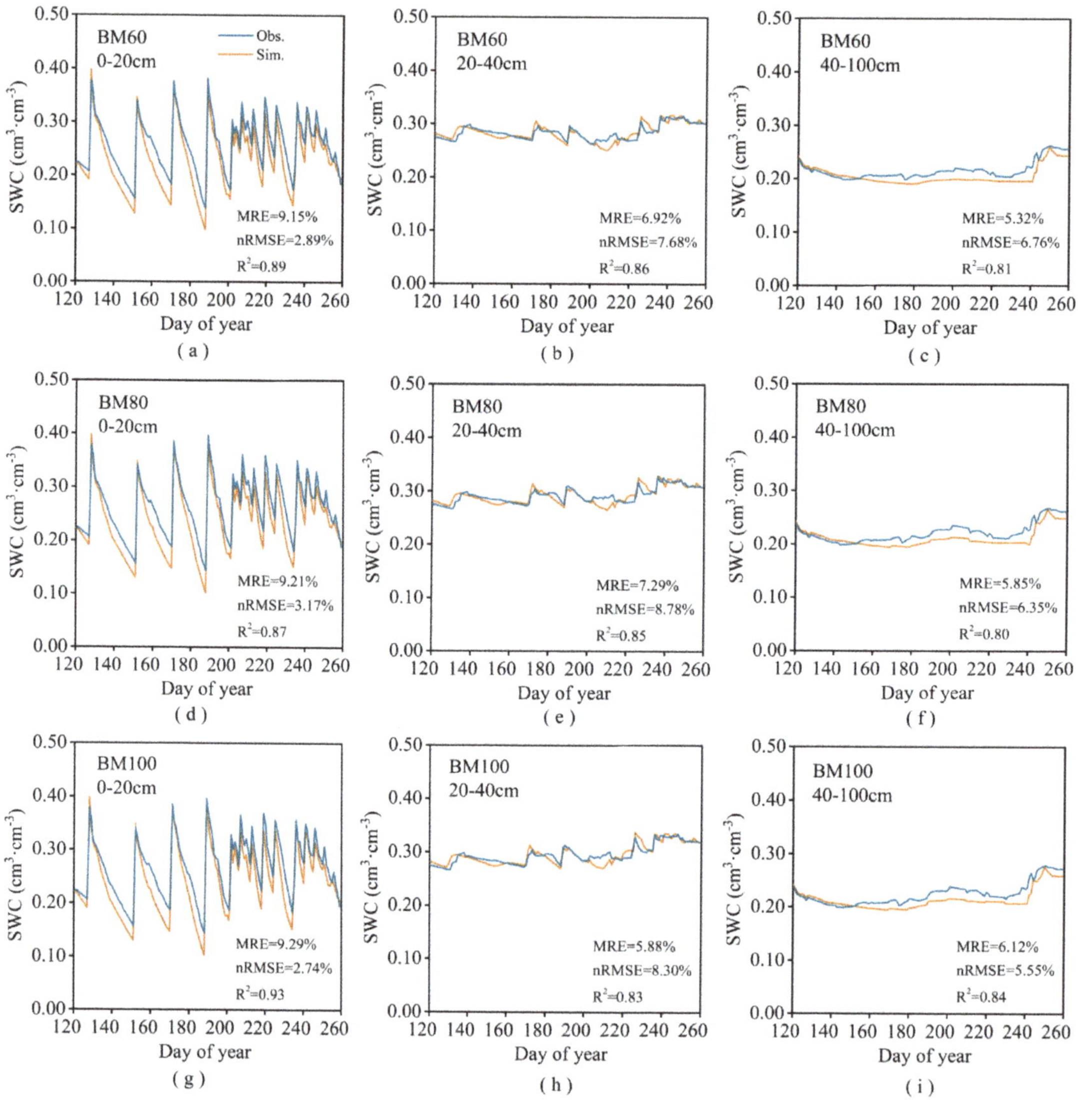


**Figure 5.** Measured soil water contents in the 0–20 cm (**a**,**d**,**g**), 20–40 cm (**b**,**e**,**h**), and 4–100 cm (**c**,**f**,**i**). Soil layer and simulated values using the improved DSSAT model for BM60, BM80, and BM100 in 2016.

### *2.4. Model Validation Under Fully Biodegradable Mulch*

The accuracy of the calibrated model was validated using observational data from experiments conducted in 2017 and 2018. The relevant statistical indicators for the phenological stages, yields, and ET are presented in Table 3. The measured and simulated values of the LAI were very similar (Figure 6). The correlation between the simulated and observed values for the soil moisture content was slightly lower than that for the LAI (Figure 7). This indicates that the improved model demonstrates high simulation accuracy for the LAI, phenological stages, yield, soil moisture content, and ET of spring corn. The coefficients of determination ($R^2$) for the relationship between the simulated and observed values were greater than 80%, and the mean relative errors (MRE) were less than 10%.

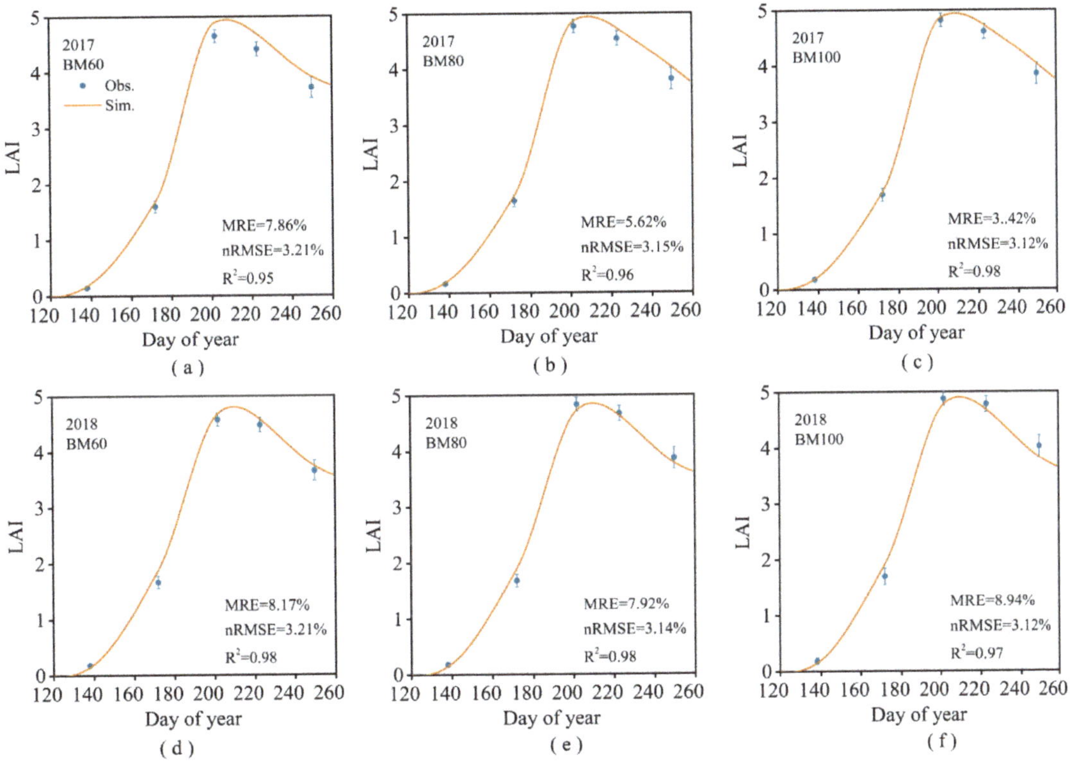


**Figure 6.** Measured values of LAI and simulated values using the improved DSSAT model for 2017BM60 (**a**), 2017BM80 (**b**), 2017BM100 (**c**), 2018BM60 (**d**), 2018BM80 (**e**), and 2018BM100 (**f**).

## *2.5. Yield Prediction Under Fully Biodegradable Mulch*

To further evaluate the impacts of different induction periods of fully biodegradable mulch on the yield and WUE of spring maize, the DSSAT model was employed along with meteorological data from 2019 to 2023. Maize yields under various induction periods of biodegradable mulch were predicted. Based on the recommended irrigation volume for maize drip irrigation in the Xiliao River Plain region, as well as rainfall data for the study area, the model utilized irrigation volumes of 183 mm for normal water years and 105 mm for wet years. The model inputs for the different induction periods, including rainfall, yield, and WUE, are summarized in Table 4. From 2019 to 2020, classified as normal water years, the simulated yield ranged from 12,140.05 to 13,692.62 kg·ha$^{-1}$, with WUE values between 26.94 and 29.85 kg·mm$^{-1}$·ha$^{-1}$, showing no significant difference compared with the measured yield of 12,109.01 to 13,301.33 kg·ha$^{-1}$ in 2016. In 2023, which was classified as a wet year, the yield was substantially lower than that in normal years. This decline can be attributed to the rainfall in July (291 mm), which accounted for 70% of the precipitation during the growing season. Excess rainfall can lead to overly wet soil conditions, which negatively affect root respiration and nutrient absorption. Additionally, continuous heavy rainfall during the tasseling stage affects maize pollination, further affecting yield. Based on the monitored yield data for 2016 to 2018 and the model predictions for 2019 to 2023, in the Xiliao River Plain and similar climatic regions, a biodegradable mulch with an induction period of approximately 80 days is recommended to ensure high yields across various hydrological conditions.

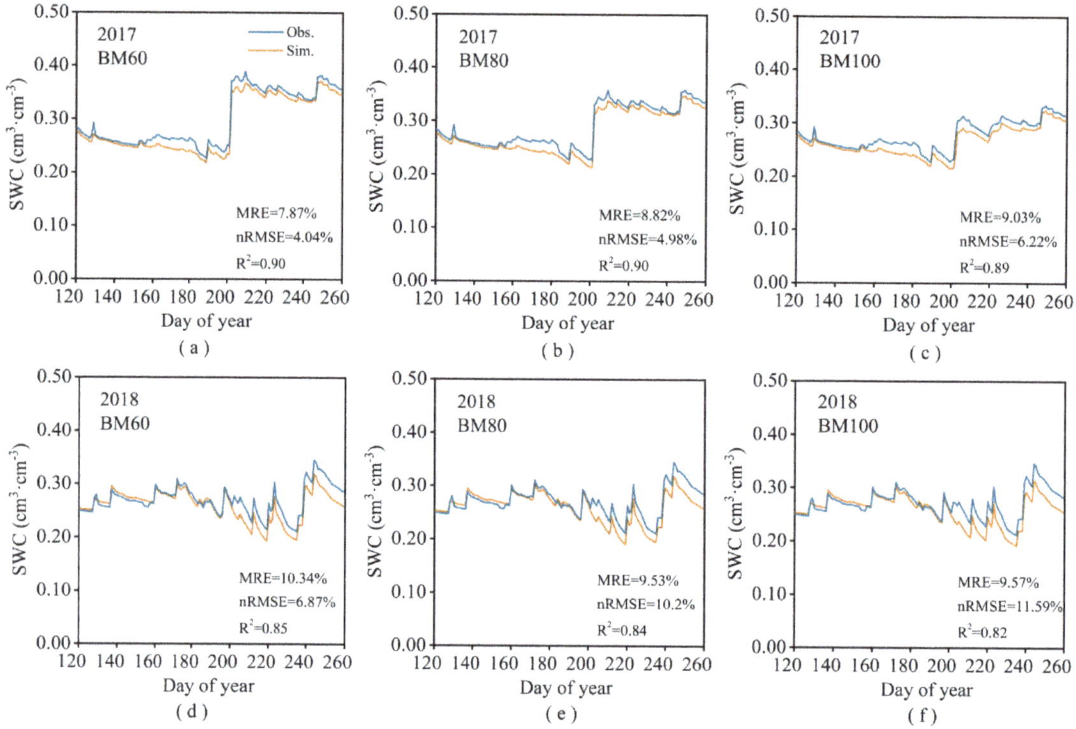


**Figure 7.** Measured values of average soil moisture in the 0–1 m soil layer and simulated values based on the improved DSSAT model for 2017BM60 (**a**), 2017BM80 (**b**), 2017BM100 (**c**), 2018BM60 (**d**), 2018BM80 (**e**), and 2018BM100 (**f**).

**Table 4.** Prediction results for rainfall, yield, and water use efficiency for different induction periods under fully biodegradable film mulching treatments.

| Treatment | | Precipitation (mm) | ETc (mm) | Yield (kg ha$^{-1}$) | WUE (kg mm$^{-1}$ ha$^{-1}$) |
|---|---|---|---|---|---|
| 2019 | BM60 | 192.53 | 449.74 | 12,659.15 | 28.15 |
| | BM80 | 177.81 | 434.68 | 12,471.10 | 28.69 |
| | BM100 | 168.97 | 423.74 | 12,223.73 | 28.85 |
| 2020 | BM60 | 203.98 | 459.85 | 13,692.62 | 29.78 |
| | BM80 | 203.98 | 455.12 | 13,531.60 | 29.73 |
| | BM100 | 184.86 | 439.28 | 13,114.28 | 29.85 |
| 2021 | BM60 | 242.34 | 449.90 | 12,903.12 | 28.68 |
| | BM80 | 226.56 | 435.48 | 12,715.48 | 29.20 |
| | BM100 | 204.58 | 412.70 | 12,140.05 | 29.42 |
| 2022 | BM60 | 209.05 | 457.10 | 12,316.49 | 26.94 |
| | BM80 | 192.87 | 442.48 | 12,284.17 | 27.76 |
| | BM100 | 183.42 | 434.80 | 12,179.04 | 28.01 |
| 2023 | BM60 | 382.13 | 498.03 | 11,819.63 | 23.73 |
| | BM80 | 364.80 | 480.30 | 11,442.29 | 23.82 |
| | BM100 | 352.64 | 467.84 | 11,163.77 | 23.86 |

## 3. Discussion

Soil temperature is a key factor determining maize growth and is influenced by various factors, such as solar radiation, soil cover, air temperature, irrigation, and rainfall. The warming effect of mulch shows substantial spatiotemporal variation. Chen et al. [16] first reported that mulching increased the accumulated soil temperature in the plowed layer during the early growth stages of cotton, compensating for insufficient air temperature. The warming effect of mulch on the plowed layer was significant, although it diminished after July. Wang et al. [17] indicated that plastic film coverage created favorable conditions for maize emergence and early growth, resulting in an average yield increase of 4.9% over the entire growth period compared with that for bare soil and significantly increasing soil temperatures. Air temperature directly affected soil temperature at a depth of 5 cm under mulched conditions, demonstrating a linear positive correlation. This relationship has been validated in field experiments using various maize, such as maize, wheat, and cotton [14,18,19]. Extensive research has shown that fully biodegradable film exhibits the same warming effect as that of conventional plastic film before degradation [20,21]. Yang et al. [18] showed that in mulched maize, the most pronounced effects occurred just before the maximum canopy cover (prior to tasseling), providing a compensatory coefficient for air temperature in mulched maize in western Liaoning. Understanding the relationship between soil temperature under mulch and air temperature, as well as quantifying the compensatory effect, is crucial for converting air temperature data into soil temperature calculations for mulched maize. Similarly, this study utilized a linear equation for air temperature and the 5 cm soil temperature alongside temperature data for 2019 to 2023 to predict the air temperature compensation coefficients for different growth stages of maize under various induction periods of a fully biodegradable film. The soil temperature under fully biodegradable film was influenced by the rate of film degradation; once the film deteriorated, the warming effect decreased gradually.

Soil moisture is another critical factor affecting maize yield and is primarily influenced by irrigation, rainfall, soil cover, and maize growth. During the seedling stage, maize requires relatively low water, with peak demand occurring from the jointing to grain-filling stages. Water requirements decrease significantly after grain filling [22]. Therefore, in the Xiliao River Plain, where rainfall is concentrated in July and August, it is essential for fully biodegradable film to remain intact during the early growth stages to reduce soil evaporation, whereas post-degradation, they should enhance rainwater utilization efficiency [23]. Li et al. [24] modeled mulched cotton in Xinjiang, which receives minimal rainfall; however, in eastern Xinjiang, once rainfall reaches moderate levels, it can move laterally to areas covered by mulch. Consequently, this study improved the rainfall input data for the model. Fu et al. [25] pointed out that because drip irrigation is a localized irrigation method, using actual irrigation amounts in the model may reduce the soil moisture content near the drip irrigation zone. Therefore, the actual amount of irrigation was adjusted based on the depth of the soil moisture layer. This study also improved the water balance equations for the soil moisture content before and after irrigation. The newly developed model accurately simulates soil moisture dynamics by refining the input values for rainfall and irrigation.

Because the management module of the DSSAT model does not account for soil cover, this study adopted the method developed by Shen et al. [14] to incorporate evaporation ratio coefficients. After the modifications, the model was calibrated using data from 2016 for spring maize genetic parameters and validated against experimental data from 2017 and 2018. The model showed high simulation accuracies for the LAI, phenology, yield, soil moisture, and ET of spring maize, with correlation coefficients ($R^2$) exceeding 80% and an MRE below 10%. During the prediction phase, the model yield forecasts were closely aligned with the observed experimental yields. Thus, the calibrated post-improvement model can effectively simulate the growth and development of spring maize under fully biodegradable film coverage, providing a reliable analytical tool for assessing water consumption and field moisture management in maize grown under such conditions.

## 4. Materials and Methods

### 4.1. Field Experiment

A spring maize drip irrigation experiment under fully biodegradable mulch coverage was conducted in the core experimental area with subsurface drip irrigation in Kezuozhongqi, Tongliao City, Inner Mongolia [44°6′ N, 122°21′ E, 173.5 m above sea level (m.a.s.l.)]. The average annual rainfall in the study area from 1951 to 2018 was 324 mm, with 80% of the rainfall concentrated during the spring maize growth period (May to September). The long-term average temperature was 20.2 °C, and the average annual sunshine duration was 3254.1 h. The groundwater depth in the experimental area was approximately 8.5 m, and the soil texture and hydraulic parameters for the 0–1 m depth in the experimental field are shown in Table 5.

**Table 5.** Soil texture and hydraulic parameters for a depth of 0–1 m at the test site.

| Soil Layers | Soil Types (%) | | | Field Capacity | Saturated Water Content | Wilting Point | Bulk Density |
|---|---|---|---|---|---|---|---|
| (cm) | Clay | Silt | Sand | ($cm^3$ $cm^{-3}$) | ($cm^3$ $cm^{-3}$) | ($cm^3$ $cm^{-3}$) | (g $cm^{-3}$) |
| 0–20 | 10.54 | 52.7 | 36.76 | 0.26 | 0.40 | 0.08 | 1.39 |
| 20–40 | 29.54 | 48.81 | 21.65 | 0.34 | 0.45 | 0.08 | 1.38 |
| 40–70 | 40.67 | 39.15 | 20.18 | 0.45 | 0.53 | 0.10 | 1.41 |
| 70–100 | 1.41 | 25.58 | 73.01 | 0.17 | 0.38 | 0.05 | 1.52 |

The planting pattern followed "one film, one drip irrigation with two rows of corn" (Figure 8). The experiment included three induction periods, 60 days (BM60), 80 days (BM80), and 100 days (BM100), with a non-mulched treatment serving as a control (CK). White, transparent, fully biodegradable mulch with a width of 70 cm and thickness of 0.008 mm was provided by Shandong Tianzhuang Eco-Benign Plastics Technology Co., Ltd. (Jinan, Shandong Province, China). The spacing between adjacent drip irrigation tapes was 1.2 m, with a distance of 35 cm between the two rows (narrow rows), resulting in a planting density of 67,500 plants per hectare. The drip emitters were spaced 25 cm apart, with an irrigation rate of 1.8 L·$h^{-1}$, and each plot was equipped with a rotary digital water meter to record the amount of water applied. The base nitrogen fertilizer (63 kg $ha^{-1}$) was applied using integrated agricultural machinery. For the application of top-dressing nitrogen, urea was fully dissolved in a fertilizer tank to obtain an amide nitrogen solution, which was then applied under the mulch to the root zone via the water pressure difference. The top-dressing was added in three applications: 82.8 kg $ha^{-1}$ at the jointing stage, 82.8 kg $ha^{-1}$ at the tasseling stage, and 41.4 kg $ha^{-1}$ during the grain-filling stage. The spring maize variety used in this study was Nonghua 106. Maize was planted on 29 April 2016, 27 April 2017, and 27 April 2018 and harvested on 26 September 2016, 24 September 2017, and 25 September 2018. The start and end dates of the key growth stages were recorded, and pest and weed management, along with agricultural machinery and agronomic practices, were conducted according to conventional local methods.

### 4.2. Data Measurement and Methods

#### 4.2.1. Meteorological Data

Field climate data were collected using a weather station (HOBO U30 Onset; Onset Computer Corporation, Bourne, MA, USA) installed at the experimental site. Reference maize evapotranspiration ($ET_0$) was calculated using the Penman–Monteith equation (Figure 9). Precipitation estimates recorded during the growing seasons from 2016 to 2023 were as follows: 272.03, 290.42, 212.56, 244.37, 262.99, 257.46, 252.56, and 414.89 mm. Notably, 2018 was classified as a low-flow year, whereas 2023 experienced above-average precipitation, making it a wet-flow year. The remaining years were classified as normal-flow years [26].

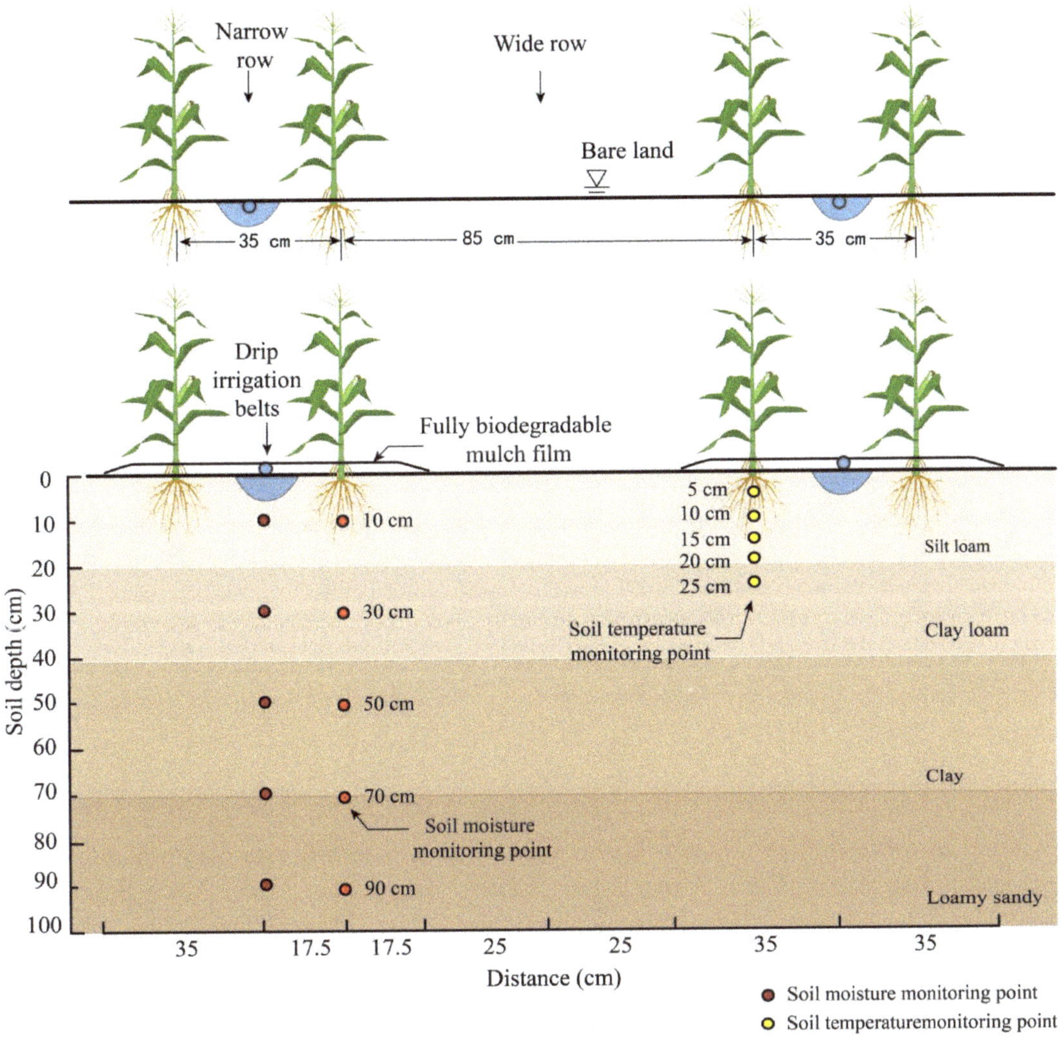


**Figure 8.** Planting patterns and soil moisture observation points.

4.2.2. Soil Water Content

The soil water content was monitored using an automatic soil water recorder (THL-TS-15; Beijing Kunlun Taiheng Technology Co., Ltd., Beijing, China.), and validation was conducted using the drying method. The automatic recorder collected data hourly, with monitoring points located directly beneath the drip irrigation tape and at depths of 10, 30, 50, 70, and 90 cm between the corn rows (Figure 8).

4.2.3. Soil Temperature

The soil temperature was monitored using an automatic soil temperature recorder (THL-TWS-14; Beijing Kunlun Taiheng Technology Co., Ltd.). Soil temperature sensors were installed at depths of 5, 10, 15, 20, and 25 cm below the surface between the corn rows to continuously track changes in soil temperature in the plowed layer, with data recorded every hour.

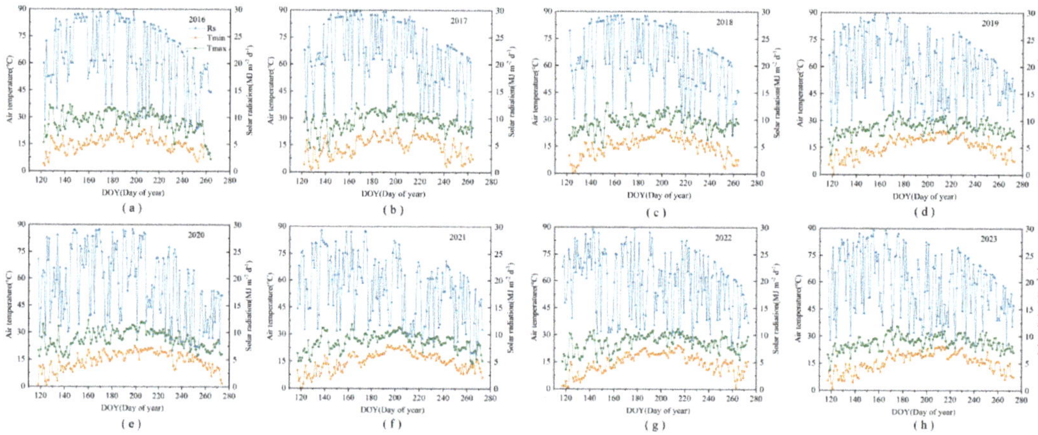

**Figure 9.** Maximum temperature ($T_{max}$), minimum temperature ($T_{min}$), and solar radiation (SR) during 2016, 2017, 2018, 2019, 2020, 2021, 2022, and 2023 (**a–h**).

4.2.4. Fully Biodegradable Mulch Film Damage Ratio Statistical Method

After sowing, three observation points were randomly selected from the plots treated with fully biodegradable mulch. Observations were recorded every 10 days, and photos were taken 30 cm above the observation area. After geometric correction and background removal, the images were imported into CAD, and the damaged areas were outlined. As most of the damaged areas are circular or elliptical, the radius of each damaged area is statistically measured to simplify the assessment by considering half of the surface area of the sphere with the damaged hole [14]. The statistical formula for fully biodegradable mulch film damage ratio ($F_d$) is as follows:

$$F_d = \frac{\sum_{i=1}^{n} A_i}{A_d} \tag{1}$$

$$A_i = 2\pi r^2 \tag{2}$$

where $A_d$ is the total area of the region between the narrow rows of corn ($cm^2$), $A_i$ is the surface area of the damaged biodegradable film within the region ($cm^2$) for $i$ ranging from 1 to $n$, and $r$ is the radius of the small hole (cm).

4.2.5. Leaf Area Index, Yield, and WUE

For each plot, five representative plants with consistent growth were selected and marked with tags for continuous monitoring of the corn leaf area throughout the growth stages. The leaf area index (LAI) was calculated using the formula proposed by Watson [27].

$$\text{LAI} = S_1/S_2 \tag{3}$$

where $S_1$ is the measured leaf area ($m^2$) and $S_2$ is the corresponding ground area ($m^2$).

At harvest, three 10 m long strips of double rows of corn were taken parallel to the drip irrigation tape from each plot. After harvesting all of the grains and allowing them to air-dry, the total weight was recorded and converted to yield per hectare.

The measured groundwater depth in the experimental area was 8.5 m. Therefore, groundwater replenishment was neglected. Drip irrigation was used, and the land surface was leveled, allowing surface runoff to be neglected. The soil water balance equation and formula for calculating WUE were as follows:

$$ET_c = I + P - \Delta W - D \tag{4}$$

$$WUE = \frac{Y}{ET_c} \quad (5)$$

where $ET_c$ represents the total evapotranspiration (mm) and $I$ and $P$ denote the irrigation amount (mm) and effective rainfall (mm), respectively. $\Delta W$ is the change in water storage during the period used to calculate $ET_c$, (mm). $D$ refers to the deep drainage or rising capillary flow (mm). $WUE$ stands for water use efficiency (kg $mm^{-1}$ $ha^{-1}$), and $Y$ represents grain yield (kg $ha^{-1}$).

### 4.3. Model Improvement

The DSSAT model includes modules for weather, soil, maize, and field management. In DSSAT, ET is typically estimated using the Priestley–Taylor model, which is based on solar radiation, rainfall, maximum temperature ($T_{\max}$), and minimum temperature ($T_{\min}$). Biodegradable film covers can increase the soil surface temperature and reduce soil evaporation during the early stages. However, this can also lead to a decrease in rainfall. As the film degrades, its insulating effect diminishes, resulting in increased soil evaporation, although some rainfall can still be utilized. The DSSAT model does not include a film cover module and assumes uniform irrigation across the entire field. In this study, we used drip irrigation, which is a form of localized irrigation. Therefore, the input values for irrigation, rainfall, and temperature were adjusted to improve the soil evaporation modulus.

#### 4.3.1. Irrigation Amount

The DSSAT model simulates soil moisture dynamics based on the assumption of uniform irrigation across the entire field. For instance, if the actual irrigation amount is 3 $m^3$ $ha^{-1}$, the DSSAT model assumes that the irrigation depth at any point in the field is 30 mm. In reality, the irrigation depth near the drip tape was greater than 30 mm. If we input a 30 mm irrigation depth into the model, it will underestimate the soil moisture near the maize roots under drip irrigation conditions. To improve the accuracy of the model, the irrigation data input was adjusted. Additionally, considering that plastic mulch is a significant influencing factor, we focused only on the soil moisture status in the area near the drip tape, specifically in the regions between the narrow rows of corn. The actual amount of irrigation, $Q_{real}$ (mm), was calculated using soil moisture data collected using automatic recording devices before and after irrigation using the following formula:

$$Q_{real} = 1000 \times \left(\theta_{after} - \theta_{before}\right) \times H \quad (6)$$

In the equation, $\theta_{before}$ represents the soil moisture content before irrigation ($cm^3$ $cm^{-3}$), $\theta_{after}$ denotes the soil moisture content after irrigation ($cm^3$ $cm^{-3}$), and $H$ is the depth of wet soil in the area between the narrow rows of corn (m). The actual irrigation amounts and model input values for 2016, 2017, and 2018 are listed in Table 6.

**Table 6.** Actual value (Act) and input model value (Inp) of the irrigation depth in 2016, 2017, and 2018.

| Year | Irrigation Depth (mm) | | | | | | | | |
|---|---|---|---|---|---|---|---|---|---|
| 2016 | Date | 9 May | 2 June | 9 July | 22 July | 2 August | 8 August | 14 August | 25 August |
| | Act | 30 | 30 | 30 | 15 | 10 | 20 | 20 | 20 |
| | Inp | 51 | 53 | 44 | 27 | 31 | 42 | 46 | 39 |
| 2017 | Date | 8 May | 11 June | 27 June | 2 July | 17 July | 22 July | 2 August | |
| | Act | 30 | 30 | 30 | 30 | 20 | 10 | 10 | |
| | Inp | 52 | 56 | 50 | 46 | 39 | 18 | 24 | |
| 2018 | Date | 9 May | 18 May | 10 June | 22 June | 17 July | 1 August | 9 August | |
| | Act | 15 | 30 | 30 | 30 | 30 | 30 | 30 | |
| | Inp | 33 | 59 | 58 | 52 | 42 | 37 | 40 | |

#### 4.3.2. Temperature

The warming effect also changes after the degradation of the biodegradable mulch. Therefore, the DSSAT model should account for the effect of soil temperature variation on maize growth following mulch degradation. Temperature compensation under biodegradable film coverage can be calculated using the following formula [28]:

$$T_{max-f} = T_{max} + \Delta T_a \tag{7}$$

$$T_{min-f} = T_{min} + \Delta T_a \tag{8}$$

$$\Delta T_a = C\Delta T_s \frac{T_a - T_b}{T_s - T_b} \tag{9}$$

$T_{max-f}$ is the maximum air temperature after compensation under biodegradable film mulching and $T_{min-f}$ is minimum air temperature (°C), $\Delta T_a$ is the air temperature difference between biodegradable film mulching and no-film mulching (°C), $C$ is the compensation coefficient, $\Delta T_s$ is the soil temperature difference between biodegradable film mulching and no-film mulching (°C), $T_a$ is the average air temperature, $T_b$ is the biological lower limit base point temperature of maize (8 °C), and $T_s$ is the daily soil temperature under no-film mulching (°C).

The compensation coefficient (C) was determined using the following equation [15]:

$$C = \frac{T_{cum-a-s} - T_{cum-a-f}}{T_{cum-s-f} - T_{cum-s-s}} \tag{10}$$

$$T_{cum} = \sum_{i=1}^{n} (T_{ai} - T_b) \tag{11}$$

where $T_{cum-a-s}$ is the effective air accumulated temperature under no-film mulching (°C), $T_{cum-a-f}$ is the effective air accumulated temperature under biodegradable film mulching (°C), $T_{cum-s-f}$ is the soil accumulated temperature under no-film mulching (°C), $T_{cum-s-s}$ is the effective soil accumulated temperature under biodegradable film mulching (°C), and $T_{ai}$ is the average air temperature on day $i$.

#### 4.3.3. Soil Evaporation

In the DSSAT model, $E_s$ is divided into two stages: stage I (constant-rate stage) and stage II (falling-rate stage). In stage I, $E_s$ was not affected by the variation in SWC, and $E_s$ in this stage was computed using the following equation:

$$E_{si} = \sum E_{s0} \tag{12}$$

where $E_{si}$ is the cumulative soil evaporation (mm d$^{-1}$) and $E_{s0}$ is the actual soil evaporation on day $i$.

In stage II, $E_s$ depended on the water vapor flux through the shallow soil layer, which was calculated using the following equation:

$$E_{si} = \alpha\sqrt{t} \tag{13}$$

where $\alpha$ is the soil hydraulic characteristic parameter (mm d$^{-0.5}$) and $t$ is the evaporation time (d).

The actual soil evaporation was related to the potential soil evaporation, and the potential soil evaporation was calculated from the potential $ET$, calculated using the default Priestley–Taylor formula of the DSSAT model [29]. The potential evapotranspiration ($ET_p$) and ($ET_c$) under the biodegradable film mulching were obtained using the following equations:

$$ET_p = ET_c \times 1.1 \tag{14}$$

$$ET_c = SLANG \times (2.04 \times 10^{-4} - 1.83 \times 10^{-4} \times ALBEDO) \times (TD + 29) \quad (15)$$

$$TD = 0.6 \times T_{max-f} + 0.4T_{min-f} \quad (16)$$

where $ET_p$ is the potential evapotranspiration rate (mm/d); $ET_c$ is the equilibrium evaporation rate, mm/d; $SLANG$ is the solar radiation (MJ m$^{-2}$ d$^{-1}$); $ALBEDO$ is the reflectance of the soil–maize surface, set to 0.23 in DSSAT 4.6; and $TD$ is the approximation of average daily temperature (°C).

Because the DSSAT model does not account for the impact of biodegradable film coverage on soil evaporation, a small amount of soil moisture evaporates from the damaged areas of the corn-planting holes before the film fully degrades. As degradation progressed, the soil evaporation rate was positively correlated with the rate of film damage. Therefore, the improved evaporation rate $E_{si}$ was determined using the following equation:

$$E_{si} = E_s \times F_d \quad (17)$$

4.3.4. Precipitation

Before the degradation of the fully biodegradable film, rainfall in the area near the drip tape could be neglected because of the obstruction caused by the film. After degradation, the utilization of rainwater is related to the damage rate of the film. The model input for precipitation $P_e$ was calculated using the formula described by Feng et al. [23].

$$P_e = P \times \lambda \quad (18)$$

$$\lambda = 54.7167 - 2.258P + 1.4297F_d + 0.0310P^2 - 0.0238F_d^2 + 0.045P \cdot F_d \quad (19)$$

In this context, $P_e$ represents the amount of rainfall accumulation in the area under the film (mm), $P$ denotes the total amount of subsequent rainfall (mm), $\lambda$ is the effective infiltration rate of the rainfall, and $F_d$ is the film damage ratio (%).

*4.4. Model Evaluation*

The prediction performance of the model was quantitatively evaluated based on the mean relative error ($MRE$, %), normalized root mean square error ($nRMSE$, %), and coefficient of determination ($R^2$).

$$\text{MRE} = \frac{1}{n}\sum_{i=1}^{n} \frac{|S_i - O_i|}{S_i} \times 100\% \quad (20)$$

$$nRMSE = \frac{\sqrt{\frac{1}{n}\sum_{i=1}^{n}(S_i - O_i)^2}}{O_{max} - O_{min}} \times 100\% \quad (21)$$

$$R^2 = \left[\frac{\sum_{i=1}^{n}(S_i - \overline{S})(O_i - \overline{O})}{\sqrt{\sum_{i=1}^{n}(S_i - \overline{S})^2}\sqrt{\sum_{i=1}^{n}(O_i - \overline{O})^2}}\right]^2 \quad (22)$$

where $S_i$ and $O_i$ represent the simulated and measured values, respectively, and $n$ represents the number of data points.

## 5. Conclusions

To accurately simulate maize growth under fully biodegradable mulch, this study utilized the evapotranspiration redistribution characteristics associated with such mulch. Moreover, the soil evaporation module was improved, the irrigation and rainfall input values were corrected using the one-dimensional water balance principle, and the effective air temperature compensation coefficient was calculated based on the warming effects of the mulch. The model was also optimized for genetic parameters of spring corn using field test data. This calibrated and verified DSSAT model effectively simulates the phenological period, leaf area index (LAI), yield, and soil moisture content of spring corn covered with biodegradable mulch across different coverage periods. Its simulation accuracy is characterized by an MRE < 10%, an nRMSE < 12%, and an $R^2 \geq 0.80$. This study offers enhanced methods for applying the DSSAT model to simulate maize under fully biodegradable mulch in a drip irrigation system.

In addition, based on experiments conducted from 2016 to 2018 and model predictions of yield and water use efficiency from 2019 to 2023, this study recommends that the Xiliaohe Plain and similar climatic areas should adopt fully biodegradable mulch with an induction period of approximately 80 days to ensure high yields across different hydrological years. The results of this study provide a scientific foundation and technical support for predicting the productivity of maize covered with fully biodegradable mulch and for optimizing the management of agricultural water and heat resources in the semi-arid regions of Northeast China in the context of climate change. In future research, continuous field experiments will be conducted in arid and semi-arid regions of China (across different temperature zones), considering the impacts of water stress and extreme weather on maize to further enhance the DSSAT model.

**Author Contributions:** Conceptualization, Y.J.; data curation, Y.F.; writing—original draft preparation, Y.J.; writing—review and editing, H.S. and X.S.; visualization, Y.J.; project administration, Q.M. and Y.F.; supervision, H.S. and Y.F.; funding acquisition, Y.F. and H.S. All authors have read and agreed to the published version of the manuscript.

**Funding:** Central Public-interest Scientific Institution Basal Research Fund (No. IFI2023-08), Sub-task of Major Scientific and Technological Tasks of Chinese Academy of Agricultural Sciences (CAAS-ZDRW20241805), The Inner Mongolia Autonomous Region Science and Technology Leading Talent Team (2022LJRC0007).

**Data Availability Statement:** Data will be made available on request.

**Conflicts of Interest:** The authors declare no conflicts of interest related to this article.

## References

1. Hengshan, Y.; Ruifu, Z.; Yuanyuan, L. Shallow-buried drip irrigation promoted the enrichment of beneficial microorganisms in surface soil. *Rhizosphere* **2023**, *28*, 100776.
2. Sun, S.; Zhang, W.; Liu, C.; Zhou, J.; Zhu, K. Degradation property of oxo-biodegradable plastic film and its mulching effect on soil moisture, soil temperature and maize growth in rainfed Northeast China. *Chin. J. Eco-Agric.* **2019**, *27*, 72–80.
3. Wang, Y.; Xu, M.; Xin, M.; Luo, X.; Zhang, T.; Pan, X.; Ji, Y.; Yin, H.; Zhao, Y.; Sui, M. Dicator construction and spatial distribution characteristics of maize spring waterlogging in Northeast China. *Trans. Chin. Soc. Agric. Eng.* **2022**, *38*, 101–110.
4. Li, R.; Zhao, J.; Shi, H.; Wang, N.; Qi, Y.; Feng, Y.; Jia, Q. Simulation of evapotranspiration of corn under mulched drip irrigation based on SIMDual_Kc model in Tongliao area of Inner Mongolia. *Trans. Chin. Soc. Agric. Eng.* **2018**, *34*, 127–134.
5. Gu, X.; Li, Y.; Du, Y. Biodegradable film mulching improves soil temperature, moisture and seed yield of winter oilseed rape (*Brassica napus* L.). *Soil Tillage Res.* **2017**, *171*, 42–50. [CrossRef]
6. Liang, S.; Li, Y.; Rashid, M.A.R.; He, S.; Wang, P.; Yin, Z.; Lu, L.; Zhan, K.; Wang, S.; Wang, X. Effects of planting patterns on rainwater use efficiency and potato tuber yield in seasonally arid areas of southwestern China. *Soil Tillage Res.* **2020**, *197*, 104502. [CrossRef]
7. Saglam, M.; Sintim, H.Y.; Bary, A.I.; Miles, C.A.; Ghimire, S.; Inglis, D.A.; Flury, M. Modeling the effect of biodegradable paper and plastic mulch on soil moisture dynamics. *Agric. Water Manag.* **2017**, *193*, 240–250. [CrossRef]
8. Sintim, H.Y.; Bary, A.I.; Hayes, D.G.; Wadsworth, L.C.; Anunciado, M.B.; English, M.E.; Bandopadhyay, S.; Schaeffer, S.M.; DeBruyn, J.M.; Miles, C.A. In situ degradation of biodegradable plastic mulch film in compost and agricultural soils. *Sci. Total Environ.* **2020**, *727*, 138668. [CrossRef]
9. Marí, A.; Pardo, G.; Aibar, J.; Cirujeda, A. Purple nutsedge (*Cyperus rotundus* L.) control with biodegradable mulches and its effect on fresh pepper production. *Sci. Hortic.* **2020**, *263*, 109111. [CrossRef]
10. Feng, Y.; Shi, H.; Cao, X.; Miao, Q.; Jia, Q.; Li, J.; Wang, N.; Zhu, K. Numerical modeling of soil water–heat transport under oxo-biodegradable film mulch and the optimal mulching period. *J. Irrig. Drain. Eng.* **2021**, *147*, 04021035. [CrossRef]
11. Chen, N.; Li, X.; Shi, H.; Yan, J.; Hu, Q.; Zhang, Y. Modeling maize evapotranspiration and associated processes under biodegradable film mulching in an arid dripped field. *Agric. For. Meteorol.* **2021**, *297*, 108247. [CrossRef]

12. Gao, X.; Xie, D.; Yang, C. Effects of a PLA/PBAT biodegradable film mulch as a replacement of polyethylene film and their residues on maize and soil environment. *Agric. Water Manag.* **2021**, *255*, 107053. [CrossRef]
13. Yin, M.; Li, Y.; Fang, H.; Chen, P. Biodegradable mulching film with an optimum degradation rate improves soil environment and enhances maize growth. *Agric. Water Manag.* **2019**, *216*, 127–137. [CrossRef]
14. Shen, H.; Gao, Y.; Guo, F.; Wang, Y.; Ma, X. A modified DSSAT-CERES model for simulating summer maize growth under film mulching. *Agron. J.* **2021**, *113*, 4819–4831. [CrossRef]
15. Chen, N.; Li, X.; Shi, H.; Zhang, Y.; Hu, Q.; Sun, Y.N. Modeling effects of biodegradable film mulching on evapotranspiration and maize yields in Inner Mongolia. *Agric. Water Manag.* **2023**, *275*, 107996. [CrossRef]
16. Chen, Q.; Nan, D. Preliminary Report on the Relationship between the Accumulated Temperature in Plow Layer of Soil and Cotton Growing Status. *Acta Pedol. Sin.* **1983**, *1*, 92–96.
17. Wang, H.; Gong, D.; Mei, X.; Hao, W. Dynamics comparison of rain-fed spring maize growth and evapotranspiration in plastic mulching and un-mulching fields. *Trans. Chin. Soc. Agric. Eng.* **2012**, *28*, 88–94.
18. Ning, Y.; Zhanxiang, S.; Zhe, Z.; Jiaming, Z.; Liangshan, F.; Kaiyu, L.; Chen, F.; Lizhen, Z. Simulation of water use process by film mulched cultivated maize, based on improved AquaMaize model and its verification. *Trans. Chin. Soc. Agric. Eng.* **2015**, *31*, 122–132.
19. Yuan, X.; Wu, P.; Wang, Y. Study on the effect of irrigation under plastic preservation of soil moisture on soil and maize. *J. Irrig. Drain.* **2006**, *25*, 25–29.
20. Sun, T.; Li, G.; Ning, T.Y. Suitability of mulching with biodegradable film to moderate soil temperature and moisture and to increase photosynthesis and yield in peanut. *Agric. Water Manag.* **2018**, *208*, 214–223. [CrossRef]
21. Wu, Q.; Wang, Z.; Zheng, X.; Zhang, J.; Li, W. Effects of biodegradation film mulching on soil temperature, moisture and yield of cotton under drip irrigation in typical oasis area. *Trans. Chin. Soc. Agric. Eng.* **2017**, *33*, 135–143.
22. Jia, Q.; Shi, H.; Li, R.; Miao, Q.; Feng, Y.; Wang, N.; Li, J. Evaporation of maize maize under mulch film and soil covered drip irrigation: Field assessment and modelling on West Liaohe Plain, China. *Agric. Water Manag.* **2021**, *253*, 106894. [CrossRef]
23. Feng, Y.; Shi, H.; Jia, Y.; Miao, Q.; Jia, Q.; Wang, N. Infiltration and water use efficiency of maize fields with drip irrigation and biodegradable mulches in the west liaohe plain, China. *Plants* **2023**, *12*, 975. [CrossRef]
24. Li, M.; Du, Y.; Zhang, F.; Bai, Y.; Fan, J.; Zhang, J.; Chen, S. Simulation of cotton growth and soil water content under film-mulched drip irrigation using modified CSM-MAIZEGRO-cotton model. *Agric. Water Manag.* **2019**, *218*, 124–138. [CrossRef]
25. Fu, C.; Wang, J.; Gong, S.; Zhang, Y.; Wang, C.; Mo, Y. Optimization of irrigation and fertilization of drip-irrigated corn in the chernozem area of north-east China based on the CERES-Maize model. *Irrig. Drain.* **2020**, *69*, 714–731. [CrossRef]
26. Jia, Q.; Shi, H.; Li, R.; Feng, Y.; Li, H.; Zheng, J. Drip irrigation schedules of maize in Tongliao. *Drain. Irrig. Mach. Eng.* **2018**, *36*, 897–902.
27. Watson, D.J. Comparative physiological studies on the growth of field maize: I. Variation in net assimilation rate and leaf area between species and varieties, and within and between years. *Ann. Bot.* **1947**, *11*, 41–76. [CrossRef]
28. Zhang, L.; Cao, W.; Zhang, S.; Luo, W. Simulation Model for Cotton Development Stages Based on Physiological Development Time. *Cotton Sci.* **2003**, *15*, 97–103.
29. Priestley, C.H.B.; Taylor, R.J. On the assessment of surface heat flux and evaporation using large-scale parameters. *Mon. Weather Rev.* **1972**, *100*, 81–92. [CrossRef]

plants

MDPI

Article

# Optimizing Irrigation Strategies to Improve Yield and Water Use Efficiency of Drip-Irrigated Maize in Southern Xinjiang

Qingyong Bian [1,2,†], Zhiduo Dong [1,†], Yanbo Fu [1,2,*], Yupeng Zhao [1], Yaozu Feng [1,2,3], Zhiguo Wang [1] and Jingquan Zhu [2]

1 Institute of Soil Fertilizer, Agricultural Water Saving, Xinjiang Academy of Agricultural Sciences, Urumqi 830091, China; 18116873795@126.com (Q.B.); dzd1281228561@163.com (Z.D.); 320223547@xjau.edu.cn (Y.Z.); guotaiminan7@163.com (Y.F.); wangzhiguo@xaas.ac.cn (Z.W.)

2 National Soil Quality Aksu Observation Experimental Station, Aksu 843000, China; ayg13779155706@163.com

3 Scientific and Technological Achievement Transformation Center, Xinjiang Academy of Agricultural Sciences, Urumqi 830091, China

* Correspondence: fuyanbo@xaas.ac.cn

† These authors contributed equally to this work.

**Abstract:** The contradiction between increased irrigation demand and water scarcity in arid regions has become more acute for crops as a result of global climate change. This highlights the urgent need to improve crop water use efficiency. In this study, four irrigation volumes were established for drip-irrigated maize under plastic mulch: 2145 $m^3$ $ha^{-1}$ (W1), 2685 $m^3$ $ha^{-1}$ (W2), 3360 $m^3$ $ha^{-1}$ (W3), and 4200 $m^3$ $ha^{-1}$ (W4). The effects of these volumes on soil moisture, maize growth, water consumption, crop coefficients, and yield were analyzed. The results showed that increasing the irrigation volume led to a 2.86% to 8.71% increase in soil moisture content, a 24.56% to 47.41% increase in water consumption, and a 3.43% to 35% increase in the crop coefficient. Maize plant height increased by 16.34% to 42.38%, ear height by 16.85% to 51.01%, ear length by 2.43% to 28.13%, and yield by 16.96% to 39.24%. Additionally, soil temperature was reduced by 1.67% to 5.67%, and the maize bald tip length decreased by 6.62% to 48%. The irrigation water use efficiency improved by 6.57% to 28.89%. A comprehensive evaluation using the TOPSIS method demonstrated that 3360 $m^3$ $ha^{-1}$ of irrigation water was an effective irrigation strategy for increasing maize yield under drip irrigation with plastic mulch in the southern border area. Compared to 4200 $m^3$ $ha^{-1}$, this strategy saved 840 $m^3$ $ha^{-1}$ of irrigation water, increased the irrigation water use efficiency by 23.96%, and resulted in only a 0.84% decrease in yield. The findings of this study provide a theoretical foundation for optimizing production benefits in the context of limited water resources.

**Keywords:** maize; irrigation; crop coefficient; yield; irrigation water use efficiency

## 1. Introduction

The importance of irrigation in ensuring food security has been well documented [1]. It has been estimated that between 70% and 90% of the world's freshwater resources were used for agricultural production [2–4]. However, the challenges of water scarcity driven by climate change, population growth, and agricultural activities have become critical issues in many parts of the world, particularly in arid and semi-arid climates [5]. At the same time, the area of irrigated farmland has been projected to continue expanding in tandem with population growth and increasing food demand [5,6], which would undoubtedly intensify the pressure on agricultural water use. In the face of escalating water scarcity and the uncertainties posed by climate change, the judicious use of limited water resources to achieve sustainable increases in crop yields has become a pivotal aspect of ensuring food security [7]. The deployment of advanced irrigation technology and the formulation of optimal irrigation strategies has facilitated the rational utilization of available water resources, thereby improving overall irrigation management [8]. Studies show that ineffective water consumption in farmlands can be reduced, and soil water retention can be increased by using methods such as film covering [9], straw covering [10], subsurface drip irrigation [11], deficit irrigation [12], and applying anti-transpirant agents [13]. These practices can subsequently improve the efficiency of agricultural water use and enhance productivity [14].

Maize (*Zea mays* L.) is a significant food crop and the most consumed cereal crop globally [15–17]. It has the second highest total production and productivity in the world, with 260 million tons total and 6.3 tons per hectare, respectively [18]. It has been anticipated that crop production will need to increase by over 70% by 2050 to meet the food demands of the growing global population [19]. However, the future growth of maize production through an expansion in planted area is limited due to the reduction in arable land and natural resources. Consequently, increasing maize productivity per unit area has become more important, and irrigation management plays a significant role [20,21]. Therefore, improving irrigation methods represents a potential solution for achieving higher yield levels in water-scarce areas while enhancing the resilience of cropping systems to climate change [22,23].

Maize is a crop with high water consumption, and identifying optimal agronomic practices is crucial for maximizing yields while conserving water and efficiently utilizing the limited irrigation water available [24]. To alleviate water scarcity and improve irrigation water use efficiency, the adoption of mulch film drip irrigation technology represents a viable solution [25]. Mulch drip irrigation integrates advanced drip irrigation technology with cover cropping technology [26], a strategy that has been demonstrated to be an effective means of conserving water, regulating salt levels, maintaining soil moisture, enhancing crop yields, and improving water use efficiency [27]. Previous research demonstrated that maize is highly susceptible to irrigation deficiencies, which was attributed to its elevated water consumption [12], increased plant density, and other changes in crop management practices [28]. Over-irrigation was shown to reduce crop yield and irrigation water productivity, whereas deficit irrigation increased yield and reduced field crop evapotranspiration [29,30]. The latter was a significant factor in determining irrigation intensity, given that crop transpiration represents a major pathway for on-farm water depletion [31,32]. Furthermore, crop evapotranspirational water consumption is a crucial consideration in this context. As a pivotal point in the transformation of water movement and energy flux within the soil–plant–atmosphere continuum, this process integrates the physics of water movement and plant physiology. It is influenced by a range of factors, including soil moisture status, meteorological conditions, living environment, and the plant's own growth and development [33].

The unique climatic conditions of Xinjiang support maize growth but are limited by water scarcity and soil salinization. To ensure food security, it is essential to reduce irrigation water use while maintaining crop yield. This study aimed to optimize irrigation strategies to enhance the yield and water use efficiency of drip-irrigated maize under membrane conditions in southern Xinjiang. We employed a weighing evapotranspiration meter to analyze changes in soil moisture, maize growth, water consumption, crop coefficients, and yield. The goal was to determine the optimal irrigation practices for maximizing yield and efficiency under these conditions. These findings provide valuable insights for guiding irrigation practices and maximizing production benefits in the context of limited water resources.

## 2. Results

### *2.1. Changes in the Soil Environment*

The average soil moisture ranged from 16.59% to 23.05%. Soil moisture was relatively low before irrigation and significantly increased after irrigation. Soil moisture exhibited an increasing trend with the rise in irrigation volume, with the W2, W3, and W4 treatments showing increases of 2.86%, 7.51%, and 8.71%, respectively, compared to W1 (Figure 1A).

The average soil temperature ranged from 14.01 °C to 26.95 °C, showing an initial increase followed by a decline as maize growth progressed. Compared to W1, soil temperature decreased by 1.67%, 3.48%, and 5.67%, respectively (Figure 1B).

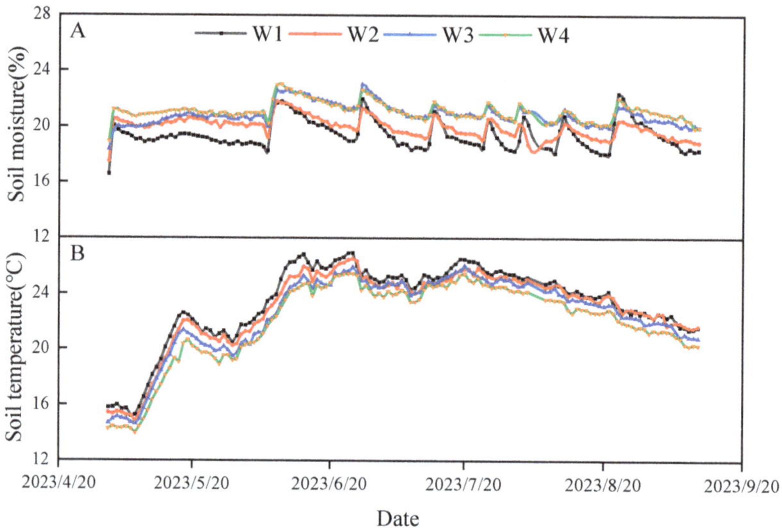


**Figure 1.** Changes in soil moisture and temperature during the maize growing period. (**A**) represents the fluctuation of soil moisture, while (**B**) represents the alteration of soil temperature.

### 2.2. Maize Growth

The plant height, ear height, and ear length of maize all exhibited a significant increasing trend ($p < 0.05$) with the increase in irrigation volume. Compared to W1, the plant height in W2, W3, and W4 increased by 16.34%, 19.69%, and 42.38%, respectively (Figure 2A), while ear height increased by 16.85%, 22.10%, and 51.01%, respectively (Figure 2B). Similarly, ear length increased by 2.43%, 8.62%, and 28.13%, respectively (Figure 2C). The bald tip length of maize decreased with increasing irrigation, with W2, W3, and W4 treatments showing reductions of 6.62%, 23.85%, and 48.55%, respectively, compared to W1 (Figure 2D).

### 2.3. Soil Water Consumption and Crop Coefficient

Soil water consumption exhibited an "N" trend with the advancement of maize growth stages, with the maximum water consumption occurring during the tasseling and flowering stages, ranging from 88 mm to 132.16 mm. The average daily soil water consumption and crop coefficient initially increased and then decreased as the growth stages progressed, with the maize filling stage reaching its peak at 3.24–5.01 mm and 0.54–0.84 (Figure 3). Soil water consumption, average daily water consumption, and crop coefficient (Kc) all showed an increasing trend with increasing irrigation volume. Compared to W1, soil water consumption in the W2, W3, and W4 treatments increased by 24.56%, 33.57%, and 47.41%, respectively; average daily water consumption increased by 26.24%, 35.12%, and 49.45%, respectively; and crop coefficients increased by 13.43%, 21.16%, and 35.80%, respectively. Compared with ET, the total ET (W4KC) calculated by W1, W2 and W3 decreased by 119.70, 58.32 and 35.53, respectively (Table 1).

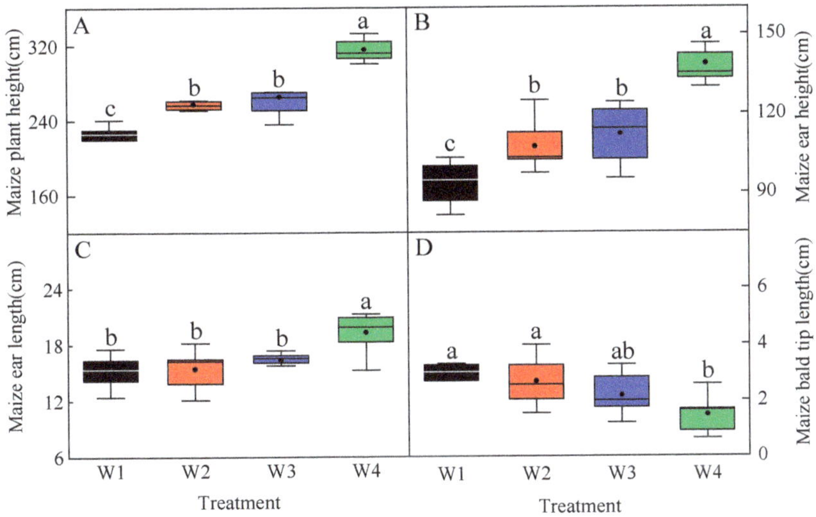


**Figure 2.** Effects of irrigation volume on maize plant height, ear height, ear length, and bald tip length. (**A**) represents the fluctuation of maize plant height, while (**B**) represents ear height. (**C**) represents ear height, and (**D**) represents bald tip length. Different lowercase letters indicate a significant difference in the mean value of different treatments at the probability level of 0.05 ($p < 0.05$), determined by a one-way analysis of variance (ANOVA) and Duncan's post hoc test. The data are presented as means ± standard deviation (SD) calculated from nine maize samples.

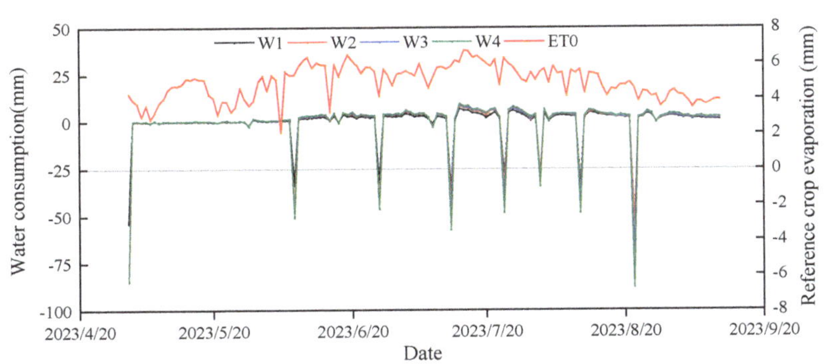


**Figure 3.** Soil water consumption during maize development.

*2.4. Maize Yield and Irrigation Water Use Efficiency*

Irrigation significantly increased maize yield and reduced the irrigation water use coefficient ($p < 0.05$). Compared to W1, maize yield under W2, W3, and W4 treatments increased by 16.96%, 38.08%, and 39.24%, respectively, while the irrigation water use coefficient decreased by 6.57%, 11.85%, and 28.89%, respectively (Figure 4).

**Table 1.** Effects of different irrigation treatment on ET during maize growth period (with Kc obtained from complete irrigation W4 as a reference).

| Treatment | Index | Jointing Stage | Bellbottom Stage | Heading–Flowering Stage | Filling Stage | Maturity Stage |
|---|---|---|---|---|---|---|
| W1 | ET | 3.54 | 17.09 | 88.82 | 55.00 | 86.00 |
| | ET0 | 88.50 | 106.81 | 188.98 | 101.85 | 156.36 |
| | ET($W_4K_C$) | 6.20 | 28.84 | 132.29 | 85.56 | 117.27 |
| | ETd | 0.18 | 0.78 | 2.69 | 3.24 | 2.61 |
| | Kc | 0.04 | 0.16 | 0.47 | 0.54 | 0.55 |
| W2 | ET | 4.72 | 21.85 | 114.24 | 73.59 | 97.54 |
| | ET0 | 78.67 | 109.25 | 190.40 | 102.21 | 154.83 |
| | ET($W_4K_C$) | 5.51 | 29.50 | 133.28 | 85.86 | 116.12 |
| | ETd | 0.24 | 0.99 | 3.46 | 4.33 | 2.96 |
| | Kc | 0.06 | 0.20 | 0.60 | 0.72 | 0.63 |
| W3 | ET | 5.58 | 24.00 | 123.34 | 77.40 | 104.20 |
| | ET0 | 79.71 | 109.09 | 189.75 | 101.84 | 155.52 |
| | ET($W_4K_C$) | 5.58 | 29.45 | 132.83 | 85.55 | 116.64 |
| | ETd | 0.28 | 1.09 | 3.74 | 4.55 | 3.16 |
| | Kc | 0.07 | 0.22 | 0.65 | 0.76 | 0.67 |
| W4 | ET | 6.36 | 28.64 | 132.16 | 85.24 | 116.78 |
| | ET0 | 90.86 | 106.07 | 188.80 | 101.48 | 155.71 |
| | ETd | 0.32 | 1.30 | 4.00 | 5.01 | 3.54 |
| | Kc | 0.07 | 0.27 | 0.70 | 0.84 | 0.75 |

Note: ET, ETd, and Kc denote water consumption, average daily water consumption, and crop coefficient, respectively.

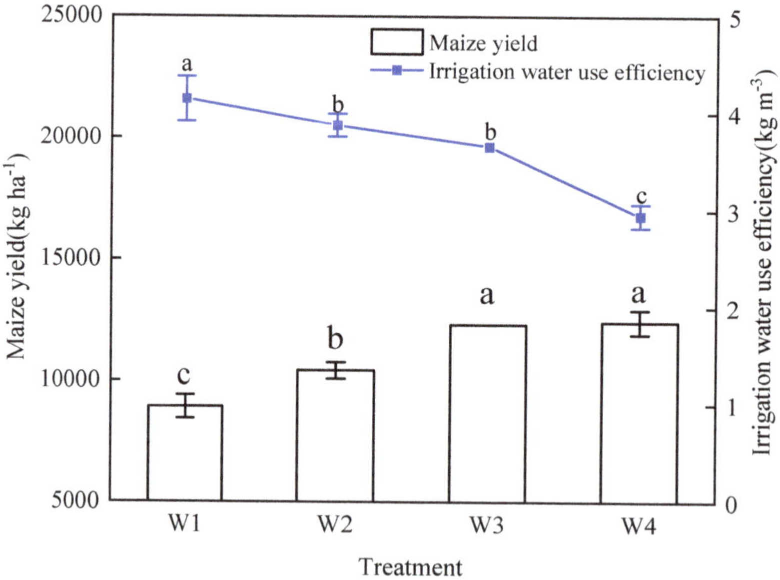


**Figure 4.** Impact of irrigation on maize yield and water use efficiency. Different lowercase letters indicate a significant difference in the mean value of different treatments at the probability level of 0.05 ($p < 0.05$), determined by one-way analysis of variance (ANOVA) and Duncan's post hoc test. The data are presented as means ± standard deviation (SD) calculated from nine maize samples.

### 2.5. Complex Evaluation Based on TOPSIS

The comprehensive evaluation of TOPSIS based on maize yield, water consumption ET, and irrigation water use efficiency is shown in Table 2. The weight of ET is 32.97%, the weight of yield is 35.37%, and the weight of irrigation water use efficiency is 31.66%. The maximum value of index weight is yield (35.37%), and the minimum value is water use irrigation efficiency (31.66%). The TOPSIS score and grade under different irrigation conditions were the best (ranked first) for the W3 treatment, with a comprehensive score of 0.73, followed by the W4 treatment at 0.59, followed by the W2 treatment at 0.56, and finally the W1 treatment at 0.41. The significant score of the W3 treatment was higher than that of other treatments, indicating that the best treatment was the irrigation amount W3.

**Table 2.** TOPSIS analysis of maize yield, soil water consumption, and irrigation water use efficiency.

| Treatment | Euclidean Distances | | Comprehensive Score Index | TOPSIS Rank |
|---|---|---|---|---|
| | d+ | d− | | |
| W1 | 0.83 | 0.56 | 0.41 | 4 |
| W2 | 0.45 | 0.59 | 0.56 | 3 |
| W3 | 0.29 | 0.78 | 0.73 | 1 |
| W4 | 0.56 | 0.83 | 0.59 | 2 |

## 3. Discussion

Drought is the result of an imbalance between evapotranspiration flux (water loss from evaporation and plant transpiration) and water intake from the soil [34]. The results of this study indicate that the soil water content increased in conjunction with the increase in irrigation water. The W2, W3, and W4 treatments exhibited an increase of 2.86%, 7.51%, and 8.71%, respectively, in comparison to W1. This suggests that the soil moisture experienced cyclic changes due to irrigation, rainfall, and evapotranspiration, with irrigation and rainfall serving to supplement soil moisture and evaporation consuming soil water [35]. Consequently, increasing the irrigation amount can compensate for the water lost through plant transpiration to a certain extent and enhance the soil water storage capacity within the root area [36], which is consistent with the results of the studies of Nxumalo et al. [37] and Brar et al. [38]. The soil water content initially increases and subsequently declines with the progression of the growth period, reaching a relatively low value during the corn filling period. This is due to the fact that the corn filling period entails a transition from vegetative growth to reproductive growth, necessitating the provision of water for both these growth stages simultaneously.

Soil temperature initially increased and then decreased as soil fertility progressed. This pattern was attributed to soil temperature being influenced by energy absorption or release in response to variations in solar radiation and air temperature [39]. The high transmittance of the film allowed most of the film area to receive solar energy, leading to surface soil warming [40]. Additionally, water droplets on the film and water vapor in the air below the film facilitated the absorption of long-wave radiation, contributing to soil warming through the greenhouse effect [41]. As the maize canopy matured during later growth stages, it influenced the heat absorbed by the soil from solar radiation, which in turn affected soil temperature. An observed decrease in soil temperature with increasing irrigation volume indicated that higher soil water content reduced soil temperature. This effect occurred because soil temperature changes were primarily influenced by the thermal properties of the soil, including specific heat capacity, thermal conductivity, and thermal diffusivity. The soil's specific heat capacity, which is influenced by its moisture content, increased with rising moisture levels, leading to a slower rate of temperature increase [42].

Strategies designed to regulate crop evapotranspiration have the potential to positively impact crop water use efficiency, thereby enhancing sustainable crop production [43]. The most prevalent method for calculating crop evapotranspiration involves multiplying the reference evapotranspiration ($ET_0$) by the crop coefficient ($K_c$), a process commonly referred to as the FAO56 method [44]. An alternative approach is to directly measure crop evapotranspiration using a gravimetric evapotranspiration meter [45]. The findings of this study indicate that maize water consumption and average daily water consumption increased with higher irrigation volumes. This suggests that increased irrigation effectively enhances soil water storage, which provides the necessary conditions for soil evaporation and plant transpiration [46].

Moreover, higher irrigation volumes promote maize growth and increase the surface area of maize leaves. A larger leaf area allows for more sunlight absorption, which subsequently increases the amount of energy converted into evapotranspiration (ET) [47]. Thus, the total evapotranspiration amount increases. Understanding the variability and magnitude of crop coefficients ($K_c$) is crucial for accurately determining crop evapotranspiration (ET) and optimizing irrigation scheduling [48]. Crop biophysical parameters, such as leaf area index, ground cover, and canopy cover, are directly correlated with crop coefficients and can be used to estimate single crop coefficients or base crop coefficients for various crops [49]. In this study, it was observed that the maize crop coefficient ($K_c$) and evapotranspiration both exhibited an increasing trend with higher irrigation volumes. This was because the crop coefficient was directly proportional to evapotranspiration, following the determination of reference crop evapotranspiration ($ET_0$) through meteorological data analysis [50,51].

Maize proved to be highly sensitive to water deficit, which significantly impacted its growth and both its physical and chemical properties [52]. Water stress could severely impair maize growth and diminish its yield potential [53]. Conversely, over-irrigation was associated with yield loss, water wastage, soil salinization, and a decline in soil fertility [54]. The findings of this study revealed that maize plant height, ear height, and ear length all increased significantly with higher irrigation volumes, while bald tip length exhibited an opposite trend. These results suggested that a favorable soil hydrothermal environment promoted maize nutrient growth, with suitable soil moisture serving as a crucial resource for photosynthesis. Additionally, optimal soil temperature was found to enhance maize root respiration, which facilitated the transport of soil moisture and nutrients, thereby improving maize photosynthesis [55].

The results indicate that maize yield tends to increase with higher irrigation volumes. However, water use efficiency demonstrates the opposite trend, suggesting that excessive irrigation could lead to a substantial decrease in yield [56,57]. Proper irrigation was shown to support the accumulation of aboveground biomass, thereby improving both crop yield and water use efficiency [58]. Nonetheless, excessive irrigation could lead to fertilizer leaching, resulting in inefficient fertilizer uptake and reduced plant nutritional and reproductive growth, ultimately affecting crop yield [59]. The TOPSIS comprehensive evaluation confirmed this conclusion, indicating that while higher irrigation volumes led to increased maize yield, they also resulted in a notable reduction in irrigation water use efficiency. The data suggest that an irrigation volume of 3360 $m^3$ $ha^{-1}$ is a highly effective strategy for enhancing maize yield under membrane drip irrigation in the southern border area.

## 4. Materials and Methods

### 4.1. Overview of the Experiment Site

The test area was situated at the base of the National Soil Quality Aksu Observation and Experiment Station in Baicheng County, Aksu Region, Xinjiang (81°54′22.6″, 41°47′37.2″; elevation 1232.1 m) (Figure 5). This region has a temperate continental arid climate, with an average annual temperature of 6.0 °C, an extreme maximum temperature of 38.3 °C, and an extreme minimum temperature of −28.0 °C. The frost-free period lasts between 133 and 163 days, and the mean annual sunshine duration is 2789.7 h. The average annual precipitation is 171.13 mm. The soil in the test area is classify as sandy loam, with a bulk density of 1.36 g $cm^{-3}$. The soil exhibits medium fertility, with an organic matter content of 23.59 g $kg^{-1}$, a total nitrogen concentration of 0.89 g $kg^{-1}$, an alkaline hydrolyzable nitrogen level of 34.93 mg $kg^{-1}$, an available phosphorus concentration of 21.81 mg $kg^{-1}$, and an available potassium concentration of 120.92 mg $kg^{-1}$. The pH value of the soil is 7.89, which is consistent with the average annual rainfall pH value [60].

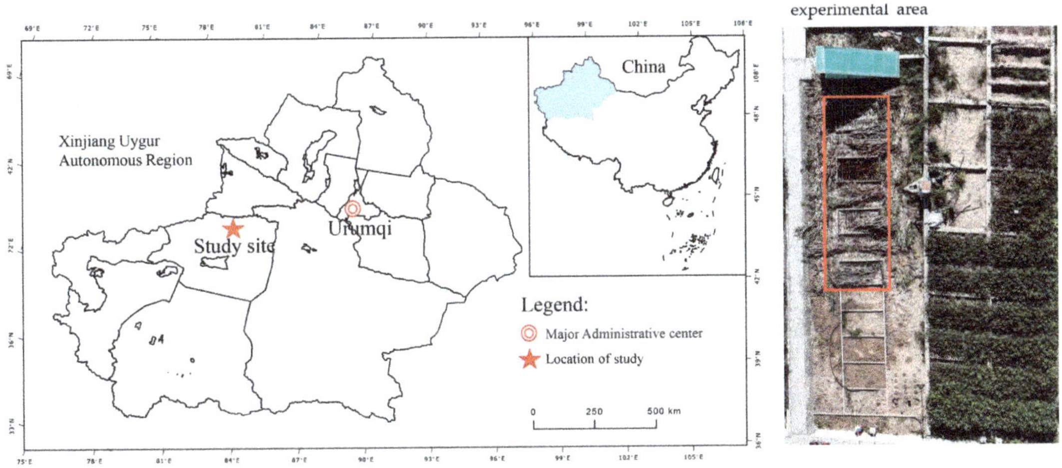


**Figure 5.** Location and layout of the study area.

## 4.2. *Field Management and Experimental Design*

In this experiment, a one-way completely randomized design was employed to establish four irrigation volumes of 2145 $m^3$ $ha^{-1}$ (W1), 2685 $m^3$ $ha^{-1}$ (W2), 3360 $m^3$ $ha^{-1}$ (W3), and 4200 $m^3$ $ha^{-1}$ (W4) for drip-irrigated maize under plastic mulch, based on the study conducted by Xiao et al. [61]. Fertilizers were applied via drip irrigation at rates of 241.5 kg $ha^{-1}$ for nitrogen (N), 172.5 kg $ha^{-1}$ for phosphorus (P), and 75 kg $ha^{-1}$ for potassium (K). The remaining field management practices were consistent with local standards.

Four treatments were implemented, with specifications of 3.3 m × 2 m × 2.3 m (length × width × depth), using a weighing soil evapotranspiration meter (XHZ-ZS201). The maize variety used in the experiment was Xinyu 31. It was sown on April 20, 2023, irrigated for the first time on May 1, and harvested on September 10. The planting was conducted with equal spacing between plants, with a row spacing of 40 cm and a plant spacing of 15 cm (Figure 6). An inlaid patch drip irrigation tape with a drip head spacing of 30 cm, a rated flow rate of 2.8 L $h^{-1}$, and an operating pressure of 0.1 MPa was selected, according to the specifications outlined in the relevant literature. The source of irrigation water was groundwater, and the flow rate of the wells was sufficient to meet the required volume of water for irrigation purposes.

## 4.3. *Measurement Items and Methods*

### 4.3.1. Soil Moisture and Temperature

An evapotranspiration meter (PH-30, Produced by Beijing Tanghua Technology Co., Ltd. Beijing, China) was installed with a soil water temperature sensor (recording frequency of once per hour) to monitor soil moisture and temperature at depths of 10 cm, 20 cm, 40 cm, 60 cm, 80 cm, 100 cm, and 130 cm during the reproductive period of maize.

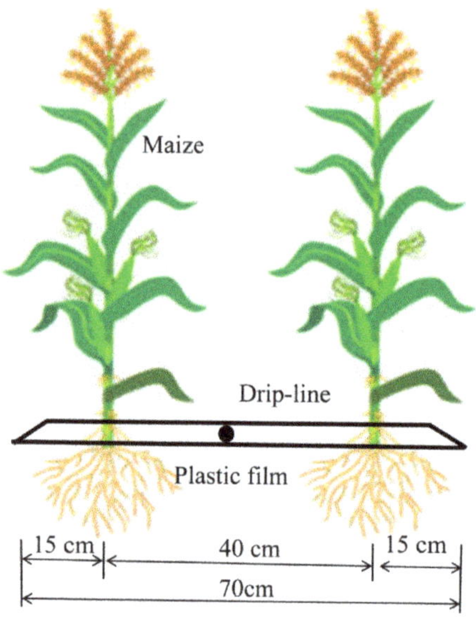


**Figure 6.** Diagram of maize planting arrangement and drip irrigation system.

4.3.2. Maize Growth Indicators

During the maize harvest period, nine maize plants with uniform growth were randomly selected and labeled for each treatment. The maize plant height, ear height, ear length, and bald tip length were measured using a steel ruler with 1 mm graduations.

4.3.3. Water Consumption and Crop Coefficient

The calculation of crop water requirements (water consumption) for various irrigation activities was achieved using methods based on crop coefficients [62]. As outlined by Allen et al. [44], multiplying the reference evapotranspiration ($ET_0$) by the crop coefficient ($K_c$) provides the crop water requirement. $K_c$ varies depending on crop type, phenological stage, and location. In this study, maize water consumption (ET) was measured using a lysimeter, and parameters such as daily maximum temperature ($T_{max}$), daily minimum temperature ($T_{min}$), average relative humidity ($RH_{mean}$), wind speed, and precipitation (P) were obtained from a meteorological station (Figure 7). Maize $K_c$ was calculated using Equation (1), and ET0 was calculated according to Equation (2) [63].

$$K_c = \frac{ET}{ET_0} \tag{1}$$

$$ET_0 = \frac{0.408 \times \Delta \times (R_n - G) + \gamma \frac{900}{T+273} \mu_2 (e_s - e_a)}{\Delta + \gamma(1 + 0.34\mu_2)} \tag{2}$$

where $ET_0$ is the reference crop evapotranspiration (mm $d^{-1}$); Rn is the net radiation [MJ $(m^2\ d)^{-1}$]; G is the soil heat flux [MJ $(m^2\ d)^{-1}$]; $\gamma$ is the psychrometric constant (kPa $°C^{-1}$); T is the daily average air temperature (°C); $\mu_2$ is the wind speed at a height of 2 m above the ground (m $s^{-1}$); $e_s$ is the saturated water vapor pressure (kPa); $e_a$ is the actual water vapor pressure (kPa); and $\Delta$ is the slope of the vapor pressure curve at temperature T (kPa $°C^{-1}$).

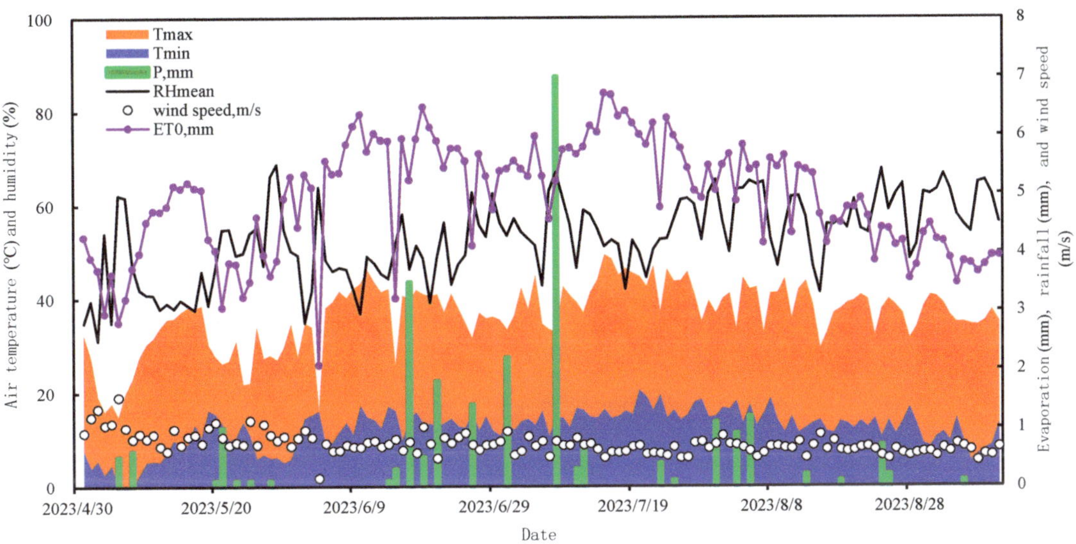


**Figure 7.** The 2023 meteorological data for the study area.

### 4.3.4. Maize Yield and Irrigation Water Use Efficiency Indicators

Maize yield was determined by harvesting all maize cobs in the evapotranspiration chamber at the point of maturity. The irrigation water use efficiency (IWUE) was calculated based on the methodology proposed by Wang et al. [64], as outlined in Equation (3):

$$IWUE = \frac{Y}{I} \quad (3)$$

where Y is the maize yield (kg $ha^{-1}$) and I is the irrigation water applied to the maize during the reproductive period ($m^3$ $ha^{-1}$).

### 4.3.5. TOPSIS Integrated Evaluation

To evaluate the optimal strategy for efficient irrigation, we employed a multi-objective optimization method, the Technique for Order of Preference by Similarity to Ideal Solution (TOPSIS). This method reflects the current status by measuring the distance from the optimal and worst solutions. The computational process is based on the study by Abdoul Kader et al. [65].

TOPSIS was calculated as follows:

(1) Establishment of the contribution matrix of the evaluation indices:

$$X = (Xij)n \times m \quad (4)$$

(2) Calculation of the normalized matrix:

$$\overline{X} = \frac{X_{ij}}{\sqrt{\sum_{i=0}^{n} X_{ij}^2}} \quad (5)$$

(3) Calculation of the weighted normalized matrix:

$$V_{ij} = \overline{X_{ij}} \times W_j \quad (6)$$

(4) Calculation of the Euclidean distances:

$$d_i^+ = \sqrt{\sum_j^m \left(V_{ij} - V_J^+\right)^2} \quad (7)$$

$$d_i^- = \sqrt{\sum_j^m \left(V_{ij} - V_J^-\right)^2} \quad (8)$$

(5) Calculation of the performance score of the treatments:

$$P_i = \frac{d_i^-}{d_i^+ + d_i^-} \quad (9)$$

where Xij is the value of the ith treatment and the jth indicator; n and m are the number of evaluation indicators and the number of treatments, respectively; Wj is the weight of each evaluation indicator; $V_J^+$ and $J_J^-$ are the positive and negative ideal solutions, respectively; and $d_i^+$ and $d_i^-$ are the distances between the positive and negative ideal solutions, respectively.

### 4.4. Statistical Analysis

Experimental data were managed using WPS version 12.1.0.17147 (Kingsoft Office Software Co., Ltd., Beijing, China). A single-factor analysis of variance (ANOVA) was performed using DPS version 16.05 (Hangzhou Ruifeng Information Technology Co., Ltd., Hangzhou, China), followed by Duncan's post hoc test to determine the significance of differences ($p < 0.05$ was considered significant). All graphs were generated using Origin 2017.

## 5. Conclusions

This study demonstrated that increasing irrigation water volume significantly improved soil moisture content, water consumption, and crop coefficients. Additionally, maize plant height, ear height, ear length, and yield all showed significant increases, ranging from 2.43% to 28.13%. Conversely, soil temperature, maize bald tip length, and irrigation water use efficiency showed notable declines. The comprehensive evaluation using TOPSIS revealed that an irrigation volume of 3360 $m^3$ $ha^{-1}$ was an effective strategy for enhancing maize yield in southern Xinjiang under membrane drip irrigation. This strategy resulted in a 23.96% reduction in irrigation water usage compared to the 4200 $m^3$ $ha^{-1}$ irrigation volume, while improving irrigation water utilization efficiency and reducing yield by only 0.84%. The findings of this study provide a theoretical foundation for optimizing production benefits under limited water resources.

**Author Contributions:** Z.W. designed the experiment, Y.Z. and J.Z. conducted the experiment. Q.B. and Z.D. wrote the manuscript. Y.F. (Yanbo Fu) provided laboratory and other technical support. Y.F. (Yanbo Fu) and Y.F. (Yaozu Feng) critically reviewed and edited the manuscript. All authors have read and agreed to the published version of the manuscript.

**Funding:** This study received funding from the Basic Scientific Research Business Fund Project of Public Welfare Research Institutes in the Autonomous Region (KY2022001) and Xinjiang Academy of Agricultural Sciences Agricultural Science and Technology Innovation Stable Support Project (XJNKYWDZC-2023002-2) and Long-Term Scientific Research Observation and Monitoring Tasks (NAES098SQ37).

**Data Availability Statement:** The original contributions presented in the study are included in the article; further inquiries can be directed to the corresponding authors.

**Acknowledgments:** We appreciate and thank the anonymous reviewers for helpful comments that led to the overall improvement of the manuscript. We also thank the journal's Editorial Board for their help and patience throughout the review process.

**Conflicts of Interest:** The authors declare no conflicts of interest.

## References


1. Wang, X.H.; Müller, C.; Elliot, J.; Mueller, N.D.; Ciais, P.; Jägermeyr, J.; Gerber, J.; Dumas, P.; Wang, C.Z.; Yang, H. Global irrigation contribution to wheat and maize yield. *Nat. Commun.* **2021**, *12*, 1235. [CrossRef] [PubMed]
2. Liu, K.; Bo, Y.; Li, X.K.; Wang, S.D.; Zhou, G.S. Uncovering current and future variations of irrigation water use across China using machine learning. *Earths Future* **2024**, *12*, e2023EF003562. [CrossRef]
3. Ju, H.R.; Liu, Y.X.; Zhang, S.R. Interprovincial agricultural water footprint in China: Spatial pattern, driving forces and implications for water resource management. *Sustain. Prod. Consum.* **2023**, *43*, 264–277. [CrossRef]
4. Wu, B.F.; Tian, F.Y.; Zhang, M.; Piao, S.L.; Zeng, H.W.; Zhu, W.W.; Liu, J.G.; Elnashar, A.; Lu, Y.M. Quantifying global agricultural water appropriation with data derived from earth observations. *J. Clean. Prod.* **2022**, *358*, 131891. [CrossRef]
5. Berrang-Ford, L.; Siders, A.; Lesnikowski, A.; Fischer, A.P.; Callaghan, M.W.; Haddaway, N.R.; Mach, K.J.; Araos, M.; Shah, M.A.R.; Wannewitz, M. A systematic global stocktake of evidence on human adaptation to climate change. *Nat. Clim. Chang.* **2021**, *11*, 989–1000. [CrossRef]
6. Rosa, L.; Chiarelli, D.D.; Sangiorgio, M.; Beltran-Peña, A.A.; Rulli, M.C.; D'Odorico, P.; Fung, I. Potential for sustainable irrigation expansion in a 3 C warmer climate. *Proc. Natl. Acad. Sci. USA* **2020**, *117*, 29526–29534. [CrossRef] [PubMed]
7. Guo, H.; Wang, X.H.; Wang, Y.H.; Li, S.E. Effect of mulched drip irrigation on crop biomass and carbon fluxes in maize field. *Agric. Water Manag.* **2024**, *303*, 109016. [CrossRef]
8. Li, F.W.; Yu, D.; Zhao, Y. Irrigation scheduling optimization for cotton based on the AquaCrop model. *Water Resour. Manag.* **2019**, *33*, 39–55. [CrossRef]
9. Wang, Z.; Chen, R.; Li, W.; Zhang, J.; Zhang, J.; Song, L.; Guo, L. Mulched drip irrigation: A promising practice for sustainable agriculture in China's arid region. *npj Sustain. Agric.* **2024**, *2*, 17. [CrossRef]
10. Zhao, H.; Liu, G.; Dou, Y.; Yang, H.; Wang, T.; Wang, Z.; Khan, A.A. Plastic mulch increases dryland wheat yield and water-use productivity, while straw mulch increases soil water storage. *J. Integr. Agric.* **2024**, *23*, 3174–3185. [CrossRef]
11. Namdarian, D.; Boroomand-Nasab, S.; Gorooei, A.; Gaiser, T.; Solymani, A.; Naseri, A.; dos Santos Vianna, M. Determination of the optimum depth for subsurface dripping irrigation of sugarcane under crop residue management. *Agric. Water Manag.* **2024**, *303*, 109026. [CrossRef]
12. Comas, L.H.; Trout, T.J.; DeJonge, K.C.; Zhang, H.; Gleason, S.M. Water productivity under strategic growth stage-based deficit irrigation in maize. *Agric. Water Manag.* **2019**, *212*, 433–440. [CrossRef]
13. Dass, A.; Bhattacharyya, R. Wheat residue mulch and anti-transpirants improve productivity and quality of rainfed soybean in semi-arid north-Indian plains. *Field Crops Res.* **2017**, *210*, 9–19. [CrossRef]
14. Alvar-Beltrán, J.; Heureux, A.; Soldan, R.; Manzanas, R.; Khan, B.; Dalla Marta, A. Assessing the impact of climate change on wheat and sugarcane with the AquaCrop model along the Indus River Basin, Pakistan. *Agric. Water Manag.* **2021**, *253*, 106909. [CrossRef]
15. Bantchina, B.B.; Qaswar, M.; Arslan, S.; Ulusoy, Y.; Gündoğdu, K.S.; Tekin, Y.; Mouazen, A.M. Corn yield prediction in site-specific management zones using proximal soil sensing, remote sensing, and machine learning approach. *Comput. Electron. Agric.* **2024**, *225*, 109329. [CrossRef]
16. Singh, S.C.; Nenavath, M.D.; Raj, T.; Singh, V. Comparison of Indian corn with The US and South American commodity corn for wet milling starch yields. *Ind. Crops Prod.* **2024**, *221*, 119444. [CrossRef]
17. Chen, F.J.; Liu, J.C.; Liu, Z.G.; Chen, Z.; Ren, W.; Gong, X.P.; Wang, L.F.; Cai, H.G.; Pan, Q.C.; Yuan, L.X. Breeding for high-yield and nitrogen use efficiency in maize: Lessons from comparison between Chinese and US cultivars. *Adv. Agron.* **2021**, *166*, 251–275. [CrossRef]
18. FAO. *A History of Farming Systems Research*; FAO: Rome, Italy, 2018; p. 403.
19. Tian, X.Y.; Engel, B.A.; Qian, H.Y.; Hua, E.; Sun, S.K.; Wang, Y.B. Will reaching the maximum achievable yield potential meet future global food demand? *J. Clean. Prod.* **2021**, *294*, 126285. [CrossRef]
20. Luo, Y.C.; Zhang, Z.; Li, Z.Y.; Chen, Y.; Zhang, L.L.; Cao, J.; Tao, F.L. Identifying the spatiotemporal changes of annual harvesting areas for three staple crops in China by integrating multi-data sources. *Environ. Res. Lett.* **2020**, *15*, 074003. [CrossRef]
21. Gao, J.; Liu, Z.; Wang, P.; Huang, S.B. Drip irrigation coupled with appropriate N input increased maize (*Zea mays* L.) yield and lodging resistance via optimizing root and stem trait. *Eur. J. Agron.* **2024**, *160*, 127298. [CrossRef]
22. Schauberger, B.; Archontoulis, S.; Arneth, A.; Balkovic, J.; Ciais, P.; Deryng, D.; Elliott, J.; Folberth, C.; Khabarov, N.; Müller, C. Consistent negative response of US crops to high temperatures in observations and crop models. *Nat. Commun.* **2017**, *8*, 13931. [CrossRef]
23. Jägermeyr, J.; Pastor, A.; Biemans, H.; Gerten, D. Reconciling irrigated food production with environmental flows for sustainable gevelopment goals implementation. *Nat. Commun.* **2017**, *8*, 15900. [CrossRef]
24. Hamoud, Y.A.; Guo, X.P.; Wang, Z.C.; Shaghaleh, H.; Chen, S.; Hassan, A.; Bakour, A. Effects of irrigation regime and soil clay content and their interaction on the biological yield, nitrogen uptake and nitrogen-use efficiency of rice grown in southern China. *Agric. Water Manag.* **2019**, *213*, 934–946. [CrossRef]
25. Zhang, W.; Dong, A.; Liu, F.; Niu, W.; Siddique, K.H. Effect of film mulching on crop yield and water use efficiency in drip irrigation systems: A meta-analysis. *Soil Tillage Res.* **2022**, *221*, 105392. [CrossRef]

26. Li, H.; Mei, X.; Wang, J.; Huang, F.; Hao, W.; Li, B. Drip fertigation significantly increased crop yield, water productivity and nitrogen use efficiency with respect to traditional irrigation and fertilization practices: A meta-analysis in China. *Agric. Water Manag.* **2021**, *244*, 106534. [CrossRef]
27. Wang, J.; Du, G.; Tian, J.; Jiang, C.; Zhang, Y.; Zhang, W. Mulched drip irrigation increases cotton yield and water use efficiency via improving fine root plasticity. *Agric. Water Manag.* **2021**, *255*, 106992. [CrossRef]
28. Han, S.; Jiang, S.; Xiong, R.; Shafique, K.; Zahid, K.; Wang, Y. Response and tolerance mechanism of food crops under high temperature stress: A review. *Braz. J. Biol.* **2022**, *82*, e253898. [CrossRef] [PubMed]
29. Tolimir, M.; Gajić, B.; Kresović, B.; Životić, L.; Gajić, K.; Brankov, M.; Todorovic, M. Impact of deficit irrigation and planting density on grain yield and water productivity of maize grown under temperate continental climatic conditions. *Agric. Water Manag.* **2024**, *302*, 109009. [CrossRef]
30. Jia, Q.M.; Sun, L.F.; Ali, S.; Liu, D.H.; Zhang, Y.; Ren, X.L.; Zhang, P.; Jia, Z.K. Deficit irrigation and planting patterns strategies to improve maize yield and water productivity at different plant densities in semi-arid regions. *Sci. Rep.* **2017**, *7*, 13881. [CrossRef] [PubMed]
31. Fu, Z.; Ciais, P.; Wigneron, J.P.; Gentine, P.; Feldman, A.F.; Makowski, D.; Viovy, N.; Kemanian, A.R.; Goll, D.S.; Stoy, P.C. Global critical soil moisture thresholds of plant water stress. *Nat. Commun.* **2024**, *15*, 4826. [CrossRef]
32. Castro-Valdecantos, P.; Puértolas, J.; Albacete, A.; Dodd, I.C. Girdling changes root and shoot hormonal balance but does not alter drought-induced stomatal closure in soybean. *Environ. Exp. Bot.* **2021**, *192*, 104657. [CrossRef]
33. Tognetti, R.; Giovannelli, A.; Lavini, A.; Morelli, G.; Fragnito, F.; d'Andria, R. Assessing environmental controls over conductances through the soil-plant-atmosphere continuum in an experimental olive tree plantation of southern Italy. *Agric. Meteorol.* **2009**, *149*, 1229–1243. [CrossRef]
34. Mumithrakamatchi, A.K.; Alagarswamy, S.; Anitha, K.; Djanaguiraman, M.; Kalarani, M.K.; Swarnapriya, R.; Marimuthu, S.; Vellaikumar, S.; Kanagarajan, S. Melatonin imparts tolerance to combined drought and high-temperature stresses in tomato through osmotic adjustment and ABA accumulation. *Front. Plant Sci.* **2024**, *15*, 1382914. [CrossRef] [PubMed]
35. Wang, H.B.; Cao, H.; Jiang, F.C.; Wang, X.P.; Gao, Y. Analysis of soil moisture, temperature, and salinity in cotton field under non-mulched drip irrigation in South Xinjiang. *Agriculture* **2022**, *12*, 1589. [CrossRef]
36. Li, Y.; Huang, G.H.; Chen, Z.J.; Xiong, Y.W.; Huang, Q.Z.; Xu, X.; Huo, Z.L. Effects of irrigation and fertilization on grain yield, water and nitrogen dynamics and their use efficiency of spring wheat farmland in an arid agricultural watershed of Northwest China. *Agric. Water Manag.* **2022**, *260*, 107277. [CrossRef]
37. Siphiwe, N.G.; Magyar, T.; Tamás, J.; Nagy, A.; Tamás, M.; János, T.; Attila, N. Modelling Soil Moisture Content with Hydrus 2D in a Continental Climate for Effective Maize Irrigation Planning. *Agronomy* **2024**, *14*, 1340. [CrossRef]
38. Brar, H.S.; Singh, P. Pre-and post-sowing irrigation scheduling impacts on crop phenology and water productivity of c-otton (*Gossypium hirsutum* L.) in sub-tropical north-western India. *Agric. Water Manag.* **2022**, *274*, 107982. [CrossRef]
39. Feng, W.; Sheng, S.J.; Gui, L.Z.; Feng, N.H.; Man, Q.X.; Jun, S.H. Effect of different irrigation scheduling on salt distribution and leaching in cotton field. *Trans. Chin. Soc. Agric. Mach.* **2013**, *44*, 120–127. [CrossRef]
40. Wang, H.; Fan, J.; Fu, W.; Du, M.; Zhou, G.; Zhou, M.; Hao, M. Good harvests of winter wheat from stored soil water and improved temperature during fallow period by plastic film mulching. *Agric. Water Manag.* **2022**, *274*, 107910. [CrossRef]
41. Wang, P.; Deng, Y.J.; Li, X.Y.; Wei, Z.W.; Hu, X.; Tian, F.; Wu, X.C.; Huang, Y.M.; Ma, Y.J.; Zhang, C.C. Dynamical effects of plastic mulch on evapotranspiration partitioning in a mulched agriculture ecosystem: Measurement with numerical modeling. *Agric. Meteorol.* **2019**, *268*, 98–108. [CrossRef]
42. Sun, T.; Li, G.; Ning, T.Y.; Zhang, Z.M.; Mi, Q.H.; Lal, R. Suitability of mulching with biodegradable film to moderate soil temperature and moisture and to increase photosynthesis and yield in peanut. *Agric. Water Manag.* **2018**, *208*, 214–223. [CrossRef]
43. Busari, M.A.; Bankole, G.O.; Adiamo, I.A.; Abiodun, R.O.; Ologunde, O.H. Influence of mulch and poultry manure application on soil temperature, evapotranspiration and water use efficiency of dry season cultivated okra. *Int. Soil Water Conserv. Res.* **2023**, *11*, 382–392. [CrossRef]
44. Allen, R.G.; Pereira, L.S.; Raes, D.; Smith, M.J.F. Crop evapotranspiration-Guidelines for computing crop water requirements-FAO Irrigation and drainage paper 56. *Food Agric. Organ. United Nations* **1998**, *300*, D05109.
45. Zhang, X.M.; Wang, N.; Cao, L.S.; Ran, B.; Wang, W.K.; Xiao, Y.; Zhang, Z.Y.; Xu, D.; Wang, Z.F. Analysis of the contribution of rainfall to recharge in the Mu Us Desert (China) based on lysimeter data. *Hydrogeol. J.* **2024**, *32*, 279–288. [CrossRef]
46. Fan, Z.L.; Chai, Q.; Huang, G.B.; Yu, A.Z.; Huang, P.; Yang, C.H.; Tao, Z.Q.; Liu, H.L. Yield and water consumption characteristics of wheat/maize intercropping with reduced tillage in an Oasis region. *Eur. J. Agron.* **2013**, *45*, 52–58. [CrossRef]
47. Liu, M.H.; Shi, H.B.; Paredes, P.; Ramos, T.B.; Dai, L.P.; Feng, Z.Z.; Pereira, L.S. Estimating and partitioning maize evapotranspiration as affected by salinity using weighing lysimeters and the SIMDualKc model. *Agric. Water Manag.* **2022**, *261*, 107362. [CrossRef]
48. Liu, B.; Han, H.H.; Liu, X.C.; Li, C.; Chen, X.P.; Wu, H.; Luo, Y.F.; Cui, Y.L. Quantifying the effects of advection on single crop coefficients over a humid paddy field for sustainable irrigation. *J. Hydrol.* **2022**, *614*, 128552. [CrossRef]
49. Pereira, L.S.; Paredes, P.; Melton, F.; Johnson, L.; Wang, T.; López-Urrea, R.; Cancela, J.J.; Allen, R.G. Prediction of crop coefficients from fraction of ground cover and height. Background and validation using ground and remote sensing data. *Agric. Water Manag.* **2020**, *241*, 106197. [CrossRef]

50. Chakilu, G.G.; Sándor, S.; Zoltán, T.; Phinzi, K. The patterns of potential evapotranspiration and seasonal aridity under the change in climate in the upper Blue Nile basin, Ethiopia. *J. Hydrol.* **2024**, *641*, 131841. [CrossRef]
51. Sharma, A.; Surkar, P.P.; Khare, R.; Choudhary, M.K.; Prasad, V. Quantifying the irrigation requirements for major crops under the influence of climate change in a Semi-Arid Region. *Water Resour. Manag.* **2024**, *38*, 5645–5660. [CrossRef]
52. Liu, Y.Q.; Gu, J.; Ma, N.N.; Li, X.; Yin, G.H.; Sun, S.J. Optimizing spring maize growth and yield through balanced irrigation and nitrogen application: A TOPSIS method approach. *Agronomy* **2024**, *14*, 1825. [CrossRef]
53. Aydinsakir, K.; Erdal, S.; Buyuktas, D.; Bastug, R.; Toker, R. The influence of regular deficit irrigation applications on water use, yield, and quality components of two corn (*Zea mays* L.) genotypes. *Agric. Water Manag.* **2013**, *128*, 65–71. [CrossRef]
54. Yan, F.L.; Zhang, F.C.; Fan, X.K.; Fan, J.L.; Wang, Y.; Zou, H.Y.; Wang, H.D.; Li, G.D. Determining irrigation amount and fertilization rate to simultaneously optimize grain yield, grain nitrogen accumulation and economic benefit of drip-fertigated spring maize in northwest China. *Agric. Water Manag.* **2021**, *243*, 106440. [CrossRef]
55. Li, Z.J.; Zou, H.Y.; Lai, Z.L.; Zhang, F.C.; Fan, J.L. Optimal drip fertigation regimes improved soil micro-environment, root growth and grain yield of spring maize in arid northwest China. *Agronomy* **2023**, *13*, 227. [CrossRef]
56. Li, C.; Luo, X.Q.; Wang, N.J.; Wu, W.J.; Li, Y.; Quan, H.; Zhang, T.B.; Ding, D.Y.; Feng, H. Transparent plastic film combined with deficit irrigation improves hydrothermal status of the soil-crop system and spring maize growth in arid areas. *Agric. Water Manag.* **2022**, *265*, 107536. [CrossRef]
57. Huang, X.; Zhao, Y.; Guo, T.K.; Mao, X.M. Enhancing SWAP simulation accuracy via assimilation of leaf area index and soil moisture under different irrigation, film mulching and maize varieties conditions. *Comput. Electron. Agric.* **2024**, *218*, 108625. [CrossRef]
58. Yaghi, T.; Arslan, A.; Naoum, F. Cucumber (*Cucumis sativus* L.) water use efficiency (WUE) under plastic mulch and drip irrigation. *Agric. Water Manag.* **2013**, *128*, 149–157. [CrossRef]
59. Wang, H.; Yin, Z.; Zhang, L.; Zhao, F.; Huang, W.; Wang, X.; Gao, Y. Irrigation modulates the effect of increasing temperatures under climate change on cotton production of drip irrigation under plastic film mulching in southern Xinjiang. *Front. Plant Sci.* **2022**, *13*, 1069190. [CrossRef] [PubMed]
60. Qian, Z.Q.; Hong, L.G.; Mei, X.Y.; Yu, Y.J.; Hong, Z.Y. Effect of water and nitrogen on the yield and quality of forage rape grown after wheat in South Xinjiang. *Chin. J. Eco-Agric.* **2019**, *27*, 1033–1041. [CrossRef]
61. Xiao, C.; Zou, H.Y.; Fan, J.L.; Zhang, F.C.; Li, Y.; Sun, S.K.; Pulatov, A. Optimizing irrigation amount and fertilization rate of drip-fertigated spring maize in northwest China based on multi-level fuzzy comprehensive evaluation model. *Agric. Water Manag.* **2021**, *257*, 107157. [CrossRef]
62. Ahmed, A.M.; Wang, Q.J.; Western, A.W.; Graham, T.D.; Wu, W.Y. Monthly disaggregation of annual irrigation water demand in the southern Murray Darling Basin. *Agric. Water Manag.* **2024**, *302*, 108961. [CrossRef]
63. Bonachela Castaño, S.; Fernández Fernández, M.D.; Hernández Rodríguez, J.; Karaca, C. Computing air temperature and humidity for reference crop evapotranspiration calculation in passive Mediterranean greenhouses. *Agric. Water Manag.* **2024**, *302*, 108991. [CrossRef]
64. Wang, H.B.; Li, G.H.; Huang, W.X.; Li, Z.Y.; Wang, X.P.; Gao, Y. Compensation of cotton yield by nitrogen fertilizer in non-mulched fields with deficit drip irrigation. *Agric. Water Manag.* **2024**, *298*, 108850. [CrossRef]
65. Hamani, A.K.M.; Abubakar, S.A.; Wang, G.S.; Kama, R.; Kurihara, T.; Liu, K.; Gao, Y.; Duan, A.W. Optimized application of combined nitrogen and microbial decomposing inoculants increases wheat (*Triticum aestivum* L.) physiological growth and mitigates global warming potential under different water regimes. *Environ. Exp. Bot.* **2023**, *206*, 105170. [CrossRef]

*plants*

MDPI

*Article*

# Growth, Evapotranspiration, Gas Exchange and *Chl a* Fluorescence of Ipê-Rosa Seedlings at Different Levels of Water Replacement

**Kalisto Natam Carneiro Silva [1], Andréa Carvalho da Silva [1,2,*], Daniela Roberta Borella [3], Samuel Silva Carneiro [1], Leonardo Martins Moura dos Santos [1], Matheus Caneles Batista Jorge [1], Beatriz Feltrin Magosso [1], Mariana Pizzatto [1] and Adilson Pacheco de Souza [1,3]**

[1] Institute of Agrarian and Environmental Sciences, Federal University of Mato Grosso, Sinop 78550-728, MT, Brazil; kalisto.silva@sou.ufmt.br (K.N.C.S.); samuel.carneiro@sou.ufmt.br (S.S.C.); leonardo.santos@sou.ufmt.br (L.M.M.d.S.); matheus.jorge@sou.ufmt.br (M.C.B.J.); beatriz.magosso@sou.ufmt.br (B.F.M.); mariana.pizzatto@ufmt.br (M.P.); adilson.souza@ufmt.br (A.P.d.S.)

[2] Postgraduate Program in Agronomy, Federal University of Mato Grosso, Sinop 78550-728, MT, Brazil

[3] Postgraduate Program in Environmental Sciences, Federal University of Mato Grosso, Sinop 78550-728, MT, Brazil; daniela.borella@sou.ufmt.br

* Correspondence: andrea.silva@ufmt.br; Tel.: +55-(66)-98105-4422

**Abstract:** In general, young plants in the establishment phase demonstrate sensitivity to changes in environmental conditions, especially regarding water availability. The effects of the seasonality of biophysical processes on plant physiology can trigger differential responses, even within the same region, making it necessary to conduct studies that characterize the physiological performance of the species at different spatial and temporal scales, making it possible to understand their needs and growth limits under water stress conditions. This paper aimed to evaluate the growth, gas exchange and *Chl a* fluorescence in ipê-rosa seedlings subjected to levels of water replacement (LWRs) of 100, 75, 50 and 25% in a greenhouse. The morphometric variables of plant height, diameter at stem height, numbers of leaves and leaflets, root length and volume, plant dry mass and leaf area were evaluated. The potential evapotranspiration of seedlings (ETc) was obtained using direct weighing, considering the water replacement of 100% of the mass variation between subsequent days as a reference; the cultivation coefficients (kc) were obtained using the ratio between ETc and the reference evapotranspiration (ETo) obtained by the Penman–Monteith FAO-56 method. Biomass and evapotranspiration data were combined to determine water sensitivity. Diurnal fluxes of gas exchange (net photosynthesis rate, transpiration rate, stomatal conductance, internal and atmospheric carbon ratio, water use efficiency and leaf temperature) and *Chl a* fluorescence (Fv/Fm, $\Phi_{PSII}$, ETR, Fv′/Fm′, NPQ and qL) were evaluated. Water restriction caused reductions of 90.9 and 84.7% in the increase in height and diameter of seedlings subjected to 25% water replacement when compared to seedlings with 100% water replacement. In comparison, biomass accumulation was reduced by 96.9%. The kc values increased throughout the seedling production cycle, ranging from 0.59 to 2.86. Maximum water sensitivity occurred at 50% water replacement, with Ky = 1.62. Maximum carbon assimilation rates occurred in the morning, ranging from 6.11 to 12.50 μmol $m^{-2}$ $s^{-1}$. Ipê-rosa seedlings regulate the physiology of growth, gas exchange and *Chl a* fluorescence depending on the amount of water available, and only 25% of the water replacement in the substrate allows the seedlings to survive.

**Keywords:** ipê-rosa - *Handroanthus impetiginosus* (Mart. ex DC.) Mattos; native tree; irrigation management; water sensitivity; photosynthesis; photochemical stress

## 1. Introduction

*Handroanthus impetiginosus* (Mart. ex DC.) Mattos, known as ipê-rosa, is a species that belongs to the Bignoniaceae family; it has a wide geographic distribution throughout the Brazilian tropical forests (Amazon Rainforest, Atlantic Forest, Caatinga, Cerrado and Pantanal) and the transitions between biomes [1–3]. This forest species is native to Brazil and has phenotypic and genotypic characteristics that are favorable for sustainable development in terms of environmental, economic and social aspects, which makes the conservation of this and many other forest species essential.

In the silvicultural field, *H. impetiginosus* is indicated for the recovery of degraded areas and insertion in the landscaping of urban areas [4–6] and in synergistic processes, such as the cycling of nutrients and carbon in natural ecosystems [7,8]. Furthermore, this tree's importance is determined by its stem's favorable characteristics and the average size of its trunk, with heights ranging from 8 to 30 m [1]. In addition, due to the high density of the wood (specific mass of 0.96 g $cm^{-3}$), it is resistant to attack by insects and other pathogens, and it is used in the furniture industry and civil construction [1,2,9]. Studies claim that *H. impetiginosus* is also used in folk medicine in Brazil and other countries in South and Central America for the treatment of edema, arthritis, diuretics and infections [10,11], as the extracts and compounds isolated from its leaves have multiple pharmacological activities [12]. Furthermore, extracts and compounds based on ipê-rosa leaves are widely used to treat cancer, as they have anti-inflammatory, antioxidant and anticancer properties [13,14].

From another perspective, *H. impetiginosus* seedlings show tolerance to water deficit, a recurring environmental characteristic in degraded environments, expressing tolerance mechanisms to maintain high water potential in their tissues, such as leaf senescence and reductions in stem growth (diameter and height) and total biomass production. Furthermore, over time, they adjust their morphophysiology to stress by emitting leaves with reduced area and a greater allocation of photoassimilates to the root system [14,15].

However, the effects of the seasonality of biophysical processes on plant physiology can trigger differential responses, even within the same region [16]. In general, young plants in the establishment phase demonstrate sensitivity to changes in their growth environment; discrepancies in solar radiation levels, water availability [17,18], salinity, temperature and relative humidity of the air imprint differential responses in the plant metabolic rhythm, which can alter the photoassimilate regime (photochemical potential and biomass accumulation) [17,19,20]. Furthermore, depending on the level of environmental stress, these individuals tend to modify their biomolecular, anatomical and morphological structure beyond survival [18,21,22]. Therefore, studies that characterize the physiological performance of species at different spatiotemporal scales are essential, making it possible to understand their needs and growth limits under water stress conditions.

In this context of water availability, there is the Cerrado–Amazon transition region (transition between savanna and forest formations), which occupies an area of approximately 152,180 $km^2$ and is appreciated for its wealth of natural resources, in addition to housing several important springs for the formation of large Brazilian hydrographic basins [23,24]. The resilience of the natural vegetation of this region (Cerrado–Amazon transition) to abiotic factors has sparked interest in understanding the morphophysiological behavior of native forest plants in response to stresses caused by water and salinity stress [25], irrigation cycles [15,23], nutrient availability [25–27], high temperatures and qualitative and quantitative variations in solar radiation [28–31], among other studies [3,32]. Despite the importance of this region of vegetation and climate transition for environmental and socioeconomic aspects, human actions without planning and adequate management of natural resources associated with climate change have generated negative impacts on these environments, such as the qualitative and quantitative loss of biodiversity and, consequently, an imbalance of plant functions in natural ecosystems. Therefore, it is necessary to carry out ecophysiological studies under abiotic stresses, such as water availability for plants, which will allow the production of information to support planning and strategic actions for the conservation of flora biodiversity in natural ecosystems.

In this sense, in this paper, we evaluated ipê-rosa seedlings in a greenhouse in the Cerrado–Amazon transition, establishing the following objectives: (i) to identify the effect of water restriction on the growth rate and final accumulation of biomass of the seedlings; (ii) to determine the accumulated crop evapotranspiration (ETc) and the cultivation coefficients (kc); (iii) to compare the water sensitivity of this species to the reduction in the amount of water available in the substrate; (iv) to verify how the gas exchange of ipê-rosa leaves responds to variations in the daily flux; (v) to determine the availability of water necessary to regulate the leaves of ipê-rosa seedlings outside the condition of photochemical stress; (vi) to verify how the *Chl a* fluorescence of ipê-rosa leaves responds to variations in the daily flux of global radiation in protected plant production environments.

## 2. Material and Methods

### 2.1. Experimental Site

The experiment was conducted for 107 days between August and December 2019, in a greenhouse in Sinop (Cerrado–Amazon transition), Mato Grosso, Brazil (11°51″51.453′ S, 55°29″6.608′ W; average altitude of 380 m) (Figure 1). This plant production structure (greenhouse) was oriented to the northeast, formed by a curved roof, with a 120-micron LDPE transparent plastic covering, and, on the sides, black polyethylene screens with 50% attenuation of solar radiation; the dimensions were 12.0 × 6.0 × 4.5 m (length, width and height), with three wooden benches inside, measuring 8.0 × 0.5 × 1.3 m (length, width and height).

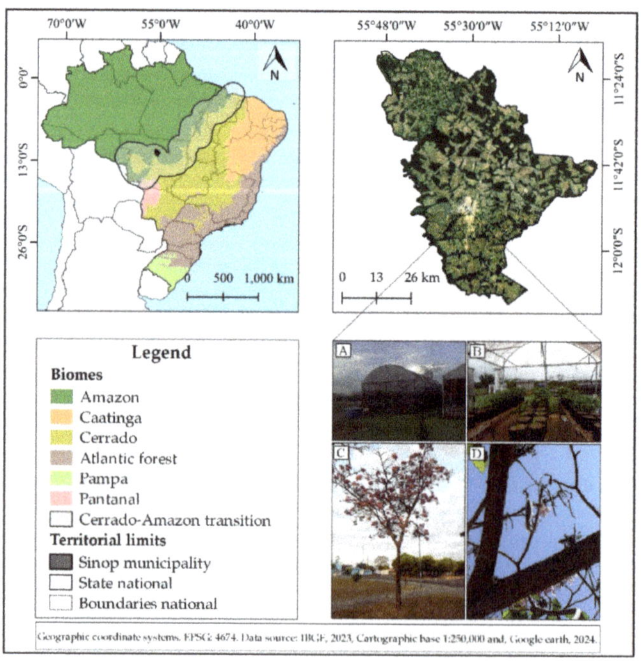


**Figure 1.** Location map of the collection region of *Handroanthus impetiginosus* seeds, Sinop, Mato Grosso state, Brazil. (**A**) plastic greenhouse for producing forestry agricultural seedlings; (**B**) view of seedlings in pots; (**C**) adult ipê-rosa tree in bloom; (**D**) ipê-rosa pods and seeds.

According to the Köppen classification, the region's climate is Aw, a hot and humid tropical savanna. Two water regimes are established during the year: dry (May to September) and rainy (October to April). Average monthly temperatures vary from 24 to 27 °C. Annual precipitation and reference evapotranspiration are 1970 and 1330 mm, respectively [33].

### 2.2. Obtaining and Germinating Seeds of Handroanthus Impetiginosus

The ipê-rosa pods were collected from urban afforestation matrix trees and dispersed in different points of public areas in the urban perimeter of Sinop, Mato Grosso state (Figure 1). Trees with similar dendrometric characteristics, with straight stems and globose crowns, were selected; the pods (fruits) were collected when ripe and with minimal incidence of pathogens. Under the region's environmental conditions, this species presents the flowering and seed dispersal phenophases between July and August. After collection, the pods were processed, and the seeds with good phytosanitary aspects were stored in a natural Kraft paper bag and kept in the refrigerator (average temperature of 12 °C, relative humidity of 10%) until the beginning of the experiment.

The ipê-rosa seeds underwent aseptic treatment in a 2% sodium hypochlorite (NaClO) solution for five minutes; subsequently, they were distributed in sets of 100 units within ten transparent plastic boxes measuring 20 × 20 cm (adapted gearboxes) with Germitest® paper moistened with distilled water; these seeds were sprayed with a 2% Protreat® fungicide solution. For 23 days, the plastic boxes were kept in a BOD germination chamber at a constant temperature of 30 °C and a 12 h photoperiod to enable the standardization of the germination process. To ensure the germination viability of the seed batch, the number of germinated seeds (those with radicle protrusion) was counted daily for 21 days, thus allowing the determination of the following variables: germination speed index (GSI) of 10.12 ± 1.3%; average seedling height of 3.08 ± 0.87 cm; stem basal diameter average of the seedling of 1.20 ± 0.09 mm; average length of the seedling root system of 3.1 ± 1.21 cm; average volume of the seedling root system of 0.49 ± 0.09 $cm^3$; total dry mass of seedlings of 0.027 ± 0.01 g.

### 2.3. Transplanting and Substrates for Seedling Production

After seedling formation (complete expansion of the pair of cotyledons leaves), a screening was performed to ensure the standardization of seedling size; subsequently, these seedlings were transplanted into 8 L plastic pots (with an effective measurement of 7.2 L of soil). The containers were filled with a mixture composed of a 2:1 ratio (forest soil/commercial substrate—the forest soil used was considered to be the surface layer (up to 0.10 m deep) of a dystrophic Red-Yellow Latosol occupied by native vegetation; it was sieved and disinfected, while the substrate consisted of a mixture of biostabilized pine bark, vermiculite, charcoal powder, water, phenolic foam and additives (fertilizers and soil improvers). Following the nutritional recommendations of Borella et al. [30] for this same substrate composition, 40, 380 and 80 g of N, $P_2O_5$ and $K_2O$ (in the forms of urea, simple superphosphate and potassium chloride) were incorporated per $m^3$ of the substrate, respectively.

### 2.4. Substrate Moisture and Application of Treatments

Before the application of treatments related to water replacement of ipê-rosa seedlings, the water retention capacity of the substrate was determined, according to Borella et al. [30]. In this case, the masses of empty 820 $cm^3$ tubes filled with the soil mixture above were determined; subsequently, the tubes were distributed in trays with water to saturate the soil through the capillary rise process, and the upper opening of the tubes was covered with plastic film to prevent water evaporation; after 48 h, the tubes were subjected to drainage of excess water, indicating that the substrate had maximum water retention capacity); then, the substrate samples were removed from the tubes, placed in aluminum foil trays, weighed (obtaining the wet weight), and kept in forced circulation ovens at 100 °C until a constant mass was obtained, to obtain the dry weight; the samples were weighed using a thousandths analytical balance (0.001 g) (BEL Engineering Ltda. company, Monza, Italy).

for computing ETo were obtained according to the procedure described by Allen et al. [34]. The daily ETc and $ET_0$ values ratio was used to obtain the crop coefficients (kc). Water sensitivity was evaluated through the coefficient "Ky" for seedlings [29] at 25, 50 and 75% LWRs (Equation (2)); for this purpose, they were considered as references for maximum evapotranspiration (ETm) and potential yield (Ym)—corresponding to the total dry mass of seedlings—in the 100% LWR.

$$Ky = \left[\left(1 - \frac{Yr}{Ym}\right) / \left(1 - \frac{ETr}{ETm}\right)\right] \tag{2}$$

where Ky = water sensitivity; Yr = actual dry mass yield (g $m^{-2}$); Ym = maximum dry mass yield in the treatment with 100% water replacement (g $m^{-2}$); ETr = actual accumulated evapotranspiration (mm); ETm = maximum accumulated evapotranspiration in the treatment with 100% LWR (mm, representing $ET_C$).

### 2.7. Seedling Growth Analysis

Non-destructive growth analyses were performed from 28/09/2019 to 13/12/2019, corresponding to 31, 59, 71, 87 and 107 days after transplanting (DATs); all plants in each experimental unit were measured in each evaluation. On these occasions, the variables analyzed were as follows: (i) seedling height—measured from the plant basal to the insertion of the apical bud with the aid of a graduated ruler (0.1 cm); (ii) basal stem diameter (BSD)—obtained at the height of the substrate with the aid of a centesimal digital caliper (0.01 mm); (iii) number of leaves; (iv) number of leaflets per seedling.

After the end of the experimental period, at 107 DATs, destructive analysis of the plants mentioned above was performed to obtain the following: (i) leaf area in $cm^2$—with a photoelectric meter (LI-3100C) (LI-COR Ltda. company, Lincoln City, OR, USA); (ii) root length—with the aid of a graduated ruler (0.1 cm); (iii) root volume—obtained with the aid of a graduated cylinder (0.1 mL), by immersing the washed roots in a known volume of water; (iv) dry matter masses of roots (DMR), stems (DMS), leaves (DML) and total (sum of partitions)—determined on an analytical balance with an accuracy of 0.0001 g (Mettler-Toledo Ltda. company, Barueri, Brazil) after the plant material was dried in a forced air circulation oven at a temperature of 65 °C ± 5 °C until a constant mass was obtained.

### 2.8. Gas Exchange and Fluorescence

The gas exchange evaluations in the leaves of ipê-rosa seedlings were performed on only one date, 100 days after transplanting, using a portable photosynthesis analyzer (IRGA), model LCi-SD from ADC BioScientific Ltda. company, Hoddesdon, UK. Three plants were selected for each treatment of 25, 50, 75 and 100% water replacement, and three readings were taken from each treatment using the median portion of the second or third adult leaf from the apical bud.

The readings were taken throughout the day at the following solar times: 09:30, 10:30 and 11:30 (a.m.); 12:30, 13:30, 14:30 and 16:00 h (p.m.). The leaves were adjusted inside the IRGA analyzer chamber, occupying an area of 6.25 $cm^2$. They were subjected to an effective pulse of 1839 μmol $m^{-2}$ $s^{-1}$ of light intensity to obtain the following variables: net photosynthesis rate—A (μmol $m^{-2}$ $s^{-1}$); transpiration rate—E (mol $m^{-2}$ $s^{-1}$); stomatal conductance—gs (mol $m^{-2}$ $s^{-1}$); internal carbon concentration—Ci (μmol $mol^{-1}$); carbon concentration in the external environment—Cref (μmol $mol^{-1}$); carbon variation between the external environment and analysis chamber—ΔC (μmol $mol^{-1}$); leaf temperature—TF (°C). Water use efficiency, EUA (μmol $CO_2$ $mol^{-1}$ $H_2O$), was obtained using the ratio between A and E; the relationship between internal carbon and atmospheric carbon—Ci/Ca—was given by the ratio between Ci and the difference between Cref and ΔC.

The fluorescence of *Chl a* in leaves of ipê-rosa seedlings was determined with an Os5p modulated light fluorometer (Brand: Opti-Sciences Ltda, Hudson City, NY, USA, Fluorpen, Model: FP-100). The protocols for the analyses were the Fv/Fm Protocol (Fv/Fm Protocol, measurements in the dark-adapted state) and Yield Protocol (Yield Protocol, measurements in the light-adapted state). The fluorescence evaluations occurred at 28, 66 and 90 DATs, for 5 plants per experimental unit, regardless of the treatment. The second or third adult leaf from the apical bud on the respective date was used to perform the readings.

*Chl a* fluorescence reading before dawn (predawn) was performed between 4:30 and 5:10 h using the Fv/Fm Protocol to obtain the photochemical performance of PSII after the complete regeneration of protein complexes during the night, as recommended by Elsheery and Cao [36]. The fiber optic bundle was placed against the leaf tissue using plastic clips to emit a pulse of modulated light incapable of triggering photochemical chains (0.05 μmol $m^{-2}$ $s^{-1}$) to measure the minimum fluorescence before dawn ($Fo_{predawn}$). At an interval of approximately 0.8 s, a pulse of saturating actinian light with an intensity greater than 5000 μmol $m^{-2}$ $s^{-1}$ was emitted to measure the maximum fluorescence before dawn ($Fm_{predawn}$). The interval between these two points provided the variable predawn fluorescence ($Fv_{predawn}$), used in calculating the maximum quantum yield of PSII before dawn ($Fv/Fm_{predawn}$), according to Equation (3).

$$\mathrm{Fv/Fm} = \frac{\mathrm{Fm} - \mathrm{Fo}}{\mathrm{Fm}} \tag{3}$$

where Fv/Fm = maximum quantum yield of PSII; Fm = maximum fluorescence in the dark-adapted state; Fo = minimum fluorescence in the dark-adapted state.

After that, *Chl a* fluorescence analyses were performed throughout the day. The first assessment was performed between 6:30 a.m. and 5:30 p.m. at sequential one-hour intervals. In the two subsequent assessments, readings were interrupted at 1:30 p.m. These analyses, on the same plants and leaves as the predawn sampling, were performed according to the equipment protocols: dark-adapted and light-adapted states.

For the "dark-adapted state" protocol, the leaves were acclimated to the dark for 30 min. After the interval, the Fv/Fm Protocol was repeated using the same procedures described for predawn. Thus, using Equation (3), the maximum quantum yield of PSII (Fv/Fm) was determined on the hourly scale for the treatments in which each plant was a replicate.

The Yield Protocol was used in the "room light-adapted state" condition. For this purpose, the leaf tissue was exposed to a pulse of actinia light with a photosynthetically active photon flux (FFF) of approximately 560 μmol $m^{-2}$ $s^{-1}$ for 5 s to obtain the variable $F'$ (fluorescence in the light-adapted state). Subsequently, the leaf was subjected to progressive saturating actinia light pulses (5000, 10,000 and 15,000 μmol $m^{-2}$ $s^{-1}$), estimating the maximum fluorescence of the light-adapted state ($Fm'$) in a curve at infinity. Through this projection, the fluorometer provided the effective quantum yield value of PSII ($\Phi_{PSII}$), according to Equation (4).

$$\Phi_{\mathrm{PSII}} = \frac{\mathrm{Fm}' - \mathrm{F}'}{\mathrm{Fm}'} \tag{4}$$

where $\Phi_{PSII}$ = effective quantum yield of PSII; $F'$ = fluorescence in the light-adapted state; $Fm'$ = maximum fluorescence in the light-adapted state.

For fluorescence in the light-adapted state, three readings were taken on each leaf per plant at each hour of analysis. Using the variables above and the equations described by Oxborough and Baker [37], Lichtenthaler et al. [38] and Baker [39], the other variables related to Chl a fluorescence were estimated. The minimum fluorescence in the light-adapted state ($Fo'$), considering the complete oxidation of QA, in this case, through the transfer of electrons from PSII to PSI, was obtained according to Equation (5).

$$\mathrm{Fo}' = \frac{\mathrm{Fo}}{(\mathrm{Fv/Fm}) + (\mathrm{Fo/Fm}')} \tag{5}$$

where $Fo'$ = minimum fluorescence in the light-adapted state; Fo = minimum fluorescence in the dark-adapted state; Fv = variable fluorescence in the dark-adapted state (difference between Fm and Fo); Fm = maximum fluorescence in the dark-adapted state; Fm′ = maximum fluorescence in the light-adapted state.

Therefore, the following parameters were also obtained: (i) the maximum efficiency of PSII (Fv′/Fm′), which provides the potential capacity of PSII, with tissues acclimated to light conditions (Equation (6)); (ii) non-photochemical dissipation (NPQ), which characterizes the portion of photochemical energy extinguished by heat loss (Equation (7)); (iii) the open fraction of PSII reaction centers (qL), which provides the capacity of the photochemical apparatus to receive electrons considering the complete oxidation state of QA (Equation (8)).

$$\mathrm{Fv'/Fm'} = \frac{\mathrm{Fm'} - \mathrm{Fo'}}{\mathrm{Fm'}} \tag{6}$$

where Fv′/Fm′ = maximum efficiency of PSII; Fm′ = maximum fluorescence in the light-adapted state; Fo′ = minimum fluorescence in the light-adapted state.

$$\mathrm{NPQ} = \frac{\mathrm{Fm} - \mathrm{Fm'}}{\mathrm{Fm}} \tag{7}$$

where NPQ = non-photochemical quenching; Fm = maximum fluorescence in the dark-adapted state; Fm′ = maximum fluorescence in the light-adapted state.

$$\mathrm{qL} = \frac{\Delta\mathrm{F}}{(\mathrm{Fm'} - \mathrm{Fo'})} \times \frac{\mathrm{Fo'}}{\mathrm{F'}} \tag{8}$$

where qL = open fraction of PSII reaction centers; ΔF = difference between Fm′ and F′; Fm′ = maximum fluorescence in the light-adapted state; Fo′ = minimum fluorescence in the light-adapted state; F′ = fluorescence in the light-adapted state.

Finally, the electron transport rate (ETR) was estimated based on the photosynthetically active photon flux (FFF) of 560 μmol $m^{-2}$ $s^{-1}$, which provides the portion of the incident flux that is effectively carried to photochemical activity in the electron transport chain, according to Equation (9).

$$\mathrm{ETR} = 0.84 \times 0.50 \times \mathrm{FFF} \times \Phi_{\mathrm{PSII}} \tag{9}$$

where ETR = electron transport rate (μmol $m^{-2}$ $s^{-1}$); FFF = photosynthetically active photon flux of 560 μmol $m^{-2}$ $s^{-1}$; ΦPSII = effective quantum yield of PSII.

Concurrent with the readings with the fluorometer and portable photosynthesis analyzer (IRGA), instantaneous measurements of global radiation (with MP-200), photosynthetically active radiation (with MQ-200) and luminance (Instrutherm lux meter LD200) were performed.

### 2.9. Basal Temperatures and Thermal Sum

To evaluate the variables studied in this study, the thermal sum, in accumulated degree days (DDAs), was adopted as the independent variable to ensure greater applicability of the results to other regions where ipê-rosa occurs. To this end, estimates of the upper (TB) and lower (Tb) basal temperatures were made using the database belonging to the Plant and Environment Interactions Research Group, collected by Monteiro [40] for seedlings of this same species subjected to different shading conditions in the region's dry season (April to August); in this case, the ipê-rosa seedlings were produced in full sunlight conditions, covered with polyethylene screens in the colors blue, red and green with respective percentages of 53, 57 and 57% of attenuation of global radiation and black screens with 38, 52 and 83% of attenuation; these shading conditions generated different microclimates that interfered with seedling growth, allowing basal temperatures to be adjusted by relationships with leaf area and dry mass. Using daily data on maximum and minimum air temperatures throughout the experimental period and in each cultivation condition, TB and Tb values were obtained by polynomial regressions (second degree), whose independent variables (x) were, respectively, maximum and minimum air temperatures of each treatment (shading screens), associated in isolation with the dependent variables total average leaf area per seedling (y1) and total average dry mass (y2), resulting in four polynomial adjustments (Figure 3). The first derivative of the adjusted equations was used to establish TB and Tb, obtaining the respective maximum points of the functions and their corresponding TB and Tb estimated at 40.6 and 16.9 °C, respectively. After TB and Tb were defined, the thermal sum of the ipê-rosa seedlings in this study was determined using the methodology of Ometto [41], according to the recommendations of Souza et al. [42].

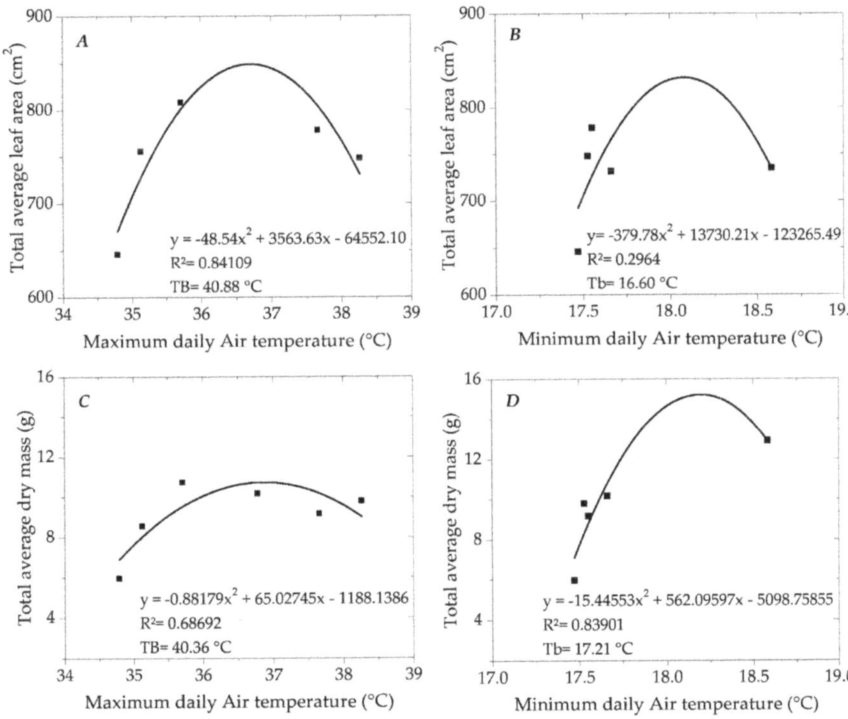


**Figure 3.** Correlation between maximum and minimum air temperatures and the total average leaf area (**A** and **B**, respectively) and total average dry mass (**C** and **D**, respectively) per plant to determine upper basal temperature (TB) and lower basal temperature (Tb), in Sinop, Mato Grosso state, Brazil. Prepared from the database assembled by Monteiro [40].

### *2.10. Data Analysis*

The data were subjected to a normality test and subsequently to an analysis of variance; when significant, the means were compared by the Tukey test at 5% probability in the Sisvar 5.6 software [43], with graphical guidelines and polynomial adjustments performed in the Origin 2018 software. The analyses of the growth, evapotranspiration, gas exchange and fluorescence variables were performed based on the accumulated thermal sum, to minimize the effects of the production environment on the growth rates and cell expansion of the plant.

To evaluate the non-destructive growth variables (height, diameter at the height of the stem, number of leaves and number of leaflets), to adjust the significant polynomials resulting from the regression analysis, the accumulated thermal sum at 31, 59, 71, 87 and 107 DATs corresponded to 508, 935, 1125, 1355 and 1655 DDAs, respectively.

Data were collected at 107 DATs (1655 DDAs) to evaluate the destructive growth variables (root length, root volume, leaf area and total dry mass). They were subjected to an analysis of variance (F test). If significant differences were found, the means were compared using the Tukey test at a 5% probability level.

The reference evapotranspiration and the crop under 100% water replacement conditions were analyzed in a grouped (cumulative) and isolated manner as a function of the thermal sum of the ipê-rosa seedlings in accumulated degree days (DDAs) on the dates of the above-mentioned non-destructive growth analyses. For evapotranspiration at 25, 50, and 75% LWRs, the data were analyzed only in a grouped manner, as well as a function of the thermal accumulation of the plants under ETc conditions. The crop coefficients (kc) were obtained daily and subjected to simple linear regression adjustments with the thermal sum as the independent variable. No applied statistical tests were performed regarding the water sensitivity coefficients (Ky).

For gas exchange, a completely randomized design (CRD) was implemented on an hourly scale, with measurements at 09:30, 10:30, 11:30, 12:30, 13:30, 14:30 and 16:00 h. Finally, the gas exchange variables above, mean air temperature and dPv (water vapor pressure deficit) were subjected to Pearson correlation analysis (r) separately within each fluid replacement level; the coefficients of these correlations were analyzed for significance using the T-test at 1 and 5% probability levels. *Chl a* fluorescence analyses were performed at 28, 66 and 90 DATs, corresponding to 464, 1046 and 1400 DDAs, respectively. For the predawn evaluation of *Chl* a fluorescence, a CRD design (completely randomized, with 5 replicates per experimental unit) was adopted. The analyses of *Chl* a fluorescence during the day is arranged in graphical form, with the average values obtained every two hours. Therefore, the results of 06:30 and 07:30 h were grouped in the averages of 07:00, the following averages of 09:00 h were obtained with the results of 08:30 and 09:30 h, and so on. Thus, in the first evaluation, at 28 DATs (accumulation of 463.6 DDAs), the average values were ordered in the following time scale: 07:00, 09:00, 11:00, 13:00, 15:00 and 17:00 h, while for the other dates, 66 and 90 DATs (accumulations of 1046.1 and 1400.0 DDAs, respectively), the points stopped at 13:00 h.

Only for the diurnal flows of the evaluations at 66 and 90 DATs, analogous to gas exchange, the *Chl* a fluorescence, mean air temperature and dPv variables were also subjected to Pearson correlation analysis (r). Furthermore, the Fv/Fm information was extracted to establish the simple linear regression adjustments as a function of the mean air temperature and dPv, following the recommendations of Marenco et al. [44]. For this purpose, Fv/Fm was the dependent variable (y) while, in isolation, mean air temperature and dPv were the independent variables (x). Eight equations were evaluated, given the division into two independent variables—TMair and dPv—and four water replacement levels, considering significance by analysis of variance (F test) at the 1 and 5% probability levels.

## 3. Results

### 3.1. Microclimate Dynamics Throughout the Experimental Period

The microclimatic dynamics throughout the experimental period in the greenhouse are arranged according to the accumulated thermal sum of the ipê-rosa seedlings (Figure 4), starting with transplanting (28 August 2019) and the last destructive growth analysis (13 December 2019), equivalent to 0.0 and 1655 accumulated degree days (DDAs). The experiment was carried out during the dry–rainy and rainy transition seasons of the mid-northern region of Mato Grosso (between August and December). During this period, the average values of temperature and relative humidity inside the greenhouse were 29.19 °C and 66.76%, with minimums and maximums of 18.60 to 47.00 °C and 16.00 to 98.60%, respectively (Figure 4A,B).

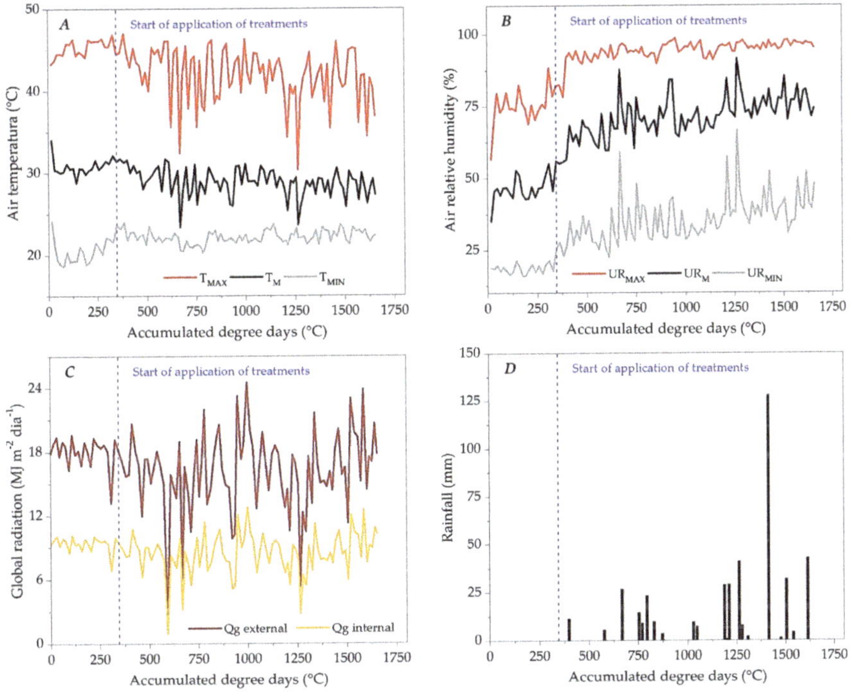


**Figure 4.** Daily variations in maximum ($T_{MAX}$), average ($T_M$) and minimum ($T_{MIN}$) air temperature (**A**) and maximum ($RH_{MAX}$), average ($RH_M$) and minimum ($RH_{MIN}$) relative air humidity (**B**) outside the greenhouse; global radiation outside and inside the greenhouse (**C**); and rainfall (**D**) as a function of the accumulated degree days of the ipê-rosa seedlings, between 28 August 2019 and 13 December 2019, in Sinop, Mato Grosso state, Brazil.

Inside the greenhouse, global radiation ranged from 0.95 to 11.81 MJ $m^{-2}$ $day^{-1}$, while in the outdoor environment (weather station), the variations were 1.83 and 24.52 MJ $m^{-2}$ $day^{-1}$. There was an accumulation of 438 mm of rainfall, concentrated in the final period of the experiment. The plastic covering of the greenhouse does not allow rainfall to occur inside the facility; however, variations in air temperature, relative humidity and internal global radiation show relationships with the influence of water vapor and cloudiness. At the beginning of the rainfall (at 29 DATs), inside the greenhouse, there was a reduction in the daily thermal amplitude of 4.95 °C, in addition a decrease of 2.11 °C in the average air temperature (Figure 4A); in the average daily relative humidity, there was an increase of 24.7% after the beginning of the rains (Figure 4B).

### 3.2. Growth of Ipê-Rosa Seedlings

The combination of water replacement levels and accumulated thermal sum (associated with the time after transplanting—DATs) significantly influenced the increase in height, diameter at stem height (DSH) and numbers of leaves and leaflets of ipê-rosa seedlings in the greenhouse (Figure 5). These findings have practical implications for seedling production, inspiring us to optimize our methods and improve the quality of our seedlings. Based on the adjusted equations (regressions), at the end of the cycle (thermal accumulation of 1655 DDAs (107 DATs)), the respective heights of 10.25, 30.43, 45.45 and 53.12 cm were estimated for ipê-rosa seedlings subjected to 25, 50, 75 and 100% LWRs; as for the diameter at stem height (DSH), in the same thermal sum, the values of 4.05, 11.24, 14.52 and 16.17 mm were obtained, at the abovementioned levels of water replacement.

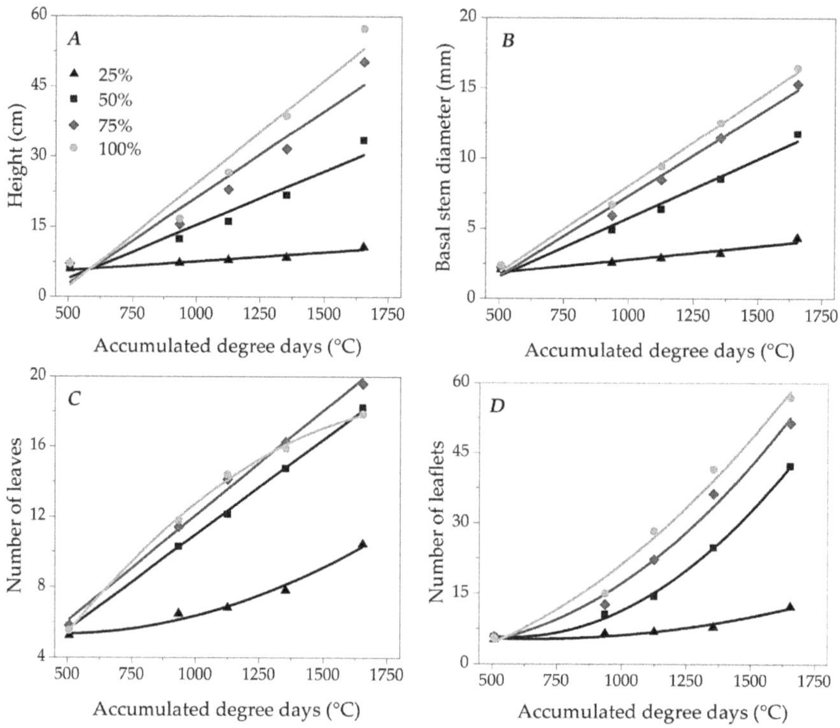


**Figure 5.** Adjusted regression curves for the non-destructive growth variables height (**A**), basal stem diameter (**B**), number of leaves (**C**) and number of leaflets (**D**) of ipê-rosa seedlings at 25, 50, 75 and 100% levels of water replacement (LWRs), as a function of the accumulated degree days, in a greenhouse.

These variables (seedling height and DAC) in all water levels presented adjustments by increasing linear equations (Figure 5 and Supplementary Table S1). The growth rates decreased with the increase in water restriction, evidenced by the angular coefficients (B1) of the regressions presented in Supplementary Table S1. Based on these regressions, at 100% LWR, the seedlings presented growth rates of 0.0441 cm $°C^{-1}$ and 0.0124 mm $°C^{-1}$ for height and DAC, respectively, whereas, with 25% LWR, the seedlings already showed rates of 0.0040 cm $°C^{-1}$ and 0.0019 mm $°C^{-1}$ for the same variables, resulting in reductions of 90.9 and 84.7% in the growth rate for each accumulated degree day.

The numbers of leaves and leaflets showed different behaviors depending on the thermal sum at different LWRs (Figure 5C,D). The seedlings subjected to 50 and 75% LWRs in the leaf count showed a linear increase. At 100% and 25% LWRs, the regressions of these morphometric variables showed quadratic behaviors with concavities downwards (number of leaves) and upwards (number of leaflets).

The gradual distance between the LWR responses led to mean differences between the curves of 25–50%, 50–75%, and 75–100% of 8.69, 6.46 and 3.29 cm for height, 3.07, 1.40 and 0.71 mm for diameter at stem height and 8.25, 6.40 and 3.89 for the number of leaflets, respectively (Figure 5A,B,D).

The leaf area and the allocation of aerial and root biomass were influenced by the levels of water replacement at the end of the production cycle of ipê-rosa seedlings [at 107 DATs, with an accumulation of 1655 DDAs] (Figure 6).

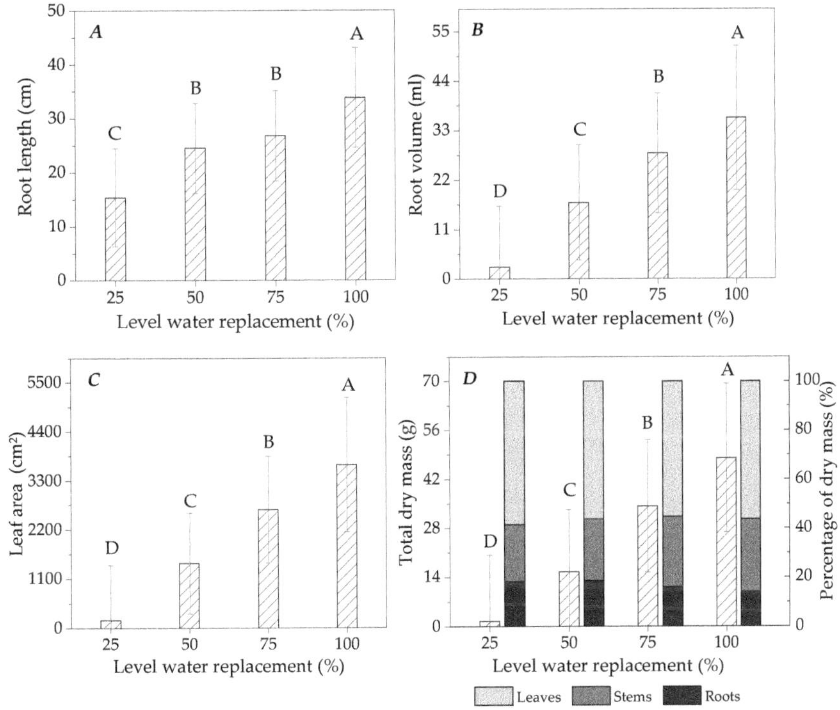


**Figure 6.** Mean values of root length (**A**), root volume (**B**), leaf area (**C**), total dry mass and percentages of dry mass partition of leaves, stems and roots (**D**) of ipê-rosa seedlings at 25, 50, 75 and 100% levels of water replacement (LWRs), 107 days after transplanting, in a greenhouse. Means followed by the same capital letter between water replacement levels do not differ by Tukey's test at the 5% probability level.

Among seedlings subjected to 100% and 25% LWRs, 54.4, 92.5, 95.2 and 96.9% reductions were observed in root length, root volume, leaf area and total dry mass, respectively.

The ipê-rosa seedlings showed average percentages of 56.6, 26.7 and 16.8% for the allocation of carbohydrates in the dry mass of leaves, stems and roots, respectively, with variations of up to 9.06% between the levels of water replacement (Figure 6).

### *3.3. Evapotranspiration, Crop Coefficients and Water Sensitivity (Ky)*

Daily evapotranspiration oscillated outside the microclimatic behavior shown in Figure 4. Before the onset of rainfall, at 24 DATs (398 DDAs), daily evapotranspiration of ipê-rosa seedlings ($ET_C$) varied between 2.55 and 6.79 mm day$^{-1}$; after this period, ETc oscillated between 1.64 and 9.89 mm day$^{-1}$; similar changes were also observed in reference evapotranspiration ($ET_0$) before and after the onset of rainfall (Figure 7A).

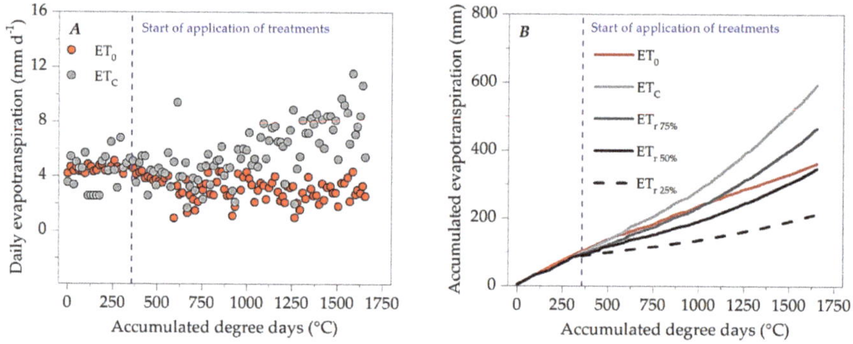


**Figure 7.** Daily evapotranspiration (**A**) and accumulated evapotranspiration (**B**) for the crop ($ET_C$: ipê-rosa seedlings, in a greenhouse and potential; $ET_0$: Penman–Monteith FAO-56 method [34]) as a function of the accumulated degree days of the seedlings, between 28 August 2019 and 13 December 2019, in Sinop, Mato Grosso state, Brazil. Real accumulated crop evapotranspiration (ETr) under 25, 50 and 75% levels of water replacement (LWRs) is represented by $ETr_{25\%}$, $ETr_{50\%}$ and $ETr_{75\%}$, respectively.

Young ipê-rosa plants subjected to 100% LWR accumulated 593.2 mm of ETc over 1655 DDAs; this water depth was 68.84% above the ETo in the same period—362.1 mm (Figure 7B). Until 33 DATs (542 DDAs), the ETc and $ET_0$ curves were close to each other; however, there was a gradual separation of ETc due to leaf expansion. At a water replacement level of 75%, the accumulation of water depth was 28.5% of $ET_0$, with 465.4 mm at 107 DATs. However, it was only after 68 DATs that the accumulated water replacement depth in this treatment was higher than the accumulation of ETo. The 25 and 50% LWR presented accumulated water depths of 213 and 346.2 mm at 107 DATs.

The cultivation coefficients (kc) of ipê-rosa seedlings in a protected environment were represented by an increasing linear regression dependent on accumulated degree days; in this sense, the minimum and maximum coefficients were estimated at 0.59 and 2.86 for the beginning (0 DATs—0 DDAs) and the end (107 DATs—1655 DDAs), respectively. The average kc throughout the seedling cycle (107 DATs) was 1.75 (Figure 8). This average value is higher than that described by Monteiro et al. [28], who, when evaluating seedlings of this species kept in polyethylene bags with a volume of 4.7 L, obtained a gradual reduction in kc according to the increase in shading levels, with a kc of 1.03 in the condition of full exposure to the sun and of 0.54 under 80% attenuation of solar radiation.

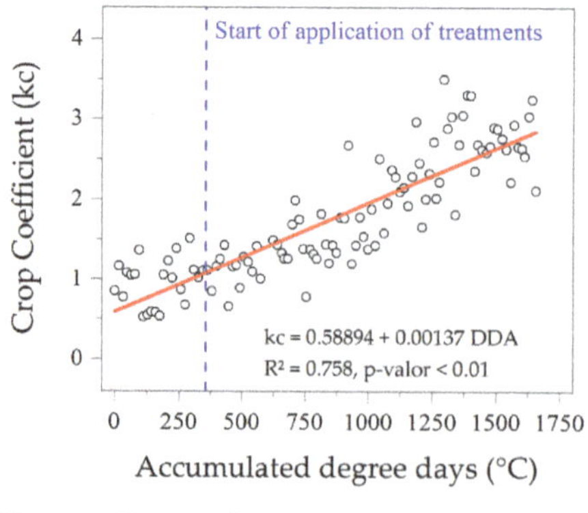


**Figure 8.** Crop coefficient as a function of accumulated degree days of the ipê-rosa seedlings, in a greenhouse ($p$-value < 0.01 indicates the significance of the adjustment at the 1% probability level; the red line represents the line of the fitted linear equation).

The averages of the parameters that make up the water sensitivity coefficient (Ky) are shown in Table 1. The dry mass yield was obtained for each level of water replacement, with the potential yield based on seedlings subjected to 100% water replacement. The Ky value represents the degree of tolerance of plants to water stress conditions; when this index is low, it demonstrates the resistance of the seedlings to the lack of water, and in the opposite case, the increase in Ky implies a greater degree of sensitivity of the seedlings to the irrigation deficit [29]. By default, the biomass yield tends to decrease as the intensity of the water restriction increases; thus, the sensitivity to water stress is classified into four categories: Ky < 0.85—low; 0.85 < Ky < 1.00—low/medium; 1.00 < Ky < 1.15—medium/high; and Ky > 1.15—high [29].

**Table 1.** Water sensitivity of ipê-rosa seedlings at 75, 50 and 25% levels of water replacement (LWRs), after 107 days after transplanting, in a greenhouse, Sinop, Mato Grosso state, Brazil.

| LWR | Etr [1] | Etm [2] | (1 − ETr/ETp) | Yr [3] | Ym [4] | (1 − Yr/Yp) | Ky [5] |
|---|---|---|---|---|---|---|---|
| | (mm) | | | (g $m^{-2}$) | | | |
| 75% | 465.41 | 593.2 | 0.22 | 802.47 | 1121.32 | 0.28 | 1.32 |
| 50% | 346.15 | 593.2 | 0.42 | 364.27 | 1121.32 | 0.68 | 1.62 |
| 25% | 212.99 | 593.2 | 0.64 | 34.4 | 1121.32 | 0.97 | 1.51 |

[1] Etr—real accumulated evapotranspiration; [2] ETm—maximum accumulated evapotranspiration in the treatment with 100% LWR, representing $ET_C$; [3] Yr—actual dry mass yield; [4] Ym—maximum dry mass yield in the treatment with 100% level of water replacement; [5] Ky—water sensitivity coefficient.

### 3.4. Diurnal Variations in Gas Exchange in Ipê-Rosa Seedlings

At 100 DATs (1558 DDAs), gas exchange assessments of ipê-rosa seedlings were carried out inside the greenhouse. Regarding the micrometeorological behavior of that day, significant differences were observed between the morning and afternoon periods for the hourly averages of the mean air temperature, dPv and PAR; in this case, the maximum values were 42.9 °C, 5.4 kPa and 937.6 µmol $m^{-2}$ $s^{-1}$, respectively (Supplementary Figure S1).

This micrometeorological dynamic influenced the gas exchange of ipê-rosa seedlings (Table 2). The net photosynthesis rates (A) of seedlings at 25, 50 and 75% LWRs showed maximums at 9:00 h, being 6.11, 12.50 and 11.12 µmol $m^{-2}$ $s^{-1}$, respectively. However, for the leaves of seedlings subjected to 100% LWR (reference condition), the maximum A values (10.41 µmol $m^{-2}$ $s^{-1}$) were recorded at 12:30 h; the minimum A values were observed in the afternoon period regardless of the water replacement level. Water use efficiency (WUE) and stomatal conductance (gs) showed lower values close to solar noon (smaller zenith angle and higher incidence of global radiation).

The ratio between internal carbon and atmospheric carbon (Ci/Ca) of ipê-rosa leaves was above 0.50 at all levels of water replacement, reaching 0.78 at 2:30 p.m. for seedlings subjected to 100% LWR; in this case, the lowest values were also obtained at times close to solar noon. However, the degree of stomatal opening remained regular at 12:30 p.m. compared to other times and was concomitant with the highest A values for seedlings with 100% LWR.

**Table 2.** Net photosynthesis rate, transpiration rate, water use efficiency, stomatal conductance, internal carbon to atmospheric carbon ratio and leaf temperature at 09:30, 12:30, 13:30 and 16:00 h in leaves of ipê-rosa seedlings at 100, 75, 50 and 25% levels of water replacement, 107 days after transplanting, the thermal sum of 1557.9 DDAs, in a greenhouse.

| Hour | Level of Water Replacement | | | |
|---|---|---|---|---|
| | 25% | 50% | 75% | 100% |
| | **Net photosynthesis rate—μmol $m^{-2}$ $s^{-1}$** | | | |
| 09:30 | 6.11 Ab | 12.54 Aa | 11.12 Aa | 10.1 Aa |
| 12:30 | 5.36 Ab | 6.92 Bb | 10.30 Aa | 10.41 Aa |
| 13:30 | 4.92 Aa | 7.15 Ba | 6.75 Ba | 8.02 Aa |
| 16:00 | 3.50 Ab | 6.28 Bab | 6.56 Bab | 8.05 Aa |
| | **Transpiration rate—mol $m^{-2}$ $s^{-1}$** | | | |
| 09:30 | 2.95 Ab | 5.47 Aa | 4.88 Aa | 5.03 Aa |
| 12:30 | 2.65 ABc | 3.93 BAcb | 4.92 Aab | 5.81 Aa |
| 13:30 | 2.43 ABb | 3.48 BCab | 3.28 Bab | 4.51 ABa |
| 16:00 | 1.26 Bb | 2.22 Cab | 2.23 Bab | 3.09 Ba |
| | **Water use efficiency—μmol $CO_2$ $mol^{-1}$ $H_2O$** | | | |
| 09:30 | 2.21 Ba | 2.29 Ba | 2.29 Ba | 1.94 Ba |
| 12:30 | 2.26 ABa | 1.82 Ba | 2.16 Ba | 1.81 Ba |
| 13:30 | 1.97 Ba | 2.12 Ba | 2.06 Ba | 1.78 Ba |
| 16:00 | 2.75 Aa | 2.83 Aa | 3.15 Aa | 2.77 Aa |
| | **Stomatal conductance—mol $m^{-2}$ $s^{-1}$** | | | |
| 09:30 | 0.10 Ab | 0.23 Aa | 0.19 Aa | 0.21 Aa |
| 12:30 | 0.07 Ab | 0.10 Bb | 0.14 ABab | 0.19 Aa |
| 13:30 | 0.06 Ab | 0.09 Bab | 0.09 Bab | 0.16 Aa |
| 16:00 | 0.06 Ab | 0.13 Bb | 0.13 ABb | 0.24 Aa |
| | **Internal carbon to atmospheric carbon ratio** | | | |
| 09:30 | 0.64 Bb | 0.71 Aab | 0.68 ABab | 0.72 ABa |
| 12:30 | 0.50 Cc | 0.64 Bab | 0.57 Cb | 0.66 Ba |
| 13:30 | 0.62 Bb | 0.59 Bb | 0.63 BCb | 0.71 ABa |
| 16:00 | 0.69 Aa | 0.67 Aa | 0.76 Aa | 0.78 Aa |
| | **Leaf temperature—°C** | | | |
| 09:30 | 40.37 Ba | 40.03 Ba | 40.40 Ba | 40.81 Aa |
| 12:30 | 41.94 Aa | 42.02 Aa | 41.97 Aa | 41.90 Aa |
| 13:30 | 39.53 Ba | 40.41 Ba | 39.47 Ba | 39.43 Ba |
| 16:00 | 33.63 Ca | 33.47 Ca | 33.83 Ca | 34.19 Ca |

Means followed by the same capital letter within columns and lowercase letter within rows do not differ by Tukey's test at the 5% probability level.

Gas exchanges were influenced by the interaction between water replacement levels and analysis times (Table 2). The leaves of seedlings subjected to 25% water replacement showed a lower net photosynthesis rate than the other treatments at the four times evaluated. In addition, it was also noted that at 25 and 100% water replacement, no differences were observed in the A value throughout the day; however, assimilation was enhanced during the morning period. Stomatal conductance showed patterns similar to the average values of A and E. In contrast, the EUA and temperatures of the leaves of ipê-rosa seedlings did not differ between the water replacement levels. Due to the microclimatic characteristics of the growth environment at 100 DATs, the analyzed gas exchange variables were associated with each other and with the mean air temperature and dPv in a distinct manner between the water replacement levels, as can be observed by the Pearson correlation coefficients (r) with value coefficients and their degrees of significance (Table 3).

**Table 3.** Pearson correlation (r) between gas exchange variables net photosynthesis rate (A), transpiration rate (E), water use efficiency (EUA), stomatal conductance (gs), relationship between internal carbon and atmospheric carbon (Ci/Ca) and leaf temperature (TF) and the microclimate factors mean air temperature ($T_M$) and water vapor pressure deficit of the air (dPv) in leaves of ipê-rosa seedlings at 100, 75, 50 and 25% levels of water replacement (LWRs), 107 days after transplanting, the thermal sum of 1556 DDAs, in a greenhouse.

| | $T_M$ | dPv | A | E | EUA | gs | Ci/Ca | Tf |
|---|---|---|---|---|---|---|---|---|
| **100% Level of Water Replacement** | | | | | | | | |
| $T_M$ | 1.00 | 1.00 ** | 0.19 NS | 0.37 ** | −0.52 ** | −0.08 NS | −0.51 ** | 0.78 ** |
| dPv | | 1.00 | 0.19 NS | 0.37 ** | −0.52 ** | −0.06 NS | −0.50 ** | 0.77 ** |
| A | | | 1.00 | 0.93 ** | 0.07 NS | 0.86 ** | 0.13 NS | 0.28 * |
| E | | | | 1.00 | −0.28 * | 0.75 ** | 0.02 NS | 0.56 ** |
| EUA | | | | | 1.00 | 0.15 NS | 0.24 NS | −0.78 ** |
| gs | | | | | | 1.00 | 0.55 ** | −0.03 NS |
| Ci/Ca | | | | | | | 1.00 | −0.61 ** |
| Tf | | | | | | | | 1.00 |
| **75% Level of Water Replacement** | | | | | | | | |
| | $T_M$ | dPv | A | E | EUA | gs | Ci/Ca | Tf |
| $T_M$ | 1.00 | 1.00 ** | 0.51 ** | 0.61 ** | −0.39 ** | 0.24 NS | −0.45 ** | 0.76 ** |
| dPv | | 1.00 | 0.52 ** | 0.61 ** | −0.37 ** | 0.25 NS | −0.45 ** | 0.75 ** |
| A | | | 1.00 | 0.85 ** | 0.00 NS | 0.80 ** | −0.22 NS | 0.41 ** |
| E | | | | 1.00 | −0.48 ** | 0.73 ** | −0.19 NS | 0.70 ** |
| EUA | | | | | 1.00 | −0.09 NS | 0.05 NS | −0.68 ** |
| gs | | | | | | 1.00 | 0.28 * | 0.09 NS |
| Ci/Ca | | | | | | | 1.00 | −0.58 ** |
| Tf | | | | | | | | 1.00 |
| **50% Level of Water Replacement** | | | | | | | | |
| | $T_M$ | dPv | A | E | EUA | gs | Ci/Ca | Tf |
| $T_M$ | 1.00 | 1.00 ** | 0.57 ** | 0.67 ** | −0.65 ** | 0.43 ** | −0.15 NS | 0.75 ** |
| dPv | | 1.00 | 0.55 ** | 0.66 ** | −0.63 ** | 0.41 ** | −0.16 NS | 0.74 ** |
| A | | | 1.00 | 0.83 ** | −0.24 NS | 0.86 ** | 0.13 NS | 0.41 ** |
| E | | | | 1.00 | −0.68 ** | 0.77 ** | 0.11 NS | 0.71 ** |
| EUA | | | | | 1.00 | −0.22 NS | 0.13 NS | −0.87 ** |
| gs | | | | | | 1.00 | 0.57 ** | 0.17 NS |
| Ci/Ca | | | | | | | 1.00 | −0.49 ** |
| Tf | | | | | | | | 1.00 |
| **25% Level of Water Replacement** | | | | | | | | |
| | $T_M$ | dPv | A | E | EUA | gs | Ci/Ca | Tf |
| $T_M$ | 1.00 | 1.00 ** | 0.13 NS | 0.28 * | −0.37 ** | 0.14 NS | −0.29 * | 0.74 ** |
| dPv | | 1.00 | 0.13 NS | 0.26 * | −0.36 ** | 0.13 NS | −0.30 * | 0.73 ** |
| A | | | 1.00 | 0.94 ** | 0.11 NS | 0.95 ** | −0.01 NS | 0.26 * |
| E | | | | 1.00 | −0.21 NS | 0.94 ** | 0.10 NS | 0.45 ** |
| EUA | | | | | 1.00 | −0.05 NS | −0.42 ** | −0.57 ** |
| gs | | | | | | 1.00 | 0.23 NS | 0.19 NS |
| Ci/Ca | | | | | | | 1.00 | −0.38 ** |
| Tf | | | | | | | | 1.00 |

Pearson correlation coefficients (r) were submitted to the *t*-test; where cells in green and red colors correspond to **, significant at the 1% probability level, and *, significant at the 5% probability level, respectively, and NS indicates a non-significant coefficient.

The leaves acclimated to intermediate water levels, 50 and 75% LWRs, presented the highest significant correlations, while Ci/Ca presented the lowest. Regardless of the water replacement, leaf temperature stood out for correlating with the highest number of variables (except for "gs"), and, of these, the correlations were negative only with WUE and Ci/Ca (Table 3). The highest correlation coefficients occurred for the interactions between

A, E and gs (all above 0.70). However, when confronted with the characteristics of the growth environment ($T_M$ and dPv), these variables showed little linearity, with 37.5% of the coefficients not having any degree of significance and the "r" values not exceeding 0.67 (r for E and $T_M$ at 50% LWR).

### 3.5. *Chl a Fluorescence: Seasonality and Diurnal Variations*

The *predawn Chl a* fluorescence of ipê-rosa plants was affected by water replacement (WR) levels and the time of the seedling cycle (associated with DDAs). At 28 DATs (464 DDAs), the maximum photochemical quantum yield of PSII before dawn ($Fv/Fm_{predawn}$) was similar among treatments, with an average of 0.80. In later evaluations (1046 and 1400 DDAs), only in plants maintained at the 25% LWR was a significant decrease in $Fv/Fm_{predawn}$ observed at 0.73 and 0.76, respectively (Figure 9).

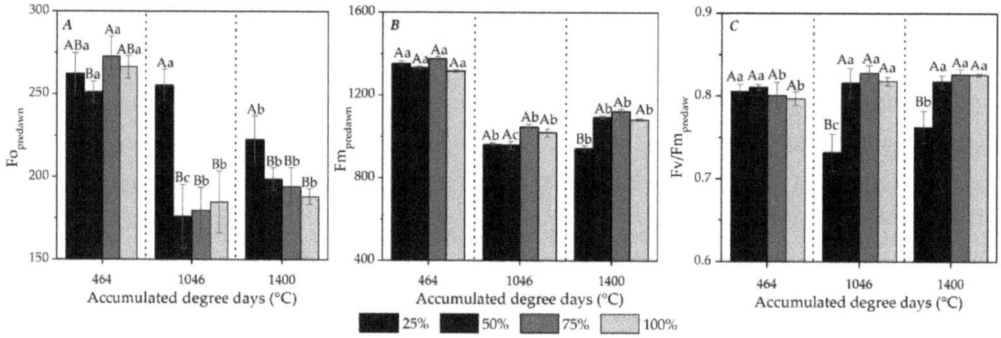


**Figure 9.** Mean values and standard deviation of minimum fluorescence ($Fo_{predawn}$) (**A**), maximum fluorescence ($Fm_{predawn}$) (**B**) and maximum quantum yield of PSII ($Fv/Fm_{predawn}$) before dawn (**C**), in ipê-rosa seedlings as a function of water replacement levels, at 28, 66 and 90 DATs (464, 1046 and 1400 DDAs). The red lines represent the standard deviation values. Means followed by the same capital letter between levels of water replacement (LWRs) and lowercase letter between DDAs do not differ by Tukey's test at the 5% probability level.

The decrease in $Fv/Fm_{predawn}$ values of plants subjected to 25% LWR with increasing experimental period and accumulation of 936 DDAs (72 DATs, time between the first and last analysis of *Chl a* fluorescence) was 5.30%. Similar dynamics were observed for minimum ($Fo_{predawn}$) and maximum ($Fm_{predawn}$) fluorescence before dawn, with the same treatment patterns. With development, plants grown at 50, 75 and 100% LWRs showed decreases in $Fo_{predawn}$, values of 180 (reduction of 31.56%) and 194 (reduction of 26.24%) in the second and third evaluations, respectively; those maintained at 25% LWR remained with a high value (255). However, the last analysis showed a relative reduction of 11.41% to 233, associated with the thermal sum of 1400 DDAs (Figure 9A). For all water treatments, the highest values of maximum predawn fluorescence ($Fm_{predawn}$) occurred at 28 DATs, when the plants accumulated 464 DDAs, an average of 1343, with a generalized decrease in subsequent evaluations (Figure 9B). However, only at 90 DATs (thermal sum of 1400 DDAs) was a difference recorded in $Fm_{predawn}$ for the water treatments, restricted to the lower level of 25% LWR compared to the others, which remained equal among themselves.

The internal diurnal micrometeorological fluctuations of the greenhouse, which occurred in the seedling growth environment in the three evaluations above, indicated thermal amplitudes of 12.02, 10.35 and 6.22 °C and dPv with variations of 4.12, 3.59 and 1.64 kPa for the evaluations at 464, 1046 and 1400 DDAs, respectively. On these days, the insolation was 7.1, 7.4 and 9.9 h, with a maximum daily PAR radiation of around 1016 $\mu mol\ m^{-2}\ s^{-1}$), at 11:00h solar time. At 28 DATs (464 DDAs), there was a similarity in the maximum quantum yield of PSII (Fv/Fm) at the water replacement levels (Table 4), with average differences of 0.42%. In the second evaluation, when the plants had accumulated

1046 DDAs, similar Fv/Fm averages were recorded until 09:00h, except for the seedlings subjected to 25% LWR. The results were similar to those of the plants with the minimum water replacement (Fv/Fm of 0.72), which maintained low values as the seedlings developed. Due to the favorable condition of the micrometeorological components in the last evaluation (1400 DDAs) (Supplementary Figure S1), mainly due to the air temperature and dPv levels below the respective monthly extremes, the Fv/Fm values were stable in all treatments during the day, with an average variation of 1.24%. In this case, the only thing that stands out is the inferiority of the plants at 25% LWR compared to the overlap of the other curves (Table 4).

Table 4 also presents the average values of the diurnal variations of *Chl a* fluorescence in the light-adapted state. When compared with the average values of the Fv/Fm variable, there was less ordering of the diurnal fluxes in the water replacement levels on the three dates analyzed for the effective quantum yield of PSII ($\Phi_{PSII}$), electron transport rate (ETR); maximum efficiency of PSII (Fv′/Fm′) and non-photochemical dissipation (NPQ), characterized mainly by inversions and overlaps between the average points in the water treatments. Nevertheless, similarly to the Fv/Fm variable, at 28 DATs, when plant development reached the accumulation of 464 DDAs, *Chl a* fluorescence in the light-adapted state of the ipê-rosa seedlings showed little distinction between the water regimes evaluated (Table 4). Furthermore, in the following evaluations, there was a distancing of the responses of the ipê-rosa seedlings subjected to 25% LWR in comparison to the other treatments, which, in turn, maintained a certain proximity during the day. A priori, although the amount of water available did not influence the current operation of the PSII, at 11:00 and 13:00 h, more accentuated favorable responses were identified in the plants at 100% LWR compared to the others, with a relative superiority of 7.59, 7.59 and 9.00% for $\Phi_{PSII}$, ETR and Fv′/Fm′, respectively, and relative inferiority of 22.92% for NPQ; however, at the next moment (15:00 h), the curves met again.

In the second evaluation, $\Phi_{PSII}$ and ETR showed the maximum point in the early morning at 25, 50 and 75% LWRs, respectively, 0.19, 0.44 and 0.50 for $\Phi_{PSII}$ and 44.32, 104.86 and 118.90 μmol $m^{-2}$ $s^{-1}$ for ETR, with a subsequent decrease at 09:00 h. Although the individuals at 100% LWR demonstrated behavior equivalent to that described, their maximum points occurred only after the decay between 9:00 and 11:00 h, at 13:00 h, with $\Phi_{PSII}$ of 0.51 and ETR of 121.21 μmol $m^{-2}$ $s^{-1}$ (Table 4).

According to Table 4, in the last evaluation with the accumulation of 1400 DDAs, within the hourly course, the dynamics of ΦPSII and ETR were approximately linear, not regular, decreasing for the seedlings maintained at 25% LWR and increasing in those with 50, 75 and 100% LWRs. On that date, despite the decreasing $\Phi_{PSII}$ and ETR values for the seedlings under water stress, 25% LWR, the results were superior to those obtained by these plants when accumulating 1046 DDAs, an average percentage of 88.4% above, yet below the other treatments. Only when the development of the seedlings reached the accumulation of 1046 DDAs, a behavioral pattern was observed in the Fv′/Fm′ curves, an asymmetric parabola with an upward concavity, with maximum values at 7:00 a.m. and minimum values close to noon local time (Table 4).

On the other dates, the ipê-rosa seedlings subjected to 100% LWR showed multiple inverse Fv′/Fm′ trends in the other measured flows when the average values were compared to the other treatments. At 90 DATs, with the accumulation of 1400 DDAs, for example, while individuals at 25, 50 and 75% of RA expressed a generalized decrease in response to the sudden gain of PAR at 13:00 h (Table 4), those subjected to full water replacement presented a maximum mean value (0.70) within the three occasions studied (Table 4). At 66 DATs, with the accumulation of 1046 DDAs, non-photochemical dissipation (NPQ) in the leaves of ipê-rosa seedlings presented a performance inverse to ΦPSII. The maximum values occurred at 09:00 h for the leaves of individuals at 50, 75 and 100% of LWR, with averages of 2.77, 2.27 and 2.07, respectively, while for the leaves growing at 25% LWR, the maximum value was concomitant with the moment of maximum daily PAR, at 11:00 h, an average of 3.39 (Table 4). However, in the third evaluation, 90 DATs, the maximums and minimums were dispersed throughout the evaluated courses in the treatments, with low comparability to $\Phi_{PSII}$; the inverse pattern, in this case, an increasing linear trend, was observed only for the 25% LWR flow (Table 4).

**Table 4.** Diurnal variations in the average maximum quantum yield of PSII (Fv/Fm), in the fluorescence of *Chl a* in the light-adapted state and in the average fraction of open reaction centers of PSII (qL) of ipê-rosa seedlings at the 25, 50, 75 and 100% levels of water replacement at 28, 66 and 90 days after transplanting (DATs), equivalent to the thermal sums of 464, 1046 and 1400 accumulated degree days (DDAs), respectively, in a greenhouse.

| Solar Time | 28 DATs (464 DDAs) | | | | 66 DATs (1046 DDAs) | | | | 90 DATs (1400 DDAs) | | | |
|---|---|---|---|---|---|---|---|---|---|---|---|---|
| | 25% | 50% | 75% | 100% | 25% | 50% | 75% | 100% | 25% | 50% | 75% | 100% |
| | **Maximum quantum yield of PSII (Fv/Fm)** | | | | | | | | | | | |
| 07:00 | 0.799 | 0.797 | 0.800 | 0.800 | 0.722 | 0.799 | 0.809 | 0.810 | 0.761 | 0.819 | 0.825 | 0.828 |
| 09:00 | 0.759 | 0.763 | 0.753 | 0.755 | 0.663 | 0.765 | 0.767 | 0.770 | 0.731 | 0.800 | 0.804 | 0.813 |
| 11:00 | 0.735 | 0.742 | 0.736 | 0.736 | 0.634 | 0.684 | 0.729 | 0.748 | 0.734 | 0.805 | 0.810 | 0.819 |
| 13:00 | 0.723 | 0.740 | 0.740 | 0.734 | 0.658 | 0.749 | 0.765 | 0.781 | 0.730 | 0.805 | 0.806 | 0.815 |
| 15:00 | 0.757 | 0.763 | 0.762 | 0.749 | - | - | - | - | - | - | - | - |
| 17:00 | 0.799 | 0.799 | 0.792 | 0.789 | - | - | - | - | - | - | - | - |
| | **Effective quantum yield of PSII ($\Phi_{PSII}$)** | | | | | | | | | | | |
| 07:00 | 0.378 | 0.423 | 0.423 | 0.417 | 0.188 | 0.444 | 0.503 | 0.405 | 0.298 | 0.357 | 0.366 | 0.356 |
| 09:00 | 0.281 | 0.352 | 0.328 | 0.284 | 0.159 | 0.273 | 0.351 | 0.385 | 0.281 | 0.426 | 0.453 | 0.497 |
| 11:00 | 0.327 | 0.335 | 0.335 | 0.364 | 0.106 | 0.294 | 0.366 | 0.385 | 0.288 | 0.438 | 0.410 | 0.396 |
| 13:00 | 0.338 | 0.358 | 0.366 | 0.375 | 0.155 | 0.410 | 0.430 | 0.513 | 0.278 | 0.468 | 0.480 | 0.506 |
| 15:00 | 0.383 | 0.352 | 0.382 | 0.349 | - | - | - | - | - | - | - | - |
| 17:00 | 0.235 | 0.228 | 0.211 | 0.185 | - | - | - | - | - | - | - | - |
| | **Electron transport rate (μmol $m^{-2}$ $s^{-1}$)** | | | | | | | | | | | |
| 07:00 | 89.27 | 100.01 | 99.84 | 98.40 | 44.32 | 104.86 | 118.90 | 95.57 | 70.48 | 84.28 | 86.45 | 84.14 |
| 09:00 | 66.50 | 83.10 | 77.47 | 67.09 | 37.67 | 64.45 | 83.034 | 90.91 | 66.43 | 100.69 | 107.10 | 117.42 |
| 11:00 | 77.36 | 79.13 | 79.24 | 85.90 | 25.12 | 69.46 | 86.499 | 91.00 | 68.13 | 103.48 | 96.76 | 93.59 |
| 13:00 | 79.90 | 84.51 | 86.42 | 88.59 | 36.56 | 96.97 | 101.67 | 121.21 | 65.69 | 110.58 | 113.49 | 119.46 |
| 15:00 | 90.47 | 83.05 | 90.19 | 82.37 | - | - | - | - | - | - | - | - |
| 17:00 | 55.48 | 53.85 | 49.92 | 43.64 | - | - | - | - | - | - | - | - |
| | **Maximum efficiency of PSII (Fv′/Fm′)** | | | | | | | | | | | |
| 07:00 | 0.511 | 0.496 | 0.501 | 0.545 | 0.422 | 0.644 | 0.719 | 0.623 | 0.575 | 0.637 | 0.733 | 0.674 |
| 09:00 | 0.484 | 0.509 | 0.479 | 0.469 | 0.341 | 0.465 | 0.508 | 0.529 | 0.475 | 0.599 | 0.685 | 0.678 |
| 11:00 | 0.479 | 0.492 | 0.495 | 0.524 | 0.296 | 0.428 | 0.479 | 0.497 | 0.497 | 0.697 | 0.672 | 0.635 |
| 13:00 | 0.463 | 0.475 | 0.490 | 0.528 | 0.339 | 0.526 | 0.561 | 0.611 | 0.466 | 0.656 | 0.660 | 0.702 |
| 15:00 | 0.525 | 0.497 | 0.527 | 0.505 | - | - | - | - | - | - | - | - |
| 17:00 | 0.584 | 0.606 | 0.581 | 0.605 | - | - | - | - | - | - | - | - |
| | **Non-photochemical dissipation (NPQ)** | | | | | | | | | | | |
| 07:00 | 3.242 | 3.281 | 3.172 | 2.383 | 2.637 | 1.200 | 0.67431 | 1.638 | 1.432 | 1.611 | 0.742 | 1.356 |
| 09:00 | 2.379 | 2.155 | 2.366 | 2.523 | 2.939 | 2.774 | 2.268 | 2.070 | 2.109 | 1.698 | 0.967 | 1.086 |
| 11:00 | 2.069 | 2.015 | 1.897 | 1.646 | 3.390 | 2.050 | 2.005 | 2.046 | 1.873 | 0.859 | 1.123 | 1.681 |
| 13:00 | 2.116 | 2.208 | 2.017 | 1.519 | 2.892 | 1.932 | 1.601 | 1.346 | 2.274 | 1.188 | 1.206 | 0.874 |
| 15:00 | 1.903 | 2.285 | 1.965 | 1.998 | - | - | - | - | - | - | - | - |
| 17:00 | 1.876 | 1.642 | 1.807 | 1.494 | - | - | - | - | - | - | - | - |
| | **Fraction of open reaction centers of the PSII (qL)** | | | | | | | | | | | |
| 07:00 | 0.600 | 0.764 | 0.757 | 0.608 | 0.317 | 0.444 | 0.401 | 0.416 | 0.326 | 0.324 | 0.22 | 0.280 |
| 09:00 | 0.418 | 0.525 | 0.533 | 0.449 | 0.367 | 0.439 | 0.530 | 0.558 | 0.425 | 0.496 | 0.378 | 0.474 |
| 11:00 | 0.535 | 0.52 | 0.518 | 0.515 | 0.296 | 0.551 | 0.630 | 0.636 | 0.410 | 0.346 | 0.342 | 0.377 |
| 13:00 | 0.604 | 0.618 | 0.597 | 0.542 | 0.356 | 0.629 | 0.597 | 0.670 | 0.440 | 0.460 | 0.471 | 0.438 |
| 15:00 | 0.566 | 0.552 | 0.554 | 0.544 | - | - | - | - | - | - | - | - |
| 17:00 | 0.239 | 0.232 | 0.233 | 0.167 | - | - | - | - | - | - | - | - |

The seedlings acclimated to 25% LWR demonstrated a lower percentage range for NPQ, with a variation relative to the average of daily courses of 25.4 and 43.8% at 66 and 90 DATs, with accumulations of 1046 and 1400 DDAs, respectively. In contrast, the other levels of water replacement maintained a percentage range between 40.1% (100% LWR at the thermal sum of 1046 DDAs) and 97.3% (75% LWR also at the thermal sum of 1046 DDAs). Regarding the diurnal fluxes of the fraction of open reaction centers of PSII (qL), in the first evaluation, there was little differentiation between the leaves of the ipê-rosa seedlings in the LWR. In this case, with the occurrence of the maximum point at 07:00 h (average of 0.68), there was a subsequent decrease and stability during the hours, until an abrupt reduction to the minimum point (average of 0.22) at 17:30 h (Table 4). At 28 DATs, with the plants accumulating 464 DDAs, at the end of the day, the regeneration of the photochemical capacity of PSII was observed, arranged by Fv/Fm and Fv′/Fm′, along with a reduction in the effective PSII activity according to $\Phi_{PSII}$ and ETR. This qL dynamic was not repeated in the other evaluations, in which the minimum points were concentrated in the early morning, between 7:00 and 9:00 h, in general, with an increasing trend until the cessation of readings at 13:00 h (Table 4).

*Chl a* fluorescence and microclimatic components—mean air temperature and dPv—showed distinct comparability in water replacement levels with the grouping of readings at 66 and 90 DATs, with accumulations of 1046 and 1400 DDAs, dates when the plants were already acclimated to water availability. Measurements were given based on Pearson's correlation coefficients (r), which showed different significance indices and values (Table 5). The maximum efficiency of PSII (Fv′/Fm′) showed significant "r" coefficients in almost all interactions analyzed, except with qL at 25% LWR, fluctuating in modular values from 0.55 to 0.91, positively associated with Fv/Fm, $\Phi_{PSII}$ and ETR. On the other hand, it was negatively linked to mean air temperature, dPv, NPQ and qL. Excluding the non-significant interaction coefficient (qL), the highest r value of the relationship with Fv′/Fm′ occurred with the minimum water supply level, a modular mean of 0.83, followed by 0.77, 0.74 and 0.71 at 50, 75 and 100% LWRs, respectively.

Non-photochemical dissipation (NPQ) showed a similar significance pattern to Fv′/Fm′. However, it showed more unsatisfactory coefficients, mainly regarding mean air temperature, dpv, Fv/Fm and qL, excluding the responses for the leaves of individuals at 75% LWR, whose coefficients were mostly above 0.50. An inversion was observed in the interactions of $\Phi_{PSII}$ and qL with mean air temperature and dPv in the treatments. The correlation coefficients between qL and the microclimate were significant for the leaves of ipê-rosa plants subjected to 50, 75 and 100% LWRs, which did not occur for the plants at 25% LWR. In the interaction of ΦPSII with the microclimate, the opposite occurred to what was described; however, in this case, only at the 5% probability level, the r value was also significant at 50% LWR. As expected, the ETR values matched the $\Phi_{PSII}$ results.

As an alternative to Fv′/Fm′, the Pearson correlations of the maximum quantum yield of PSII (Fv/Fm) with the average air temperature and dPv were, in the module, decreasing in the water treatments, with the maximum average for 100% LWR (−0.89) and minimum at 25% LWR (−0.72). The interactions between the maximum quantum yield of PSII (Fv/Fm) and the microclimatic variables were translated into linear regression fits at each LWR, as shown in Table 5.

**Table 5.** Pearson correlation (r) between *Chl a* fluorescence variables maximum PSII quantum yield (Fv/Fm), effective PSII quantum yield ($\Phi_{PSII}$), non-photochemical dissipation (NPQ), electron transport rate (ETR), maximum PSII efficiency (Fv′/Fm′) and fraction of PSII open reaction centers (qL) and microclimate factors mean air temperature ($T_M$) and air water vapor pressure deficit (dPv) in leaves of ipê-rosa seedlings at levels of water replacement (LWRs) of 100, 75, 50 and 25%. Grouping of readings at 66 and 90 days after transplanting.

| | | | | | | | | |
|---|---|---|---|---|---|---|---|---|
| **100% Level of Water Replacement** | | | | | | | | |
| | $T_M$ | DPV | Fv/Fm | $\Phi_{PSII}$ | NPQ | ETR | Fv′/Fm′ | qL |
| $T_M$ | 1.00 | 0.99 ** | −0.87 ** | −0.04 NS | 0.46 ** | −0.04 NS | −0.72 ** | 0.82 ** |
| DPV | | 1.00 | −0.91 ** | −0.10 NS | 0.45 ** | −0.10 NS | −0.76 ** | 0.83 ** |
| Fv/Fm | | | 1.00 | 0.17 NS | −0.50 ** | 0.17 NS | 0.80 ** | −0.84 ** |
| $\Phi_{PSII}$ | | | | 1.00 | −0.68 ** | 1.00 ** | 0.55 ** | 0.18 NS |
| NPQ | | | | | 1.00 | −0.68 ** | −0.91 ** | 0.48 ** |
| ETR | | | | | | 1.00 | 0.55 ** | 0.18 NS |
| Fv′/Fm′ | | | | | | | 1.00 | −0.71 ** |
| qL | | | | | | | | 1.00 |
| **75% Level of Water Replacement** | | | | | | | | |
| | $T_M$ | DPV | Fv/Fm | $\Phi_{PSII}$ | NPQ | ETR | Fv′/Fm′ | qL |
| $T_M$ | 1.00 | 0.99 ** | −0.82 ** | −0.12 NS | 0.57 ** | −0.12 NS | −0.74 ** | 0.84 ** |
| DPV | | 1.00 | −0.85 ** | −0.17 NS | 0.59 ** | −0.17 NS | −0.77 ** | 0.84 ** |
| Fv/Fm | | | 1.00 | 0.27 NS | −0.56 ** | 0.27 NS | 0.82 ** | −0.81 ** |
| $\Phi_{PSII}$ | | | | 1.00 | −0.70 ** | 1.00 ** | 0.60 ** | 0.08 NS |
| NPQ | | | | | 1.00 | −0.70 ** | −0.93 ** | 0.54 ** |
| ETR | | | | | | 1.00 | 0.60 ** | 0.08 NS |
| Fv′/Fm′ | | | | | | | 1.00 | −0.73 ** |
| qL | | | | | | | | 1.00 |
| **50% Level of Water Replacement** | | | | | | | | |
| | $T_M$ | DPV | Fv/Fm | $\Phi_{PSII}$ | NPQ | ETR | Fv′/Fm′ | qL |
| $T_M$ | 1.00 | 0.98 ** | −0.79 ** | −0.36 * | 0.42 ** | −0.36 * | −0.70 ** | 0.62 ** |
| DPV | | 1.00 | −0.83 ** | −0.39 * | 0.42 ** | −0.39 * | −0.72 ** | 0.62 ** |
| Fv/Fm | | | 1.00 | 0.50 ** | −0.38 * | 0.50 ** | 0.79 ** | −0.57 ** |
| $\Phi_{PSII}$ | | | | 1.00 | −0.76 ** | 1.00 ** | 0.78 ** | 0.07 NS |
| NPQ | | | | | 1.00 | −0.76 ** | −0.86 ** | 0.37 * |
| ETR | | | | | | 1.00 | 0.78 ** | 0.07 NS |
| Fv′/Fm′ | | | | | | | 1.00 | −0.55 ** |
| qL | | | | | | | | 1.00 |
| **25% Level of Water Replacement** | | | | | | | | |
| | $T_M$ | DPV | Fv/Fm | $\Phi_{PSII}$ | NPQ | ETR | Fv′/Fm′ | qL |
| $T_M$ | 1.00 | 0.99 ** | −0.72 ** | −0.77 ** | 0.58 ** | −0.77 ** | −0.80 ** | −0.20 NS |
| DPV | | 1.00 | −0.73 ** | −0.78 ** | 0.57 ** | −0.78 ** | −0.79 ** | −0.24 NS |
| Fv/Fm | | | 1.00 | 0.79 ** | −0.35 * | 0.79 ** | 0.79 ** | 0.34 ** |
| $\Phi_{PSII}$ | | | | 1.00 | −0.65 ** | 1.00 ** | 0.88 ** | 0.51 ** |
| NPQ | | | | | 1.0 | −0.65 ** | −0.84 ** | 0.20 NS |
| ETR | | | | | | 1.00 | 0.88 ** | 0.51 ** |
| Fv′/Fm′ | | | | | | | 1.00 | 0.06 NS |
| qL | | | | | | | | 1.00 |

Pearson correlation coefficients (r) submitted to the *t*-test, where cells in green and red colors correspond to **, significant at the 1% probability level, and *, significant at the 5% probability level, respectively, and NS indicates a non-significant coefficient.

## 4. Discussion

### 4.1. Morphometric Growth of Ipê-Rosa Seedlings

During the experimental period, global radiation and air temperature variations related to cloud cover and seasonal rainfall in the region were observed [45]. The dynamics of the climatic elements recorded during the experimental period result from the region's water seasonality, independent of the degree of shading resulting from the protected environments [29,46,47].

The application of treatments associated with levels of water replacement (LWRs) began at 21 DATs when the seedlings had already accumulated 345 DDAs. This timing is crucial, as it allows us to experiment in a protected seedling production environment, minimizing the effects of rainfall. Such effects could potentially homogenize the humidity of the tube substrates and thus minimize the effects of water replacement treatments. The period of execution of the experiment was carefully chosen to align with the seed dispersal season of the species and the production systems of native and exotic seedlings in Brazilian forest nurseries. These facilities have a planning basis for producing and hardening seedlings during the transition from dry to rainy seasons, ensuring higher survival rates in the field.

The adjustments to the mathematical models of the growth variables as a function of the thermal sum served a crucial purpose in our experiment. They allow us to predict DDA values so that the forest seedlings reach the ideal morphometric level to be taken to the field. These models consider the seasonality of air temperature within the production cycle and the role of this micrometeorological variable in plant physiology and cycle length, which does not necessarily include seasons favorable to good vegetative performance [48]. The mathematical adjustments showed linear seedling height and stem diameter growth, evidencing that the seedlings maintained linear development and expressed regularity in the face of environmental conditions. However, growth rates decreased with increased water restriction (Supplementary Table S1).

Lima et al. [49] found a 38% reduction in the growth of the aerial part of ipê-rosa seedlings when increasing the number of days between waterings during plant hardening. In addition, these authors observed effects on the metabolic pattern of seedlings subjected to water restriction, with substantial phenotypic plasticity during acclimatization to field conditions and with preferential redirection of photoassimilates to the root regions of the plant. Plant growth is generally represented by a sigmoid curve, modeled by phases with more or less accentuated productivity according to the plant's phenophase and the environment in which it is inserted [50,51]. However, forest seedlings present little phenological distinction in nursery conditions, given the short permanence period and growth environment with restrictive characteristics [52]. Thus, there was constant growth at the same level of water replenishment, with the rate of increase being influenced by the availability of water in the substrate.

Growth assessments were carried out after the beginning of the rainy season in the region when the temperature and relative humidity were at levels that were biologically more favorable to plant development than the dry season (Figure 4). This environmental condition, associated with the protected environment of plant production, may also have favored the regular growth in height and DAC of the ipê-rosa seedlings. Excessive temperature increases and relative humidity reductions antagonistically affect the assimilation of $CO_2$ used in biomass formation by plant species, as they alter the activity of enzymes that act in the photosynthetic pathways and the electron transport chain. In this case, there may be a reduction in the degree of opening of the stomata and an increase in the $O_2/CO_2$ ratio dissolved inside the *Chl*oroplasts, which in turn increases the photorespiration rates (a route that occurs due to the ability of the Rubisco enzyme—a carboxylation agent—to use oxygen as a substrate); in addition, there may be an increase in respiratory activity, with the use of assimilated organic compounds to maintain cellular biochemical integrity [53,54]. Vieira et al. [5], when evaluating the growth of ipê-rosa in its juvenile phase, grown at different levels of base saturation in the soil in a greenhouse for 90 days, indicated a height of 23.3 cm as ideal for transplanting in the field, also considering technical and financial perspectives. Under the conditions of the present study, thermal accumulations of 1344, 1056 and 979 DDAs would be required for the ipê-rosa seedlings to become suitable in height to leave the nursery conditions for 50, 75 and 100% LWRs. Notably, for seedlings with the minimum LWR (25%), around 4940 DDAs would be necessary to reach the aforementioned morphometric standard.

The numbers of leaves and leaflets showed different behaviors depending on the thermal sum, indicating the existence of accumulated thermal sums that provide maximum and minimum numbers of leaflets, respectively. Leaf tissues represent the basis of the photosynthetic process and are generally the first parts of the plant body to respond to the appearance of unfavorable conditions in the growth environment, demonstrating regulated ontogeny and morphophysiology to minimize irreversible damage to the metabolic integrity of the plant, while also seeking to maximize photoassimilative processes [21]. The ipê-rosa seedlings did not begin the growth period at 25% LWR since all treatments presented the same level of water replacement (irrigation) until the thermal accumulation of 345 DDAs (21 DATs). In this case, these plants initially captured the condition of water restriction and adjusted their physiology to a lower LWR to assume a new growth rhythm, demonstrated by the number of leaves and leaflets. For the other morphometric components of these same seedlings, it would probably be necessary to extend the analyses in time for growth adjustments to occur as long as the other environmental conditions were maintained. The gradual separation in the curves of the responses to LWR for seedling height, stem diameter at height and number of leaflets showed us the negative effect of restricted water supply on the growth of young ipê-rosa plants. Thus, we can state that juvenile ipê-rosa can tolerate short-term water stress and maintain its growth rate. However, if this unfavorable water condition is sustained, the species tends to regulate its morphological structure to the margin of survival; that is, it reduces stem growth and leaf senescence to maintain the physiological stability of the plant until water is re-established in the environment [15]. Furthermore, water plays a fundamental role in plant growth because it is associated with the activity of signaling hormones and protein groups in the cell wall, enabling the process of cell expansion. Its absence gradually makes cell elongation impossible and partially stops the tissue growth cycle [55,56].

The stagnation in the gain in leaf number of seedlings subjected to 100% LWR was not qualified by the "turnover" of the photosynthetic apparatus. According to Monteiro [21], this moment is defined by the emergence and senescence of equivalent leaves, renewing the photoassimilative capacity of forest seedlings. In this case, leaflets were more pronounced at this 100% LWR, indicating the anticipation of the leaf ontogeny of ipê-rosa cultivated at maximum water replacement.

At the end of the experiment (at 107 DATs, with an accumulation of 1655 DDAs), the influence of water restriction on the accumulation of seedling mass was evident, with a significant reduction for the 25% replacement level. Lima et al. [49] found a 50% loss in biomass input in ipê-rosa seedlings subjected to water deficiency compared to irrigated plants of the species. Water restriction reduces the carbon fixation capacity of plants by affecting multiple pathways that make up the photosynthetic chains dependent on the presence of water, such as changing the balance between cyclic and acyclic photophosphorylation, in addition to inducing the occurrence of alternative processes of stress tolerance, maintenance of internal water content, and cellular integrity and homeostasis; these effects are also transmitted to the morphological structure of the plant through the relationship with the allocation of carbon in the aerial and root parts [57–59]. The change in leaf area due to water availability, in turn, is a mechanism associated with reducing the leaf surface in contact with the atmospheric stratum to reduce water loss through transpiration, which is recurrent in different higher plants [60]. However, the marked loss in leaf area observed in 25% water replacement was probably more related to insufficient carboxylation than to the phenotypic expression of tolerance of ipê-rosa seedlings under water stress. This finding is reinforced by the patterns of dry mass partition between parts at LWRs (Figure 6), since regardless of the amount of water in the substrate, there was regularity in the proportions of area and root dry mass contribution after 107 DATs. Also, the ipê-rosa seedlings demonstrated a single phenotypic category for the distribution of photoassimilates among the plant organs, unaltered by the water status of the growth environment, which was based on leaf production.

Plants in the Amazon biome acclimate to drought, with adjustments observed throughout the metabolic chain to minimize permanent damage to the photosynthetic apparatus. Once the water regime is resumed, vegetative performance tends to reach the "optimal" level, indicating resilience to extreme climate events [61]. The ipê-rosa seedlings demonstrated the ability to adjust their morphological structure to the survival margin of water deficit; consequently, they expressed a gradual reduction in growth variables with the reduction in water availability in the substrate.

### 4.2. Water Requirements and Sensitivity of the Ipê-Rosa Seedlings

The climatic seasonality during the experiment caused changes in the crop evapotranspiration (ETc) and reference (ETo) patterns; with the onset of rains, there was a greater variation in daily values. Biophysical processes of plant interaction with the growth environment regulate daily water demand. Thus, the plant's seasonality depends on water energy balances, since plants can adjust their morphophysiology to microclimatic seasonal patterns [16,62]. A continuous increase in the water demand of the seedlings of this species was observed, accompanying the increase in the transpirational surface and morphometric variables in the seedling cycle. Although forest seedlings do not present defined phenological phases, water requirements are regulated by the plant's growth stage, culminating in an increase in potential evapotranspiration with increased allocation to plant organs, thus requiring more and more water to maintain good plant performance and homeostasis [63].

The size of seedling growth containers (tubes, pots, plastic bags, among others) can generate different growth rates for ipê-rosa seedlings, which can interfere with potential evapotranspiration. According to Cunha et al. [64], in the volumetric range of 2.7 to 11.5 L, there are gains in height, stem diameter, and aerial and root biomass of ipê-rosa seedlings, since larger containers, associated with other components of forest seedling management, tend to improve the vegetative performance of this species by up to 76%. The crop coefficient is an index that facilitates irrigation management of agricultural and forestry crops, associating it with easily obtainable climate variables [65]. However, few studies in the silvicultural field quantify kc values as a function of the thermal requirements for seedlings of native Amazon biome species, reinforcing the present study's importance. However, there is still a demand for research that associates these responses with other components essential to the success of production in forest nurseries, such as container size, substrate quality and level of shading, among others.

Regarding the degree of tolerance of seedlings to water conditions (Ky), although no linear behavior of Ky was observed as a function of water replacement, ipê-rosa seedlings were sensitive to water shortage. Pessoa et al. [66] characterized a rapid and progressive reduction in the physiological performance of ipê-rosa seedlings after irrigation was suppressed at 10-day intervals; however, the plants maintained regular water use efficiency. Keffer et al. [29] described the sensitivity to water restriction of young *Handroanthus serratifolius* plants under different shading levels, with Ky values lower than those obtained in the present study, ranging from 0.89 to 1.28. In a similar analysis of Ky, Bueno et al. [67] measured the plant growth factor (Gpf) of *Schizolobium parahyba*, *Cytharexylum myrianthum* and *Ceiba speciosa* at the seedling stage as a variable that estimates water sensitivity based on non-destructive growth parameters (stem height and diameter); for this purpose, they obtained a Gpf below 0.5, classifying the seedlings of these species as having low sensitivity to water restriction. In addition to the degree of incident solar radiation and water availability, other technical and ecophysiological components, such as container size and substrate quality, can regulate the biomass yield of forest seedlings and, consequently, their demands on the environment [68,69]. In general, water demand is expected to increase with the vegetative development of the seedlings [63]. Plants subjected to 25% LWR regulated dry matter production to a restricted margin from the beginning of treatment application, while those at 50% LWR possibly had sufficient supply to regulate growth for a longer cycle period. When the growth stage requires greater water availability, it is characterized by greater water deficiency in this condition, reflected in Ky.

Studies that combine irrigation regimes, growth factors, and performance indices, such as Ky for forest nurseries, are gradually being requested. However, they are still scarce in qualifying the best cultivation conditions for conserving and managing environmental and financial resources. From another perspective, Amazonian ecosystems have demonstrated differential behaviors in recent years in the face of climate change, mainly regarding the extension of dry days, resulting in changes at the level of individuals and entire plant communities, through the regulation of organisms already established to the dynamics of the environment, as well as the recruitment of taxa more related to drought conditions [70]. Despite the expressive water sensitivity observed in the present study, Ky has few literary parameters for Amazonian tree species. Thus, ipê-rosa plants expressed a certain resistance to water shortages, with vegetative performance regulated to the survival margin, verified by the inversion of Ky between 50 and 25% LWRs.

### 4.3. *Gas Exchange of Ipê-Rosa Seedlings*

The region where the experiment was conducted is characterized by variations in cloudiness during the dry–rainy and rainy transition periods [45], and these atmospheric dynamics influenced the micrometeorological variables of the protected production environment and, consequently, the gas exchange of the ipê-rosa seedlings throughout the day. The seasonality and diurnal variation of the growth environment affect the physiological responses of evergreen species in different production systems, with the circadian rhythm of the plant's water status indicators responding together with the micrometeorological variables, which enhance the occurrence of maximum gas exchange values at solar noon, associated with maximum air temperature, water vapor pressure deficit and PAR radiation [71].

Marenco et al. [72] evaluated the diurnal photosynthetic fluxes of leaves of seedlings of ten Amazonian Forest species under the forest canopy in consideration of regional rainfall seasonality (dry and rainy). They observed that, regardless of the season, the net photosynthesis rate in saturating light increased until local noon, with a reduction in the afternoon. According to these authors, in the shaded condition under the canopy, the photosynthetic activity of these emerging plants is only favored when a greater amount of light enters the environment, associated with higher temperatures that allow better enzymatic dynamics, since, in both seasons (dry and rainy), water availability does not become a restrictive element in native forests.

In the ipê-rosa seedlings evaluated, stomatal conductance did not present significant differences between the levels of water replacement at solar noon. The degree of stomatal opening of Amazonian species is expected to respond directly to changes in leaf photosynthesis and transpiration rates [72]. However, the leaf anatomy of the ipê-rosa may have contributed to the control of stomatal activity since this species is classified as hypostomatic, i.e., stomata are mostly dispersed in the abaxial portion of the leaf organs, thus minimizing their responses to the influence of direct solar radiation [73]. However, for the leaves of seedlings with 100% water replacement, at solar noon, stomatal conductance showed maximum values concurrent with peaks of A and minimum values of Ci/Ca. Therefore, internal carbon reduction was assumed to be closely associated with $CO_2$ consumption in photoassimilative processes. For plants with 25% LWR, the minimum Ci/Ca ratio at local noon may be associated with the stability of gs and A; in this case, $CO_2$ was probably released through respiration and photorespiration processes due to the high temperature and low dPv.

Despite the antagonistic nature of respiration and photorespiration to the effective assimilation of inorganic $CO_2$, given the increase in the proportion of sugars destined for metabolic regulation and, consequently, the reduction in their availability for growth [74], both routes are essential to avoid cellular collapse, enabling acclimatization and maximum physiological integrity of the plant during stress [54]. The gas exchanges observed for ipê-rosa seedlings in a greenhouse provide a central alternative based on diurnal micrometeorological dynamics: C3 plants, such as ipê-rosa, regulate photosynthesis according to the combination of extremes related to air temperature and water vapor since values below or above the "optimum" induce restrictions in four crucial processes of inorganic $CO_2$ assimilation. In this case, the following can be highlighted: (i) the capacity of the enzyme ribulose 1,5-bisphosphate carboxylase/oxygenase (Rubisco) to consume ribulose-bisphosphate (RuBP) molecules; (ii) the regenerative potential of RuBP in reactions adjacent to the Calvin Cycle and thylakoid membrane; (iii) the capacity to synthesize sugars for the consumption of ATP and regeneration of inorganic phosphate; (iv) and the transport of electrons in the conversion of light into chemical energy [53]. Furthermore, in plant organisms, the thermal index of the mesophyll defines the cellular metabolic rhythm; within certain limits, the increase in leaf temperature implies the favoring of carboxylation processes, which, on the other hand, is associated with greater water loss for regulation of physiological integrity [53]. However, studies are still needed to characterize this species' ideal environment for photosynthesis based on other micrometeorological conditions.

Another perspective to be discussed refers to the depletion of water available for the metabolism of the plants evaluated. Despite the generalized reduction in E and A in all water regimes after 12:30 h, the responses of the leaves subjected to 100% LWR remained above the other treatments throughout the afternoon period, indicating that the availability of water in the substrate was consistent with that required by the evapotranspiration demand of the seedlings. Thus, in this reference condition (100% LWR), there were no restrictions on the amount of water available, and the seedlings presented losses linked to "potential evapotranspiration". On the other hand, the other treatments enhanced photosynthetic activity during the morning period. Gas exchange is extremely sensitive to water availability since the lack of this resource stimulates the regulation of plant physiology to prevent water loss and the occurrence of successive damage to its biomolecular, anatomical and, ultimately, morphological structure and cellular homeostasis [18].

Dombroski et al. [75], when suppressing irrigation of ipê-rosa seedlings, found a combined negative effect on A, E and gs from the third day in the restrictive environment; however, photosynthetic activity was quickly able to recover after the re-establishment of the irrigation regime. In the present study, throughout the day (hours), the lack of statistical differences for A within the extremes of water supply (25 and 100% LWRs) indicates that environmental changes in the daily flow were insufficient to affect the assimilative stability of these plants. However, under conditions of 50 and 75% LWRs, photoassimilation was favored by the water availability of the morning period; even so, the EUA remained unchanged. This species has a high capacity to use water in the photoassimilative process, modifying the morphophysiological structure and increasing its efficiency according to the restrictive impositions of the growth environment [27]. In this sense, treatments with lower water replacement were expected to demonstrate higher average WUE values in consideration of the complete compensation of daily evapotranspiration. Thus, it is assumed that, at the genotypic level, the evaluated ipê-rosa seedlings had an ideal margin for WUE, regardless of water replacement, associated with adjustment based on water depletion in the substrate at the end of the day. Different progenies tend to react differently when subjected to environmental stress, altering the phenotype of their morphophysiology according to the genetic structure previously adjusted to the native place of their growth and reproduction [19]. This effect can be even more pronounced when we deal with forest species cultivated in a seed form, subject to intraspecific genetic variation; starting from the same mother tree, there are multiple genetic groups [76].

Seedlings of Amazonian tree species show a high capacity to adjust gas exchange in leaf organs in response to environmental stress [77]. When subjected to drought, these plants demonstrate a gradual reduction in photosynthesis, accompanied by alternative mechanisms to carbon and water flows throughout the plant and the development of survival and vegetative resilience strategies. Once regularly rehydrated, they maximize the carbon assimilation process and regulate other indices close to the "optimum", as long as the other components of the environment are not affected [77]. The leaves of ipê-rosa seedlings reacted similarly at metabolic extremes in comparing the gas exchange of seedlings with 25 and 100% LWRs. Given the diurnal micrometeorological changes, the leaves at 25% water replacement leveled their photosynthetic activity downwards, differently from what was observed for the seedlings at 100% water replacement. On the other hand, the leaf organs acclimated to 50 and 75% LWRs showed greater environmental sensitivity as mentioned above; assimilation was intensified during the morning period.

Finally, the species' genetic set possibly showed the capacity to adjust its leaves' photosynthetic physiology to water deficiency since the variables evaluated concerning adaptability to the growth condition, EUA and TF, remained similar in the diurnal flow, even though the amounts of water replacement were different throughout the experimental period.

### 4.4. Chl a Fluorescence of Ipê-Rosa Seedlings

The literature describes that higher plant organisms, under non-stress conditions (biotic or abiotic), present an Fv/Fm close to 0.83 [78]. This behavior can be observed in the evaluated ipê-rosa seedlings since, initially, at 28 DATs (437 DDAs), the variations in the amount of water replaced in the substrate did not alter the integrity of the photochemical apparatus of the leaves. With the evolution of the plant cycle (increase in thermal sum), it was noted that only the plants subjected to 25% LWR underwent photochemical stress. At the same time, the other treatments maintained the integrity of the PSII reaction centers. It was expected that the reduction in water replacement levels would cause a gradual reduction in $Fv/Fm_{predawn}$ after the plants acclimated to the respective treatments [36]; however, this was not observed. Yin et al. [19] also did not observe a decrease in Fv/Fm in *Jatropha curcas* L. seedlings grown in soil at 30% of field capacity compared to those treated at 50 and 80% of field capacity; these authors attributed the differences, not classified as photochemical stress, to the origin of the genetic materials used, which, in turn, demonstrated different phenotypic expressions as an effect of the respective growth environments.

The biochemical functioning of the structure of the dimeric protein supercomplex with multiple subunits of PSII in higher plants is resistant to water stress when this is not associated with other factors that limit good vegetative performance (such as high levels of light and salinity), given the nature of photoprotective mechanisms capable of dissipating excess energy through alternative non-photochemical extinction pathways [79]. Furthermore, Elsheery and Cao [36] suggest that the reaction centers may not be closed and that the accumulation of irreparable damage to the protein complexes associated with the conversion of solar energy into chemical energy may be avoided, presumably through the de-epoxidation of the pigments of the xanthophyll cycle. Therefore, given that photochemical stress was characterized only in the 25% LWR, biochemical and physiological acclimatization of the photochemical apparatus of the plants occurred in water replacement of 50% of that required by daily evapotranspiration or higher, which is sufficient to maintain the functioning of the protein complexes that make up the photochemical chains (PSI, PSII, Cytochrome B6F and ATP synthase).

In addition, the protective potential of plant tissues against damage to the photochemical apparatus, preceding the effective closure of the reaction centers responsible for regulating the maximum support of photochemical energy to ideal levels [80], may have triggered the late impact of water restriction (25% LWR) on $Fm_{predawn}$, which was observed only at 90 DATs (1400 DDAs).

The diurnal flow of Fv/Fm, with acclimation to darkness for 30 min, exposed at 28 DATs (463 DDAs), showed little variation as a function of water availability, probably affected by the genetic variability inherent in seedlings from seed propagation, before water replacement. Despite the influence of water treatments on subsequent evaluations, as well as the distinct variations in the growth environment, the daily dynamics of Fv/Fm for ipê-rosa were related to micrometeorological patterns, outlining curves with a downward concavity, whose minimum points occurred with the maximum daily photosynthetically active radiation, between 11:00 and 13:00 h. Marenco et al. [74] observed a similar pattern for Fv/Fm in leaves in the upper canopy of the species *Coussapoa orthoneura* and *Protium opacum*, with a reduction from 0.80 to 0.70 at midday and subsequent recovery in the late afternoon. The recovery of the Fv/Fm variable values after the midday depression in Amazonian plants is associated with the adaptive potential of these organisms in the face of diurnal climate extremes, characterizing the existence of dynamic photoinhibition [54]. Specifically, there is an increase in energy dissipation in the form of heat and the occurrence of reversible damage to the photochemical apparatus during the stress period; after the reduction in radiation levels and air temperature, there is a gradual temporary closure of the xanthophyll cycle and reactivation of damaged PSII proteins [81]. Analogous to gas exchange, due to the influence of the micrometeorological dynamics during the day inherent to the period in which the experiment was conducted, there was less ordering in the values of $\Phi_{PSII}$, ETR, Fv′/Fm′, NPQ and qL in consideration of Fv/Fm throughout the diurnal evolution.

The variables that make *Chl a* fluorescence in the light-adapted state provide efficiency in coordinating the operation of the photochemical apparatus with protein complexes acclimated to the current growth environment without the "turnover" of the chains and cycles of the photosynthetic machinery or complete oxidation of the primary electron acceptor—QA—and are regulated by momentary fluctuations in the microclimate in a more pronounced way than Fv/Fm, which in turn is determined after 30 min of dark adaptation [39]. However, at 90 DATs (1400 DDAs), the $\Phi_{PSII}$ and ETR of the ipê-rosa seedlings with water availability of 50% water replacement or higher demonstrated an increasing linear trend since the levels of air temperature, dPv and PAR being lower than those of the other evaluation dates gradually made the environment more favorable to the photochemical stage of photosynthesis. The seedlings under 25% LWR showed a linear decreasing trend in the variables mentioned on the same date, with results superior to those obtained in previous evaluations in this water condition. These plants obtained better photochemical performance in the milder environment; even so, the increase in air temperature, dPv and PAR after 11:00 h was also associated with the depletion of water available in the substrate, causing a reduction in $\Phi_{PSII}$ and ETR.

Plant organisms tend to present different phenotypes depending on the interaction between air temperature, available water, dPv and solar irradiance, altering their morpho-physiological structure to survive stress, accompanied by photoprotective mechanisms and the partitioning of the flow of absorbed electrons into assimilative and non-assimilative processes of chemical energy [21]. Furthermore, the functioning of the photosynthetic apparatus is regulated under the optimum limits of environmental factors, air temperature and irradiance. Quantities below these horizons demonstrate insufficiency for the perpetuation of photochemical activity. On the other hand, exceeding them triggers the occurrence of photoinhibition, in principle, in a non-lethal way to the tissues under the condition [82]. Reinforcing this discussion, it is clear that enzymatic activity is regulated by the balance between the components necessary for the occurrence of biochemical reactions, that is, substrate availability and other factors and cofactors that characterize the good performance of a given enzyme [57]. Within the electron transport chain, PSII works in three stages: capturing light energy, converting it into chemical energy and oxidizing the water molecule, triggering the formation of NADPH molecules and ATP synthesis [83].

Thus, the relationship between water availability and irradiance levels, associated with other environmental patterns, qualifies the performance and balance between the energy dissipation pathways of PSII [17]. Concomitant with the micrometeorological variations at 90 DATs, the leaves of the ipê-rosa seedlings maintained at 100% water replacement had the potential amount of water based on the daily evapotranspiration demand. Thus, the passage of the flow of photosynthetic photons through the electron transport chain, continuously generating NADPH and ATP, controlled the demand for the photochemical activity of PSII to the maximum. When already acclimated to the level of 25% LWR, at 66 and 90 DATs, the ipê-rosa seedlings exhibited the lowest percentage variability for NPQ compared to the other treatments.

The physiological expression observed in the leaves of these seedlings provides two lines of analysis that suggest substantial plasticity in the photosynthetic apparatus of ipê-rosa plants in the juvenile phase: (i) seedlings under water stress maintained greater stability in their photochemical activity due to the lower need to change their physiology for survival within micrometeorological extremes; (ii) on the contrary, individuals grown at regular to optimal levels of water replacement responded to environmental limitations to a greater extent, demonstrating tolerance and high photochemical recovery capacity [61,84]. In the first analysis, the fraction of open reaction centers of PSII (qL), which estimates the capacity of the photochemical apparatus to receive electrons considering the complete oxidation state of QA, classifies the degree of impairment of this protein complex [39].

Phytochrome interconversion, regulated by red light wavelengths, directly influences the photochemical activity of PSII. In contrast, the 600 to 700 nm (red) range, which is prominent at the beginning and end of the day, induces activation of the photosynthetic machinery. In comparison, the far-red range (from 730 nm onwards) induces activation of the photosynthetic machinery; in this case, in the final moments of sunlight, PSII "turnover" occurs until the beginning of the night, characterized by a decrease in the action of the electron acceptor and donor fractions of PSII [85]. However, despite the "closure" of the reaction centers, the wavelengths that make up the far-red are essential for maintaining the integrity of PSII; exposure of leaf tissues to these wavelengths drives the regeneration of the proteins in the complex, with subsequent improvement in photochemical capacity [86]. Depending on the times and dates of subsequent assessments, qL values increased until 1:00 p.m. Ouzounis et al. [87] observed opposite behavior for qL in young *Lactuca sativa* plants, with a reduction in qL due to the gain in radiation incident on the leaf surface. This confirms the ability of young ipê-rosa plants to regulate their photochemical activity in the different water replacement treatments according to the characteristics of the growth environment. The perpetuation of photoprotective mechanisms ensured the regulation of the degree of opening of the photoreceptor channels based on the integrity of PSII as a whole, controlling the fractionation between photoassimilative or non-photoassimilative dissipation routes [21,36].

Under general environmental conditions, the variables that make up the *Chl a* fluorescence in the light-adapted state of ipê-rosa seedlings responded partially to microclimate variations during the diurnal periods evaluated, depending on water replacement. For the Cerrado tree species *Miconia fallax*, *Didymopanax macrocarpum*, *Qualea grandiflora*, *Ouratea hexasperma* and *Roupala montana*, Franco and Lüttge [88] described a direct relationship between photochemical performance and photosynthetic photon flux density, high ΦPSII in the early morning, decreasing with increasing irradiance until the maximum depression, prolonged between 11:00 and 15:00 h, with values of 0.09 to 0.26, and upcoming regeneration at the end of the afternoon period. On the other hand, Bacarin et al. [89] demonstrated greater stability for $\Phi_{PSII}$ and $Fv'/Fm'$ in *Gallesia integrifolia* seedlings in the face of diurnal variations in the environment, with averages showing a small increase between 7:30 and 11:30 a.m., a sharp reduction during local noon (time of maximum air temperature and irradiance) and a successive increase until the end of the daylight hours.

The transition of the environment (end of the dry season to the beginning of the rainy season) provided distinct responses for the *Chl a* fluorescence variables in the light-adapted state of ipê-rosa seedlings at 464, 1046 and 1400 DDAs. This also indicates the adaptive degree of the species in consideration of the combination of air temperature, dPv and PAR radiation associated with water availability in consideration of photosynthetic integrity and performance. The physiological patterns of higher plants respond to the combination of environmental factors once adjusted to a certain stress circumstance, regulating their metabolic activity to the margin of survival, being susceptible to the occurrence of processes antagonistic to cellular homeostasis due to the prolongation of unfavorable conditions [90].

Ipê-rosa is a species dispersed across multiple Brazilian biomes, and as such, it demonstrates the ability to adjust its phenotypic expression according to local environmental variations [2]. In its juvenile phase, it requires a substantial period in a given environment to effectively show effects on its photochemical potential and activity [91]. Given the results, only the submission of the ipê-rosa seedlings to 25% water replacement (compared to the daily potential evapotranspiration) resulted in damage to the PSII reaction centers, which, although not very high, characterized the existence of photochemical stress and chronic photoinhibition. Therefore, there were metabolic adjustments, as well as gene expression that conferred a certain degree of tolerance to water restriction on the part of these seedlings in the face of changes in the microclimate, with the continuous adjustment between the dissipation routes of solar energy intercepted by the leaf surface being combined with the balance with photoprotective chains to regulate the metabolic rhythm for survival.

Phylogenetic characteristics are essential in discerning the ideal environment for photosynthetic integrity in Amazonian species, not only defining the ideal environmental levels to guarantee good vegetative performance but also influencing the ability to resist environmental adversities, mainly regarding the combination of irradiance, air temperature and water availability [92,93]. Multiple studies attest to the effectiveness of using *Chl a* fluorescence as a tool for monitoring water stress in plants, allowing the identification of management practices that lead to good silvicultural and agricultural results [94,95]. In this sense, the water supply to ipê-rosa seedlings grown in a protected environment with replacements greater than 50% of the daily potential evapotranspiration was sufficient to keep them out of the state of water stress.

## 5. Conclusions

Ipê-rosa seedlings regulate the physiology of growth, gas exchange and *Chl a* fluorescence according to the amount of water available. Only a 25% level of water replacement (LWR) in the substrate allows the seedlings to survive.

The growth in height and diameter at the collar decreases according to the water restriction, with 90.9 and 84.7% decreases in the increase in these variables in ipê-rosa seedlings with 25% LWR compared to those with 100% LWR. The biomass accumulation of seedlings responds with a reduction of 96.9% compared to the same water levels.

Ipê-rosa, in a greenhouse, with a thermal accumulation of 1655 degree days, presents an accumulated crop evapotranspiration of 593.20 mm and a crop coefficient (kc) that increases over time, varying from 0.59 to 2.86 during the seedling cycle, with an average kc of 1.75. The water sensitivity of young ipê-rosa plants is inverted, with the highest Ky occurring at 50% LWR.

The gas exchange variables (net photosynthesis, transpiration and stomatal conductance) decrease throughout the day in 50% and 75% LWR conditions. However, during the diurnal flow, the lowest values of these variables occur in the treatment of 25% LWR, and the highest at 100% LWR.

A water supply of 50% of daily evapotranspiration keeps the leaves of ipê-rosa seedlings out of photochemical stress. In the long term, only 25% LWR of this quantity regulates them for survival, with the occurrence of non-photoassimilative energy dissipation pathways.

Regardless of water availability, the maximum quantum yield of PSII in ipê-rosa leaves responds inversely to diurnal variations in the microclimate, reaching minimum values at times of maximum daily photosynthetically active radiation.

The variables that make up the *Chl a* fluorescence of ipê-rosa seedlings in the light-adapted state ($\Phi_{PSII}$, ETR, Fv′/Fm′, NPQ and qL) demonstrate different physiological responses to the combination of microclimatic factors and available water.

**Supplementary Materials:** The following supporting information can be downloaded at https://www.mdpi.com/article/10.3390/plants13202850/s1, Figure S1: Hourly variations of mean air temperature—$T_M$ (A), water vapor pressure deficit—dPv (B) and photosynthetically active radiation—PAR (C), at 100 DAT (1558 DDA), in a greenhouse (internal measurements), Sinop City, Mato Grosso state, Brazil; Table S1: Adjusted equations for non-destructive growth variables of ipê-rosa seedlings at different levels of water replacement (LWR), as a function of accumulated degree days (DDA), in a greenhouse.

**Author Contributions:** Conceptualization, K.N.C.S. and A.C.d.S.; methodology, K.N.C.S., A.C.d.S., D.R.B., M.P. and A.P.d.S.; software, K.N.C.S. and A.P.d.S.; validation, K.N.C.S., A.C.d.S., D.R.B., M.P. and A.P.d.S.; formal analysis, K.N.C.S., A.C.d.S., D.R.B., M.P. and A.P.d.S.; investigation, K.N.C.S., S.S.C., L.M.M.d.S., M.C.B.J., B.F.M., D.R.B., M.P. and A.C.d.S.; resources, A.C.d.S. and A.P.d.S.; data curation, K.N.C.S., A.C.d.S., D.R.B., M.P. and A.P.d.S.; writing—original draft preparation, K.N.C.S.; writing—review and editing, K.N.C.S., D.R.B., M.P. and A.P.d.S.; visualization, K.N.C.S., A.C.d.S., D.R.B., M.P. and A.P.d.S.; supervision, A.C.d.S. and A.P.d.S.; project administration, A.C.d.S.; funding acquisition, A.C.d.S. and A.P.d.S. All authors have read and agreed to the published version of the manuscript.

**Funding:** This research was funded by Coordenação de Aperfeiçoamento de Pessoal de Nível Superior—Brazil (CAPES)—Finance Code 001; the National Council for Scientific and Technological Development (CNPq)—Process 308784/2019-7 and undergraduate scientific initiation scholarships (PIBIC); and Federal University of Mato Grosso, specifically for supporting Process 23108.072552/2024-01.

**Data Availability Statement:** Study data can be obtained upon request from the corresponding author or the first author via e-mail. The data are unavailable on the website as the research project is still being developed.

**Conflicts of Interest:** The authors declare no conflicts of interest.

## References

1. Lorenzi, H. *Brazilian Trees: Manual for Identification and Cultivation of Tree Plants in Brazil*, 4th ed.; Instituto Plantarum: Nova Odessa, Brazil, 2002; Volume 2, 368p.
2. Pereira, M.S. *Technical Manual Knowing and Producing Seeds and Seedlings from the Caatinga;* Association Caatinga: Fortaleza, Brazil, 2011; 60p.
3. Mendonça, A.M.C.; Lira, J.M.S.; Vilela, A.L.O.; Vieira, D.A.; Melo, N.C.; Barbosa, J.P.R.A.D. High aluminum concentration and initial establishment of *Handroanthus impetiginosus*: Clues about an Al non-resistant species in Brazilian Cerrado. *J. For. Res.* **2020**, *31*, 2075–2082. [CrossRef]
4. Benevides, D.S.; Carvalho, F.G. Survey of bee flora present in Caatinga areas in the municipality of Caraúbas-RN. *Soc. Territ.* **2009**, *21*, 44–54.
5. Vieira, C.R.; Araujo, M.M.V.; Ferreira, A.F. Base saturation in the initial growth of *Tabebuia impetiginosa* seedlings. *Rev. Estud. Ambient.* **2020**, *22*, 6–14. [CrossRef]
6. Pimenta, J.M.A.; Souza, W.M.A.T.; Ferrari, C.S.; Vieira, F.A.; Fajardo, C.G.; Pacheco, M.V. Selection of *Handroanthus impetiginosus* mother trees to support seed collection areas. *Rev. Árvore* **2023**, *47*, e4706. [CrossRef]
7. Lisboa, M.A.N.; Silva, L.V.A.; Nascimento, A.S.; Silva, A.O.; Teixeira, M.R.A.; Ferreira, M.F.R.; Ferreira, S.C.; Silva, A.C.V.; Colares, A.V.; Calixto Júnior, J.T. Diversity, structure, and carbon sequestration potential of the woody flora of urban squares in the Brazilian semiarid region. *Trees For. People* **2024**, *16*, 100561. [CrossRef]
8. Morichetti, M.; Vangi, E.; Collalti, A. Predicted Future Changes in the Mean Seasonal Carbon Cycle Due to Climate Change. *Forests* **2024**, *15*, 1124. [CrossRef]
9. Santos, M.E.C.D.; Melo, R.R.D.; Correia, D.; Sousa, J.A.d.; Santos, A.M.; Silva, A.K.V.D.; Paula, E.A.D.O.; Alves, A.R.; Scatolino, M.V.; Rusch, F.; et al. Variation in the Basic Density of Woods Produced in the Brazilian Semiarid Region Subjected to Different Irrigation Regimes. *Forests* **2023**, *14*, 2168. [CrossRef]
10. Castellanos, J.R.G.; Prito, J.M.; Heinrich, M. Red Lapacho (*Tabebuia impetiginosa*)—A global ethnopharmacological commodity? *J. Ethnopharmacol.* **2009**, *121*, 1–13. [CrossRef]
11. Lourenço, J.A.; Pitangui, C.P.; Jordão, A.A.; Vannucchi, H.; Cecchi, A.O. Absence of mutagenicity and antimutagenicity of the extract obtained from the flowers of ipê-roxo *Tabebuia Impetiginosa* (Mart. ex DC.) Standl. *Rev. Bras. Plantas Med.* **2010**, *12*, 414–420. [CrossRef]
12. Zhang, J.; Hunto, S.T.; Yang, Y.; Lee, J.; Cho, J.Y. *Tabebuia impetiginosa*: A Comprehensive Review on Traditional Uses, Phytochemistry, and Immunopharmacological Properties. *Molecules* **2020**, *25*, 4294. [CrossRef]
13. Nahar, J.; Morshed, M.N.; Rupa, E.J.; Lee, J.H.; Kariyarath Valappil, A.; Awais, M.; Hun, K.J.; Sook, L.J.; Al-Amin, M.; Ahn, J.C.; et al. Roasting Extract of *Handroanthus impetiginosus* Enhances Its Anticancer Activity in A549 Lung Cancer Cells and Improves Its Antioxidant and Anti-Inflammatory Effects in Normal Cells. *Appl. Sci.* **2023**, *13*, 13171. [CrossRef]
14. Santos, L.C.D.; Azevedo, L.S.; Siqueira, E.P.; Castro, A.H.F.; Lima, L.A.R.S. Chemical characterization, antioxidant activity, and cytotoxicity of fatty acids methyl esters from *Handroanthus impetiginosus* (Mart. Ex DC.) Mattos (Bignoniaceae) seeds. *Nat. Prod. Res.* **2024**, *38*, 619–623. [CrossRef] [PubMed]
15. Oliveira, N.P.; Nascimento, J.W.S.; Madalena, N.S., Jr.; Serafim, E.O.; Leandro, B.S.; Pereira, L.S.; Santos, M.C.C.; Nascimento, H.H.C. Ecophysiology of *Handroanthus impetiginosus* seedlings subjected to different irrigation cycles. *Braz. J. Dev.* **2020**, *6*, 36563–36574. [CrossRef]

16. Juárez, R.I.N.; Hodnett, M.G.; Fu, R.; Goulden, M.L.; Randow, C.V. Control of Dry Season Evapotranspiration over the Amazonian Forest as Inferred from Observations at a Southern Amazon Forest Site. *J. Clim.* **2007**, *20*, 2827–2839. [CrossRef]
17. Costa, A.C.; Rezende-Silva, S.L.; Megguer, C.A.; Moura, L.M.F.; Rosa, M.; Silva, A.A. The effect of irradiance and water restriction on photosynthesis in young jatobá-do-cerrado (*Hymenea stigonocarpa*) plants. *Phoyosynthetica* **2015**, *53*, 118–127. [CrossRef]
18. Seleiman, M.F.; Al-Suhaibani, N.; Ali, N.; Akmal, M.; Alotaibi, M.; Refay, Y.; Dindaroglu, T.; Abdul-Wajid, H.H.; Battaglia, M.L. Drought Stress Impacts on Plants and Different Approaches to Alleviate Its Adverse Effects. *Plants* **2021**, *10*, 259. [CrossRef]
19. Yin, C.Y.; Pang, X.Y.; Peuke, A.D.; Wang, X.; Chen, K.; Gong, R.G. Growth and photosynthetic responses in *Jatropha curcas* L. seedlings of different provenances to watering regimes. *Phoyosynthetica* **2016**, *54*, 367–373. [CrossRef]
20. Liu, L.; Li, S.; Tang, F.; Li, P.; Liu, J.; Fu, R.; Zheng, L.; Zhang, J.; Chao, N. MaMYBR30, a Novel 1R-MYB, Plays Important Roles in Plant Development and Abiotic Stress Resistance. *Plants* **2024**, *13*, 1794. [CrossRef]
21. Shirke, P.A.; Pathre, U.V. Diurnal and Seasonal Changes in Photosynthesis and Photosystem 2 Photochemical Efficiency in *Prosopis juliflora* Leaves Subjected to Natural Environmental Stress. *Phoyosynthetica* **2003**, *41*, 83–89. [CrossRef]
22. Bandurska, H. Drought Stress Responses: Coping Strategy and Resistance. *Plants* **2022**, *11*, 922. [CrossRef]
23. Oliveira, B.; Marimon Junior, B.H.; Mews, H.A.; Valadão, M.B.X.; Marimon, B.S. Unraveling the ecosystem functions in the Amazonia–Cerrado transition: Evidence of hyperdynamic nutrient cycling. *Plant Ecol.* **2017**, *218*, 225–239. [CrossRef]
24. Marques, E.Q.; Marimon-Junior, B.H.; Marimon, B.S.; Matricardi, E.A.; Mews, H.A.; Colli, G.R. Redefining the Cerrado–Amazonia transition: Implications for conservation. *Biodivers. Conserv.* **2019**, *29*, 1501–1517. [CrossRef]
25. Santos, W.R.; Souza, L.S.B.; Jardim, A.M.R.F.; Morais, J.E.F.; Santos, M.M.P.; Souza, C.A.A.; Silva, T.G.F. How is the water footprint of the species *Vachellia farnesiana*, *Amburana cearensis*, and *Handroanthus impetiginosus* influenced by abiotic stresses as water deficit and salinity? *Int. J. Phytoremediat.* **2024**, *26*, 784–792. [CrossRef] [PubMed]
26. de Oliveira, I.P.; Schaaf, C.; de Setta, N. Drought Responses in Poaceae: Exploring the Core Components of the ABA Signaling Pathway in Setaria italica and Setaria viridis. *Plants* **2024**, *13*, 1451. [CrossRef]
27. Scarpa, A.L.M.; Cruz, Y.C.; Duarte, V.P.; Castro, E.M.; Pasqual, M.; Oliveira, J.P.V.; Pereira, F.J. Growth Response, Gas Exchange, and Leaf Anatomy of *Handroanthus* spp. Seedlings in Mine Tailings Enriched with Nutrient Solution. *J. Soil Sci. Plant Nutr.* **2022**, *22*, 3774–3787. [CrossRef]
28. Monteiro, E.B.; Silva, A.C.; Souza, A.P.; Tanaka, A.A.; Ferneda, B.G.; Martim, C.C. Water requirements and crop coefficients of tropical forest seedlings in different shading conditions. *Rev. Bras. Eng. Agrícola Ambient.* **2016**, *20*, 709–715. [CrossRef]
29. Keffer, J.F.; Silva, C.C.; Souza, A.P.; Silva, A.C.; Bouvié, L.; Dias, T.K.R. Evapotranspiration and water sensitivity of Amazonian yellow ipe seedlings under different shading conditions. *Rev. Bras. Eng. Agrícola Ambient.* **2019**, *23*, 733–740. [CrossRef]
30. Borella, D.R.; Souza, A.P.; Silva, A.C.; Pizzatto, M.; Keffer, J.F.; Lima, D.C. Water requirements of *Dipteryx alata* Vogel Seedlings at different solar radiation levels in Cerrado-Amazon transition. *Trop. Subtrop. Agroecosyst.* **2020**, *23*, 1–13. [CrossRef]
31. Sabino, M.; Ferneda, B.G.; Martim, C.C.; Bouvié, L.; Da Silva, C.C.; Souza, A.P.; Silva, A.C.; Felipe, R.T.A. Initial growth of amazonian and brazilian Cerrado yellow ipe cultivated under different shading intensities and spectral wavelength. *Interciência* **2020**, *45*, 183–191.
32. Su, X.; Yang, Z.; Zhou, C.; Geng, S.; Chen, S.; Cai, N.; Tang, J.; Chen, L.; Xu, Y. The Response and Evaluation of Morphology, Physiology, and Biochemistry Traits in Triploid Passiflora edulis Sims 'Mantianxing' to Drought Stress. *Plants* **2024**, *13*, 1685. [CrossRef]
33. Souza, A.P.; Mota, L.L.; Zamadei, T.; Martim, C.C.; Almeida, F.T.; Paulino, J. Climate classification and climatic water balance in Mato Grosso state, Brazil. *Nativa* **2013**, *1*, 34–43. [CrossRef]
34. Allen, R.G.; Pereira, L.S.; Raes, D.; Smith, M. Crop Evapotranspiration: Guidelines for Computing Crop Water Requirements. In *FAO Irrigation and Drainage Paper, 56*; FAO: Rome, Italy, 1998; 300p.
35. Martim, C.C.; Zamadei, T.; Souza, A.P.; Almeida, F.T.; Zolin, C.A. Angström-Prescott coefficients and reference evapotranspiration in the Cerrado-Amazon transition region of Mato Grosso. *Rev. Bras. Climatol.* **2020**, *26*, 579–594. [CrossRef]
36. Elsheery, N.; Cao, K.F. Gas exchange, *Chl*orophyll fluorescence, and osmotic adjustment in two mango cultivars under drought stress. *Acta Physiol. Plant.* **2008**, *30*, 769–777. [CrossRef]
37. Oxborough, K.; Baker, N.R. Resolving *Chl*orophyll a fluorescence images of photosynthetic efficiency into photochemical and non-photochemical components—Calculation of qP and Fv′/Fm′ without measuring Fo′. *Photosynth. Res.* **1997**, *54*, 135–142. [CrossRef]
38. Lichtenthaler, H.K.; Buschmann, C.; Knapp, M. How to correctly determine the different *Chl*orophyll fluorescence parameters and the *Chl*orophyll fluorescence decrease ratio RFd of leaves with the PAM fluorometer. *Phoyosynthetica* **2005**, *43*, 379–393. [CrossRef]
39. Baker, N.R. *Chl*orophyll Fluorescence: A Probe of Photosynthesis In Vivo. *Annu. Rev. Plant Biol.* **2008**, *59*, 89–113. [CrossRef]
40. Monteiro, E.B. Growth and Water Requirements of Forest Seedlings under Different Shading Conditions. Master's Thesis, Federal University of Mato Grosso, Sinop, Turkey, 2015; 218p.
41. Ometto, J.C. *Plant Bioclimatology*; Agronômica Ceres: São Paulo, Brazil, 1981; 425p.
42. Souza, A.P.; Leonel, S.; Silva, A.C. Basal temperature and thermal sum in phenological phases of nectarine and peach cultivars. *Pesqui. Agropecuária Bras.* **2011**, *46*, 1588–1596. [CrossRef]
43. Ferreira, D.F. Sisvar: A computer statistical analysis system. *Ciência Agrotecnologia* **2011**, *35*, 1039–1042. [CrossRef]
44. Marenco, R.A.; Neves, T.S.; Camargo, M.A.B.; Dias, D.P.; Costa, G.F.; Rodrigues, J.C. Dynamic photoinhibition of photosynthesis in canopy trees from the Central Amazon. *Rev. Bras. Biociências* **2007**, *5*, 150–152.

45. Souza, A.P.; Zamadei, T.; Monteiro, E.B.; Casavecchia, B.H. Atmospheric Transmissivity of the Global Radiation in the Amazonic Region of Mato Grosso. *Rev. Bras. Meteorol.* **2016**, *31*, 639–648. [CrossRef]
46. Borella, D.R.; Souza, A.P.; Silva, K.N.C.; Santos, L.M.M.; Ximenes, E.S.O.C.; Anjos, A.M. Dynamics and estimates of air temperature and relative humidity in nurseries protected with different shading. *Nativa* **2021**, *9*, 62–75. [CrossRef]
47. Monteiro, E.B.; Silva, C.C.; Silva, A.C.; Souza, A.P. Estimating Emission of Leaves Seedlings Forest in Different Shading Levels, at Conditions of Transition Amazon-Cerrado, Brazil. *Am. J. Plant Sci.* **2014**, *5*, 2330–2341. [CrossRef]
48. Oliveira, A.S.; Ribeiro, A.; Silva, C.R.A.; Xavier, A.; Freitas, A.F. Modeling the growth of eucalyptus seedlings based on thermal sum. *Rev. Árvore* **2017**, *41*, e410212. [CrossRef]
49. Lima, P.R.; Horbach, M.A.; Dranski, J.A.L.; Ecco, M.; Malavasi, M.M.; Malavasi, U.C. Morphophysiological evaluation of *Handroanthus impetiginosus* (Mart. ex DC.) Mattos seedlings during hardening. *Floresta Ambiente* **2014**, *21*, 316–326. [CrossRef]
50. Augostinho, L.M.D.; Prado, R.M.; Rozane, D.E.; Freitas, N. Accumulation of dry mass and nutrient absorption rate in 'Pedro Sato' guava seedlings. *Bragantia* **2008**, *67*, 577–585. [CrossRef]
51. Nogueira, R.C.; Paiva, R.; Lima, E.C.; Soares, G.A.; Oliveira, L.M.; Santos, B.R.; Emrich, E.B.; Castro, A.H.F. Curva de crescimento e análises bioquímicas de calos de murici-pequeno (*Byrsonima intermedia* A. Juss.). *Rev. Bras. Pl. Med.* **2008**, *10*, 44–48.
52. César, F.R.C.F.; Matsumoto, S.N.; Viana, A.E.S.; Bonfim, J.A. Initial growth and quality of *Pterogyne nitens* Tull. seedling under artificial shading gradient. *Ciência Florest.* **2014**, *24*, 357–366. [CrossRef]
53. Sage, R.F.; Kubien, D.S. The temperature response of $C_3$ and $C_4$ photosynthesis. *Plant Cell Environ.* **2007**, *30*, 1086–1106. [CrossRef]
54. Marenco, R.A.; Nascimento, H.C.S.; Magalhães, N.S. Stomatal conductance in Amazonian tree saplings in response to variations in the physical environment. *Phoyosynthetica* **2014**, *52*, 493–500. [CrossRef]
55. Kutschera, U. Review Cell Expansion in Plant Development. *Braz. J. Plant Physiol.* **2000**, *12*, 65–98.
56. Bashline, L.; Lei, L.; Li, S.; Gu, Y. Cell wall, cytoskeleton, and cell expansion in higher plants. *Mol. Plant* **2014**, *7*, 586–600. [CrossRef] [PubMed]
57. Johnson, G.N. Physiology of PSI cyclic electron transport in higher plants. *Biochim. Biophys. Acta—Bioenerg.* **2011**, *1807*, 384–389. [CrossRef] [PubMed]
58. Duarte, D.M.; Rocha, G.T.F.B.L.; Matos, F.S.; Rodrigues, F. Responses of paricá seedlings to water stress. *Rev. Floresta* **2016**, *46*, 405–412. [CrossRef]
59. Yamori, W.; Shikanai, T. Physiological functions of cyclic electron transport around photosystem I in sustaining photosynthesis and plant growth. *Annu. Rev. Plant Physiol.* **2016**, *67*, 81–106. [CrossRef]
60. Zhou, H.; Zhou, G.; He, Q.; Zhou, L.; Ji, Y.; Zhou, M. Environmental explanation of maize specific leaf area under varying water stress regimes. *Environ. Exp.* **2020**, *171*, 103932. [CrossRef]
61. Santos, V.A.H.F.; Ferreira, M.J.; Rodrigues, J.V.F.C.; Garica, M.N.; Ceron, J.V.B.; Nelson, B.W.; Saleska, S.R. Causes of reduced leaf-level photosynthesis during strong El Niño drought in a Central Amazon forest. *Glob. Chang. Biol.* **2018**, *24*, 4266–4279. [CrossRef]
62. Maeda, E.E.; Ma, X.; Wagner, F.H.; Kim, H.; Oki, T.; Eamus, D.; Huete, A. Evapotranspiration seasonality across the Amazon Basin. *Earth Syst. Dynam.* **2017**, *8*, 439–454. [CrossRef]
63. Sabonaro, D.Z.; Galbiatti, J.A. Effect of irrigation levels in substrates for "ipê-roxo" seedlings production. *Sci. For.* **2007**, *1*, 95–102.
64. Cunha, A.O.; Andrade, L.A.; Bruno, R.L.A.; Silva, J.A.L.; Souza, V.C. Efeitos de substratos e das dimensões dos recipientes na qualidade das mudas de *Tabebuia impetiginosa* (Mart. Ex D.C.) Standl. *Rev. Árvore* **2005**, *29*, 507–516. [CrossRef]
65. Carvalho, D.F.; Lima, M.E.; Oliveira, A.D.; Rocha, H.S.; Guerra, J.G.M. Crop coefficiente and water consumption of eggplant in no-tillage system and conventional soil preparation. *Eng. Agrícola* **2012**, *32*, 784–793. [CrossRef]
66. Pessoa, J.L.; Freire, A.L.O.; Costa, A.S. Gas exchange of *Handroanthus impetiginosus* (Mart. ex DC) Mattos plants under water stress and rehydration. *Rev. Ciências Agroveterinárias* **2017**, *16*, 269–276. [CrossRef]
67. Bueno, M.M.; Leles, P.S.S.; Abreu, J.F.G.; Santos, J.J.S.; Carvalho, D.F. Water requirement and growth indicators of forest tree species seedlings produced with automated irrigation management. *PLoS ONE* **2020**, *15*, e0238677. [CrossRef] [PubMed]
68. Melo, L.A.; Abreu, A.H.M.; Leles, P.S.S.; Oliveira, R.R.; Silva, D.T. Quality and initial growth of seedlings *Mimosa caesalpiniifolia* Benth. produced in different volumes of containers. *Ciência Florest.* **2018**, *28*, 47–55. [CrossRef]
69. Silva, B.L.B.; Costa, E.; Binotti, F.F.S.; Benett, C.G.S.; Silva, A.G. Quality and growth of achachairu seedlings depending on substrate and shading. *Pesqui. Agropecuária Trop.* **2018**, *48*, 407–413. [CrossRef]
70. Esquivel-Muelbert, A.; Baker, T.R.; Dexter, K.G.; Lewis, S.L.; Brienen, R.J.; Feldpausch, T.R.; Lloyd, J.; Monteagudo-Mendoza, A.; Arroyo, L.; Álvarez-Dávila, E.; et al. Compositional response of Amazon forests to climate change. *Glob. Chang. Biol.* **2018**, *25*, 39–56. [CrossRef]
71. Mira-García, A.B.; Romero-Triguero, C.; Gambín, J.M.B.; Sáchez-Iglesias, M.P.; Tortosa, P.A.N.; Nicolás, E.N. Estimation of stomatal conductance by infra-red thermometry in citrus trees cultivated under regulated deficit irrigation and reclaimed water. *Agr. Water Manag.* **2020**, *276*, e108057. [CrossRef]
72. Marenco, R.A.; Antezana-Vera, S.A.; Gouvêa, R.S.; Camargo, M.A.B.; Oliveira, M.F.; Santos, J.K.S. Physiology of Amazon tree species: Photosynthesis, respiration and water relations. *Ceres* **2014**, *61*, 786–799. [CrossRef]
73. Silva, A.M.L.; Costa, M.F.B.; Leite, V.G.; Rezende, A.A.; Teixeira, S.P. Leaf anatomy with taxonomic implications in ipe species. *Hoehnea* **2009**, *36*, 329–338. [CrossRef]

74. Marenco, R.A.; Lopes, N.F. *Fisiologia Vegetal: Fotossíntese, Respiração, Relações Hídricas e Nutrição Mineral*, 3rd ed.; Editora UFV: Viçosa, Brazil, 2009; 486p.
75. Dombroski, J.L.D.; Freitas, R.M.O.; Tomczak, V.E.; Pinto, J.R.S.; Farias, R.M. Ecophysiology of water stressed *Handroanthus impetiginosus* (Mart. Ex. DC) Mattos) Seedlings. *Sci. For.* **2014**, *42*, 155–163.
76. Souza, A.F.; Rocha Junior, E.O.; Laura, V.A. Early development and efficiency in water and nitrogen use by seedlings of *Calophyllum brasiliense, Eucalyptus urograndis, Tabebuia impetiginosa* and *Toona ciliata. Ciência Florest.* **2018**, *28*, 1465–1477. [CrossRef]
77. Shimpl, F.C.; Ferreira, M.J.; Jaquetti, R.K.; Martins, S.C.V.; Gonçalves, J.F.C. Physiological responses of young Brazil nut (*Bertholletia excelsa*) plants to drought stress and subsequent rewatering. *Flora* **2019**, *252*, 10–17. [CrossRef]
78. Björkman, O.; Demmig, B. Photon yield of $O_2$ evolution and *Chl*orophyll fluorescence characteristics at 77 K among vascular plants of diverse origins. *Planta* **1987**, *170*, 489–504. [CrossRef] [PubMed]
79. Melo, H.F.; Souza, E.R.; Cunha, J.C. Fluorescence of *Chl*orophyll a and photosynthetic pigments in *Atriplex nummularia* under abiotic stresses. *Rev. Bras. De Eng. Agrícola E Ambient.* **2017**, *21*, 232–237. [CrossRef]
80. Guo, Y.Y.; Yu, H.Y.; Kong, D.S.; Yan, F.; Zhang, Y.J. Effects of drought stress on growth and *Chl*orophyll fluorescence of *Lycium ruthenicum* Murr. Seedlings. *Phoyosynthetica* **2016**, *54*, 524–531. [CrossRef]
81. Dias, D.P.; Marenco, R.A. Photoinhibition of photosynthesis in *Minquartia guianensis* and *Swietenia macrophylla* inferred by monitoring the initial fluorescence. *Phoyosynthetica* **2006**, *44*, 235–240. [CrossRef]
82. Dongsansuk, A.; Lütz, C.; Neuner, G. Effects of temperature and irradiance on quantum yield of PSII photochemistry and xanthophyll cycle in a tropical and a temperate species. *Phoyosynthetica* **2013**, *51*, 118–127. [CrossRef]
83. Nelson, N.; Yocum, C.F. Structure and function of photosystems I and II. *Annu. Rev. Plant Physiol.* **2006**, *57*, 521–565. [CrossRef]
84. Liberato, M.A.R.; Gonçalves, J.F.C.; Chevreuil, L.R.; Nina, A.R., Jr.; Fernandes, A.V.; Santos, U.M., Jr. Leaf water potential, gas exchange and *Chl*orophyll a fluorescence in acariquara seedlings (*Minquartia guianensis* Aubl.) under water stress and recovery. *Braz. J. Plant Physiol.* **2006**, *18*, 315–323. [CrossRef]
85. Taiz, L.; Zeiger, E.; Moller, I.M.; Murphy, A. *Fisiologia e Desenvolvimento Vegetal*, 6th ed.; Artmed: Porto Alegre, Brazil, 2017; 888p.
86. Thapper, A.; Mamedov, F.; Mokvist, F.; Hammarström, L.; Styring, S. Defining the Far-Red Limit of Photosystem II in Spinach. *Plant Cell* **2009**, *21*, 2391–2401. [CrossRef]
87. Ouzounis, T.; Parjikolaei, B.R.; Fretté, X.; Rosenqvist, E.; Ottosen, C.O. Predawn and high intensity application of supplemental blue light decreases the quantum yield of PSII and enhances the amount of phenolic acids, flavonoids, and pigments in *Lactuca sativa. Front. Plant Sci.* **2015**, *6*, 19. [CrossRef]
88. Franco, A.; Lüttge, U. Midday depression in savanna trees: Coordinated adjustments in photochemical efficiency, photorespiration, $CO_2$ assimilation and water use efficiency. *Oecologia* **2002**, *31*, 356–365. [CrossRef] [PubMed]
89. Bacarin, M.A.; Martinazzo, E.G.; Cassol, D.; Falqueto, A.R.; Silva, D.M. Daytime variations of *Chl*orophyll *a* fluorescence in Pau d'alho seedlings. *Rev. Árvore* **2016**, *40*, 1023–1030. [CrossRef]
90. Weng, J.H.; Chen, Y.N.; Liao, T.S. Relationships between *Chl*orophyll fluorescence parameters and photochemical reflectance index of tree species adapted to different temperature regimes. *Plant Funct. Evol. Biol.* **2006**, *33*, 241–246. [CrossRef]
91. Silva, L.K.S.; Alves, M.C.J.L.; Costa, R.N.; Silva, D.M.R.; Santos, J.C.C.S.; Moura, F.B.P.; Silva, J.M., Jr.; Silva, J.V. Gas exchange and photochemical efficiency of Caatinga plants submitted to different water management strategies. *J. Agric. Sci.* **2019**, *11*, 53–67. [CrossRef]
92. Morais, R.R.; Gonçalvez, J.F.C.; Santos, U.M., Jr.; Santos, A.L.W. *Chl*oroplastid pigment contents and *Chl*orophyll a fluorescence in Amazonian tropical three species. *Rev. Árvore* **2007**, *31*, 959–966. [CrossRef]
93. Ortiz, D.; Moreno, F.; Díez, M.C. Photosynthesis, growth, and survival in seedlings of four tropical fruit-tree species under intense radiation. *Acta Amaz.* **2021**, *51*, 1–9. [CrossRef]
94. Reis, L.C.; Scalon, S.P.Q.; Dresch, D.M.; Foresti, A.C.; Santos, C.C.; Pereira, Z.V. *Chl*orophyll a fluorescence as an indicator of water stress in *Calophyllum brasiliense. Not. Bot. Horti Agrobot. Cluj-Napoca* **2020**, *48*, 210–2020. [CrossRef]
95. Reis, L.A.C.; Oliveira, J.A.; Farnese, F.S.; Rosado, A.M.; Reis, L.A.C. *Chl*orophyll fluorescence and water content parameters are good biomarkers for selecting drought tolerant eucalyptus clones. *For. Ecol. Manag.* **2021**, *481*, 118682. [CrossRef]

plants

MDPI

Article

# Water and Fertilizer Management Is an Important Way to Synergistically Enhance the Yield, Rice Quality and Lodging Resistance of Hybrid Rice

Haijun Zhu [1,†], Lingli Nie [1,2,†], Xiaoe He [1], Xuehua Wang [1,*], Pan Long [1,*] and Hongyi Chen [2,*]

[1] Key Laboratory of Ministry of Education for Crop Physiology and Molecular Biology, Hunan Agricultural University, Changsha 410128, China; 15211104718@163.com (H.Z.); linglinie@163.com (L.N.); hexiaoe2022@163.com (X.H.)

[2] State Key Laboratory of Hybrid Rice, Hunan Hybrid Rice Research Center, Changsha 410125, China

* Correspondence: 13873160151@163.com (X.W.); longpan_lp@126.com (P.L.); chy7960@163.com (H.C.)

† These authors contributed equally to this work.

**Abstract:** This study comprehensively investigated the synergistic effects and underlying mechanisms of optimized water and fertilizer management on the yield, quality, and lodging resistance of hybrid rice (*Oryza sativa*), through a two-year field experiment. Two hybrid rice varieties, Xinxiangliangyou 1751 (XXLY1751) and Yueliangyou Meixiang Xinzhan (YLYMXXZ), were subjected to three irrigation methods (W1: wet irrigation, W2: flooding irrigation, W3: shallow-wet-dry irrigation) and four nitrogen fertilizer treatments (F1 to F4 with application rates of 0, 180, 225, and 270 kg ha$^{-1}$, respectively). Our results revealed that the W1F3 treatment significantly enhanced photosynthetic efficiency and non-structural carbohydrate (NSC) accumulation, laying a robust foundation for high yield and quality. NSC accumulation not only supported rice growth but also directly influenced starch and protein synthesis, ensuring smooth grain filling and significantly improving yield and quality. Moreover, NSC strengthened stem fullness and thickness, converting them into structural carbohydrates such as cellulose and lignin, which substantially increased stem mechanical strength and lodging resistance. Statistical analysis demonstrated that water and fertilizer treatments had significant main and interactive effects on photosynthetic rate, dry matter accumulation, yield, quality parameters, NSC, cellulose, lignin, and stem bending resistance. This study reveals the intricate relationship between water and fertilizer management and NSC dynamics, providing valuable theoretical and practical insights for high-yield and high-quality cultivation of hybrid rice, significantly contributing to the sustainable development of modern agriculture.

**Keywords:** yield; quality; lodging resistance; NSC; fertilizer management; irrigation methods

## 1. Introduction

Water and fertilizer management has always occupied a crucial position in the agricultural field, especially for hybrid rice (*Oryza sativa*), and its significance is even more far-reaching and significant [1]. In the current context where the global food security situation is becoming increasingly severe, continuously increasing the yield of rice to meet the huge demand for food from the continuously growing population has become an urgent and crucial task [2]. For hybrid rice, as a crop type with significant yield-increasing potential, the significance of effectively enhancing its yield to ensure an adequate supply of food goes without saying [3,4]. Numerous past research results clearly indicate that a reasonable and precise water and fertilizer management model can have a significant positive impact on the yield of hybrid rice [5–7]. Through in-depth research, Zhu [1] found that precisely regulating the irrigation water volume and the time, type, and dosage of fertilizer application can create the most suitable growth environment for hybrid rice, thereby greatly promoting the efficiency of photosynthesis and the effect of nutrient absorption, laying a solid foundation for a substantial increase in yield.

However, the goal we pursue is by no means limited solely to the simple increase in yield. With the development of modern society and the significant improvement of people's living standards, strict requirements for the quality of rice have also gradually emerged [8]. High-quality rice is not only closely related to the enjoyment of taste and the level of nutritional value, but also directly related to people's health status [9]. Water and fertilizer management play a vital role in influencing the quality of rice [10]. The just-right water and fertilizer supply model can finely regulate many aspects of rice, including appearance quality, processing quality, and cooking and taste quality [11,12]. The research of Sun [13] clearly pointed out that a specific water management method can effectively reduce the chalkiness degree and thereby significantly improve the rate of whole milled rice, while Ning [14], through a large number of experiments and data analysis found that a reasonable fertilizer operation plan can effectively improve the protein content and starch characteristics of rice, thereby optimizing its eating quality. At the same time, the lodging of hybrid rice has long been one of the troublesome problems that have plagued the majority of growers [15]. Lodging will not only directly lead to a serious loss of yield, but also bring great difficulties and cost increases to the harvesting work, and have a serious impact on the economic benefits of rice production [16]. Furthermore, the scientific and reasonable water and fertilizer management strategy can effectively enhance the stem strength and lodging resistance ability of rice plants. By rationally regulating moisture and avoiding long-term excessive water logging or severe drought conditions, it can strongly promote the good development of the root system and the robust growth of the stem [17]. In terms of fertilizer management, achieving a balanced supply of various nutritional elements such as nitrogen, phosphorus, and potassium, especially strictly controlling the excessive application of nitrogen fertilizer, plays a crucial role in enhancing the mechanical strength and toughness of the stem [18]. Many related research works such as Zhang's [19] have deeply revealed the close internal connection existing between water and fertilizer management and lodging resistance.

In the past abundant research process, many scholars have carried out extensive and in-depth explorations on the relationship between water and fertilizer management and the yield, quality, and lodging resistance of hybrid rice [20–22]. However, despite the fruitful research results that have been achieved, there is still a series of key problems that urgently need to be further deeply studied and properly solved. How to accurately formulate personalized water and fertilizer management plans for different ecological environments and planting conditions to achieve the best synergistic effect of yield, quality, and lodging resistance is still a highly challenging topic. In addition, regarding the internal mechanism of how water and fertilizer management affects the physiological and biochemical processes of hybrid rice, our current understanding still has certain limitations and requires more in-depth and comprehensive research to fully reveal its internal laws.

This study aims to deeply and systematically explore the relationship between water and fertilizer management and the yield, quality, and lodging resistance of hybrid rice. By comprehensively using diversified research methods such as field experiments and physiological and biochemical analyses, it strives to fully reveal the internal mechanism and synergistic effect. We earnestly aim that through this study, we can provide a solid scientific basis and practical operational guidance for optimizing the water and fertilizer management of hybrid rice, strongly promoting the sustainable development of the hybrid rice industry, and making due contributions to effectively ensuring global food security and significantly improving people's quality of life. Meanwhile, our goal is also that this study can attract more scholars' high attention and in-depth discussion in this field, thereby promoting the continuous progress and development of relevant research.

## 2. Results

### 2.1. Photosynthetic Rate

Water and fertilizer management has a significant impact on the net photosynthetic rate of leaves during different growth stages of hybrid rice (Figure 1). Among them, different irrigation treatments had a significant effect ($p < 0.05$) on the transpiration rate of rice leaves during the tillering and grain filling stages in 2022, as well as the tillering, booting, and grain filling stages in 2023. There was no significant effect during the booting stage in 2022, but there was a highly significant effect ($p < 0.01$) during other growth stages. Fertilizer had a highly significant effect ($p < 0.01$) on the photosynthetic rate during all examined periods. Varietal differences had no significant effect on the net photosynthetic rate during the tillering stage in both 2022 and 2023, but had a highly significant effect ($p < 0.01$) during other growth stages. The interaction between water and fertilizer only had a significant effect ($p < 0.05$) during the grain filling stage in 2023.

As rice grows, the photosynthetic rate shows a downward trend, with the highest rate during the tillering stage and the lowest during the grain filling stage. The water treatment strategy varies at different growth stages: from tillering to booting, W3 is the most optimal; from full heading to grain filling, W1 is better. The effect of fertilizer application also varies by stage: from tillering to booting, the photosynthetic rate increases with increasing fertilizer application, i.e., F4 > F3 > F2 > F1; from full heading to grain filling, it first increases and then decreases, i.e., F3 > F4 > F2 > F1.

There are significant differences in the net photosynthetic rate of rice leaves under different water, fertilizer, and cultivar treatments. Taking the full heading stage as an example, Xinxiangliangyou 1751 had the highest net photosynthetic rate under W1F4V1 treatment in both 2022 and 2023, with rates of 17.01 $\mu$mol $m^{-2}$ $s^{-1}$ and 16.89 $\mu$mol $m^{-2}$ $s^{-1}$, respectively. For Yueliangyou Meixiangxinzhan, the highest net photosynthetic rate during the full heading stage in both 2022 and 2023 was achieved under W1F3V1 treatment, with rates of 16.29 $\mu$mol $m^{-2}$ $s^{-1}$ and 16.18 $\mu$mol $m^{-2}$ $s^{-1}$, respectively.

### 2.2. Total Dry Weight (TDW)

Distinct differences in TDW accumulation were observed across rice growth stages under diverse water and fertilizer treatments (Figure 2). Treatments W1F3V1 and W1F3V2 were identified as the most effective in promoting TDW accumulation, achieving above-ground dry matter levels of 1465.19 g $m^{-2}$ and 1377.51 g $m^{-2}$ at maturity, respectively, significantly surpassing other treatment groups. These findings unequivocally demonstrate that appropriate water and fertilizer management strategies can significantly enhance the dry matter accumulation of rice, potentially leading to positive impacts on its yield and quality.

### 2.3. Yield

The yield of rice was significantly influenced by water and fertilizer management strategies (Figure 3). Treatments W1F3V1 and W1F3V2 demonstrated the highest yields, averaging 9.64 t $ha^{-1}$ and 8.99 t $ha^{-1}$, respectively, while W2F1V1 and W2F1V2 resulted in the lowest yields. The significant differences in yields among treatment groups highlight the effectiveness of optimized water and fertilizer management in enhancing rice productivity.

### 2.4. Appearance Quality

Water and fertilizer management strategies had significant effects on the appearance quality of rice, as evidenced by variations in the head rice rate and chalky grain rate (Figure 4). Treatments W1F3V1 and W1F3V2 resulted in higher head rice rates and lower chalky grain rates compared to other treatments, indicating improvements in both processing and appearance qualities. Specifically, the head rice rate reached 70.4% under treatment W1F3V1, while treatment W1F3V2 achieved an impressive 88.3%, both of which were significantly higher than other treatment groups. However, in terms of chalky grain rate, the W1F3V1 and W1F3V2 treatments showed relatively lower performance. The data reveal that the chalky grain rate was 3.2% for W1F3V1 and 4.8% for W1F3V2, both significantly lower than other treatments. The finding indicates that through careful water and fertilizer management, the head rice rate can be effectively increased while reducing the chalky grain rate, thereby enhancing the overall quality of rice.

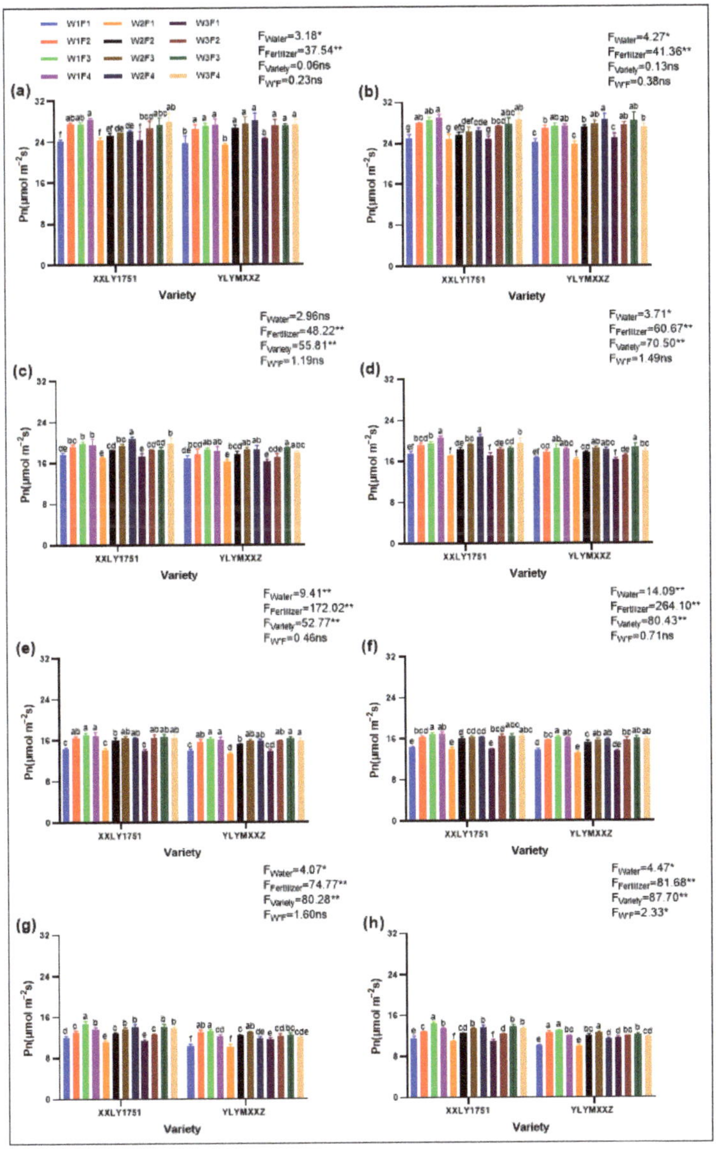

**Figure 1.** (**a**,**b**) represent the tillering stages in 2022 and 2023, respectively; (**c**,**d**) represent the booting stages in 2022 and 2023, respectively; (**e**,**f**) represent the full heading stages in 2022 and 2023, respectively; (**g**,**h**) represent the grain filling stages in 2022 and 2023, respectively. Different lowercase letters on the error bars denote statistical differences (at the 0.05 level) among treatments of various varieties based on the LSD test. Significant differences within the same treatment are denoted by ns ($p > 0.05$), * ($0.01 < p \leq 0.05$), and ** ($p \leq 0.01$).

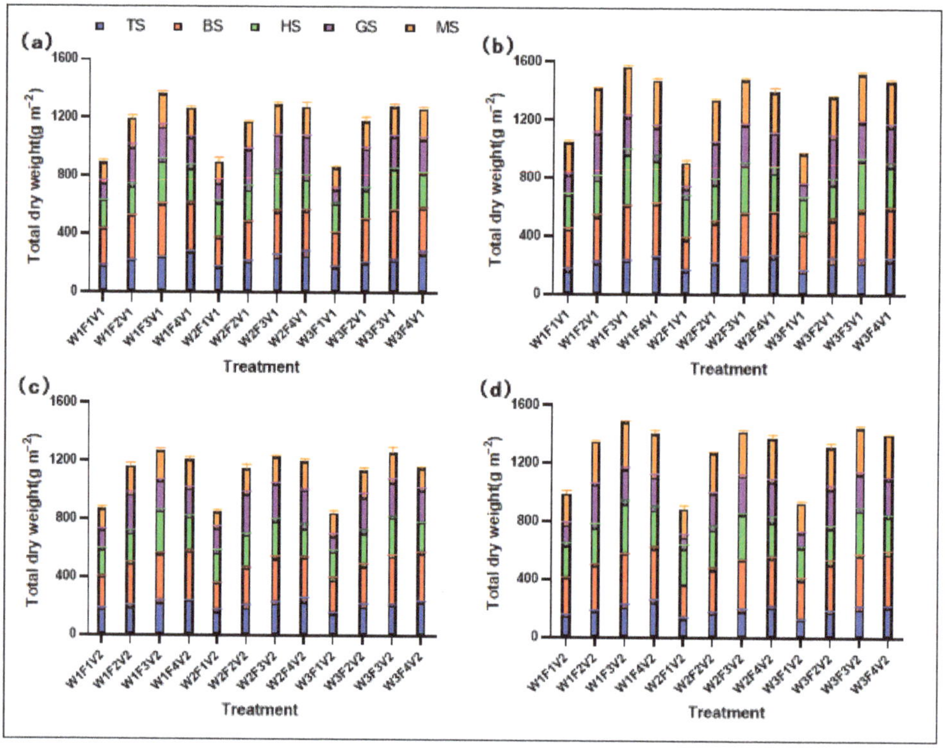


**Figure 2.** TDW of XXLY1751 (**a**) and YLYMXXZ (**b**) in 2022, and XXLY1751 (**c**) and YLYMXXZ (**d**) in 2023.

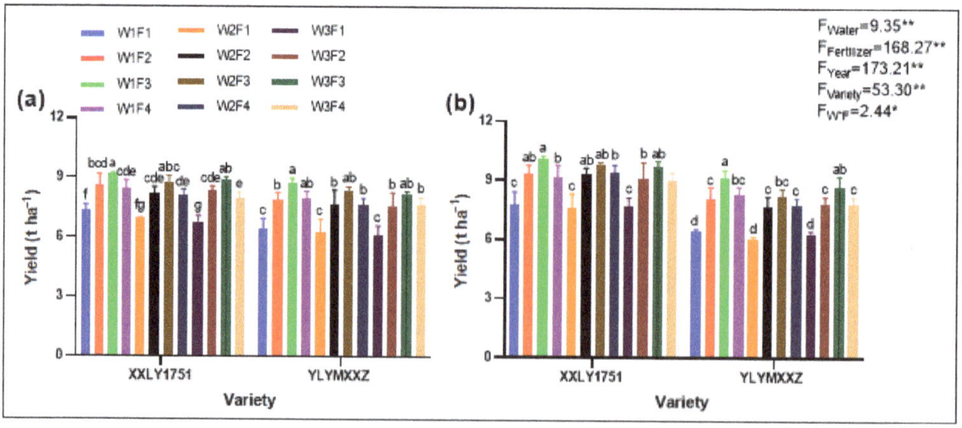


**Figure 3.** Figures (**a**,**b**) represent the yields of Xinxiangliangyou 1751 and Yueliangyou Meixiangxinzhan in 2022 and 2023, respectively. Different lowercase letters on the error bars indicate statistical differences (at a significance level of 0.05) between treatments of various cultivars in the LSD test. Significant differences within the same treatment are denoted by * ($0.01 < p \leq 0.05$), and ** ($p \leq 0.01$).

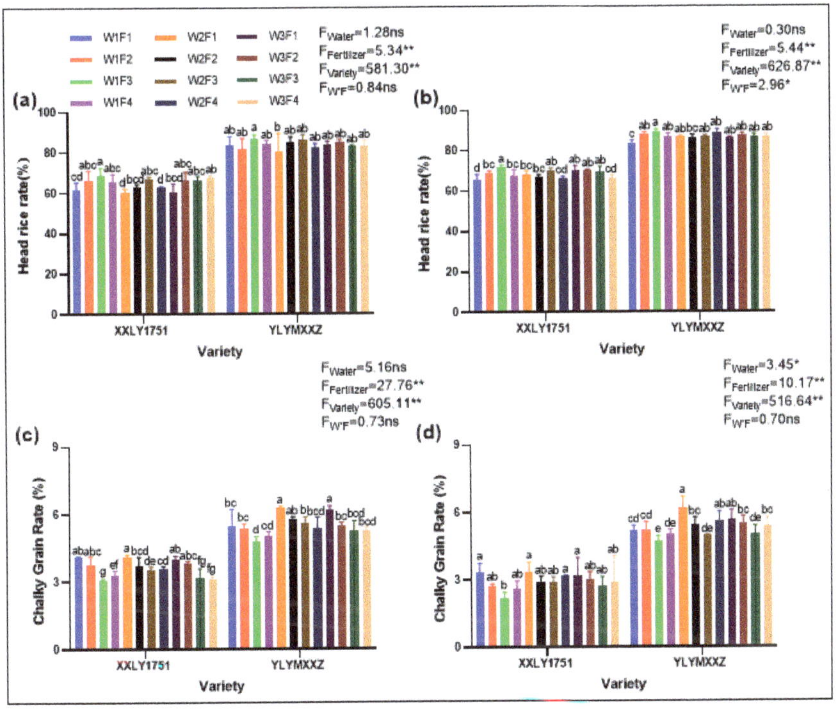


**Figure 4.** Head rice rate of XXLY1751 and YLYMXXZ in 2022 (**a**) and 2023 (**b**), and chalky grain rate of XXLY1751and YLYMXXZ in 2022 (**c**) and in 2023 (**d**). Different lowercase letters denote statistical differences between treatments of each season according to the LSD test (0.05). Significant differences within the same treatment are denoted by ns ($p > 0.05$), * ($0.01 < p \leq 0.05$), and ** ($p \leq 0.01$).

### *2.5. Nutritional Quality*

The effects of water and fertilizer management on protein and amylose content in different high-quality hybrid rice varieties are shown in the accompanying figures (Figure 5). After thorough analysis, we have drawn the following conclusions: Regarding protein content, irrigation treatments did not significantly impact it in 2022 and 2023. However, fertilizer treatments had a highly significant effect ($p < 0.01$), showing an increasing trend with higher fertilizer application, where the F4 treatment resulted in the highest protein content. The cultivar treatment had a significant influence on protein in 2023 ($p < 0.05$), but no notable difference was observed in 2022. It is worth mentioning that the interaction between water and fertilizer did not significantly affect protein content. In terms of specific cultivar responses, Xinxiangliangyou 1751 had the highest protein content under the W2F4V1 treatment, while Yueliangyou Meixiangxinzhan led under the W3F4V2 treatment.

As for amylose content, irrigation, fertilizer, and cultivar treatments all had a highly significant impact ($p < 0.01$) in both years. Wet irrigation (W1 treatment) generally effectively reduced amylose content. The effect of fertilizer treatments showed varying patterns, particularly in the case of Xinxiangliangyou 1751 in 2022, where the amylose content decreased initially with increasing fertilizer application and then rose, with the F3 treatment resulting in the lowest content. Similarly, the interaction between water and fertilizer did not significantly affect amylose content. Regarding cultivar responses, Xinxiangliangyou 1751 had the highest amylose content under the W1F2V1 treatment, while Yueliangyou Meixiangxinzhan led under the W1F2V2 treatment.

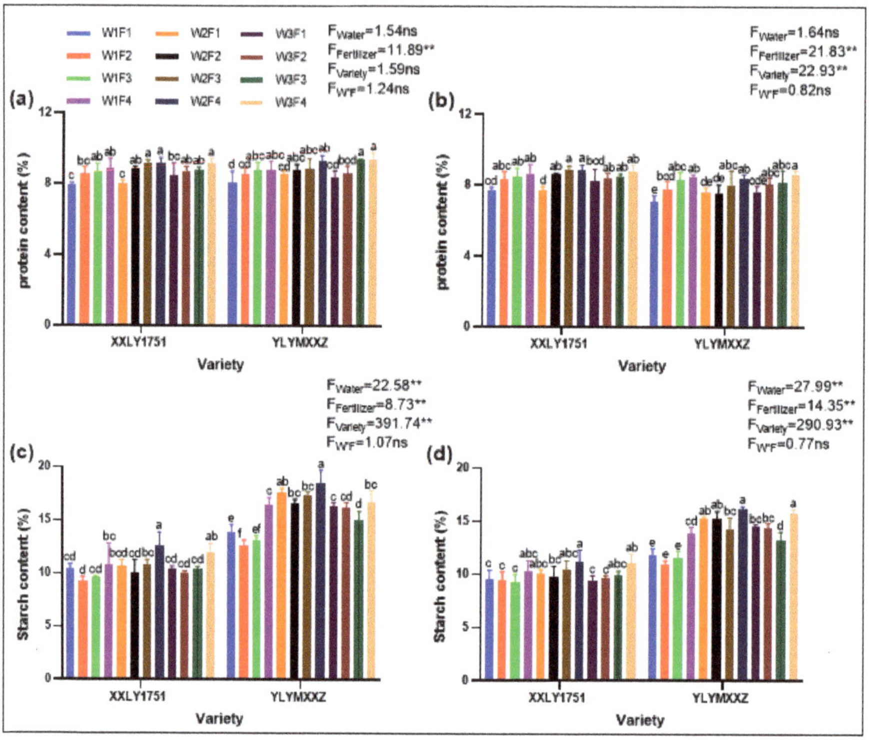


**Figure 5.** Protein content of XXLY1751 and YLYMXXZ in 2022 (**a**) and 2023 (**b**), and amylose content of XXLY1751and YLYMXXZ in 2022 (**c**) and 2023 (**d**). Different lowercase letters denote statistical differences between treatments of each season according to the LSD test (0.05). Significant differences within the same treatment are denoted by ns ($p > 0.05$), and ** ($p \leq 0.01$).

## 2.6. Plant Height at Center of Gravity and Panicle Length

Irrigation and fertilization strategies significantly affected plant height at the center of gravity and panicle length. W1F3V1 and W1F3V2 treatments resulted in longer panicles, indicating the potential of optimized water and fertilizer management in promoting rice panicle growth (Figure 6). Regarding plant height at the center of gravity, distinct differences were observed among different irrigation treatments, with an overall trend of W2 > W3 > W1. Meanwhile, various fertilization treatments also exhibited a certain pattern of influence on plant height at the center of gravity, specifically F4 > F2 > F3 > N1. These results indicate that irrigation and fertilization strategies have a significant regulatory effect on the plant height at the center of gravity of rice plants. In terms of panicle length, the W1F3V1 and W1F3V2 treatments demonstrated notable advantages. Specifically, the panicle length of the W1F3V1 treatment reached 31.46 cm, while that of the W1F3V2 treatment was 27.70 cm.

## 2.7. Lodging Index

Irrigation and fertilization treatments demonstrated significant effects on rice lodging resistance (Table 1). The F3 treatment was identified as the most effective in enhancing stem bending resistance and lodging index, while the W1 treatment also showed a positive impact on lodging resistance. Regarding the influence of irrigation treatments on bending resistance, we observed a general trend of W1 > W3 > W2. However, this trend is reversed when considering the lodging index. This finding indicates that wet irrigation treatment (i.e., W1 treatment) has a significant effect on enhancing rice lodging resistance. Therefore, we can conclude that both W1F3 and W1F2 treatments are effective regulatory measures to improve rice lodging resistance.

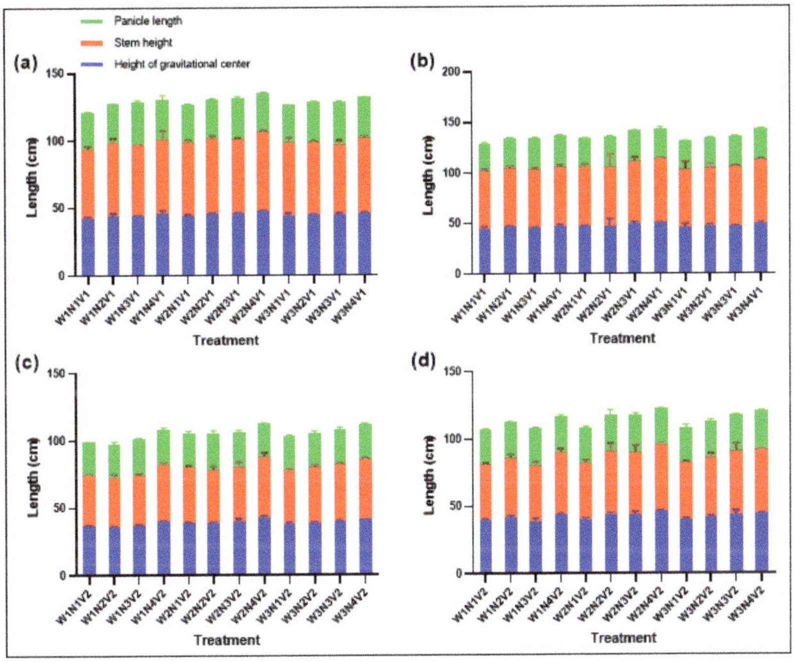


**Figure 6.** Plant height at center of gravity and panicle length of XXLY1751 (**a**) and YLYMXXZ (**b**) in 2022, and XXLY1751 (**c**) and YLYMXXZ (**d**) in 2023.

**Table 1.** Irrigation and fertilization treatments effect on the lodging resistance of different rice stem internodes in 2022 and 2023. J3 and J4 represent the third and fourth basal stem internodes, respectively.

| Year | Variety | Irrigation | Fertilizer | M | | WP | | LI | |
|---|---|---|---|---|---|---|---|---|---|
| | | | | J3 | J4 | J3 | J4 | J3 | J4 |
| 2022 | V1 | W1 | F1 | 861.99 e | 833.07 g | 1159.26 bcd | 923.69 def | 0.74 de | 0.9 f |
| | | | F2 | 955.67 c | 932.54 de | 1475.44 a | 1174.63 a | 0.65 g | 0.79 g |
| | | | F3 | 1008.96 b | 1055.43 a | 1545.99 a | 1176.86 a | 0.65 g | 0.9 f |
| | | | F4 | 999.54 b | 761.54 h | 1069.37 de | 1125.94 ab | 0.93 b | 0.68 h |
| | | W2 | F1 | 717.56 g | 757.67 h | 1024.27 de | 769.23 gh | 0.7 f | 0.98 e |
| | | | F2 | 742.9 fg | 947.31 cd | 1218.76 bc | 915.86 def | 0.61 h | 1.03 d |
| | | | F3 | 982.38 bc | 1005.86 b | 1120.87 cde | 834.78 fg | 0.88 c | 1.2 b |
| | | | F4 | 1005.08 b | 983.44 bc | 1013.36 e | 699.08 h | 0.99 a | 1.41 a |
| | | W3 | F1 | 769.54 f | 880.81 f | 1064.27 de | 990.99 cd | 0.72 ef | 0.89 f |
| | | | F2 | 981.74 bc | 895.71 ef | 1290.56 b | 1038.12 bc | 0.76 d | 0.86 f |
| | | | F3 | 904.99 d | 884.15 f | 1222.8 bc | 932.08 de | 0.74 de | 0.95 e |
| | | | F4 | 1053.27 a | 927.14 def | 1029.34 de | 852.12 efg | 1.02 a | 1.09 c |
| | | Mean | | 915.30 | 905.39 | 1186.19 | 952.78 | 0.78 | 0.97 |

**Table 1.** *Cont.*

| Year | Variety | Irrigation | Fertilizer | M | | WP | | LI | |
|---|---|---|---|---|---|---|---|---|---|
| | | | | J3 | J4 | J3 | J4 | J3 | J4 |
| 2022 | V2 | W1 | F1 | 460.05 e | 623.43 d | 1134.24 f | 882.62 de | 0.41 d | 0.71 cd |
| | | | F2 | 465.03 e | 598.32 d | 1520.39 a | 1130.49 a | 0.31 g | 0.53 g |
| | | | F3 | 551.55 b | 708.63 b | 1422.99 ab | 1073.58 ab | 0.39 d | 0.66 f |
| | | | F4 | 599.2 a | 758.58 a | 1212.07 def | 1030.63 abc | 0.49 a | 0.74 c |
| | | W2 | F1 | 454.6 e | 627.8 d | 1008.7 g | 764.52 f | 0.45 c | 0.82 b |
| | | | F2 | 460.53 e | 697.12 bc | 1373.79 bc | 1027.06 abc | 0.34 f | 0.68 ef |
| | | | F3 | 463.35 e | 679.84 bc | 1249.3 de | 928.85 cd | 0.37 e | 0.73 c |
| | | | F4 | 515.49 d | 745.11 a | 1168.81 ef | 820.59 ef | 0.44 c | 0.91 a |
| | | W3 | F1 | 451.34 e | 625.83 d | 1027.19 g | 784.05 ef | 0.44 c | 0.8 b |
| | | | F2 | 529.81 cd | 674.48 c | 1417.64 b | 1043.46 ab | 0.37 e | 0.65 f |
| | | | F3 | 541.7 bc | 749.42 a | 1280.45 cd | 1016.42 bc | 0.42 cd | 0.74 c |
| | | | F4 | 558.12 b | 620.01 d | 1174.3 ef | 886.62 de | 0.48 b | 0.7 de |
| | | Mean | | 504.23 | 675.71 | 1249.16 | 949.07 | 0.41 | 0.72 |
| 2023 | V1 | W1 | F1 | 895.28 g | 1078.83 f | 1401.65 bc | 1070.58 de | 0.64 g | 1.01 d |
| | | | F2 | 1082.66 e | 1271.38 bc | 1577.03 a | 1206.32 b | 0.69 fg | 1.05 d |
| | | | F3 | 1239.04 bcd | 1327.3 abc | 1573.71 a | 1299.14 a | 0.79 de | 1.02 d |
| | | | F4 | 1317.46 ab | 1301.19 abc | 1382.95 bc | 1161.1 bc | 0.95 c | 1.12 d |
| | | W2 | F1 | 867.32 g | 1134.01 def | 1095.14 d | 867.17 g | 0.79 de | 1.31 c |
| | | | F2 | 884.6 g | 1359 ab | 1302.8 c | 990.76 e | 0.68 fg | 1.37 bc |
| | | | F3 | 1202.37 cd | 1415.9 a | 1267.95 c | 954.06 ef | 0.95 c | 1.48 ab |
| | | | F4 | 1263.2 abc | 1223.3 cde | 1110.95 d | 895.99 f | 1.14 a | 1.37 bc |
| | | W3 | F1 | 994.43 f | 1115.41 ef | 1340.03 bc | 984.98 e | 0.74 ef | 1.13 d |
| | | | F2 | 1040.69 ef | 1259.89 bcd | 1473.39 ab | 1108.15 cd | 0.71 f | 1.14 d |
| | | | F3 | 1173.33 d | 1373.67 ab | 1403.52 bc | 1006.83 e | 0.84 d | 1.36 bc |
| | | | F4 | 1338.62 a | 1415.28 a | 1279.68 c | 901.6 f | 1.05 b | 1.57 a |
| | | Mean | | 1108.25 | 1272.93 | 1350.73 | 1037.22 | 0.83 | 1.24 |
| | V2 | W1 | F1 | 599.5 e | 773.81 d | 1240.13 de | 822.6 g | 0.48 bc | 0.94 ab |
| | | | F2 | 682.76 cd | 841.56 bc | 1616.17 a | 1153.71 a | 0.42 e | 0.73 ef |
| | | | F3 | 701.6 bc | 877.17 bc | 1600.53 a | 1107.21 ab | 0.44 de | 0.79 d |
| | | | F4 | 757.63 a | 944.24 a | 1473.88 b | 1050.86 bcd | 0.51 b | 0.9 bc |
| | | W2 | F1 | 524.99 f | 705.23 e | 1128.8 e | 940.1 ef | 0.47 cd | 0.75 def |
| | | | F2 | 640.9 de | 851.91 bc | 1380.83 bc | 1101.69 abc | 0.46 cd | 0.77 de |
| | | | F3 | 663.18 cd | 992.58 a | 1319.2 cd | 1015.98 cde | 0.5 b | 0.98 a |
| | | | F4 | 740.67 ab | 831 bcd | 1298.16 cd | 921.32 f | 0.57 a | 0.9 bc |
| | | W3 | F1 | 543.63 f | 699.8 e | 1298.25 cd | 1007.25 def | 0.42 e | 0.69 f |
| | | | F2 | 696.26 bc | 881.03 b | 1606.12 a | 1122.51 ab | 0.43 de | 0.78 de |
| | | | F3 | 732.91 ab | 1000.65 a | 1458.66 b | 1016.52 cde | 0.5 b | 0.98 a |
| | | | F4 | 770.8 a | 815.84 cd | 1304.82 cd | 936.19 ef | 0.59 a | 0.87 c |
| | | Mean | | 671.24 | 851.24 | 1393.80 | 1016.33 | 0.48 | 0.84 |

Within a column for each year, means followed by the same letters are not significantly different according to LSD (0.05). M, breaking bending moment; WP, bending moment; LI, lodging index.

### 2.8. Non-Structural Carbohydrate (NSC)

NSC content in rice stems was influenced by water and fertilizer management (Figure 7). The W1F3V1 and W1F3V2 treatments significantly increased NSC content during the grain-filling period, suggesting a positive effect of optimized water and fertilizer management on NSC accumulation. From the heading stage to maturity, we observed a continuous decline in stem NSC content across all treatment groups. This indicates that NSC in the stems is gradually transported to the panicle to support its growth and grain filling. Taking the grain-filling period as an example, we further analyzed the impact of different treatments on stem NSC content. The results showed that the stem NSC content of the W1F3V1 and W1F3V2 treatments was significantly higher than that of other treatment groups, reaching 230.93 mg g$^{-1}$ and 222.20 mg g$^{-1}$, respectively. This suggests that water and fertilizer management can effectively increase the stem NSC content during the grain-filling period of rice, which may positively impact rice yield and quality.

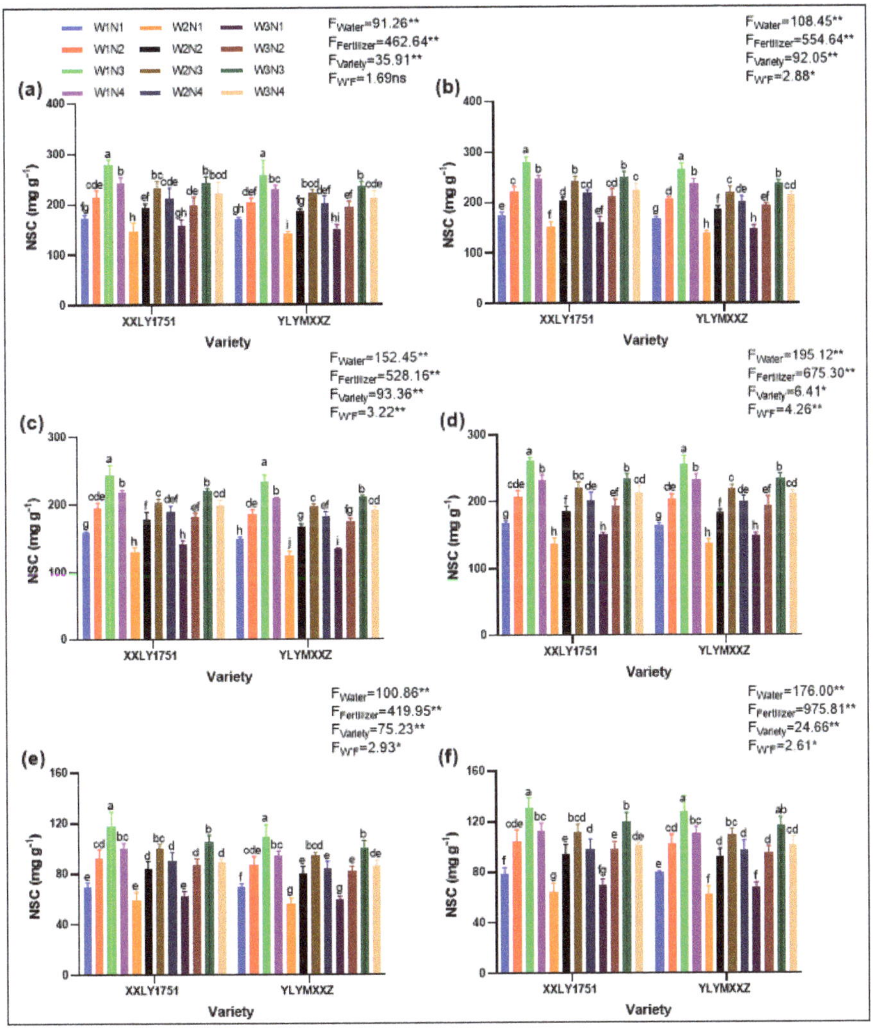


**Figure 7.** (**a**,**b**) represent the full heading stages in 2022 and 2023, respectively; (**c**,**d**) represent the grain filling stages in 2022 and 2023, respectively; (**e**,**f**) represent the mature stage in 2022 and 2023, respectively. Different lowercase letters on the error bars denote statistical differences (at the 0.05 level) among treatments of various varieties based on the LSD test. Significant differences within the same treatment are denoted by ns ($p > 0.05$), * ($0.01 < p \leq 0.05$), and ** ($p \leq 0.01$).

## *2.9. Lignin and Cellulose*

Water and fertilizer management significantly affected the structural carbohydrate content of rice stems (Figure 8). According to comprehensive data from both years, the W1F3V1 and W1F3V2 treatments exhibited clear advantages in stem cellulose content. Specifically, the stem cellulose content in the W1F3V1 treatment was 398.11 mg g$^{-1}$, which was 15.01 to

99.4 mg g$^{-1}$ higher than other treatment groups. Similarly, the content in the W1F3V2 treatment was 378.74 mg g$^{-1}$, exceeding other treatments by 19.28 to 89.68 mg g$^{-1}$. Likewise, in terms of stem lignin content, the W1F3V1 and W1F3V2 treatments also led significantly. The lignin content in W1F3V1 was 191.58 mg g$^{-1}$, which was 5.42 to 44.96 mg g$^{-1}$ higher than other treatments. For W1F3V2, the content was 181.85 mg g$^{-1}$, surpassing other treatment groups by 8.23 to 44.09 mg g$^{-1}$. These findings suggest that specific water and fertilizer management strategies, such as the W1F3V1 and W1F3V2 treatments, can significantly increase the structural carbohydrate content in rice stems. This may positively impact the mechanical strength and lodging resistance of the stems.

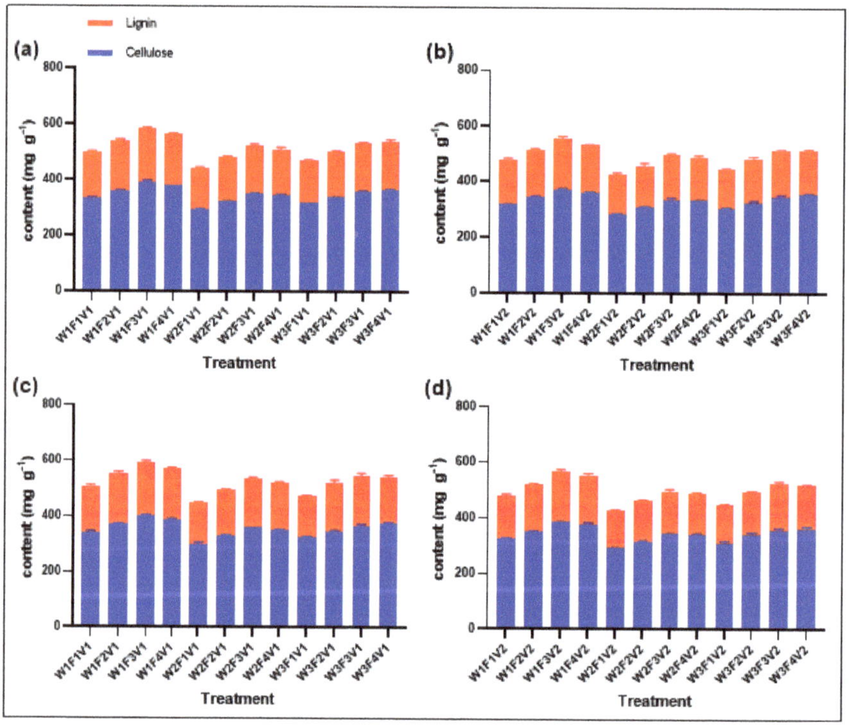


**Figure 8.** Lignin and cellulose of XXLY1751 (**a**) and YLYMXXZ (**b**) in 2022, and XXLY1751 (**c**) and YLYMXXZ (**d**) in 2023. Different lowercase letters denote statistical differences between treatments of each season according to an LSD test (0.05).

## *2.10. Amylase Activity*

The activities of SPS enzyme, α-amylase, and β-amylase in rice stems were influenced by water and fertilizer management (Figure 9). Upon comprehensive comparison of the experimental outcomes from both years, it was observed that during the grain filling period, the contents of SPS enzyme, α-amylase, and β-amylase in the stems under the W1F3V1 and W1F3V2 treatments were notably higher than those in other treatment groups. Specifically, the contents of SPS enzyme, α-amylase, and β-amylase in the W1F3V1 treatment were 45.12 mg g$^{-1}$, 12.16 mg g$^{-1}$, and 6.11 mg g$^{-1}$, respectively, while those in the W1F3V2 treatment were 41.03 mg g$^{-1}$, 11.20 mg g$^{-1}$, and 5.63 mg g$^{-1}$, respectively. The W1F3V1 and W1F3V2 treatments were associated with higher enzyme activities, suggesting a positive impact on carbohydrate metabolism.

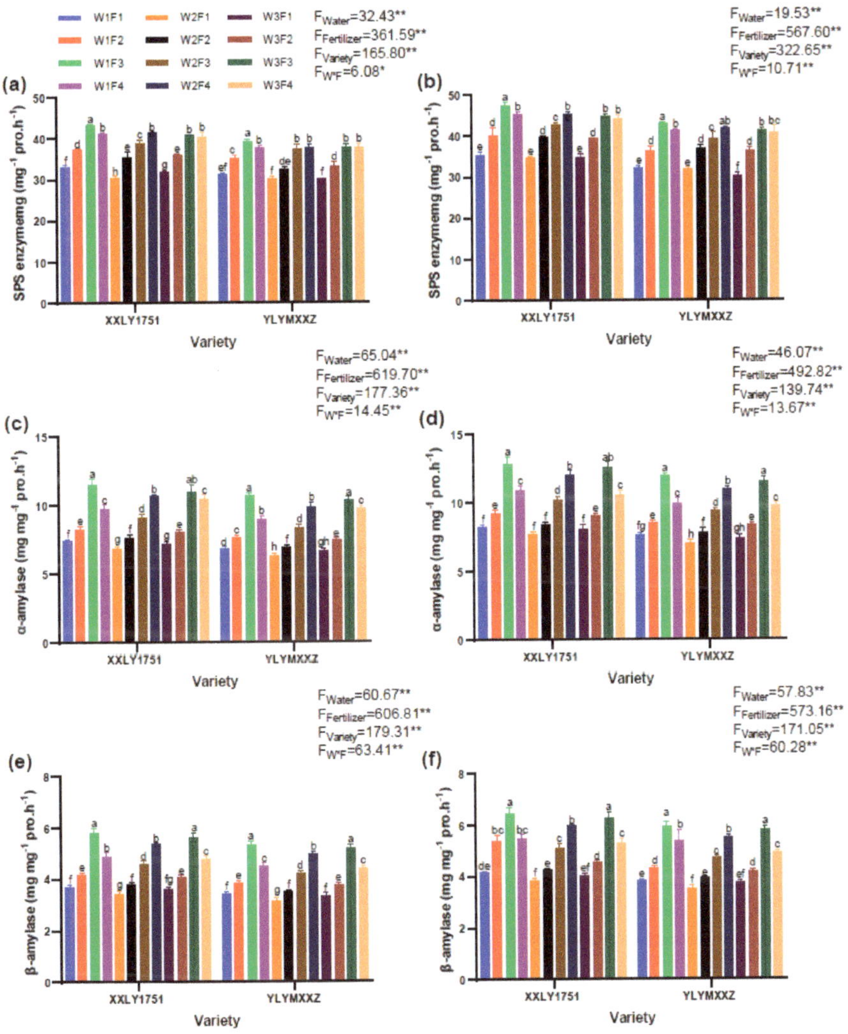


**Figure 9.** SPS enzyme activity of XXLY1751 and YLYMXXZ in 2022 (**a**) and in 2023 (**b**), α-amylase activity of XXLY1751 and YLYMXXZ in 2022 (**c**) and in 2023 (**d**), and β-amylase activity of XXLY1751 and YLYMXXZ in 2022 (**e**) and in 2023 (**f**). Different lowercase letters denote statistical differences between treatments of each season according to an LSD test (0.05). Significant differences within the same treatment are denoted by * ($0.01 < p \leq 0.05$), and ** ($p \leq 0.01$).

### *2.11. Variance Analysis*

We have discovered that both water and fertilizer treatments exert significant influences on the physiological indicators, yield, and quality of rice (Table 2). More specifically, these two treatments serve as primary effects, effectively enhancing the photosynthetic rate, dry matter accumulation during maturity, yield, head rice rate, as well as the protein and amylose content in rice grains. Concurrently, there is a notable increase in the content of NSC, cellulose, lignin, and the bending resistance of the stems.

**Table 2.** Fertilizer, irrigation, variety, and year affect Pn (net photosynthetic rate), TDW (total dry weight of the ground part), yield, HRY (head rice rate), protein, amylose, NSC, cellulose, lignin, and BR (bending resistance force).

| SOV | Pn | TDW | Yield | HRY | Protein | Amylose | NSC | Cellulose | Lignin | BR |
|---|---|---|---|---|---|---|---|---|---|---|
| F | ** | ** | ** | ** | ** | ** | ** | ** | ** | ** |
| V | ** | ** | ** | ** | ** | ** | ** | ** | ** | ** |
| W | ** | ** | * | ** | ** | ** | ** | ** | ** | ** |
| Y | ** | ** | ** | * | ** | ** | ** | ** | NS | ** |
| F*W | * | ** | ** | NS | NS | ** | ** | ** | NS | * |
| F*W*V | NS | ** | NS | NS | NS | NS | NS | NS | NS | NS |
| F*W*Y | NS | NS | NS | NS | NS | ** | NS | NS | NS | NS |
| F*V*W*Y | NS | ** | NS | NS | NS | NS | NS | NS | NS | NS |

Significant treatment effects within a main category are denoted by NS ($p > 0.05$) or * ($0.01 < p \leq 0.05$) or ** ($p \leq 0.01$). Pn, net photosynthetic rate; TDW, total dry weight of the ground part; HRY, head rice rate; BR, bending resistance force.

Further data analysis has unveiled significant interactive effects between water and fertilizer on various aspects of rice, including the photosynthetic rate, dry matter accumulation during maturity, yield, head rice rate, amylose content in rice grains, cellulose content in stems, and stem bending resistance. It is worth noting that this interaction demonstrates consistency in its impact on two rice varieties across two years of experimentation, indicating that there is no significant interaction between water, fertilizer, and rice varieties. Similarly, no apparent interaction was observed between water, fertilizer, and the experimental year. These findings strongly suggest that when water management and fertilizer application strategies are appropriately combined, they can lead to a more substantial improvement in the aforementioned physiological, yield, and quality indicators.

## 3. Discussion

### 3.1. Effect of Water and Fertilizer on the Yield

The application of optimized water and fertilizer management strategies in this study, specifically the W1F3 treatment, has demonstrated significant effectiveness in boosting rice yield. This approach notably improved photosynthetic efficiency, which is critical for biomass accumulation and grain production in rice [1]. By enhancing photosynthesis through precise regulation of water and nitrogen, the vegetative organs of the rice were strengthened, the leaf area was increased, and leaf senescence was delayed. These physiological changes led to improved carboxylation efficiency of $CO_2$ in mesophyll cells, thereby increasing the rate of $CO_2$ assimilation [4]. This series of physiological responses not only established a solid material foundation for high rice yields but also reflected the excellent efficacy of the W1F3 treatment in optimizing rice's physiological functions.

The elevated photosynthetic rates resulting from the W1F3 treatment supported healthy growth in rice plants, leading to improved dry matter production efficiency. This increase in dry matter accumulation directly contributed to building a productive rice population [23]. Additionally, efficient transfer of NSC stored in rice stems and sheaths during the grain-filling phase is essential for determining the final yield. By promoting NSC accumulation and refining their transfer mechanism, the W1F3 treatment further optimized the yield formation process [24,25].

In conclusion, the W1F3 treatment enhanced yield through various mechanisms: improving photosynthetic efficiency, facilitating dry matter accumulation, and optimizing NSC dynamics. These results offer valuable insights into the potential for increasing rice yield and provide new perspectives for achieving high-yield rice cultivation in the future.

### 3.2. Effect of Water and Fertilizer on Quality

The application of refined water and fertilizer management tactics, especially the W1F3 treatment, has exhibited notable impacts on elevating the quality of hybrid rice.

Through meticulous regulation of the growth conditions, this method not only augments photosynthetic efficiency but also substantially affects carbohydrate metabolism and nutritional makeup.

The boosted photosynthetic efficiency resulting from the W1F3 treatment leads to an amplified accumulation of NSC, encompassing soluble sugars and starches. These NSC constitute a vital material foundation for starch synthesis and protein formation in rice grains, thereby fostering enhanced rice quality [26]. Specifically, the W1F3 treatment guarantees seamless grain filling, giving rise to fuller grains and an overall elevation in the aesthetic quality of rice. This improvement is manifested in the considerable augmentation of the head rice rate and grain length–width ratio, signifying advancements in both processing and visual attributes [24,27].

Furthermore, the W1F3 treatment brings about a diminished chalkiness degree and chalky grain rate, further underscoring its proficiency in upgrading rice quality. Chalkiness, a pivotal aspect influencing rice quality, is intimately linked to carbohydrate metabolism and nutrient absorption [28]. The optimized water and fertilizer management inherent in the W1F3 treatment adeptly regulates these procedures, consequently minimizing chalkiness and elevating rice quality.

In essence, the W1F3 treatment, via its thorough optimization of the rice growth milieu, considerably boosts photosynthetic efficiency and carbohydrate metabolism. These repercussions not solely advance the accumulation of NSC but also refine nutritional composition and ultimately upgrade rice quality. This investigation offers invaluable perceptions into the modulation of rice quality through optimized water and fertilizer management, contributing to the sustainable advancement of rice production and accommodating the escalating demand for high-quality rice in the marketplace.

### 3.3. Effect of Water and Fertilizer on Lodging Resistance

The integration of water and fertilizer management is crucial in shaping the growth conditions of hybrid rice, thereby significantly influencing its lodging resistance. By ensuring an optimal supply of water and nutrients, this management approach effectively stimulates photosynthesis and carbohydrate biosynthetic processes [29]. The augmented photosynthesis leads to the accumulation of additional NSC, specifically soluble sugars and starches, which form the material basis for stem strength and durability [30,31].

The elevated NSC content not only enhances the fullness and thickness of the rice stem, resulting in a stronger morphological appearance, but also significantly boosts its physical resilience [32]. Furthermore, these NSC can potentially be converted to structural carbohydrates, such as cellulose and lignin. This conversion notably increases the mechanical strength of the stem, effectively improving its lodging resistance, a key factor in preventing lodging caused by natural elements like wind and rain [31,33].

Additionally, the combined approach of water and fertilizer management favorably impacts the morphological structure of rice stems [17]. Through meticulous regulation, it promotes thicker stem nodes and judicious adjustments in internode length [34]. These morphological enhancements significantly contribute to lodging resistance, allowing rice plants to maintain upright growth even in challenging environmental conditions, thus ensuring normal physiological functions and yield production [35,36].

In conclusion, the coordinated regulation of water and fertilizer management optimizes the growth environment for hybrid rice, boosting photosynthetic efficiency and carbohydrate accumulation. Moreover, it significantly improves the morphological structure and mechanical strength of the stem, laying a firm foundation for enhanced lodging resistance and consistent high yields. This strategy is vital in satisfying the demand for high-yield, high-quality, and stress-resistant rice varieties in contemporary agriculture, further promoting the sustainable progress of rice production.

### 3.4. Synergistic Regulation of Yield, Quality, and Lodging Resistance by Water and Fertilizer Management

The implementation of a carefully crafted water and fertilizer management strategy, notably the W1F3 treatment, has demonstrated remarkable efficacy in optimizing the growth environment for hybrid rice. As clearly depicted in Figure 10 created by Figdraw, this sophisticated strategy comprehensively regulates rice yield, quality, and lodging resistance via multiple mechanisms, wherein NSC assumes a pivotal function.

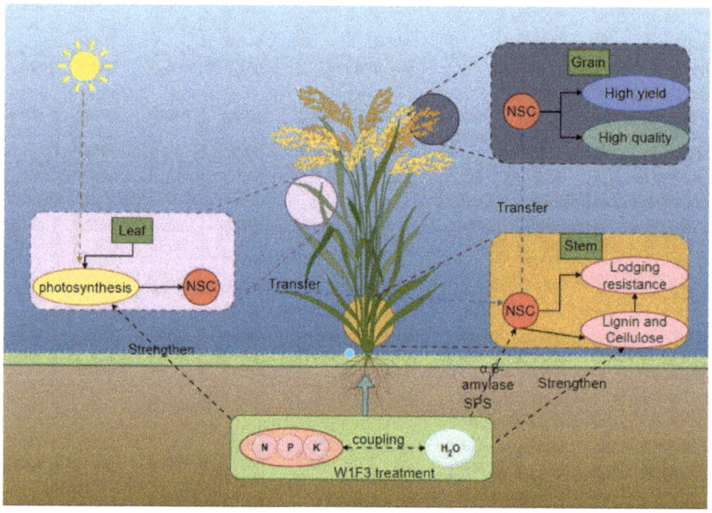


**Figure 10.** Synergistic regulation of yield, quality, and lodging resistance by water and fertilizer management.

Initially, the optimized water and fertilizer management notably boosts the photosynthetic efficiency of leaves, facilitating significant NSC synthesis and accumulation. These NSC, including soluble sugars and starches, furnish essential energy for rice growth and development [37]. Simultaneously, they directly contribute to starch accumulation and protein synthesis within rice grains, laying a robust material foundation for high-yielding rice [38,39].

Furthermore, regarding yield formation, the ample NSC reserve expedites the rate of dry matter accumulation [24]. Augmented photosynthesis enables rice plants to more efficiently convert light energy into chemical energy, storing it in the form of NSC [40,41]. The precise regulation of carbohydrate production and transport further advances the grain filling process, thereby contributing to elevated yield [42,43].

Moreover, in terms of rice quality, the timely transfer and judicious distribution of NSC ensure uniform grain filling. This not only leads to fuller rice grains but also enhances the head rice rate and length–width ratio while diminishing chalkiness and chalky grain rate, significantly elevating the overall quality of rice [24,27].

Additionally, NSC accumulation plays a pivotal role in lodging resistance. The buildup of NSC augments the stem's fullness and thickness, bolstering its visual sturdiness [30]. Furthermore, optimized water and fertilizer management fosters the conversion of certain NSC into structural carbohydrates, such as cellulose and lignin, through amplified activity of enzymes like SPS, α-amylase, and β-amylase [24,31]. These components considerably strengthen the mechanical robustness of the stem, thereby effectively enhancing its lodging resistance [44–46].

In conclusion, through the implementation of an optimized water and fertilizer management strategy, particularly the W1F3 treatment, comprehensive modulation of rice growth has been attained, with NSC functioning as the core regulator. This approach not only elevates photosynthetic efficiency but also promotes NSC accumulation and transport processes. More significantly, it exhibits remarkable impacts in synergistically optimizing rice yield, quality, and lodging resistance. This investigation not only reveals the intricate relationship between water and fertilizer management and NSC dynamics but also offers invaluable theoretical and practical insights for attaining high-yield and high-quality cultivation of hybrid rice, significantly contributing to the sustainable advancement of rice production.

## 4. Materials and Methods

### 4.1. Climatic Condition

In 2022, the average daily minimum and maximum temperatures were notably higher than those recorded in 2023. Conversely, the average daily rainfall was significantly lower in 2022 compared to 2023. While temperatures exhibited an upward trajectory from the transplanting to flowering stages in both years, rainfall patterns were distinct. Specifically, rainfall was markedly more concentrated during the Transplanting to Panicle Initiation phase in 2022, and from Panicle Initiation to Flowering in 2023, as detailed in Table 3.

**Table 3.** The climate conditions by rice growth stages in 2022 and 2023.

| Year | Min T | Max T | ADR |
|---|---|---|---|
| | | Transplanting to Panicle initiation | |
| 2022 | 26.3 | 34.2 | 11.2 |
| 2023 | 25.9 | 32.1 | 5.1 |
| | | Panicle initiation to Flowering | |
| 2022 | 28.2 | 37.7 | 0.1 |
| 2023 | 26.1 | 34.5 | 6.9 |
| | | Flowering to Maturity | |
| 2022 | 22.6 | 31.6 | 0.1 |
| 2023 | 21.2 | 29.8 | 0.4 |
| | | Transplanting to Maturity | |
| 2022 | 25.8 | 34.6 | 3.3 |
| 2023 | 24.4 | 32.2 | 4.2 |

Average daily minimum temperature (Min T, °C), average daily maximum temperature (Max T, °C), and average daily rainfall (ADR, mm day$^{-1}$) for each growing period.

### 4.2. Experiment Design and Plant Materials

The experiments were conducted in rice paddies located in Oujiangcha Village (geographic coordinates: 28°29′55″ N, 112°35′59″ E; altitude: 12 m above sea level), Yiyang City, Hunan Province, during 2022 and 2023. The experimental design employed a split-split plot design, focusing on three primary factors: rice variety (V), water management (W), and nitrogen fertilizer application (F). The study encompassed 72 split plots, each with an area of 20 m$^2$, totaling an experimental area of 1440 m$^2$ (excluding water channels). The soil had the following properties: pH 5.1, 34.8 g kg$^{-1}$ organic matter, 1.6 g kg$^{-1}$ total nitrogen (N), 13.8 mg kg$^{-1}$ Olsen phosphorus (P), and 103.4 mg kg$^{-1}$ exchangeable potassium (K). The soil test was based on samples taken from the upper 20 cm of the soil before the application of basal fertilizers in 2022.

The main plots were assigned different water treatments (W), categorized into three levels: W1—continuous wet-shallow irrigation throughout the entire growth cycle, W2—conventional flooded irrigation, and W3—thin, shallow, wet, and dry irrigation. The specific water layer control standards for these irrigation methods are summarized in Table 4. All irrigation treatments were manually administered, with precise measurements of the irrigation volume using water meters (refer to Figure 11). The split plots were designated for evaluating different rice varieties (V), specifically as follows.

Table 4. Thresholds of water level for different rice irrigation modes (mm).

| Irrigation | RS | ES | LS | JS | HS | GS | MS |
|---|---|---|---|---|---|---|---|
| W1 | 0–20–30 | 0–20–30 | SF | 0–20–30 | 0–20–30 | 0–20–30 | NF |
| W2 | 20–50–70 | 20–50–70 | SF | 20–50–70 | 20–50–70 | 20–50–70 | NF |
| W3 | 10–30–50 | 10–30–50 | SF | 10–40–60 | 10–40–60 | 10–20–40 | NF |

Three continuous data in the table (e.g., 0–20–30, mm) represent the lower and upper limits of irrigation and the maximum limit after rain, respectively; RS represents the returning green stage, ES represents the early tillering stage, LS represents the late tillering stage, JS represents the jointing and booting stage, and HS represents the heading stage. GS represents the grain-filling stage, and MS represents the maturity stage. SF represents sun-dried field, NF represents natural drying.

Figure 11. Water pipes and water meters in the community.

This study employed a split-split plot experimental design, involving three factors: cultivar, water, and nitrogen fertilizer. The main plots focused on water treatment (W), with three levels: W1 represents continuous wet-shallow irrigation throughout the entire growth cycle, W2 stands for flooded irrigation (the conventional irrigation method used by large-scale grain farmers), and W3 signifies thin, shallow, wet, and dry irrigation (the typical irrigation approach adopted by small-scale grain farmers). Detailed water layer control standards for these three irrigation modes are outlined in Table 4. The split plots concentrated on cultivar (V), comprising two levels: Xinxiangliangyou 1751 (V1, XXLY1751), sourced from Hunan Jinse Nonghua Seed Industry Technology Co., Ltd. (Changsha, China), and Yueliangyou Meixiangxinzhan (V2, YLYMXXZ), provided by Yuan Longping High-Tech Agriculture Co., Ltd. (Changsha, China)—both widely grown rice in the middle and lower Yangtze River. The split-split plots investigated nitrogen application rates (N), divided into four levels: F1, F2, F3, and F4, corresponding to pure nitrogen applications of 0, 180, 225, and 270 kg ha$^{-1}$, respectively. The fertilizer adopted the multiple fertilization method, with the nitrogen fertilizer (urea) application ratio of base fertilizer: tiller fertilizer: ear fertilizer being 5:3:2. The base fertilizer, tiller fertilizer, and ear fertilizer were applied 1–2 days before transplanting, 10 days after transplanting, and at the beginning of young panicle differentiation (jointing stage), respectively. The fertilization ratio uniformly adopted an NPK ratio of N:$P_2O_5$:$K_2O$ = 1:0.6:1.2. Phosphate fertilizer ($P_2O_5$, 16%) was entirely used as base fertilizer, while potassium fertilizer ($K_2O$, 60%) was split evenly between the base fertilizer and the tiller fertilizer. Before transplanting, equal amounts of phosphate and potassium fertilizers were applied as base fertilizers to supplement the nutritional elements required by rice across all treatments. The experiment comprised 24 treatments, with separate isolation between plots to ensure independent drainage and irrigation. There were three replicates, arranged randomly according to a split-plot design, resulting in a total of 72 split plots. Each split plot covered an area of

20 $m^2$, and the total experimental area was 1440 $m^2$ (excluding water channels). Disease, pest, and weed control were strictly managed according to high-yield cultivation practices.

### 4.3. Sampling and Measurement

**Photosynthetic rate** Photosynthetic rates were measured using a portable photosynthesis measurement system LI-6800 (LI-COR, Inc., Lincoln, NE, USA) at the tillering stage (selecting the second fully expanded leaf from the top), booting stage, full heading stage, and milk-ripe stage (selecting the flag leaf). Measurements were conducted on clear, cloudless, and windless days between 9:00 and 12:00. Two representative plants were selected from each plot, and each plant was measured three times. The average value of these measurements was taken as the final result [47].

**Total dry weight (TDW)** Sampling was conducted during the tillering stage, booting stage, full heading stage, and filling stage. Ten clusters were consecutively selected from each plot, and the stems, sheaths, leaves, and panicles were separated and packaged. Samples were oven-dried at 105 °C for 30 min to stop enzyme activity and then dried to a constant weight at 80 °C [48].

**Yield** Yield was determined by harvesting and drying grains from each plot separately. The actual yield was calculated based on a moisture content of 14% in the rice grains [49].

**Appearance Quality** Appearance quality measurements, including chalkiness degree and grain length–width ratio, were conducted using a Wanshen SC-E rice appearance quality detector (Wanshen Testing Technology Co., Ltd., Hangzhou, China) and its analytical system [50].

**Nutritional Quality Amylose** content was determined using the iodine-blue colorimetric method following the Ministry of Agriculture standard NY/T 2639-2014. Protein content was measured by digesting polished rice flour with $H_2SO_4$-$H_2O_2$ and using Nessler's colorimetric method to determine nitrogen content. Protein content was calculated by multiplying nitrogen content by a conversion factor of 5.95 [51].

**Lodging index** Twenty uniformly growing main stalks were randomly selected from each plot 25 days after heading. The third, fourth, and fifth internodes from the base were cut off, and their lengths were measured. Bending resistance at the midpoint of these internodes was determined using a YYD-1 stalk strength tester (Tuopu Instruments Co., Ltd., Hangzhou, China), with a support span of 5 cm between the two fulcrums. Various mechanical parameters were calculated using established formulas [52].

$$\text{The breaking bending moment (M, g·cm)} = F \times L/4$$

where F is the force applied when the measured basal section is broken, with the unit of kg; L is the distance between the two supports, with the unit of cm.

$$\text{The bending moment (WP, g·cm)} = SL \times FW$$

where SL is the distance from the base of the internode to the top of the panicle, with the unit of cm; FW is the fresh weight from the base of the internode to the top of the panicle, with the unit of g.

$$\text{The lodging index (LI)} = WP/M \times 100\%$$

**NSC** Concentrations of NSC (soluble sugars and starch) in straw were determined at the heading stage, grain filling stage, and maturity stage, following the method outlined by [53]. Aliquots from the soluble sugars and starch extracts were assayed using anthrone reagent colorimetry, employing a spectrophotometer (L8, Shanghai Yidian Analytical Instrument Co., Ltd., Shanghai, China) set at a wavelength of 620 nm. Starch content was calculated by multiplying the glucose level by a conversion factor of 0.9. NSC concentration was estimated by summing the concentrations of soluble sugars and starch.

**Amylase** The detection of α-amylase, β-amylase, and sucrose phosphate synthase (SPS) was conducted using ELISA kits. The detection results were measured in terms of absorbance (OD value) at a wavelength of 450 nm using the DNM-9606 microplate reader (Beijing Perlong New Technology Co., Ltd., Beijing, China). With the OD value plotted on the vertical axis (Y) and the corresponding standard concentration of the analyte on the horizontal axis (X), a standard curve was constructed. By referencing the OD value of the sample against this standard curve, the concentration of the analyte in the sample could be determined.

**Lignin and Cellulose** The lignin and cellulose content measurements were described by [54].

*4.4. Statistical Analysis*

The field experiment employed a 3 × 4 × 2 factorial design, comprising three water levels, four fertilizer levels, and two varieties for each treatment combination within three blocks. We utilized a mixed-effects analysis of variance for all variables in our study. The fixed effects included water, fertilizer, variety, year, and the interaction between water and fertilizer. We regarded blocks as a random effect. The data were analyzed using analysis of variance (ANOVA, SAS Version 9.1.2, Cary, NC, USA). The means of the water and fertilizer treatments for each variety were compared using the least significant difference test (LSD) at a 0.05 probability level. Graphs were constructed using Microsoft Excel 2017 (Microsoft Corp., Redmond, USA).

**5. Conclusions**

This study found that the W1F3 treatment (wet-shallow irrigation and appropriate fertilization (N-$P_2O_5$-$K_2O$: 225–135–270 kg $ha^{-1}$)) had a significant positive impact on hybrid rice. Specifically, the effects are manifested in the following ways: (1) The yield of hybrid rice was significantly enhanced by improving photosynthetic efficiency, promoting dry matter accumulation, and optimizing the dynamics of non-structural carbohydrates (NSC). (2) The effective increase in NSC accumulation and improvement in carbohydrate metabolism resulted in an increased rate of head rice, an increased grain length/width ratio, and a reduction in chalkiness and the chalky grain rate. (3) By promoting photosynthesis, increasing NSC accumulation, enhancing stem fullness and thickness, and facilitating the conversion of NSC to structural carbohydrates such as cellulose and lignin, the mechanical strength of the stems was significantly improved, thereby enhancing the lodging resistance of hybrid rice. In summary, the W1F3 treatment comprehensively regulated the yield, quality, and lodging resistance of hybrid rice through the key role of NSC. This reveals a close connection between water and fertilizer management and NSC dynamics, providing an important basis for achieving high-yield and high-quality cultivation of hybrid rice. It has positive implications for promoting the sustainable development of rice production.

**Author Contributions:** H.C. and X.W. conceived the project and designed the experiments. H.Z. and L.N. conducted the experiments. H.Z. and P.L. designed and conducted computational and statistical analyses. H.Z. and X.W. wrote the manuscript. X.H. and X.W. revised the manuscript. All authors have read and agreed to the published version of the manuscript.

**Funding:** This research was funded by the National Key R&D Program of China (2017YFD0301500, 2022YFD2300301, 2023YFD1523019).

**Data Availability Statement:** The data presented in this study are available upon request from the corresponding author.

**Acknowledgments:** I am indebted to fieldworkers Manwen Sun (Bureau of Agriculture and Rural Affairs of Heshan District) and Yanheng Zeng (Hunan Biological and Electromechanical Polytechnic) for their dedication and hard work in collecting the data. Additionally, I am thankful to Weijian Tan (Bureau of Agriculture and Rural Affairs of Heshan District) for providing the necessary facilities and resources for the field experiments.

**Conflicts of Interest:** The authors declare no competing interests.

## References


1. Zhu, H.; Zhang, T.; Zhang, C.; He, X.; Shi, A.; Tan, W.; Yi, Z.; Wang, X. Optimizing Irrigation and Nitrogen Management to Increase Yield and Nitrogen Recovery Efficiency in Double-Cropping Rice. *Agronomy* **2022**, *12*, 1190. [CrossRef]
2. He, X.; Zhu, H.; Shi, A.; Wang, X. Optimizing Nitrogen Fertilizer Management Enhances Rice Yield, Dry Matter, and Nitrogen Use Efficiency. *Agronomy* **2024**, *14*, 919. [CrossRef]
3. Huang, M.; Hu, L.; Cao, J.; Zhang, R.; Chen, J.; Cao, F.; Liu, L.; Fang, S.; Zhang, M. Texture and digestion properties of hybrid rice: A comparison between two cultivars released 18 years apart. *Food Chem. X.* **2022**, *13*, 100215. [CrossRef] [PubMed]
4. Ma, G.; Yuan, L. Hybrid rice achievements, development and prospect in China. *J. Integr. Agric.* **2015**, *14*, 197–205. [CrossRef]
5. Djaman, K.; Mel, V.C.; Diop, L.; Sow, A.; El-Namaky, R.; Manneh, B.; Saito, K.; Futakuchi, K.; Irmak, S. Effects of Alternate Wetting and Drying Irrigation Regime and Nitrogen Fertilizer on Yield and Nitrogen Use Efficiency of Irrigated Rice in the Sahel. *Water* **2018**, *10*, 711. [CrossRef]
6. Pan, J.; Liu, Y.; Zhong, X.; Lampayan, R.M.; Singleton, G.R.; Huang, N.; Liang, K.; Peng, B.; Tian, K. Grain yield, water productivity and nitrogen use efficiency of rice under different water management and fertilizer-N inputs in South China. *Agric. Water Manag.* **2017**, *184*, 191–200. [CrossRef]
7. Sun, Y.; Yan, F.; Sun, Y.; Xu, H.; Guo, X.; Yang, Z.; Yin, Y.; Guo, C.; Ma, J. Effects of different water regimes and nitrogen application strategies on grain filling characteristics and grain yield in hybrid rice. *Arch. Agron. Soil Sci.* **2018**, *64*, 1152–1171. [CrossRef]
8. Zhang, Y.; Zhang, S.; Zhang, J.; Wei, W.; Zhu, T.; Qu, H.; Liu, Y.; Xu, G. Improving rice eating and cooking quality by enhancing endogenous expression of a nitrogen-dependent floral regulator. *Plant Biotechnol. J.* **2023**, *21*, 2654–2670. [CrossRef]
9. Wang, S.; Li, S.; Liu, Q.; Wu, K.; Zhang, J.; Wang, S.; Wang, Y.; Chen, X.; Zhang, Y.; Gao, C.; et al. The *OsSPL16-GW7* regulatory module determines grain shape and simultaneously improves rice yield and grain quality. *Nat. Genet.* **2015**, *47*, 949–954. [CrossRef]
10. Guo, L.; Liu, M.; Tao, Y.; Zhang, Y.; Li, G.; Lin, S.; Dittert, K. Innovative water-saving ground cover rice production system increases yield with slight reduction in grain quality. *Agric. Syst.* **2020**, *180*, 102795. [CrossRef]
11. Qi, D.; Yan, J.; Zhu, J. Effect of a reduced fertilizer rate on the water quality of paddy fields and rice yields under fishpond effluent irrigation. *Agric. Water Manag.* **2020**, *231*, 105999. [CrossRef]
12. Liu, X.; Li, M.; Guo, P.; Zhang, Z. Optimization of water and fertilizer coupling system based on rice grain quality. *Agric. Water Manag.* **2019**, *221*, 34–46. [CrossRef]
13. Sun, Y.; Wu, Y.; Sun, Y.; Luo, Y.; Guo, C.; Li, B.; Li, F.; Xing, M.; Yang, Z.; Ma, J. Effects of Water and Nitrogen on Grain Filling Characteristics, Canopy Microclimate with Chalkiness of Directly Seeded Rice. *Agriculture* **2022**, *12*, 122. [CrossRef]
14. Ning, H.; Sun, J.; Liu, H.; Gao, Y.; Shen, X.; Wang, G.; Zhang, K. The effects of nitrogen application rate on the grain physicochemical properties of<i>japonica</i>rice under controlled and flooding irrigation. *J. Sci. Food Agric.* **2021**, *101*, 2428–2438.
15. Huang, M.; Tao, Z.; Lei, T.; Cao, F.; Chen, J.; Yin, X.; Zou, Y.; Liang, T. Improving lodging resistance while maintaining high grain yield by promoting pre-heading growth in rice. *Field Crops Res.* **2021**, *270*, 108212. [CrossRef]
16. Tao, Z.; Lei, T.; Cao, F.; Chen, J.; Yin, X.; Liang, T.; Huang, M. Contrasting Characteristics of Lodging Resistance in Two Super-Rice Hybrids Differing in Harvest Index. *Phyton-Int. J. Exp. Bot.* **2022**, *91*, 429–437. [CrossRef]
17. Wu, D.-H.; Chen, C.-T.; Yang, M.-D.; Wu, Y.-C.; Lin, C.-Y.; Lai, M.-H.; Yang, C.-Y. Controlling the lodging risk of rice based on a plant height dynamic model. *Bot. Stud.* **2022**, *63*, 25. [CrossRef]
18. Gao, H.; Dou, Z.; Chen, L.; Lu, Y.; Sun, D.; Xu, Q.; Sun, R.; Chen, X. Effects of semi-deep water irrigation on hybrid indica rice lodging resistance. *Front. Plant Sci.* **2022**, *13*, 1038129. [CrossRef]
19. Zhang, M.; Mo, Z.; Liao, J.; Pan, S.; Chen, X.; Zheng, L.; Luo, X.; Wang, Z. Lodging Resistance Related to Root Traits for Mechanized Wet-Seeding of Two Super Rice Cultivars. *Rice Sci.* **2021**, *28*, 200–208.
20. Lampayan, R.M.; Bouman, B.A.M.; de Dios, J.L.; Espiritu, A.J.; Soriano, J.B.; Lactaoen, A.T.; Faronilo, J.E.; Thant, K.M. Yield of aerobic rice in rainfed lowlands of the Philippines as affected by nitrogen management and row spacing. *Field Crops Res.* **2010**, *116*, 165–174. [CrossRef]
21. Jing, Q.; Van Keulen, H.; Hengsdijk, H.; Cao, W.; Bindraban, P.S.; Dai, T.; Jiang, D. Quantifying N response and N use efficiency in rice-wheat (RW) cropping systems under different water management. *J. Agr. Sci.* **2009**, *147*, 303–312. [CrossRef]
22. Li, Y.; Shao, X.; Li, D.; Xiao, M.; Hu, X.; He, J. Effects of water and nitrogen coupling on growth, physiology and yield of rice. *Int. J. Agr. Biol. Eng.* **2019**, *12*, 60–66. [CrossRef]
23. Bhuiyan, K.A.; Bhuiya, S.U.; Saleque, M.A.; Khatun, A. Grain yield, growth response, and water use efficiency of direct wet-seeded rice as affected by nitrogen rates under alternate wetting and drying irrigation system. *Commun. Soil Sci. Plant Anal.* **2018**, *49*, 2527–2545. [CrossRef]
24. Li, G.; Hu, Q.; Shi, Y.; Cui, K.; Nie, L.; Huang, J.; Peng, S. Low Nitrogen Application Enhances Starch-Metabolizing Enzyme Activity and Improves Accumulation and Translocation of Non-structural Carbohydrates in Rice Stems. *Front. Plant Sci.* **2018**, *9*, 1128. [CrossRef] [PubMed]
25. Xu, G.-W.; Song, K.-J.; Lu, D.-K.; Wang, H.-Z.; Chen, M.-C. Influence of Water Management and Nitrogen Application on Rice Root and Shoot Traits. *Agron. J.* **2019**, *111*, 2232–2244. [CrossRef]
26. Fu, J.; Huang, Z.; Wang, Z.; Yang, J.; Zhang, J. Pre-anthesis non-structural carbohydrate reserve in the stem enhances the sink strength of inferior spikelets during grain filling of rice. *Field Crops Res.* **2011**, *123*, 170–182. [CrossRef]

27. Wei, H.; Ge, J.; Zhang, X.; Zhu, W.; Deng, F.; Ren, W.; Chen, Y.; Meng, T.-; Dai, Q. Decreased panicle N application alleviates the negative effects of shading on rice grain yield and grain quality. *J. Integr. Agric.* **2023**, *22*, 2041–2053. [CrossRef]
28. Gao, Y.; Liu, C.; Wang, J.; Liu, X.; Zhang, X.; Zhou, J.; Li, X.; Wang, Y.; Dong, G.; Huang, J.; et al. Modified TAL expression in rice plant regulates yield components and grain quality in a N-rate dependent manner. *Field Crops Res.* **2024**, *306*, 109219. [CrossRef]
29. Zhang, Q.; Liu, X.; Yu, G.; Duan, B.; Wang, H.; Zhao, H.; Feng, D.; Gu, M.; Hu, Y.; Chen, Y.; et al. Alternate wetting and moderate soil drying could increase grain yields in both main and ratoon rice crops. *Crop Sci.* **2022**, *62*, 2413–2427. [CrossRef]
30. Liu, W.; Fan, X.; Liu, Y.; Bao, S.; Lu, Y.; Gai, D.; Fu, X.; Du, J.; Guo, L.; Zhang, Q.; et al. Relationship between characteristics of basal internodes and lodging and its physiological mechanism in direct-seeded rice. *J. Agron. Crop Sci.* **2023**, *209*, 632–650. [CrossRef]
31. Zhang, J.; Li, G.; Huang, Q.; Liu, Z.; Ding, C.; Tang, S.; Chen, L.; Wang, S.; Ding, Y.; Zhang, W. Effects of culm carbohydrate partitioning on basal stem strength in a high-yielding rice population. *Crop J.* **2017**, *5*, 478–487. [CrossRef]
32. Zheng, Y.; Ding, Y.; Liu, Z.; Wang, S. Effects of Panicle Nitrogen Fertilization on Non-Structural Carbohydrate and Grain Filling in Indica Rice. *Agric. Sci. China* **2010**, *9*, 1630–1640. [CrossRef]
33. Cao, P.; Sun, W.; Huang, Y.; Yang, J.; Yang, K.; Lv, C.; Wang, Y.; Yu, L.; Hu, Z. Effects of Elevated $CO_2$ Concentration and Nitrogen Application Levels on the Accumulation and Translocation of Non-Structural Carbohydrates in Japonica Rice. *Sustainability* **2020**, *12*, 5386. [CrossRef]
34. Li, W.; Han, M.; Pang, D.; Chen, J.; Wang, Y.; Dong, H.; Chang, Y.; Jin, M.; Luo, Y.; Li, Y.; et al. Characteristics of lodging resistance of high-yield winter wheat as affected by nitrogen rate and irrigation managements. *J. Integr. Agric.* **2022**, *21*, 1290–1309. [CrossRef]
35. Liu, Q.; Ma, J.; Zhao, Q.; Zhou, X. Physical Traits Related to Rice Lodging Resistance under Different Simplified-Cultivation Methods. *Agron. J.* **2018**, *110*, 127–132. [CrossRef]
36. Zhang, W.; Wu, L.; Wu, X.; Ding, Y.; Li, G.; Li, J.; Weng, F.; Liu, Z.; Tang, S.; Ding, C.; et al. Lodging Resistance of Japonica Rice (*Oryza Sativa* L.): Morphological and Anatomical Traits due to top-Dressing Nitrogen Application Rates. *Rice* **2016**, *9*, 31. [CrossRef]
37. Hu, Y.; Lin, Y.; Xia, Y.; Xu, X.; Wang, Z.; Cui, X.; Han, L.; Li, J.; Zhang, R.; Ding, Y.; et al. Overexpression of *OsSnRK1a* through a green tissue-specific promoter improves rice yield by accelerating sheath-to-panicle transport of nonstructural carbohydrates and increasing leaf photosynthesis. *Plant Physiol. Bioch.* **2023**, *203*, 108048. [CrossRef]
38. Dong, C.; Xu, N.; Ding, C.; Gu, H. Rapid evaluation method for rice straw feeding quality. *Field Crops Res.* **2018**, *228*, 204–209. [CrossRef]
39. Dong, C.F.; Liu, X.B.; Qu, H.; Shen, Y.X. Dynamical partition of photosynthates in tillers of rice during late growth period and its correlation with feeding value of rice straw at harvest. *Field Crops Res.* **2011**, *123*, 273–280. [CrossRef]
40. Ishikawa, J.; Okamura, M.; Arai-Sanoh, Y.; Baba, K.; Furuya, M.; Maeda, H.; Kondo, M. Close relationship between potassium accumulation in internodes and translocation of non-structural carbohydrate to grains in rice cultivars with different yields. *Field Crops Res.* **2023**, *300*, 109007. [CrossRef]
41. Nakano, H.; Tanaka, R.; Hakata, M. Nonstructural carbohydrate content in the stubble per unit area regulates grain yield of the second crop in rice ratooning. *Crop Sci.* **2022**, *62*, 1603–1613. [CrossRef]
42. Wei, H.; Meng, T.; Ge, J.; Zhang, X.; Shi, T.; Ding, E.; Lu, Y.; Li, X.; Tao, Y.; Chen, Y.; et al. Reduced nitrogen application rate with dense planting improves rice grain yield and nitrogen use efficiency: A case study in east China. *Crop J.* **2021**, *9*, 954–961. [CrossRef]
43. Zhu, T.; Yin, C.; Zhu, T.; Zhou, P.; Wu, L.; Wang, G.; He, H.; You, C.; Zhang, Q.; Ke, J. Delaying panicle nitrogen application to emergence of 3rd leaf from flag leaf increases the grain-filling ability and yield of large-panicle rice by increasing stem nonstructural carbohydrates at heading. *Field Crops Res.* **2024**, *312*, 109405. [CrossRef]
44. Zhu, H.; He, X.; Wang, X.; Long, P. Increasing Hybrid Rice Yield, Water Productivity, and Nitrogen Use Efficiency: Optimization Strategies for Irrigation and Fertilizer Management. *Plants* **2024**, *13*, 1717. [CrossRef] [PubMed]
45. Wang, D.R.; Wolfrum, E.J.; Virk, P.; Ismail, A.; Greenberg, A.J.; McCouch, S.R. Robust phenotyping strategies for evaluation of stem non-structural carbohydrates (NSC) in rice. *J. Exp. Bot.* **2016**, *67*, 6125–6138. [CrossRef]
46. Zhang, R.; Jia, Z.; Ma, X.; Ma, H.; Zhao, Y. Characterising the morphological characters and carbohydrate metabolism of oat culms and their association with lodging resistance. *Plant Biol.* **2020**, *22*, 267–276. [CrossRef]
47. Du, T.; Meng, P.; Huang, J.; Peng, S.; Xiong, D. Fast photosynthesis measurements for phenotyping photosynthetic capacity of rice. *Plant Methods* **2020**, *16*, 6. [CrossRef]
48. Zhu, H.; Wen, T.; Sun, M.; Ali, I.; Sheteiwy, M.S.; Wahab, A.; Tan, W.; Wen, C.; He, X.; Wang, X. Enhancing Rice Yield and Nitrogen Utilization Efficiency through Optimal Planting Density and Reduced Nitrogen Rates. *Agronomy* **2023**, *13*, 1387. [CrossRef]
49. Huang, M.; Chen, J.; Cao, F.; Zou, Y. Increased hill density can compensate for yield loss from reduced nitrogen input in machine-transplanted double-cropped rice. *Field Crops Res.* **2018**, *221*, 333–338. [CrossRef]
50. Ali, I.; Iqbal, A.; Ullah, S.; Muhammad, I.; Yuan, P.; Zhao, Q.; Yang, M.; Zhang, H.; Huang, M.; Liang, H.; et al. Effects of Biochar Amendment and Nitrogen Fertilizer on RVA Profile and Rice Grain Quality Attributes. *Foods* **2022**, *11*, 625. [CrossRef]
51. Li, X.; Zhang, M.; Xiao, Z.; Liu, L.; Cao, F.; Chen, J.; Huang, M. Relationships between texture properties of cooked rice with grain amylose and protein content in high eating quality *indica* rice. *Cereal Chem.* **2024**, *101*, 577–582. [CrossRef]

52. Huang, M.; Lei, T.; Cao, J.; Tao, Z.; Cao, F.; Chen, J.; Yin, X.; Zou, Y. Linking grain yield and lodging resistance with growth patterns in rice. *Exp. Agric.* **2022**, *58*, e24. [CrossRef]
53. Li, X.; Zhang, R.; Chen, G.; Xie, J.; Xiao, Z.; Cao, F.; Ali, I.; Iqbal, A.; Wahab, A.; Huang, M.; et al. Increasing grain weight and yield stability by increasing pre-heading non-structural carbohydrate reserves per spikelet in short-growth duration rice. *Crop J.* **2023**, *11*, 1912–1920. [CrossRef]
54. Zhang, W.; Wu, L.; Ding, Y.; Yao, X.; Wu, X.; Weng, F.; Li, G.; Liu, Z.; Tang, S.; Ding, C.; et al. Nitrogen fertilizer application affects lodging resistance by altering secondary cell wall synthesis in japonica rice (*Oryza sativa*). *J. Plant Res.* **2017**, *130*, 859–871. [CrossRef] [PubMed]

plants

MDPI

*Article*

# Soybean Yield Simulation and Sustainability Assessment Based on the DSSAT-CROPGRO-Soybean Model

**Lei Zhang [1], Zhenxi Cao [1,2,3], Yang Gao [1,3,4], Weixiong Huang [5], Zhuanyun Si [4], Yuanhang Guo [1], Hongbo Wang [1,2,3,*] and Xingpeng Wang [1,2,3,6,*]**

1 Modern Agricultural Engineering Key Laboratory at Universities of Education Department of Xinjiang Uygur Autonomous Region, College of Water Hydraulic and Architectural Engineering, Tarim University, Alar 843300, China; zhanglei101598@163.com (L.Z.); cy211119@163.com (Z.C.); gaoyang@caas.cn (Y.G.); vooyagee@163.com (Y.G.)

2 Key Laboratory of Tarim Oasis Agriculture, Ministry of Education, Tarim University, Alar 843300, China

3 Western Agricultural Research Center, Chinese Academy of Agricultural Sciences, Changji 831100, China

4 Institute of Farmland Irrigation, Chinese Academy of Agricultural Sciences, Xinxiang 453002, China; sizhuanyun@caas.cn

5 Hubei Key Laboratory of Yangtze Catchment Environmental Aquatic Science, School of Environmental Studies, China University of Geosciences, Wuhan 430078, China; huangwx@cug.edu.cn

6 Key Laboratory of Northwest Oasis Water-Saving Agriculture, Ministry of Agriculture and Rural Affairs, Shihezi 832000, China

* Correspondence: 18083915561@163.com (H.W.); 13999068354@163.com (X.W.); Tel.: +86-18083915561 (H.W.); +86-13999068354 (X.W.)

**Abstract:** In order to ensure national grain and oil security, it is imperative to expand the soybean planting area in the Xinjiang region. However, the scarcity of water resources in southern Xinjiang, the relatively backward soybean planting technology, and the lack of a supporting irrigation system have negatively impacted soybean planting and yield. In 2022 and 2023, we conducted an experiment which included three irrigation amounts of 27 mm, 36 mm, and 45 mm and analyzed the changes in dry mass and yield. Additionally, we simulated the potential yield using the corrected DSSAT-CROPGRO-Soybean model and biomass based on the meteorological data from 1994 to 2023. The results demonstrated that the model was capable of accurately predicting soybean emergence (the relative root mean square error (nRMSE) = 0, the absolute relative error (ARE) = 0), flowering (nRMSE = 0, ARE = 2.78%), maturity (nRMSE = 0, ARE = 3.21%). The model demonstrated high levels of accuracy in predicting soybean biomass ($R^2$ = 0.98, nRMSE = 20.50%, ARE = 20.63%), 0–80 cm soil water storage ($R^2$ = 0.64, nRMSE = 7.78%, ARE = 3.24%), and yield ($R^2$ = 0.81, nRMSE = 10.83%, ARE = 8.79%). The biomass of soybean plants increases with the increase in irrigation amount. The highest biomass of 63 mm is 9379.19 kg·hm$^{-2}$. When the irrigation yield is 36–45 mm ($p < 0.05$), the maximum yield can reach 4984.73 kg·hm$^{-2}$; the maximum efficiency of soybean irrigation water was 33–36 mm. In light of the impact of soybean yield and irrigation water use efficiency, the optimal irrigation amount for soybean cultivation in southern Xinjiang is estimated to be between 36 and 42 mm. The simulation results provide a theoretical foundation for soybean cultivation in southern Xinjiang.

**Keywords:** soybean; DSSAT model; yield; biomass; irrigation water use efficiency

## 1. Introduction

Soybeans are a significant edible oil crop in China [1]. As a consequence of insufficient domestic production, China has become the world's first importer of soybeans [2]. It is of great importance to increase the expansion of soybean oil crops and further promote national grain and oil security. Xinjiang is situated in the arid and semi-arid inland region of China, which is rich in land, light, and heat resources. It is therefore an ideal location for the cultivation of soybeans [3]. However, the scarcity of water represents the principal constraint factor limiting agricultural production in this region [4]. The key challenge in ensuring the sustainable development of agriculture in southern Xinjiang is to improve the efficiency with which water resources are utilized and to implement a rational irrigation system.

The continuous expansion of agricultural production has led to an urgent need to address the issue of water shortage. Different degrees of water shortage have been shown to have a significant impact on agriculture [5,6]. One of the most effective ways to ensure high and stable crop yields and save water is through the development of appropriate irrigation systems. This is particularly important in the case of soybean yield. The yield of soybeans was found to be significantly affected by different irrigation amounts. It has been demonstrated that soybeans have high water requirements during the flowering and grain-filling stages. Short-term water shortage does not significantly affect soybean yield, whereas long-term water shortage has a detrimental impact on yield, particularly during the grain-filling period. This is evidenced by the fact that water deficit has a significant impact on soil moisture content [7–9]. In a series of field experiments spanning three years, Rajanna and colleagues observed that high soil moisture content extended the growth period of soybean plants. They also demonstrated that an optimal irrigation regimen could enhance soybean yield [10].

Given the complexity of the farmland ecosystem and the limitations of field experiments, the use of crop models to simulate crop growth processes and yields represents an important approach to addressing the impact of climate change and human activities as well as achieving water savings and yield increases in agriculture [11]. A crop growth model is a quantitative and dynamic representation of crop production, informed by the knowledge and research findings of plant physiology, ecology, agro-meteorology, soil science, and other related disciplines. It has been employed extensively in the prediction of crop production potential and the guidance of the irrigation, fertilization, and tillage of agricultural land [12]. The DSSAT-CROPGRO-Soybean model is capable of simulating soybean growth, development, and yield based on genetic parameters interacting with the environment and crop management. Jiang et al. [13] conducted a simulation of the irrigation amount and sowing date of maize in northwestern arid areas. Their findings indicated that the optimized irrigation system could reduce the current irrigation amount by 50%. Guo et al. [14] applied the DSSAT-CROPGRO-Soybean model to correct and validate the genetic parameters of the vegetative phase, reproductive phase, and yield formation process of soybeans at different maturity stages. Additionally, they predicted the soybean yield from 2021 to 2100 under climate change and various technological improvements. Some scholars have also observed that the model accurately simulates crop growth and soil moisture change, but the accuracy of the results decreases with increasing drought [15].

The arid and semi-arid regions of southern Xinjiang are typified by a scarcity of water, high rates of evaporation, and a limited capacity for soil fertility, all of which constrain the potential for agricultural development [16], In the South Xinjiang region, crop cultivation often adopts the planting mode of drip irrigation under the membrane, which not only reduces soil evaporation, maintains soil moisture, and raises soil temperature [17–19], but also realizes precise water and fertilizer management, thus increasing crop yields and slowing down the negative impacts of soil evaporation and temperature on the sustainable production of agriculture. Soybeans are a crop with high water requirements, and their production is significantly impacted by water stress. The implementation of rational and effective irrigation strategies is therefore essential to ensure stable yields and enhance irrigation water use efficiency [20,21]. Currently, soybean planting in Xinjiang is primarily focused on the sowing period [22], planting density [23,24], and rhizobia application [25], with relatively few studies addressing the soybean irrigation strategy. Furthermore, there are distinct differences in the management of water and fertilizers in Xinjiang when compared to other regions.

Therefore, in this study, the applicability of DSSAT-CROPGRO-Soybean was calibrated and verified using two-year field trial data from 2022 and 2023. Furthermore, the potential yield and biomass of soybeans were simulated based on meteorological data from 1994 to 2023, and the study analyzed the sustainability and stability of soybean yields under different irrigation strategies for drip-irrigated soybean cultivation under mulch in southern Xinjiang.

The aim was to identify suitable irrigation strategies for improving the sustainability and profitability of soybean production and promoting the large-scale production of drip-irrigated soybeans under mulch in arid and semi-arid regions.

## 2. Results

### 2.1. Calibration and Validation of the DSSAT-CROPGRO-Soybean Model

The simulated and measured values of soybean biomass in 2022 and 2023 are shown in Figure 1 and Table 1. Under different irrigation and fertilization treatments, the soybean biomass normalization index (d) and coefficient of determination ($R^2$) were greater than 0.9, indicating that the model has a good simulation effect on soybean biomass. The correction and validation results of soybean phenological period, 0–80 cm soil water storage, yield, and biomass are shown in Table 1. Using the measured data in 2023, we found that the absolute relative error (ARE), relative root mean square error (nRMSE) and $R^2$ were 0~22.35%, 0~23.47%, and 0.85~0.99, respectively, indicating that the model simulation was good (ARE < 30%, nRMSE < 30%). The verification of the model in 2022 showed that the simulated values of soybean phenology, 0–80 cm soil water storage, yield, and biomass were in agreement with the measured values (all less than 30%), and ARE, nRMSE, and $R^2$ were 0~27.70%, 0~26.34%, and 0.81~0.99, respectively, which further showed that the model had a good simulation effect on soybean growth and yield.

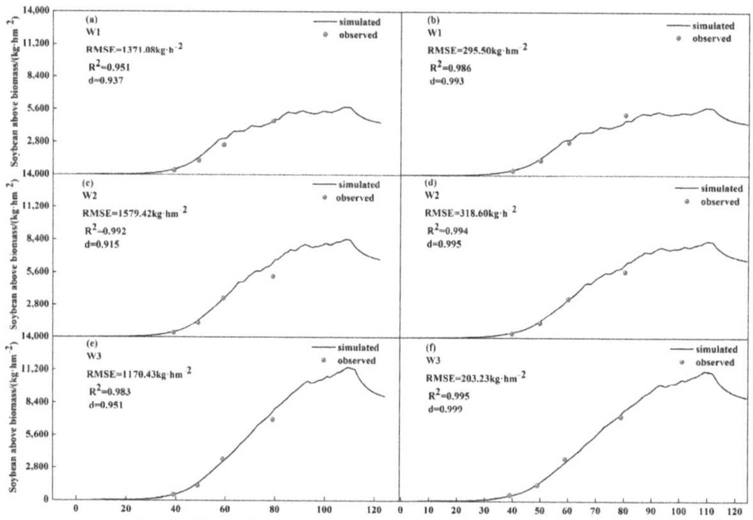


**Figure 1.** Simulated vs. measured biomass in 2022 and 2023. Note: (**a**,**c**,**e**) refer to the years 2022. (**b**,**d**,**f**) refer to the years 2023.

**Table 1.** DSSAT-CROPGRO-Soybean model simulates soybean phenology, biomass, and yield assessment results.

| Indicator | Treatment | 2022 | | 2023 | | RMSE | | nRMSE/% | | ARE/% | | $R^2$ | |
|---|---|---|---|---|---|---|---|---|---|---|---|---|---|
| | | Sim. | Obs. | Sim. | Obs. | 2022 | 2023 | 2022 | 2023 | 2022 | 2023 | 2022 | 2023 |
| Seedling stage | W1 | 7 | 7 | 7 | 7 | | | | | 0 | 0 | | |
| | W2 | 7 | 7 | 7 | 7 | 0 | 0 | 0 | 0 | 0 | 0 | | |
| | W3 | 7 | 7 | 7 | 7 | | | | | 0 | 0 | | |
| | Average | 7 | 7 | 7 | 7 | 0 | 0 | 0 | 0 | 0 | 0 | | |
| Anthesis | W1 | 35 | 36 | 46 | 44 | | | | | 2.78 | 4.55 | | |
| | W2 | 35 | 36 | 46 | 44 | 1 | 2 | 2.79 | 4.55 | 2.78 | 4.55 | | |
| | W3 | 35 | 36 | 46 | 44 | | | | | 2.78 | 4.55 | | |
| | Average | 35 | 36 | 46 | 44 | | | | | 2.78 | 4.55 | | |

**Table 1.** *Cont.*

| Indicator | Treatment | 2022 | | 2023 | | RMSE | | nRMSE/% | | ARE/% | | $R^2$ | |
|---|---|---|---|---|---|---|---|---|---|---|---|---|---|
| | | Sim. | Obs. | Sim. | Obs. | 2022 | 2023 | 2022 | 2023 | 2022 | 2023 | 2022 | 2023 |
| Maturity | W1 | 117 | 114 | 122 | 118 | | | | | 2.63 | 3.39 | | |
| | W2 | 118 | 114 | 122 | 118 | 3.7 | 4.61 | 3.24 | 4.04 | 3.51 | 3.39 | | |
| | W3 | 118 | 114 | 124 | 118 | | | | | 3.51 | 5.08 | | |
| | Average | 118 | 114 | 123 | 118 | 3.7 | 4.61 | 3.24 | 4.04 | 3.22 | 3.95 | | |
| Biomass | W1 | 4104 | 4938 | 2423 | 2421 | 1371.08 | 295.5 | 27.76 | 12.21 | 28.46 | 9.56 | 0.951 | 0.986 |
| | W2 | 4155 | 5591 | 2901 | 2701 | 1579.42 | 318.60 | 28.25 | 11.80 | 31.49 | 10.63 | 0.992 | 0.994 |
| | W3 | 4146 | 5086 | 3178 | 3185 | 1170.43 | 203.23 | 23.01 | 6.38 | 23.14 | 3.34 | 0.983 | 0.995 |
| | Average | 4135 | 5205 | 2834 | 2769 | 1373.64 | 272.44 | 26.34 | 10.13 | 27.70 | 7.84 | 0.975 | 0.992 |
| 0–80 cm soil water storage | W1 | 150 | 155 | 123 | 102 | 15.68 | 29.66 | 8.61 | 23.91 | 3.22 | 20.59 | 0.635 | 0.721 |
| | W2 | 186 | 188 | 134 | 133 | 12.78 | 15.86 | 7.02 | 12.79 | 1.06 | 0.75 | 0.639 | 0.770 |
| | W3 | 190 | 202 | 148 | 137 | 14.02 | 13.79 | 7.70 | 11.12 | 5.45 | 8.03 | 0.650 | 0.859 |
| | Average | 175 | 182 | 135 | 124 | 14.16 | 19.77 | 7.78 | 15.94 | 3.24 | 9.79 | 0.641 | 0.783 |
| Yield | W1 | 3525 | 3665 | 1866 | 2920 | | | | | 3.82 | 36.1 | | |
| | W2 | 3906 | 4210 | 3085 | 4236 | 450.57 | 906.2 | 10.83 | 23.47 | 7.22 | 27.17 | 0.81 | 0.85 |
| | W3 | 3893 | 4598 | 4261 | 4428 | | | | | 15.33 | 3.77 | | |
| | Average | 3775 | 4158 | 3071 | 3861 | 450.57 | 906.2 | 10.83 | 23.47 | 8.79 | 22.35 | 0.81 | 0.85 |

Note: Maturity refers to the number of days until soybeans reach harvest; above biomass refers to leaves, stems, and pods; Yield refers to the yield of the seed at the time of the soybean harvest; Sim refers to simulated value; Obs refers to measured values.

## 2.2. Soybean Biomass, Yield, and Irrigation Water Use Efficiency

The modified DSSAT-CROPGRO-Soybean model was employed to simulate the average soybean biomass, yield, and irrigation water use efficiency under 14 irrigation amount scenarios from 1994 to 2023 (Figure 2). The results indicated that the soybean yield exhibited a fluctuating pattern with the increase in irrigation amount. The average soybean yield was observed to range from 2529.10 kg·hm$^{-2}$ to 4984.73 kg·hm$^{-2}$. The soybean yield was found to be significantly higher when the irrigation amount was greater than or equal to 36 mm ($p < 0.05$). The highest yield observed was 4984.73 kg·hm$^{-2}$. The biomass of the soybeans increased with the increase in irrigation amount. The average soybean biomass was 5346.97 kg·hm$^{-2}$~9379.19 kg·hm$^{-2}$. The increasing trend of soybean biomass when the irrigation amount was higher than 36 mm was significantly higher than the low irrigation amount treatment (<36 mm). The biomass of soybeans grown with an irrigation amount of 63 mm was the highest at 9379.19 kg·hm$^{-2}$. In a variety of simulation scenarios, the efficiency of soybean irrigation water use is found to be closely correlated with yield change. The mean irrigation water use efficiency for soybeans is 0.79 to 1.30, with the highest efficiency observed at an irrigation amount of 30 to 39 mm (Figure 2b).

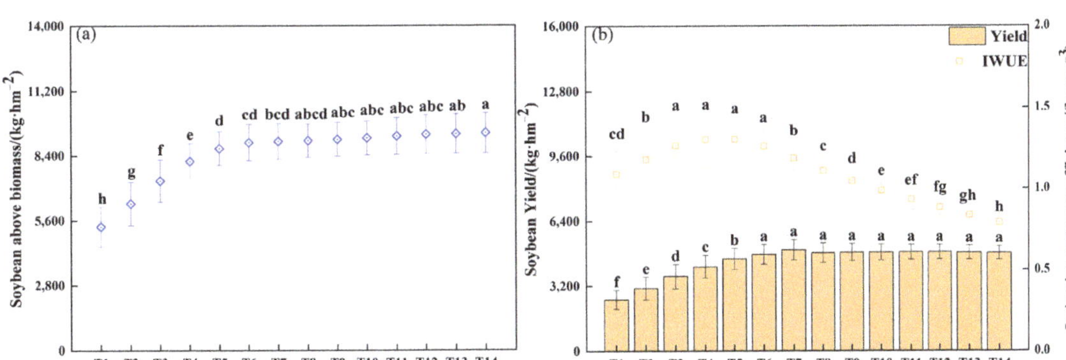


**Figure 2.** Simulated values of average soybean yield, average biomass, and irrigation water use efficiency from 1994 to 2023. (**a**) soybean above biomass. (**b**) soybean yield and irrigation water use efficiency. Note: T1~T14 represent the irrigation amounts of 24 to 63 mm, respectively. The lowercase letters indicate the difference in significance among treatments at the 0.05 level. The short line represents the standard deviation.

### 2.3. Evaluation of Simulated Soybean Yield Sustainability and Stability under Different Scenarios

The inter-annual variation, stability, and sustainability index of soybean yields from 1994 to 2023 is presented in Table 2 and Figure 3. The inter-annual variation amplitude of soybean yields exhibits a decreasing trend with an increase in irrigation amount, while the stability index displays an increasing trend with the same increase in irrigation amount. The change trend of the sustainability index is consistent with that of stability. The sustainability index of the T5–T14 yields was greater than 0.7, indicating a high degree of sustainability in soybean production. The variation coefficient of T4–T14 was less than 15%, indicating a high degree of stability in soybean yield change. Comprehensive analysis indicates that the soybean yields under the T5–T7 treatments exhibit high stability and good sustainability.

**Table 2.** Soybean yield coefficients of variation and sustainability indices from 1994 to 2023.

| Treatment | CV | SYI | Treatment | CV | SYI |
|---|---|---|---|---|---|
| T1 | 0.60 | 18.25 | T8 | 0.72 | 9.81 |
| T2 | 0.56 | 18.34 | T9 | 0.73 | 8.75 |
| T3 | 0.61 | 16.61 | T10 | 0.74 | 8.00 |
| T4 | 0.69 | 13.67 | T11 | 0.75 | 7.39 |
| T5 | 0.73 | 11.35 | T12 | 0.76 | 7.30 |
| T6 | 0.75 | 10.15 | T13 | 0.78 | 6.97 |
| T7 | 0.73 | 10.02 | T14 | 0.79 | 6.78 |

Note: CV and SYI represent the variation coefficient and sustainability index, respectively.

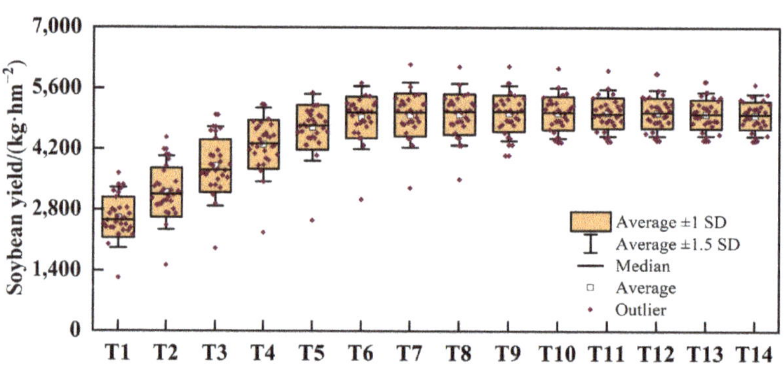


**Figure 3.** Soybean yield simulation, yield coefficients of variation, and sustainability indices from 1994 to 2023. Note: T1~T14 represent the irrigation amounts of 24 to 63 mm, respectively.

## 3. Discussion

The DSSAT crop model has been employed extensively to assess the impact of climate change and management strategies on crop growth and yield [26]. In this study, the model was calibrated and verified using soybean phenology, biomass, and yield data from 2022 to 2023. The ARE, nRMSE, and $R^2$ were introduced to evaluate the model simulation results. The results demonstrated that the corrected and validated DSSAT model provided a highly accurate simulation of soybean biomass, phenology, and yield. The nRMSE of the phenology, 0–80 cm soil water storage, and biomass was less than 30%, while the nRMSE of the soybean yield was less than 30%. Additionally, the $R^2$ values were 0. 81 and 0.85. The mean ARE of the different treatments was less than 30%, which is consistent with the research conclusions of Mulazzani [27], Wang Xingpeng [28], and Du Jiangtao in [29]. This indicates that the model constructed in this paper is a viable approach for simulating soybean growth and yield in southern Xinjiang.

Southern Xinjiang is a region that is typified by irrigated agriculture, and the implementation of moderate irrigation is a fundamental requirement for the achievement of high crop yields. The findings of this study indicate that the response of soybean yield and biomass to irrigation amount differs. The yield exhibits a downward parabolic change trend with an increase in irrigation amount, whereas biomass demonstrates linear growth. This suggests that excessive or inadequate irrigation is detrimental to soybean dry matter accumulation and nutrient absorption and utilization [30]. The yield decreased as a result of the contribution rate of the soybean yield after an irrigation amount exceeding a certain amount. This resulted in more water directly promoting the development of soybeans' nutritional organs, inhibiting reproductive growth, and ultimately leading to a decrease in yield [31]. Studies have demonstrated that a high irrigation amount of soybean shoot biomass and yield inhibition [32,33], which differs from this study, is largely influenced by regional environmental conditions. The southern Xinjiang area, for instance, is prone to high evaporation and transpiration rates, which have not yet reached the upper limit required to inhibit soybean growth. Consequently, subsequent tests will moderately increase the irrigation amount to verify the aforementioned conclusion. The irrigation water use efficiency is a crucial indicator for assessing the effective utilization of agricultural water resources. The utilization efficiency of irrigation water is negatively impacted by both high and low irrigation amounts, as evidenced by research [34]. The research of Li [35] found that although increasing the irrigation amount can effectively alleviate salinity stress caused by production, it can also cause a significant loss of water and nutrients, which in turn can lead to a reduction in irrigation water use efficiency. Zhang [36] set up 625 different simulation scenarios to examine the impact of increasing and reducing irrigation water on winter wheat. This paper reaches the same conclusion as Zhang [36] regarding the impact of irrigation water on winter wheat.

The sustainability index (SYI) and coefficient of variation (CV) are crucial indicators for measuring the sustainable development and stable yield of crops [37]. This study revealed that the irrigation volume had a significant impact on the sustainability and stability of soybean yields. The low irrigation amount was found to have a detrimental effect on the sustainability and low stability of soybeans. The sustainability and stability of soybean production increases with an increase in irrigation amount. Wang [38] analyzed the results of cotton yields, but too high an irrigation amount inhibits cotton production and increases the waste of water resources. However, increased irrigation under drought conditions promotes increased yields [39]. To ensure the sustainable development of soybeans, it is necessary to choose an appropriate irrigation amount. In addition to the irrigation amount of water, the amount of fertilizer is also an important factor affecting the growth and yield of soybeans. Consequently, the subsequent study will investigate the coupling relationship between irrigation amount and fertilizer amount, with the objective of further promoting the growth and yield of soybeans [40,41]. Meanwhile, only two years of pilot study were considered. In agriculture, trials of more than 3 years may be more convincing, and we will continue to conduct relevant experimental studies. This will provide a theoretical basis and technical support for the appropriate planting of soybeans in southern Xinjiang.

## 4. Materials and Methods

### 4.1. Overview of the Experimental Site

The field experiment was conducted in 2022 and 2023 at the Modern Agriculture Field Experimental Base in Xinjiang (81.17′56.52″ E, 40.32′36.90″ N, elevation 1, 100 m) (Figure 4). The area exhibits a typical temperate continental climate, with an average annual temperature of 12.04 °C, annual precipitation of about 46 mm, annual evaporation of approximately 2100 mm, and a groundwater depth of approximately 4 m.

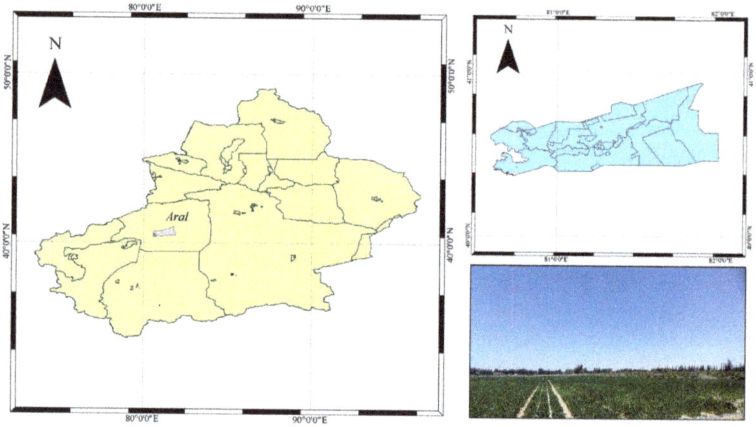


**Figure 4.** Study area.

### 4.2. Field Management and Experimental Design

The soybean variety utilized in the field trial was designated as "Tianyou 2986". Prior to planting, the residential area was levelled. The planting method employed was one film, six lines, and three bands (Figure 5). The wide behavior was 40 cm, the narrow behavior was 20 cm, the plant spacing was 10 cm, and the drip head spacing was 25 cm. The seeding density in 2022 and 2023 was $2.6 \times 10^5$ plants per hectare on 20 April and 17 April, respectively. The flowering dates were 29 May and 28 May, and the mature soybean harvest occurred in mid-August. The soil texture in the experimental area is sandy loam, and the average bulk capacity of 0–100 cm is 1.58 g·cm$^{-3}$, and the water holding in the field is 0.24 g·g$^{-1}$; the nitrogen, phosphorus, and potassium levels are 225 kg m$^{-2}$, 180 kg·m$^{-2}$, and 180 kg·m$^{-2}$, respectively.

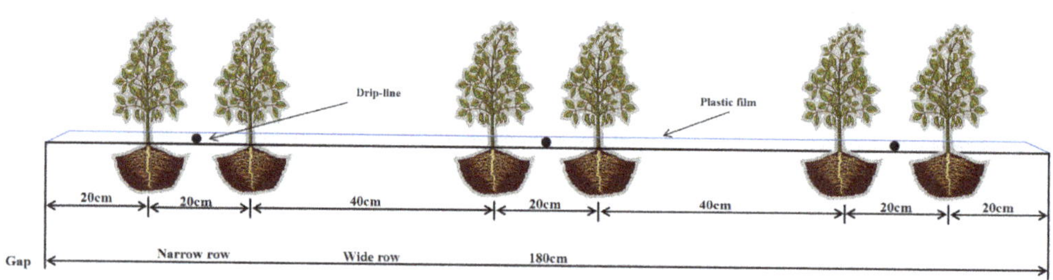


**Figure 5.** Soybean planting pattern map.

It was found that the optimal irrigation volume for soybeans in northern Xinjiang was 337–420 mm [42,43], and the optimal irrigation volume for cotton in southern Xinjiang was 406–462 mm [44], whereas less research has been conducted on the irrigation volume of soybean in the southern Xinjiang. For this reason, in the field experiment, three irrigation amounts were applied: 27 mm (W1), 36 mm (W2), and 45 mm (W3). Ten irrigation dates gave a total of 270, 360, and 450 mm of water (Table 3). The amount of nitrogen applied was 225 kg·hm$^{-2}$. Each treatment was replicated three times, with 18 cells in total. The small interval was reserved for shelter rows, and the cell size was 2.33 m by 15 m. Since the soybean with film drip irrigation in southern Xinjiang is still in the preliminary study stage and the *Kc* value has not been determined, this study used meteorological data to calculate the daily water loss, and irrigation was conducted when the accumulated water loss reached 36 mm ($ET_0 - P = 36$ mm) [35,36]. In the soybean seedling stage, one irrigation was applied to the soybean seedlings, two irrigations were applied during the flowering stage, three irrigations were applied during the pod stage, three irrigations were applied during the grain stage, and one irrigation was applied during the maturity stage, making ten irrigations in total throughout the growth period. Fertilizer was applied with water intervals. The irrigation and fertilization systems for soybeans in 2022 and 2023 are shown in Table 3.

**Table 3.** Irrigation and fertilization regimes for soybean fertility in 2022 and 2023.

| Planting Date | | Date of Irrigation | | Irrigation Amount/mm | | | Nitrogen Application/kg·hm$^{-2}$ | Planting Density/Plants·hm$^{-2}$ | Planting Depth/cm |
|---|---|---|---|---|---|---|---|---|---|
| 2022 | 2023 | 2022 | 2023 | W1 | W2 | W3 | | | |
| 20-April | 17-April | 12-May | 18-May | 27 | 36 | 45 | 0 | 2.6 × 10$^5$ | 2 |
| | | 31-May | 29-May | 27 | 36 | 45 | 45 | | |
| | | 8-June | 9-June | 27 | 36 | 45 | 0 | | |
| | | 15-June | 16-June | 27 | 36 | 45 | 45 | | |
| | | 23-June | 23-June | 27 | 36 | 45 | 0 | | |
| | | 1-July | 30-June | 27 | 36 | 45 | 45 | | |
| | | 8-July | 7-July | 27 | 36 | 45 | 0 | | |
| | | 16-July | 14-July | 27 | 36 | 45 | 45 | | |
| | | 22-July | 21-July | 27 | 36 | 45 | 0 | | |
| | | 28-July | 28-July | 27 | 36 | 45 | 45 | | |
| | | Total amount/mm | | 270 | 360 | 450 | 225 | | |

### 4.3. *DSSAT-CROPGRO-Soybean Model*

The DSSAT model is used to simulate crop growth and development in various environments [45]. The model effectively takes into account all of the parameters in the simulation that directly or indirectly affect soybean seed yield, which mainly includes 18 parameters, and is calibrated and validated using two years of experimental data. The measured data were entered into the DSSAT soybean file, run with the weather and soil files, and the process was repeated until the observed and simulated phenological periods and yields were optimized [46].

#### 4.3.1. Overview of the Model

The DSSAT model includes many plant growth modules, especially on soybeans [47,48], maize [49], and wheat [50]. The DSSAT (v4.8.2) soybean model (DSSAT-CROPGRO-Soybean) was employed to simulate soybean biomass and yield under membrane drip irrigation in southern Xinjiang. The applicability of the model in simulating soybean growth and development and yield in southern Xinjiang was then evaluated. The CROPGRO-Soybean module is employed to simulate the growth and development of soybeans under a range of climatic, soil, and field management conditions. The model input data encompass weather, crop, soil, and field management data [51–53]. The soil input data comprise the physical and chemical properties of different deep soil layers. The field management information includes crop varieties, planting date, planting depth, and irrigation and fertilization. Experimental data are entered via the Experimental Data module. It is necessary to create T-files (leaf area index and biomass, etc.) and A-files (phenological period and yield, etc.) in accordance with the standards and formats set out in the DSSAT model.

#### 4.3.2. Meteorological Data

The meteorological data for the period from 2022 to 2023 were primarily obtained from the HOBO meteorological monitoring station installed at the test site. The meteorological data for the period from 1994 to 2021 were obtained from the China Meteorological Science Data Sharing Service Network (http://www.cma.gov.cn/) (accessed on 1 March 2024). The meteorological data required by the CROPGRO-Soybean model encompass a range of variables, including daily maximum temperature (in degrees Celsius), minimum temperature (in degrees Celsius), solar radiation (in megajoules per square meter), rainfall (in millimeters), and others. The data were utilized in conjunction with the WeatherMan program to generate the requisite meteorological documents in the form of WHT files for the model.

#### 4.3.3. Soil Data

Prior to commencing the field test, it is necessary to sample the initial physical properties of the soil at four depths: 0–20 cm, 20–40 cm, 40–60 cm, and 60–80 cm (Table 4). The SBuild module should be employed to generate the SOL soil files that are necessary for the DSSAT-CROPGRO-Soybean model.

**Table 4.** Physical properties of the soil in the test area.

| Depth/cm | Clay/% | Silt/% | Sand/% | Wilting Point ($g \cdot g^{-1}$) | Field Capacity ($g \cdot g^{-1}$) | Saturated Water Content ($g \cdot g^{-1}$) | Bulk Density ($g \cdot cm^{-3}$) |
|---|---|---|---|---|---|---|---|
| 0~20 | 2.43 | 41.49 | 56.08 | 0.10 | 0.21 | 0.24 | 1.60 |
| 20~40 | 2.56 | 41.40 | 56.04 | 0.10 | 0.24 | 0.30 | 1.55 |
| 40~60 | 2.88 | 42.82 | 54.30 | 0.12 | 0.25 | 0.33 | 1.58 |
| 60~80 | 2.60 | 41.40 | 56.00 | 0.13 | 0.25 | 0.32 | 1.59 |

#### 4.3.4. Field Management Data

The field management data primarily comprise seeding date, planting density and depth, and irrigation and fertilization strategies (Table 3). The model field management database was established with the measured field data in 2022 and 2023. The localization application of the DSSAT-CROPGRO-Soybean model necessitates the parameter correction of the varieties. In this paper, the maximum likelihood estimation module (GLUE) in the DSSAT model is employed to validate the phenological period (emergence period, flowering period), biomass, and yield. The output includes the emergence period, flowering period, maturity period, yield, and harvest period, as well as the $R^2$ values for the crop varieties (Table 5).

**Table 5.** Plant Genetic Parameters.

| Crop Genetic Parameters | Parameter Description | Value | Value Range |
|---|---|---|---|
| CSDL | Critical short-day length below which reproductive development progresses with no daylength effect (for short-day plants) (hour) | 14.01 | 11.78~14.60 |
| PPSEN | Slope of the relative response of development to photoperiod with time (positive for short-day plants) (1/hour) | 0.262 | 0.129~0.385 |
| EM-FL | Time between plant emergence and flower appearance (R1) (photothermal days) | 17.83 | 9~28.9 |
| FL-SH | Time between first flower and first pod (R3) (photothermal days) | 5.5 | 5~10 |
| FL-SD | Time between first flower and first seed (R5) (photothermal days) | 13.34 | 11~22 |
| SD-PM | Time between first seed (R5) and physiological maturity (R7) (photothermal days) | 31.86 | 22~37 |
| FL-LF | Time between first flower (R1) and end of leaf expansion (photothermal days) | 13 | 11~19 |
| LFMAX | Maximum leaf photosynthesis rate at 30 C, 350 vpm $CO_2$, and high light (mg $CO_2 \cdot m^2 \cdot s^{-1}$) | 1.054 | 1~1.4 |
| SLAVR | Specific leaf area of crop under standard growth conditions ($cm^2 \cdot g^{-1}$) | 370.6 | 300~400 |
| SIZLF | Maximum size of full leaf (three leaflets) ($cm^2$) | 199 | 137~230 |
| XFRT | Maximum fraction of daily growth that is partitioned to seed + shell | 1 | 1 |
| WTPSD | Maximum weight per seed (g) | 0.169 | 0.15~0.19 |
| SFDUR | Seed filling duration for pod cohort at standard growth conditions (photothermal days) | 25.28 | 17~25.5 |
| SDPDV | Average seed per pod under standard growing conditions (grain/pod) | 1.82 | 1.7~2.44 |
| PODUR | Time required for crop to reach final pod load under optimal conditions (photothermal days) | 10.5 | 10~12 |
| THRSH | Threshing percentage. The maximum ratio of (seed/(seed + shell)) at maturity. Causes seeds to stop growing as their dry weight increases until shells are filled in a cohort. | 77 | 77~78 |
| SDPRO | Fraction protein in seeds (g(protein)/g(seed)) | 0.405 | 0.400~0.425 |
| SDLIP | Fraction oil in seeds (g(oil)/g(seed)) | 0.205 | 0.200~0.225 |

#### 4.3.5. Setting Simulated Scenarios

The quantity of irrigation water applied is a significant factor influencing the growth and yield of soybeans. In order to develop an irrigation schedule that is optimal for the growth requirements of soybeans in southern Xinjiang, a series of 14 simulation scenarios were employed in this study to determine the irrigation amount. The following irrigation amounts were considered: 24 mm, 27 mm, 30 mm, 33 mm, 36 mm, 39 mm, 42 mm, 45 mm, 48 mm, 51 mm, 54 mm, 57 mm, 60 mm, and 63 mm (Table 6). The impact of irrigation amount on potential soybean yield and biomass was evaluated through the use of meteorological data spanning the period from 1994 to 2023. The remaining parameters were aligned with the management protocol employed in the 2023 field experiment.

**Table 6.** Scenario simulation processing.

| Treatment | Irrigation Amount mm | Treatment | Irrigation Amount mm |
|---|---|---|---|
| T1 | 24 | T8 | 45 |
| T2 | 27 | T9 | 48 |
| T3 | 30 | T10 | 51 |
| T4 | 33 | T11 | 54 |
| T5 | 36 | T12 | 57 |
| T6 | 39 | T13 | 60 |
| T7 | 42 | T14 | 63 |

### *4.4. Aboveground Biomass and Seed Yield of Soybean*

From the beginning of the soybean seedling stage, 9 soybean plants were randomly selected in each cell at 15 d intervals, and then disassembled into stems, leaves, and pods. All samples were subjected to drying in an oven at 105 °C for 30 min, after which they were dried in an oven at 65 °C until they reached a constant weight. The dry material weight of each sample was quantified using an electronic balance with a precision of 0.1 g. During the soybean harvest, 2.33 $m^2$ quadrats were taken for each treatment, with each treatment being repeated three times in order to determine the seed yield of soybeans and to calculate the yield per unit area.

### *4.5. Irrigation Water Use Efficiency*

The calculation of irrigation water use efficiency (IWUE) is as shown in Equation (1).

$$\mathrm{IWUE} = \mathrm{Y}/\mathrm{I} \tag{1}$$

where Y is the soybean yield ($\mathrm{kg \cdot hm^{-2}}$) and I is the total amount of soybean irrigation water ($\mathrm{m^3 \cdot hm^{-2}}$) during the growth period.

### *4.6. Calculation of Soil Water Storage*

Soil water storage refers to the volumetric water content at a given soil depth, which can be calculated using the following equation:

$$\mathrm{SW} = \sum_{i=1}^{5} \theta_i h_i \tag{2}$$

where SW is the soil water storage from 0 to 80 cm (mm), $\theta_i$ is the soil water content of each layer, and $h_i$ is the corresponding soil depth (mm).

### *4.7. Model and Simulation Evaluation Method*

The GLUE (generalized likelihood uncertainty estimation) parameter debugging program package was used to determine the parameter rate of crop varieties, and the parameter adjustment number was set to 10,000 times. In this study, the parameters were corrected using data from the 2023 soybean field experiment, and the model was verified and evaluated using data from the 2022 field experiment. In both the model correction and validation processes, the root mean square error between the simulated and measured values (root mean square error, RMSE), nRMSE, ARE, $R^2$, and the homing index (normalization index and d) were used for evaluation, and the calculation formula is shown below:

$$\text{RMSE} = \sqrt{\frac{\sum_{i=1}^{n} (S_i - M)^2}{n}} \tag{3}$$

$$\text{nRMSE} = \frac{\sqrt{\frac{\sum_{i=1}^{n} (S_i - M)^2}{n}}}{\overline{M}} \times 100 \tag{4}$$

$$\text{ARE} = \frac{|S_i - M_i|}{M_i} \times 100\% \tag{5}$$

$$R^2 = \frac{\left[\sum_{i=1}^{n} (M_i - \overline{M}) \times (S_i - \overline{S})\right]}{\sum_{i=1}^{n} (M_i - \overline{M})^2 \times \sum_{i=1}^{n} (S_i - \overline{S})^2} \tag{6}$$

$$d = 1 - \frac{\sum_{i=1}^{n} (S_i - M_i)^2}{\sum_{i=1}^{n} \left(|S_i - \overline{M}| + |M_i - \overline{M}|\right)^2} \tag{7}$$

where S is the simulated value, M is the measured value, $\overline{S}$ is the average of the simulated value, and $\overline{M}$ is the average of the measured value. The accuracy of the simulated value is excellent when the nRMSE is <10%, 10% nRMSE is 20%, and 20% nRMSE is 30%. When the nRMSE is >30%, the simulated value is less accurate. The model performs well when nRMSE is 20%, ARE is 30%, and $R^2$ and d are close to 1 [26].

### *4.8. Evaluation of Production Sustainability*

Simulations were conducted using the DSSAT model. The yield of soybeans was simulated under different irrigation amounts. The sustainability of local soybean production is evaluated by the sustainability index (SYI), while yield stability is expressed by the coefficient of variation (CV). The calculations are as follows:

$$\text{SYI} = \frac{Y_{men} - \sigma}{Y_{max}} \tag{8}$$

$$\text{CV} = \frac{\sigma}{Y_{men}} \times 100 \tag{9}$$

where $Y_{men}$ is the mean yield (kg·hm$^{-2}$), $\sigma$ is the standard deviation of the yield, and $Y_{max}$ is the maximum yield (kg·hm$^{-2}$).

### *4.9. Statistical Analysis*

SPSS 22.0 software was used to analyze the difference in the measured data ($p < 0.05$), and Origin 2021 software was used for plotting.

## 5. Conclusions

This study demonstrates the efficacy of the DSSAT-CROPGRO-Soybean model in accurately predicting the phenological period, biomass, and yield of soybeans. These findings suggest that the model can be utilized to simulate the irrigation system of soybeans in southern Xinjiang. The simulation results indicate that when the irrigation amount is 42 mm, the highest soybean average yield is 4984.73 kg·hm$^{-2}$, the average biomass of soybeans is 5346.97 kg·hm$^{-2}$ to 9379.19 kg·hm$^{-2}$, and the average irrigation water use efficiency for soybeans is 0.79 to 1.30. The irrigation water use efficiency at the irrigation amount of 33 and 36 mm is 1.30. When the irrigation amount exceeds 33 mm, it contributes to the sustainability and stability of the soybean yield. In light of the impact of the irrigation amount on soybean biomass, yield, and irrigation water use efficiency, it is recommended that the optimal irrigation amount for soybean planting in southern Xinjiang be set at 36–42 mm, Irrigate ten times throughout the reproductive period: one time in the seedling stage, two times in the flowering stage, three times in the podding stage, three times in the bulging stage, and one time in the ripening stage, with a total irrigation volume of 360–420 mm.

**Author Contributions:** Conceptualization, methodology, software, writing—original draft preparation, L.Z.; conceptualization, methodology, formal analysis, investigation, Z.C.; conceptualization, methodology, resources, Y.G. (Yang Gao); writing—review and editing, W.H.; software, Z.S.; investigation, Y.G. (Yuanhang Guo); conceptualization, methodology, validation, writing—review and editing, H.W.; supervision, project administration, funding acquisition, X.W. All authors have read and agreed to the published version of the manuscript.

**Funding:** This research was supported by the Tarim University President's Fund Project (TDZKYS202301), the Major Science and Technology Project of Xinjiang Production and Construction Corps (2021AA003), and the Tarim University Graduate Student Research and Innovation Program (TDGRI202363).

**Data Availability Statement:** Data will be made available upon request.

**Conflicts of Interest:** The authors declare no conflicts of interest.

## References

1. Wang, X.; Feng, J.; Ao, G.; Qin, W.; Han, M.; Shen, Y. Globally nitrogen addition alters soil microbial community structure, but has minor effects on soil microbial diversity and richness. *Soil Biol. Biochem.* **2023**, *179*, 108982. [CrossRef]
2. Hu, M.; Wiatrak, P. Effect of planting date on soybean growth, yield, and grain quality. *Agron. J.* **2012**, *104*, 785–790. [CrossRef]
3. Ding, Y.; Yang, J.; Li, L.; Zhang, H.; Zeng, J. Effects of deficit irrigation and film mulching on biomass and production of Cyperus esculentus in the southern Xinjiang Basin. *Arid Zone Res.* **2022**, *39*, 883–892. [CrossRef]
4. Jueraiti, W.; Anwaer, M.; Xue, D. Coupling and Coordinated Development of Urbanization and Intensive Utilization of Cultivated Land: A Case Study in Aksu City, Xinjiang. *Arid. Zone Res.* **2019**, *36*, 1333–1343. [CrossRef]
5. Rosegrant, M.W.; Ringler, C.; Zhu, T. Water for agriculture: Maintaining food security under growing scarcity. *Annu. Rev. Environ. Resour.* **2009**, *34*, 205–222. [CrossRef]
6. Rosa, L.; Chiarelli, D.D.; Rulli, M.C.; Dellangelo, J.; Dodorico, P. Global agricultural economic water scarcity. *Sci. Adv.* **2020**, *6*, eaaz6031. [CrossRef]
7. Morales-Santos, A.; García-Vila, M.; Nolz, R. Assessment of the impact of irrigation management on soybean yield and water productivity in a subhumid environment. *Agric. Water Manag.* **2023**, *284*, 108356. [CrossRef]
8. Sandhu, R.; Irmak, S. Effects of subsurface drip-irrigated soybean seeding rates on grain yield, evapotranspiration and water productivity under limited and full irrigation and rainfed conditions. *Agric. Water Manag.* **2022**, *267*, 107614. [CrossRef]
9. Xiong, R.; Liu, S.; Considine, M.J.; Siddique, K.H.; Lam, H.M.; Lan, H.; Chen, Y. Root system architecture, physiological and transcriptional traits of soybean (*Glycine max* L.) in response to water deficit: A review. *Physiol. Plant.* **2021**, *172*, 405–418. [CrossRef]
10. Rajanna, G.A.; Dass, A.; Suman, A.; Babu, S.; Venkatesh, P.; Singh, V.; Upadhyay, P.K.; Sudhishri, S. Co-implementation of tillage, irrigation, and fertilizers in soybean: Impact on crop productivity, soil moisture, and soil microbial dynamics. *Field Crops Res.* **2022**, *288*, 108672. [CrossRef]
11. Gao, F.; Luan, X.; Yin, Y.; Sun, S.; Li, Y.; Mo, F.; Wang, J. Exploring long-term impacts of different crop rotation systems on sustainable use of groundwater resources using DSSAT model. *J. Clean. Prod.* **2022**, *336*, 130377. [CrossRef]
12. Wang, H.; Cheng, M.; Liao, Z.; Gou, J.; Zhang, F.; Fan, J.; Feng, H.; Yang, Q.; Wu, L.; Wang, X. Performance evaluation of AquaCrop and DSSAT-SUBSTOR-Potato models in simulating potato growth, yield and water productivity under various drip fertigation regimes. *Agric. Water Manag.* **2023**, *276*, 108076. [CrossRef]
13. Jiang, Y.; Zhang, L.; Zhang, B.; He, C.; Jin, X.; Bai, X. Modeling irrigation management for water conservation by DSSAT-maize model in arid northwestern China. *Agric. Water Manag.* **2016**, *177*, 37–45. [CrossRef]
14. Guo, S.; Zhang, Z.; Guo, E.; Fu, Z.; Gong, J.; Yang, X. Historical and projected impacts of climate change and technology on soybean yield in China. *Agric. Syst.* **2022**, *203*, 103522. [CrossRef]
15. Ge, J.; Yu, Z.; Gong, X.; Ping, Y.; Luo, J.; Li, B. Evaluation of Irrigation Modes for Greenhouse Drip Irrigation Tomatoes Based on AquaCrop and DSSAT Models. *Plants* **2023**, *12*, 3863. [CrossRef]
16. Zhang, Q.; Shi, R.; Singh, V.P.; Xu, C.; Yu, H.; Fan, K.; Wu, Z. Droughts across China: Drought factors, prediction and impacts. *Sci. Total Environ.* **2022**, *803*, 150018. [CrossRef] [PubMed]

17. Wang, H.; Yin, Z.; Zhang, L.; Zhao, F.; Huang, W.; Wang, X.; Gao, Y. Irrigation modulates the effect of increasing temperatures under climate change on cotton production of drip irrigation under plastic film mulching in southern Xinjiang. *Front. Plant Sci.* **2022**, *13*, 1069190. [CrossRef]
18. Yu, Q.; Kang, S.; Zhang, L.; Hu, S.; Li, Y.; Parsons, D. Incorporating new functions into the WAVES model, to better simulate cotton production under film mulching and severe salinity. *Agric. Water Manag.* **2023**, *288*, 108470. [CrossRef]
19. Lv, S.; Li, J.; Yang, Z.; Yang, T.; Li, H.; Wang, X.; Peng, Y.; Zhou, C.; Wang, L.; Abdo, A. The field mulching could improve sustainability of spring maize production on the Loess Plateau. *Agric. Water Manag.* **2023**, *279*, 108156. [CrossRef]
20. El Amine, B.; Mosseddaq, F.; Houssa, A.A.; Bouaziz, A.; Moughli, L.; Oukarroum, A. How far can the interactive effects of continuous deficit irrigation and foliar iron fertilization improve the physiological and agronomic status of soybeans grown in calcareous soils under arid climate conditions? *Agric. Water Manag.* **2024**, *300*, 108926. [CrossRef]
21. Valliyodan, B.; Ye, H.; Song, L.; Murphy, M.K.; Shannon, J.G.; Nguyen, H.T. Genetic diversity and genomic strategies for improving drought and waterlogging tolerance in soybeans. *J. Exp. Bot.* **2017**, *68*, 1835–1849. [CrossRef]
22. Wu, S.; Ran, X.; Huang, X.; Zhang, L.; Chen, G.; Wu, Q.; Zhai, Y. Effects of Plant and Row Configuration on Growing and Yield of Multiple Planting Soybean in Southern Xinjiang. *Soybean Sci.* **2023**, *42*, 194–203.
23. Ran, X.; Wu, S.; Huang, X.; Chen, G.; Wu, Q.; Wang, D.; Zhai, Y. Effects of Plant-Row Configuration on Grain Filling Characteristics and Yield of Reseeding Soybean in Southern Xinjiang. *Shandong Agric. Sci.* **2022**, *54*, 57–62. [CrossRef]
24. Huang, X.; Ran, X.; Wu, S.; Chen, G.; Wang, D.; Wu, Q.; Chen, G.; Zhai, Y. Effects of Sowing Dates and Densities on Photosynthetic Characteristics and Yield of Spring Soybean in Southern Xinjiang. *Soybean Sci.* **2022**, *41*, 546–556.
25. Xu, Y.; Xu, Y.; Li, L.; Huang, X.; Wang, D.; Wu, Q.; Zhai, Y. Effect of Nitrogen Fertilizer and Rhizobium lnoculation on Nodulation and Growth of Southern Xinjiang Spring Soybean. *Soybean Sci.* **2021**, *40*, 241–248. [CrossRef]
26. Sun, Y.; Sun, S.; Li, M.; Chang, J.; Wang, L.; Gao, X. Study on High and Stable Yield Areas of Soybean in Northeast Chinabased on Dssat Model. *Chin. J. Agric. Resour. Reg. Plan.* **2024**, *45*, 106–120.
27. Mulazzani, R.P.; Gubiani, P.I.; Zanon, A.J.; Drescher, M.S. Impact of soil compaction on 30-year soybean yield simulated with CROPGRO-DSSAT. *Agric. Syst.* **2022**, *203*, 103523. [CrossRef]
28. Wang, X.; Xin, L.; Du, J.; Li, M. Simulation of Cotton Growth and Yield under Film Drip lrrigation Condition Basedon DSSAT Model in Southern Xinjiang. *Trans. Chin. Soc. Agric. Mach.* **2022**, *53*, 314–321. [CrossRef]
29. Du, J.; Zhang, N.; Gong, K.; Du, M.; Yang, Y.; Wang, X. Optimization of cotton irrigation schedule under mulch drip irrigation in southern Xinjiang based on DSSAT model. *Chin. J. Ecol.* **2021**, *40*, 3760–3768. [CrossRef]
30. Kubar, M.S.; Shar, A.H.; Kubar, K.A.; Rind, N.A.; Ullah, H.; Kalhoro, S.A.; Wang, C.; Feng, M.; Gujar, A. Optimizing nitrogen supply promotes biomass, physiological characteristics and yield components of soybean (*Glycine max* L. Merr.). *J. Saudi J. Biolog. Sci.* **2021**, *28*, 6209–6217. [CrossRef]
31. Si, Z.; Zain, M.; Mehmood, F.; Wang, G.; Gao, Y.; Duan, A. Effects of nitrogen application rate and irrigation regime on growth, yield, and water-nitrogen use efficiency of drip-irrigated winter wheat in the North China Plain. *Agric. Water Manag.* **2020**, *231*, 106002. [CrossRef]
32. Singh, B.; Kaur, G.; Quintana-Ashwell, N.E.; Singh, G.; Lo, T.H.; Nelson, K.A. Row spacing and irrigation management affect soybean yield, water use efficiency and economics. *Agric. Water Manag.* **2023**, *277*, 108087. [CrossRef]
33. Zhu, B.; Zhang, C.; Jia, H.; Meng, Q.; Wang, N.; Kuang, E.; Zhang, L.; Shen, X.; Gao, X. Effect of Different Drip lrrigation on Spring Soybean Growth and lts Soil Physical Properties. *J. Nucl. Agric. Sci.* **2017**, *31*, 2250–2257. [CrossRef]
34. Dar, E.A.; Brar, A.S.; Mishra, S.K. Simulating response of wheat to timing and depth of irrigation water in drip irrigation system using CERES-Wheat model. *Field Crops Res.* **2017**, *214*, 149–163. [CrossRef]
35. Li, D. Quantifying water use and groundwater recharge under flood irrigation in an arid oasis of northwestern China. *Agric. Water Manag.* **2020**, *240*, 106326. [CrossRef]
36. Zhang, C.; Xie, Z.; Wang, Q.; Tang, M.; Feng, S.; Cai, H. AquaCrop modeling to explore optimal irrigation of winter wheat for improving grain yield and water productivity. *Agric. Water Manag.* **2022**, *266*, 107580. [CrossRef]
37. Liu, C.; He, C.; Wang, F.; Hang, J. Co-incorporation of rice stubble and Chinese milk vetch saves fertilizer input and increases production efficiency of rice. *J. Plant Nutr. Fertil.* **2024**, *30*, 279–288. [CrossRef]
38. Wang, H.; Li, G.; Xu, X.; Hang, W.; Zhao, Z.; Gao, Y.; Wang, X. Assessing the Sustainability of Cotton Production under Climate Change Based on the AquaCrop Model. *Chin. J. Agrometeorol.* **2023**, *44*, 588–598. [CrossRef]
39. Grassini, P.; Torrion, J.A.; Yang, H.S.; Ress, J.; Andersen, D.; Cassman, K.G.; Specht, J.E. Soybean yield gaps and water productivity in the western US Corn Belt. *Field Crops Res.* **2015**, *179*, 150–163. [CrossRef]
40. Liu, X.; Meng, L.; Yin, T.; Wang, X.; Zhang, S.; Ogundeji, A.O.; Li, S. Maize/soybean intercrop over time has higher yield stability relative to matched monoculture under different nitrogen-application rates. *Field Crops Res.* **2023**, *301*, 109015. [CrossRef]
41. Wang, H.; Li, G.; Huang, W.; Gao, Y. Compensation of cotton yield by nitrogen fertilizer in non-mulched fields with deficit drip irrigation. *Agric. Water Manag.* **2024**, *298*, 108850. [CrossRef]
42. Zhang, Y.; Xu, W.; Li, Y.; Zhang, N.; Peng, J. Suitable drip irrigation amount for summer soybean sownafter wheat in Xinjiang. *J. Plant Nutr. Fertil.* **2016**, *22*, 1133–1140. [CrossRef]
43. Bai, P.; Ren, J.; Zhang, J.; Guo, T.; Xue, L.; Wang, Z.; Sun, S.; Zheng, W. Effects of Different Water and Nitrogen Treatments on RootGrowth and Yield of Spring. *Soybean Mulched Drip Lrrigation. Soybean Sci.* **2023**, *42*, 326–334.

44. Tan, S.; Wang, Q.; Zhang, J.; Chen, Y.; Shan, Y.; Xu, D. Performance of AquaCrop model for cotton growth simulation under film-mulched drip irrigation in southern Xinjiang, China. *Agric. Water Manag.* **2018**, *196*, 99–113. [CrossRef]
45. Chisanga, C.B.; Phiri, E.; Shepande, C.; Sichingabula, H. Evaluating CERES-Maize model using planting dates and nitrogen fertilizer in Zambia. *J. Agric. Sci* **2015**, *7*, 79–97. [CrossRef]
46. Choudhury, A.K.; Molla, M.S.H.; Zahan, T.; Sen, R.; Biswas, J.C.; Akhter, S.; Lshtiaque, S.; Ahmed, F.; Maniruzaman, M.; Hosain, M.B.; et al. Optimum sowing window and yield forecasting for maize in northern and western Bangladesh using CERES Maize Model. *Agronomy* **2021**, *11*, 635. [CrossRef]
47. Mota, M.C.; Candido, L.A.; Cuadra, S.V.; Marenco, R.A.; Souza, E.V.; Tome, A.M.; Andrade Lopes, A.B.; Lima, F.L.; Reis, J.; Brizolla, R.M. CROPGRO-soybean model–Validation and application for the southern Amazon, Brazil. *Comput. Electron. Agric.* **2024**, *216*, 108478. [CrossRef]
48. Dias, H.B.; Cuadra, S.V.; Boote, K.J.; Lamparelli, R.A.C.; Figueiredo, G.K.A.; Suyker, A.E.; Magalhaes, P.S.G.; Hoogenboom, G. Coupling the CSM-CROPGRO-Soybean crop model with the ECOSMOS Ecosystem Model–An evaluation with data from an AmeriFlux site. *Agric. For. Meteorol.* **2023**, *342*, 109697. [CrossRef]
49. Wang, Y.; Jiang, K.; Shen, H.; Wang, N.; Liu, R.; Wu, J.; Ma, X. Decision-making method for maize irrigation in supplementary irrigation areas based on the DSSAT model and a genetic algorithm. *Agric. Water Manag.* **2023**, *280*, 108231. [CrossRef]
50. Ishaque, W.; Osman, R.; Hafiza, B.S.; Malghani, S.; Zhao, B.; Xu, M.; Ata-UI-Karim, S.T. Quantifying the impacts of climate change on wheat phenology, yield, and evapotranspiration under irrigated and rainfed conditions. *Agric. Water Manag.* **2023**, *275*, 108017. [CrossRef]
51. Jones, J.W.; Hoogenboom, G.; Porter, C.H.; Botte, K.J.; Batchelor, W.D.; Hunt, L.A.; Wilkens, P.W.; Singh, U.; Gijsman, A.J.; Ritchie, J.T. The DSSAT crop system model. *Eur. J. Agron.* **2003**, *18*, 235–265. [CrossRef]
52. Ritchie, J.T. Soil water balance and plant water stress. *Underst. Options Agric. Prod.* **1998**, 41–54. [CrossRef]
53. Mehrabi, F.; Sepaskhah, A.R. Winter wheat yield and DSSAT model evaluation in a diverse semi-arid climate and agronomic practices. *Int. J. Plant Prod.* **2020**, *14*, 221–222. [CrossRef]

plants

MDPI

Article

# Effect of Delayed Irrigation at the Jointing Stage on Nitrogen, Silicon Nutrition and Grain Yield of Winter Wheat in the North China Plain

Hao Zheng [1,2,†], Jinyang Sun [1,2,†], Yueping Liang [3], Caiyun Cao [2], Yang Gao [3], Junpeng Zhang [1], Hongkai Dang [2,*] and Chunlian Zheng [2,*]

[1] Shandong Key Laboratory of Agricultural Water-Saving Technology and Equipment, College of Water Conservancy and Civil Engineering, Shandong Agricultural University, Tai'an 271018, China; 17866702681@163.com (H.Z.); sunjinyang123456@126.com (J.S.); jpengzhang@163.com (J.Z.)

[2] Hebei Provincial Key Laboratory of Crop Drought Resistance Research, Institute of Dryland Farming, Hebei Academy of Agriculture and Forestry Sciences, Hengshui 053000, China; cycao1234@126.com

[3] Institute of Farmland Irrigation, Chinese Academy of Agricultural Sciences, Xinxiang 453000, China; yueping0520@163.com (Y.L.); gaoyang@caas.cn (Y.G.)

* Correspondence: wheatcrop@126.com (H.D.); nkzheng@126.com (C.Z.)

† These authors contributed equally to this work.

**Abstract:** Water scarcity is a key limitation to winter wheat production in the North China Plain, and it is essential to explore the optimal timing of spring irrigation to optimize N and Si uptake as well as to safeguard yields. The aim of this study was to systematically study the effect mechanism of nitrogen and silicon absorption of winter wheat on yield under spring irrigation and to provide a scientific basis for optimizing irrigation strategy during the growth period of winter wheat. In this experiment, the winter wheat 'Heng 4399' was used. Five irrigation periods, i.e., 0 d (CK), 5 d (AJ5), 10 d (AJ10), 15 d (AJ15), and 20 d (AJ20) after the jointing stage, were set up to evaluate the nitrogen (N) and silicon (Si) absorption and grain yield (GY). The results showed that delayed irrigation for 5–10 days at the jointing stage had increased the GY. With the delay of irrigation time, the N/Si content of the entire plant at the maturity period increased first and then decreased; among that, the maximum N contents appeared in AJ15 and AJ5 in 2015 and 2020, respectively, while the Si concentrations appeared in AJ5 and AJ10 in sequence. Compared with AJ15 and AJ20, the N accumulation of vegetative organs in AJ5 increased by 3.05~23.13% at the flowering stage, 14.12~40.12% after the flowering stage, and a 1.76~6.45% increase in the N distribution rate at maturity stage. A correlation analysis revealed that the GY was significantly and positively correlated with the N/Si accumulation at the anthesis and N translocation after the anthesis stage. In conclusion, under limited irrigation conditions, delaying watering for 5 to 10 days at the jointing stage can improve the nitrogen and silicon absorption and nutrient status of wheat plants and increase wheat yield.

**Keywords:** wheat; grain yield; nitrogen; silicon; water-saving irrigation

## 1. Introduction

Winter wheat is one of the main grain crops in the North China Plain (NCP) [1], which accounts for about 70% of the total wheat production in China [2]. In the NCP, the maize-in-summer/wheat-in-winter rotation system is common under the constraint of precipitation deficit, which is far less than the amount required for normal growth and development of winter wheat during the growing season [3–5], where two irrigations in spring after sowing are needed for high yield [6]. Since the 1970s, groundwater as the main source for crop irrigation has been increasingly depleted from the groundwater table, causing a series of geological disasters [7]. Therefore, implementing a more efficient irrigation strategy and technology in the region is imperative to conserve water while achieving high and stable yield.

Deficit irrigation is a water-saving irrigation strategy in which less than the full water requirement of the crop is applied only during critical growth stages. Extensive research has shown that an appropriate reduction of irrigation can increase grain yield and water use efficiency [8]. Zeng et al. [2] showed that the water productivity of winter wheat increased significantly when irrigation was reduced by 50%, while the difference in yield was minor. The jointing stage is a critical period of water demand for wheat, and sufficient water at this stage can significantly increase yield [9]. However, a mild water deficit before the jointing stage did not negatively affect wheat yield but slightly increased it [10]. Fan et al. [11] reported that delayed irrigation during the jointing stage could effectively improve photosynthetic capacity and water use efficiency, resulting in increased yield, while Li et al. [12] showed that the high yield of winter wheat can be maintained by irrigating once in spring.

Deficit irrigation will cause drought stress to wheat, and it is a crucial action for wheat to take measures to reduce the adverse effects of water shortage. Nitrogen (N) is one of the main factors limiting the productivity of various cereal crops, including wheat [13–15]. Nitrogen is involved in various key processes, such as leaf expansion, biomass accumulation, and plant height [16]. The application of a high nitrogen supply promoted the transport of carbohydrates from vegetative organs to panicles at the late growth stage [17]. More and more studies have shown that nitrogen nutrition is essential to improving plant stress resistance. Reasonable nitrogen management can alleviate the negative effects of drought on wheat yield formation and resource utilization efficiency. Yan et al. [18] found that with the increase in water stress, the total nitrogen uptake decreased, but there was no significant difference between normal irrigation and mild deficit irrigation. Even Yong et al. [19] thought that a certain degree of water stress could promote the absorption of nitrogen. The potential of silicon (Si) to enhance environmental stress tolerance has been confirmed [20]. In the mineral elements found in different crops, silicon content reached 0.1%~10.0% of plant dry weight [21]. Silicon can reduce the destructive effects of biotic [22] and abiotic [23] stresses in wheat plants. It plays a vital role in plant growth by enhancing mineral nutrition and mechanical strength and improving plant resistance to abiotic stresses [24]. Nadeem et al. [25] found that silicon can enhance the performance of wheat in reverse environments such as drought, saline-alkali, and nutrient deficit conditions. Silicon-induced plant resistance to stress may be attributed to the formation of plant rocks and the polymerization of insoluble silicon around cell walls and vascular bundles, reducing water loss by controlling evaporation [26]. Therefore, silicon is one of the important elements required for wheat to maintain a high yield [27]. However, Salem [28] showed that a decrease in irrigation water leads to a decrease in plant silicon content.

At present, the growth [29] and photosynthetic characteristics [30] of winter wheat under different stages of spring irrigation have been well understood through previous research. However, the research on the accumulation and distribution of N and Si in winter wheat plants varied according to geographical location, climatic conditions, and other factors. Therefore, this paper aimed to compare the effects of spring irrigation at the critical jointing stage on the accumulation and distribution of N and Si and the GY of winter wheat between a normal (2014–2015) and a dry (2019–2020) year in the NCP. It will provide a theoretical basis for the water-saving and high-yield cropping strategy.

## 2. Materials and Methods

### 2.1. Experimental Sites

The field experiment was conducted during the 2014–2015 and 2019–2020 wheat seasons in a site located in the Hengshui Experimental Station of Dry-land Farming Research Institute, Hebei Academy of Agricultural and Forest Sciences (115°72′ E, 37°91′ N), Shenzhou County, Hengshui, Hebei Province in the NCP. The experimental field has a semi-humid and semi-arid climate with an annual average temperature of 12.8 °C, an annual evaporation rate of 1785 mm, an annual sunshine duration of 2509.4 h, an annual frost-free period of 188 days, and an annual precipitation of 510 mm. The soil of the experimental site was loamy with a soil bulk density of 1.46 g·cm$^{-3}$. The nutrient contents in topsoil (20 cm) had 15.66~17.42 g·kg$^{-1}$ organic matter, 1.48~1.54 g·kg$^{-1}$ total N, 127.2~134.68 mg·kg$^{-1}$ available N, 21.95~24.13 mg·kg$^{-1}$ available P, and 113.68~128.42 mg·kg$^{-1}$ available K.

The climatic year type was divided by the classification method of annual crop climate types: $\Delta R = \frac{X - X_p}{s}$, in which $\Delta R$ is the anomaly percentage distance, $X$ is the rainfall during the whole growth period of crops, $X_p$ is the average value of rainfall elements, and $s$ is the standard deviation. When $\Delta R \geq 0.3$, it is a wet year; when $\Delta R \leq -0.3$, it is a dry year; and when $-0.3 \geq \Delta R \geq 0.3$, it is a normal year. According to the calculation, the values of ΔR in 2014–2015 and 2019–2020 are −0.08 and −1.09, respectively, which belong to the normal year and the dry year. The meteorological data during two wheat-growing seasons are presented in Figure 1 and Table 1.

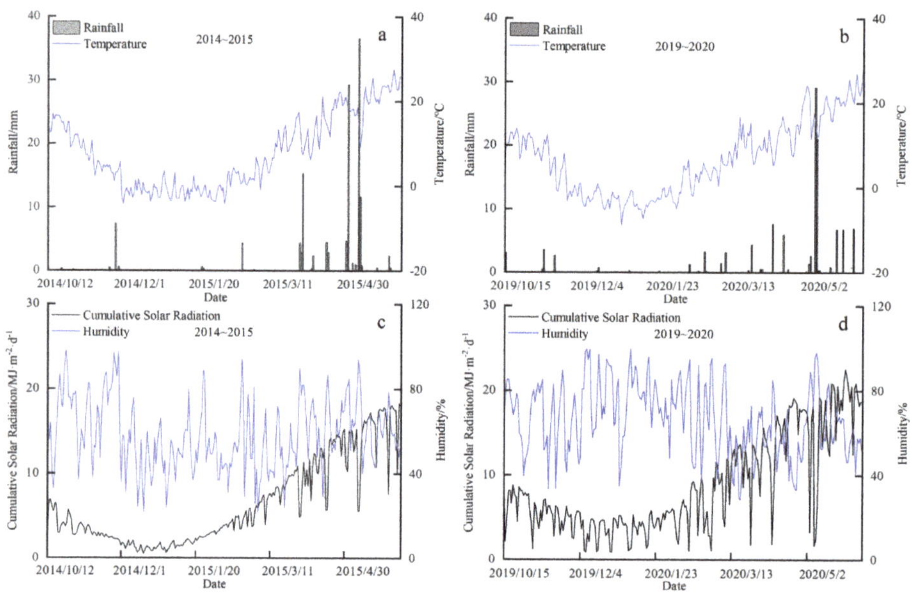


**Figure 1.** Meteorological map of winter wheat growth period. Note: The subfigures (**a**,**b**) represent the daily rainfall and average temperature in 2014–2015 and 2019–20202, respectively. The subfigures (**c**,**d**) represent the daily cumulative solar radiation and relative humidity in 2014–2015 and 2019–2020, respectively.

**Table 1.** Precipitation during the winter wheat growing seasons (mm) in 2014–2015 and 2019–2020.

| Year | Month | | | | | | | | | During the Whole Growing Season |
|---|---|---|---|---|---|---|---|---|---|---|
| | 10 | 11 | 12 | 1 | 2 | 3 | 4 | 5 | 6 | |
| 2014–2015 | 4.3 | 15.6 | 0.0 | 1.2 | 9.5 | 4.5 | 54.9 | 64.0 | 0.0 | 154 |
| 2019–2020 | 4.2 | 5.0 | 2.4 | 2.0 | 7.8 | 6.4 | 15.8 | 68.4 | 6.4 | 118.4 |

## 2.2. Experimental Design

The winter wheat variety planted in the field experiment was 'Heng 4399'. The winter wheat-summer maize double cropping system was adopted in one year. After the maize was harvested, the straw was all crushed and returned to the field. Winter wheat was sown on 12 October 2014 and 15 October 2019, respectively, and harvested on 6 June 2015 and 7 June 2020, respectively. The sowing rate was 225 kg·hm$^{-2}$ with a row spacing of 15 cm. The plots were irrigated with 75 mm of water before sowing to ensure seed germination and emergence. Urea, diammonium hydrogen phosphate, and potassium chloride were applied at 375 $kg \cdot hm^{-2}$ and 150 $kg \cdot hm^{-2}$, respectively, equivalent to 267 $kg \cdot hm^{-2}$ N, 241.5 $kg \cdot hm^{-2}$ $P_2O_5$, and 90 $kg \cdot hm^{-2}$ $K_2O$. The field experiment was arranged in a randomized complete block design with three replications for each treatment. During the critical jointing period, 5 treatments of delayed irrigation days were set as follows: 0d (CK, AJ0), 5d (AJ5), 10d (AJ10), 15d (AJ15), and 20d (AJ20) after the start of jointing. In AJ0, 75 mm irrigation was used as a control. The individual plot was 40 $m^2$ (5 m $\times$ 8 m) with a 0.5 m buffer between plots.

### 2.3. Date Collection and Calculation

#### 2.3.1. Plant Biomass

Aboveground plant samples (n = 30) were randomly collected at the jointing stage (JS), heading stage (HS), anthesis stage (AS), filling stage (FS), and maturity stage (MS) of winter wheat. Plant samples at anthesis and maturity were separated into three components (stem/sheath, leaf, and spike). The samples were dried in an oven at 80 °C for 72 h after heating at 105 °C for 30 min and then ground. Finally, N and Si contents were measured in different parts of the winter wheat.

#### 2.3.2. N Content Determination

Dried samples from all growth stages were ground and the dry matter was measured, and the N content of each sample was calculated by digestion of a 0.25 g sample in sulfuric acid ($H_2SO_4$) and hydrogen peroxide ($H_2O_2$) and measured using the Kjeldahl method [31]. The N content in different parts of the wheat plant was determined by multiplying the total dry matter of each plant organ by the N content and calculated in $kg \cdot hm^{-2}$.

$$\text{NAA} = \text{aboveground dry matter accumulation} \times \text{N content} \quad (1)$$

$$\text{NTA}_{\text{stem}} = \text{NAAA}_{\text{stem}} - \text{NAAM}_{\text{stem}} \quad (2)$$

$$\text{NTA}_{\text{leaf}} = \text{NAAA}_{\text{leaf}} - \text{NAAM}_{\text{leaf}} \quad (3)$$

$$\text{NTA}_{\text{spike}} = \text{NAAA}_{\text{spike}} - \text{NAAM}_{\text{spike}} \quad (4)$$

$$\text{N distribution rate in each organ} = \frac{\text{N accumulation in each organ}}{\text{total N accumulation}} \times 100\% \quad (5)$$

#### 2.3.3. Si Content Determination

The incinerated samples and their extraction were performed according to the method described by Hogendorp [32], and the colorimetric determination was calculated according to the method of Yang [33]:

$$\text{SiO}_2\% = \frac{\text{Microgram number of solution} \times \text{Color volume} \times \text{Dividing multiple}}{\text{Sample weight} \times 10^4} \times 100 \quad (6)$$

$$\text{SiAA} = \text{aboveground dry matter accumulation} \times \text{Si content} \quad (7)$$

$$\text{SiAAAM}_{\text{stem}} = \text{SiAAA}_{\text{stem}} - \text{SiAAM}_{\text{stem}} \quad (8)$$

$$\text{SiAAAM}_{\text{leaf}} = \text{SiAAA}_{\text{leaf}} - \text{SiAAM}_{\text{leaf}} \quad (9)$$

$$\text{SiAAAM}_{\text{spike}} = \text{SiAAA}_{\text{spike}} - \text{SiAAM}_{\text{spike}} \quad (10)$$

$$\text{Si distribution rate in each organ} = \frac{\text{Si accumulation in each organ}}{\text{total Si accumulation}} \times 100\% \quad (11)$$

Formula: NAA: N accumulation amount; $NTA_{stem}$: N translocation amount from stem and sheath; $NAAA_{stem}$: N accumulation amount in stem and sheath at anthesis; $NAAM_{stem}$: N accumulation amount in stem and sheath at maturity; $NTA_{leaf}$: N translocation amount from leaf; $NAAA_{leaf}$: N accumulation amount in leaf at anthesis; $NAAM_{leaf}$: N accumulation amount in leaf at maturity; $NTA_{spike}$: N translocation and accumulation amount to the ear; $NAAA_{spike}$: N accumulation amount in ear at anthesis; $NAAM_{spike}$: N accumulation amount in ear at maturity; SiAA: Si accumulation amount; $SiAAAM_{stem}$: Si accumulation amount in stem and sheath from anthesis to maturity; $SiAAA_{stem}$: Si accumulation amount in stem and sheath at anthesis; $SiAAM_{stem}$: Si accumulation amount in stem and sheath at maturity; $SiAAAM_{leaf}$: Si accumulation amount in leaf from anthesis to maturity; $SiAAA_{leaf}$: Si accumulation amount in leaf at anthesis; $SiAAM_{leaf}$: Si accumulation amount in leaf at maturity; $SiAAAM_{spike}$: Si accumulation amount in ear from anthesis to maturity; $SiAAA_{spike}$: Si accumulation amount in ear at anthesis; $SiAAM_{spike}$: Si accumulation amount in ear at maturity.

#### 2.3.4. Grain Yield

Grain yield (GY) was determined by harvesting the wheat ears in the middle rows of each treatment plot. Winter wheat plants in 1 $m^2$ were harvested as one of three replicates of each treatment for the calculation of GY.

### *2.4. Statistical Analyses*

The experimental data were collected and analyzed using Excel 2021 and plotted using the R-4.3.3 program and Origin 2021. Multiple comparisons were performed using SPSS 26.0 software, and the significance testing was performed using LSD with 0.05 as the level of significant difference.

## 3. Results

### *3.1. Changes of N and Si Contents in Winter Wheat and Their Distribution in Various Organs*

#### 3.1.1. Dynamics of the N and Si Content in Winter Wheat

The N contents at different growth stages were 12.41~34.48 $mg \cdot g^{-1}$ (Table 2), and the contents in different treatments in the same growing season decreased continuously. At the maturity stage, the N contents in 2015 and 2020 were 14.21~15.98 $mg \cdot g^{-1}$ and 12.54~13.43 $mg \cdot g^{-1}$, respectively. There was no significant difference in N content between treatments at maturity in 2015. The treatment with the highest N content in 2020 was AJ5, and there was little difference between AJ10 and AJ20, but significantly higher than CK and AJ20. In the same year, the treatments with the highest N content required for 100 kg grain construction were CK and AJ15 in 2015 and 2020, respectively, and the treatment with the lowest N content was AJ10. The Si content in winter wheat with different treatments was 7.83~19.27 $mg \cdot g^{-1}$. At maturity, the Si content was highest in AJ5 in 2015, 7.80%~10.40% higher than that in AJ10, AJ15, and AJ20. The highest Si content was recorded in AJ10 in 2020, which was not significantly different from AJ15, but significantly higher than the other treatments. The treatment with the greatest Si accumulation in 2015 was CK, which was significantly greater than AJ15 and AJ20 with an increase of 10.67~11.48%. AJ10 with the highest Si accumulation in 2020 was significantly greater than the other treatments with an increase of 15.03~39.19%.

**Table 2.** N/Si contents and its accumulations in winter wheat at different stages.

| Year | | Treatment | Content (mg·g$^{-1}$) | | | | | Accumulation (kg·hm$^{-2}$) | | | | | Accumulation per 100 kg GY (kg) |
|---|---|---|---|---|---|---|---|---|---|---|---|---|---|
| | | | JS | HS | AS | FS | MS | JS | HS | AS | FS | MS | |
| Nitrogen | 2015 | CK | 31.15 b | 23.58 b | 17.77 a | 15.51 ab | 15.28 a | 166.21 a | 215.37 a | 226.63 a | 255.61 a | 297.61 a | 3.67 |
| | | AJ5 | 31.80 b | 20.83 c | 46.69 b | 15.01 b | 15.04 a | 158.39 a | 204.8 a | 229.69 a | 250.09 a | 285.37 a | 3.44 |
| | | AJ10 | 34.48 a | 26.78 a | 15.41 c | 15.77 ab | 14.21 a | 165.95 a | 204.16 a | 229.95 a | 255.84 a | 290.03 a | 3.35 |
| | | AJ15 | 31.76 b | 23.37 b | 18.36 a | 17.34 a | 15.98 a | 128.7 b | 175.92 b | 226.05 a | 252.08 a | 305.96 a | 3.59 |
| | | AJ20 | 30.94 b | 22.73 bc | 16.29 b | 15.87 ab | 14.83 a | 131.35 b | 187.14 b | 199.59 b | 229.17 a | 279.22 a | 3.4 |
| | 2020 | CK | 18.36 d | 14.87 a | 13.03 a | 10.83 c | 12.54 b | 82.86 a | 133.68 a | 165.23 b | 168.52 b | 208.18 b | 4.28 |
| | | AJ5 | 18.88 cd | 14.62 b | 13.91 a | 12.74 b | 13.43 a | 82.12 a | 126.23 b | 185.89 a | 202.88 a | 241.73 a | 4.07 |
| | | AJ10 | 19.60 b | 14.34 c | 13.00 a | 12.97 b | 13.29 ab | 84.06 a | 107.88 c | 164.60 b | 197.12 a | 234.89 a | 4.01 |
| | | AJ15 | 20.90 a | 12.91 d | 13.28 a | 14.26 a | 12.50 b | 82.82 a | 90.86 d | 147.33 b | 166.94 b | 201.15 b | 4.31 |
| | | AJ20 | 19.36 bc | 12.41 e | 13.20 a | 13.32 ab | 13.15 ab | 75.07 b | 74.3 e | 152.64 b | 164.14 b | 201.57 b | 4.27 |

**Table 2.** *Cont.*

| Year | | Treatment | Content (mg·g$^{-1}$) | | | | | Accumulation (kg·hm$^{-2}$) | | | | | Accumulation per 100 kg GY (kg) |
|---|---|---|---|---|---|---|---|---|---|---|---|---|---|
| | | | JS | HS | AS | FS | MS | JS | HS | AS | FS | MS | |
| Silicon | 2015 | CK | 8.92 a | 15.05 a | 16.23 a | 18.45 a | 18.29 a | 47.55 a | 137.51 a | 206.98 b | 304.09 a | 356.35 a | 4.39 |
| | | AJ5 | 8.37 a | 11.08 a | 15.27 b | 18.14 a | 18.43 a | 41.71 ab | 109.01 a | 210.13 b | 302.24 a | 349.67 ab | 4.22 |
| | | AJ10 | 9.73 a | 11.37 a | 15.51 ab | 18.58 a | 16.81 a | 46.84 a | 86.69 a | 231.33 a | 301.36 a | 342.95 ab | 3.96 |
| | | AJ15 | 8.44 a | 12.38 a | 15.27 ab | 18.14 a | 16.70 b | 34.19 b | 93.17 a | 188.81 c | 263.66 a | 319.64 b | 3.75 |
| | | AJ20 | 7.83 a | 13.94 a | 15.27 b | 18.12 a | 17.10 a | 33.22 b | 114.82 a | 187.03 c | 261.71 a | 321.98 ab | 3.92 |
| | 2020 | CK | 7.90 c | 8.40 c | 9.92 c | 12.23 c | 13.70 b | 35.66 b | 75.51 d | 125.77 c | 190.33 d | 227.44 d | 4.67 |
| | | AJ5 | 9.03 b | 11.08 b | 10.83 bc | 13.97 bc | 14.55 b | 39.29 b | 95.68 a | 144.74 b | 222.50 bc | 261.78 bc | 4.40 |
| | | AJ10 | 10.27 a | 10.65 b | 13.40 a | 18.01 a | 17.91 a | 44.03 a | 80.14 c | 169.63 a | 273.77 a | 316.55 a | 5.41 |
| | | AJ15 | 11.03 a | 12.27 a | 11.95 ab | 19.27 a | 17.10 a | 43.73 a | 86.34 b | 132.56 bc | 225.60 b | 275.19 b | 5.89 |
| | | AJ20 | 8.17 bc | 8.15 c | 10.52 bc | 15.92 b | 15.21 b | 31.61 c | 48.79 e | 121.63 c | 196.09 cd | 233.16 cd | 4.94 |

Note: JS: jointing stage of winter wheat, HS: heading stage of winter wheat, AS: flowering stage of winter wheat, FS: filling stage of winter wheat, MS: maturity stage of winter wheat. Values within columns followed by the same letter are statistically insignificant at the 0.05 level.

### 3.1.2. Content and Accumulation of N and Si in Different Parts

In different parts of wheat at anthesis, the N content is shown as ear > leaf > stem, while the N accumulation showed leaf > stem > ear with more delayed irrigation time (Figure 2). The N accumulation in each part showed an increasing and then decreasing trend. At the anthesis stage, the stem and leaf obtained the maximum N accumulation in AJ10 and AJ5 in 2015, respectively. In 2020, the N accumulation of stem and leaf reached the maximum in AJ5 with an increase of 9.63~40.14% and 5.04~18.84%, respectively. At the maturity stage, the N contents and accumulations in different parts of wheat showed a trend of spike > leaf > stem, and the N accumulation in spike reached the maximum in AJ15 and AJ5 in the two years. Between them, the difference in the N accumulation was not significant in 2015, but in 2020, the difference between AJ5 and AJ10 was not significant but significantly greater than other treatments.

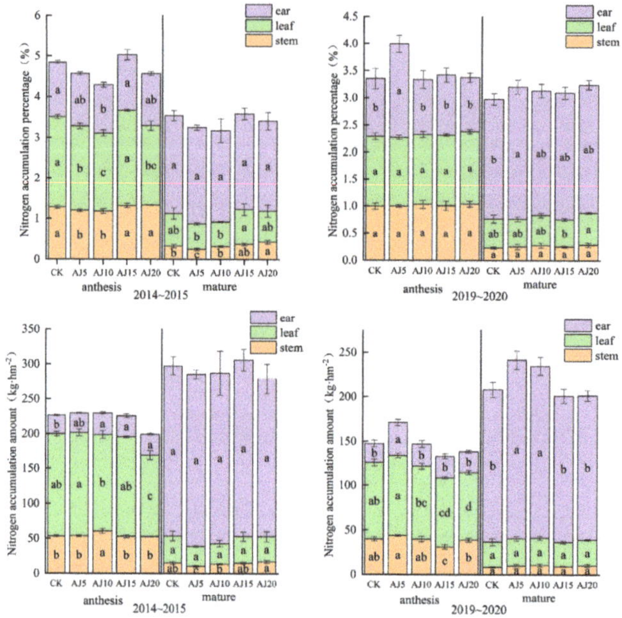


**Figure 2.** The N contents and accumulations in different wheat parts at the anthesis and maturity stage. Note: Values within columns followed by the same letter are statistically insignificant at the 0.05 level.

The contents of Si in different parts at anthesis were higher in leaf and spike than in stems, and the trend of Si accumulation is manifested as leaf > stem > spike (Figure 3). With more delayed irrigation time, the Si accumulation in each wheat part showed a trend of first increasing and then decreasing. At the anthesis stage, Si accumulation in stems reached the maximum in AJ5 in both years, and Si accumulation in leaves and spikes reached the maximum in AJ10. At the maturity stage with the delayed irrigation time, the Si accumulation in stems showed a trend of first decreasing and then increasing in 2015, and the maximum value was obtained in CK. The difference in Si accumulation between leaves and spikes was not significant. In 2020, all the wheat parts showed a trend of increasing and then decreasing. The Si contents in stems, leaves, and spikes reached the maximum in AJ10, AJ10, and AJ15 with an increase of 13.13~23.43%, 13.18~30.93%, and 10.42~64.83%, respectively, compared to other treatments.

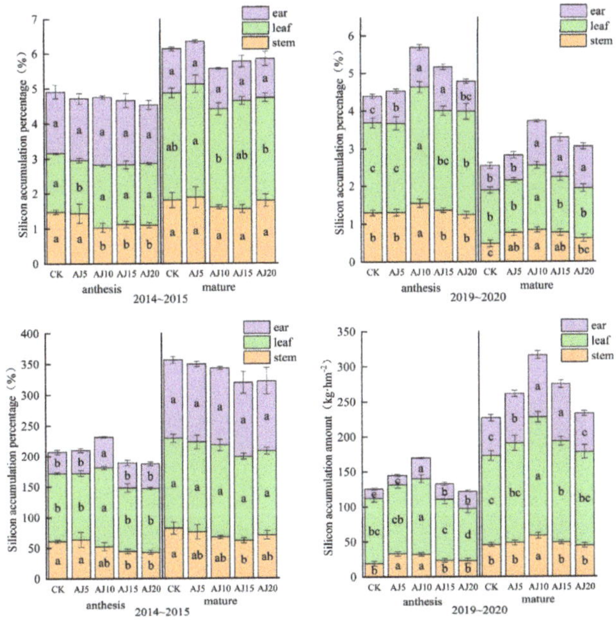


**Figure 3.** The Si contents and accumulations in different wheat parts at the anthesis and maturity stage. Note: Values within columns followed by the same letter are statistically insignificant at the 0.05 level.

### 3.1.3. Distribution of N and Si in Different Wheat Parts

The N distribution rate of N in leaves was 58.4~64.8% at anthesis and decreased to 10.1~14.4% at maturity (Table 3). The distribution rate of N in stems decreased from 23.4~28.4% at the anthesis stage to 3.5~6.1% at the maturity stage. The N distribution rate in ears gradually increased from 11.8~19.8% at the anthesis stage to 80.6~86.4% at the maturity stage. The N distribution rates in stems and leaves of different treatments in the two-year maturity period showed a trend of first decreasing and then increasing, and the N distribution rate in spikes showed a trend of first increasing and then decreasing. In 2015 and 2020, the N distribution rates in spikes reached the maximum in AJ5 with an increase of 2.73~6.40% and 0.97~3.35% compared to other treatments, respectively.

At the anthesis stage, Si distribution rates were 51.5~74.3%, 15~30.2%, and 9~21.9% in leaves, stems, and spikes, respectively. At the maturity stage, the Si distribution rate decreased to 41.3~57.3% in leaves, remained unchanged at 17.6~23% in stems, and increased to 23.6~37.6% in spikes. At the maturity stage, there was no significant difference in the Si distribution rates between different treatments in 2015. In 2020, the Si distribution rate showed a trend of first increasing and then decreasing with a maximum value in AJ15 that was significantly greater than other treatments.

**Table 3.** Distribution of N and Si in different wheat parts.

| Year | Stage | Treatment | Percentages of Nitrogen (%) | | | Percentages of Silicon (%) | | |
|---|---|---|---|---|---|---|---|---|
| | | | Stem | Leaf | Ear | Stem | Leaf | Ear |
| 2015 | anthesis | CK | 23.45 b | 64.77 a | 11.78 d | 29.32 a | 53.63 ab | 17.04 b |
| | | AJ5 | 23.39 b | 64.39 a | 12.22 cd | 30.20 a | 51.46 b | 18.34 b |
| | | AJ10 | 26.26 a | 60.24 b | 13.50 b | 22.57 b | 55.52 a | 21.91 a |
| | | AJ15 | 23.41 b | 63.26 a | 13.32 bc | 23.61 b | 54.79 a | 21.59 a |
| | | AJ20 | 26.28 a | 58.38 b | 15.34 a | 22.90 b | 55.75 a | 21.35 a |
| | mature | CK | 4.98 ab | 12.90 a | 82.12 b | 22.96 a | 41.28 b | 35.76 a |
| | | AJ5 | 3.52 c | 10.05 a | 86.43 a | 21.56 ab | 42.12 ab | 36.32 a |
| | | AJ10 | 4.58 bc | 11.32 a | 84.10 ab | 19.56 ab | 43.89 a | 36.55 a |
| | | AJ15 | 4.85 ab | 12.42 a | 82.73 ab | 19.20 b | 43.21 a | 37.60 a |
| | | AJ20 | 6.10 a | 12.70 a | 81.20 b | 21.87 ab | 42.95 ab | 35.17 a |
| 2020 | anthesis | CK | 26.97 ab | 58.18 ab | 14.84 a | 14.96 c | 74.35 a | 10.70 d |
| | | AJ5 | 26.18 b | 54.01 b | 19.81 a | 22.81 a | 68.17 b | 9.02 e |
| | | AJ10 | 27.04 ab | 55.75 ab | 17.22 a | 18.96b | 63.59 d | 17.46 b |
| | | AJ15 | 23.60 c | 58.94 a | 17.47 a | 17.60 b | 65.84 c | 16.55 c |
| | | AJ20 | 28.40 a | 55.61 ab | 15.99 a | 18.60 b | 61.28 e | 20.12 a |
| | mature | CK | 3.96 a | 13.59 a | 82.45 ab | 20.12 a | 55.85 b | 24.04 c |
| | | AJ5 | 3.93 a | 12.74 a | 83.32 a | 18.50 bc | 54.43 c | 27.07 b |
| | | AJ10 | 4.47 a | 13.02 a | 82.50 ab | 18.44 bc | 53.67 c | 27.89 b |
| | | AJ15 | 4.51 a | 13.61 a | 81.88 ab | 17.77 c | 52.61 d | 29.62 a |
| | | AJ20 | 5.04 a | 14.36 a | 80.60 b | 19.09 b | 57.32 a | 23.58 c |

Note: Values within columns followed by the same letter are statistically insignificant at the 0.05 level.

#### 3.1.4. Net Uptake and Transfer of N and Si in Different Organs

The Table 4 shows that stems and leaves started to transfer N after flowering, and the transferred amounts were 22.02~47.10 kg·hm$^{-2}$ and 46.83~119.34 kg·hm$^{-2}$, respectively. The amount transferred by leaves was the highest in both years. In 2015 and 2020, the N contribution rate from stems and leaves to ears reached about 50%, indicating that N uptake and accumulation largely depended on the N redistribution from vegetative organs. Between the different irrigation treatments and with delayed irrigation at the jointing stage, the total amounts of N transferred from stems and leaves showed a trend of first increasing and then decreasing, reaching a peak amount at AJ5 or AJ10 in both years. The Si content of each wheat part increased from the flowering to the maturing stage, and the trend of Si accumulation in different wheat parts was ear > leaf > stem. The amount of Si absorbed by spikelet in 2015 was much higher in 2015 than in 2020. In 2015, the Si uptake of each wheat part was not significantly different. In 2020, Si uptake in stems, leaves, and ears was maximum in CK, AJ10, and AJ15, respectively.

**Table 4.** The net absorption and transfer of nitrogen and silicon in different parts of wheat from anthesis to maturity stage.

| Year | Treatment | NTA (kg·hm$^{-2}$) | | | SiAAAM (kg·hm$^{-2}$) | | |
|---|---|---|---|---|---|---|---|
| | | Stem | Leaf | Ear | Stem | Leaf | Ear |
| 2015 | CK | −38.15 bc | −108.38 ab | 217.46 a | 21.28 a | 36.15 a | 91.95 a |
| | AJ5 | −43.70 ab | −119.34 a | 218.54 a | 12.08 a | 39.24 a | 88.23 a |
| | AJ10 | −47.10 a | −106.19 ab | 213.27 a | 14.75 a | 22.24 b | 74.63 a |
| | AJ15 | −38.10 bc | −104.76 b | 222.87 a | 16.59 a | 34.33 a | 79.9 a |
| | AJ20 | −35.44 c | −80.91 c | 195.95 a | 27.13 a | 33.95 a | 73.85 a |

**Table 4.** *Cont.*

| Year | Treatment | NTA (kg·hm$^{-2}$) | | | SiAAAM (kg·hm$^{-2}$) | | |
|---|---|---|---|---|---|---|---|
| | | Stem | Leaf | Ear | Stem | Leaf | Ear |
| 2020 | CK | −31.5 ab | −57.47 ab | 149.69 ab | 26.85 a | 33.68 b | 41.25 b |
| | AJ5 | −34.07 a | −59.30 a | 167.92 a | 15.41 b | 44.02 ab | 57.75 a |
| | AJ10 | −29.22 b | −51.24 bc | 168.47 a | 26.22 a | 62.01 a | 58.75 a |
| | AJ15 | −22.02 c | −50.12 c | 141.48 b | 25.49 ab | 57.42 ab | 59.58 a |
| | AJ20 | −28.56 b | −46.83 c | 140.53 b | 21.76 ab | 59.15 ab | 30.51 b |

Note: Values within columns followed by the same letter are statistically insignificant at the 0.05 level. '−' represents the decrease in dry matter quality from the anthesis stage to the maturity stage, and no '−' represents the increase in dry matter quality from the anthesis stage to the maturity stage. NTA: N translocation amount from anthesis to maturity; SiAAAM: Si accumulation amount from anthesis to maturity.

### 3.2. GY and Its Correlation with the N and Si Accumulation

#### 3.2.1. Impact of Irrigation Duration and Different Years on the GY and Irrigation Water Use Efficiency of Winter Wheat

In the two experimental years, the GY and irrigation water use efficiency (IWUE) of winter wheat first increased and then decreased with delayed irrigation time (Figures 4 and 5). Among them, the GY of winter wheat reached the maximum in AJ10 with an increase of 1.37~6.57% compared with other treatments in 2015. Compared with CK and other treatments, the GY in AJ5 and AJ10 increased significantly by 22.12~27.20% in 2020, and the difference between AJ5 and AJ10 was not significant. The GY of winter wheat in 2020 was much lower than that in 2015. Compared with 2020, the GY of each treatment decreased by 28.25~45.25% than that in 2015. The IWUE was consistent with the change rule of yield.

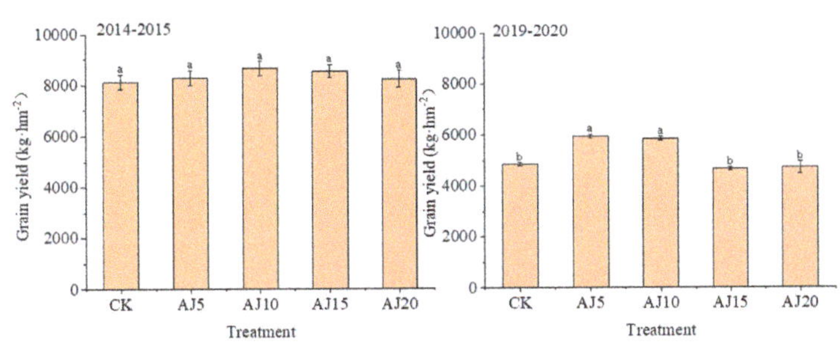


**Figure 4.** Grain yield of winter wheat. Mean values in a separate column followed by similar letters were not significantly different at $p < 0.05$. The values are the means ± SE (standard error).

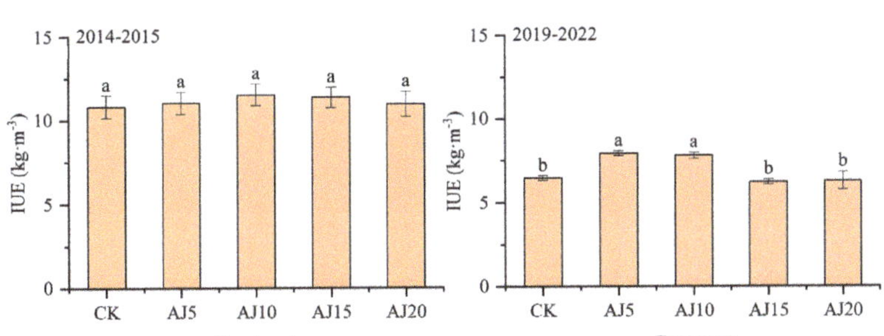


**Figure 5.** IUE of winter wheat. Mean values in a separate column followed by similar letters were not significantly different at $p < 0.05$. The values are the means ± SE (standard error).

The F-value test (Table 5) showed that irrigation time (I) had a very significant effect on GY, NAAA, $NTA_{stem}$, $NTA_{leaf}$, SiAAA, SiAAM, while the different year (Y) had a very significant effect on GY, NAAA, NAAM, $NTA_{stem}$, $NTA_{leaf}$, SiAAA, SiAAM. The interaction between them (I × Y) had a significant effect on GY, NAAM, $NTA_{leaf}$, and a most significant effect on NAAA, $NTA_{stem}$, and SiAAM.

**Table 5.** The significance of the F-value from the analysis of variance of various parameters of winter wheat under different irrigation times and years at various developmental stages.

| Parameter | GY | NAAA | NAAM | $NTA_{stem}$ | $NTA_{leaf}$ | SiAAA | SiAAM |
|---|---|---|---|---|---|---|---|
| Irrigation duration (I) | 5.423 ** | 13.580 ** | 1.744 NS | 9.912 ** | 13.577 ** | 26.583 ** | 6.844 ** |
| Years (Y) | 494.287 ** | 417.794 ** | 137.851 ** | 110.919 ** | 500.729 ** | 441.056 ** | 129.995 ** |
| I × Y | 3.526 * | 4.829 ** | 3.108 * | 4.681 ** | 3.841 * | 1.750 NS | 7.521 ** |

Note: * and ** denote the significance levels at alpha 0.05 and 0.01 obtained by morality. Significant difference (HSD) test. 'NS' denotes non-significance. GY: grain yield; NAAA: N accumulation amount at anthesis; NAAM: N accumulation amount at maturity; $NTA_{stem}$: N translocation amount from stem and sheath; $NTA_{leaf}$: N translocation amount from leaf; SiAAA: Si accumulation amount at anthesis; SiAAM: Si accumulation amount at maturity.

### 3.2.2. Relationship between the Studied Parameters

Correlation analysis of GY with N and Si accumulation and translocation (Figure 6) showed that it was significantly and positively correlated with NAAA, $NAAA_{leaf}$, $NAA_{meaf}$, $NAAA_{stem}$, $NAAN_{stem}$, $NAAA_{stem+leaf}$, $NAAA_{spike}$, $NAAM_{spike}$, $NTA_{stem}$, $NTA_{leaf}$, $NTA_{stem+leaf}$, and $NTA_{spike}$, but negatively correlated with NAA for 100 kg GY. At the same time, $NAAA_{stem}$ and $NAAA_{leaf}$ were positively correlated with the amount of N translocated from different parts of the wheat after flowering. In this experiment, delayed irrigation during the jointing stage increased the N accumulation in vegetative organs at the anthesis stage, increased the N transfer from vegetative organs after the anthesis stage, and increased the N accumulation in grains for higher GY. GY was found to be positively correlated with $SiAAA_{leaf}$, $SiAAM_{leaf}$, $SiAAA_{stem}$, $SiAAA_{spike}$, $SiAAM_{spike}$, and $SiAAAM_{spike}$, and negatively correlated with $SiAAAM_{leaf}$ and SiAA for 100 kg GY, indicating that increasing Si content in stems and ears promotes GY to some extent.

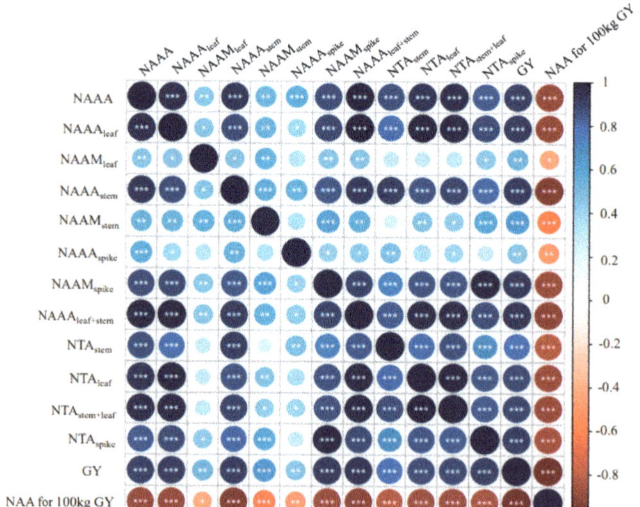


(**a**)

**Figure 6.** *Cont.*

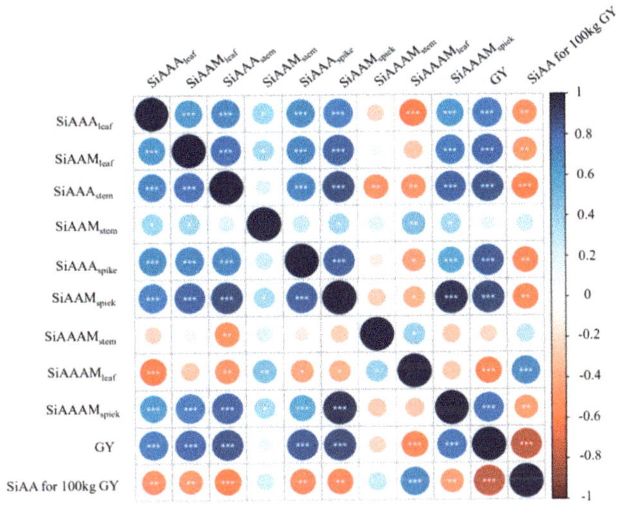

(**b**)

**Figure 6.** The relationship between the yield and other studied parameters. (**a**) Correlation analysis between grain yield and N accumulation and translocation; (**b**) Correlation analysis between grain yield and silicon accumulation and translocation. Note: The intensity of color represents the significance of a variable. Blue represents a positive correlation, and red represents a negative correlation. *, **, *** significant at the 0.05, 0.01, and 0.001 probability levels, respectively. GY: grain yield; NAAA: N accumulation amount at anthesis; NAAM: N accumulation amount at maturity; $NAAA_{leaf}$: N accumulation amount in leaf at anthesis; $NAAM_{leaf}$: N accumulation amount in leaf at maturity; NAAAstem: N accumulation amount in stem and sheath at anthesis; $NAAM_{stem}$: N accumulation amount in stem and sheath at maturity; $NAAA_{spike}$: N accumulation amount in ear at anthesis; $NAAM_{spike}$: N accumulation amount in ear at maturity; $NAAA_{stem+ leaf}$: N accumulation amount in stem, sheath, and leaf at anthesis; $NTA_{stem}$: N translocation amount from stem and sheath; NTAleaf: N translocation amount from leaf; NTAspike: N translocation and accumulation amount to the ear; NAA for 100 GY: N accumulation amount for per 100 kg grain yield construction; SiAAA: Si accumulation amount at anthesis; SiAAM: Si accumulation amount at maturity; $SiAAA_{leaf}$: Si accumulation amount in leaf at anthesis; $SiAAM_{leaf}$: Si accumulation amount in leaf at maturity; $SiAAA_{stem}$: Si accumulation amount in stem and sheath at anthesis; $SiAAM_{stem}$: Si accumulation amount in stem and sheath at maturity; SiAAAspike: Si accumulation amount in ear at anthesis; SiAAMspike: Si accumulation amount in ear at maturity; $SiAAAM_{stem}$: Si accumulation amount in stem and sheath from anthesis to maturity; $SiAAAM_{leaf}$: Si accumulation amount in leaf from anthesis to maturity; $SiAAAM_{spike}$: Si accumulation amount in ear from anthesis to maturity; SiAA for 100 GY: Si accumulation amount for per 100 kg grain yield construction.

## 4. Discussion

### *4.1. The Relationship between the N and Si Content, Accumulation, and Yield Level of Winter Wheat*

Our research showed that the N contents in the whole plant at the maturity stage of winter wheat were 1.42~1.6% and 1.25~1.34% in 2015 and 2020, respectively, which was slightly lower than the results (2.10~3.24%) of Li [34]. The Si content was 1.67~1.84% and 1.37~1.79% in both years, respectively, which was lower than that determined by Li [35] in wheat straw (2.5~5%). At the yield level of this study, the highest N accumulation in both years was 201.15~305.97 kg·hm$^{-2}$, which was similar to 200~400 kg·hm$^{-2}$ determined by Liang [36] at the yield level of 7881~9588 kg·hm$^{-2}$ at the maturity stage. The highest Si accumulation was 227.44~356.35 kg·hm$^{-2}$, which was higher than 115.42~159.24 kg·hm$^{-2}$ determined by He [37] at the yield level of 3673.47~5102.04 kg·hm$^{-2}$. The difference may be attributed to the light quality and quantity, cultivation pattern, soil texture, and other parameters.

In 2015 and 2020, the N uptake per 100 kg grain was 3.35~3.67 kg and 4.01~4.31 kg, respectively. He et al. [38] showed that under the yield level of 7387~10,414 kg·hm$^{-2}$, the total N accumulation was 151~256 kg·hm$^{-2}$, and the N uptake per 100 kg grain was 3.35~5.15 kg. Li et al. [39] proved that under the yield level of 5161~7671 kg·hm$^{-2}$, the total N accumulation was 193~290 kg·hm$^{-2}$, and the N uptake per 100 kg grain was 3.53~3.78 kg. With the increase of GY per unit area, the total N uptake of wheat increased significantly (r = 0.950, $p < 0.001$), while the N uptake per unit GY decreased (r = −0.959, $p < 0.001$), which was consistent with the result of previous research [40]. The Si uptake per 100 kg grain ranged from 3.75 to 4.39 kg and 4.40 to 5.89 kg in 2015 and 2020, respectively. With the increase in yield, the total Si uptake had an increasing trend (r = 0.773, $p < 0.001$), and the Si uptake per 100 kg of grains had a decreasing trend (r = 0.800, $p < 0.001$). In conclusion, we found that under the condition with different delayed spring irrigation times, the amount of the N and Si uptake per 100 kg of grains decreased, and the N/Si utilization efficiency for the grain production was improved to increase the GY.

### 4.2. N Nutritional Characteristics of Winter Wheat and Its Effect on GY under Different Times of Spring Irrigation

Water affects N accumulation and transport in crops [41]. Studies have shown that N accumulation in wheat increased with more irrigation water [42]. Water deficiency from the jointing to anthesis stage has the greatest impact on N absorption [43,44]. However, slight water stress can be beneficial to increase the N accumulation in wheat [19]. The jointing stage is a key period for water demand in winter wheat [9]. Delayed jointing irrigation has been shown to promote N accumulation in vegetative organs and its transport to grains, and to increase N accumulation after the anthesis stage and N partitioning ratio in grains at the maturity stage [45]. With a delay in irrigation time, our data showed that N accumulation in each wheat part tended to first increase and then decrease. At the anthesis stage, N accumulation in stems and leaves reached the maximum amount in AJ10 and AJ5 in 2015, respectively. In 2020, N accumulation in both leaves and stems reached the maximum amount in AJ5, while at the maturity stage, N accumulation in ears reached the maximum amount in AJ15 and AJ5 in both years. Therefore, the timing of irrigation may affect the efficiency of water and available N uptake by roots [46,47]. Early irrigation may lead to insufficient water supply later, resulting in reduced N uptake from the soil due to reduced nutrient diffusion and massive leakage in the soil [28]. The Lu et al. [48] study found that water shortage would reduce the nitrogen accumulation of winter wheat by 23.2%. However, if irrigation is applied too late, it can lead to a water shortage at the critical stage, and water cannot be fully utilized. Combined with previous studies on water use [29,34], we found that whole-plant N was highly correlated with water use (r = 0.820, $p < 0.001$). In addition, in this experiment, supplemental N fertilization with irrigation met the second N requirement of wheat and promoted GY [49]. During the two years, the yield reached the maximum in AJ10 and AJ5, respectively, which increased by 6.57%, 1.37%, 5.27%, and 22.12%, 27.21%, and 25.86% compared with CK, AJ15, and AJ20, respectively. However, much delayed irrigation may be affected by more rainfall in the late growth stage, and some N fertilizer may be lost through guttation, root secretion, and rain leaching [50], which in turn inhibits N accumulation during flowering. The statistical analysis of this study showed that GY was significantly and positively correlated with N accumulation in vegetative parts at anthesis, N translocation after anthesis, and N accumulation in spikes at maturity. It suggested that appropriate delayed irrigation at the jointing stage could promote N accumulation after anthesis and the N partitioning ratio in the ear at the maturing stage by promoting N accumulation in vegetative organs and its transport to grains for high yield.

### 4.3. Si Nutritional Characteristics of Winter Wheat and Its Effect on the GY under Different Irrigation Times in Spring

Si can increase the tolerance of plants to environmental stresses [25]. Previous studies have shown that exogenous Si could significantly improve the growth and development of winter wheat under drought stress [51], and it was generally believed that there were differences in the distribution of Si in plants, which generally followed the end-distribution [52], where the highest Si content was in leaves and the lowest in roots. This experiment confirmed that Si in wheat followed the end distribution and that the highest accumulation occurred mainly in leaves. The accumulation of silicon in leaves is beneficial to the growth and yield of winter wheat. The results showed that Si had a beneficial effect on the greening of leaves and could improve the chlorophyll content. In addition, Si could reduce the stomatal conductance and water loss of leaves [53,54]. In this experiment, compared with CK, AJ15, and AJ20 treatments, the leaf silicon accumulation of AJ10 treatment at the flowering stage increased by 15.56~44.67%, and the yield increased by 1.37~23.93%, which was consistent with the above conclusions. In addition, this study has shown that the Si content in the whole plant in 2015 and 2020 showed a trend of first increasing and then decreasing, and the variation range of silicon accumulation in each treatment was 319.64~356.35 $kg \cdot hm^{-2}$ and 227.44~346.56 $kg \cdot hm^{-2}$, respectively. The highest amount is in AJ5 and AJ10, respectively. There was also a significant difference in Si content between years ($F = 10.038$, $p = 0.004$), as Si absorption is water dependent [28]. Appropriately delayed irrigation brings soil moisture in line with the water demand and promotes water uptake by wheat [29]. Combined with previous studies on water use [28,33], we found that whole-plant Si accumulation was highly correlated with water use ($r = 0.762$, $p < 0.001$). With delayed irrigation time, Si partitioning rates of winter wheat in leaves and ears first increased and then decreased. Correlation analysis showed that GY was significantly and positively correlated with Si content in each of the wheat organs at the anthesis stage and its content in ears at the maturity stage. These results indicate that suitable delayed irrigation at the jointing stage can promote the proportion of silicon in the growth center of leaves and ears, improve stress resistance, and then affect the higher GY.

### 4.4. Effect of Different Spring Irrigation Times on the IUE of Winter Wheat

Moderate water stress can improve the irrigation water utilization efficiency of wheat [55]. Through the analysis of winter wheat IUE, it was found that IUE first increased and then decreased with the change in spring irrigation time, which was consistent with the research results of Liu et al. [56]. The reason was that properly postponing spring irrigation time could promote wheat roots to tie down and improve their absorption and utilization of deep soil water. Optimizing population indexes and canopy structure of winter wheat can promote kernel development [11], thereby increasing yield and water use efficiency. However, if spring irrigation time is too late, the decline rate of tillering will be significantly reduced, and the number of panicles will be reduced, thereby reducing winter wheat yield and leading to the decrease of IWUE [57]. In addition, compared with conventional spring irrigation with second water, wheat yield will be reduced, but IWUE can be increased [58,59]. Wang et al. [60] found that spring irrigation with one water significantly increased IWUE by 72.65~75.71% compared with spring irrigation with second water, which was also confirmed in our previous report [12,61]. In contrast, this study can not only stabilize wheat yield but also improve irrigation water utilization efficiency by adjusting irrigation time under the condition of spring irrigation with one water, which is a water-saving irrigation model suitable for promotion in this region.

## 5. Conclusions

Delayed irrigation for 5 to 10 days at the jointing stage could increase the nitrogen concentration, nitrogen accumulation, silicon concentration, and silicon distribution in the whole plant or different organs, optimize the distribution ratio of nitrogen and silicon in different organs, and realize the improvement of GY. Compared with AJ15 and AJ20, the accumulation of N in the vegetative organs of AJ5 increased by 3.05–23.11% in the flowering stage, the transfer volume of N in the post-flowering vegetative organs increased by 14.09–40.10%, and the N distribution rate in the ear of the mature stage increased by 1.71–6.40%. In the past two years, the GY reached the maximum in AJ10 and AJ5, respectively, and increased by 1.37–6.57% and 22.12–27.21% compared with CK, AJ15, and AJ20, respectively. Therefore, under the condition of spring irrigation, it is suggested that the irrigation time should be 5–10 days after the jointing stage, and the topdressing time should be synchronized with irrigation.

**Author Contributions:** Conceptualization, methodology and project administration, H.D. and C.Z.; data curation, formal analysis, investigation and writing—original draft, H.Z. and J.S.; funding acquisition, H.D., C.Z. and J.Z.; supervision, Y.L. and C.C.; Validation, Y.L. and Y.G.; Writing—review and editing, Y.L., C.C., Y.G. and J.Z. All authors have read and agreed to the published version of the manuscript.

**Funding:** This research was funded by Dang Hongkai, project leader of the National Key Research and Development Program (2023YFD2301500), Zhang Junpeng, project leader of the Shandong Key Research and Development Program (2023TZXD011), and Dang Hongkai, project leader of the Hebei Key Research and Development Program (21327003D).

**Data Availability Statement:** Data will be made available on request.

**Conflicts of Interest:** The authors declare no conflicts of interest.

## References


1. Li, X.; Wang, S.W.; Chen, Y.F.; Zhang, D.W.; Yang, S.S.; Wang, J.W.; Zhang, J.H.; Bai, Y.; Zhang, S. Improved simulation of winter wheat yield in North China Plain by using PRYM-Wheat integrated dry matter distribution coefficient. *J. Integr. Agric.* **2024**, *23*, 1381–1392. [CrossRef]
2. Zeng, R.Y.; Yao, F.M.; Zhang, S.; Yang, S.S.; Bai, Y.; Zhang, J.H.; Wang, J.W.; Wang, X. Assessing the effects of precipitation and irrigation on winter wheat yield and water productivity in North China Plain. *Agric. Water Manag.* **2021**, *256*, 107063. [CrossRef]
3. Cao, G.L.; Han, D.M.; Song, X.F. Evaluating actual evapotranspiration and impacts of groundwater storage change in the North China Plain. *Hydrol. Process.* **2014**, *28*, 1797–1808. [CrossRef]
4. Zhou, L.L.; Liao, S.H.; Wang, Z.M.; Wang, P.; Zhang, Y.H.; Yan, H.J.; Gao, Z.; Shen, S.; Liang, X.G.; Wang, J.H.; et al. A simulation of winter wheat crop responses to irrigation management using CERES-Wheat model in the North China Plain. *J. Integr. Agric.* **2017**, *17*, 1181–1193. [CrossRef]
5. Fang, Q.X.; Ma, L.; Green, T.R.; Yu, Q.; Wang, T.D.; Ahuja, L.R. Water resources and water use efficiency in the North China Plain: Current status and agronomic management options. *Agric. Water Manag.* **2010**, *97*, 1102–1116. [CrossRef]
6. Lv, L.H.; Yao, Y.R.; Zhang, L.H.; Dong, Z.Q.; Jia, X.L.; Liang, S.B.; Ji, J.J. Winter wheat grain yield and its components in the North China Plain: Irrigation management, cultivation, and climate. *Chil. J. Agric. Res.* **2013**, *73*, 233–242. [CrossRef]
7. Wu, X.F.; Qi, Y.Q.; Shen, Y.J.; Yang, W.; Zhang, Y.C.; Kondoh, A. Change of winter wheat planting area and its impacts on groundwater depletion in the North China Plain. *J. Geogr. Sci.* **2019**, *29*, 891–908. [CrossRef]
8. Zhao, D.D.; Shen, J.Y.; Lang, K.; Liu, Q.R.; Li, Q.Q. Effects of irrigation and wide-precision planting on water use, radiation interception, and grain yield of winter wheat in the North China Plain. *Agric. Water Manag.* **2013**, *118*, 87–92.
9. Yang, Y.H.; Ding, J.L.; Zhang, Y.H.; Wu, J.C.; Zhang, J.M.; Pan, X.Y.; Gao, C.M.; Wang, Y.; He, F. Effects of tillage and mulching measures on soil moisture and temperature, photosynthetic characteristics and yield of winter wheat. *Agric. Water Manag.* **2018**, *201*, 299–308. [CrossRef]
10. Meng, Z.J.; Jia, D.L.; Liu, A.N.; Pang, H.B.; Wang, H.Z.; Chen, J.P. Effect of regulated deficit irrigation on physiological mechanism and water use efficiency of winter wheat. *Trans. CSAE* **2003**, *19*, 66–69.
11. Fan, Y.L.; Liu, J.M.; Zhao, J.T.; Ma, Y.Z.; Li, Q.Q. Effects of delayed irrigation during the jointing stage on the photosynthetic characteristics and yield of winter wheat under different planting patterns. *Agric. Water Manag.* **2019**, *221*, 371–376. [CrossRef]
12. Li, X.S.; Dang, H.K.; Song, N.; Shen, X.J.; Ma, F.J.; Sun, J.S. Effects of irrigation once at different time on water consumption and dry matter production of winter wheat in spring. *Trans. Agric. Mach.* **2019**, *50*, 280–289.
13. Gayatri; Venkatesh, K.; Sinha, S.K.; Roy, P.; Mandal, P.K. Molecular Characterization of GS2 and Fd-GOGAT Homeologues and Their Biased Response to Nitrogen Stress in Bread Wheat (*Triticum aestivum* L.). *J. Plant Growth Regul.* **2021**, *41*, 2555–2569. [CrossRef]
14. Ma, Q.; Wang, M.Y.; Zheng, G.L.; Yao, Y.; Tao, R.R.; Zhu, M.; Ding, J.F.; Li, C.Y.; Guo, W.S.; Zhu, X.K. Field Crops Research. Twice-split application of controlled-release nitrogen fertilizer met the nitrogen demand of winter wheat. *Field Crop. Res.* **2021**, *267*, 108163. [CrossRef]
15. Ferreira, L.A.R.; Silva, S.R.; Lollato, R.P.; Ferreira, E.B.; Kolln, O.T. Wheat nitrogen utilization efficiency and yield as affected by nitrogen management and environmental conditions. *Emir. J. Food Agric.* **2021**, *33*, 944–957. [CrossRef]

16. Leghari, S.J.; Wahocho, N.A.; Laghari, G.M.; Laghari, A.H.; Bhabhan, G.M.; Talpur, K.H.; Bhutto, T.A.; Wahocho, S.A.; Lashari, A.A. Role of Nitrogen for Plant Growth and Development: A review. *Adv. Environ. Biol.* **2016**, *10*, 209–218.
17. Yang, J.C.; Zhang, J.H.; Wang, Z.Q.; Zhu, Q.S.; Liu, L.J. Water Deficit–Induced Senescence and Its Relationship to the Remobilization of Pre-Stored Carbon in Wheat during Grain Filling. *Agron. J.* **2001**, *93*, 196–206. [CrossRef]
18. Yan, S.C.; Wu, Y.; Fan, J.L.; Zhang, F.C.; Guo, J.J.; Zheng, J.; Wu, L.F. Quantifying grain yield, protein, nutrient uptake and utilization of winter wheat under various drip fertigation regimes. *Agric. Water Manag.* **2022**, *261*, 107380. [CrossRef]
19. Yu, Y.; Qu, Z.Y.; Liu, Y.J. Experimental Research on the Influence of Limited Irrigation on Wheat Growth under the Condition of Intercropping Plantation. *Trans. China Rural Water Hydropower* **2010**, *3*, 11–13.
20. Rodrigues, F.A.; Resende, R.S.; Dallagnol, L.J.; Datnoff, L.E. Silicon Potentiates Host Defense Mechanisms Against Infection by Plant Pathogens. In *Silicon and Plant Diseases*; Springer International Publishing: New York, NY, USA, 2015; pp. 109–138.
21. Epstein, E. Silicon. *Annu. Rev. Plant Physiol. Plant Mol. Biol.* **1999**, *50*, 641–664. [CrossRef]
22. Moldes, C.A.; Filho, O.F.d.L.; Merini, L.J.; Tsai, S.M.; Plantarum, J.M.C.J.A.P. Occurrence of powdery mildew disease in wheat fertilized with increasing silicon doses: A chemometric analysis of antioxidant response. *Acta Physiol. Plant.* **2016**, *38*, 206. [CrossRef]
23. Ahmed, M.; Kamran, A.; Asif, M.; Qadeer, U.; Ahmed, Z.I.; Goyal, A. Silicon priming: A potential source to impart abiotic stress tolerance in wheat: A review. *Aust. J. Crop Sci.* **2013**, *7*, 484–491.
24. Hamayun, M.; Sohn, E.Y.; Khan, S.A.; Shinwari, Z.K.; Lee, I.J. Silicon alleviates the adverse effects of salinity and drought stress on growth and endogenous plant growth hormones of soybean (*Glycine max* L.). *Pak. J. Bot.* **2010**, *42*, 1713–1722.
25. Nadeem, M.; Ul Haq, M.A.; Saqib, M.; Maqsood, M.; Iftikhar, I.; Ali, T.; Awais, M.; Ulah, R.; He, Z.L. Nutrients, Osmotic and Oxidative Stress Management in Bread Wheat (*Triticum aestivum* L.) by Exogenously Applied Silicon Fertilization Under Water Deficit Natural Saline Conditions. *Silicon* **2022**, *14*, 11869–11880. [CrossRef]
26. Mecfel, J.; Hinke, S.; Goedel, W.A.; Marx, G.; Fehlhaber, R.; Bäucker, E.; Wienhaus, O. Effect of silicon fertilizers on silicon accumulation in wheat. *J. Plant Nutr. Soil Sci.* **2007**, *170*, 769–772. [CrossRef]
27. Khan, A.L. Silicon: A valuable soil element for improving plant growth and $CO_2$ sequestration. *J. Adv. Res.* **2024**. [CrossRef] [PubMed]
28. Salem, E.M.M.; Kenawey, M.K.M.; Saudy, H.S.; Mubarak, M. Influence of Silicon Forms on Nutrients Accumulation and Grain Yield of Wheat Under Water Deficit Conditions. *Gesunde Pflanz.* **2022**, *74*, 539–548. [CrossRef]
29. Liu, Z.L. Effects of Irrigation Once in Spring on Different Dates on Growth and Water Use Efficiency for Winter Wheat. Master's Thesis, Shandong Agricultural University, Tai'an, China, 2021.
30. Xu, J.K.; Shi, Y.; Yu, Z.W.; Zhao, J.Y. Irrigation methods affect wheat flag leaf senescence and chlorophyll fluorescence in the North China Plain. *Int. J. Plant Prod.* **2017**, *11*, 361–377. [CrossRef]
31. Bao, S.D. *Soil and Agricultural Chemistry Analysis*; China Agriculture Press: Beijing, China, 2000.
32. Hogendorp, B.K.; Cloyd, R.A.; Swiader, J.M. Determination of Silicon Concentration in Some Horticultural Plants. *HortScience* **2012**, *47*, 1593–1595. [CrossRef]
33. Yang, Y.J. A method for the determination of silicon in rice plants by alkali-dissolution digestion with nickel crucible. *Biotechnology* **1995**, *3*, 42–44. [CrossRef]
34. Li, X.J. A Study on Physiological Basis and Optimized Agronomic Measures for Water-Saving and High-Yield of Spring Wheat in Hetao Irrigation District. Master's Thesis, Inner Mongolia Agricultural University, Hohhot, China, 2014.
35. Li, H. Effect of Rice Straw Returning on Si and Zn Absorption and Utilization of Soil and Crop in Rice-Wheat Rotation System. Master's Thesis, Southwest University, Chongqing, China, 2017.
36. Liang, X.Q. Effects of Plant-Row Configuration on Growth and Plant Nutrients of Winter Wheat in South Xinjiang. Master's Thesis, Tarim University, Alaer, China, 2022.
37. He, R. Effects of Straw and Its Ash Returning on Soil Nutrient, Crop Nutrient Uptake and Crop Yield. Master's Thesis, Sichuan Agricultural University, Chengdu, China, 2020.
38. He, Z.R. Effects of Nitrogen and Phosphorus Reduction on Plant Physiology, Soil Biological Characteristics and Yield of Strong Gluten Wheat Varieties. Master's Thesis, Henan Agricultural University, Zhengzhou, China, 2024.
39. Li, R.Q.; Li, Y.Q.; He, J.X.; Li, G.C.; Hao, X.C.; Wang, F. Effect of Nitrogen Application Rate on Nitrogen Utilization and Grain Yield of Winter Wheat. *J. Triticeae Crops* **2011**, *42*, 1713–1722.
40. Dang, H.K.; Li, R.Q.; Sun, Y.H.; Zhang, X.W.; Li, Y.M. Absorption, accumulation and distribution of potassium in super highly-yielding winter wheat. *Sci. Direct* **2013**, *19*, 1037–1047.
41. Li, W.; Zhou, X.N.; Yu, K.; Zhang, Z.; Liu, Y.; Hu, N.Y.; Liu, Y.; Yao, C.S.; Yang, X.G.; Wang, Z.M.; et al. Spectroscopic Estimation of N Concentration in Wheat Organs for Assessing N Remobilization Under Different Irrigation Regimes. *Front. Plant Sci.* **2021**, *12*, 657578. [CrossRef] [PubMed]
42. Moradi, L.; Siosemardeh, A.; Sohrabi, Y.; Bahramnejad, B.; Hosseinpanahi, F. Effects of supplemental irrigation on the accumulation, partitioning, and remobilization of nitrogen, yield and water use efficiency of wheat cultivars. *J. Plant Nutr.* **2022**, *45*, 2758–2773. [CrossRef]
43. Liu, E.K.; Mei, X.R.; Gong, D.Z.; Yan, C.R.; Zhuang, Y. Effects of drought on N absorption and utilization in winter wheat at different developmental stages. *Chin. J. Plant Ecol.* **2010**, *34*, 555–562.

44. Li, L. Physiological and Biochemical Mechanisms of Wheat Response to Water and Nitrogen Stress and Elevated $CO_2$ Concentration. Ph.D. Thesis, Chinese Academy of Agricultural Sciences, Beijing, China, 2021.
45. Sun, W.H.; Hamani, A.K.M.; Si, Z.Y.; Abubakar, S.A.; Liang, Y.P.; Liu, K.; Gao, Y. Effects of Timing in Irrigation and Fertilization on Soil $NO_3^-$-N Distribution, Grain Yield and Water–Nitrogen Use Efficiency of Drip-Fertigated Winter Wheat in the North China Plain. *Water* **2022**, *14*, 1780. [CrossRef]
46. Hou, R.; Xu, X.; Ouyang, Z. Effect of experimental warming on nitrogen uptake by winter wheat under conventional tillage versus no-till systems. *Soil Tillage Res.* **2018**, *180*, 116–125. [CrossRef]
47. Gao, Z.Z.; Wang, Y.X.; Feng, Y.P.; Laraib, I.; Wang, C.; Zhao, J.C.; Chu, Q.Q. Optimized irrigation management mitigates reactive nitrogen loss and shapes nitrogen fate in wheat production. *Field Crops Res.* **2024**, *308*, 109269. [CrossRef]
48. Lu, J.S.; Hu, T.T.; Li, Y.; Cui, X.L.; Cheng, M.H.; Yan, S.C.; Xiang, Y.Z.; Fan, J.L.; Zhang, F.C.; Xiong, Y.C. Effects of different water management strategies on critical nitrogen concentration dilution curves, nitrogen accumulation, and grain yield in winter wheat. *Agric. Commun.* **2024**, *2*, 2949–7981. [CrossRef]
49. Nie, W.T. Effects of Nitrogen Application on Yield and Quality of Different Wheat Varieties. Master's Thesis, Northwest Agriculture and Forestry Technology University, Shanxi, China, 2018.
50. Chen, X.; Tong, Y.A.; Kang, H.H.; Yu, J.B.; Wang, Z.H.; Yang, J.F. Effect of postponing N application on the yield, apparent N recovery and N absorption of winter wheat. *Plant Nutr. Fertil. Sci.* **2008**, *14*, 450–455.
51. Ahmad, M.; El-Saeid, M.H.; Akram, M.A.; Ahmad, H.R.; Haroon, H.; Hussain, A. Silicon fertilization—A tool to boost up drought tolerance in wheat (*Triticum aestivum* L.) crop for better yield. *J. Plant Nutr.* **2016**, *39*, 1283–1291. [CrossRef]
52. Zhan, H. Morphological Characteristics of Phytolith and Effects of Exogenous Silicon on Cold Tolerance of Bamboo Seedlings in 10 Clumped Bamboo Species. Ph.D. Thesis, Southwest Forestry University, Kunming, China, 2017.
53. Silva, O.N.; Lobato, A.K.S.; Ávila, F.W.; Costa, R.L.C.; Oliveira Neto, C.F.; Santos Filho, B.G.; Martins Filho, A.M.; Lemos, R.; Pinho, J.M.; Medeiros, M.B.C.L.; et al. Silicon induced increase in chlorophyll is modulated by the leaf water potential in two water-deficient tomato cultivars. *Plant Soil Environ.* **2012**, *58*, 481–486. [CrossRef]
54. Radkowski, A.; Radkowska, I. Effects of silicate fertilizer on seed yield in timothy-grass (*Phleum pratense* L.). *Ecol. Chem. Eng. S-Chem. I Inz. Ekol. S* **2018**, *25*, 169–180. [CrossRef]
55. Liu, B.H.; Wang, B.Y.; Chen, Z.Y.; Zhang, W.Y. Effects of irrigation time on yield and growth characteristics of winter wheat in spring. *Trans. J. Irrig. Drain.* **2021**, *40*, 122–128.
56. Liu, R.; Zhou, W.W.; Jin, S.A.; Shang, M.X.; Wang, M.; Li, J.L. Effects of soil moisture supplemental irrigation on farmland water and nitrogen and wheat yield under a new irrigation model. *Trans. J. Irrig. Drain.* **2023**, *42*, 39–46.
57. Xu, X.X.; Zhang, Y.H.; Li, J.P.; Zhang, M.; Zhou, S.L.; Wang, Z.M. Optimizing single irrigation scheme to improve water use efficiency by manipulating winter wheat sink-source relationships in Northern China Plain. *PLoS ONE* **2018**, *13*, e0193895. [CrossRef]
58. Feng, Y.P.; Yang, M.; Shang, M.F.; Jia, H.; Chu, Q.Q.; Chen, F. Improving the annual yield of a wheat and maize system through irrigation at the maize sowing stage. *Irrig. Drain.* **2018**, *67*, 755–761. [CrossRef]
59. Wang, Y.Q.; Zhang, Y.H.; Zhang, R.; Li, J.P.; Zhang, M.; Zhou, S.L.; Wang, Z.M. Reduced irrigation increases the water use efficiency and productivity of winter wheat-summer maize rotation on the north China plain. *Sci. Total Environ.* **2017**, *618*, 112–120. [CrossRef]
60. Wang, D.; Yu, Z.W. Effects of irrigation amount and irrigation period on water consumption characteristics and yield of wheat. *Trans. J. Appl. Ecol.* **2008**, *9*, 1965–1970.
61. Liu, Z.L.; Li, X.S.; Cao, C.Y.; Zheng, C.L.; Ma, J.Y.; Li, K.J.; Yang, H.Y.; Li, S.N.; Dang, H.K. Effects of time of spring one irrigation on the grain filling characteristics and water use efficiency of winter wheat. *Chin. J. Eco-Agric.* **2021**, *29*, 1296–1304.

*plants*

MDPI

*Article*

# Dry Matter Accumulation, Water Productivity and Quality of Potato in Response to Regulated Deficit Irrigation in a Desert Oasis Region

**Hengjia Zhang [1,*,†], Xietian Chen [1,2,†], Daoxin Xue [2], Wanheng Zhang [2], Fuqiang Li [2], Anguo Teng [3], Changlong Zhang [3], Lian Lei [3] and Yuchun Ba [3]**

1 College of Agronomy and Agricultural Engineering, Liaocheng University, Liaocheng 252059, China; gsaucxt@163.com
2 College of Water Conservancy and Hydropower Engineering, Gansu Agricultural University, Lanzhou 730070, China; tl2024202405@126.com (D.X.); 15214097901@163.com (W.Z.); lifq@gsau.edu.cn (F.L.)
3 Yimin Irrigation Experimental Station, Hongshui River Management Office, Minle, Zhangye 734500, China; tenganguo@126.com (A.T.); zcl13830689195@163.com (C.Z.); cxt19952024@126.com (L.L.); bayuchun@163.com (Y.B.)
* Correspondence: zhanghengjia@lcu.edu.cn
† These authors contributed equally to this work.

**Abstract:** As one of the most important food crops, the potato is widely planted in the oasis agricultural region of Northwest China. To ascertain the impact of regulated deficit irrigation (RDI) on various facets including dry matter accumulation, tuber yield, quality and water use efficiency (WUE) of potato plants, a two-growth season field experiment under mulched drip irrigation was conducted in the desert oasis region of Northwest China. Water deficits, applied at the seedling, tuber formation, tuber expansion and starch accumulation stages, encompassed two distinctive levels: mild (55–65% of field capacity, $F_C$) and moderate (45–55% $F_C$) deficit, with full irrigation (65–75% $F_C$) throughout the growing season as the control (CK). The results showed that water deficit significantly reduced ($p < 0.05$) above-ground dry matter, water consumption and tuber yield compared to CK, and the reduction increased with the increasing water deficit. A mild water deficit at the tuber formation stage, without significantly reducing ($p > 0.05$) yield, could significantly increase WUE and irrigation water use efficiency (IWUE), with two-year average increases of 25.55% and 32.33%, respectively, compared to CK. Water deficit at the tuber formation stage increased starch content, whereas water deficit at tuber expansion stage significantly reduced starch, protein and reducing sugar content. Additionally, a comprehensive evaluation showed that a mild water deficit at the tuber formation stage is the optimal RDI strategy for potato production, providing a good balance between yield, quality and WUE. The results of this study can provide theoretical support for efficient and sustainable potato production in the desert oasis regions of Northwest China.

**Keywords:** water deficit; sub-membrane drip irrigation; water consumption; water use efficiency; tuber quality; comprehensive evaluation

## 1. Introduction

Agriculture is the most basic sector of material production and the foundation of the national economy, and water resources are essential to ensure agricultural production. In China, more than 60% of water consumption is used for agriculture, of which about 90% is used for irrigation [1–3]. In recent years, with the increase in population and water demand as well as climate change, water scarcity has become a major factor restricting agricultural production, seriously threatening the sustainable development of China's agriculture [4–6]. Especially in the arid desert regions of Northwest China, irrigation is the main way to provide water consumption for crop growth [7]. However, the increasing expansion of irrigated cropland and high water-consuming crop species has further aggravated the conflicts over agricultural water use in this region [8]. Therefore, the development of efficient water-saving irrigation technologies to improve water productivity is an inevitable choice for ensuring sustainable agricultural production in the arid regions.

As one of the world's most important food crops, potato (*Solanum tuberosum* L.) plays an irreplaceable role in ensuring food security [9]. China is the world's largest producer of potato, with the world's largest area under cultivation and total production [10]. The Hexi Oasis region of Northwest China has a cold climate and abundant light, which is suitable for potato growth, and is one of the main growing areas for potato [11]. Previous studies have shown that the potato plant is more sensitive to soil moisture than other food crops due to its shallow root system and smaller root mass [6,12]. Typically, the water requirement of potato plants ranges from 400 to 800 mm throughout the growing season [13,14]. Rainfall is scarce ($\leq$300 mm) and inter-annual variability is great in the Hexi Oasis region of Northwest China, so that in almost all years the rainfall is insufficient to meet the normal growth requirements of potato plants. Therefore, irrigation becomes a necessary measure to ensure the safe production of potatoes [15]. However, the vast majority of potato growing areas in the region still use flood and furrow irrigation to deliver water to the crop. Such crude irrigation practices result in significant water wastage and scarcity, which can easily lead to reduced yields and crop failure. Improving water productivity by optimizing the irrigation mode is therefore essential for efficient potato production in arid oasis region of Northwest China.

Among several common irrigation strategies, regulated deficit irrigation (RDI) is considered as a promising water-saving technology [16–18]. This technique is based on the principle of applying a certain degree of water stress during crop growth, thereby regulating the nutritional and reproductive growth of the crop, and ultimately achieving the objectives of increasing water productivity and improving crop fruit quality without significantly affecting yield [19,20]. As there are differences in crop sensitivity to soil moisture at different growth stages, a clear understanding of crop response to irrigation periods and levels is crucial for successful implementation of RDI [21]. Currently, many studies have applied RDI to potato cultivation in arid areas. However, the results of these studies are not consistent due to the differences in climatic conditions and soil environments in the experimental areas. Several studies have demonstrated that deficit irrigation significantly reduces potato yield despite increasing water use efficiency [22–24]. However, some studies have indicated that an appropriate deficit irrigation can significantly increase water use efficiency and improve the tuber nutritional quality while maintaining potato yield [25,26]. Jensen et al. [27] showed that 30% was the limit of water saving for RDI compared to full irrigation during the whole growth phase of potato. The optimal RDI strategy to maintain potato yield was 70% of full irrigation at the end of tuber formation and 50% of full irrigation in the first two weeks at the end of the growing season. Wagg et al. [12] revealed that water deficit at any growth stage in potato negatively affected biomass, yield and tuber quality. Kifle et al. [28] indicated that potato tuber yield significantly decreased with increasing water deficit, but applying 75% of the crop water requirement during the entire growing season showed better water use efficiency. In summary, the effect of RDI on potato growth and yield formation varies considerably in different research settings. This means that RDI strategies in one region are unlikely to be transferable to other regions, and that efficient and appropriate RDI strategies need to be identified through local field trials over many years [29].

Several previous studies have reported the response mechanism of potato production to RDI; however, it has not been possible to identify the optimal RDI strategy for potato plants in the desert oasis region of Northwest China based on the current scarce research, especially under the condition of drip irrigation under a membrane. We hypothesized that excessive and insufficient irrigation is detrimental to efficient potato production in arid regions, whereas an appropriate water deficit at a certain stage of potato plant growth not only maintains yield but also improves water use efficiency (WUE) and tuber quality.

Therefore, this study quantified the response of potato plant dry matter accumulation, water productivity and tuber quality to regulated deficit irrigation based on a field experiment with drip irrigation under mulch. Specifically, the precise objectives were to determine the following: (1) the characteristics of above-ground dry matter accumulation under different irrigation levels using logistic regression equations; (2) the effects of water deficit on potato water consumption (ET), yield, tuber quality and WUE at different growth stages and (3) the optimal RDI strategy based on entropy weight and technique for order preference by similarity to an ideal solution (TOPSIS) methods. The results of this study may provide useful information for researchers and farmers to implement water-saving irrigation in potato cultivation in the desert oasis region of Northwest China.

## 2. Results

### 2.1. Above-Ground Dry Matter Accumulation Characteristics

#### 2.1.1. Above-Ground Dry Matter at Different Stages

As shown in Figure 1, above-ground dry matter accumulation of potato plants under different water deficit treatments showed an increasing trend during the growing season. At the seedling stage, above-ground dry matter of the potato plant was generally low, ranging from 0.5 to 1.2 t ha$^{-1}$. Above-ground dry matter in S-D1 (mild water deficit at the seedling stage) and S-D2 (moderate water deficit at the seedling stage) was significantly reduced ($p < 0.05$) by an average of 26.52% and 28.51%, respectively, compared to CK (full irrigation), while the rest of the water deficit treatments were not significantly different ($p > 0.05$) from CK. At the tuber formation stage, above-ground dry matter was significantly decreased by an average of 18.14% and 22.46% in F-D1 (mild water deficit at the tuber formation stage) and F-D2 (moderate water deficit at the tuber formation stage), respectively, compared to CK. At tuber expansion stage, E-D1 (mild water deficit at the tuber expansion stage) and E-D2 (moderate water deficit at the tuber expansion stage) showed a significant reduction in above-ground dry matter, averaging 19.54% and 28.77%, respectively, compared to CK. Meanwhile, above-ground dry matter in S-D1, S-D2, F-D1, and F-D2 treatments remained significantly lower than that in CK, suggesting that the effect of water deficit at the seedling and tuber formation stages on above-ground dry matter accumulation would persist even if full irrigation was resumed at later stages. At the starch accumulation stage, the above-ground dry matter was significantly lower than CK in all water deficit treatments, with decreases ranging from 4.57 to 26.74% in 2016 and 6.66 to 27.83% in 2018. In conclusion, water deficit at any stage significantly reduced above-ground dry matter accumulation in potato plants, especially at moderate water deficits.

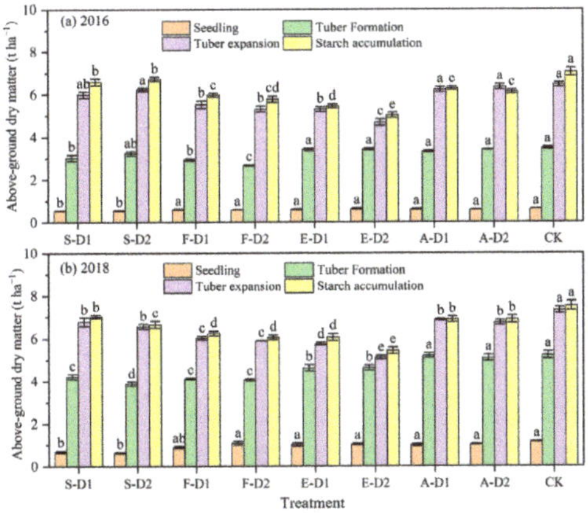


**Figure 1.** Above-ground dry matter accumulation characteristics of potato plants at different stages under each water deficit treatment in 2016 (**a**) and 2018 (**b**). Different lowercase letters indicate significant differences between treatments at the $p < 0.05$ level. The bars represent the standard deviation.

#### 2.1.2. Above-Ground Dry Matter Accumulation Rate and Its Characteristic Parameters

The dynamic changes in above-ground dry matter accumulation in potato plants under different deficit irrigation treatments were fitted using a logistic model with days after planting as the independent variable (Table 1). It can be seen that the determination coefficients ($R^2$) of the logistic model equations under all treatments were all greater than 0.99, indicating that the logistic model has a high simulation accuracy. In 2016 and 2018, the time at which the maximum rate of above-ground dry matter accumulation in potato plants occurred varied from 52 to 79 d after planting under the different water deficit treatments, the duration of rapid growth ranged from 32 to 43 d, the maximum accumulation rate ranged from 0.083 to 0.165 t ha$^{-1}$ d$^{-1}$, and the average accumulation rate ranged from 0.036 to 0.055 t ha$^{-1}$ d$^{-1}$, indicating that water deficit at different growth periods had a significant effect on the accumulation rate of above-ground dry matter. Specifically, the start time of the rapid growth period of above-ground dry matter accumulation in potato plants was earliest in the E-D2 group in both years, 11.6 d and 4.5 d earlier than CK, respectively. A-D2 (moderate water deficit at the starch accumulation stage) had the highest maximum accumulation rate, which increased by 10.34% (2016) and 28.91% (2018), respectively, compared to CK. The mean rates of potato dry matter accumulation under water deficit treatments were all lower than CK, with the greatest decrease seen in E-D2, which was 28.00% (2016) and 27.27% (2018), respectively. Furthermore, the mean accumulation rate of the moderate deficit treatments was lower than that of the mild deficit treatments over the same period, indicating that the potato plant above-ground biomass decreased more significantly with the increasing water deficit.

**Table 1.** Parameters and eigenvalues of the logistic model for above-ground dry matter accumulation in potato plants under different water deficit treatments.

| Year | Treatment | Regression Equation | $R^2$ | $T_1$ | $T_2$ | $T_{max}$ | $\Delta T$ | $V_{max}$ | $V_{mean}$ |
|---|---|---|---|---|---|---|---|---|---|
| 2016 | S-D1 | $y = 6.57/(1 + 202.270e^{-0.067x})$ | 0.997 | 59.59 | 98.90 | 79.25 | 39.31 | 0.110 | 0.047 |
| | S-D2 | $y = 6.72/(1 + 211.388e^{-0.068x})$ | 0.998 | 59.36 | 98.10 | 78.73 | 38.73 | 0.114 | 0.048 |
| | F-D1 | $y = 5.96/(1 + 168.382e^{-0.066x})$ | 0.997 | 57.72 | 97.62 | 77.67 | 39.91 | 0.098 | 0.042 |
| | F-D2 | $y = 5.77/(1 + 195.330e^{-0.067x})$ | 0.996 | 59.07 | 98.38 | 78.73 | 39.31 | 0.097 | 0.041 |
| | E-D1 | $y = 5.45/(1 + 145.365e^{-0.071x})$ | 0.999 | 51.58 | 88.68 | 70.13 | 37.10 | 0.097 | 0.039 |
| | E-D2 | $y = 5.01/(1 + 78.941e^{-0.066x})$ | 0.999 | 46.24 | 86.15 | 66.19 | 39.91 | 0.083 | 0.036 |
| | A-D1 | $y = 6.27/(1 + 470.230e^{-0.081x})$ | 0.995 | 59.71 | 92.22 | 75.97 | 32.52 | 0.127 | 0.044 |
| | A-D2 | $y = 6.33/(1 + 428.301e^{-0.081x})$ | 0.993 | 58.55 | 91.07 | 74.81 | 32.52 | 0.128 | 0.045 |
| | CK | $y = 7.04/(1 + 169.513e^{-0.066x})$ | 0.999 | 57.82 | 97.73 | 77.77 | 39.91 | 0.116 | 0.050 |
| 2018 | S-D1 | $y = 7.01/(1 + 113.236e^{-0.070x})$ | 0.999 | 48.75 | 86.38 | 67.56 | 37.63 | 0.123 | 0.052 |
| | S-D2 | $y = 6.64/(1 + 139.822e^{-0.073x})$ | 0.999 | 49.64 | 85.72 | 67.68 | 36.08 | 0.121 | 0.049 |
| | F-D1 | $y = 6.24/(1 + 62.181e^{-0.066x})$ | 0.998 | 42.62 | 82.53 | 62.58 | 39.91 | 0.103 | 0.046 |
| | F-D2 | $y = 6.05/(1 + 43.133e^{-0.062x})$ | 0.999 | 39.47 | 81.96 | 60.71 | 42.48 | 0.094 | 0.044 |
| | E-D1 | $y = 6.06/(1 + 63.009e^{-0.072x})$ | 0.998 | 39.25 | 75.84 | 57.55 | 36.58 | 0.109 | 0.045 |
| | E-D2 | $y = 5.42/(1 + 82.067e^{-0.084x})$ | 0.997 | 36.79 | 68.15 | 52.47 | 31.36 | 0.114 | 0.040 |
| | A-D1 | $y = 6.98/(1 + 88.792e^{-0.076x})$ | 0.999 | 41.70 | 76.36 | 59.03 | 34.66 | 0.133 | 0.051 |
| | A-D2 | $y = 6.88/(1 + 77.639e^{-0.074x})$ | 0.999 | 41.02 | 76.61 | 58.81 | 35.59 | 0.165 | 0.051 |
| | CK | $y = 7.51/(1 + 62.0229e^{-0.068x})$ | 0.999 | 41.33 | 80.07 | 60.70 | 38.73 | 0.128 | 0.055 |

$R^2$, determination coefficient; $T_1$, the start time of the rapid growth period, d; $T_2$, the termination time of the rapid growth period, d; $T_{max}$, the time corresponding to the maximum accumulation rate, d; $\Delta T$, duration of the rapid growth period, d; $V_{max}$, maximum accumulation rate, t ha$^{-1}$ d$^{-1}$; $V_{mean}$, mean accumulation rate, t ha$^{-1}$ d$^{-1}$.

### 2.2. Water Consumption

The ET characteristics of potato plants at different growth stages under each water deficit treatments are shown in Figure 2a,b. The ET of potato plants during the whole reproductive period showed a unimodal trend, and the peaks of ET in the two years occurred at the tuber formation and tuber expansion stages, respectively. In general, the total ET of potato plants during the whole reproductive period was 424.50–610.62 mm and 411.08–567.01 mm for all treatments in 2016 and 2018, respectively (Figure 2c,d). Compared to full irrigation, the ET of the deficit treatments was significantly reduced by 11.62–30.48% and 7.36–25.53% in the two years, respectively.

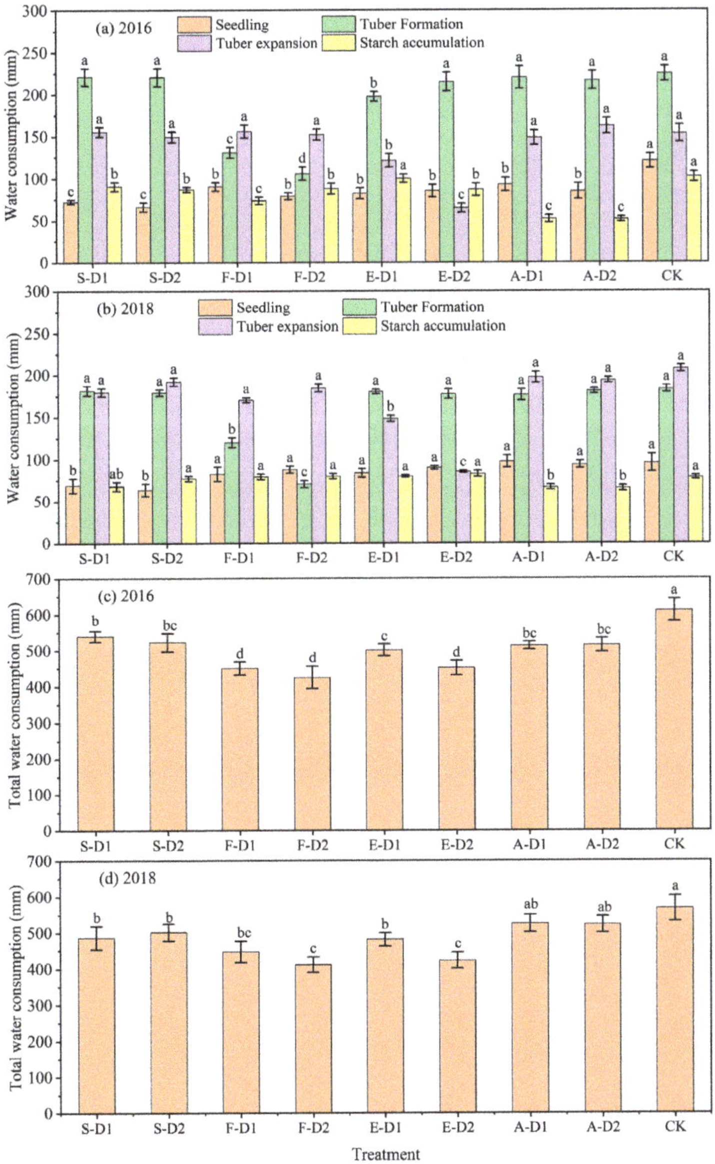


**Figure 2.** Water consumption characteristics of potato plants at different growth stages under each water deficit treatment in 2016 (**a**,**c**) and 2018 (**b**,**d**). Different lowercase letters indicate significant differences between treatments at the $p < 0.05$ level. The bars represent the standard deviation.

Specifically, the ET at the seedling stage was 69.62–120.26 mm (2016) and 63.66–97.16 mm (2018) for all treatments, with S-D1 and S-D2 having significantly lower ($p < 0.05$) ET than CK and other deficit treatments (Figure 2a,b). The ET at the tuber formation stage was significantly increased compared to the seedling stage, accounting for 24.90–47.61% and

17.14–41.92% of the total ET in 2016 and 2018, respectively. The ET in F-D1 and F-D2 was significantly lower than that of CK at the tuber formation stage, with a significant reduction of 38.09 and 57.17% on average in both years, respectively, while no significant difference ($p > 0.05$) was observed between the ET of the other water deficit treatments and that of CK. At the tuber expansion stage, the ET in all treatments was 64.64–161.84 mm and 84.94–207.17 mm in 2016 and 2018, respectively, accounting for approximately 14.33–35.83% and 20.09–44.88% of the total ET. The ET in E-D1 and E-D2 was significantly lower compared to other deficit treatments, with an average reduction of 24.74% and 58.38%, respectively, compared to CK. During the starch accumulation stage, the ET was significantly reduced in all treatments, accounting for 9.99–20.70% (2016) and 12.53–19.39% (2018) of the total ET. A-D1 (mild water deficit at starch accumulation stage) and A-D2 had significantly lower ET than the other treatments at the starch accumulation stage, with an average reduction of 31.77% and 32.76% compared to CK in both years, respectively. These results indicated that the degree of water deficit had a significant effect on ET at all growth stages. At the same stage, ET increased with increasing water deficit. Additionally, due to the large proportion of ET in the tuber formation and tuber expansion stages, deficit irrigation during these two periods can significantly reduce the total ET of the whole growth period.

### 2.3. Tuber Yield, Dry Matter Content and Harvest Index

Table 2 shows the effect of each water deficit treatment on potato tuber yield indexes. It can be seen that potato tuber yield varied greatly among different deficit treatments. CK had the highest yield of 36,037.23 kg ha$^{-1}$ and 35,317.71 kg ha$^{-1}$ in 2016 and 2018, respectively. Water deficit treatments at the seedling stage (S-D1 and S-D2) led to a reduction in tuber yield, but the difference was not significant ($p > 0.05$) compared to CK. Water deficit at the tuber formation stage (F-D1 and F-D2) significantly reduced ($p < 0.05$) potato yield compared to CK, especially by 17.58% for F-D2 treatment in 2016. The most significant effect on yield was observed for water deficit at the tuber expansion stage, which decreased by an average of 22.03% (E-D1) and 31.62% (E-D2) in 2016 and 2018, respectively, compared to CK. Potato yield was significantly lower under water deficit at the starch accumulation stage (A-D1 and A-D2) compared to CK in 2016, but the decrease was not significant in 2018. Furthermore, dry matter content of potato tubers under water deficit treatments was 20.86–22.64% (2016) and 22.25–25.84% (2018). In 2016, the dry matter content of F-D1, F-D2 and A-D1 was significantly increased by 6.59%, 4.28% and 6.40% compared with CK, respectively, and the dry matter content of other water deficit treatments was at the same level as CK. In 2018, the dry matter content of A-D2 was significantly reduced by 13.19% compared to CK, and the remaining water deficit treatments were not significantly different from CK.

As shown in Table 2, different stages and degrees of water deficit had significant effects on total potato plant biomass. Total dry biomass was highest in CK in both 2016 and 2018. Total dry biomass in S-D1 and SD2 did not show a significant difference ($p > 0.05$) compared to CK in 2016 but showed a significant decrease ($p < 0.05$) of 6.90% and 8.26% in 2018, respectively. Water deficit during the tuber formation and tuber expansion stages resulted in a significant decrease in total dry biomass, with the highest decrease in E-D2, which was significantly reduced by 39.21% (2016) and 24.11% (2018) compared to CK. In 2018, total dry biomass for water deficit during the starch accumulation stage was not significantly different from CK, while it was significantly reduced by 13.26% (A-D1) and 21.30% (A-D2) in 2016. In addition, the HI of potato plants showed significant differences between different water deficit treatments. Specifically, among all water deficit treatments, F-D1 treatment had the highest HI in both 2016 and 2018, which was significantly increased by 8.16% and 5.68% compared to CK, respectively. The difference between the HI of S-D1 and S-D2 and CK was not significant, whereas the HI of E-D1 and E-D2 was significantly reduced compared to CK, with an average decrease of 7.91% and 8.34% in both years, respectively. Water deficit during the starch accumulation stage (A-D1 and A-D2) caused a decrease in HI in both years, especially in A-D2, which was significantly reduced by 7.45% and 3.35% in 2016 and 2018 compared to CK, respectively.

**Table 2.** Effect of water deficit on potato yield, tuber dry matter content, total dry biomass and harvest index. Different lowercase letters indicate significant differences between treatments at the $p < 0.05$ level.

| Year | Treatment | Fresh Tuber Yield (kg ha$^{-1}$) | Tuber Dry Matter Content (%) | Total Dry Biomass (kg ha$^{-1}$) | Harvest Index |
|---|---|---|---|---|---|
| 2016 | S-D1 | 35,271.74 ab | 21.37 b | 13,312.03 a | 0.566 b |
| | S-D2 | 35,834.08 a | 21.15 b | 13,293.98 a | 0.570 b |
| | F-D1 | 34,276.71 b | 22.64 a | 12,726.34 b | 0.610 a |
| | F-D2 | 29,701.00 c | 22.15 ab | 11,357.03 c | 0.579 b |
| | E-D1 | 25,839.44 d | 20.80 b | 10,430.18 d | 0.515 c |
| | E-D2 | 20,066.11 e | 21.02 b | 8255.62 e | 0.511 c |
| | A-D1 | 28,784.62 c | 22.60 a | 11,780.24 c | 0.552 b |
| | A-D2 | 25,975.15 d | 21.47 b | 10,688.34 d | 0.522 c |
| | CK | 36,037.23 a | 21.24 b | 13,581.47 a | 0.564 b |
| 2018 | S-D1 | 31,913.54 ab | 25.84 a | 12,257.50 b | 0.673 bc |
| | S-D2 | 32,508.33 ab | 25.67 a | 12,083.72 b | 0.691 b |
| | F-D1 | 34,852.08 a | 24.66 ab | 11,841.06 b | 0.726 a |
| | F-D2 | 32,879.17 ab | 23.98 ab | 11,733.67 bc | 0.672 bc |
| | E-D1 | 29,747.92 b | 24.01 ab | 11,197.44 c | 0.638 d |
| | E-D2 | 28,633.33 b | 22.25 b | 9996.05 d | 0.637 d |
| | A-D1 | 33,619.79 a | 25.19 a | 13,145.88 a | 0.644 d |
| | A-D2 | 34,050.00 a | 24.89 ab | 12,755.40 a | 0.664 c |
| | CK | 35,317.71 a | 25.63 a | 13,171.44 a | 0.687 b |

### *2.4. Water Use Efficiency (WUE) and Irrigation Water Use Efficiency (IWUE)*

Figure 3 shows the effect of different water deficits on the WUE and IWUE of potato plants. In 2016, water deficit at both the seedling and tuber formation stages significantly increased ($p < 0.05$) WUE and IWUE. Among them, WUE and IWUE were the highest in F-D1 and F-D2, with a significant increase of 29.04% and 18.59% in WUE and 55.30% and 49.75% in IWUE, respectively, compared to CK. Water deficit at the tuber expansion and starch accumulation stages did not significantly increase ($p > 0.05$) WUE and IWUE, and may even lead to a decrease, such as for the WUE and IWUE in F-D2 which were decreased by 24.60% and 10.89% compared to CK, respectively. In 2018, WUE in F-D1 and F-D2 was significantly higher than CK with 22.07% and 25.03%, respectively, while the WUE in the other water deficit treatments was at the same level as CK. Additionally, the IWUE in all moderate deficit treatments was significantly higher than CK in 2018, with F-D2 having the highest IWUE with a significant increase of 23.83% over CK. Conversely, E-D1 had the lowest IWUE, which was significantly lower than CK by 14.22%.

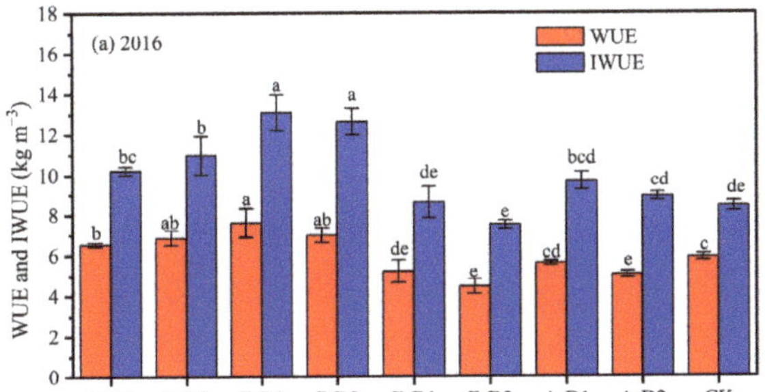


**Figure 3.** *Cont.*

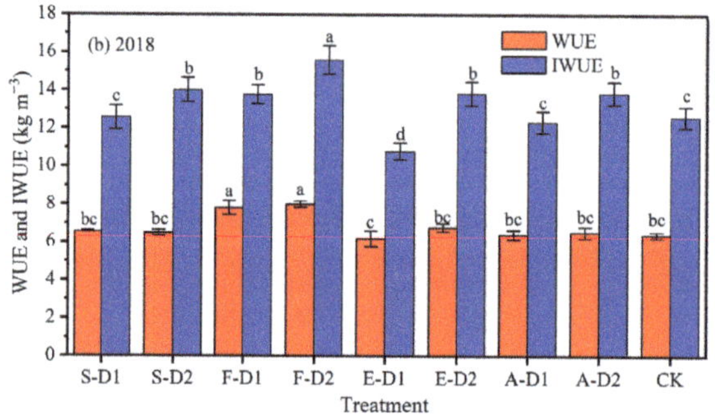

**Figure 3.** Effect of different water deficits on potato plant WUE and IWUE in 2016 (**a**) and 2018 (**b**). Different lowercase letters indicate significant differences between treatments at the $p < 0.05$ level. The bars represent the standard deviation.

*2.5. Tuber Quality*

The effect of each water deficit treatment on potato starch content (SC), protein content (PC) and reducing sugar content (RSC) is shown in Figure 4. In 2016, SC for water deficit treatments at the tuber formation stage (F-D1 and F-D2) was significantly higher ($p < 0.05$) than CK and other treatments, while SC for water deficit treatments at the seedling (S-D1 and S-D2) and tuber expansion (E-D1 and E-D2) stages was significantly lower than CK, with decreases ranging from 17.63% to 27.63%. In 2018, there was no significant difference ($p > 0.05$) in SC between all water deficit treatments and CK. In 2016, water deficit at the tuber expansion stage resulted in a significant average reduction in PC and RSC of 14.13% and 25.64%, respectively, compared to CK, while the remaining water deficit treatments did not differ significantly from CK. In 2018, PC in S-D1, F-D1 and A-D1 was not significantly different from CK, while PC in the remaining treatments was significantly lower than CK by 12.95–28.50%. Furthermore, A-D1 exhibited the highest RSC, which was significantly higher by 6.25% compared to CK in 2018. RSC of S-D1, E-D1 and E-D2 was significantly lower compared to CK, with decreases of 31.25%, 34.38% and 46.88%, respectively.

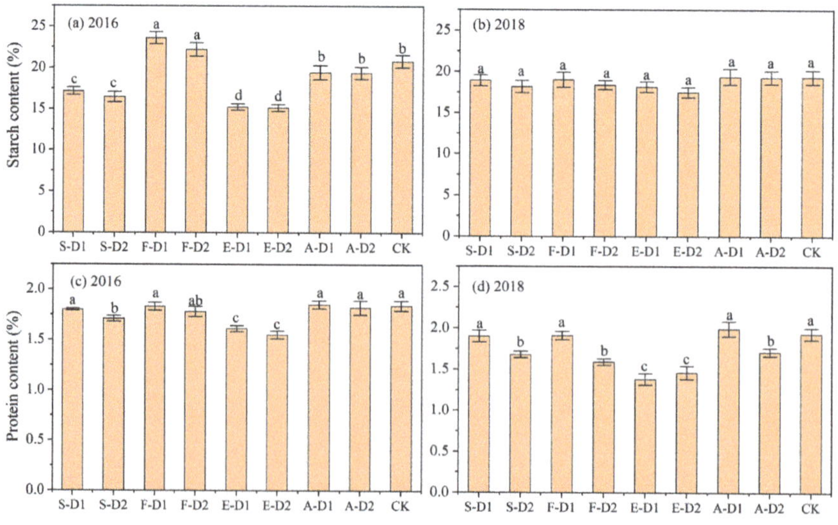

**Figure 4.** *Cont.*

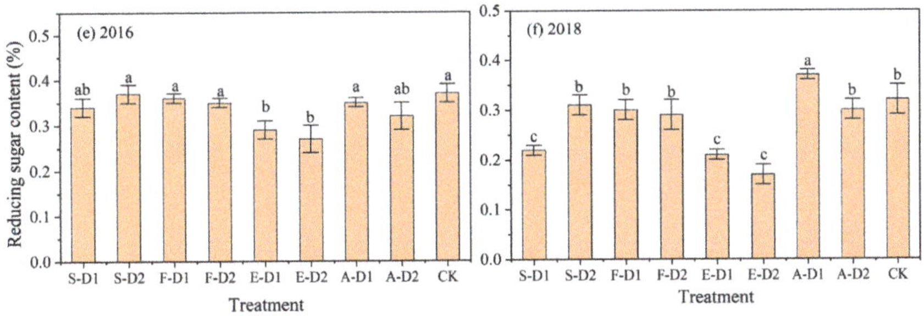


**Figure 4.** Effect of different water deficits on starch (**a**,**b**), protein (**c**,**d**) and reducing sugar (**e**,**f**) content of potato tubers in 2016 and 2018. Different lowercase letters indicate significant differences between treatments at the $p < 0.05$ level. The bars represent the standard deviation.

### *2.6. Optimization of Potato Water Deficit Strategy Based on Entropy Weight and TOPSIS*

Using the nine water deficit treatments undertaken in this experiment as feasible scenarios, eight indicators—yield (X1), HI (X2), WUE (X3), IWUE (X4), above-ground dry matter accumulation (X5), SC (X6), PC (X7) and RSC (X8)—were selected for the comprehensive evaluation of the multi-objectives. Firstly, the original evaluation matrix was constructed using the participating indicators and normalized. Subsequently, the weights of each of the indicators were obtained utilizing the entropy weighting method (the weights of X1–X8 were, in order, 0.129, 0.120, 0.125, 0.122, 0.129, 0.117, 0.129 and 0.13 in 2016 and 0.127, 0.115, 0.111, 0.129, 0.130, 0.128, 0.128 and 0.133 in 2018). Finally, the comprehensive score for each treatment was calculated using TOPSIS as shown in Table 3. In both 2016 and 2018, the F-D1 treatment had the highest comprehensive scores of 0.897 and 0.738, respectively. Consequently, F-D1 treatment emerges as the optimal deficit irrigation strategy that can balance multiple objectives.

**Table 3.** Comprehensive score and ranking of different water deficit treatments based on entropy weight and TOPSIS.

| Treatment | 2016 | | | | 2018 | | | |
|---|---|---|---|---|---|---|---|---|
| | $D^+$ | $D^-$ | $C_i$ | Ranking | $D^+$ | $D^-$ | $C_i$ | Ranking |
| S-D1 | 0.021 | 0.051 | 0.711 | 4 | 0.025 | 0.020 | 0.446 | 7 |
| S-D2 | 0.019 | 0.055 | 0.749 | 3 | 0.015 | 0.027 | 0.641 | 6 |
| F-D1 | 0.007 | 0.060 | 0.897 | 1 | 0.011 | 0.031 | 0.738 | 1 |
| F-D2 | 0.014 | 0.054 | 0.794 | 2 | 0.015 | 0.029 | 0.654 | 4 |
| E-D1 | 0.037 | 0.039 | 0.514 | 8 | 0.035 | 0.008 | 0.184 | 9 |
| E-D2 | 0.048 | 0.030 | 0.382 | 9 | 0.037 | 0.010 | 0.222 | 8 |
| A-D1 | 0.025 | 0.047 | 0.654 | 6 | 0.015 | 0.034 | 0.695 | 3 |
| A-D2 | 0.032 | 0.041 | 0.564 | 7 | 0.015 | 0.027 | 0.644 | 5 |
| CK | 0.023 | 0.056 | 0.710 | 5 | 0.014 | 0.033 | 0.699 | 2 |

$D^+/D^-$, Euclidean distance between the evaluation object and the positive/negative ideal value. $C_i$, the relative closeness of each evaluation object.

## 3. Discussion

### *3.1. Effects of Water Deficit on Potato Plant Above-Ground Dry Matter, Yield and Water Use Efficiency*

In arid farming areas, water is the most critical constraint to efficient crop production. Achieving higher crop yields, economic efficiency and water productivity have become the main objectives of modern agriculture [30]. Numerous studies have shown that drip irrigation techniques, especially drip irrigation under mulch, can achieve higher potato tuber yields and crop water use efficiency [31,32]. In the present study, it was found that with plastic film-covered drip irrigation, water deficit treatments caused a reduction in above-ground dry matter and its accumulation rate and tuber yield in potato plants compared to fully irrigated plants, and the reduction increased with worsening water deficit. Consistent results were also obtained from many previous studies, which revealed that water deficit had a negative effect on above-ground biomass and tuber fresh and dry yields [24,28,33,34]. The reduction in potato yield due to water deficit can be mainly attributed to its inhibition of above-ground nutrient growth. When a potato plant experiences water stress, its leaf chloroplasts are damaged and net photosynthetic rate decreases, and photosynthesis provides the basic source of material for tuber yield formation. Consequently, the decline in above-ground biomass in turn limits potato production [12,35]. Moreover, the degree of water deficit is an important factor affecting tuber yield. Martínez-Romero et al. [33] observed that a 20% reduction in water input from the normal water supply resulted in a 5% reduction in tuber yield. Greenwood et al. [36] revealed that when the water deficit exceeded 35% of the maximum allowable value, the growth rate of the potato was significantly reduced, ultimately leading to lower yields.

Improving WUE in potato plants is key to ensuring high yields, reducing production costs and minimizing negative environmental impacts [23,37]. Previous studies have shown that although water deficit leads to lower potato yields, it increases WUE due to reduced water inputs [24]. In this study, water deficit at the seedling and tuber formation stages significantly increased ($p < 0.05$) WUE and IWUE without significantly reducing yield. However, water deficit at the tuber expansion stage significantly reduced yield, resulting in a significant decrease in WUE and IWUE. This is in agreement with the findings of Li et al. [26] who found that potato plants have a high water requirement during the tuber expansion stage and water deficit at this stage had a significant effect on tuber yield, causing a significant decrease in WUE and IWUE. Furthermore, a previous study recommended that in order to improve water productivity in arid areas, water deficits should be applied during the nutritive growth and maturity stages of the potato plant and that water deficits should be avoided during the tuber expansion stage [33].

### 3.2. Effect of Water Deficit on Tuber Quality

With the continuous development of the potato processing industry, the quality of potato tubers has been allocated increasingly more attention. The main component in the dry matter of potato tubers is starch, with additional small amounts of sugar and protein [24]. Previous studies have shown that water supply has a significant effect on potato tuber quality [38]. In this study, water deficit at the tuber formation stage did not reduce potato tuber SC, and even mild water deficit at this stage can significantly increase ($p < 0.05$) SC. This result is similar to that of Carli et al. [39], who showed that potato tuber SC increased with decreasing irrigation volume. Additionally, the present study identified that water deficit at the seedling and tuber expansion stages led to a decrease in tuber SC, and the decrease with water deficit at the tuber expansion stage was particularly significant, whereas water deficit at the starch accumulation stage had no significant effect on potato tuber quality. Consistent results were obtained by Xue et al. [40] who indicated that reduced water supply at the seedling and tuber expansion stages had a significant negative effect on tuber SC.

The present study found a gradual decrease in tuber PC with increasing water deficit, with the greatest decrease seen with a moderate water deficit at the tuber expansion stage. Similarly, previous studies have found that a water deficit generally causes a decrease in plant N concentration and the decrease increases with decreasing irrigation levels [41]. Furthermore, sugar content is one of the important indicators of potato tuber quality, which is generally regulated by variety, growing environment and water management [42]. In the present study, it was found that water deficit at the tuber expansion stage resulted in a significant decrease ($p < 0.05$) in RSC, whereas the effect of water deficit at other stages was not significant ($p > 0.05$). Similar results were reported by Wegener et al. [43] who found that the total sugar, glucose and fructose contents of potato tubers under water stress were lower than those of fully irrigated tubers. Contrarily, Eldredge et al. [44] showed that water deficit increased sugar content in tubers. This divergence from the present study's results may be attributed to the inconsistent response of different genotypes to the same treatments, that is, the response of potato tubers to water stress is not singular and uniform [45,46].

### 3.3. Comprehensive Evaluation Based on Entropy Weight and TOPSIS

With a single analysis it is difficult to identify a water management strategy that simultaneously balances yield, quality and resource use efficiency. Therefore, in the current study, a comprehensive multi-objective evaluation was performed using TOPSIS with different water deficit treatments as evaluation objects. In the process of determining the optimal water supply strategy by using TOPSIS, it is extremely critical to determine the weights of the participating indicators. In several previous studies, the weights of the evaluation indicators were set equal, which may affect the accuracy of the evaluation results [47,48]. Therefore, in order to make the evaluation results more reasonable and reliable, the entropy weight method is used in this study to objectively assign weights to the evaluation indicators. This would go some way towards removing the bias of subjective factors from the evaluation results. The results of the combined evaluation utilizing the entropy weight method and TOPSIS showed that the comprehensive score of F-D1 was optimal in both years, indicating that a mild water deficit at the tuber formation stage could better achieve the goals of stable yield, quality improvement and higher water productivity. Additionally, the comprehensive scores of E-D1 and E-D2 were in the bottom two in both years, indicating that water deficit at the tuber expansion stage would lead to a decrease in tuber yield, quality and WUE, thus suggesting that water deficit at this stage should be avoided.

## 4. Materials and Methods

### 4.1. Experimental Site Description

A field experiment, spanning two growing seasons (2016 and 2018), was executed at the Yimin Irrigation Experimental Station (100°47′ E, 38°35′ N, 1977 m a.s.l.) in Minle County, Zhangye City, Northwest China (Figure 5). The region has a cool climate and sufficient light, creating favorable conditions for photosynthesis and organic matter accumulation in crops. The multi-year average total precipitation in this site is less than 200 mm, which is far from meeting the crop's water requirements, thus irrigation is a necessary measure to ensure normal crop production. In addition, the average annual temperature at the test site was about 6.0 °C, the frost-free period was in the range of 109–174 d and the number of sunshine hours was about 3000 h per year. The soil texture was loam, with a field capacity ($F_C$) of 24.0% in the tillage layer (0–60 cm), an average soil bulk density of 1.48 t m$^{-3}$, a pH of 8.2 and a water table greater than 20 m. The content of ammonium nitrogen, available phosphorus and available potassium in the 0–20 cm soil layer was 3.56 mg kg$^{-1}$, 17.12 mg kg$^{-1}$ and 81.41 mg kg$^{-1}$, respectively.

Rainfall, evaporation and temperature during the two potato growing seasons are shown in Figure 6. The total rainfall during the potato growing season was 182.7 mm and 197.4 mm in 2016 and 2018, respectively. The mean daily evaporation was 6.64 mm d$^{-1}$ in 2016 and 6.42 mm d$^{-1}$ in 2018. The mean air temperature in 2016 was 14.02 °C, with a mean maximum of 20.23 °C and a mean minimum of 8.77 °C. The mean air temperature in 2018 was 16.32 °C, with a mean maximum temperature of 23.58 °C and a mean minimum temperature of 9.55 °C.

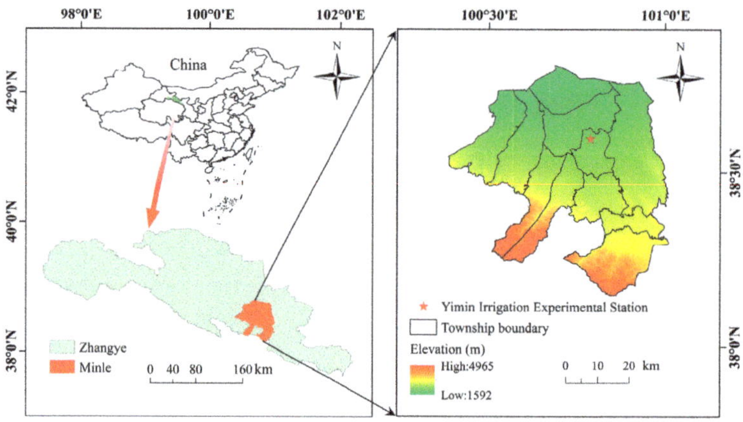


**Figure 5.** Location of the experiment station.

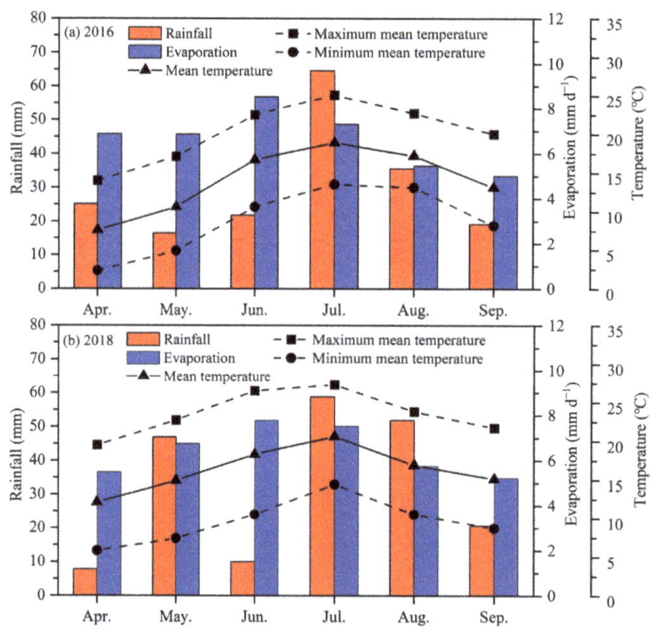


**Figure 6.** Monthly rainfall, mean evaporation and mean temperature during the potato growing season at the experimental site in 2016 (**a**) and 2018 (**b**).

## 4.2. Experimental Design and Field Management

The potato cultivar "Qingshu168" was selected for this experiment, which has a long fertility cycle, strong adaptability, disease resistance, storage resistance and is mainly used for making starch and food processing. In 2016 and 2018, potatoes lasted 172 d and 170 d from planting (8 April 2016 and 11 April 2018) to harvesting (27 September 2016 and 28 September 2018), respectively. In this experiment, the planting mode was a combination of ridging and covering with white plastic film, in which the film width was 120 cm, the thickness was 0.01 mm, the width of the ridges was 80 cm, the height of the ridges was 20 cm, and the width of the furrow between the two ridges was 40 cm (Figure 7). Two rows of potatoes were planted on each ridge, with a spacing of 40 cm × 20 cm (plants × rows). The irrigation method used was drip irrigation with drip tape placed under the film and in the middle of the two rows of potatoes, with drippers spaced 30 cm apart and a nominal flow rate of 2.0 L $h^{-1}$. To ensure an accurate water supply, a ball valve and water meter were installed at the water inlet of each plot. In addition, the same amount of diamine phosphate (90 kg $ha^{-1}$), urea (120 kg $ha^{-1}$) and potassium fertilizer (120 kg $ha^{-1}$) were evenly spread in each plot at planting.

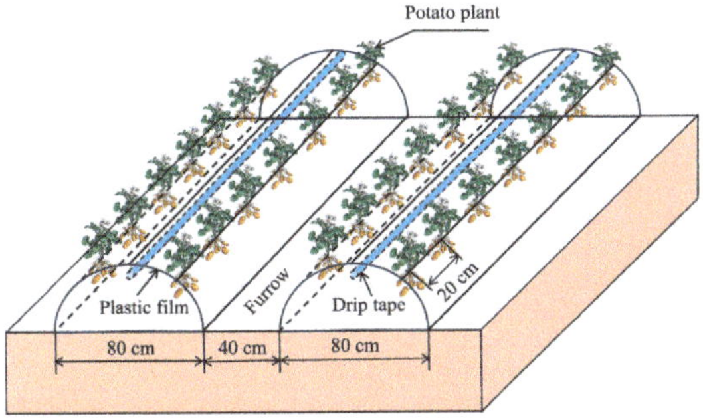


**Figure 7.** Schematic layout of drip irrigation under film for potato plants.

The entire growth period of the potato plant, dictated by its growth characteristics, was segmented into four distinct stages: seedling (S), tuber formation (F), tuber expansion (E) and starch accumulation (A). Three irrigation levels were defined in this experiment, specifically fully irrigation (65–75% $F_C$), mild deficit (55–65% Fc, D1) and moderate deficit (45–55% $F_C$, D2). Water deficit was applied at each growth stage, for a total of eight water deficit treatments, and full irrigation was applied during the whole growth stage as a control treatment (CK), as shown in Table 4 for the specific scheme. The experiment was performed in a single factor randomized complete block design with three replicates per treatment for a total of 27 plots. Each plot was 24 $m^2$ (2.4 m × 10 m) in size, with a 1.0 m wide isolation zone between neighboring plots. For each irrigation event, the amount of water applied is based on the soil moisture content at the depth of the planned wetted layer (40 cm at seedling stage and 60 cm at other stages). Specifically, the moisture content of the soil in the planned wetted layer is measured using the dry method every approximately 7 days, and if the observed value is close to or below the designed lower limit, irrigation is immediately applied to increase the soil moisture content to the designed upper limit.

**Table 4.** Potato deficit irrigation design scheme.

| Treatment | Relative Soil Water Content (% of the Field Capacity) | | | |
|---|---|---|---|---|
| | **Seedling** | **Tuber Formation** | **Tuber Expansion** | **Starch Accumulation** |
| S-D1 | 55–65% | 65–75% | 65–75% | 65–75% |
| S-D2 | 45–55% | 65–75% | 65–75% | 65–75% |
| F-D1 | 65–75% | 55–65% | 65–75% | 65–75% |
| F-D2 | 65–75% | 45–55% | 65–75% | 65–75% |
| E-D1 | 65–75% | 65–75% | 55–65% | 65–75% |
| E-D2 | 65–75% | 65–75% | 45–55% | 65–75% |
| A-D1 | 65–75% | 65–75% | 65–75% | 55–65% |
| A-D2 | 65–75% | 65–75% | 65–75% | 45–55% |
| CK | 65–75% | 65–75% | 65–75% | 65–75% |

S-D1, mild water deficit at seedling stage; S-D2, moderate water deficit at seedling stage; F-D1, mild water deficit at tuber formation stage; F-D2, moderate water deficit at tuber formation stage; E-D1, mild water deficit at tuber expansion stage; E-D2, moderate water deficit at tuber expansion stage; A-D1, mild water deficit at starch accumulation stage; A-D2, moderate water deficit at starch accumulation stage; CK, full irrigation throughout the growing season.

### *4.3. Measurements, Calculations and Methodologies*

#### 4.3.1. Dry Matter Accumulation

Three potato plants were randomly sampled from each plot at the end of each growth stage. After cleaning the attached soil, the above-ground part (stems and leaves) and the underground part (roots and tubers) were placed in an oven at 105 °C for 30 min, followed by drying each organ to a constant weight at 75 °C. Dry matter accumulation per hectare was the average dry weight of individual plants multiplied by the planting density and the emergence rate.

In order to explore the accumulation characteristics of potato dry matter under different water deficits, the logistic model was used to nonlinearly fit the changes in above-ground dry matter during the growing period. The equation of the logistic model is as follows [49]:

$$Y = \frac{K_m}{1 + ae^{-bt}} \tag{1}$$

where $Y$ is the observed value of potato plant above-ground biomass at a given stage (t ha$^{-1}$); $K_m$ is the theoretical maximum value of potato plant above-ground biomass (t ha$^{-1}$); $t$ is the number of days after planting (d) and $a$ and $b$ are the specific fitting parameters of the equation.

The maximum accumulation rate of potato plant dry matter ($V_{max}$) (t ha$^{-1}$ d$^{-1}$) and its corresponding time ($T_{max}$) can be obtained by taking the first order derivation of the logistic equation and setting it to 0. Then, by taking the derivative of the second order and setting it to 0, the start time ($T_1$) and the end time ($T_2$) of the rapid growth period can be determined. The detailed equations are as follows:

$$V_{max} = \frac{K_m \times b}{4} \tag{2}$$

$$T_{max} = \frac{ln\ a}{b} \tag{3}$$

$$T_1 = \frac{1}{b} \times ln\left(\frac{a}{2 + \sqrt{3}}\right) \tag{4}$$

$$T_2 = \frac{1}{b} \times ln\left(\frac{a}{2 - \sqrt{3}}\right) \tag{5}$$

#### 4.3.2. Tuber Yield

At harvest, fresh potatoes from an area of 2.4 m × 3 m were selected from each plot and weighed. The tuber yield was then calculated for each plot and per hectare. Harvest index (HI) was calculated by dividing the dry tuber mass by the sum of the dry tuber mass and the dry above-ground mass [50].

#### 4.3.3. Tuber Quality

After ripening, five potato tubers from each plot were sampled for quality determination. Three replicates were made for each sample. The starch content was determined using iodine colorimetry, the reducing sugar content was determined using 3,5-dinitrosalicylic acid colorimetry, and the protein content was determined using the Coomassie Brilliant Blue G-250 method [51].

#### 4.3.4. Water Consumption

The ET of potato plants at each growth stage is determined by the following water balance equation [52]:

$$ET = P + I + K - C + \Delta W \tag{6}$$

$$\Delta W = 10 \times \sum_{i=1}^{n} \gamma_i H_i (W_{i1} - W_{i2}) \tag{7}$$

where $P$ is the effective rainfall (mm); if the individual rainfall is less than 5 mm, $P$ is ignored. $I$ is the irrigation amount (mm). $K$ and $C$ are groundwater recharge and deep seepage, respectively. Since drip irrigation was used for water supply in the test, and the groundwater depth exceeded 30 m, seepage loss and groundwater recharge could not be generated, so the values of $K$ and $C$ were 0. $\Delta W$ indicates the change in soil water storage over a period of time. $n$ is the number of soil layers. In this study, the drying method was used to measure the soil water content at the depth of 0–100 cm at the interval of 20 cm soil layers, so the value of $n$ was 5. $\gamma_i$ is the soil bulk density (g cm$^{-3}$). $H_i$ is the thickness of the layer (20 cm). $W_{i1}$ and $W_{i2}$ are the soil water content (mass water content) at the beginning and end of a certain stage, respectively.

#### 4.3.5. Water Use Efficiency and Irrigation Water Use Efficiency

*WUE* and *IWUE* are calculated using the following formula [47]:

$$WUE = \frac{Y}{ET} \tag{8}$$

$$IWUE = \frac{Y}{I} \tag{9}$$

where $Y$ is the potato tuber yield (kg ha$^{-1}$); $ET$ is crop water consumption (mm) and $I$ is the total irrigation amount during the potato growing season (mm).

#### 4.3.6. Multi-Objective Optimization Based on Entropy Weight Method and TOPSIS

Entropy weighting is an objective weighting method, which determines the weight of each index according to the information provided by the observed value of each index. TOPSIS is a commonly used multi-objective comprehensive optimization method, and its decision-making basis is to compare the distance (i.e., the closeness degree) between the alternative scheme and the optimal target and the worst target. In this study, a comprehensive evaluation system was established by considering potato plant growth, yield, quality and water use efficiency indicators to obtain a multi-objective optimal deficit irrigation scheme. The specific steps are as follows [53]:

(1) Construct the original matrix.

$$\boldsymbol{R} = \left(r_{ij}\right)_{m\times n} = \begin{bmatrix} r_{11} & r_{12} & \cdots & r_{1n} \\ r_{21} & r_{22} & \cdots & r_{2n} \\ \vdots & \vdots & \ddots & \vdots \\ r_{m1} & r_{m2} & \cdots & r_{mn} \end{bmatrix} \tag{10}$$

where $r_{ij}$ denotes the observed value of the evaluation indicator, and $m$ and $n$ are the number of treatments and indicators, respectively;

(2) Determination of weights using entropy weighting.

The dimensionless transformation of the original matrix is carried out by using the extreme value normalization method, and the matrix $\boldsymbol{A}$ is obtained. Subsequently, the entropy weight $W_j$ of each index is determined.

$$A_{ij} = \frac{r_{ij} - min\,(r_i)}{max\,(r_i) - min\,(r_i)} \tag{11}$$

$$P_{ij} = \frac{A_{ij}}{\sum_{i=1}^{m} A_{ij}} \tag{12}$$

$$E_j = -Ln(m)^{-1} \sum_{i=1}^{m} P_{ij} Ln(P_{ij}) \tag{13}$$

$$W_j = \frac{1 - E_j}{\sum_{j=1}^{n} E_j} \tag{14}$$

where $P_{ij}$ is the characteristic proportion of the $i$th evaluation object on the $j$th index; $E_j$ is the entropy of the $j$th index;

(3) Multi-objective optimization using TOPSIS.

The original matrix $\boldsymbol{R}$ was normalized to obtain matrix $\boldsymbol{B} = [b_{ij}]_{\mathrm{m \times n}}$, and then the weighted matrix $\boldsymbol{Z}$ was obtained by multiplying the entropy weight.

$$b_{ij} = \frac{r_{ij}}{\sqrt{\sum_{i=1}^{m} r_{ij}^2}} \tag{15}$$

$$Z_{ij} = W_j \times b_{ij} \tag{16}$$

Calculation of the Euclidean distance $D^+$ and $D^-$ and the relative proximity ($C_i$) between each evaluation object and the ideal solution is then performed.

$$D_i^+ = \sqrt{\sum_{j=1}^{n} (z_j^+ - z_{ij})^2} \tag{17}$$

$$D_i^- = \sqrt{\sum_{j=1}^{n} (z_{ij} - z_j^-)^2} \tag{18}$$

$$C_i = \frac{D_i^-}{D_i^+ + D_i^-} \tag{19}$$

where $z_j^+$ is defined as a positive ideal solution, $z^+ = [max(z_1), max(z_2), \cdots, max(z_n)]$, $z_j^-$ is defined as a negative ideal solution and $z^- = [min(z_1), min(z_2), \cdots, min(z_n)]$. $C_i$ is the decision basis for ranking the schemes. A value of $C_i$ closer to 1 indicates a better comprehensive evaluation result.

### *4.4. Statistical Analysis*

Microsoft Excel 2010 was utilized for data processing and computation. One-way analysis of variance (ANOVA) was performed using SPSS 23.0 software (SPSS Inc., Chicago, IL, USA). Mean differences were examined via the Duncan's multiple range tests, adopting a significance level of $p < 0.05$. Origin 2021 software (OriginLab Inc., Northampton, MA, USA) was used to generate the figures.

## 5. Conclusions

Water deficit decreased above-ground biomass and its accumulation rate, tuber yield and water consumption in potato plants, and the decrease was more significant with increasing water deficit. Among all deficit treatments, water deficit at the tuber expansion stage had the greatest negative effect on potato growth, resulting in the most significant reductions ($p < 0.05$) in yield and harvest index. Water deficit at the seedling and tuber formation stages can significantly increase water use efficiency and irrigation water use efficiency while slightly reducing tuber yield. Moreover, water deficit at the tuber formation stage was beneficial for increasing starch content but had no significant effect ($p > 0.05$) on protein and reducing sugar content. On the contrary, water deficit at the tuber expansion stage significantly reduced the content of starch, protein and reducing sugar in the potato. The evaluation results using entropy weighting and TOPSIS methods showed that a mild water deficit at the tuber formation stage ranked first in the overall score, which effectively balanced multiple objectives such as yield, quality and water use efficiency. Therefore, mild water deficit at the tuber formation stage can be recommended as the optimal RDI strategy for potato cultivation in the Hexi desert oasis of northwest China. Nonetheless, the optimal RDI strategy identified in this study may not be applicable to other arid agricultural regions due to differences in soil properties and climatic conditions. Consequently, future research should be conducted in multiple regions to elucidate how different climatic variations and soil textures affect the development of optimal RDI strategies.

**Author Contributions:** Conceptualization, H.Z., D.X. and X.C.; methodology, X.C.; software, X.C. and F.L.; validation, X.C. and W.Z.; formal analysis, D.X.; investigation, A.T., C.Z., L.L. and Y.B.; resources, H.Z.; data curation, D.X. and W.Z.; writing—original draft preparation, D.X. and X.C.; writing—review and editing, H.Z., F.L. and X.C.; visualization, X.C.; supervision, H.Z.; project administration, H.Z.; funding acquisition, H.Z. All authors have read and agreed to the published version of the manuscript.

**Funding:** This work was supported by the National Natural Science Foundation of China (No. 52269008, 51669001), the Industrial Support Plan Project of Gansu Provincial Department of Education (No. 2022CYZC-51), the Key Research and Planning Projects of Gansu Province (No. 18YF1NA073), the Scientific Research Foundation for High-level Talented Scholars of Liaocheng Universtiy (No. 318042401) and the Open Project of Liaocheng Universtiy Landscape Architecture Discipline (No. 31946221236).

**Data Availability Statement:** The datasets used and/or analyzed during the current study are available from the corresponding author upon reasonable request and the approval of the data owner.

**Acknowledgments:** We thank everyone who helped during the manuscript writing. We also thank the reviewers for their useful comments and suggestions.

**Conflicts of Interest:** The authors declare no conflicts of interest.

## References

1. Cheng, M.H.; Wang, H.D.; Fan, J.L.; Zhang, S.H.; Wang, Y.L.; Li, Y.P.; Sun, X.; Yang, L.; Zhang, F.C. Water productivity and seed cotton yield in response to deficit irrigation: A global meta-analysis. *Agric. Water Manag.* **2021**, *255*, 12. [CrossRef]
2. Wang, Y.B.; Wu, P.T.; Zhao, X.N.; Li, J.L. Development tendency of agricultural water structure in China. *Chin. J. Eco-Agric.* **2010**, *18*, 399–404. [CrossRef]
3. Yawson, D.O.; Adu, M.O.; Armah, F.A.; Chiroro, C. Virtual water and phosphorus gains through rice imports to Ghana: Implications for food security policy. *Int. J. Agric. Resour. Gov. Ecol.* **2014**, *10*, 374–393. [CrossRef]
4. Chai, Q.; Gan, Y.T.; Zhao, C.; Xu, H.L.; Waskom, R.M.; Niu, Y.N.; Siddique, K.H.M. Regulated deficit irrigation for crop production under drought stress. A review. *Agron. Sustain. Dev.* **2016**, *36*, 21. [CrossRef]
5. Li, H.; Wu, Y.Z.; Huang, X.J.; Sloan, M.; Skitmore, M. Spatial-temporal evolution and classification of marginalization of cultivated land in the process of urbanization. *Habitat Int.* **2017**, *61*, 1–8. [CrossRef]
6. Wang, H.D.; Wu, L.F.; Cheng, M.H.; Fan, J.L.; Zhang, F.C.; Zou, Y.F.; Chau, H.W.; Gao, Z.J.; Wang, X.K. Coupling effects of water and fertilizer on yield, water and fertilizer use efficiency of drip-fertigated cotton in northern Xinjiang, China. *Field Crops Res.* **2018**, *219*, 169–179. [CrossRef]
7. Tang, J.Z.; Wang, J.; Fang, Q.X.; Dayananda, B.; Yu, Q.; Zhao, P.Y.; Yin, H.; Pan, X.B. Identifying agronomic options for better potato production and conserving water resources in the agro-pastoral ecotone in North China. *Agric. For. Meteorol.* **2019**, *272*, 91–101. [CrossRef]
8. Cao, X.C.; Liu, Z.; Wu, M.Y.; Guo, X.P.; Wang, W.G. Temporal-spatial distribution and driving mechanism of arable land water scarcity index in China from water footprint perspective. *Trans. Chin. Soc. Agric. Eng.* **2019**, *35*, 94–100.
9. Tang, J.Z.; Xiao, D.P.; Wang, J.; Wang, R.D.; Bai, H.Z.; Guo, F.H.; Liu, J.F. Optimizing irrigation and nitrogen management for potato production under multi-objective production conditions. *Trans. Chin. Soc. Agric. Eng.* **2021**, *37*, 108–116.

10. Devaux, A.; Goffart, J.P.; Kromann, P.; Andrade-Piedra, J.; Polar, V.; Hareau, G. The Potato of the Future: Opportunities and Challenges in Sustainable Agri-Food Systems. *Potato Res.* **2022**, *65*, 211. [CrossRef]
11. Yang, Y.D.; Duan, D.D.; Ju, Z.H.; Du, Y.T.; Luo, Q.Y.; Zhang, Q. Optimization analysis of China's potato spatial layout at county level based on FAO-GAEZ model. *Agric. For. Meteorol.* **2022**, *41*, 3352–3363. [CrossRef]
12. Wagg, C.; Hann, S.; Kupriyanovich, Y.; Li, S. Timing of short period water stress determines potato plant growth, yield and tuber quality. *Agric. Water Manag.* **2021**, *247*, 10. [CrossRef]
13. Haverkort, A.J.; Van de Waart, M.; Bodlaender, K.B.A. The effect of early drought stress on numbers of tubers and stolons of potato in controlled and field conditions. *Potato Res.* **1990**, *33*, 89–96. [CrossRef]
14. Tang, J.Z.; Wang, J.; Wang, E.L.; Yu, Q.; Yin, H.; He, D.; Pan, X.B. Identifying key meteorological factors to yield variation of potato and the optimal planting date in the agro-pastoral ecotone in North China. *Agric. For. Meteorol.* **2018**, *256*, 283–291. [CrossRef]
15. Wang, Z.Y.; Yu, S.C.; Zhang, H.J.; Lei, L.; Liang, C.; Chen, L.L.; Su, D.D.; Li, X. Deficit mulched drip irrigation improves yield, quality, and water use efficiency of watermelon in a desert oasis region. *Agric. Water Manag.* **2023**, *277*, 16. [CrossRef]
16. Dong, X.C.; Tang, H.X.; Zhang, Q.; Zhang, C.M.; Wang, Z.T. Transcriptomic analyses provide new insights into jujube fruit quality affected by water deficit stress. *Sci. Hortic.* **2022**, *291*, 11. [CrossRef]
17. Du, T.S.; Kang, S.Z.; Zhang, J.H.; Davies, W.J. Deficit irrigation and sustainable water-resource strategies in agriculture for China's food security. *J. Exp. Bot.* **2015**, *66*, 2253–2269. [CrossRef] [PubMed]
18. Dingre, S.K.; Gorantiwar, S.D. Soil moisture based deficit irrigation management for sugarcane (*Saccharum officinarum* L.) in semiarid environment. *Agric. Water Manag.* **2021**, *245*, 14. [CrossRef]
19. Cantore, V.; Lechkar, O.; Karabulut, E.; Sellami, M.H.; Albrizio, R.; Boari, F.; Stellacci, A.M.; Todorovic, M. Combined effect of deficit irrigation and strobilurin application on yield, fruit quality and water use efficiency of "cherry" tomato (*Solanum lycopersicum* L.). *Agric. Water Manag.* **2016**, *167*, 53–61. [CrossRef]
20. Nikanorova, A.D.; Milanova, E.V.; Dronin, N.M.; Telnova, N.O. Estimation of water deficit under climate change and irrigation conditions in the Fergana Valley of Central Asia. *Arid. Ecosyst.* **2016**, *6*, 260–267. [CrossRef]
21. Istanbulluoglu, A. Effects of irrigation regimes on yield and water productivity of safflower (*Carthamus tinctorius* L.) under Mediterranean climatic conditions. *Agric. Water Manag.* **2009**, *96*, 1792–1798. [CrossRef]
22. Woli, P.; Hoogenboom, G.; Alva, A. Simulation of potato yield, nitrate leaching, and profit margins as influenced by irrigation and nitrogen management in different soils and production regions. *Agric. Water Manag.* **2016**, *171*, 120–130. [CrossRef]
23. Onder, S.; Caliskan, M.E.; Onder, D.; Caliskan, S. Different irrigation methods and water stress effects on potato yield and yield components. *Agric. Water Manag.* **2005**, *73*, 73–86. [CrossRef]
24. Elhani, S.; Haddadi, M.; Csákvári, E.; Zantar, S.; Hamim, A.; Villányi, V.; Douaik, A.; Bánfalvi, Z. Effects of partial root-zone drying and deficit irrigation on yield, irrigation water-use efficiency and some potato (*Solanum tuberosum* L.) quality traits under glasshouse conditions. *Agric. Water Manag.* **2019**, *224*, 10. [CrossRef]
25. Zhang, F.; Chen, M.R.; Xing, Y.Y.; Dang, F.F.; LI, Y.; WANG, X.K. Optimization of fertilizer and drip irrigation levels for efficient potato production based on entropy weight method and TOPSIS. *J. Plant Nutr. Fertil.* **2023**, *29*, 732–744.
26. Li, F.Q.; Deng, H.L.; Wang, Y.C.; Li, X.; Chen, X.T.; Liu, L.T.; Zhang, H.J. Potato growth, photosynthesis, yield, and quality response to regulated deficit drip irrigation under film mulching in a cold and arid environment. *Sci. Rep.* **2021**, *11*, 16. [CrossRef] [PubMed]
27. Jensen, C.R.; Battilani, A.; Plauborg, F.; Psarras, G.; Chartzoulakis, K.; Janowiak, F.; Stikic, R.; Jovanovic, Z.; Li, G.T.; Qi, X.B.; et al. Deficit irrigation based on drought tolerance and root signalling in potatoes and tomatoes. *Agric. Water Manag.* **2010**, *98*, 403–413. [CrossRef]
28. Kifle, M.; Gebretsadikan, T.G. Yield and water use efficiency of furrow irrigated potato under regulated deficit irrigation, Atsibi-Wemberta, North Ethiopia. *Agric. Water Manag.* **2016**, *170*, 133–139. [CrossRef]
29. Patanè, C.; Tringali, S.; Sortino, O. Effects of deficit irrigation on biomass, yield, water productivity and fruit quality of processing tomato under semi-arid Mediterranean climate conditions. *Sci. Hortic.* **2011**, *129*, 590–596. [CrossRef]
30. Ledoigt, G.; Griffaut, B.; Debiton, E.; Vian, C.; Mustel, A.; Evray, G.; Maurizis, J.C.; Madelmont, J.C. Analysis of secreted protease inhibitors after water stress in potato tubers. *Int. J. Biol. Macromol.* **2006**, *38*, 268–271. [CrossRef]
31. Erdem, T.; Halim, O.A.; Erdem, Y.; Okursoy, H. Crop water stress index for potato under furrow and drip irrigation systems. *Potato Res.* **2005**, *48*, 49–58. [CrossRef]
32. Dalla Costa, L.; Delle Vedove, G.; Gianquinto, G.; Giovanardi, R.; Peressotti, A. Yield, water use efficiency and nitrogen uptake in potato: Influence of drought stress. *Potato Res.* **1997**, *40*, 19–34. [CrossRef]
33. Martínez-Romero, A.; Domínguez, A.; Landeras, G. Regulated deficit irrigation strategies for different potato cultivars under continental Mediterranean-Atlantic conditions. *Agric. Water Manag.* **2019**, *216*, 164–176. [CrossRef]
34. Bélanger, G.; Walsh, J.R.; Richards, J.E.; Milburn, P.H.; Ziadi, N. Yield response of two potato culivars to supplemental irrigation and N fertilization in New Brunswick. *Am. J. Potato Res.* **2000**, *77*, 11–21. [CrossRef]
35. Zhang, L.L.; Shi, Y.; Qi, X.; Wang, Q.X.; Cui, L. Effects of drought stress on the ultrastructure and physiological indexes of leaf cells in three potato varieties. *Agric. Res. Arid. Areas* **2015**, *33*, 75–80.
36. Greenwood, D.J.; Zhang, K.; Hilton, H.W.; Thompson, A.J. Opportunities for improving irrigation efficiency with quantitative models, soil water sensors and wireless technology. *J. Agric. Sci.* **2010**, *148*, 1–16. [CrossRef]

37. Badr, M.A.; El-Tohamy, W.A.; Zaghloul, A.M. Yield and water use efficiency of potato grown under different irrigation and nitrogen levels in an arid region. *Agric. Water Manag.* **2012**, *110*, 9–15. [CrossRef]
38. Wang, X.K.; Guo, T.; Wang, Y.; Xing, Y.Y.; Wang, Y.F.; He, X.L. Exploring the optimization of water and fertilizer management practices for potato production in the sandy loam soils of Northwest China based on PCA. *Agric. Water Manag.* **2020**, *237*, 14. [CrossRef]
39. Carli, C.; Yuldashev, F.; Khalikov, D.; Condori, B.; Mares, V.; Monneveux, P. Effect of different irrigation regimes on yield, water use efficiency and quality of potato (*Solanum tuberosum* L.) in the lowlands of Tashkent, Uzbekistan: A field and modeling perspective. *Field Crops Res.* **2014**, *163*, 90–99. [CrossRef]
40. Xue, D.X.; Zhang, H.J.; Ba, Y.C.; Zhang, M.; Wang, S.J. Effects of regulated deficit irrigation on soil environment and yield of potato under drip irrigation. *Acta Agric. Boreali-Sin.* **2017**, *3*, 229–238.
41. He, M.Z.; Dijkstra, F.A. Drought effect on plant nitrogen and phosphorus: A metaanalysis. *New Phytol.* **2014**, *204*, 924–931. [CrossRef] [PubMed]
42. Kumar, D.; Singh, B.P.; Kumar, P. An overview of the factors affecting sugar content of potatoes. *Ann. Appl. Biol.* **2004**, *145*, 247–256. [CrossRef]
43. Wegener, C.B.; Jürgens, H.U.; Jansen, G. Drought stress affects nutritional and bioactive compounds in potatoes (*Solanum tuberosum* L.) relevant to human health. *Funct. Foods Health Dis.* **2017**, *7*, 17–35. [CrossRef]
44. Eldredge, E.P.; Holmes, Z.A.; Mosley, A.R.; Shock, C.C.; Stieber, T.D. Effects of transitory water stress on potato tuber stem-end reducing sugar and fry color. *Am. Potato J.* **1996**, *73*, 517–530. [CrossRef]
45. Andre, C.M.; Schafleitner, R.; Legay, S.; Lefevre, I.; Aliaga, C.A.A.; Nomberto, G.; Hoffmann, L.; Hausman, J.-F.; Larondelle, Y.; Evers, D. Gene expression changes related to the production of phenolic compounds in potato tubers grown under drought stress. *Phytochemistry* **2009**, *70*, 1107–1116. [CrossRef] [PubMed]
46. Muttucumaru, N.; Powers, S.J.; Elmore, J.S.; Mottram, D.S.; Haford, N.G. Effects of Water Availability on Free Amino Acids, Sugars, and Acrylamide-Forming Potential in Potato. *J. Agric. Food Chem.* **2015**, *63*, 2566–2575. [CrossRef]
47. Wang, H.D.; Wang, X.K.; Bi, L.F.; Wang, Y.; Fan, J.L.; Zhang, F.C.; Hou, X.H.; Cheng, M.H.; Hu, W.H.; Wu, L.F.; et al. Multi-objective optimization of water and fertilizer management for potato production in sandy areas of northern China based on TOPSIS. *Field Crops Res.* **2019**, *240*, 55–68. [CrossRef]
48. Rasool, G.; Guo, X.P.; Wang, Z.C.; Ali, M.U.; Chen, S.; Zhang, S.X.; Wu, Q.J.; Ullah, M.S. Coupling fertigation and buried straw layer improves fertilizer use efficiency, fruit yield, and quality of greenhouse tomato. *Agric. Water Manag.* **2020**, *239*, 9. [CrossRef]
49. Mahbod, M.; Sepaskhah, A.R.; Zand-Parsa, S. Estimation of yield and dry matter of winter wheat using logistic model under different irrigation water regimes and nitrogen application rates. *Arch. Agron. Soil Sci.* **2014**, *60*, 1661–1676. [CrossRef]
50. Feng, Y.Y.; Shi, H.B.; Jia, Y.H.; Li, R.P.; Miao, Q.F.; Jia, Q. Multi-Objective Optimization Water-Nitrogen Coupling Zones of Maize under Mulched Drip Irrigation: A Case Study of West Liaohe Plain, China. *Agronomy* **2023**, *13*, 486. [CrossRef]
51. Zhang, Y.; Tian, F. *Potato Experiment Research Method*; China Agricultural Science and Technology Press: Beijing, China, 2007.
52. Wang, N.; Zhang, T.H.; Cong, A.Q.; Lian, J. Integrated application of fertilization and reduced irrigation improved maize (*Zea mays* L.) yield, crop water productivity and nitrogen use efficiency in a semi-arid region. *Agric. Water Manag.* **2023**, *289*, 14. [CrossRef]
53. Wang, Z.Y.; Zhang, H.J.; Wang, Y.C.; Zhou, C.L. Integrated Evaluation of the Water Deficit Irrigation Scheme of Indigowoad Root under Mulched Drip Irrigation in Arid Regions of Northwest China Based on the Improved TOPSIS Method. *Water* **2021**, *13*, 1532. [CrossRef]

*plants*

MDPI

*Article*

# Increasing Hybrid Rice Yield, Water Productivity, and Nitrogen Use Efficiency: Optimization Strategies for Irrigation and Fertilizer Management

**Haijun Zhu, Xiaoe He, Xuehua Wang * and Pan Long ***

Key Laboratory of Ministry of Education for Crop Physiology and Molecular Biology, Hunan Agricultural University, Changsha 410128, China; 15211104718@163.com (H.Z.); hexiaoe2022@163.com (X.H.)

* Correspondence: 13873160151@163.com (X.W.); longpan_lp@126.com (P.L.)

**Abstract:** Water and fertilizer are crucial in rice growth, with irrigation and fertilizer management exhibiting synergies. In a two-year field study conducted in Yiyang City, Hunan Province, we examined the impact of three irrigation strategies—wet-shallow irrigation (W1), flooding irrigation (W2), and the "thin, shallow, wet, dry irrigation" method (W3)—in combination with distinct fertilizer treatments (labeled F1, F2, F3, and F4, with nitrogen application rates of 0, 180, 225, and 270 kg ha$^{-1}$, respectively) on rice yield generation and water–fertilizer utilization patterns. The study employed Hybrid Rice Xin Xiang Liang you 1751 (XXLY1751) and Yue Liang you Mei Xiang Xin Zhan (YLYMXXZ) as representative rice cultivars. Key findings from the research include water, fertilizer, variety, and year treatments, which all significantly influenced the yield components of rice. Compared to W2, W1 in 2022 reduced the amount of irrigation water by 35.2%, resulting in a 42.0~42.8% increase in irrigation water productivity and a 25.7~25.9% increase in total water productivity. In 2023, similar improvements were seen. Specifically, compared with other treatments, the W1F3 treatment increased nitrogen uptake and harvest index by 1.4–7.7% and 5.9–7.7%, respectively. Phosphorus and potassium uptake also improved. The W1 treatment enhanced the uptake, accumulation, and translocation of nitrogen, phosphorus, and potassium nutrients throughout the rice growth cycle, increasing nutrient levels in the grains. When paired with the F3 fertilization approach, W1 treatment boosted yields and improved nutrient use efficiency. Consequently, combining W1 and F3 treatment emerged as this study's optimal water–fertilizer management approach. By harnessing the combined effects of water and fertilizer management, we can ensure efficient resource utilization and maximize the productive potential of rice.

**Keywords:** irrigation strategies; fertilizer management; rice yield; water productivity; nutrient utilization

## 1. Introduction

With the growth of the population and the decrease in arable land, food security has emerged as a global concern. As the world's largest producer and consumer of rice, China undertakes a crucial responsibility in rice production [1]. Hybrid rice, a significant achievement of Chinese agricultural technology, boasts a yield far higher than traditional rice varieties, which is pivotal in ensuring China's food security [2,3]. However, the high yield of hybrid rice also raises a series of issues, such as high water consumption and low fertilizer utilization efficiency [4]. Therefore, improving water and fertilizer utilization efficiency while maintaining the high yield of hybrid rice and reducing the environmental burden has become an important topic in current agricultural scientific research [5].

Water resources are an indispensable condition for agricultural production [6]. In 2021, Chinese agricultural water consumption amounted to 364.43 billion cubic meters, accounting for 61.5% of the country's total water consumption [7]. Rice, in particular, is the most water-consuming crop in agricultural production, utilizing 70% of the agricultural water consumption [8]. Over 95% of rice irrigation in China relies primarily on traditional flooded irrigation methods. However, this irrigation method consumes significant water, with approximately half lost through various means [9]. According to research, the current irrigation water consumption for paddy fields in southern China exceeds 9000 $m^3$ per hectare. In contrast, due to low rainfall in some northern regions, the irrigation water amount surpasses 15,000 $m^3$ per hectare [10]. This is far higher than the irrigation water required for local rice production, indicating a severe waste of water resources. Due to the lack of effective water-saving irrigation techniques and management, a significant portion of irrigation water is lost during field production [11]. Therefore, developing water-saving agriculture and researching high-yield and efficient water-saving irrigation techniques for rice is a significant strategic need in China. Approximately 95% of China's rice paddies are suitable for water-saving irrigation [12]. By fully adopting water-saving irrigation methods, China's rice production could increase by 5.4% to 6.9% while saving 22.1% to 26.4% of irrigation water [13]. Researching water-saving irrigation technologies to grow rice with less water and improve water productivity, thereby achieving sustainable irrigation, is significant for stabilizing China's rice production safety and optimizing water resource utilization.

Notably, fertilizer is also an essential nutrient for the growth of hybrid rice [14]. China occupies only 9% of the world's cultivated land area. Yet, its fertilizer application accounts for one-third of the global total [15]. The fertilizer application per hectare of crops in China reaches 506.11 kg, which is 2.05 times that of the UK, 3.69 times that of the US, and 9.45 times that of Australia, significantly higher than the levels seen in developed countries worldwide [16]. This high usage is primarily attributed to the pressure to increase rice production, low soil fertility, high utilization intensity of cultivated land, large-scale agricultural production, and outdated fertilization techniques. Due to outdated fertilization techniques and a limited range of fertilizer types, the utilization rate of fertilizers remains low in practical agricultural production, resulting in significant fertilizer runoff into the environment. This leads to a waste of resources and causes environmental pollution [17]. In order to overcome these problems, several nitrogen-efficient management strategies, such as the balanced application of nitrogen fertilizer, optimized and reduced nitrogen management, organic fertilizers, and the adoption of controlled-release fertilizers, have been widely carried out [18,19].

Water and fertilizer are two fundamental factors for rice growth, which significantly impact crops and the environment and interact and restrict each other [20]. Studies have shown that the application effect of suitable water and fertilizer coupling technology is far greater than the single application effect of the two factors. Reasonable water and fertilizer regulation can promote the release of soil nutrients in paddy fields, enhance root activity, and enable more soil water and nutrients to enter the roots, which are then transported to the aboveground parts [21,22]. This helps to increase the transpiration rate, enhance net photosynthetic production, and facilitate the accumulation of rice dry matter, thereby increasing yield. Meanwhile, due to enhanced root activity, nutrients' absorption and transformation capacity is relatively increased, improving the utilization rate of elements such as nitrogen, phosphorus, and potassium [23].

The middle and lower reaches of the Yangtze River are important rice cultivation areas, constituting the most intensively rice-farmed region in China and one of the most crucial rice production regions globally [9]. Upon investigation, it is discovered that rice growers in the middle and lower reaches of the Yangtze River have a tradition of cultivating rice under flood conditions and applying excessive fertilization [24,25]. This study aims to evaluate the interaction effects of three irrigation systems (wet-shallow irrigation, flooding irrigation, and "thin, shallow, wet, dry" irrigation) and four fertilizer management strategies (F1, F2, F3, and F4, corresponding to nitrogen application rates of 0, 180, 225, and 270 kg $ha^{-1}$, respectively) on rice yield, fertilizer utilization efficiency, and WP, and to elucidate the possible reasons for any observed differences. This is crucial for rice production, water-conserving irrigation in the rice planting area in the middle and lower Yangtze River basin, and promoting the sustainable development of paddy fields in China.

## 2. Results

### 2.1. Grain Yield and Yield Components in Response to Irrigation Water and Fertilizer

Both water and fertilizer treatments significantly increased rice yield as primary effects (Table 1). We found a notable interaction between water and fertilizer. This interaction had a similar impact on the two rice varieties tested over two years (i.e., there were no significant interactions among water, fertilizer, and variety, neither were there significant interactions among water, fertilizer, and year). This suggests that when water and fertilizer treatments are combined, the increase in yield is even more substantial. Additionally, there were differences in rice yield between years, which can be attributed to significant interactions between variety and water treatment, as well as variety and fertilizer. Indeed, water, fertilizer, variety, and year treatments significantly influenced rice yield components (including panicles, spikelet per panicle, spikelet filling, and grain weight).

**Table 1.** Fertilizer, irrigation, variety, and year affect grain yield and yield components. No significant four-way interactions were observed, and only two three-way interactions (F × V × Y and V × W × Y) were significant for rice yield properties. Variables for which the model simplification process eliminated the non-significant (NS) three-way interactions are marked as NS. Significant treatment effects within a main category are denoted by * ($0.01 < p \leq 0.05$) or ** ($p \leq 0.01$).

| SOV | Yield | Panicles ($\times 10^4$ ha$^{-1}$) | Spikelet's Panicle$^{-1}$ | Spikelet Filling (%) | Grain Weight (mg) |
|---|---|---|---|---|---|
| F | ** | ** | ** | ** | ** |
| V | ** | ** | ** | ** | ** |
| W | ** | ** | ** | ** | * |
| Y | ** | ** | ** | ** | ** |
| F × V | ** | ** | ** | ** | NS |
| F × W | ** | NS | NS | NS | NS |
| F × Y | ** | NS | NS | ** | NS |
| V × W | NS | ** | NS | ** | NS |
| V × Y | ** | NS | NS | NS | NS |
| W × Y | NS | NS | NS | ** | NS |
| F × V × Y | NS | NS | NS | ** | NS |
| V × W × Y | NS | NS | NS | ** | NS |

## 2.2. The Accumulation of N, P, and K

Irrigation methods and nitrogen fertilizer management have significant or highly significant effects on the accumulation of nitrogen, phosphorus, and potassium in vegetative organs, ear parts, and mature plants during the heading and maturity stages (Tables 2–4). Additionally, both factors show significant or highly significant interaction effects on the accumulation of nitrogen, phosphorus, and potassium in various organs during the heading and maturity stages. Focusing on nitrogen accumulation in plants as an illustrative example, since the accumulation of nitrogen, phosphorus, and potassium exhibits similarity across different treatments, Table 1 indicates notable differences in nitrogen accumulation among varieties and between years for straw, grain, and the total plant. Specifically, XXLY1751 shows significantly higher nitrogen accumulation compared to YLYMXXZ. Regarding yearly variation, 2023 demonstrates significantly higher nitrogen accumulation than 2022. Regarding the distribution of nitrogen among organs, straw accounts for an average of 57.86%, and grain accounts for 42.14%. Significant differences in nitrogen accumulation are observed among various water and fertilizer treatments.

**Table 2.** The absorption capacity of nitrogen by the aboveground part of hybrid rice plants (kg ha$^{-1}$).

| Item | Treatment | Heading | | Mature | | |
|---|---|---|---|---|---|---|
| | | Straw | Grain | Straw | Grain | Total |
| Year | 2022 | 125.8 b | 19.96 a | 54.84 b | 111.73 b | 166.57 b |
| | 2023 | 132.72 a | 21.66 a | 58.51 a | 115.58 a | 174.53 a |
| Variety | XXLY1751 | 129.95 a | 22 a | 57.22 a | 117.34 a | 174.56 a |
| | YLYMXXZ | 127.37 a | 19.78 b | 55.75 b | 110.62 b | 166.37 b |
| Strategies | W1F1 | 75 f | 17.48 ef | 36.35 e | 74.47 g | 110.81 f |
| | W1F2 | 127.05 d | 19.61 cd | 57.03 d | 123.03 d | 180.06 d |
| | W1F3 | 149.52 c | 23.34 b | 62.22 bcd | 143.89 a | 206.11 a |
| | W1F4 | 165.86 b | 26.59 a | 67.57 abc | 136.63 b | 204.2 ab |
| | W2F1 | 72.24 f | 15.54 g | 33.58 e | 68.3 h | 101.88 g |
| | W2F2 | 118.74 e | 18.28 def | 58.99 d | 110.9 f | 169.89 e |
| | W2F3 | 144.53 c | 20.42 c | 62.56 bcd | 128.27 cd | 190.83 c |
| | W2F4 | 176.71 a | 25.17 a | 68.26 ab | 125.18 cd | 193.44 c |
| | W3F1 | 75.93 f | 16.67 fg | 36.97 e | 72.82 gh | 109.79 f |
| | W3F2 | 126.78 d | 18.95 cde | 59.47 d | 117.03 e | 176.5 d |

**Table 2.** *Cont.*

| Item | Treatment | Heading | | Mature | | |
|---|---|---|---|---|---|---|
| | | Straw | Grain | Straw | Grain | Total |
| | W3F3 | 146.86 f | 22.42 b | 65.62 abc | 136.89 b | 202.5 ab |
| | W3F4 | 164.66 f | 26.23 a | 69.23 a | 130.37 c | 199.6 b |
| SOV | F | ** | ** | ** | ** | ** |
| | W | ** | ** | ** | ** | ** |
| | V | ** | ** | ** | ** | ** |
| | Y | ** | ** | ** | ** | ** |
| | F × W | ** | NS | ** | ** | ** |
| | F × W × V | NS | NS | NS | NS | NS |
| | F × W × Y | * | NS | NS | NS | NS |
| | F × W × V × Y | NS | NS | NS | NS | NS |

Within a column for each year, means followed by the same letters are not significantly different according to LSD (0.05). SOV, source of variation, and significant treatment effects within a main category are denoted by * ($0.01 < p \leq 0.05$) or ** ($p \leq 0.01$).

**Table 3.** The absorption capacity of phosphorus by the aboveground part of hybrid rice plants (kg ha$^{-1}$).

| Item | Treatment | Heading | | Mature | | |
|---|---|---|---|---|---|---|
| | | Straw | Grain | Straw | Grain | Total |
| Year | 2022 | 20.06 b | 6.27 b | 8.26 b | 27.9 b | 36.15 b |
| | 2023 | 23.16 a | 6.66 a | 9.77 a | 30.6 a | 40.37 a |
| Variety | XXLY1751 | 21.72 a | 6.84 a | 9.08 a | 30.58 a | 39.66 a |
| | YLYMXXZ | 21.44 a | 6.12 b | 8.89 a | 28.29 b | 37.17 b |
| Strategies | W1F1 | 14.51 g | 4.4 g | 5.72 e | 24.61 f | 30.33 a |
| | W1F2 | 20.99 ef | 6.35 d | 8.62 d | 27.45 e | 36.06 a |
| | W1F3 | 24.67 bc | 7.78 a | 9.69 bcd | 37.63 a | 47.32 b |
| | W1F4 | 25.67 b | 7.53 bc | 10.89 ab | 31.96 c | 42.85 c |
| | W2F1 | 14.23 g | 4.6 f | 6.06 e | 22.78 g | 28.84 c |
| | W2F2 | 19.8 f | 6.22 de | 8.7 cd | 25.15 f | 33.85 d |
| | W2F3 | 23.09 cd | 7.64 ab | 9.6 bcd | 35.04 b | 44.65 e |
| | W2F4 | 27.42 a | 7.57 bc | 11.22 a | 29.98 d | 41.2 e |
| | W3F1 | 15.18 g | 4.54 fg | 6.13 e | 24.01 fg | 30.14 f |
| | W3F2 | 21.72 de | 6.12 e | 9.24 cd | 26.77 e | 36.01 g |
| | W3F3 | 25.28 b | 7.53 bc | 10.48 ab | 36.68 a | 47.16 gh |
| | W3F4 | 26.39 ab | 7.45 c | 11.45 a | 31.16 cd | 42.61 h |
| SOV | F | ** | ** | ** | ** | ** |
| | W | ** | ** | ** | ** | ** |
| | V | * | ** | ** | ** | ** |
| | Y | ** | ** | ** | ** | ** |
| | F × W | ** | ** | ** | NS | NS |
| | F × W × V | NS | ** | NS | NS | NS |
| | F × W × Y | NS | * | NS | NS | ** |
| | F × W × V × Y | NS | NS | NS | NS | NS |

Within a column for each year, means followed by the same letters are not significantly different according to LSD (0.05). SOV, source of variation, and significant treatment effects within a main category are denoted by * ($0.01 < p \leq 0.05$) or ** ($p \leq 0.01$).

**Table 4.** The absorption capacity of potassium by the aboveground part of hybrid rice plants (kg ha$^{-1}$).

| Item | Treatment | Heading | | Mature | | |
|---|---|---|---|---|---|---|
| | | Straw | Grain | Straw | Grain | Total |
| Year | 2022 | 172.96 b | 6.27 b | 140.22 b | 27.9 b | 36.15 b |
| | 2023 | 193.46 a | 6.66 a | 150.90 a | 30.6 a | 40.37 a |
| Variety | XXLY1751 | 188.68 a | 6.84 a | 155.01 a | 30.58 a | 39.66 a |
| | YLYMXXZ | 178.49 b | 6.12 b | 136.36 b | 28.29 b | 37.17 b |
| Strategies | W1F1 | 132.05 g | 9.51 g | 96.98 h | 15.45 g | 112.44 h |
| | W1F2 | 187.2 e | 13.46 de | 141.83 e | 23.43 d | 165.26 e |
| | W1F3 | 207.26 c | 16.9 c | 168.36 b | 29.26 a | 197.63 b |
| | W1F4 | 222.09 a | 21.2 b | 169.76 b | 28.26 bc | 198.02 b |
| | W2F1 | 127.06 h | 9.98 f | 105.85 f | 16.31 f | 122.16 f |
| | W2F2 | 180.58 f | 13.78 d | 145.88 d | 22.88 de | 168.76 d |
| | W2F3 | 201.97 d | 16.74 c | 166.94 b | 28.69 ab | 195.63 b |
| | W2F4 | 215.13 b | 21.78 a | 174.57 a | 28.41 bc | 202.98 a |
| | W3F1 | 128.26 h | 9.38 g | 102.23 g | 16.03 fg | 118.26 g |
| | W3F2 | 182.56 f | 13.27 e | 142.9 e | 22.5 e | 165.4 e |
| | W3F3 | 203.75 d | 16.66 c | 163.28 c | 28.25 bc | 191.53 c |
| | W3F4 | 215.11 b | 20.91 b | 169.62 b | 27.92 c | 197.54 b |

**Table 4.** *Cont.*

| Item | Treatment | Heading | | Mature | | |
|---|---|---|---|---|---|---|
| | | Straw | Grain | Straw | Grain | Total |
| SOV | F | ** | ** | ** | ** | ** |
| | W | ** | ** | ** | ** | ** |
| | V | ** | ** | ** | ** | ** |
| | Y | ** | ** | ** | ** | ** |
| | F × W | ** | ** | ** | ** | ** |
| | F × W × V | ** | ** | ** | ** | ** |
| | F × W × Y | ** | ** | ** | * | ** |
| | F × W × V × Y | ** | NS | ** | NS | ** |

Within a column for each year, means followed by the same letters are not significantly different according to LSD (0.05). SOV, source of variation, and significant treatment effects within a main category are denoted by * (0.01 < $p \leq 0.05$) or ** ($p \leq 0.01$).

### 2.3. N Harvest Index and Nitrogen Use Efficiency

Water and fertilizer treatments significantly affected the N harvest index (NHI) and N use efficiency (NAE), partial factor productivity of applied N (NPFP), and apparent recovery efficiency of N (NRE). At the same time, varietal differences and yearly variations also had a notable impact on these indicators, except NAE for varieties and NRE for yearly variations (Table 5). A significant interaction was observed between F×W for the NHI, but this interaction was not significant for NAE, NPFP, and NRE. Moreover, nitrogen application had a more pronounced effect on nitrogen use efficiency than irrigation methods. Compared to W2, W1 resulted in significant increases in NHI and NAE, NPFP, and NRE, with increases of 3.99%, 4.66%, 4.15%, and 3.03% in 2022, and 3.42%, 8.75%, 4.08%, and 3.58% in 2023, respectively. Among fertilizer treatments, F3 significantly increased NHI, NAE, NPFP, and NRE compared to F4, with notable increases of 3.96%, 71.14%, 30.86%, and 21.28% in 2022, and 3.86%, 59.90%, 29.84%, and 21.11% in 2023.

**Table 5.** Agronomic N use efficiency (NAE), partial factor productivity of applied N (NPFP), apparent recovery efficiency of N (NRE), and N harvest index (NHI) under different irrigation and fertilizer treatments in 2022 and 2023.

| Item | Treatment | AE | PFP | RE | HI |
|---|---|---|---|---|---|
| Year | 2022 | 8.94 a | 37.67 b | 36.12 a | 67.02 a |
| | 2023 | 10.39 a | 39.14 a | 36.45 a | 66.37 a |
| Variety | XXLY1751 | 9.85 a | 40.99 a | 36.58 a | 67.25 a |
| | YLYMXXZ | 9.78 a | 36.92 b | 36.24 a | 66.4 b |
| Strategies | W1F1 | | | | 67.12 bcd |
| | W1F2 | 10.55 a | 47.13 a | 34.72 b | 68.32 ab |
| | W1F3 | 12.15 a | 41.4 c | 42.35 a | 69.82 a |
| | W1F4 | 6.96 b | 31.35 e | 34.74 b | 66.93 bcd |
| | W2F1 | | | | 67.01 bcd |
| | W2F2 | 10.83 a | 45.77 ab | 34.5 b | 65.27 de |
| | W2F3 | 11.08 a | 39.02 d | 39.57 a | 67.21 bcd |
| | W2F4 | 7.31 b | 30.6 e | 33.6 b | 64.69 e |
| | W3F1 | | | | 66.29 cde |
| | W3F2 | 10.82 a | 45.63 b | 33.72 b | 66.29 cde |
| | W3F3 | 11.71 a | 39.56 d | 41.21 a | 67.6 bc |
| | W3F4 | 6.91 b | 30.12 e | 33.26 b | 65.32 de |
| SOV | F | ** | ** | ** | ** |
| | W | ** | ** | ** | ** |
| | V | NS | ** | NS | ** |
| | Y | NS | ** | NS | NS |
| | F × W | ** | ** | ** | ** |
| | F × W × V | NS | NS | NS | NS |
| | F × W × Y | NS | NS | NS | NS |
| | F × W × V × Y | NS | NS | NS | NS |

Within a column for each year, means followed by the same letters are not significantly different according to LSD (0.05). SOV, source of variation, and significant treatment effects within a main category are denoted by ** ($p \leq 0.01$).

### 2.4. Water Productivity

The water productivity (WP) of rice under different water and fertilizer management in 2022 and 2023 is illustrated in the accompanying figure (Figure 1). Compared to W2, W1 significantly enhanced the irrigation water productivity (IWP) and the overall WP. Specifically, in 2022, there was an increase of 24.04% to 40.99% for IWP and 15.61% to 4.58% for WP. Similarly, in 2023, the improvements ranged from 24.22% to 42.41% for IWP and 13.33% to 22.39% for WP. Under the same irrigation treatment, when compared to the N1 treatment, the IWP and WP were elevated in the N2, N3, and N4 treatments. Moreover, W1F2 and W1F3 treatments contributed to higher IWP and WP, with W1F3 being the most effective. In 2022, under W1F3, the IWP and WP were, respectively, higher than other treatments by 3.91% to 78.93% and 2.83% to 64.56%. Similarly, in 2023, the increases ranged from 4.47% to 7.11% for IWP and a remarkable 105.61% to 85.19% for WP.

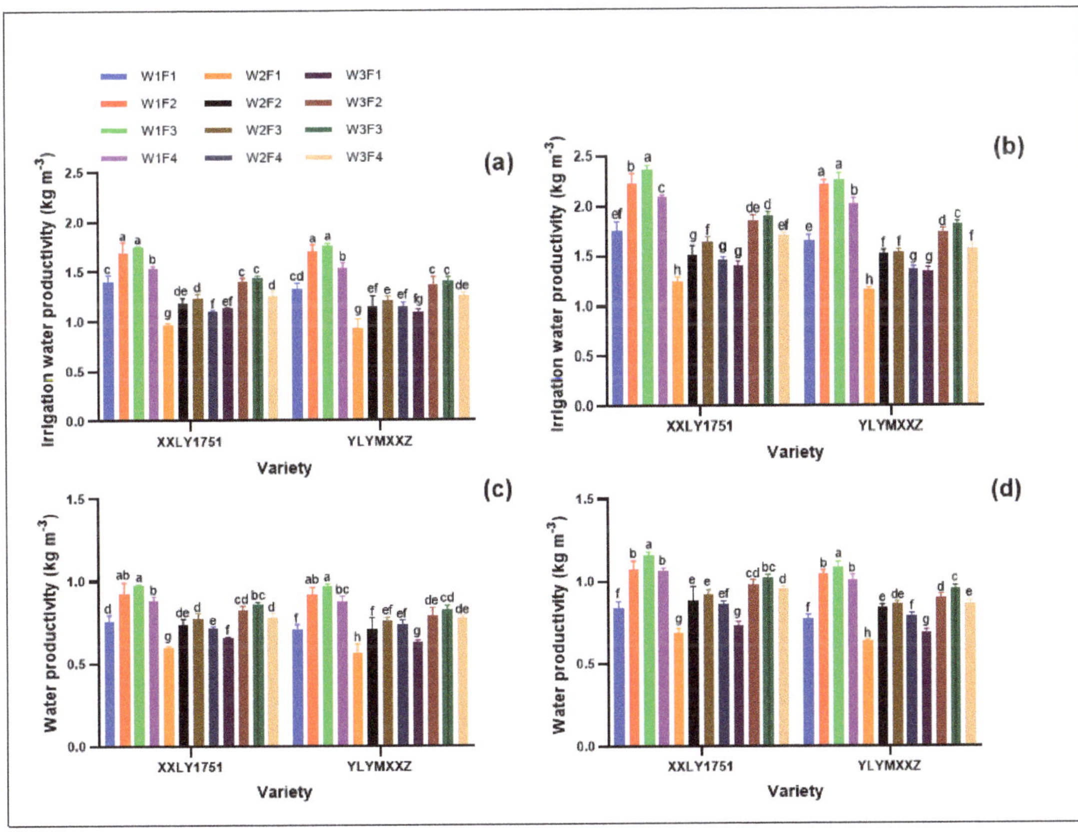


**Figure 1.** Irrigation water productivity in 2022 (**a**) and 2023 (**b**), and water productivity in 2022 (**c**) and 2023 (**d**). Different lowercase letters denote statistical differences between treatments of each variety according to the LSD test (0.05).

### 2.5. Optimization of Water and Fertilizer Management Based on Yield

We constructed a bubble chart to determine the optimal combination (Figure 2). As illustrated in the chart, the objective was achieved by applying 4322 $m^3$ $ha^{-1}$ of irrigation water and 225–135–270 kg N: $P_2O_5$: $K_2O$ $ha^{-1}$.

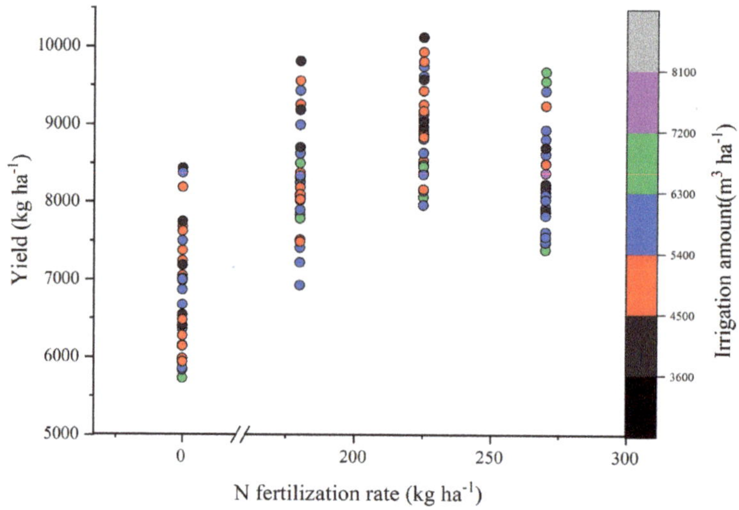


**Figure 2.** Relationship between irrigation amount, fertilizer rate, and yield.

### *2.6. Water and Fertilizer Management Based on Yield Using Grain Nitrogen Accumulation Correlation Analysis*

We constructed a correlation matrix to compare the correlations between all indicators (Figure 3). As illustrated in the figure, NHI, TAN, NAE, NPFP, NRE, WP, and IWP exhibit significant correlations with yield. NHI, TAN, NAE, NPFP, NRE, and yield significantly correlate with WP.

**Figure 3.** Water and fertilizer management based on yield using grain nitrogen accumulation correlation analysis., and significant treatment effects within a main category are denoted by * ($0.01 < p \leq 0.05$), ** ($p \leq 0.01$) or *** ($p \leq 0.001$).

## 3. Discussion

### 3.1. The Relationship between Rice Yield and the Absorption of Nitrogen, Phosphorus, and Potassium in Rice Plants under the Interaction of Irrigation and Fertilizer Rice Plants under the Interaction of Irrigation and Fertilizer

Sun [26] reported that there is a significant interaction between water and nitrogen that affects the accumulation, translocation, allocation, and yield of nitrogen, phosphorus, and potassium during the main growth stages of rice. This water–nitrogen interaction demonstrates a significant or highly significant positive correlation with the absorption, and translocation of nitrogen, phosphorus, and potassium, as well as with yield, across various growth stages. We discovered that the alterations in fertilizer application rates beneath diverse irrigation methods continuously restricted the overall accumulation of nitrogen, phosphorus, and potassium in rice plants, as well as the translocation of these nutrients from vegetative organs during the fruiting period. Remarkably, there was a notable consistency in the changes noticed for nitrogen, phosphorus, and potassium throughout different growth stages under various water and fertilizer management practices in both the experimental grains. In addition, the correlation analyses of nitrogen, phosphorus, and potassium during the two-year trial period disclosed significant or highly significant correlations between different irrigation methods and fertilizer application rates regarding the absorption and accumulation of these nutrients in each growth stage. This implies that the interaction between water and fertilizer can also facilitate the synergistic uptake of nutrients in rice. This finding validates and supplements previous research, further highlighting that capitalizing on the advantages of the water–fertilizer interaction is fundamental for high-yield cultivation [27,28]. Some studies [29–31] reported that the total amount of nitrogen accumulation is different among different types and among different rice varieties of the same type, which in turn affects the yield. Sun [26] also clearly stated that under the interaction of different water management and nitrogen application rates, there is an extremely significant positive correlation between the total cumulative amounts of nitrogen, phosphorus, and potassium in rice plants and the yield. Our correlation analysis revealed a highly significant relationship between the total accumulation of nitrogen, phosphorus, and potassium in rice plants and the yield under different irrigation methods and fertilizer application rates. During the two-year experimental period, the W1 treatment enhanced the absorption, accumulation, and translocation of nitrogen, phosphorus, and potassium in various growth stages of rice, increasing the nutrient content in the grains. When combined with the F3 fertilizer application rate, it achieved the dual goals of increasing the yield and enhancing the nutrient use efficiency, establishing it as the optimal water–fertilizer coupling strategy in this study.

### 3.2. The Effect of Different Irrigation and Fertilizer Interactions on Rice Yield and Water–Fertilizer Use Efficiency

The water-saving irrigation techniques for rice, such as wet-shallow irrigation, alternate wetting and drying irrigation, controlled irrigation, and dry cultivation, have significantly increased water productivity [32–34]. However, the impact on yield varies depending on soil texture, soil drying level, and rainfall during the rice growing season [35]. Some studies have shown a decrease or no significant change, while others have demonstrated a decrease or no significant change [36,37]. The water-saving management technique employed in this study aligns with previous research findings regarding increasing rice yield. Its effect on water productivity is primarily achieved by reducing irrigation water input while enhancing yield. Research has indicated that wet-shallow irrigation can significantly enhance water productivity while reducing water consumption [38]. Xie [39] found that compared to conventional irrigation, wet-shallow irrigation led to a notable increase in water productivity by 11.3% despite a 24.3% reduction in irrigation water. Similarly, Zhang [40] reported that wet-shallow irrigation decreased irrigation water usage in paddy fields by 11.3–17.2% compared to shallow water irrigation while also increasing water productivity in rice irrigation by 3.2–30.0%. In our study, wet-shallow irrigation reduced irrigation water by 35.2% in 2022 compared to flooded irrigation, resulting in improvements of 42.0~42.8% and 25.7~25.9% in irrigation water productivity and total water productivity, respectively. In 2023, Similar results were also observed. These improvements might be attributed to the reduced irrigation volume and frequency associated with wet-shallow irrigation, which decreased water depth in the paddy fields, thus minimizing water leakage and evaporation [41]. Additionally, wet-shallow irrigation subjects rice to optimal stress during its growth cycle, promoting healthier growth and production, and ultimately enhancing rice yield and water productivity [42].

Within the fertilizer application range set in this study, the effect of increased fertilizer application on rice grain yield varies under different water management conditions. The trend of rice yield change is consistent in two years under water-saving conditions, with the highest yield achieved at the F3 treatment. In this study, the W1F3 treatment improved rice nitrogen (phosphorus, potassium) uptake, harvest index, and fertilizer use efficiency compared to other treatments. This indicates that wet-shallow irrigation improves rice fertilizer utilization efficiency, consistent with Sun's [26] research results. This may be because wet-shallow irrigation reduces the water level in the paddy field, promotes root respiration, and enhances the absorption of nutrients, thereby improving rice fertilizer utilization efficiency [43]. Thus, high-quality hybrid rice varieties can achieve a synergistic efficiency of yield and water and fertilizer efficiency in intensive rice systems under a cultivation model combining lower fertilizer application (F3) with wet-shallow irrigation water management (W1).

## 4. Materials and Methods

### 4.1. Overview of Study Site

The experiment took place in the rice fields of farmers located in Oujiangcha village (coordinates: 28°29′55″ N 112°35′59″ E, 12 m a.s.l.), Oujiangcha town, Yiyang city, during the years 2022 and 2023. Table 6 presents the chemical properties of the soil in the 0 to 20 cm layer of the rice field prior to transplanting in 2022. This study conducted a split-split plot experiment with three factors: variety, water, and nitrogen fertilizer. The experiment comprised 72 split plots, each covering an area of 20 $m^2$, resulting in a total experimental area of 1440 $m^2$ (excluding water channels). The main plots were assigned to water treatments (W), with three levels: W1—continuous wet-shallow irrigation throughout the growth period, W2—flooded irrigation, and W3—thin, shallow, wet, dry irrigation. The water layer control standards for these three irrigation modes are summarized in Table 7. All three irrigation treatments involved manual watering, with a precise calculation of the irrigation volume using water meters (Figure 4). The split plots were designated for variety (V), featuring two levels: Xin Xiang Liang You 1751 (V1), provided by Hunan Jinse Nonghua Seed Industry Technology Co., Ltd. (Changsha, China), and Yue Liang You Meixiang Xin Zhan (V2), supplied by Yuan Longping High-Tech Agriculture Co., Ltd. (Changsha, China).

**Table 6.** Soil pH, organic matter, and total nitrogen and available nutrients (N, P, K) of the experimental field.

| PH | Organic Matter (g $kg^{-1}$) | Total Nitrogen (g $kg^{-1}$) | Olsen Phosphorus (mg $kg^{-1}$) | Alkali-Hydrolysis Nitrogen (mg $kg^{-1}$) | Exchangeable Potassium (mg $kg^{-1}$) |
|---|---|---|---|---|---|
| 5.1 | 34.8 | 1.6 | 13.8 | 166.8 | 106.4 |

**Table 7.** Thresholds of water level for different rice irrigation modes.

| Irrigation | RS | ES | LS | JS | HS | GS | MS |
|---|---|---|---|---|---|---|---|
| W1 | 0–20–30 | 0–20–30 | SF | 0–20–30 | 0–20–30 | 0–20–30 | DF |
| W2 | 20–50–70 | 20–50–70 | SF | 20–50–70 | 20–50–70 | 20–50–70 | DF |
| W3 | 10–30–50 | 10–30–50 | SF | 10–40–60 | 10–40–60 | 10–20–40 | DF |

**Figure 4.** Water pipes and water meters in the community during the early stage of the experiment.

The split-split plots focused on nitrogen application (N) with four levels, N1, N2, N3, and N4, corresponding to pure nitrogen applications of 0, 180, 225, and 270 kg $ha^{-1}$. The basic seedling rate was set at three seedlings, with a transplanting density of 225,000 hills per hectare. Nitrogen fertilizer (urea) was applied in a ratio of base fertilizer/tiller fertilizer/ear fertilizer = 5:3:2. The base fertilizer, tiller fertilizer, and ear fertilizer were applied 1–2 days before transplanting, 10 days after transplanting, and at the beginning of young ear differentiation (jointing stage), respectively. A consistent NPK ratio of N: $P_2O_5$: $K_2O$ = 1: 0.6: 1.2 was used for all fertilizer applications, with all of the phosphorus fertilizer ($P_2O_5$, 16%) applied as the base fertilizer and potassium fertilizer ($K_2O$, 60%) split evenly between the base and tiller fertilizers. The rigorous management of pests, diseases, and weeds follows the principles of high-yield cultivation.

Three continuous data in the table (e.g., 0–20–30) represent the lower and upper limits of irrigation and the maximum limit after rain, respectively; RS represents the returning green stage, ES represents the early tillering stage, LS represents the late tillering stage, JS represents the jointing and booting stage, and HS represents the heading stage. GS represents the grain-filling stage, and MS represents the maturity stage.

### 4.2. Sampling and Measurement

The meteorological data, including minimum temperature, maximum temperature, and precipitation, was sourced from the local meteorological bureau (Figure 5). Irrigation water productivity (IWP) is the amount of dry matter produced per unit volume of water ($m^3$) irrigated in a 1-hectare rice field, representing the production efficiency of irrigation water. Total water productivity (WP), on the other hand, refers to the amount of dry matter produced per unit volume of water ($m^3$) consumed in a 1-hectare rice field, indicating the overall efficiency of water utilization [44].

At the heading and maturity stages, 10 representative rice plants were selected based on the average number of panicles per plot. These plants were thoroughly washed with water, their roots were trimmed, and they were manually threshed. The plant materials were then separated into three parts: straw, filled grains, and empty grains. The samples were first killed at 105 °C for 30 min and then dried to a constant weight at 80 °C before being weighed. A plant grinder was used to pulverize the plant samples to determine nitrogen, phosphorus, and potassium contents in the stems, leaves, and grains. Nitrogen was measured using the Kjeldahl method, phosphorus was determined by the molybdenum–antimony colorimetric method, and potassium was analyzed with a flame photometer [45].

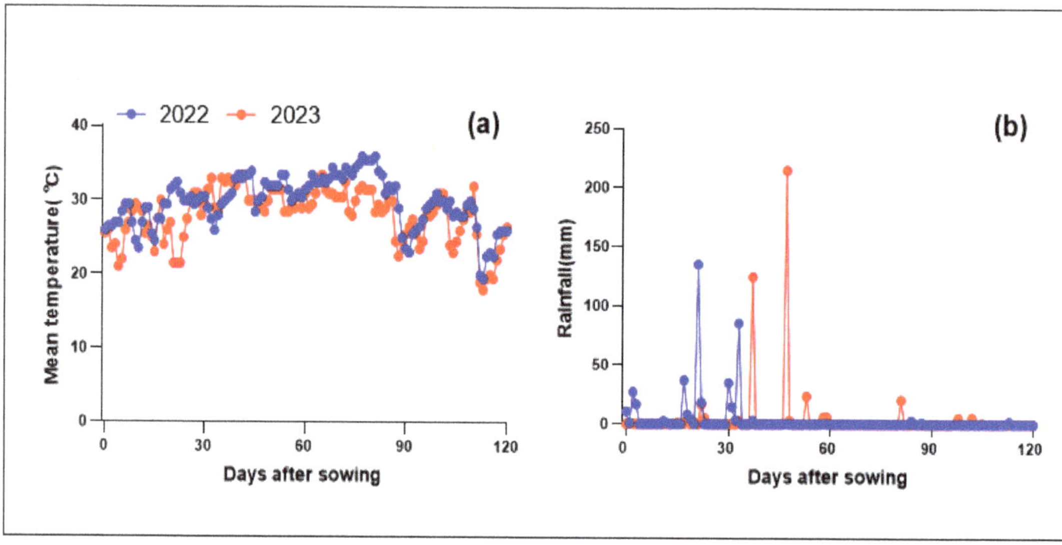


**Figure 5.** Daily mean temperature (**a**) and daily cumulative rainfall (**b**) during rice-growing season in 2022 and 2023.

At the maturity stage, one hundred effective rice spikes were meticulously examined in each plot. Subsequently, twelve hills were randomly chosen from each plot, taking into account the average number of effective spikes. These samples were then transported to the laboratory to investigate several key parameters, including panicle number × 104 $ha^{-1}$, spikelet per panicle, spikelet filling, and grain weight. The entire plot of plants was harvested and dried separately for each plot, and the actual yield was calculated after adjusting the grain moisture content to 14% [46].

The plant's total nitrogen (phosphorus, potassium) uptake is calculated as the sum of the nitrogen (phosphorus, potassium) absorbed by the stem, leaves, and grains. The nitrogen (phosphorus, potassium) harvest index (HI) is calculated as the ratio of grain nitrogen (phosphorus, potassium) uptake at maturity to the total nitrogen (phosphorus, potassium) uptake of the plant. Nitrogen (phosphorus, potassium) recovery efficiency (RE) represents the percentage difference between the accumulation of nitrogen (phosphorus, potassium) in plants from treated and untreated areas relative to the amount of nitrogen (phosphorus, potassium) applied. Nitrogen (phosphorus, potassium) partial factor productivity (PFP) is determined by dividing the yield in the treated area by the amount of nitrogen (phosphorus, potassium) applied. Agronomic efficiency (AE) of nitrogen (phosphorus, potassium) fertilizer refers to the increase in rice yield per unit of nitrogen (phosphorus, potassium) applied, calculated as the difference in rice yield between treated and untreated areas divided by the level of nitrogen (phosphorus, potassium) applied [47].

### 4.3. Data Analysis

The data were analyzed using analysis of variance (ANOVA) conducted in SAS Version 9.1.2. Means of fertilizer level, irrigation amount, and variety treatments were compared using the least significant difference test (LSD) at a 0.05 probability level. Graphs were constructed using Microsoft Excel 2017 (Microsoft Corp., Redmond, WA, USA).

## 5. Conclusions

In a two-year field study, we have determined the interactive effects of irrigation and fertilization on rice yield. Across different varieties, irrigation and fertilization significantly enhanced grain yield and the absorption of nitrogen, phosphorus, and potassium absorption. By combining reasonable irrigation methods with nitrogen management, we observed improvements in yield, water productivity, and fertilizer utilization efficiency. Therefore, we recommend using wet-shallow irrigation and appropriate fertilization (N- $P_2O_5$-$K_2O$: 225–135–270 kg $ha^{-1}$) in rice cultivation, especially in the middle and lower reaches of the Yangtze River, to enhance rice yield and water and fertilizer utilization efficiency. The combined application of wet-shallow irrigation and adequate fertilization achieves "compensating water with fertilizer" and "regulating water with fertilizer", potentially contributing to the synergistic efficiency of yield, water, and fertilizer in intensive rice systems. In the future, further research can develop real-time sensors to measure the soil moisture and nutrient levels, so as to achieve more adaptive and targeted irrigation and fertilization.

**Author Contributions:** P.L. and X.W. conceived the project and designed the experiments. H.Z. and X.H. conducted the experiments. H.Z. designed and conducted computational and statistical analyses. H.Z. and X.W. wrote the manuscript. H.Z. and X.W. revised the manuscript. All authors have read and agreed to the published version of the manuscript.

**Funding:** This research was funded by the National Key R&D Program of China (2022YFD2300301, 2023YFD1523019).

**Data Availability Statement:** The data presented in this study are available upon request from the corresponding author.

**Acknowledgments:** I would like to extend my heartfelt gratitude to Abdul Wahab (University of Chinese Academy of Sciences) for their invaluable contributions to the conceptualization and writing of this paper. I am also indebted to the fieldworkers, Manwen Sun (Bureau of Agriculture and Rural Affairs of Heshan District) and Yanheng Zeng (Hunan Biological and Electromechanical Polytechnic) for their dedication and hard work in collecting the data. Additionally, I am thankful to Weijian Tan (Bureau of Agriculture and Rural Affairs of Heshan District) for providing the necessary facilities and resources for the field experiments.

**Conflicts of Interest:** The authors declare no competing interests.

## References

1. Huang, M. The decreasing area of hybrid rice production in China: Causes and potential effects on Chinese rice self-sufficiency. *Food Secur.* **2022**, *14*, 267–272. [CrossRef]
2. Yuan, L. Progress in super-hybrid rice breeding. *Crop. J.* **2017**, *5*, 100–102. [CrossRef]
3. Huang, M.; Tang, Q.-Y.; Ao, H.-J.; Zou, Y.-B. Yield potential and stability in super hybrid rice and its production strategies. *J. Integr. Agric.* **2017**, *16*, 1009–1017. [CrossRef]
4. Pan, J.; Liu, Y.; Zhong, X.; Lampayan, R.M.; Singleton, G.R.; Huang, N.; Liang, K.; Peng, B.; Tian, K. Grain yield, water productivity and nitrogen use efficiency of rice under different water management and fertilizer-N inputs in South China. *Agric. Water Manag.* **2017**, *184*, 191–200. [CrossRef]
5. Luo, W.Q.; Chen, M.T.; Kang, Y.H.; Li, W.P.; Li, D.; Cui, Y.L.; Khan, S.; Luo, Y.F. Analysis of crop water requirements and irrigation demands for rice: Implications for increasing effective rainfall. *Agric. Water Manag.* **2022**, *260*, 107285. [CrossRef]
6. Gautam, N.K.; Sarkar, S. Estimating Crop Water Requirements and Assessing Irrigation Water Distribution Cost for Rice Production Using a Pipe Network. *Pol. J. Environ. Stud.* **2018**, *27*, 1071–1083. [CrossRef]
7. Yan, Q.; Zhang, Y.; Wang, Q. Analysis of Demand and Structural Changes in China's Water Resources over the Next 30 Years. *Water* **2023**, *15*, 745. [CrossRef]

8. Zheng, J.; Wang, W.; Liu, G.; Ding, Y.; Cao, X.; Chen, D.; Engel, B. Towards quantification of the national water footprint in rice production of China: A first assessment from the perspectives of single-double rice. *Sci. Total Environ.* **2020**, *739*, 140032. [CrossRef] [PubMed]
9. Zhu, H.; Zhang, T.; Zhang, C.; He, X.; Shi, A.; Tan, W.; Yi, Z.; Wang, X. Optimizing Irrigation and Nitrogen Management to Increase Yield and Nitrogen Recovery Efficiency in Double-Cropping Rice. *Agronomy* **2022**, *12*, 1190. [CrossRef]
10. Faiz-ul Islam, S.; Sander, B.O.; Quilty, J.R.; de Neergaard, A.; van Groenigen, J.W.; Jensen, L.S. Mitigation of greenhouse gas emissions and reduced irrigation water use in rice production through water-saving irrigation scheduling, reduced tillage and fertiliser application strategies. *Sci. Total Environ.* **2020**, *739*, 140215. [CrossRef]
11. Zoffoli, G.; Gangi, F.; Ferretti, G.; Masseroni, D. The potential of a coordinated system of gates for flood irrigation management in paddy rice farm. *Agric. Water Manag.* **2023**, *289*, 108536. [CrossRef]
12. Lv, Y.; Gu, L.; Xu, J.; Liu, X. A coupled hourly water-carbon flux model at plot and field scales for water-saving irrigated rice paddy. *Agric. Water Manag.* **2024**, *293*, 108706. [CrossRef]
13. Liu, L.; Lian, Z.; Ouyang, W.; Yan, L.; Liu, H.; Hao, F. Potential of optimizing irrigation and fertilization management for sustainable rice production in China. *J. Clean. Prod.* **2023**, *432*, 139738. [CrossRef]
14. Xu, L.; Yuan, S.; Wang, X.; Yu, X.; Peng, S. High yields of hybrid rice do not require more nitrogen fertilizer than inbred rice: A meta-analysis. *Food Energy Secur.* **2021**, *10*, 341–350. [CrossRef]
15. Zhu, X.; Ros, G.H.; Xu, M.; Cai, Z.; Sun, N.; Duan, Y.; de Vries, W. Long-term impacts of mineral and organic fertilizer inputs on nitrogen use efficiency for different cropping systems and site conditions in Southern China. *Eur. J. Agron.* **2023**, *146*, 126797. [CrossRef]
16. Xu, Y.; Xu, X.; Li, J.; Guo, X.; Gong, H.; Ouyang, Z.; Zhang, L.; Mathijs, E. Excessive synthetic fertilizers elevate greenhouse gas emissions of smallholder-scale staple grain production in China. *J. Clean. Prod.* **2023**, *430*, 139720. [CrossRef]
17. Chen, J.; Abou-Elwafa, S.F.; Huang, M. Dynamic changes in the fossil energy use efficiency in crop production: A case study from Hunan province of China. *J. Clean. Prod.* **2022**, *371*, 133627. [CrossRef]
18. Amirahmadi, E.; Moudrý, J.; Konvalina, P.; Hörtenhuber, S.J.; Ghorbani, M.; Neugschwandtner, R.W.; Jiang, Z.; Krexner, T.; Kopecký, M. Environmental Life Cycle Assessment in Organic and Conventional Rice Farming Systems: Using a Cradle to Farm Gate Approach. *Sustainability* **2022**, *14*, 15870. [CrossRef]
19. Ghorbani, M.; Neugschwandtner, R.W.; Konvalina, P.; Asadi, H.; Kopecký, M.; Amirahmadi, E. Comparative effects of biochar and compost applications on water holding capacity and crop yield of rice under evaporation stress: A two-years field study. *Paddy Water Environ.* **2023**, *21*, 47–58. [CrossRef]
20. Duan, J.; Liu, H.; Zhang, X.; Ren, C.; Wang, C.; Cheng, L.; Xu, J.; Gu, B. Agricultural management practices in China enhance nitrogen sustainability and benefit human health. *Nat. Food* **2024**, *5*, 378–389. [CrossRef] [PubMed]
21. Nkebiwe, P.M.; Weinmann, M.; Bar-Tal, A.; Müller, T. Fertilizer placement to improve crop nutrient acquisition and yield: A review and meta-analysis. *Field Crop. Res.* **2016**, *196*, 389–401. [CrossRef]
22. Dong, B.; Mao, Z.; Cui, Y.; Luo, Y.; Li, Y. Controlled Irrigation for Paddy Rice in China. *Irrig. Drain.* **2020**, *69*, 61–74. [CrossRef]
23. Liu, S.W.; Zhang, L.; Jiang, J.Y.; Chen, N.N.; Yang, X.M.; Xiong, Z.Q.; Zou, J.W. Methane and nitrous oxide emissions from rice seedling nurseries under flooding and moist irrigation regimes in Southeast China. *Sci. Total Environ.* **2012**, *426*, 166–171. [CrossRef] [PubMed]
24. Hu, Y.M.; Wang, L.; Chen, F.X.; Ren, X.N.; Tan, Z.X. Soil carbon sequestration efficiency under continuous paddy rice cultivation and excessive nitrogen fertilization in South China. *Soil Tillage Res.* **2021**, *213*, 105108. [CrossRef]
25. Bhaduri, D.; Purakayastha, T.J.; Chakraborty, D.; Chakraborty, K.; Singh, M. Integrated tillage-water-nutrient management effects on selected soil physical properties in a rice-wheat system in the Indian subcontinent. *Arch. Agron. Soil Sci.* **2018**, *64*, 132–145. [CrossRef]
26. Sun, Y.; Ma, J.; Sun, Y.; Xu, H.; Yang, Z.; Liu, S.; Jia, X.; Zheng, H. The effects of different water and nitrogen managements on yield and nitrogen use efficiency in hybrid rice of China. *Field Crop. Res.* **2012**, *127*, 85–98. [CrossRef]
27. Sun, Y.J.; Sun, Y.Y.; Li, X.Y.; Zhang, R.P.; Guo, X.; Ma, J. Effects of water-nitrogen interaction on absorption, translocation and distribution of nitrogen, phosphorus, and potassium in rice. *Acta Agron. Sin.* **2010**, *36*, 655–664. [CrossRef]
28. Sun, Y.J.; Sun, Y.Y.; Li, X.Y.; Guo, X.; Ma, J. Relationship of nitrogen utilization and activities of key enzymes involved in nitrogen metabolism in rice under water–nitrogen interaction. *Acta Agron. Sin.* **2009**, *35*, 2055–2063. [CrossRef]
29. Su, Q.; Cang, B.; Ullah, R.; Song, Z.; Li, Y.; Cui, J.; Wei, X.; Yang, M.; Yang, X.; Wu, Z. Interaction of the coupled effects of irrigation mode and nitrogen fertilizer timing on rice yield in different regions. *Irrig. Drain.* **2023**, *72*, 999–1012. [CrossRef]
30. Yang, J.; Liu, K.; Wang, Z.; Du, Y.; Zhang, J. Water-saving and high-yielding irrigation for lowland rice by controlling limiting values of soil water potential. *J. Integr. Plant Biol.* **2007**, *49*, 1445–1454. [CrossRef]
31. Hameed, F.; Xu, J.Z.; Rahim, S.F.; Wei, Q.; Khalil, A.U.R.; Liao, Q. Optimizing Nitrogen Options for Improving Nitrogen Use Efficiency of Rice under Different Water Regimes. *Agronomy* **2019**, *9*, 39. [CrossRef]
32. Jin, Z.; Yue, R.; Ma, Z.; Cheng, S.; Khan, M.N.; Nie, L. Effect of water and nitrogen coupling on energy balance and production efficiency in rice production. *Energy* **2024**, *288*, 129739. [CrossRef]
33. Song, T.; Das, D.; Hu, Q.; Yang, F.; Zhang, J. Alternate wetting and drying irrigation and phosphorus rates affect grain yield and quality and heavy metal accumulation in rice. *Sci. Total Environ.* **2021**, *752*, 141862. [CrossRef] [PubMed]

34. Cabangon, R.J.; Castillo, E.G.; Tuong, T.P. Chlorophyll meter-based nitrogen management of rice grown under alternate wetting and drying irrigation. *Field Crop. Res.* **2011**, *121*, 136–146. [CrossRef]
35. Qiu, H.; Yang, S.; Jiang, Z.; Xu, Y.; Jiao, X. Effect of Irrigation and Fertilizer Management on Rice Yield and Nitrogen Loss: A Meta-Analysis. *Plants* **2022**, *11*, 1690. [CrossRef] [PubMed]
36. Yan, J.; Wu, Q.; Qi, D.; Zhu, J. Rice yield, water productivity, and nitrogen use efficiency responses to nitrogen management strategies under supplementary irrigation for rain-fed rice cultivation. *Agric. Water Manag.* **2022**, *263*, 107486. [CrossRef]
37. Sudhir, Y.; Evangelista, G.; Fardnilo, J.; Humphreys, E.; Henry, A.; Fernandez, L. Establishment method effects on crop performance and water productivity of irrigated rice in the tropics. *Field Crops Res.* **2014**, *166*, 112–127. [CrossRef]
38. Wu, M.Y.; Cao, X.C.; Guo, X.P.; Xiao, J.F.; Ren, J. Assessment of grey water footprint in paddy rice cultivation: Effects of field water management policies. *J. Clean. Prod.* **2021**, *313*, 127876. [CrossRef]
39. Xie, Y.; Wang, Z.; Cheng, X.; Qiu, R.; Hamoud, Y.A.; Hong, C.; Zong, X.; Wang, Y.; Agathokleous, E.; Guo, X. Dissecting the combined effects of cultivar, fertilization, and irrigation on rhizosphere bacterial communities and nitrogen productivity in rice. *Sci. Total Environ.* **2022**, *835*, 155534. [CrossRef] [PubMed]
40. Zhang, S.X.; Rasool, G.; Guo, X.P.; Sen, L.; Cao, K.W. Effects of Different Irrigation Methods on Environmental Factors, Rice Production, and Water Use Efficiency. *Water* **2020**, *12*, 2239. [CrossRef]
41. Zhao, Z.; Xia, L.; Qin, Z.; Cao, J.; Mohammed, A.A.O.; Toland, H. The environmental fate of phenanthrene in paddy field system and microbial responses in rhizosphere interface: Effect of water-saving patterns. *Chemosphere* **2021**, *269*, 128774. [CrossRef] [PubMed]
42. Cao, X.-C.; Wu, L.-L.; Lu, R.-H.; Zhu, L.-F.; Zhang, J.-H.; Jin, Q.-Y. Irrigation and fertilization management to optimize rice yield, water productivity and nitrogen recovery efficiency. *Irrig. Sci.* **2021**, *39*, 235–249. [CrossRef]
43. Xu, J.; Peng, S.; Yang, S.; Wang, W. Ammonia volatilization losses from a rice paddy with different irrigation and nitrogen managements. *Agric. Water Manag.* **2012**, *104*, 184–192. [CrossRef]
44. Zhao, L.; Wu, L.; Wu, M.; Li, Y. Nutrient uptake and water use efficiency as affected by modified rice cultivation methods with reduced irrigation. *Paddy Water Environ.* **2011**, *9*, 25–32. [CrossRef]
45. Kapoor, V.; Singh, U.; Patil, S.K.; Magre, H.; Shrivastava, L.K.; Mishra, V.N.; Das, R.O.; Samadhiya, V.K.; Sanabria, J.; Diamond, R. Rice growth, grain yield, and floodwater nutrient dynamics as affected by nutrient placement method and rate. *Agron. J.* **2008**, *100*, 526–536. [CrossRef]
46. Huang, M.; Tao, Z.; Lei, T.; Cao, F.; Chen, J.; Yin, X.; Zou, Y.; Liang, T. Improving lodging resistance while maintaining high grain yield by promoting pre-heading growth in rice. *Field Crop. Res.* **2021**, *270*, 108212. [CrossRef]
47. Meng, T.; Zhang, X.; Ge, J.; Chen, X.; Zhu, G.; Chen, Y.; Zhou, G.; Wei, H.; Dai, Q. Improvements in grain yield and nutrient utilization efficiency of japonica inbred rice released since the 1980s in eastern China. *Field Crop. Res.* **2022**, *277*, 108427. [CrossRef]

plants

Article

# Exploring the Water–Soil–Crop Dynamic Process and Water Use Efficiency of Typical Irrigation Units in the Agro-Pastoral Ecotone of Northern China

**Guoshuai Wang [1,2,†], Xiangyang Miao [1,2,3,*,†], Bing Xu [1,2,3,*], Delong Tian [1,2], Jie Ren [1,2], Zekun Li [1,2], Ruiping Li [3], Hexiang Zheng [1,2,3], Jun Wang [1,2], Pengcheng Tang [1,2], Yayang Feng [4], Jie Zhou [1,2,3] and Zhiwei Xu [5]**

1 Yinshanbeilu Grassland Eco-Hydrology National Observation and Research Station, China Institute of Water Resources and Hydropower Research, Beijing 100038, China; imau_wgs@163.com (G.W.); renj@iwhr.com (J.R.); tangpc1988@163.com (P.T.)
2 Institute of Water Resources for Pastoral Area, Ministry of Water Resources, Hohhot 010020, China
3 College of Water Conservancy and Civil Engineering, Inner Mongolia Agricultural University, Hohhot 010018, China; nmglrp@imau.edu.cn
4 Farmland Irrigation Research Institute, Chinese Academy of Agricultural Sciences (CAAS), Xinxiang 453002, China
5 Agriculture, Animal Husbandry and Water Resources Bureau of Saihan District, Hohhot 010018, China; xly001225@163.com
* Correspondence: miaoxiangyang1108@163.com (X.M.); nmxubing@163.com (B.X.)
† These authors contributed equally to this work.

**Abstract:** Groundwater resources serve as the primary source of water in the agro-pastoral ecotone of northern China, where scarcity of water resources constrains the development of agriculture and animal husbandry. As a typical rainfed agricultural area, the agro-pastoral ecotone in Inner Mongolia is entirely dependent on groundwater for agricultural irrigation. Due to the substantial groundwater consumption of irrigated farmland, groundwater levels have been progressively declining. To obtain a sustainable irrigation pattern that significantly conserves water, this study faces the challenge of unclear water transport relationships among water, soil, and crops, undefined water cycle mechanism in typical irrigation units, and water use efficiency, which was not assessed. Therefore, this paper, based on in situ experimental observations and daily meteorological data in 2022–2023, utilized the DSSAT model to explore the growth processes of *potato*, *oat*, *alfalfa*, and *sunflower*, the soil water dynamics, the water balance, and water use efficiency, analyzed over a typical irrigation area. The results indicated that the simulation accuracy of the DSSAT model was ARE < 10%, nRMSE/% < 10%, and $R^2 \geq 0.85$. The consumption of the soil moisture during the rapid growth stage for the *potatoes*, *oats*, *alfalfa*, and *sunflower* was 7–13% more than that during the other periods, and the yield was 67,170, 3345, 6529, and 4020 kg/ha, respectively. The soil evaporation of oat, potato, alfalfa, and sunflower accounted for 18–22%, 78–82%; 57–68%, and 32–43%, and transpiration accounted for 40–44%, 56–60%, 45–47%, and 53–55% of ETa (333.8 mm–369.2 mm, 375.2 mm–414.2 mm, 415.7 mm–453.7 mm, and 355.0 mm–385.6 mm), respectively. It was advised that irrigation water could be appropriately reduced to decrease ineffective water consumption. The water use efficiency and irrigation water use efficiency for *potatoes* was at the maximum amount, ranging from 16.22 to 16.62 kg/m$^3$ and 8.61 to 10.81 kg/m$^3$, respectively, followed by alfalfa, sunflowers, and oats. For the perspective of water productivity, it was recommended that *potatoes* could be extensively cultivated, alfalfa planted appropriately, and oats and sunflowers planted less. The findings of this study provided a theoretical basis for efficient water resource use in the agro-pastoral ecotone of Northern China.

**Keywords:** agro-pastoral ecotone of Northern China; the different forage crops; DSSAT model; soil water dynamics; crop growth processes; crop evapotranspiration; water productivity

## 1. Introduction

The Yinshanbeilu region, which stretches across the central part of Inner Mongolia, represents a typical agro-pastoral ecotone in the north of China. This area serves dual functions in production and ecology; however, agriculture and animal husbandry are constrained by insufficient water resources. The area generally supports the cultivation of a short-growing period, cold- and drought-resistant crops such as *oats*, *potatoes*, and *sunflowers*, and forage crops [1]. Soil erosion is severe in this region and the long-term cultivation in grasslands and improper planting practices have led to coarsening of the soil texture [2,3]; therefore, the soil water retention and fertilizer capacity have drastically declined [4,5]. The local agricultural development is facing serious threats. The region experiences a dry climate. With the increase in crop varieties and irrigation, the over-extraction of groundwater has become serious, which has led to water scarcity becoming a primary limiting factor for agricultural and pastoral production [6,7]. In the agro-pastoral ecotone of northern China, agricultural irrigation relies entirely on groundwater. The higher water consumption characteristics of crops, such as *potatoes*, *sunflowers*, and forage crops [8–10], mean their cultivation is constrained by local climate and water resources. Establishing rational planting patterns [11] and optimizing irrigation systems [12] can fully utilize space, improve soil structure, and increase water productivity and water use efficiency; however, the region currently faces challenges, including the presence of unclear relationships in water transport among soil, crops, and irrigation, and undefined water cycle processes and water use efficiency. Thus, it is important to explore the dynamic processes of water–soil–crops and subsequently improve water use efficiency and yield. Consequently, numerous scholars have researched the water resource utilization and crop yield of the agro-pastoral ecotone. Wang et al. [13], based on economic benefits, tuber quality, and soil environmental benefits, applied the Analytic Hierarchy Process and Fuzzy Comprehensive Evaluation Method to optimize irrigation and fertilization and found that the optimal pattern for irrigation is 100% $ET_C$ and for fertilization (N-P-K) is 200-34.9-248.9 kg/ha. To increase oat yields, Ma et al. [14], based on the APSIM model, studied the yield and water–nitrogen response under rainfed and irrigated conditions and found that it was recommended to apply nitrogen at 90 kg/ha for rainfed conditions and apply nitrogen at 120 kg/ha with an irrigation quota of 120 mm for irrigation conditions. Liu et al. [15] studied the relationship between water use efficiency and alfalfa yield in different irrigation methods (Flood irrigation (Figures S1 and S2), FI, and subsurface drip irrigation, SDI) and found that water use efficiency in alfalfa production for SDI was 3.0–17.8 kg/(ha·mm) higher than that for FI. Zhang et al. [16] used the Two-Factor Design Method to study the optimal planting methods and conservation tillage pattern for sunflower in arid regions. Extensive research had been conducted on the crop physiology, irrigation methods, and irrigation scheduling of *potatoes*, *oats*, *alfalfa*, and *sunflowers*. However, currently, research is mainly focused on study areas with flat terrains, long growing seasons, and higher accumulated temperatures. In contrast, there is relatively less research on crops in agro-pastoral ecotones with shorter growing periods and lower accumulated temperatures, and few studies have researched the dynamic processes of water–soil–crops, water balance or water use efficiency in agro-pastoral ecotones.

Crop growth models can quantitatively demonstrate the process of crop growth, dynamics of crop yield, and the impact of the environment on these factors [17]. These multifunctional, spatial, digital, and visual models enable the more accurate prediction of crop growth patterns, potential, and climate effects [18,19]. The Decision Support System for Agrotechnology Transfer (DSSAT) is one of the most widely used crop growth models worldwide [20,21]. Miao et al. [22] employed the DSSAT model to study the growth process of perennial legume alfalfa, revealing the feedback relationship between alfalfa yield and soil moisture. Furthermore, they identified the optimal soil moisture levels for alfalfa regreening as well as the best irrigation regime for this plant. Zhou et al. [23] integrated remote sensing into the DSSAT model to study surface and root-zone soil moisture, assessing the impact of data assimilation on agricultural drought monitoring.

Their research showed that soil moisture estimates could be improved significantly by data assimilation, enhancing drought monitoring by at least one drought level as compared to non-assimilated data. Based on the small-hole effect, Shen et al. [24] developed an estimation method for soil evaporation under plastic film mulch and integrated it into the DSSAT model, addressing the lack of a soil evaporation module for plastic film mulch in DSSAT. Meanwhile, Ma et al. [25]—aiming to rapidly evaluate crop genetic parameters for model applications and improve optimization efficiency and accuracy—coupled the Parameter Estimation (PEST) tool (an independent automatic parameter optimization tool) with the DSSAT crop growth model using the R programming language and established the DSSAT-PEST optimization method. Further, Wang et al. [26] assessed various drip irrigation and fertilization schemes for *potato* farming using the DSSAT-SUBSTOR–potato model. Accordingly, they recommended an optimal irrigation level of 100% ETc, with a fertilization rate ($N\text{-}P_2O_5\text{-}K_2O$) of 200-80-300 kg/ha, for *potato* farming in the northwest region. Together, these studies demonstrate that the DSSAT model can effectively evaluate agricultural production and help in enhancing the yield of various crops, but the effective use of the model depends on the results of the validation in each zone and with the crops of that zone. However, current research on typical crop irrigation methods and planting patterns in agro-pastoral zones mainly relies on field experiments, and the application of crop growth models in this context has thus far been limited. However, field irrigation experiments are influenced by many factors. Therefore, coupling field trials with DSSAT models can reduce labor and resource utilization, thus improving efficiency. Nevertheless, the success of DSSAT model-based simulations is highly dependent on the appropriate selection of crop genetic parameters, field management parameters, and soil parameters. Hence, it is necessary to optimize the crop genetic parameters for common plants such as *potatoes*, oats, sunflowers, and alfalfa in agro-pastoral zones and propose DSSAT model parameters suitable for irrigated farmlands in these zones.

Therefore, this study utilizes the DSSAT model to investigate the following: (1) soil water dynamics and growth processes for *potatoes*, *oats*, *alfalfa* and *sunflowers*; (2) water balance analysis; and (3) water use efficiency evaluation. This research aimed to provide a theoretical basis for the efficient utilization of agricultural water resources, scientific enhancement of crop yields, and rational optimization of cropping pattern in the agro-pastoral ecotone of northern China.

## 2. Materials and Methods

### 2.1. Study Area Overview

The study area was located in Wuchuan County (Yinshanbeilu agro-pastoral ecotone in Inner Mongolia: 41°08.344′ N, 111°17.580′ E; Altitude: 1648.5 m). Wuchuan County has a total cultivated land area of 144,667 ha, including 28,667 ha of irrigated land. The agricultural irrigation water was groundwater. Due to groundwater over-extraction, the groundwater table had been declining at a rate of 27.50 cm/a to 47.50 cm/a, with the depth of wells increasing by 50 to 120 m over the past two decades. The climate of the study area is characterized by a temperate continental monsoon climate with an average annual temperature of 2.5 °C, annual precipitation ranging from 100 to 200 mm, and a frost-free period of 90 to 120 days. The bulk density of soil was from 1.5 to 1.7 g/cm$^3$, and the soil texture is sandy loam. The main crops were *potatoes*, *oats*, *sunflowers*, and *alfalfa*. Detailed results are presented in Figure 1.

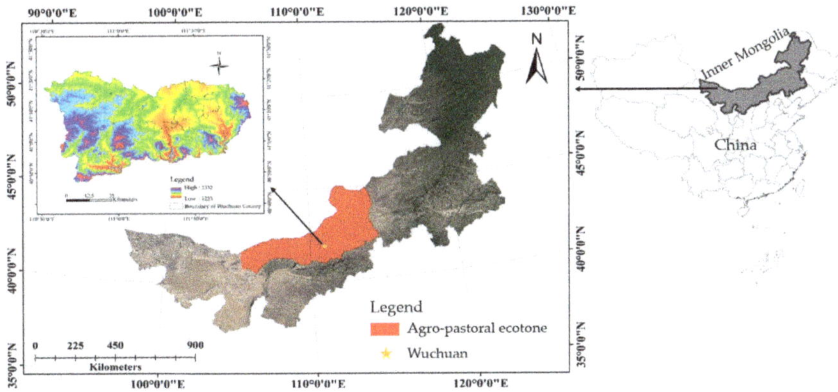


**Figure 1.** Overview of the study area.

## 2.2. *Experimental Design*

For this study, we selected a typical irrigation unit within the research area, where we established one groundwater irrigation well. This well serviced a total irrigation area of 35.3 $hm^2$, and the area of cultivation encompassed *potatoes* (15.2 $hm^2$), *oats* (10.8 $hm^2$), *alfalfa* (4.9 $hm^2$), and *sunflowers* (4.4 $hm^2$). The research site featured a land gradient, with the elevation decreasing by 39 m from the east to the west. We strategically positioned 68 soil sampling points across the research area, with a 100 m interval maintained between each point (Figure 2). Monitoring was performed from 1 May to 30 September, and soil samples were collected on the 1st and 15th of each month in 2022 and 2023. The sampling frequency was increased during the irrigation season. Taking into account crop traits and root development (Figure S3), soil samples were obtained from a depth of 100 cm, and growth indicators such as plant height and leaf area were measured for the *potato*, *oat*, *alfalfa*, and *sunflower* plants. The research area's initial planting, field management, and irrigation strategies are based on the real local circumstances. Drip irrigation was the chosen method for irrigation within the research area. We meticulously recorded data for each crop, including its growth stages, irrigation schedules, frequency of irrigation, and the allotted irrigation quota.

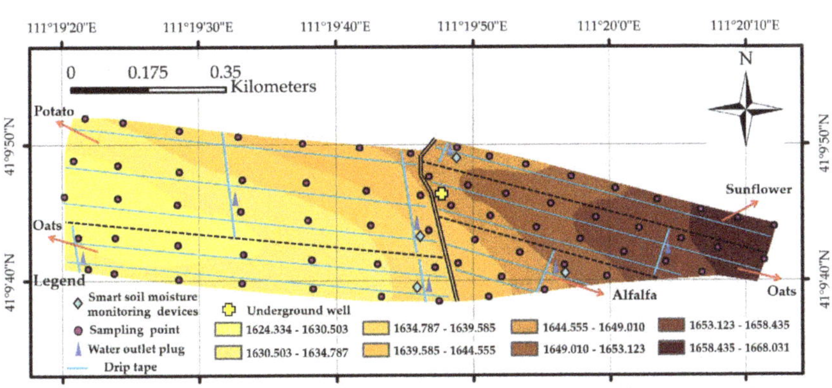


**Figure 2.** Experimental design diagram.

### 2.3. *Test Indicators and Methods*

#### 2.3.1. Physical Properties of Soil in the Study Area

We installed smart soil moisture monitoring devices to track soil moisture dynamics for each crop. These sensors were embedded at depths of 10, 20, 30, 40, 50, 60, 70, 80, 90, and 100 cm, and the data collection interval was set to 1 h. Additionally, we used the oven-drying method to determine the soil moisture content across each crop growth stage (there are four phases of growth: early, fast, medium, and late), as well as before and after irrigation to calibrate the smart soil moisture monitoring device data. As a layer, samples are obtained every 20 cm. Soil samples were manually collected from depths of 0–20 cm, 20–40 cm, 40–60 cm, 60–80 cm, and 80–100 cm, with three samples obtained from each layer. The core cutter method was employed to measure soil bulk density, field capacity, and saturated moisture content. We also analyzed the soil samples to determine the levels of available phosphorus, potassium, and organic matter content, thereby determining the fundamental physical parameters of the soil layers within the experimental area. Detailed results are presented in Tables 1–4.

**Table 1.** Basic physical and chemical parameters of *potato* soil.

| Soil Layer Depth (cm) | Bulk Density ($g/cm^3$) | | Field Moisture Capacity ($cm^3/cm^3$) | | Saturated Moisture ($cm^3/cm^3$) | | Available P ($mg \cdot kg^{-1}$) | | Available K ($mg \cdot kg^{-1}$) | | Soil Organic Matter ($g \cdot kg^{-1}$) | | pH | |
|---|---|---|---|---|---|---|---|---|---|---|---|---|---|---|
| | 2022 | 2023 | 2022 | 2023 | 2022 | 2023 | 2022 | 2023 | 2022 | 2023 | 2022 | 2023 | 2022 | 2023 |
| 0–20 | 1.34 | 1.36 | 0.32 | 0.30 | 0.35 | 0.35 | 11.59 | 12.08 | 163.88 | 158.72 | 2.17 | 2.31 | 7.29 | 7.16 |
| 20–40 | 1.57 | 1.51 | 0.25 | 0.24 | 0.29 | 0.26 | 4.56 | 5.34 | 132.75 | 128.31 | 5.61 | 5.74 | 7.11 | 6.97 |
| 40–60 | 1.58 | 1.57 | 0.23 | 0.18 | 0.23 | 0.20 | 3.09 | 2.98 | 74.56 | 71.78 | 1.49 | 1.73 | 7.08 | 7.03 |
| 60–80 | 1.62 | 1.61 | 0.17 | 0.21 | 0.21 | 0.21 | 1.84 | 1.85 | 66.37 | 67.23 | 0.69 | 0.72 | 6.96 | 6.92 |
| 80–100 | 1.65 | 1.63 | 0.13 | 0.14 | 0.17 | 0.16 | 1.57 | 1.69 | 92.32 | 88.75 | 0.52 | 0.48 | 6.93 | 6.82 |

**Table 2.** Basic physical and chemical parameters of *oat* soil.

| Soil Layer Depth (cm) | Bulk Density ($g/cm^3$) | | Field Moisture Capacity ($cm^3/cm^3$) | | Saturated Moisture ($cm^3/cm^3$) | | Available P ($mg \cdot kg^{-1}$) | | Available K ($mg \cdot kg^{-1}$) | | Soil Organic Matter ($g \cdot kg^{-1}$) | | pH | |
|---|---|---|---|---|---|---|---|---|---|---|---|---|---|---|
| | 2022 | 2023 | 2022 | 2023 | 2022 | 2023 | 2022 | 2023 | 2022 | 2023 | 2022 | 2023 | 2022 | 2023 |
| 0–20 | 1.37 | 1.39 | 0.29 | 0.28 | 0.34 | 0.33 | 11.13 | 11.60 | 155.69 | 150.78 | 2.20 | 2.34 | 7.10 | 7.01 |
| 20–40 | 1.54 | 1.54 | 0.23 | 0.22 | 0.27 | 0.24 | 4.38 | 5.13 | 126.11 | 121.89 | 5.68 | 5.81 | 6.95 | 6.79 |
| 40–60 | 1.57 | 1.59 | 0.21 | 0.17 | 0.22 | 0.19 | 2.97 | 2.86 | 70.83 | 68.19 | 1.51 | 1.75 | 6.91 | 6.81 |
| 60–80 | 1.62 | 1.63 | 0.16 | 0.19 | 0.20 | 0.20 | 1.77 | 1.78 | 63.05 | 63.87 | 0.70 | 0.73 | 6.77 | 6.73 |
| 80–100 | 1.64 | 1.65 | 0.12 | 0.13 | 0.16 | 0.15 | 1.51 | 1.62 | 87.70 | 84.31 | 0.53 | 0.49 | 6.74 | 6.63 |

**Table 3.** Basic physical and chemical parameters of *sunflower* soil.

| Soil Layer Depth (cm) | Bulk Density ($g/cm^3$) | | Field Moisture Capacity ($cm^3/cm^3$) | | Saturated Moisture ($cm^3/cm^3$) | | Available P ($mg \cdot kg^{-1}$) | | Available K ($mg \cdot kg^{-1}$) | | Soil Organic Matter ($g \cdot kg^{-1}$) | | pH | |
|---|---|---|---|---|---|---|---|---|---|---|---|---|---|---|
| | 2022 | 2023 | 2022 | 2023 | 2022 | 2023 | 2022 | 2023 | 2022 | 2023 | 2022 | 2023 | 2022 | 2023 |
| 0–20 | 1.38 | 1.33 | 0.28 | 0.27 | 0.33 | 0.31 | 10.68 | 11.13 | 155.69 | 150.78 | 2.08 | 2.14 | 7.13 | 7.04 |
| 20–40 | 1.58 | 1.56 | 0.22 | 0.21 | 0.25 | 0.22 | 4.20 | 4.92 | 126.11 | 121.89 | 5.27 | 5.40 | 7.02 | 6.86 |
| 40–60 | 1.54 | 1.56 | 0.20 | 0.16 | 0.21 | 0.17 | 2.85 | 2.75 | 70.83 | 68.19 | 1.40 | 1.63 | 6.96 | 6.87 |
| 60–80 | 1.64 | 1.62 | 0.15 | 0.18 | 0.19 | 0.19 | 1.70 | 1.70 | 63.05 | 63.87 | 0.65 | 0.68 | 6.84 | 6.79 |
| 80–100 | 1.65 | 1.64 | 0.11 | 0.12 | 0.15 | 0.14 | 1.45 | 1.56 | 87.70 | 84.31 | 0.48 | 0.42 | 6.83 | 6.71 |

Table 4. Basic physical and chemical parameters of *alfalfa* soil.

| Soil Layer Depth (cm) | Bulk Density (g/cm$^3$) | | Field Moisture Capacity (cm$^3$/cm$^3$) | | Saturated Moisture (cm$^3$/cm$^3$) | | Available P (mg·kg$^{-1}$) | | Available K (mg·kg$^{-1}$) | | Soil Organic Matter (g·kg$^{-1}$) | | pH | |
|---|---|---|---|---|---|---|---|---|---|---|---|---|---|---|
| | 2022 | 2023 | 2022 | 2023 | 2022 | 2023 | 2022 | 2023 | 2022 | 2023 | 2022 | 2023 | 2022 | 2023 |
| 0–20 | 1.36 | 1.31 | 0.29 | 0.28 | 0.31 | 0.32 | 11.16 | 11.63 | 159.95 | 154.91 | 2.19 | 2.34 | 7.21 | 7.02 |
| 20–40 | 1.57 | 1.54 | 0.23 | 0.22 | 0.27 | 0.25 | 4.39 | 5.14 | 129.56 | 125.23 | 5.54 | 5.68 | 7.03 | 6.84 |
| 40–60 | 1.59 | 1.55 | 0.21 | 0.17 | 0.22 | 0.18 | 2.98 | 2.87 | 72.77 | 70.06 | 1.42 | 1.64 | 6.96 | 6.88 |
| 60–80 | 1.63 | 1.61 | 0.16 | 0.19 | 0.20 | 0.20 | 1.77 | 1.78 | 64.78 | 65.62 | 0.67 | 0.71 | 6.86 | 6.76 |
| 80–100 | 1.65 | 1.63 | 0.12 | 0.13 | 0.16 | 0.15 | 1.51 | 1.63 | 90.10 | 86.62 | 0.49 | 0.45 | 6.81 | 6.76 |

### 2.3.2. Meteorological Data Collection

The meteorological data required for the study were obtained from an automatic weather station located within the experimental area (HOBO-U30, Onset Computer Corp., Bourne, MA, USA). The temperature and amount of rainfall during the crop growth period are shown in Figure 3. The amount of rainfall for the crop growth period in 2022 and 2023 was 213.6 mm and 257.8 mm, with the daily average temperature ranges being 0.4–29.8 °C and 1.3–26.9 °C, respectively.

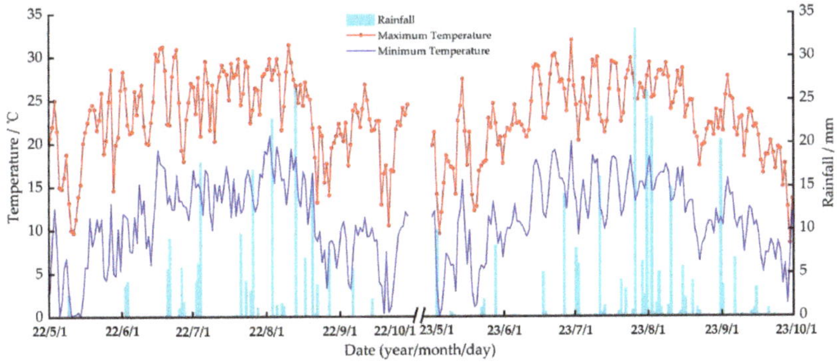


Figure 3. Temperature and rainfall during the growth period.

### 2.3.3. Crop Growth Periods

Data for the different crops were recorded from the time of seeding. We obtained seeding dates, emergence time, plant heights, leaf areas, and harvest times (Table 5).

Table 5. Growth stages of different crops.

| Crop Varieties | Growth Period | 2022 Start and End Date | 2022 Fertility Days | 2023 Start and End Date | 2023 Fertility Days |
|---|---|---|---|---|---|
| *Oat* | Sowing–emergence | 19 May~28 May | 10 d | 21 May~2 Jun. | 12 d |
| | Emergence–jointing | 29 May~12 Jul. | 45 d | 3 Jun.~14 Jul. | 41 d |
| | Jointing–grouting | 13 Jul.~6 Aug. | 25 d | 15 Jul.~8 Aug. | 25 d |
| | Grout–mature | 7 Aug.~16 Sep. | 40 d | 9 Aug.~19 Sep. | 41 d |
| | Total | 19 May~16 Sep. | 120 d | 21 May~19 Sep. | 119 d |
| *Potato* | Sowing–emergence | 2 May~27 May | 25 d | 1 May~28 May | 29 d |
| | Emergence–formation | 28 May~26 Jun. | 30 d | 29 May~29 Jun. | 32 d |
| | Form–expand | 27 Jun.~15 Aug. | 50 d | 30 Jun.~19 Aug. | 51 d |
| | Expansion–harvest | 16 Aug.~11 Sep. | 25 d | 31 Aug.~20 Sep. | 21 d |
| | Total | 2 May~11 Sep. | 133 d | 1 May~13 Sep. | 136 d |

**Table 5.** *Cont.*

| Crop Varieties | Growth Period | 2022 | | 2023 | |
|---|---|---|---|---|---|
| | | **Start and End Date** | **Fertility Days** | **Start and End Date** | **Fertility Days** |
| *Sunflower* | Sowing–emergence | 29 May~3 Jul. | 36 d | 30 May~6 Jul. | 38 d |
| | Emergence–budding | 4 Jul.~25 Jul. | 21 d | 7 Jul.~27 Jul. | 20 d |
| | Budding–blooming | 26 Jul.~13 Aug. | 18 d | 28 Jul.~17 Aug. | 20 d |
| | Bloom–ripen | 14 Aug.~26 Sep. | 43 d | 18 Aug~29 Sep. | 41 d |
| | Total | 29 May~26 Sep. | 118 d | 30 May~28 Sep. | 119 d |
| *Alfalfa* | Greening–branching | 11 May~30 May | 19 d | 10 May~25 May | 15 d |
| | Branching–budding | 31 May~22 Jun. | 23 d | 26 May~18 Jun. | 24 d |
| | Budding–blooming | 23 Jun.~10 Jul. | 17 d | 19 Jun.~5 Jul. | 16 d |
| | Blossom–harvest | 11 Jul.~24 Jul. | 13 d | 6 Jul.~21 Jul. | 15 d |
| | Greening–branching | 25 Jul.~10 Aug. | 16 d | 22 Jul.~4 Aug. | 14 d |
| | Branching–budding | 11 Aug.~28 Aug. | 17 d | 5 Aug.~24 Aug. | 19 d |
| | Budding–blooming | 29 Aug.~9 Sep. | 12 d | 25 Aug.~6 Sep. | 12 d |
| | Blossom–harvest | 10 Sep.~20 Sep. | 10 d | 7 Sep.~18 Sep. | 11 d |
| | Total | 11 May~20 Sep. | 127 d | 15 May~18 Sep. | 126 d |

### 2.3.4. Crop Irrigation Regimes

The amount of rainfall during the crop growth period for *oats*, *potatoes*, *alfalfa*, and *sunflowers* was 191.7 mm, 185.3 mm, 213.6 mm, and 198.7 mm in 2022 and 241.1 mm, 236.0 mm, 257.8 mm, and 241.2 mm in 2023, respectively. The irrigation quotas for each growth period over the two years are presented in Table 6.

**Table 6.** Irrigation quotas for different crop growth stages.

| Crop Varieties | Growth Period | Irrigation Amount (mm) | |
|---|---|---|---|
| | | **2022** | **2023** |
| *Oat* | Sowing–emergence | 18 | 15 |
| | Emergence–jointing | 41 | 36 |
| | Jointing–grouting | 52 | 45 |
| | Grout–mature | 35 | 32 |
| | Total | 146 | 128 |
| *Potato* | Sowing–emergence | 32 | 31 |
| | Emergence–formation | 55 | 50 |
| | Form–expand | 66 | 57 |
| | Expansion–harvest | 20 | 18 |
| | Total | 173 | 156 |
| *Sunflower* | Sowing–emergence | 36 | 23 |
| | Emergence–budding | 30 | 35 |
| | Budding–blooming | 46 | 39 |
| | Bloom–ripen | 25 | 21 |
| | Total | 137 | 118 |
| *Alfalfa* | Greening–branching | 16 | 16 |
| | Branching–budding | 39 | 35 |
| | Budding–blooming | 19 | 16 |
| | Blossom–harvest | 18 | 15 |
| | Greening–branching | 19 | 17 |
| | Branching–budding | 38 | 33 |
| | Budding–blooming | 18 | 18 |
| | Blossom–harvest | 15 | 14.0 |
| | Total | 182 | 164 |

### 2.4. Crop Growth and Water Consumption Models

The DSSAT model—which is one of the most widely applied crop growth models globally and includes numerous modules such as CROPGRO, CERES, SUBSTOR, CANEGRO, and more—was employed. The model has been shown to simulate the dynamics of a wide variety of crops [27,28].

#### 2.4.1. Soil Water Movement Equation

Figure 4 shows the water transport process module, taking the *potato* as an example. Crop transpiration, soil evaporation, surface runoff, and soil profile drainage make up the outputs of the system, whereas rainfall and irrigation water are the inputs of the system.

**Figure 4.** Schematic diagram of water movement process in DSSAT model.

The formula used by the DSSAT model to calculate changes in soil water content [29] is as follows:

$$ET(EP + ES) = P + I - \Delta S - R - D \quad (1)$$

where $\Delta S$ represents the change in soil water content; $P$ is the rainfall; $I$ is the irrigation amount; $EP$ is the transpiration; $ES$ is the soil evaporation; $R$ is the surface runoff; and $D$ is the soil profile drainage.

The DSSAT model calculates water stress using the following formula [30]:

$$SWDF_1 = \frac{WS_P}{RWUEP \times EP_0} \quad (2)$$

$$SWDF_2 = \frac{WS_P}{EP_0} \quad (3)$$

where $SWDF_1$ is the first water stress factor in the model; $WSP$ denotes potential root water uptake (mm); $RWUEP$ is a species-specific characteristic parameter; $EP_0$ represents the crop water demand (mm); and $SWDF_2$ is the second water stress factor in the crop model.

#### 2.4.2. Crop Dry Matter Accumulation Equation

The DSSAT model's equation for dry matter accumulation is [31]

$$\Delta TOT = 0.758 \times \left(PARCE \times 10^{-6} \times IPAR \ - 0.004 \times TOT\right) \times SWDF \quad (4)$$

where $\Delta TOT$ is the daily increment of crop dry matter (t/ha); $PARCE$ is the photosynthetically active radiation conversion efficiency (g/MJ); $IPAR$ is the intercepted photosynthetically active radiation (MJ/ha); $TOT$ is the total dry matter (t/ha); and $SWDF$ is the water stress factor affecting dry matter accumulation.

The calculation formula for $PARCE$ in the $DSSAT$ model is

$$PARCE = PARCE_{max} \times \{1 \ - exp[-0.008 \times \ (T - 8)]\} \quad (5)$$

where $PARCE_{max}$ is the maximum value for photosynthetically active radiation conversion efficiency (g/MJ), and $T$ is the daily average temperature (°C).

### *2.5. Model Development, Calibration, and Validation*

#### 2.5.1. Simulation Unit Division

Based on field data (e.g., growth stages, leaf area index [LAI], and yield), the SUBSTOR, CERES, OILCROP, and FORAGES models within DSSAT were employed to simulate the growth process and yield of typical crops like *potatoes, oats, sunflowers*, and *alfalfa* in the agro-pastoral ecotone of the northern foothills of Yinshan. The simulation periods for *oats, potatoes, sunflowers*, and *alfalfa* in 2022–2023 were set as 19 May to 1 September, 8 May to 20 September, 29 May to 28 September, and 14 May to 30 September, respectively.

#### 2.5.2. Meteorological Data

The meteorological parameters required for the DSSAT model include solar radiation ($MJ \cdot m^{-2}$), maximum temperature (°C), minimum temperature (°C), and rainfall (mm). The field management parameters required by the model include crop variety, planting method, planting date, planting density and depth, irrigation quota and timing, and fertilization amount and timing, among others. For this model, solar radiation must be calculated based on the duration of sunshine recorded by the weather station, as follows [32]:

$$R_s = R_{\max}\left(a_s - b_s\frac{n}{N}\right) \tag{6}$$

where $R_s$ is the total solar radiation ($MJ/m^2$); $a_s$ is the clear sky solar radiation constant; $b_s$ is a proportionality factor, both of which are constants related to atmospheric conditions, usually taken as 0.25 and 0.5, respectively; $n$ denotes sunshine duration (h); $N$ is the maximum possible sunshine duration (h); and $R_{max}$ is the solar radiation on a clear day ($MJ/m^2$).

$$R_{\max} = 37.586 \times d \times (W_s + \sin\varnothing\sin\delta + \cos\varnothing\cos\delta\sin w_s) \tag{7}$$

where $R_{\max}$ is the solar radiation on a clear day ($MJ/m^2$), $d$ is the distance between the earth and the sun, $W_s$ is the sunset hour angle (°), $\delta$ is the declination of the equator (°), and $\varnothing$ is the latitude of the meteorological station (°).

$$N = \frac{24}{\pi}W_s \tag{8}$$

where $N$ represents the maximum possible duration of daylight (h), and $W_s$ is the sunset hour angle (°).

$$\delta = 0.4093 \times \sin\left(\frac{2\pi}{365} \times J - 1.045\right) \tag{9}$$

where $\delta$ is the declination of the equator (°), and $J$ is the Julian day, which is the day of the year, with 1 January equal to 1.

$$W_s = arccos(-\tan\varnothing\tan\delta) \tag{10}$$

where $Ws$ is the sunset hour angle (°), $\delta$ is the declination of the equator (°), and $\varnothing$ is the latitude of the meteorological station (°).

#### 2.5.3. Soil Parameters

The soil parameters required for the DSSAT model consist of the basic physicochemical properties of the soil at different depths, including field capacity, saturated moisture content, soil bulk density, soil organic carbon content, total nitrogen content, and pH value, as shown in Tables 1–4.

2.5.4. Model Calibration and Validation

In this study, the absolute relative error (*ARE*), root mean square error (*RMSE*), normalized root mean square error (*nRMSE*), and coefficient of determination ($R^2$) [33–35] were used to measure the relative difference between simulated values from the DSSAT model and actual measurements. Generally, the closer the *ARE* is to 0, the higher the model's accuracy. An *nRMSE* < 10% is considered excellent and 10–20% is good. The $R^2$ is used to test the fit between model simulations and actual measurements, with values closer to 1 indicating a higher degree of fit. The formula for the index calculation is as follows:

$$ARE = \frac{|S_i - M_i|}{M_i} \times 100\% \tag{11}$$

$$MSE = \sqrt{\frac{\sum_{i=1}^{n}(M_i - S_i)^2}{\text{n}}} \tag{12}$$

$$nRMSE = \frac{RMSE}{M} \times 100\% \tag{13}$$

$$R^2 = \frac{\left[\sum_{i=1}^{n}(M_i - M)(S_i - S)\right]^2}{\sum_{i=1}^{n}(M_i - M)^2 \sum_{i=1}^{n}(S_i - S)^2} \tag{14}$$

where *Mi* is the observed value, *Si* is the simulated value, *M* is the average of the measured values, *S* is the average of the simulated values, and n is the number of samples.

### 2.6. Crop Irrigation Evaluation Indices

This paper uses water use efficiency (WUE, kg/m$^3$) and irrigation water use efficiency (IWUE, kg/m$^3$) as two indices [36,37] to evaluate and reflect the relationship between crop yield and irrigation water use. The calculation formula [38] is

$$WUE = \frac{Y}{10 \times ET} \tag{15}$$

where *Y* is the crop yield (kg/ha), *ET* is the actual evapotranspiration during the growth stage (mm), and 10 is a numerical conversion factor.

To study the sensitivity of crops to irrigation water use, the *IWUE* was calculated as follows:

$$IWUE = \frac{Y - Y_a}{10 \times I} \tag{16}$$

where *Y* is the crop yield (kg/ha), $Y_a$ is the crop yield without irrigation (kg/ha), *I* is the amount of irrigation water supplied during the growth stage (mm), and 10 is a numerical conversion parameter.

## 3. Results

### 3.1. Model Calibration

In this study, calibration was conducted using experimental data collected in 2022, with *ARE*, *nRMSE*, and $R^2$ serving as the metrics. According to Table 7, the *ARE* for different crop leaf area indices ranged from 3.06% to 4.18%; the *nRMSE* values ranged between 5.32% and 6.46%; and the $R^2$ values ranged from 0.87 to 0.89. The *ARE* for yield ranged from 3.47% to 5.82%, with *nRMSE* values ranging between 4.51% and 5.88% and $R^2$ values ranging from 0.87 to 0.90. Table 8 shows that the accuracy for soil moisture content at 0–60 cm was $ARE \leq 7\%$, $nRMSE \leq 10\%$, and $R^2 \geq 0.85$. The calibration of crop genetic parameters is shown in Tables 9–12. The measured values were in good agreement with the simulated values.

**Table 7.** Evaluation of accuracy indexes for crop leaf area and crop yield.

| Crop Type | Statistical Indicators | LAI/($cm^2 \cdot cm^{-2}$) | Yield/(kg/ha) |
|---|---|---|---|
| *Oat* | ARE/% | 3.06 | 5.82 |
| | nRMSE/% | 6.46 | 5.25 |
| | $R^2$ | 0.87 | 0.88 |
| *Potato* | ARE/% | 4.18 | 4.89 |
| | nRMSE/% | 5.64 | 6.57 |
| | $R^2$ | 0.88 | 0.87 |
| *Alfalfa* | ARE/% | 3.82 | 4.27 |
| | nRMSE/% | 6.31 | 5.88 |
| | $R^2$ | 0.87 | 0.89 |
| *Sunflower* | ARE/% | 3.39 | 3.47 |
| | nRMSE/% | 5.32 | 4.51 |
| | $R^2$ | 0.89 | 0.90 |

**Table 8.** Evaluation of accuracy indexes for the soil water content.

| Crop Type | Soil Depth | ARE/% | nRMSE/% | $R^2$ |
|---|---|---|---|---|
| *Oat* | 0–20 | 4.53 | 8.41 | 0.86 |
| | 20–40 | 3.96 | 7.65 | 0.87 |
| | 40–60 | 5.39 | 9.73 | 0.85 |
| *Potato* | 0–20 | 5.27 | 7.87 | 0.87 |
| | 20–40 | 6.34 | 8.65 | 0.86 |
| | 40–60 | 4.88 | 6.82 | 0.88 |
| *Alfalfa* | 0–20 | 4.38 | 6.22 | 0.87 |
| | 20–40 | 4.94 | 5.87 | 0.87 |
| | 40–60 | 5.57 | 8.06 | 0.86 |
| *Sunflower* | 0–20 | 6.36 | 7.58 | 0.86 |
| | 20–40 | 3.31 | 6.34 | 0.88 |
| | 40–60 | 5.48 | 6.97 | 0.87 |

**Table 9.** Genetic parameters of potato varieties.

| Argument | Definition | Calibrated Value |
|---|---|---|
| *G2* | Leaf area expansion rate [$cm^2 \cdot (m^2 \cdot d)^{-1}$] | 1100 |
| *G3* | Potential tuber growth rate [$g \cdot (plant \cdot d)^{-1}$] | 23.3 |
| *PD* | Tuber growth stress coefficient (%) | 0.9 |
| *P2* | Photoperiod coefficient (%) | 0.5 |
| *TC* | Upper limit critical temperature for tubers to start growing | 20 |

**Table 10.** Genetic parameters of oat varieties.

| Argument | Definition | Calibrated Value |
|---|---|---|
| *PIV* | Number of days required for vernalization under optimal temperature conditions (d) | 20 |
| *PID* | Photoperiod coefficient (%) | 30.5 |
| *P5* | Accumulated temperature during grain filling stage (°C d) | 450 |
| *G1* | Number of grains per unit canopy mass of a single plant at flowering stage ($grain \cdot g^{-1}$) | 16 |
| *G2* | Standard grain weight under optimal conditions (mg) | 24 |
| *G3* | Standard dry mass of stem and spike per plant during the maturation stage under non-stress conditions (g) | 1.9 |
| *PHINT* | Accumulated temperature required to complete the growth of one leaf (leaf thermal time) (°C d) | 99 |

**Table 11.** Genetic parameters of alfalfa varieties.

| Argument | Definition | Calibrated Value |
|---|---|---|
| *CSDL* | Critical short day duration (h) | 10.5 |
| *PPSEN* | Relative response slope to photoperiod (1/h) | 0.2 |
| *EM-FL* | Duration of light and heat from seedling emergence to first blossom appearance (d) | 21.5 |
| *FL-SH* | From the initial inflorescence blossoming to the first inflorescence fruit setting, light and heat conditions (d) | 6.7 |
| *FL-SD* | The light and heat time from the first inflorescence blooming to the first inflorescence grain production (d) | 12.6 |
| *SD-PM* | Photothermal duration from seed production to the first inflorescence's physiological ripening (d) | 33.5 |
| *FL-LF* | The photothermal time between the flowering of the first inflorescence and the cessation of leaf expansion (d) | 16 |
| *LFMAX* | Maximum photosynthetic rate of leaves (mg $CO_2/m^2 \cdot s^{-1}$) | 2.5 |
| *SLAVR* | Specific leaf area ($cm^2/g$) | 290 |
| *SIZLF* | Maximum blade size ($cm^2$) | 5 |

**Table 12.** Genetic parameters of sunflower varieties.

| Argument | Definition | Calibrated Value |
|---|---|---|
| *CSDL* | Critical short day duration (h) | 15 |
| *PPSEN* | Relative response slope to photoperiod (1/h) | −0.086 |
| *EM-FL* | Duration of light and heat from seedling emergence to first blossom appearance (d) | 22.6 |
| *FL-SH* | From the initial inflorescence blossoming to the first inflorescence fruit setting, light and heat conditions (d) | 7.2 |
| *FL-SD* | The light and heat time from the first inflorescence blooming to the first inflorescence grain production (d) | 12.5 |
| *SD-PM* | Photothermal duration from seed production to the first inflorescence's physiological ripening (d) | 32.5 |
| *FL-LF* | The photothermal time between the flowering of the first inflorescence and the cessation of leaf expansion (d) | 15 |
| *LFMAX* | Maximum photosynthetic rate of leaves (mg $CO_2/m^2 \cdot s^{-1}$) | 2.1 |
| *SLAVR* | Specific leaf area ($cm^2/g$) | 240 |
| *SIZLF* | Maximum blade size ($cm^2$) | 200 |
| *XFRT* | Maximum proportion of daily dry matter allocated to fruits (g) | 0.81 |
| *WTPSD* | Maximum weight per seed (g) | 0.1 |
| *SFDUR* | Maximum weight per seed (d) | 24 |
| *PODUR* | Optimal photothermal time required for final fruit load (d) | 4.5 |
| *THRSH* | Shattering percentage. Maximum ratio of seeds to (seeds + hulls) | 72.5 |
| *SDPRO* | Protein content in seeds (g (protein)/g (seed)) | 0.14 |
| *SDLIP* | Oil content in seeds (g (oil)/g (seed)) | 0.45 |

### 3.2. Model Validation

After calibration based on the 2022 data, the corrected genetic parameters for different crops (Tables 9–12) were added to the corresponding modules for validation. The model was then validated based on the actual data from 2023. The validation results, as shown in Figure 5, indicated that the *ARE* for soil moisture in different crops ranged from 4.23% to 6.04%. Meanwhile, the *nRMSE* values ranged from 4.97% to 5.74% and $R^2$ ranged from 0.901 to 0.912. The *LAI* in different crops was as follows: *ARE*, 2.85% to 5.37%; *nRMSE*, 4.01% to 5.62%; and $R^2$, 0.911 to 0.919. The yields were as follows: *ARE*, 3.44% to 5.09%; *nRMSE*, 4.29% to 6.05%; and $R^2$, 0.917 to 0.930. Hence, the model demonstrated good simulation accuracy, and the calibrated and validated models were deemed suitable for use in this study.

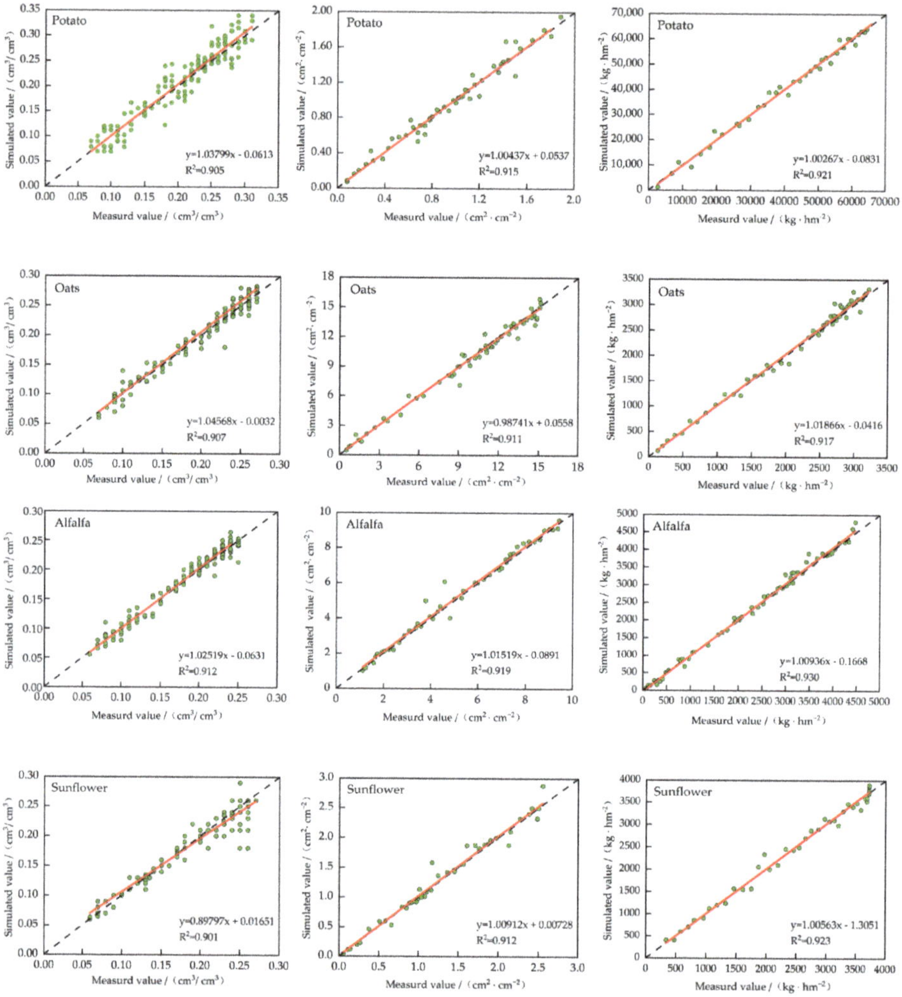

**Figure 5.** Verification of the soil moisture, leaf area index, and yield in different crops. Note: The green dots are the scatter of the simulated and measured values, and the red line is the fitted line.

*3.3. Soil Water Dynamics and Growth Processes of the Different Crops*

3.3.1. The Soil Water Content Dynamics for *Potatoes*, *Oats*, *Alfalfa*, and *Sunflowers*

Using the calibrated DSSAT model, we simulated and analyzed the impact of irrigation on soil moisture, crop growth, and yield. The changes in soil moisture content for *potatoes*, *oats*, *alfalfa*, and *sunflowers* are illustrated in Figures 6–9. The layer of cultivated soil in the study area was rather thin and had a sandy loam texture. Further, the presence of small pebbles increased with the soil depth, leading to poor water retention.

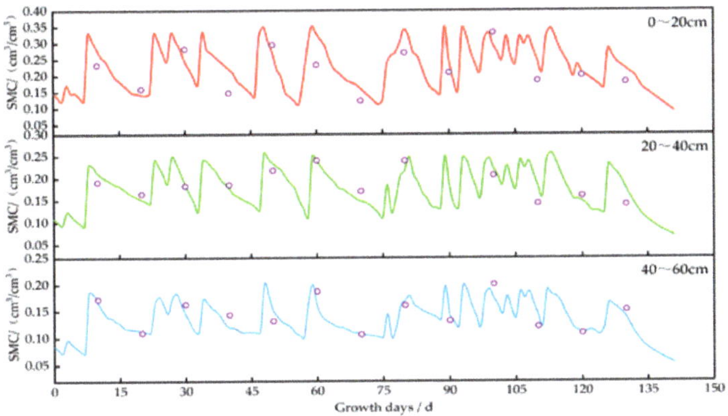


**Figure 6.** Changes in soil moisture content at a depth of 0–60 cm in potato plots. Note: the circles represent the measured values, and the lines represent the simulated values.

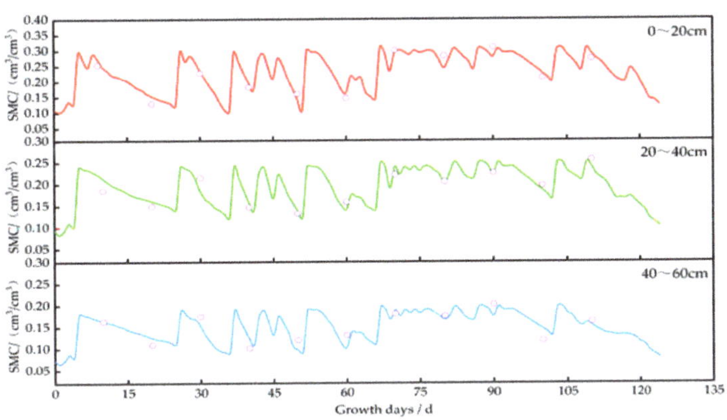


**Figure 7.** Changes in soil moisture content at a depth of 0–60 cm in oat plots. Note: the circles represent the measured values, and the lines represent the simulated values.

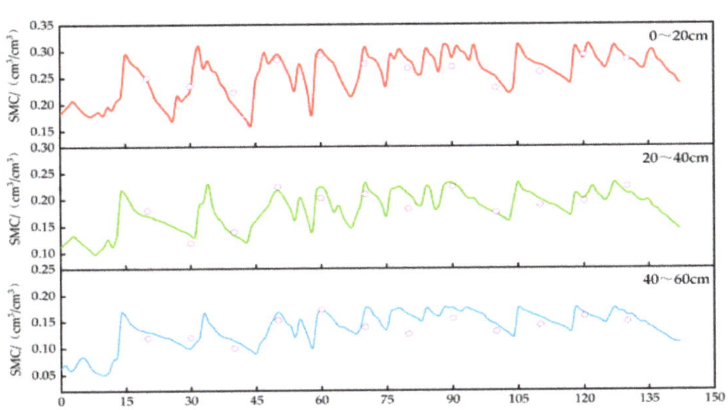


**Figure 8.** Changes in soil moisture content at a depth of 0–60 cm in alfalfa plots. Note: the circles represent the measured values, and the lines represent the simulated values.

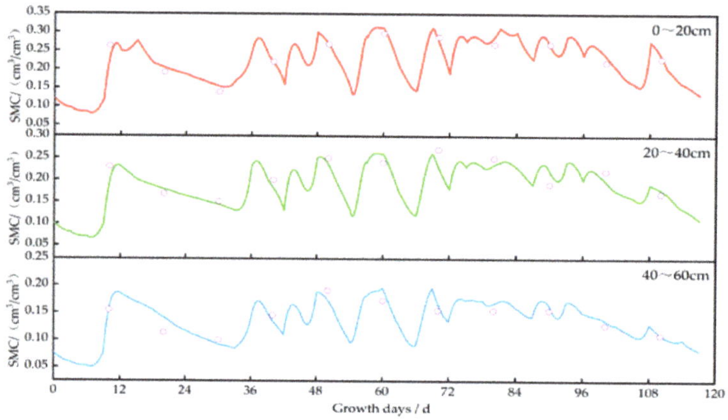


**Figure 9.** Changes in soil moisture content at a depth of 0–60 cm in sunflower plots. Note: the circles represent the measured values, and the lines represent the simulated values.

The saturated moisture contents in the 0–20 cm soil layer for the *potato*, *oat*, *alfalfa*, and *sunflower* plots were 0.35 $cm^3/cm^3$, 0.31 $cm^3/cm^3$, 0.31 $cm^3/cm^3$, and 0.32 $cm^3/cm^3$, respectively. Meanwhile, in the 20–60 cm layer, these values were 0.20–0.26 $cm^3/cm^3$, 0.19–0.24 $cm^3/cm^3$, 0.17–0.22 $cm^3/cm^3$, and 0.18–0.25 $cm^3/cm^3$, respectively. After each irrigation event, the soil moisture level quickly dropped from saturation to maximum water holding capacity.

In 2023, the *potatoes* were irrigated eight times at short intervals. During the rapid growth period (45–75 days), soil moisture decreased by about 60–66%, 52–60%, and 42–51% in the 0–20 cm, 20–40 cm, and 40–60 cm layers before the subsequent round of irrigation, respectively. During the mid-growth stage (75–95 days), soil moisture reduction was around 53–59% in the 0–20 cm layer, 43–48% in the 20–40 cm layer, and 35–40% in the 40–60 cm layer. The water consumption during the rapid growth stage was 7–12% more than that during the mid-growth period.

In 2023, the *oats* were irrigated four times, i.e., at low frequency with longer intervals. During the rapid growth period (30–50 days), soil moisture decreased by about 62–68% in the 0–20 cm layer, 44–52% in the 20–40 cm layer, and 41–49% in the 40–60 cm layer. During the mid-growth period (50–70 days), the reduction was about 51–55% in the 0–20 cm layer, 36–44% in the 20–40 cm layer, and 33–42% in the 40–60 cm layer. The water consumption during the rapid growth period was 11–13% more than that during the mid-growth period.

In 2023, the *alfalfas* were irrigated eight times, i.e., frequently and at relatively short intervals. During the rapid growth periods (30–45 days and 92–105 days), soil moisture decreased by about 50–52% in the 0–20 cm layer, 46–48% in the 20–40 cm layer, and 40–45% in the 40–60 cm layer. During the mid-to-late growth stage (45–50 days and 105–117 days), the reduction was about 43–46% in the 0–20 cm layer, 40–44% in the 20–40 cm layer, and 32–35% in the 40–60 cm layer. The water consumption during the rapid growth period was 7–10% higher than that during the mid-to-late growth stage.

In 2023, the *sunflowers* were irrigated five times, i.e., less frequently and at prolonged intervals. During the early growth period (12–48 days), soil moisture in the 0–20 cm layer decreased by about 43–50% after each irrigation; this reduction was 41–48% in the 20–40 cm layer and 39–47% in the 40–60 cm layer. During the rapid growth period (48–67 days), the reduction was about 56–60% in the 0–20 cm layer, 51–55% in the 20–40 cm layer, and 45–50% in the 40–60 cm layer. The water consumption during the initial growth stage was 10–13% lower than that during the rapid growth period.

### 3.3.2. The LAI Dynamics for Potatoes, Oats, Alfalfa, and Sunflowers

The changes in the LAI for potatoes, oats, alfalfa, and sunflowers are shown in Figure 10. The LAI of these four crops underwent significant changes during different growth stages. The LAI for *potatoes* peaked on the 45th day after planting, corresponding to substantial soil moisture consumption. To meet the growth needs of the *potatoes*, irrigation was performed four times during this stage. Thereafter, the LAI of *potatoes* began to decline. Meanwhile, the LAI for oats peaked on the 75th day after planting, with rapid growth observed between days 30 and 60, leading to a swift decrease in soil moisture and a higher water demand. During this phase, the oats were irrigated four times to meet water requirements. As oats entered the later stages of growth and development, the LAI started to decrease, and the rate of soil moisture decline started to slow down. Alfalfa exhibited multiple peaks in LAI during its growth, with rapid increases observed following each cutting, indicating a significant increase in water demand with the rapid decrease in soil moisture. As alfalfa entered its blooming stage, the rate of soil moisture decline slowed down. The LAI for sunflowers reached the maximum value on the 67th day after planting, corresponding to significant soil moisture consumption. Irrigation was primarily concentrated during this phase.

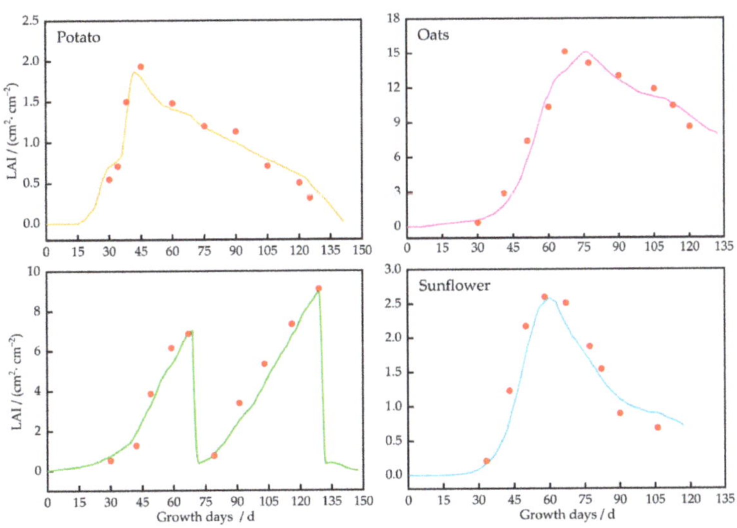


**Figure 10.** Change trends in leaf area of potatoes, oats, alfalfa, and sunflower. Note: Red dots are measured values.

### 3.3.3. The Yield Dynamics for Potatoes, Oats, Alfalfa, and Sunflowers

The yield trends for potatoes, oats, alfalfa, and sunflowers are illustrated in Figure 11. A rapid increase in potato yield occurred during the tuber formation and bulking phase (55–105 days), with significant soil moisture consumption observed at 45–90 days. Throughout the growth period, the yield increased at an average rate of 733.7 kg/ha per day, resulting in a final yield of 67,170 kg/ha. The yield of oats showed exponential growth between 75 and 95 days, and the water demand during this stage was high. The average daily yield increase across the entire growth period was 60.3 kg/ha, resulting in a total yield of 3345 kg/ha. As a perennial plant, alfalfa showed rapid increases in yield at 45–60 and 105–120 days. The average daily yield increase was 35.7 kg/ha for the first cutting and 64.2 kg/ha for the second cutting, with an annual yield of 6529 kg/ha. Sunflower yield increased linearly throughout the growth period, with an average daily yield increase of 124.9 kg/ha. This culminated in a total yield of 4020 kg/ha.

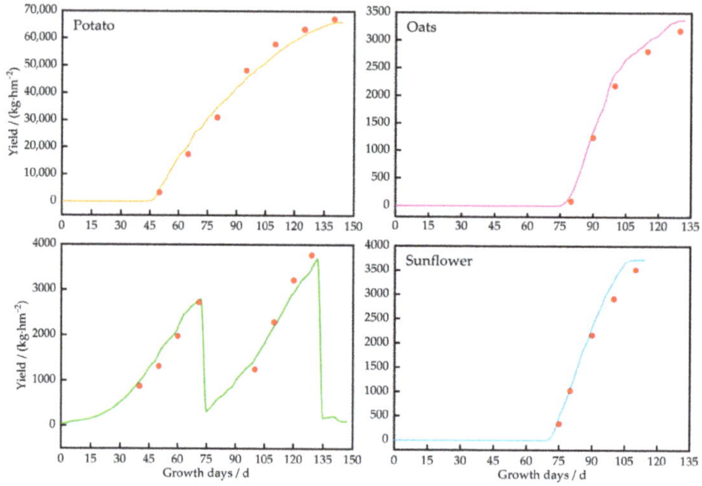


**Figure 11.** Yield trends in potatoes, oats, alfalfa, and sunflowers. Note: Red dots are measured values.

### 3.4. Water Balance Analysis

The changes in water content within the 0–60 cm layer of soil were analyzed throughout the growth season based on the simulated soil moisture levels for *potatoes, oats, alfalfa,* and *sunflowers* in 2022 and 2023, using the water balance principle. The results are presented in Table 13.

**Table 13.** Water balance in the 0–60 cm layer of soil for different crops in 2022 and 2023.

| Crop Type | P/mm | | I/mm | | ΔW/mm | | ET/mm | | E/mm | | T/mm | | D/mm | |
|---|---|---|---|---|---|---|---|---|---|---|---|---|---|---|
| | 2022 | 2023 | 2022 | 2023 | 2022 | 2023 | 2022 | 2023 | 2022 | 2023 | 2022 | 2023 | 2022 | 2023 |
| *Oat* | 191.7 | 241.1 | 146 | 128 | −16.5 | −18.7 | 333.8 | 369.2 | 73.4 | 67.1 | 260.4 | 302.1 | 20.4 | 18.6 |
| *Potato* | 185.3 | 236.0 | 173 | 156 | −41.6 | −44.9 | 375.2 | 414.2 | 254.7 | 238.0 | 120.5 | 176.2 | 24.7 | 22.7 |
| *Alfalfa* | 213.6 | 257.8 | 182 | 164 | −47.4 | −56.7 | 415.7 | 453.7 | 184.3 | 181.2 | 231.4 | 272.5 | 27.3 | 24.8 |
| *Sunflower* | 198.7 | 241.2 | 137 | 118 | −37.8 | −42.3 | 355 | 385.6 | 160.7 | 179.6 | 194.3 | 206.0 | 18.5 | 15.9 |

For *oats*, in 2022 and 2023, the evapotranspiration (ET) was 333.8 mm and 369.2 mm, respectively, with total percolation amounts of 20.4 mm and 18.6 mm. Soil evaporation and oat transpiration accounted for 22% and 78% of ET in 2022 and 18% and 82% of ET in 2023, respectively.

For *potatoes*, the ET was 375.2 mm and 414.2 mm in 2022 and 2023, respectively, and the total percolation amount was 24.7 mm and 22.7 mm, respectively. Soil evaporation and potato transpiration accounted for 68% and 32% of ET in 2022 and 57% and 43% of ET in 2023, respectively (Figure S4).

For *alfalfa*, the ET in 2022 and 2023 was 415.7 mm and 453.7 mm, and the total percolation amount was 27.3 mm and 24.8 mm, respectively. Soil evaporation and alfalfa transpiration accounted for 44% and 56% of ET in 2022 and 40% and 60% of ET in 2023, respectively.

For *sunflowers*, the ET was 355 mm and 385.6 mm and the total percolation amount was 18.5 mm and 15.9 mm in 2022 and 2023, respectively. Soil evaporation and sunflower transpiration accounted for 45% and 55% of ET in 2022 and 47% and 53% of ET in 2023, respectively. Among the four crops, alfalfa had the highest ET, percolation amount, and total amount of introduced water.

### 3.5. Water Use Efficiency Analysis

Based on the simulated yield of *potatoes, oats, alfalfa,* and *sunflowers* (2022 and 2023), and by employing the formulae for WUE and IWUE, the relationships between crop yields and irrigation water usage were analyzed. The results, as shown in Table 14, indicated that the WUE for oats was 1.04 kg/m$^3$ in 2022 and 1.01 kg/m$^3$ in 2023. This value was relatively consistent, suggesting that the unit water productivity for oat crops remained stable. However, the IWUE decreased from 0.64 kg/m$^3$ in 2022 to 0.52 kg/m$^3$ in 2023, indicating a reduction in the efficiency of irrigation water use and highlighting the potential need to modify the irrigation strategy. For *potatoes*, the WUE decreased from 16.62 kg/m$^3$ in 2022 to 16.22 kg/m$^3$ in 2023, while the IWUE dropped from 10.81 kg/m$^3$ to 8.61 kg/m$^3$. Despite the slight decrease in WUE, the reduction in IWUE was quite significant, since precipitation was 213.6 mm in 2022 and 257.8 mm in 2023, and irrigation was 173 mm in 2022 and 156 mm in 2023. Precipitation in 2023 was more than 44.2 mm, and irrigation in 2023 was less than 17 mm. Rainfall contributed significantly to potato yield. The WUE of alfalfa showed a slight increase from 1.17 kg/m$^3$ to 1.44 kg/m$^3$ between 2022 and 2023, while its IWUE decreased from 0.93 kg/m$^3$ to 0.71 kg/m$^3$. Sunflowers maintained a relatively stable WUE, only showing a slight decrease from 1.39 kg/m$^3$ to 1.34 kg/m$^3$, but their IWUE decreased from 0.85 kg/m$^3$ to 0.68 kg/m$^3$. Nevertheless, when compared with other crops, sunflowers exhibited a relatively minor decline in water use efficiency. Thus, overall, oats and sunflowers exhibited a stable WUE, with a slight downward trend for IWUE. Potatoes showed a minor decrease in WUE but a more noticeable decline in IWUE. This suggests that the irrigation regime was not effectively adapted to the water demands of the crops.

**Table 14.** Water use rates and irrigation water use rates of different crops in 2022 and 2023.

| **Crop Type** | **I/ mm** | | **ET/ mm** | | **Y/ (kg/ha)** | | **$Y_a$/ (kg/ha)** | | **WUE/ (kg/m$^3$)** | | **IWUE/ (kg/m$^3$)** | |
|---|---|---|---|---|---|---|---|---|---|---|---|---|
| | **2022** | **2023** | **2022** | **2023** | **2022** | **2023** | **2022** | **2023** | **2022** | **2023** | **2022** | **2023** |
| *Oat* | 146 | 128 | 333.8 | 369.2 | 3120 | 3345 | 2184 | 2676 | 1.04 | 1.01 | 0.64 | 0.52 |
| *Potato* | 173 | 156 | 375.2 | 414.2 | 62,355 | 67,170 | 43,649 | 53,736 | 16.62 | 16.22 | 10.81 | 8.61 |
| *Alfalfa* | 182 | 164 | 415.7 | 453.7 | 4879 | 6529 | 3185 | 5362 | 1.17 | 1.44 | 0.93 | 0.71 |
| *Sunflower* | 137 | 118 | 355 | 385.6 | 3870 | 4020 | 2709 | 3216 | 1.39 | 1.34 | 0.85 | 0.68 |

The value of water productivity and irrigation water use efficiency for *potatoes* was at the maximum amount, ranging from 16.22 to 16.62 kg/m$^3$ and 8.61 to 10.81 kg/m$^3$, respectively, followed by *alfalfa*, *sunflowers*, and *oats*. For the perspective of water productivity, it was recommended that *potatoes* could be extensively cultivated, *alfalfa* planted appropriately, and *oats* and *sunflowers* planted less.

## 4. Discussion

Using field trial data from 2022 and 2023, this study investigated the soil water dynamics and the growth processes for *potatoes, oats, Alfalfa,* and *Sunflowers* in the agro-pastoral ecotone. The adjusted model aligned well with the measured data ($ARE$ < 10%, $nRMSE/\%$ < 10%, and $R^2 \geq 0.85$), which provided effective predictions of crop growth trends, consistent with findings from Wang et al. (2022) [39], Shen et al. (2020) [40], and Jiang et al. (2019) [41]. The present study also demonstrated that soil moisture changes differed across the different layers of soil at a depth of 0–60 cm, which corresponds to the active rootzone (Figures S5 and S6). These results are in line with those from Gao et al. (2015) [42] and Guo et al. (2020) [43], who investigated soil moisture content across different soil depths under various conditions of land use. They demonstrated that the water consumption in the topsoil layer (0–20 cm) is significantly higher than that in the deeper layer (20–40 cm) when the crops have a higher water demand. Similarly, our study revealed that the reduction in soil water within the surface layer (0–20 cm) was 6–11% and

15–19% greater than that in the deeper layer (20–60 cm) for *potatoes* in 2022 and 2023, 11–18% and 13–21% higher for *oats*, 3–4% and 7–11% higher for *Alfalfa*, and 2–5% and 6–10% higher for *Sunflowers*. Additionally, the majority of roots in *potato* and *oat* plants are distributed at a depth of 0–40 cm [44,45]. The significant differences in water absorption and utilization among the four crops led to variations in leaf growth, with the peak LAI being 1.9 for *potatoes*; 16 for *oats*; and 2.7 for *Sunflowers*. Meanwhile, *Alfalfa* exhibited multiple LAI peaks (7.5 and 9). The sensitivity to water deficit across various yield accumulation stages differs among different crops [46], and these differences lead to varying rates of yield increase. For example, Song et al. [47] studied the coupling effect of water and nitrogen on *potato* yield, quality, and water productivity under subsurface drip irrigation conditions in the northwest arid area of Weiwu City. They found that P2N2 (soil moisture ratio of 70% and nitrogen application rate of 135 kg/ha) treatment had the best performance in *potato* tuber starch content and yield, with the highest yield of 54,187 kg/ha, and P1N2 (soil moisture ratio of 40% and nitrogen application rate of 135 kg/ha) treatment had the highest water productivity of 12.86 kg/m$^3$. This study found that the yield was 62,355–67,170 kg/ha. The *potato* yield in this study is higher, since the nitrogen fertilizer (N) applied in this study was 185 kg/ha, which was 50 kg/ha more than that of Song et al. [47] and the rainfall in Weiwu City was less than 100 mm.

Simulations based on the DSSAT model were conducted to track water consumption changes, ET, and yield for *potatoes*, *oats*, *Alfalfa*, and *Sunflowers* [48]. The model accurately depicted the water requirements throughout the crop growth process, highlighting the critical role of concentrated rainfall and irrigation in supporting crop growth and enhancing soil moisture content in arid regions [49]. This study identified varying water consumption levels at different stages of crop growth. The total water consumption was found to be 375.2–414.2 mm for *potatoes*, 333.8–369.2 mm for *oats*, 415.7–453.7 mm for *Alfalfa*, and 355.0–385.6 mm for *Sunflowers*. The water consumption for *oats* was consistent with the value reported by Hao et al. (2019) [50], whereas the consumption for *potatoes* was higher than the total water consumption of 216–278 mm reported by Chen et al. (2019) [51]. Conversely, the water consumption calculated for *Alfalfa* and *Sunflowers* was lower than the previously reported water consumption of 400–500 mm for *Alfalfa* [52] and 400 mm for *Sunflowers* [53] during the growth season. The lower values observed for *Alfalfa* and *Sunflowers* were likely due to factors such as shorter growth seasons, less rainfall, and insufficient heat accumulation. Additionally, the water percolation for *potatoes*, *oats*, *Alfalfa*, and *Sunflowers* was 20.1–22.7 mm, 15.9–18.6 mm, 22.7–24.8 mm, and 12.5–15.9 mm, respectively. This suggests that agricultural irrigation management practices should focus on controlling leakage during this period to reduce ineffective water consumption. The irrigation water for the four crops could be reduced to potentially decrease ineffective percolation and consumption. The WUE of *Alfalfa* in 2023 was higher than that in 2022, whereas the WUEs for *oats*, *potatoes*, and *Sunflowers* all decreased from 2022 to 2023, indicating the superior water utilization capability of *Alfalfa*.

Overall, the optimal planting patterns for crops under different water and fertilizer combinations will be examined in future studies based on the DSSAT model, providing a scientific basis for sustainable agricultural development in arid regions.

## 5. Conclusions

The consumption of the soil moisture during the rapid growth stage for the *potatoes*, *oats*, *Alfalfa*, and *Sunflower* was 7–12%, 11–13%, 7–10% and 10–13% more than that during the rapid, middle, mid-to-late, and initial growth period, and the yield was 67,170 kg/ha, 3345 kg/ha, 6529 kg/ha, and 4020 kg/ha, respectively. The water consumption of *oats*, *potatoes*, and *Sunflowers* was 333.8 mm–369.2 mm, 375.2 mm–414.2 mm, and 355.0 mm–385.6 mm, respectively. The maximum water consumption of *Alfalfa* was 415.7 mm–453.7 mm. Among the four crops, *Alfalfa* was the highest water consuming crop. The average percolation of the four crops was 12.5 mm–15.9 mm. The value of water productivity and irrigation water use efficiency for *potatoes* was at the maximum value, ranging from 16.22 to 16.62 kg/m$^3$ and 8.61 to 10.81 kg/m$^3$, respectively, followed by *Alfalfa*, *Sunflowers*, and *oats*. It was advised that irrigation water could be appropriately reduced to decrease ineffective water percolation and consumption. For the perspective of water productivity, it was recommended that *potatoes* could be extensively cultivated, *Alfalfa* planted appropriately, and *oats* and *Sunflowers* planted less.

**Supplementary Materials:** The following supporting information can be downloaded at: https://www.mdpi.com/article/10.3390/plants13141916/s1, Figure S1. Flood irrigation 1. Figure S2. Flood irrigation 2. Figure S3. Root depth of *Alfalfa*. Figure S4. Spacing between *potato* rows. Figure S5. Root depth of *oats*. Figure S6. Root depth of *potato*.

**Author Contributions:** G.W. and X.M. conceived and designed the experiments. G.W. and X.M. developed the initial and final versions of this manuscript and analyzed the data. G.W. and X.M. made the same contribution to this paper and were both the first co-authors of this paper. R.L., H.Z. and B.X. contributed their expertise and advice. D.T. and J.W. supported the writing of the manuscript. P.T. was responsible for project management and funding enhancement. Y.F. contributed to the experiments and model applications. J.R., Z.L. and J.Z. conducted the experiments. Z.X. provided the test materials. All authors have read and agreed to the published version of the manuscript.

**Funding:** This study was supported by the Water use strategy and smart control technology for farmland under limited groundwater irrigation in the Agro-pastoral Ecotone of the Yinshanbeilu in Inner Mongolia, China, N0. MKGP2024JK015; the Study on the Water Cycle Mechanism of Typical Irrigation Areas in the Agro-pastoral Ecotone of Yinshanbeilu Mountains in Inner Mongolia, NO. MK2022J02; Study and Demonstration of the Key Technologies for Efficient Water-saving in the Forage Area of the Yellow River Basin in Inner Mongolia: No. 2023JBGS0014; Inner Mongolia National Natural Science Foundation: No. 2022QN05027; Technical Demonstration of Collaborative Utilization of Water–Soil–Plant Resources System in Ecological Agriculture and Animal Husbandry Area: No. 2021CG0012.

**Data Availability Statement:** Data are contained within the article.

**Acknowledgments:** We highly appreciate the reviewers' and editors' useful suggestions on this work. This study was supported by the Yinshanbeilu Grassland Eco-hydrology National Observation and Research Station Field testing station.

**Conflicts of Interest:** The authors declare no conflicts of interest.

## References

1. Li, S.; An, P.; Pan, Z. Farmers' initiative on adaptation to climate change in the Northern Agro-pastoral Ecotone. *Int. J. Disaster Risk Reduct.* **2015**, *12*, 278–284. [CrossRef]
2. Fu, D.; Ren, X.; Wang, Y. Distribution characteristics of soil particle size in farming-pastoral ecotone: A case study of Wuchuan County in Inner Mongolia. *Arid Zone Res.* **2022**, *39*, 1322–1332.
3. Wuyun, D.; Sun, L.; Chen, Z. The spatiotemporal change of cropland and its impact on vegetation dynamics in the farming-pastoral ecotone of northern China. *Sci. Total Environ.* **2022**, *805*, 150286. [CrossRef] [PubMed]
4. Gao, T.; Zhan, R. Damage of Reclamation on Grassland Slope in Farmingpastoral Interleaving Areas, North Yinshan Mountains. *J. Agric. Sci. Technol.* **2014**, *16*, 125.
5. Wang, S.; Wang, W.; Wu, Y. Effects of a changing climate and anthropogenic impacts on net primary production in Yinshanbeilu, Inner Mongolia, China. *Environ. Impact Assess. Rev.* **2023**, *102*, 107185. [CrossRef]
6. Cao, X.; Liu, Z.; Wu, M. Temporal-spatial distribution and driving mechanism of arable land water scarcity index in China from water footprint perspective. *Trans. Chin. Soc. Agric. Eng.* **2019**, *35*, 94–100.
7. Yin, L.; Feng, X.; Fu, B. Irrigation water consumption of irrigated cropland and its dominant factor in China from 1982 to 2015. *Adv. Water Resour.* **2020**, *143*, 103661. [CrossRef]
8. Zhang, X.; Ma, Y.; Yu, X. Effects of vertically rotary sub-soiling tillage on water utilization and yield of *potato* in semi-arid area of northwest China. *Chin. J. Appl. Ecol.* **2018**, *29*, 3293–3301.
9. Wang, Y.; Lv, W.; Wang, X. Distribution Characteristics of Soil Water, Nutrients, and Roots in Rain-fed *Alfalfa* Fields in Southern Ningxia Mountain Area. *Chin. J. Grassl.* **2022**, *44*, 37–45.
10. Kang, W.; Shi, X.; Duan, H. Analysis of Water Requirement and Water Requirement Law of Main Crops in Dryland Area at The Northern Foot of Yinshan Mountain. *Tianjin Agric. Sci.* **2023**, *29*, 13–16+20.
11. Yu, H.; Liu, K.; Bai, Y. The agricultural planting structure adjustment based on water footprint and multi-objective optimisation models in China. *J. Clean. Prod.* **2021**, *297*, 126646. [CrossRef]

12. Tang, J.; Xiao, D.; Wang, J. Optimizing water and nitrogen managements for *potato* production in the agro-pastoral ecotone in North China. *Agric. Water Manag.* **2021**, *253*, 106945. [CrossRef]
13. Wang, H.; Cheng, M.; Zhang, S. Optimization of irrigation amount and fertilization rate of drip-fertigated *potato* based on Analytic Hierarchy Process and Fuzzy Comprehensive Evaluation methods. *Agric. Water Manag.* **2021**, *256*, 107130. [CrossRef]
14. Ma, Q.; Zhang, X.; Wu, Y. Optimizing water and nitrogen strategies to improve forage *oat* yield and quality on the Tibetan Plateau using APSIM. *Agronomy* **2022**, *12*, 933. [CrossRef]
15. Liu, M.; Wang, Z.; Mu, L. Effect of regulated deficit irrigation on *Alfalfa* performance under two irrigation systems in the inland arid area of midwestern China. *Agric. Water Manag.* **2021**, *248*, 106764. [CrossRef]
16. Zhang, J.; Du, L.; Xing, Z. Effects of dual mulching with wheat straw and plastic film under three irrigation regimes on soil nutrients and growth of edible *Sunflower*. *Agric. Water Manag.* **2023**, *288*, 108453. [CrossRef]
17. Rauff, K.O.; Bello, R. A review of crop growth simulation models as tools for agricultural meteorology. *Agric. Sci.* **2015**, *6*, 1098–1105. [CrossRef]
18. Lenz-Wiedemann, V.I.; Klar, C.W.; Schneider, K. Development and test of a crop growth model for application within a Global Change decision support system. *Ecol. Model.* **2010**, *221*, 314–329. [CrossRef]
19. Ge, J.; Yu, Z.; Gong, X. Evaluation of Irrigation Modes for Greenhouse Drip Irrigation Tomatoes Based on AquaCrop and DSSAT Models. *Plants* **2023**, *12*, 3863. [CrossRef]
20. Corbeels, M.; Chirat, G.; Messad, S. Performance and sensitivity of the DSSAT crop growth model in simulating maize yield under conservation agriculture. *Eur. J. Agron.* **2016**, *76*, 41–53. [CrossRef]
21. Pierre, J.F.; Singh, U.; Ruiz-Sánchez, E. Development of a cereal–legume intercrop model for DSSAT version 4.8. *Agriculture* **2023**, *13*, 845. [CrossRef]
22. Miao, X.; Wang, G.; Li, R. Study on Modeling and Evaluating *Alfalfa* Yield and Optimal Water Use Efficiency in the Agro-Pastoral Ecotone of Northern China. *Plants* **2024**, *13*, 229. [CrossRef] [PubMed]
23. Zhou, H.; Geng, G.; Yang, J. Improving Soil Moisture Estimation via Assimilation of Remote Sensing Product into the DSSAT Crop Model and Its Effect on Agricultural Drought Monitoring. *Remote Sens.* **2022**, *14*, 3187. [CrossRef]
24. Shen, H.; Gao, Y.; Guo, F. A modified DSSAT-CERES model for simulating summer maize growth under film mulching. *Agron. J.* **2021**, *113*, 4819–4831. [CrossRef]
25. Ma, H.; Malone, R.W.; Jiang, T.; Yao, N.; Chen, S.; Song, L.; Feng, H.; Yu, Q.; He, J. Estimating crop genetic parameters for DSSAT with modified PEST software. *Eur. J. Agron.* **2020**, *115*, 126017. [CrossRef]
26. Wang, H.; Cheng, M.; Liao, Z. Performance evaluation of AquaCrop and DSSAT-SUBSTOR-*Potato* models in simulating *potato* growth, yield and water productivity under various drip fertigation regimes. *Agric. Water Manag.* **2023**, *276*, 108076. [CrossRef]
27. Mubeen, M.; Ahmad, A.; Hammad, H.M. Evaluating the climate change impact on water use efficiency of cotton-wheat in semi-arid conditions using DSSAT model. *J. Water Clim. Chang.* **2020**, *11*, 1661–1675. [CrossRef]
28. Abayechaw, D. Review on decision support system for agrotechnology transfer (DSSAT) model. *Int. J. Intell. Inf. Syst.* **2021**, *10*, 117–124. [CrossRef]
29. Jiang, Y.; Zhang, L.; Zhang, B. Modeling irrigation management for water conservation by DSSAT-maize model in arid northwestern China. *Agric. Water Manag.* **2016**, *177*, 37–45. [CrossRef]
30. Wang, P. Research of Optimizing Deficit Irrigation for Sugarcane in Guangxi Province Using the DSSAT Model Based on Genetic Algorithm. Master's Thesis, Hebei University of Engineering, Handan, China, 2022.
31. Singels, A.; Bezuidenhout, C.N. A new method of simulating dry matter partitioning in the Canegro sugarcane model. *Field Crops Res.* **2002**, *78*, 151–164. [CrossRef]
32. Zhang, M. Remote Sensing for Maize Nitrogen Nutrition Diagnosis Based on Field Experiments and the DSSAT Model. Master's Thesis, Jilin University, Changchun, China, 2023.
33. Wu, C.; Anlauf, R.; Ma, Y. Application of the DSSAT model to simulate wheat growth in Eastern China. *J. Agric. Sci.* **2013**, *5*, 198–208. [CrossRef]
34. Yan, W.; Jiang, W.; Han, X. Simulating and predicting crop yield and soil fertility under climate change with fertilizer management in Northeast China based on the decision support system for agrotechnology transfer model. *Sustainability* **2020**, *12*, 2194. [CrossRef]
35. Gao, Y. Growth Simulation of Plastic Film Mulching in Summer Maize Based on Improved DSSAT-CERES Model. Master's Thesis, Northwest A & F University, Xianyang, China, 2021.
36. Blum, A. Effective use of water (EUW) and not water-use efficiency (WUE) is the target of crop yield improvement under drought stress. *Field Crops Res.* **2009**, *112*, 119–123. [CrossRef]
37. Attia, A.; Rajan, N.; Xue, Q. Application of DSSAT-CERES-Wheat model to simulate winter wheat response to irrigation management in the Texas High Plains. *Agric. Water Manag.* **2016**, *165*, 50–60. [CrossRef]
38. Yang, X.; Huang, X.; Chen, Z. Evaluation index comparison of different irrigation water use efficiency based on DSSAT simulation. *Trans. Chin. Soc. Agric. Eng.* **2015**, *31*, 95–100.
39. Wang, X.; Xin, L.; Du, J. Simulation of Cotton Growth and Yield under Film Drip Irrigation Condition Based on DSSAT Model in Southern Xinjiang. *Trans. Chin. Soc. Agric. Mach.* **2022**, *53*, 314–321.
40. Shen, H.; Xu, F.; Zhao, R. Optimization of Sowing Date, Irrigation, and Nitrogen Management of Summer Maize Using the DSSAT-CERES-Maize Model in the Guanzhong Plain, China. *Trans. ASABE* **2020**, *63*, 789–797. [CrossRef]

41. Jiang, R.; He, W.; Zhou, W. Exploring management strategies to improve maize yield and nitrogen use efficiency in northeast China using the DNDC and DSSAT models. *Comput. Electron. Agric.* **2019**, *166*, 104988. [CrossRef]
42. Gao, L.; Shao, M.; Peng, X. Spatio-temporal variability and temporal stability of water contents distributed within soil profiles at a hillslope scale. *Catena* **2015**, *132*, 29–36. [CrossRef]
43. Guo, X.; Fu, Q.; Hang, Y. Spatial variability of soil moisture in relation to land use types and topographic features on hillslopes in the black soil (mollisols) area of northeast China. *Sustainability* **2020**, *12*, 3552. [CrossRef]
44. Boguszewska-Mańkowska, D.; Zarzyńska, K.; Nosalewicz, A. Drought differentially affects root system size and architecture of *potato* cultivars with differing drought tolerance. *Am. J. Potato Res.* **2020**, *97*, 54–62. [CrossRef]
45. Osborne, S.L.; Chim, B.K.; Riedell, W.E. Root length density of cereal and grain legume crops grown in diverse rotations. *Crop Sci.* **2020**, *60*, 2611–2620. [CrossRef]
46. Wagg, C.; Hann, S.; Kupriyanovich, Y. Timing of short period water stress determines *potato* plant growth, yield and tuber quality. *Agric. Water Manag.* **2021**, *247*, 106731. [CrossRef]
47. Song, N.; Wang, F.; Yang, C.; Yang, K. Coupling effects of water and nitrogen on yield, quality and water use of *potato* with drip irrigation under plastic film mulch. *Trans. Chin. Soc. Agric. Eng. (Trans. CSAE)* **2013**, *29*, 98–105.
48. Si, Z.; Zain, M.; Li, S. Optimizing nitrogen application for drip-irrigated winter wheat using the DSSAT-CERES-Wheat model. *Agric. Water Manag.* **2021**, *244*, 106592. [CrossRef]
49. Zhang, Y.; Wang, Y.; Niu, H. Effects of temperature, precipitation and carbon dioxide concentrations on the requirements for crop irrigation water in China under future climate scenarios. *Sci. Total Environ.* **2019**, *656*, 373–387. [CrossRef]
50. Hao, J.; Zhang, T.; Yin, X. Impact of climate change on the water requirements of *oat* in Northeast and North China. *Water* **2019**, *11*, 91. [CrossRef]
51. Chen, Y.; Chai, S.; Tian, H. Straw strips mulch on furrows improves water use efficiency and yield of *potato* in a rainfed semiarid area. *Agric. Water Manag.* **2019**, *211*, 142–151. [CrossRef]
52. Zheng, H.; Wang, B.; Cao, X. Optimization of Subsurface Drip Irrigation Schedule of *Alfalfa* in Northwest China. *Water* **2023**, *15*, 2242. [CrossRef]
53. Pan, Y.; Pan, X.; Zi, T. Optimal ridge–furrow ratio for maximum drought resilience of *Sunflower* in semi-arid Region of China. *Sustainability* **2019**, *11*, 4047. [CrossRef]

*plants*

MDPI

*Article*

# Long-Term Straw Incorporation under Controlled Irrigation Improves Soil Quality of Paddy Field and Rice Yield in Northeast China

**Peng Zhang [1], Peng Chen [2], Tangzhe Nie [1,3,*], Zhongxue Zhang [3,4], Tiecheng Li [3,4], Changlei Dai [1], Lili Jiang [1], Yu Wu [1], Zhongyi Sun [5] and Shuai Yin [6]**

[1] School of Water Conservancy and Electric Power, Heilongjiang University, Harbin 150080, China; 2221975@s.hlju.edu.cn (P.Z.)
[2] College of Agricultural Science and Engineering, Hohai University, Nanjing 211100, China
[3] Key Laboratory of Effective Utilization of Agricultural Water Resources, Ministry of Agriculture and Rural Affairs, Northeast Agricultural University, Harbin 150030, China
[4] School of Water Conservancy and Civil Engineering, Northeast Agricultural University, Harbin 150030, China
[5] College of Ecology and Environment, Hainan University, Haikou 570208, China
[6] State Key Laboratory of Remote Sensing Science, Aerospace Information Research Institute, Chinese Academy of Sciences, Beijing 100101, China
* Correspondence: 2019036@hlju.edu.cn; Tel.: +86-189-4508-3525

**Abstract:** Soil quality is an indicator of the ability to ensure ecological security and sustainable soil usage. The effects of long-term straw incorporation and different irrigation regimes on the yield and soil quality of paddy fields in cold regions remain unclear. This study established four treatments: controlled irrigation + continuous straw incorporation for 3 years (C3), controlled irrigation + continuous straw incorporation for 7 years (C7), flooded irrigation + continuous straw incorporation for 3 years (F3), and flooded irrigation + continuous straw incorporation for 7 years (F7). Analysis was conducted on the impact of various irrigation regimes and straw incorporation years on the physicochemical characteristics and quality of the soil. The soil quality index (SQI) for rice fields was computed using separate datasets for each treatment. The soil nitrate nitrogen, available phosphorus, soil organic carbon, and soil organic matter contents of the C7 were 93.51%, 5.80%, 8.90%, and 8.26% higher compared to C3, respectively. In addition, the yield of the C7 treatment was 5.18%, 4.89%, and 10.32% higher than those of F3, C3, and F7, respectively. The validity of the minimum data set (MDS) was verified by correlation, $E_f$ and $E_R$, which indicated that the MDS of all treatments were able to provide a valid evaluation of soil quality. The MDS based SQI of C7 was 11.05%, 11.97%, and 27.71% higher than that of F3, C3, and F7, respectively. Overall, long-term straw incorporation combined with controlled irrigation increases yield and soil quality in paddy fields in cold regions. This study provides a thorough assessment of soil quality concerning irrigation regimes and straw incorporation years to preserve food security and the sustainability of agricultural output. Additionally, it offers a basis for soil quality diagnosis of paddy fields in the Northeast China.

**Keywords:** paddy fields; soil quality index; straw incorporation; irrigation regime; soil nutrient management

## 1. Introduction

Rice holds immense significance as a staple food for nearly 50% of the global population [1,2]. However, the productivity of rice cultivation has encountered stagnation in many rice-growing regions [3], largely due to the influence of climatic conditions, field management practices [4], and soil quality [5]. Among these factors, soil quality is shaped not only by soil genesis but also by various factors associated with soil utilization and management [6], including straw incorporation and irrigation regimes [7,8]. Enhancing soil quality via field management will promote sustainable agricultural management approaches [9], attain food security, and contribute to the preservation of agro-ecosystems [10].

In recent years, organic amendments such as manure, compost, and straw have been used to improve the environment for rice production [11]. Among them, straw incorporation is recognized as an effective method to maintain or improve rice yield and soil quality [12]. In the short term, straw incorporation offered the advantages of mitigating environmental pollution caused by straw burning and piling [13], while simultaneously improving soil structure and microbial activity [12,14]. The long-term consequences of straw incorporation on the soil quality in paddy fields have not, however, been thoroughly studied. Li et al. [15] concluded that long-term straw incorporation enhanced soil carbon and nitrogen content and facilitated seed yield enhancement by synchronizing nitrogen demand during rice growth. Yang et al. [16] concluded that long-term straw incorporation can increase rice quality. Nonetheless, the benefits of long-term straw incorporation were not universally applicable to paddy fields [17], and the stabilization and augmentation of soil organic carbon did not consistently exhibit a positive or linear correlation with the quantity of straw incorporated into the field [18]. Excessive amounts of straw in croplands may have adverse effects on crop yields due to elevated temperatures and limited oxygen availability [19]. The release of chemicals during the breakdown of straw can impede the growth and development of crops [20].

Furthermore, prolonged flood irrigation combined with straw decomposition in paddy fields often resulted in a rapid decline in soil oxygen levels, leading to a deterioration in the rice root environment [21]. This phenomenon facilitated the accumulation of ammonium nitrogen, also increased the loss of nitrogen during the nitrification-denitrification process [22]. Conversely, water-saving irrigation is currently being widely promoted [23], as it reduced the water consumption in the field [24]. Additionally, water-saving irrigation helped to diminish soil nitrogen loss [25,26]. Some studies have shown that straw incorporation under water-saving irrigation increased soil organic carbon sequestration [27]. The soil microenvironment and straw decomposition process were modified by the implementation of water-efficient irrigation, thereby impacting the productivity of paddy fields. However, the specific relationships between soil physicochemical property and soil quality under different water management and straw incorporation conditions remained unclear. Furthermore, thorough assessments of soil indicators have been lacking in previous research [28,29]. Therefore, a comprehensive evaluation of soil quality considering both straw incorporation and irrigation regimes is imperative for ensuring food security and sustaining agricultural production [30].

Soil quality can be inferred from management-induced changes in soil physicochemical properties [31]. Various methods have been developed for assessing soil quality [32,33]. The most commonly used method for assessment was based on the soil quality index (SQI), as it combined multiple indicators into a composite index using a scoring function [34–36], this method effectively handled the multivariate datasets generated by experiments and reflected the trends in soil quality changes [9]. It should be noted that the minimum data set (MDS) approach, as compared to the total data set (TDS), reduced data redundancy and considered the complex interrelationships among multiple indicators [7] and offered the advantages of adaptability and operability [37]. So far, the combined effect of straw incorporation and different irrigation practices on soil quality assessment has been limited [38,39]. Hence, it is imperative to investigate whether MDS based SQI (MDS-SQI) calculation could produce superior outcomes when subjected to varying water management and straw incorporation treatments.

The northeastern region of China holds significant importance as a grain-producing area and has emerged as a prominent rice production base in China [40,41]. However, the soil quality in this region has deteriorated rapidly as a result of long-term irrational farming practices [42]. Additionally, a considerable amount of straw has accumulated over the years of rice cultivation. Consequently, straw incorporation has become an optimal measure to improve soil quality [43,44]. Furthermore, the limited availability of irrigation resources in agriculture has expedited the promotion of local water-saving irrigation techniques [45]. Some studies have pointed out that the combination of water-saving irrigation and nitrogen fertilizer improved rice yield in Northeast China [46]; however, the studies under long-term straw incorporation conditions are not sufficient. Therefore, the objectives of this study were: (1) to analyze the effects of different straw incorporation years and irrigation regimes on soil nutrients, and (2) to screen the indicators of the MDS to calculate the SQI and thus to assess the effects of straw incorporation years and different irrigation regimes on the soil quality of paddy fields. This exploration aims to establish a foundation for future soil quality diagnosis and sustainable development of paddy fields in cold region in Northeast China.

## 2. Results

### 2.1. Soil Physical and Chemical Properties

Compared to the F3, soil pH (pH) (Figure 1a), ammonium nitrogen ($NH_4^+$-N) (Figure 1b), available phosphorus (AP) (Figure 1c), soil organic carbon (SOC) (Figure 1e), soil organic matter (SOM) (Figure 1e), total nitrogen (TN) (Figure 1d), microbial carbon (MBC) (Figure 1g), and dissolved organic nitrogen (DON) (Figure 1d) showed higher values in F7 with increases of 6.41%, 39.79%, 45.28%, 21.75%, 21.32%, 49.11%, 38.61%, and 55.14%, respectively, while nitrate nitrogen ($NO_3^-$-N) (Figure 1b), dissolved organic carbon (DOC) (Figure 1f), available potassium (AK) (Figure 1c), and microbial nitrogen (MBN) (Figure 1g) exhibited significant reductions ($p < 0.05$). $NO_3^-$-N, AP, SOC, and SOM levels in C7 were higher than those in C3 by 93.51%, 5.80%, 8.90%, and 8.26%, respectively. These findings suggested that the straw incorporation years had a positive impact on soil physicochemical properties, with a more pronounced effect observed under flooded irrigation compared to controlled irrigation.

The F3 resulted in lower AK, AP, TN, DON, and MBC compared to the C3, with reductions of 8.91%, 16.46%, 37.81%, 40%, and 23.28%, respectively, despite the same straw incorporation years. Similarly, when straw was incorporated for 7 years, the pH, $NH_4^+$-N, AP, DOC, SOC, SOM, TN, DON, MBC, and MBN of F7 were higher than C7 by 8.01%, 104.94%, 14.69%, 99.31%, 24.65%, 25.02%, 107.39%, 34.93%, and 15.34% respectively ($p < 0.05$). The largest increases were in $NH_4^+$-N, DOC, and TN.

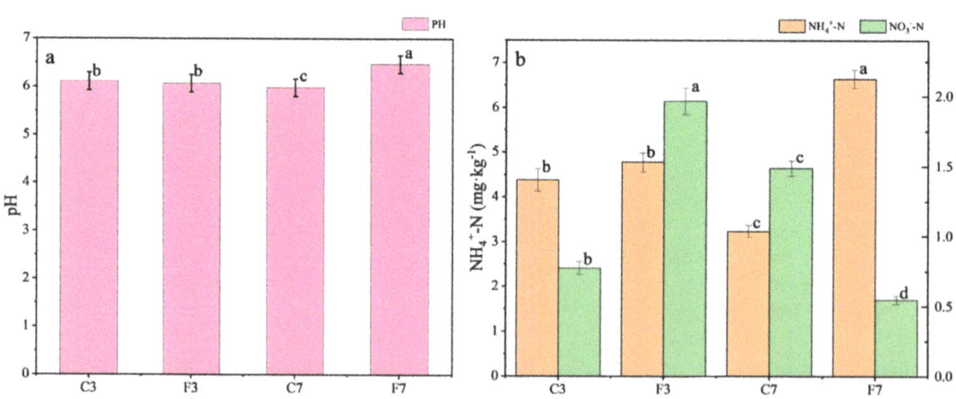


**Figure 1.** *Cont.*

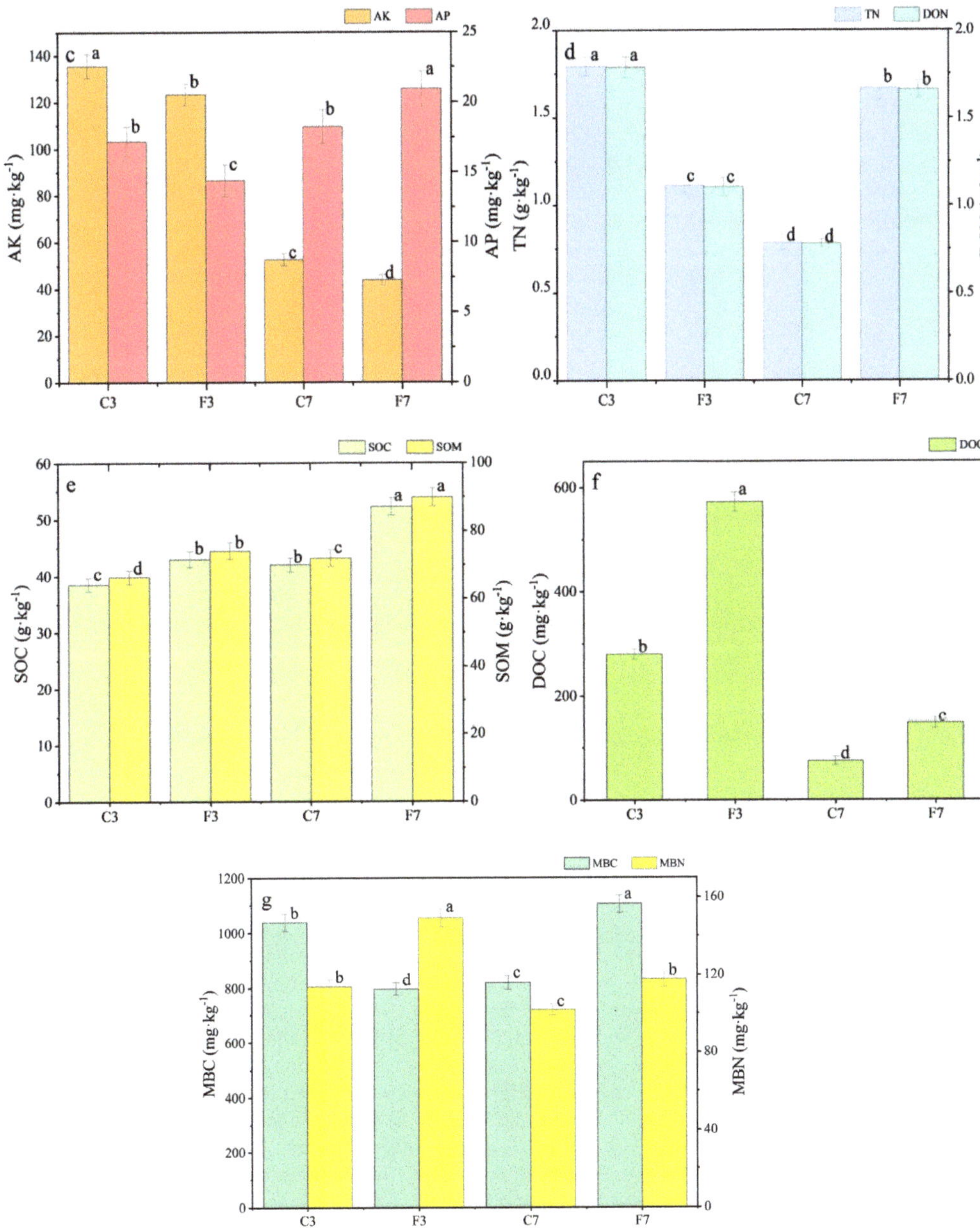


Note: Error bars represent standard deviation. Different letters in the graph indicate statistical differences of $p<0.05$ derived from the least significant difference test.

**Figure 1.** Soil (**a**) pH, (**b**) $NH_4^+$-N and $NO_3^-$-N, (**c**) AK and AP, (**d**) TN and DON, (**e**) SOC and SOM, (**f**) DOC, (**g**) MBC and MBN of paddy fields in different treatments. Abbreviations: pH: Soil pH; $NH_4^+$-N: Ammonium Nitrogen; $NO_3^-$-N: Nitrate Nitrogen; AK: Available Potassium; AP: Available Phosphorus; DOC: Dissolved Organic Carbon; SOC: Soil Organic Carbon; SOM: Soil Organic Matter; TN: Total Nitrogen; DON: Dissolved Organic Nitrogen; MBC: Microbial Carbon; MBN: Microbial Nitrogen.

### 2.2. Rice Yields in Different Treatments

Rice yields under different treatments was observed to follow the order of C7 > C3 > F3 > F7 (Figure 2). The C7 treatment showed 4.89%, 5.72%, and 11.27% more than C3, F3, and F7 respectively. In terms of the same straw incorporation years, rice yields were higher in all controlled irrigation treatments compared to flooded irrigation treatments. The rice yield was significantly affected by the irrigation regime, while the straw incorporation year had no significant effect ($p < 0.05$) (Figure 2). However, the combination of straw incorporation years and irrigation regime had a significant effect on yield ($p < 0.05$).

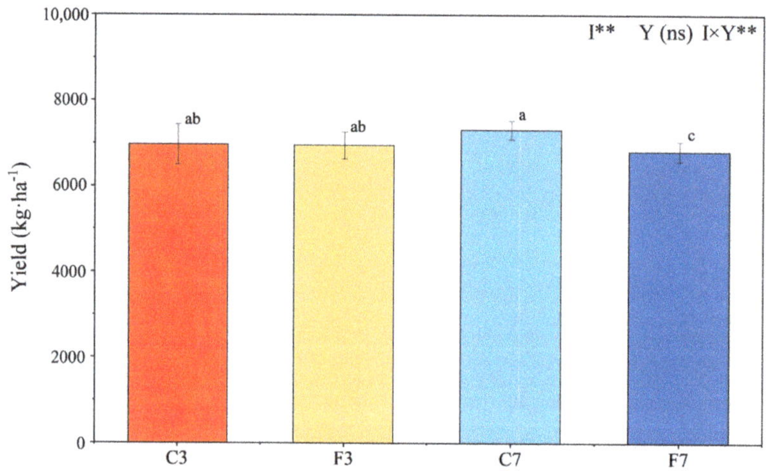


Different letters in the graph indicate statistical differences of $p$<0.05 derived from the least significant difference test. Error bars represent standard deviation.ns and ** mean no significant difference and significant difference at $p$<0.01, respectively.

**Figure 2.** Rice yields in different treatments. Note: I: irrigation regime; Y: year.

### 2.3. Minimum Data Set Filtering

Soil nutrients were analyzed separately for each treatment by principal component analysis (Figure S1). Taking F3 as an example, PC1, PC2, PC3, and PC4 (with eigenvalues greater than 1) accounted for 33.78%, 31.28%, 20.60%, and 7.60% of the variability, respectively. Cumulatively, they contributed to a total variance of 92.35%. Additionally, the metric variance of each indicator was greater than 80%, indicating that the first four principal components effectively captured the information of each indicator and the overall soil quality (Table 1). From Table 1, the high weight indicators under F3 were PC1: MBC, TN, MBC/MBN, DON; PC2: DOC, SOC, SOM; PC3: $NH_4^+$-N, $NO_3^-$-N; and PC4: C/N. Based on the Tables S1 and S2, the MDS for F3 were identified as MBC/MBN, DOC, $NO_3^-$-N, and C/N. The differences in MDS for evaluating paddy soil quality under different treatments indicated variations in the main limiting factors among the treatments (Table 2).

**Table 1.** Principal component analysis of soil quality indicators in paddy fields under different treatments.

| | F3 Principal Component | | | | C3 Principal Component | | | | C7 Principal Component | | | | F7 Principal Component | | | |
|---|---|---|---|---|---|---|---|---|---|---|---|---|---|---|---|---|
| | 1 | 2 | 3 | 4 | 1 | 2 | 3 | 4 | 1 | 2 | 3 | 4 | 1 | 2 | 3 | 4 |
| Eigenvalues | 4.73 | 4.38 | 2.88 | 1.06 | 5.12 | 4.01 | 2.66 | 1.82 | 4.12 | 3.55 | 3.38 | 2.54 | 3.96 | 3.76 | 3.31 | 2.69 |
| Contribution rate (%) | 33.78 | 31.28 | 20.60 | 7.60 | 36.56 | 28.67 | 18.97 | 12.98 | 29.42 | 25.39 | 24.11 | 18.13 | 28.30 | 26.82 | 23.66 | 19.24 |
| Cumulative contribution rate (%) | 33.78 | 65.06 | 85.66 | 93.26 | 36.56 | 65.23 | 84.20 | 97.18 | 29.42 | 54.81 | 79.92 | 97.05 | 28.30 | 55.13 | 78.79 | 98.03 |
| Soil indicators | | | | | | | | | | | | | | | | |
| pH | 0.50 | 0.36 | 0.70 | 0.01 | 0.47 | 0.22 | 0.29 | 0.77 | 0.40 | 0.11 | 0.23 | 0.87 | 0.37 | 0.26 | 0.04 | 0.88 |
| $NH_4^+$-N | −0.08 | 0.14 | 0.97 | −0.07 | 0.50 | 0.20 | 0.74 | −0.07 | 0.36 | 0.59 | 0.29 | 0.59 | 0.39 | 0.28 | 0.12 | 0.86 |
| $NO_3^-$-N | −0.03 | 0.02 | 0.97 | 0.09 | 0.04 | 0.03 | −0.35 | 0.92 | 0.32 | −0.03 | 0.11 | 0.93 | 0.21 | −0.02 | 0.62 | 0.73 |
| AK | 0.27 | 0.72 | −0.13 | 0.21 | 0.10 | 0.52 | 0.80 | 0.10 | 0.06 | −0.22 | 0.89 | 0.27 | 0.01 | 0.95 | 0.02 | 0.21 |
| AP | −0.43 | 0.76 | 0.43 | 0.09 | 0.23 | −0.07 | 0.93 | −0.17 | −0.01 | 0.97 | 0.01 | 0.10 | 0.01 | 0.96 | −0.14 | 0.17 |
| DOC | 0.12 | 0.97 | 0.15 | −0.06 | 0.17 | 0.98 | 0.02 | 0.12 | 0.16 | 0.98 | −0.02 | −0.05 | 0.29 | −0.03 | 0.94 | 0.09 |
| SOC | 0.19 | 0.95 | 0.13 | −0.10 | 0.24 | 0.96 | 0.14 | 0.06 | 0.27 | 0.16 | 0.92 | 0.11 | 0.31 | 0.46 | 0.81 | 0.13 |

Abbreviations refer to Figure 1.

**Table 2.** Minimum data set in different treatments.

| Treatments | MDS |
|---|---|
| F3 | MBC/MBN, DOC, $NO_3^-$-N, C/N |
| C3 | MBC/MBN, SOC, AP, $NO_3^-$-N |
| C7 | DON, SOM, DOC, $NO_3^-$-N |
| F7 | MBN, AK, DOC, pH |

Abbreviations refer to Figure 1.

## 2.4. Soil Quality Index

The weight value occupied by SQI was calculated using Equation (1) (Tables S3 and S4), the SQI for TDS and MDS were calculated separately (Figure 3). To verify the validity of the MDS, ANOVA was performed on the TDS-SQI and MDS-SQI values. The results showed that there was no significant difference in TDS-SQI among treatments. The TDS-SQI values were ranked as follows: C7 > C3 > F3 > F7, with C7 having a higher SQI of 2.58%, 6.94%, and 8.02% compared to C3, F3, and F7, respectively. Additionally, neither the irrigation regime nor the straw incorporation years had a significant effect on the TDS-SQI ($p < 0.05$) (Figure 3a).

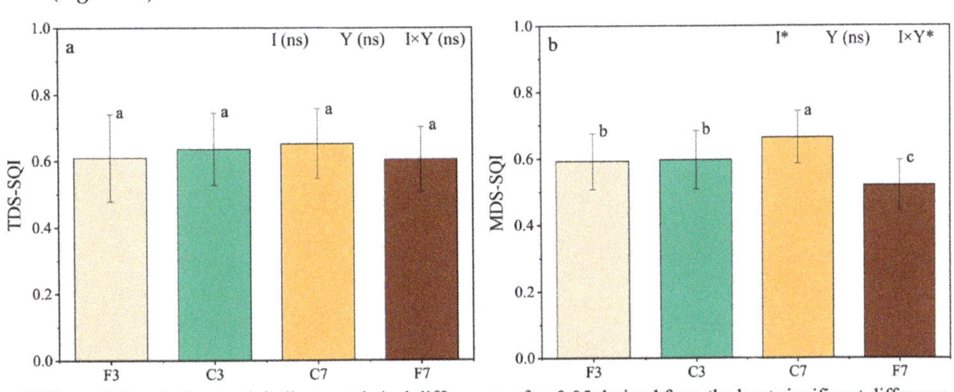


Different letters in the graph indicate statistical differences of p<0.05 derived from the least significant difference test. Error bars represent standard deviation.ns and * mean no significant difference and significant difference at *p*<0.05, respectively.

**Figure 3.** (**a**)TDS-SQI and (**b**)MDS-SQI of paddy soil in different treatments. Note: I: irrigation regime; Y: year.

The MDS-SQI values were followed the same ranking order: C7 > C3 > F3 > F7, with C7 exhibiting a higher SQI of 11.05%, 11.97%, and 27.71% compared to C3, F3, and F7, respectively. The irrigation regime had a significant effect on MDS-SQI values of paddy soils, while the straw incorporation years did not have a significant effect on MDS-SQI. However, when the two factors were interacted, the MDS-SQI showed a significant effect ($p < 0.05$) (Figure 3b).

The TDS-SQI and MDS-SQI were significantly correlated for each treatment ($p < 0.01$). Furthermore, the $E_f$ and $E_R$ were calculated using Equations (3) and (4) to validate the rationality, respectively. The $E_f$ values measured for F3, C3, C7, and F7 were 0.67, 0.63, 0.78, and 0.72, respectively, and the ER values were 0.11, 0.16, 0.08, and 0.01. These findings suggested that the MDS was a more effective evaluation method for soil quality compared to the TDS.

### *2.5. Correlation between Rice Yield and MDS-SQI*

As the MDS-SQI increased, the yields of rice for the F3 ($R^2 = 0.93$), C3 ($R^2 = 0.94$), C7 ($R^2 = 0.85$), and F7 ($R^2 = 0.90$) treatments also increased (Figure 4). The rice yield in the F3, C3, C7, and F7 were significantly and positively correlated with their corresponding MDS-SQIs ($p < 0.01$), indicating that MDS-SQI correctly evaluates the soil quality.

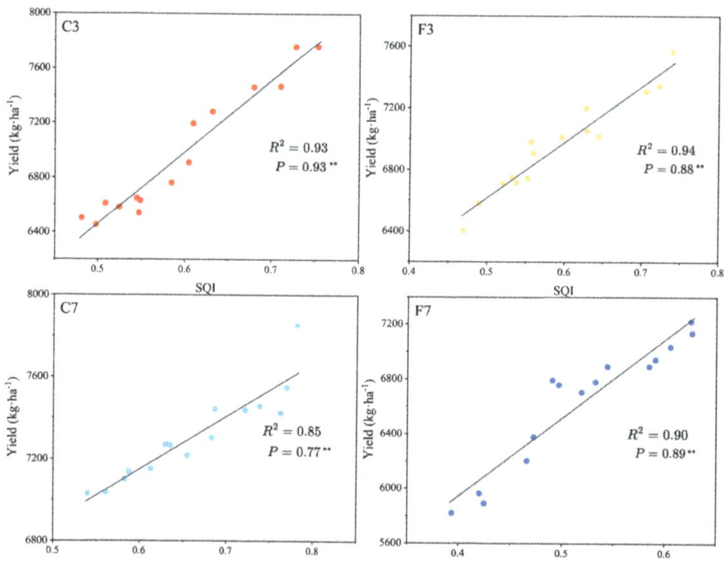


** mean significant at $p$<0.01.

**Figure 4.** Correlation analysis of rice yield and SQI under different treatments. Note: $P$ is the correlation between SQI and yield.

## 3. Discussion

### *3.1. Changes in Soil Physicochemical Properties*

In addition to being a complete indication of soil chemical characteristics, soil pH is essential for controlling microbial activity and nutrient availability. He et al. [47] concluded that controlling soil water condition increased soil pH, which was inconsistent with the conclusions of this study. Our findings revealed a significant decrease in pH after 7 years of continuous straw incorporation under controlled irrigation. The anaerobic decomposition of straw incorporated, which generated reducible organic acids, might be responsible for this reduction in pH [48]. Conversely, we observed the highest pH in the F7 treatment. This was because persistent flooding conditions led to a lowering of redox potential, which promoted pH elevation [49]. The availability of $NH_4^+$-N/$NO_3^-$-N directly affected crop nutrient uptake [50]. Our study revealed that under flooded conditions, $NO_3^-$-N/$NH_4^+$-N decreased with increasing years of straw incorporation. In contrast, under controlled irrigation conditions, we observed the opposite trend. Previous studies have demonstrated the significant impact of soil acidification on $NO_3^-$-N/$NH_4^+$-N [51,52]. The increase in pH might restrict the movement of $NO_3^-$-N through the root system [53], thereby affecting nitrogen metabolism. Moreover, anaerobic environments favored the accumulation of $NH_4^+$-N [49].

According to previous studies, incorporating straw into the field might increase the potassium content of the soil and reduce the demand for potash fertilizer [54]. However, as the straw incorporation years increased, our study revealed a considerable decline in AK content. This was especially true for the C7 and F7 treatments, where AK content dropped below the crucial threshold of 100 mg $kg^{-1}$, suggesting a serious deficiency. The long-term application of nitrogen, which enhanced crop yields and increased plant uptake of potassium, was probably responsible for the AK reduction. Conversely, long-term straw incorporation might increase the potassium content in the farmland moisture, making it susceptible to runoff losses [54]. The AP content in all treatments exceeded the critical level of 10 mg $kg^{-1}$. The increase in soil pH under water-saving irrigation could reduce the adsorption and immobilization of AP. Conversely, continuous flooded irrigation combined with straw incorporation increased the total soil reductant content, potentially inhibiting AP uptake by rice [48]. Furthermore, a decrease in pH altered the form and availability of elements in the soil solution, leading to phosphorus deficiency [49].

SOM is an essential indication for assessing the level of soil fertility because it provides plants with nutrients and energy for soil microbial activities [55,56]. A previous study indicated a positive relationship between SOM and the productivity of paddy yield [57], and a higher SOM content was associated with the soil's ability to retain fertilizers and increase yields [58], which is different from this study. The SOM of the F7 was significantly higher than that of the F3, indicating that long-term straw incorporation effectively increased the SOC content, thereby providing more nutrients for plants [59]. Enhanced SOC content could be linked to reduced pH, which inhibited SOC decomposition under controlled irrigation [60]. The DOC content generally constituted less than 2% of the total SOC [61]. This study observed a decrease in DOC content with increasing years of straw incorporation. Long-term tillage practices led to a significant depletion of DOC content, and consecutive years of straw incorporation were insufficient to compensate for this reduction [62].

Soil TN is a key factor in determining soil fertility [63]. Loss of nitrogen from paddy fields usually resulted in nitrogen deficiency [64]. We observed that increasing of incorporating straw years under flooded irrigation significantly enhanced the nitrogen content in the soil. This effect was likely due to the additional nitrogen contributed to the soil through the decomposition of straw, resulting in the formation of humus and other organic compounds. Organic nitrogen is easily stored in soil [65]. The C7 exhibited the highest C/N ratio and significantly higher crop yield ($p < 0.05$). This improvement might be attributed to the involvement of microorganisms in converting straw nitrogen during the process of straw carbon conversion, thereby influencing soil nitrogen [66].

Numerous studies have shown that the amounts of MBC and MBN represent the soil's ability to store and recycle vital nutrients, making them significant markers of soil microbial fertility [67]. Increasing SOM did not result in a corresponding increase in MBN content in this study. This observation may be attributed to long-term tillage practices that lead to soil disturbance and suppress microbial activity [68]. Furthermore, changes in soil pH due to straw incorporation could also exert an influence on soil microbial biomass [69]. MBC/MBN can effectively characterize the proportion of active carbon to nitrogen in the soil. Some studies have pointed out that straw incorporation replenished carbon and nitrogen organisms in the soil and enriched microbial biomass [70]. Our findings indicated that MBC/MBN differed among treatments, with the highest MBC/MBN observed in F7, followed by C3, C7, and F3. Factors such as nitrogen and phosphorus losses and elevated soil temperature also contributed to a decrease in microbial population, thereby influencing the MBC/MBN.

### 3.2. MDS-Based Soil Quality Evaluation

The TDS and MDS have been widely used in soil quality evaluation [9,71]. The TDS approach considers all identified variables [36], but it has limitations related to indicator redundancy and the complexity of data interpretation [7]. In our study, separate datasets were established for different treatments, and the results revealed no significant difference in TDS-SQI. This suggested that TDS might not be suitable for analyzing changes in soil quality of paddy fields in cold regions. In contrast, the MDS can accurately reflect soil quality, reduces data redundancy, and is highly adaptable [37,72]. The reliability of MDS was verified through correlation, $E_f$ and $E_R$, indicating that MDS outperformed TDS in effectively and accurately evaluating soil quality, and provided valuable information for farmland management.

Previous studies have demonstrated the positive impact of straw incorporation on soil quality and subsequent crop yield improvement [73]. In the present study, the MDS-SQI for the four treatments followed the order of C7 > C3 > F3 > F7. Furthermore, significant and positive correlations were observed between the MDS-SQI and yields of all treatments ($p< 0.01$), indicating the influence of soil quality on paddy yield across all treatments. Specifically, the SQI values decreased under the flooded irrigation with long-term straw incorporation, whereas SQI values significantly increased ($p< 0.05$) under controlled irrigation with increasing years of straw incorporation. Importantly, the rice yields followed the order of C7 > C3 > F3 > F7, with the SQI level directly determining the yield [74]. Zhu et al. [75] observed that continuous straw incorporation combined with the promotion of soil material cycling enhanced soil fertility and increased crop yields. While initial straw incorporation under flooded irrigation improved soil quality and subsequently increased yields [76], long-term straw incorporation was less beneficial than controlled irrigation in enhancing paddy yield and soil quality. This may be attributed to prolonged flooding negatively affecting rice root growth, with root systems primarily concentrating on the soil surface and inhibiting proper tillering [26]. Additionally, prolonged flooding might accumulate toxic substances in soil, cause a decrease in microbial populations [77], and diminish soil fertility, resulting in lower yields.

This study describes changes in soil nutrient characteristics. However, future studies should also consider soil microbial indicators. Soil microorganisms facilitate nutrient transformation and maintain soil ecological health [78]. Additionally, soil indicators closely related to water management, such as redox potential and soil enzyme activity, should also be included.

## 4. Materials and Methods

### 4.1. Study Area

The experiment was conducted at Qing'an National Key Station of Irrigation Experiment in Heilongjiang Province (127°40′44′ E, 46°57′29′ N). This station has a cold-temperate continental monsoon climate with low precipitation in spring, which is prone to drought, and early cold spells in autumn. From the rice re-greening stage to maturity, the average precipitation over the years is 550 mm. The soil used for the experiment was sandy clay loam, with a soil tillage thickness of 11.5 cm. Rice has been a major food crop in the study site for over forty years. To increase yields and save water, measures such as water-saving irrigation have been proposed since 2004. Figure 5 shows the maximum, minimum, and average air temperatures together with the amount of precipitation at the experimental site in 2023 during the rice growth period.

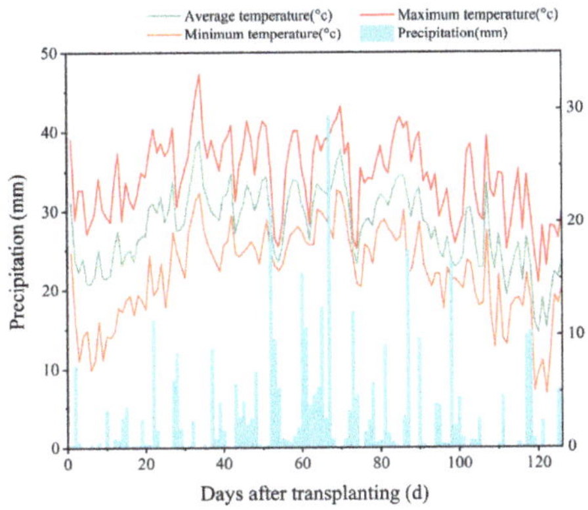


**Figure 5.** Air temperature and precipitation during the rice growth period in 2023.

### 4.2. Experimental Design

The experimental plots underwent consecutive years of straw incorporation experiments, initiated in 2016 and 2020, respectively, and have been ongoing since then. The basic properties of straw are shown in Table 3. The experiment of this study took place from May to September 2023, encompassing 7 and 3 consecutive years of straw incorporation, respectively. Two irrigation regimes—controlled irrigation (C) and flooded irrigation (F)—were used in this study. Table 4 detailed the experimental treatments and water management during the rice growth period of each treatment; there were a total of four treatments: C3, F3, C7, and F7; three randomized replications were set up for each experimental treatment. The primary local cultivar used throughout the entire experimental cycle was "Suijing18", planted at a density of 24 cm × 16 cm, with three plants per hill. The fertilization management and straw incorporation standards adopted were in line with the recommendations of Nie et al. [79].

**Table 3.** The basic properties of straw.

| Indicators | Values |
|---|---|
| pH | 7.05 |
| Total C (%) | 37.06 |
| Total N (g kg$^{-1}$) | 6.34 |
| Total P (g kg$^{-1}$) | 2.31 |
| Total K (g kg$^{-1}$) | 10.67 |

**Table 4.** Experimental treatments and water management of rice at each growth stage under different irrigation regimes.

| Treatments | Irrigation Regimes | Straw Incorporation Years | Regreening Stage | Tillering Stage | Later Tillering | Booting Stage | Flowering Stage | Milk Stage | Mature Stage |
|---|---|---|---|---|---|---|---|---|---|
| C3<br>C7 | Controlled irrigation | 3 | 0~30 mm | 70%~100% $\theta_s$ | Drainage | 80%~100% $\theta_s$ | 80%~100% $\theta_s$ | 70%~100% $\theta_s$ | Naturally drying |
| F3<br>F7 | Flooded irrigation | 7 | 0~30 mm | 20~50 mm | Drainage | 60~80 mm | 60~80 mm | 0~30 mm | |

Note: $\theta_s$ refers to soil saturated water content of root layer.

### 4.3. Sample Collection and Determination

At the mature stage of rice, to avoid marginal effects, a buffer of two rows and two columns around the plots was excluded. Yield measurements were taken from rice plants of 3 m$^2$ collected separately from each plot and manually threshed [79].

On 23 September 2023, soil samples were collected from the middle of each plot by soil auger at 5 points in the 0~20 cm soil layer and preserved according to the protocol of Zhang et al. [56]. During the soil natural air-drying process, once the soil became crushable, larger sample pieces were broken to expedite drying. The samples were regularly turned to eliminate any debris or stones and were sieved and mixed for soil physicochemical property analyses. The determination of soluble organic nitrogen (DON) could not be performed directly and therefore relied on the differential subtraction method. Table 5 shows the remaining soil physicochemical analysis methods.

**Table 5.** Methods for soil physicochemical properties analysis.

| Soil Indicators | Methods | Abbreviations | References |
|---|---|---|---|
| Soil Acidity and Alkalinity | pH meter | pH | Ran et al. [80] |
| Ammonium Nitrogen | Indophenol blue colorimetric | $NH_4^+$-N | Nie et al. [79] |
| Nitrate Nitrogen | UV dual-wavelength | $NO_3^-$-N | |
| Available Phosphorus | $NaHCO_3$-leaching—molybdenum antimony colorimetric | AP | Yang et al. [37] |
| Available Potassium | Flame photometer analysis | AK | |
| Total Nitrogen | Kjeldahl method | TN | Qi et al. [81] |
| Soil Organic Carbon | | SOC | |
| Soil Organic Matter | Dichromate oxidation | SOM | Das et al. [67] |
| Dissolved Organic Carbon | | DOC | |
| Microbial Biomass Carbon | Chloroform-fumigation technique | MBC | Marion et al. [7] |
| Microbial Biomass Nitrogen | | MBN | |

### 4.4. Establishment of TDS and MDS Metrics

The TDS considered all the identified variables (Table 5) and the MDS is developed by screening using principal component analysis (PCA) to evaluate the indicators of the quality of the tillage layer. First, the selected indicators were subjected to PCA. The magnitude of the eigenvalues indicated the extent to which the principal components represented the variability in the data. Within each group of principal components, the factor loading variables that contributed most were identified, with high factor loading indicating greater importance within that particular principal component. An indicator with a factor loading close to or reaching 90% of the maximum factor loading in that principal component was considered a high loading indicator and included in the MDS. It is important to determine their interrelationships when a principal component has several indicators with high factor loadings.

The formula for calculating the weight ($W_{Ni}$) of each indicator is as follows:

$$W_{Ni} = W_i / \sum_{i=1}^{n} W_i \quad (1)$$

where: $W_i$ is the variance contribution of all indicators in the dataset, $n$ is the number of indicators considered.

### 4.5. Soil Quality Evaluation

The main affiliation scoring functions were categorized into three types: positive S, inverse S, and parabolic. For indicators without well-defined boundaries, a simple linear scoring method [81,82] was employed, where the highest value of the measured value was scored as 1, and the ratio of the other measured values to this highest value was equal to the respective score.

The SQI was calculated using the following formula:

$$SQI_i = \sum_{i}^{n} W_{Ni} \times S_i \quad (2)$$

where: $S_i$ denotes the indicator score.

Finally, the accuracy of SQI was verified using $E_f$ and $E_R$. The closer $E_f$ is to 1, the more accurate the result, as it indicates that the MDS-SQI is closer to the baseline value.

The closer $E_R$ is to 0, the more accurate the result, as it indicates that the MDS-SQI has less deviation from the benchmark value. The calculation formula is as follows [83]:

$$E_f = 1 - \frac{\sum(R_0 - R_{cal})^2}{\sum(R_0 - \overline{R}_0)^2} \quad (3)$$

$$E_R = \frac{\left|\sum_{i=1}^{n} R_{0i} - \sum_{i=1}^{n} R_{cali}\right|}{\sum_{i=1}^{n} R_{0i}} \quad (4)$$

where: $R_0$ and $\overline{R}_0$ are the SQI and the average value of the SQI calculated based on the TDS, respectively. $R_{cal}$ is the MDS-SQI. Figure 6 illustrates the specific procedures for evaluating and analyzing soil quality (refer to Karaca et al. [84] for plotting).

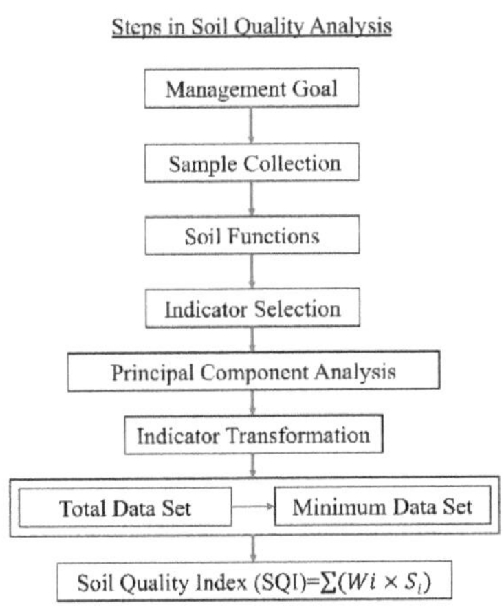


**Figure 6.** Comprehensive procedures for evaluating and analyzing soil quality.

### 4.6. Statistical Analysis

This study employed one-way analysis of variance (ANOVA) to determine the significance of differences in soil physicochemical indicators and yield among different treatments. The interactions between straw incorporation years and irrigation regimes were examined using a two-way ANOVA. The relationship between SQI and yield was assessed by linear regression. Statistical analyses were conducted using SPSS 26.0 (IBM Co., New York, NY, USA), and resulting plots were generated using Origin 2023 (Origin Lab Corporation, Northampton, MA, USA).

## 5. Conclusions

This study presents an analysis of the soil physicochemical quality conditions of rice paddies in cold region, focusing on how successive years of straw incorporation and different irrigation regimes affect soil quality. The results indicated that the C3 showed improvements in AK, AP, TN, DON, and MBC compared to the F3, with increases of 8.91%, 16.46%, 37.81%, 40%, and 23.28%, respectively. The levels of $NO_3^-$-N, AP, SOC, and SOM were higher in the C7 compared to the C3, with increases of 93.51%, 5.80%, 8.90%, and 8.26% respectively. The MDS-SQI values followed the same ranking order: C7>C3>F3>F7, with C7 exhibiting a higher SQI of 11.05%, 11.97%, and 27.71% compared to C3, F3, and F7, respectively. In conclusion, this study demonstrated that long-term straw incorporation under controlled irrigation enhanced rice yield and improved soil quality, while long-term straw incorporation under flooded irrigation did not necessarily result in increased yield and fertilizer conservation. These results help diagnose the condition of the soil quality in cold regions and provide a basis for agricultural practices in rice production in Northeast

China, which is important for preserving food security and guaranteeing the sustainability of agricultural output.

**Supplementary Materials:** The following supporting information can be downloaded at: https://www.mdpi.com/article/10.3390/plants13101357/s1, Figure S1. Principal component analysis for (a) F3, (b) C3, (c) C7 and (d) F7. Table S1. Correlation coefficients and sum of correlation coefficients of high weight indicators under different treatments. Table S2. Correlation of high factor load indicators under different treatments. Table S3. Weights of TDS evaluation indicators of paddy soil under different treatments. Table S4. Weights of MDS evaluation indicators of paddy soil under different treatments.

**Author Contributions:** P.Z.: Investigation, Methodology, Data curation, Writing—original draft. T.N. and Z.Z.: Conceptualization, Supervision, Writing—review and editing. P.C., T.L., C.D., L.J., Y.W., Z.S. and S.Y.: Writing—review and editing. All authors have read and agreed to the published version of the manuscript.

**Funding:** This research was funded by the Natural Science Foundation Project of Heilongjiang Province (LH2023E109), the National Natural Science Foundation Project of China (52079028), the Basic Scientific Research Fund of Heilongjiang Provincial Universities (2021-KYYWF-0019), the Opening Project of Key Laboratory of Efficient Use of Agricultural Water Resources, Ministry of Agriculture and Rural Affairs of the People's Republic of China (AWR2021002) and Postdoctoral Fellowship Program of CPSF (GZC20230668).

**Data Availability Statement:** Data that support the findings of this study are available from the corresponding author upon reasonable request.

**Acknowledgments:** We thank all the members in Lab of Pumping, Hydraulic Teaching and Experimental Center of Heilongjiang University. We also thank the anonymous reviewers and the editor for their suggestions, which substantially improved the manuscript.

**Conflicts of Interest:** The authors declare that they have no known competing financial interests or personal relationships that could have appeared to influence the work reported in this paper.

## References


1. Muthayya, S.; Sugimoto, J.D.; Montgomery, S.; Maberly, G.F. An overview of global rice production, supply, trade, and consumption. *Ann. N. Y. Acad. Sci.* **2014**, *1324*, 7–14. [CrossRef] [PubMed]
2. Zhao, X.; Chen, M.; Xie, H.; Luo, W.; Wei, G.; Zheng, S.; Wu, C.; Khan, S.; Cui, Y.; Luo, Y. Analysis of irrigation demands of rice: Irrigation decision-making needs to consider future rainfall. *Agric. Water Manag.* **2023**, *280*, 108196. [CrossRef]
3. Ray, D.K.; Ramankutty, N.; Mueller, N.D.; West, P.C.; Foley, J.A. Recent patterns of crop yield growth and stagnation. *Nat. Commun.* **2012**, *3*, 1293. [CrossRef] [PubMed]
4. Chen, S.; Elrys, A.S.; Zhao, C.; Cai, Z.; Zhang, J.; Müller, C. Global patterns and controls of yield and nitrogen use efficiency in rice. *Sci. Total Environ.* **2023**, *898*, 165484. [CrossRef] [PubMed]
5. Shao, X.; Gao, J.; Liu, Y.; Wang, X.; Guo, L. Effects of plastic mulching and straw incorporation on rice yield and nitrogen use efficiency in a cold region. *Eng. Agric.* **2021**, *41*, 504–516. [CrossRef]
6. Rinot, O.; Levy, G.J.; Steinberger, Y.; Svoray, T.; Eshel, G. Soil health assessment: A critical review of current methodologies and a proposed new approach. *Sci. Total Environ.* **2019**, *648*, 1484–1491. [CrossRef] [PubMed]
7. Marion, L.F.; Schneider, R.; Cherubin, M.R.; Colares, G.S.; Wiesel, P.G.; da Costa, A.B.; Lobo, E.A. Development of a soil quality index to evaluate agricultural cropping systems in southern Brazil. *Soil Tillage Res.* **2022**, *218*, 150293. [CrossRef]
8. Tang, L.; Hayashi, K.; Ohigashi, K.; Shimura, M.; Kohyama, K. Developing characterization factors to quantify management impacts on soil quality of paddy fields within life cycle assessment. *J. Clean. Prod.* **2019**, *238*, 117890. [CrossRef]
9. Li, Y.; Ma, J.; Li, Y.; Jia, Q.; Shen, X.; Xia, X. Spatiotemporal variations in the soil quality of agricultural land and its drivers in China from 1980 to 2018. *Sci. Total Environ.* **2023**, *892*, 164649. [CrossRef]
10. Dossou-Yovo, E.R.; Brüggemann, N.; Jesse, N.; Huat, J.; Ago, E.E.; Agbossou, E.K. Reducing soil $CO_2$ emission and improving upland rice yield with no-tillage, straw mulch and nitrogen fertilization in northern Benin. *Soil Tillage Res.* **2016**, *156*, 44–53. [CrossRef]
11. Amirahmadi, E.; Moudrý, J.; Konvalina, P.; Hörtenhuber, S.J.; Ghorbani, M.; Neugschwandtner, R.W.; Jiang, Z.; Krexner, T.; Kopecký, M. Environmental Life Cycle Assessment in Organic and Conventional Rice Farming Systems: Using a Cradle to Farm Gate Approach. *Sustainability* **2022**, *14*, 15870. [CrossRef]
12. Liu, C.; Lu, M.; Cui, J.; Li, B.; Fang, C. Effects of straw carbon input on carbon dynamics in agricultural soils: A meta-analysis. *Glob. Change Biol.* **2014**, *20*, 1366–1381. [CrossRef] [PubMed]

13. Yan, S.S.; Liu, C.X.; Li, J.A.; Li, J.W.; Cui, C.; Fan, J.S.; Gong, Z.P.; Zhang, Z.X.; Yan, C. Changes in the soil phosphorus supply with rice straw return in cold region. *Agronomy* **2023**, *13*, 2214. [CrossRef]
14. Zhang, X.; Wang, J.; Feng, X.; Yang, H.; Li, Y.; Yakov, K.; Liu, S.; Li, F.-M. Effects of tillage on soil organic carbon and crop yield under straw return. *Agric. Ecosyst. Environ.* **2023**, *354*, 18053. [CrossRef]
15. Li, Z.K.; Shen, Y.; Zhang, W.Y.; Zhang, H.; Liu, L.J.; Wang, Z.Q.; Gu, J.F.; Yang, J.C. Effects of long-term straw returning on rice yield and soil properties and bacterial community in a rice-wheat rotation system. *Field Crops Res.* **2023**, *291*, 108800. [CrossRef]
16. Yang, S.; Chen, L.; Xiong, R.; Jiang, J.; Liu, Y.; Tan, X.; Liu, T.; Zeng, Y.; Pan, X.; Zeng, Y. Long-term straw return improves cooked indica rice texture by altering starch structural, physicochemical properties in South China. *Food Chem. X* **2023**, *20*, 100965. [CrossRef] [PubMed]
17. Chen, K.; Ma, T.; Ding, J.; Yu, S.; Dai, Y.; He, P.; Ma, T. Effects of straw return with nitrogen fertilizer reduction on rice (*Oryza sativa* L.) morphology, photosynthetic capacity, yield and water–nitrogen use efficiency traits under different water regimes. *Agronomy* **2022**, *13*, 133. [CrossRef]
18. Zhu, X.; Xie, H.; Masters, M.D.; Rui, Y.; Luo, Y.; He, H.; Zhang, X.; Liang, C. Microorganisms, their residues, and soil carbon storage under a continuous maize cropping system with eight years of variable residue retention. *Appl. Soil Ecol.* **2023**, *187*, 104846. [CrossRef]
19. Liu, B.; Wu, Q.; Wang, F.; Zhang, B. Is straw return-to-field always beneficial? Evidence from an integrated cost-benefit analysis. *Energy* **2019**, *171*, 393–402. [CrossRef]
20. Yu, F.; Chen, Y.; Huang, X.; Shi, J.; Xu, J.; He, Y. Does straw returning affect the root rot disease of crops in soil? A systematic review and meta-analysis. *J. Environ. Manag.* **2023**, *336*, 117673. [CrossRef]
21. Chen, S.; Zheng, X.; Wang, D.; Chen, L.; Xu, C.; Zhang, X. Effect of long-term paddy-upland yearly rotations on rice (Oryza sativa) yield, soil properties, and bacteria community diversity. *Sci. World J.* **2012**, *2012*, 279641. [CrossRef]
22. Liu, S.; Qin, Y.; Zou, J.; Liu, Q. Effects of water regime during rice-growing season on annual direct $N_2O$ emission in a paddy rice–winter wheat rotation system in southeast China. *Sci. Total Environ.* **2010**, *408*, 906–913. [CrossRef] [PubMed]
23. Champness, M.; Ballester, C.; Hornbuckle, J. Effect of soil moisture deficit on aerobic rice in temperate Australia. *Agronomy* **2023**, *13*, 168. [CrossRef]
24. Nie, T.; Chen, P.; Zhang, Z.; Qi, Z.; Lin, Y.; Xu, D. Effects of Different Types of Water and Nitrogen Fertilizer Management on Greenhouse Gas Emissions, Yield, and Water Consumption of Paddy Fields in Cold Region of China. *Int. J. Environ. Res. Public Health* **2019**, *16*, 1639. [CrossRef]
25. Wei, Q.; Wei, Q.; Xu, J.; Liu, Y.; Wang, D.; Chen, S.; Qian, W.; He, M.; Chen, P.; Zhou, X.; et al. Nitrogen losses from soil as affected by water and fertilizer management under drip irrigation: Development, hotspots and future perspectives. *Agric. Water Manag.* **2024**, *296*, 108791. [CrossRef]
26. Peng, S.-Z.; Yang, S.-H.; Xu, J.-Z.; Luo, Y.-F.; Hou, H.-J. Nitrogen and phosphorus leaching losses from paddy fields with different water and nitrogen managements. *Paddy Water Environ.* **2011**, *9*, 333–342. [CrossRef]
27. Han, Y.; Zhang, Z.; Li, T.; Chen, P.; Nie, T.; Zhang, Z.; Du, S. Straw return alleviates the greenhouse effect of paddy fields by increasing soil organic carbon sequestration under water-saving irrigation. *Agric. Water Manag.* **2023**, *287*, 108434. [CrossRef]
28. Liu, L.; Cheng, M.; Yang, L.; Gu, X.; Jin, J.; Fu, M. Regulation of straw decomposition and its effect on soil function by the amount of returned straw in a cool zone rice crop system. *Sci. Rep.* **2023**, *13*, 15673. [CrossRef]
29. Zhang, J.; Hou, J.; Zhang, H.; Meng, C.; Zhang, X.; Wei, C. Low soil temperature inhibits yield of rice under drip irrigation. *Eur. J. Agron.* **2019**, *19*, 228–236. [CrossRef]
30. Bhaduri, D.; Purakayastha, T.J. Long-term tillage, water and nutrient management in rice–wheat cropping system: Assessment and response of soil quality. *Soil Tillage Res.* **2014**, *144*, 83–95. [CrossRef]
31. Biswas, S.; Hazra, G.C.; Purakayastha, T.J.; Saha, N.; Mitran, T.; Singha Roy, S.; Basak, N.; Mandal, B. Establishment of critical limits of indicators and indices of soil quality in rice-rice cropping systems under different soil orders. *Geoderma* **2017**, *292*, 34–48. [CrossRef]
32. Dengiz, O. Soil quality index for paddy fields based on standard scoring functions and weight allocation method. *Arch. Agron. Soil Sci.* **2019**, *66*, 301–315. [CrossRef]
33. Wang, X.; Shi, W.; Sun, X.; Wang, M. Comprehensive benefits evaluation and its spatial simulation for well-facilitated farmland projects in the Huang-Huai-Hai Region of China. *Land Degrad. Dev.* **2020**, *31*, 1837–1850. [CrossRef]
34. Gunasekaran, Y.; Kaliappan, S.B.; Porpavai, S. Developing soil quality indices for different crop rotations of deltaic inceptisol regions of India. *Commun. Soil Sci. Plant Anal.* **2021**, *52*, 1363–1376. [CrossRef]
35. Majhi, P.; Rout, K.K.; Nanda, G.; Singh, M. Soil quality for rice productivity and yield sustainability under long-term fertilizer and manure application. *Commun. Soil Sci. Plant Anal.* **2019**, *50*, 1330–1343. [CrossRef]
36. Paz-Ferreiro, J.; Liu, Z.; Rong, Q.; Zhou, W.; Liang, G. Effects of inorganic and organic amendment on soil chemical properties, enzyme activities, microbial community and soil quality in yellow clayey soil. *PLoS ONE* **2017**, *12*, e0172767. [CrossRef]
37. Yang, E.; Zhao, X.; Qin, W.; Jiao, J.; Han, J.; Zhang, M. Temporal impacts of dryland-to-paddy conversion on soil quality in the typical black soil region of China: Establishing the minimum data set. *Catena* **2023**, *231*, 107303. [CrossRef]
38. Dinh Thi, L.; Thi Hang, N.; Nguyen, H.; Nguyen, L. Rice growth, grain zinc, and soil properties under saline irrigation conditions. *J. Ecol. Eng.* **2021**, *22*, 58–69. [CrossRef]

39. Liu, Y.; Li, J.; Jiao, X.; Li, H.; Hu, T.; Jiang, H.; Mahmoud, A. Effects of biochar on water quality and rice productivity under straw returning condition in a rice-wheat rotation region. *Sci. Total Environ.* **2022**, *819*, 152063. [CrossRef] [PubMed]
40. Xin, F.; Xiao, X.; Dong, J.; Zhang, G.; Zhang, Y.; Wu, X.; Li, X.; Zou, Z.; Ma, J.; Du, G.; et al. Large increases of paddy rice area, gross primary production, and grain production in Northeast China during 2000–2017. *Sci. Total Environ.* **2020**, *711*, 135183. [CrossRef] [PubMed]
41. Yang, D.; Wang, Y.; Wu, Q. Impact of Tillage and Straw Management on Soil Properties and Rice Yield in a Rice-Ratoon Rice System. *Agronomy* **2023**, *13*, 1762. [CrossRef]
42. Wang, E.; Lin, X.; Tian, L.; Wang, X.; Ji, L.; Jin, F.; Tian, C. Effects of short-term rice straw return on the soil microbial community. *Agriculture* **2021**, *11*, 561. [CrossRef]
43. Hou, Y.; Xu, X.; Kong, L.; Zhang, L.; Wang, L. The combination of straw return and appropriate K fertilizer amounts enhances both soil health and rice yield in Northeast China. *Agron. J.* **2021**, *113*, 5424–5435. [CrossRef]
44. Jiao, F.; Zhang, D.; Chen, Y.; Wu, J.; Zhang, J. Effects of long-term straw returning and nitrogen fertilizer reduction on soil microbial diversity in black soil in Northeast China. *Agronomy* **2023**, *13*, 2036. [CrossRef]
45. Du, S.; Zhang, Z.; Li, T.; Wang, Z.; Zhou, X.; Gai, Z.; Qi, Z. Response of rice harvest index to different water and nitrogen management modes in the black soil region of Northeast China. *Agriculture* **2022**, *12*, 115. [CrossRef]
46. Sun, Y.; Lai, Y.; Wang, Q.; Song, Q.; Jin, L.; Zeng, X.; Feng, Y.; Lu, X. Combination of water-saving irrigation and nitrogen fertilization regulates greenhouse gas emissions and increases rice yields in high-cold regions, Northeast China. *Int. J. Environ. Res. Public Health* **2022**, *19*, 16506. [CrossRef]
47. He, H.; Li, D.; Wu, Z.; Wu, Z.; Hu, Z.; Yang, S. Assessment of the straw and biochar application on greenhouse gas emissions and yield in paddy fields under intermittent and controlled irrigation patterns. *Agric. Ecosyst. Environ.* **2024**, *359*, 108745. [CrossRef]
48. Yang, H.; Feng, J.; Weih, M.; Meng, Y.; Li, Y.; Zhai, S.; Zhang, W. Yield reduction of direct-seeded rice under returned straw can be mitigated by appropriate water management improving soil phosphorus availability. *Crop Pasture Sci.* **2020**, *71*, 134–146. [CrossRef]
49. Borin, J.B.M.; Carmona, F.d.C.; Anghinoni, I.; Martins, A.P.; Jaeger, I.R.; Marcolin, E.; Hernandes, G.C.; Camargo, E.S. Soil solution chemical attributes, rice response and water use efficiency under different flood irrigation management methods. *Agric. Water Manag.* **2016**, *176*, 9–17. [CrossRef]
50. Wang, J.; Wang, D.; Zhang, G.; Wang, Y.; Wang, C.; Teng, Y.; Christie, P. Nitrogen and phosphorus leaching losses from intensively managed paddy fields with straw retention. *Agric. Water Manag.* **2014**, *141*, 66–73. [CrossRef]
51. Shaaban, M.; Peng, Q.-a.; Bashir, S.; Wu, Y.; Younas, A.; Xu, X.; Rashti, M.R.; Abid, M.; Zafar-ul-Hye, M.; Núñez-Delgado, A.; et al. Restoring effect of soil acidity and Cu on $N_2O$ emissions from an acidic soil. *J. Environ. Manag.* **2019**, *250*, 109535. [CrossRef]
52. Shaaban, M.; Wu, Y.; Khalid, M.S.; Peng, Q.-a.; Xu, X.; Wu, L.; Younas, A.; Bashir, S.; Mo, Y.; Lin, S.; et al. Reduction in soil $N_2O$ emissions by pH manipulation and enhanced nosZ gene transcription under different water regimes. *Environ. Pollut.* **2018**, *235*, 625–631. [CrossRef]
53. Wang, H.; Ahan, J.; Wu, Z.; Shi, D.; Liu, B.; Yang, C. Alteration of nitrogen metabolism in rice variety 'Nipponbare' induced by alkali stress. *Plant Soil* **2011**, *355*, 131–147. [CrossRef]
54. Xiong, Z.; Zhu, D.; Lu, Y.; Lu, J.; Liao, Y.; Ren, T.; Li, X. Continuous potassium fertilization combined with straw return increased soil potassium availability and risk of potassium loss in rice-upland rotation systems. *Chemosphere* **2023**, *344*, 140390. [CrossRef]
55. Nath, C.P.; Kumar, N.; Das, K.; Hazra, K.K.; Praharaj, C.S.; Singh, N.P. Impact of variable tillage based residue management and legume based cropping for seven years on enzymes activity, soil quality index and crop productivity in rice ecology. *Environ. Sustain. Indic.* **2021**, *10*, 100107. [CrossRef]
56. Zhang, J.; Nie, J.; Cao, W.; Gao, Y.; Lu, Y.; Liao, Y. Long-term green manuring to substitute partial chemical fertilizer simultaneously improving crop productivity and soil quality in a double-rice cropping system. *Eur. J. Agron.* **2023**, *142*, 126641. [CrossRef]
57. Liu, Z.; Zhou, W.; Shen, J.; He, P.; Lei, Q.; Liang, G. A simple assessment on spatial variability of rice yield and selected soil chemical properties of paddy fields in South China. *Geoderma* **2014**, *235–236*, 39–47. [CrossRef]
58. Pan, G.; Smith, P.; Pan, W. The role of soil organic matter in maintaining the productivity and yield stability of cereals in China. *Agric. Ecosyst. Environ.* **2009**, *129*, 344–348. [CrossRef]
59. Wang, S.; Zhai, L.; Guo, S.; Zhang, F.; Hua, L.; Liu, H. Returned straw reduces nitrogen runoff loss by influencing nitrification process through modulating soil C:N of different paddy systems. *Agric. Ecosyst. Environ.* **2023**, *354*, 108438. [CrossRef]
60. Dong, Z.; Li, H.; Xiao, J.; Sun, J.; Liu, R.; Zhang, A. Soil multifunctionality of paddy field is explained by soil pH rather than microbial diversity after 8-years of repeated applications of biochar and nitrogen fertilizer. *Sci. Total Environ.* **2022**, *853*, 158620. [CrossRef] [PubMed]
61. Aumtong, S.; Chotamonsak, C.; Glomchinda, T. Study of the interaction of dissolved organic Carbon, available nutrients, and clay content driving soil carbon storage in the rice rotation cropping system in Northern Thailand. *Agronomy* **2023**, *13*, 142. [CrossRef]
62. Park, S.-I.; Yang, H.I.; Park, H.-J.; Seo, B.-S.; Jeong, Y.-J.; Lim, S.-S.; Kwak, J.-H.; Kim, H.-Y.; Yoon, K.-S.; Lee, S.-M.; et al. Rice straw cover decreases soil erosion and sediment-bound C, N, and P losses but increases dissolved organic C export from upland maize fields as evidenced by δ13C. *Sci. Total Environ.* **2021**, *753*, 142053. [CrossRef]
63. Al-Kaisi, M.M.; Yin, X.; Licht, M.A. Soil carbon and nitrogen changes as influenced by tillage and cropping systems in some Iowa soils. *Agric. Ecosyst. Environ.* **2005**, *105*, 635–647. [CrossRef]

64. Bashir, M.A.; Zhai, L.-m.; Wang, H.-y.; Liu, J.; Raza, Q.-U.-A.; Geng, Y.-c.; Rehim, A.; Liu, H.-b. Apparent variations in nitrogen runoff and its uptake in paddy rice under straw incorporation. *J. Integr. Agric.* **2022**, *21*, 3356–3367. [CrossRef]
65. Cui, S.; Zhu, X.; Cao, G. Effects of years of rice straw return on soil nitrogen components from rice–wheat cropped fields. *Agronomy* **2022**, *12*, 1247. [CrossRef]
66. Liu, N.; Li, Y.; Cong, P.; Wang, J.; Guo, W.; Pang, H.; Zhang, L. Depth of straw incorporation significantly alters crop yield, soil organic carbon and total nitrogen in the North China Plain. *Soil Tillage Res.* **2021**, *205*, 104772. [CrossRef]
67. Das, S.; Bhattacharyya, R.; Das, T.K.; Sharma, A.R.; Dwivedi, B.S.; Meena, M.C.; Dey, A.; Biswas, S.; Aditya, K.; Aggarwal, P.; et al. Soil quality indices in a conservation agriculture based rice-mustard cropping system in North-western Indo-Gangetic Plains. *Soil Tillage Res.* **2021**, *208*, 104914. [CrossRef]
68. Bu, R.; Ren, T.; Lei, M.; Liu, B.; Li, X.; Cong, R.; Zhang, Y.; Lu, J. Tillage and straw-returning practices effect on soil dissolved organic matter, aggregate fraction and bacteria community under rice-rice-rapeseed rotation system. *Agric. Ecosyst. Environ.* **2020**, *287*, 106681. [CrossRef]
69. Li, Y.; Feng, H.; Dong, Q.; Xia, L.; Li, J.; Li, C.; Zang, H.; Andersen, M.N.; Olesen, J.E.; Jørgensen, U.; et al. Ammoniated straw incorporation increases wheat yield, yield stability, soil organic carbon and soil total nitrogen content. *Field Crops Res.* **2022**, *284*, 108558. [CrossRef]
70. Liu, X.; Liu, H.; Zhang, Y.; Chen, G.; Li, Z.; Zhang, M. Straw return drives soil microbial community assemblage to change metabolic processes for soil quality amendment in a rice-wheat rotation system. *Soil Biol. Biochem.* **2023**, *185*, 109131. [CrossRef]
71. Rahmanipour, F.; Marzaioli, R.; Bahrami, H.A.; Fereidouni, Z.; Bandarabadi, S.R. Assessment of soil quality indices in agricultural lands of Qazvin Province, Iran. *Ecol. Indic.* **2014**, *40*, 19–26. [CrossRef]
72. Jin, H.; Shi, D.; Lou, Y.B.; Zhang, J.; Ye, Q.; Jiang, N. Evaluation of the quality of cultivated-layer soil based on different degrees of erosion in sloping farmland with purple soil in China. *Catena* **2021**, *198*, 105048. [CrossRef]
73. Wang, X.; Wang, Q.; Zhang, Y.; Zhang, J.; Xia, S.; Qin, H.; Feng, C.; Bie, S. Influence of decomposition agent application and schedule in wheat straw return practice on soil quality and crop yield. *Chem. Biol. Technol. Agric.* **2023**, 10. [CrossRef]
74. Adak, S.; Bandyopadhyay, K.; Purakayastha, T.J.; Sen, S.; Sahoo, R.N.; Shrivastava, M.; Krishnan, P. Impact of contrasting tillage, residue mulch and nitrogen management on soil quality and system productivity under maize-wheat rotation in the north-western Indo-Gangetic Plains. *Front. Sustain. Food Syst.* **2023**, *7*, 1230207. [CrossRef]
75. Zhu, D.; Lu, J.; Cong, R.; Ren, T.; Zhang, W.; Li, X. Potassium management effects on quantity/intensity relationship of soil potassium under rice-oilseed rape rotation system. *Arch. Agron. Soil Sci.* **2019**, *66*, 1274–1287. [CrossRef]
76. Chang, F.; Zhang, H.; Song, J.; Yu, R.; Zhang, X.; Li, H.; Wang, J.; Kan, Z.; Li, Y. Once-middle amount of straw interlayer enhances saline soil quality and sunflower yield in semi-arid regions of China: Evidence from a four-year experiment. *J. Environ. Manag.* **2023**, *344*, 118530. [CrossRef]
77. Moyano, F.E.; Manzoni, S.; Chenu, C. Responses of soil heterotrophic respiration to moisture availability: An exploration of processes and models. *Soil Biol. Biochem.* **2013**, *59*, 72–85. [CrossRef]
78. Mi, W.; Sun, T.; Ma, Y.; Chen, C.; Ma, Q.; Wu, L.; Wu, Q.; Xu, Q. Higher yield sustainability and soil quality by manure amendment than straw returning under a single-rice cropping system. *Field Crops Res.* **2023**, *292*, 108805. [CrossRef]
79. Nie, T.; Huang, J.; Zhang, Z.; Chen, P.; Li, T.; Dai, C. The inhibitory effect of a water-saving irrigation regime on $CH_4$ emission in Mollisols under straw incorporation for 5 consecutive years. *Agric. Water Manag.* **2023**, *278*, 108163. [CrossRef]
80. Ran, C.; Gao, D.; Liu, W.; Guo, L.; Bai, T.; Shao, X.; Geng, Y. Straw and nitrogen amendments improve soil, rice yield, and roots in a saline sodic soil. *Rhizosphere* **2022**, *24*, 100606. [CrossRef]
81. Qi, Y.; Darilek, J.L.; Huang, B.; Zhao, Y.; Sun, W.; Gu, Z. Evaluating soil quality indices in an agricultural region of Jiangsu Province, China. *Geoderma* **2009**, *149*, 325–334. [CrossRef]
82. Zhang, X.; You, Y.; Wang, D.; Zhu, L. Quality evaluation of the soil-root composites layer of *Leymus chinensis* grassland based on different degradation degrees. *Catena* **2022**, *215*, 106330. [CrossRef]
83. Lin, S.; Lei, Q.; Liu, Y.; Zhao, Y.; Su, L.; Wang, Q.; Tao, W.; Deng, M. Quantifying the impact of organic fertilizers on soil quality under varied irrigation water sources. *Water* **2023**, *15*, 3618. [CrossRef]
84. Karaca, S.; Dengiz, O.; Demirağ Turan, İ.; Özkan, B.; Dedeoğlu, M.; Gülser, F.; Sargin, B.; Demirkaya, S.; Ay, A. An assessment of pasture soils quality based on multi-indicator weighting approaches in semi-arid ecosystem. *Ecol. Indic.* **2021**, *121*, 107001. [CrossRef]

*plants*

MDPI

*Article*

# Root Zone Water Management Effects on Soil Hydrothermal Properties and Sweet Potato Yield

**Shihao Huang, Lei Zhao, Tingge Zhang, Minghui Qin, Tao Yin, Qing Liu and Huan Li ***

College of Resources and Environmental Sciences, Qingdao Agricultural University, Qingdao 266109, China; hsh25495@163.com (S.H.); zl199511231111@163.com (L.Z.); zhangtingge0929@163.com (T.Z.); 19861607932@163.com (M.Q.); yintao@qau.edu.cn (T.Y.); qy7271@163.com (Q.L.)

* Correspondence: lihuancomcomcom@163.com; Tel.: +86-132-4532-5806; Fax: +86-0532-8803-0461

**Abstract:** Sufficient soil moisture is required to ensure the successful transplantation of sweet potato seedlings. Thus, reasonable water management is essential for achieving high quality and yield in sweet potato production. We conducted field experiments in northern China, planted on 18 May and harvested on 18 October 2021, at the Nancun Experimental Base of Qingdao Agricultural University. Three water management treatments were tested for sweet potato seedlings after transplanting: hole irrigation ($W_1$), optimized drip irrigation ($W_2$), and traditional drip irrigation ($W_3$). The variation characteristics of soil volumetric water content, soil temperature, and soil $CO_2$ concentration in the root zone were monitored in situ for 0–50 days. The agronomy, root morphology, photosynthetic parameters, $^{13}C$ accumulation, yield, and yield components of sweet potato were determined. The results showed that soil VWC was maintained at 22–25% and 27–32% in the hole irrigation and combined drip irrigation treatments, respectively, from 0 to 30 days after transplanting. However, there was no significant difference between the traditional ($W_3$) and optimized ($W_2$) drip irrigation systems. From 30 to 50 days after transplanting, the VWC decreased significantly in all treatments, with significant differences among all treatments. Soil $CO_2$ concentrations were positively correlated with VWC from 0 to 30 days after transplanting but gradually increased from 30 to 50 days, with significant differences among treatments. Soil temperature varied with fluctuations in air temperature, with no significant differences among treatments. Sweet potato survival rates were significantly lower in the hole irrigation treatments than in the drip irrigation treatments, with no significant difference between $W_2$ and $W_3$. The aboveground biomass, photosynthetic parameters, and leaf area index were significantly higher under drip irrigation than under hole irrigation, and values were higher in $W_3$ than in $W_2$. However, the total root length, root volume, and $^{13}C$ partitioning rate were higher in $W_2$ than in $W_3$. These findings suggest that excessive drip irrigation can lead to an imbalance in sweet potato reservoir sources. Compared with $W_1$, the $W_2$ and $W_3$ treatments exhibited significant yield increases of 42.98% and 36.49%, respectively. The $W_2$ treatment had the lowest sweet potato deformity rate.

**Keywords:** sweet potato; soil volumetric water content; soil temperature; soil $CO_2$ concentration; root morphology; physiological characteristics

## 1. Introduction

Despite the continuous development of agricultural irrigation systems, the irrational use of water resources has led to imbalances in water supply and demand that have gradually become a key factor restricting agricultural production in China [1]. The efficient utilization of water resources has become a crucial issue for sustainable agricultural development and improvement of the ecological environment [2].

Sweet potato is a drought-resistant crop that tolerates both waterlogging and drought. Natural precipitation is a crucial water source for sweet potato, and excessive or insufficient water supply can have adverse effects on its growth [3]. Northern China is among the main sweet potato production areas in China, with harvests typically conducted in the spring and summer. Water supply is the main factor limiting sweet potato production in northern China [4]. Watering seedlings after transplanting is crucial for their rooting and branching, and sweet potato fields in northern China are generally arid or semi-arid in spring, resulting in water shortages due to insufficient precipitation. In the past, sweet potato cultivation in China has often employed hole irrigation for transplanted seedlings; however, this method is time-consuming, laborious, and susceptible to drought stress, resulting in poor and unstable yields. In recent years, there has been an attempt to introduce drip irrigation for sweet potato seedling transplanting. Drip irrigation is preferred because it does not damage the soil structure and helps to maintain soil water, gases, and heat at suitable conditions. Additionally, this approach conserves both labor and water [5]. However, sweet potato producers tend to believe that more drip irrigation promotes seedling survival, resulting in excessive irrigation due to a lack of appropriate scientific guidance on seedling water amounts.

Soil is a complex medium composed of solid, liquid, and gaseous phases, which must be maintained in appropriate proportions to form a favorable environment for crop growth [6]. Studies have shown that excessive single-application irrigation in relatively permeable soil causes water to leak into deeper layers, keeping irrigation water away from areas where roots are concentrated and hindering the efficient use of irrigation water [7–9]. In addition, when soil water levels are too high, poor aeration can inhibit root respiration and slow the belowground development of sweet potatoes. In severe cases, over-irrigation can cause root rot, which reduces seedling growth rates [10–12]. Conversely, insufficient irrigation to meet the water demand of sweet potato seedlings may increase soil oxygen content, improving soil aeration but resulting in stunted crop roots [13,14].

A previous study altered soil water content, aeration, and soil temperature through the application of different irrigation levels after sweet potato transplantation; changes to the soil environment of the root zone indirectly affected the above- and belowground growth of sweet potato seedlings [15]. Other studies have primarily conducted pot experiments to examine individual soil water, air, or heat effects, without examining interactions among multiple factors. Additionally, few researchers have explored the spatiotemporal variability of sweet potato responses to factors related to soil water, air, and heat, or the impact of the root zone soil environment on sweet potato growth in domestic and international contexts. Therefore, in the present study, we conducted field experiments and in situ monitoring of various factors to investigate the impacts of soil water, aeration, and heat on sweet potato seedling growth to reveal their intrinsic relationships with sweet potato physiology. We predicted that water had a direct effect on the growth and yield of sweet potato, while temperature had an indirect effect. Our findings provide a theoretical basis by which to guide the rational irrigation of sweet potato plants after transplantation.

## 2. Results

### 2.1. Soil Volumetric Water Content (VWC), Soil Temperature, and Soil $CO_2$ Content Dynamics

Soil VWC increased as the amount of water provided to transplanted seedlings increased; this increase was significantly higher under drip irrigation than under hole irrigation ($p < 0.05$). However, there was no significant difference between optimized and traditional drip irrigation methods ($p > 0.05$). VWC remained stable in all treatments for the first month after transplantation; from 30 to 50 days, VWC gradually decreased during the fertility period, with significant differences among treatments ($p < 0.05$).

The soil temperature in the 0–20 cm soil layer increased over time, but with non-significant differences among the three treatments ($p > 0.05$). Significant cooling occurred on days 21 and 28, but the temperature quickly recovered thereafter. By day 35, the treatments had stabilized. These results suggest that soil temperature was primarily influenced by changes in atmospheric temperature.

Soil $CO_2$ content decreased as the irrigation amount increased. In the first 30 days, drip irrigation resulted in a significantly higher $CO_2$ content than hole irrigation ($p < 0.05$), whereas no significant difference was detected between optimized and conventional irrigation methods ($p > 0.05$). From 30 to 50 days after transplantation, soil $CO_2$ content was influenced by soil moisture content and plant root respiration, decreasing in the order $W_2 > W_3 > W_1$ (Figure 1).

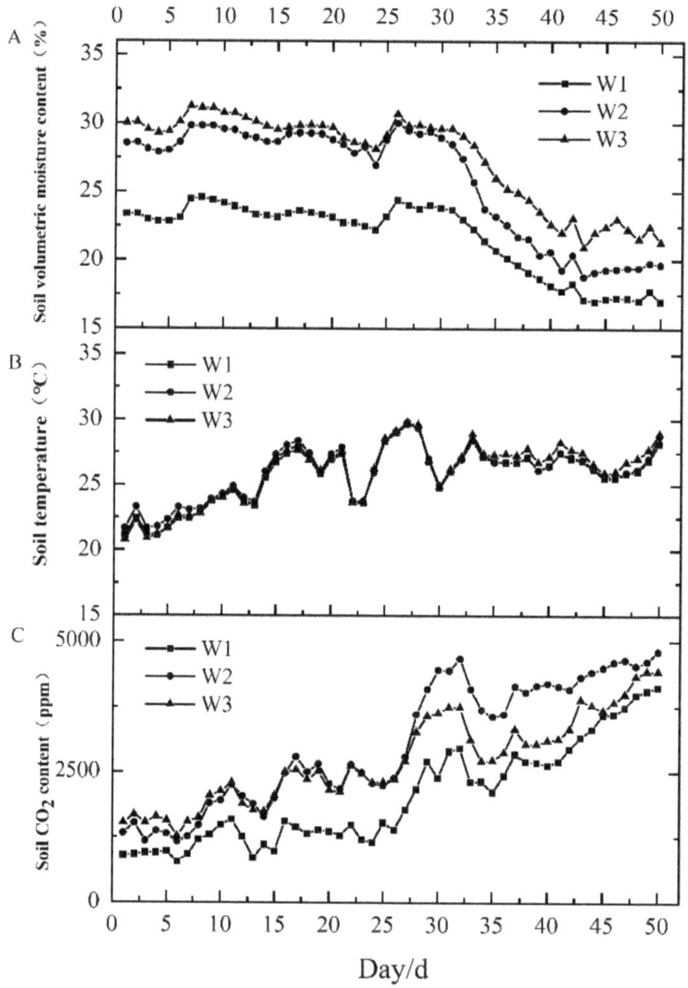


**Figure 1.** Dynamic variation in the soil volume moisture content, temperature, and $CO_2$ content at 0–50 days. Note: (**A**) soil volume moisture content (%); (**B**) soil temperature (°C); (**C**) soil $CO_2$ content (ppm). $W_1$, $W_2$, and $W_3$, respectively, represent different experimental treatments. In this study, the effect of soil water vapor heat on stratification at 15 and 30 cm depths was investigated. However, it was found that the change rule of each factor at 15 and 30 cm depths was identical; thus, the use of an average value for soil water vapor heat to express the change rule.

### 2.2. Treatment Effects on Seedling Survival Rates, Biomass, Leaf Area Index (LAI), and Photosynthetic Parameters

#### 2.2.1. Survival Rates

The survival rates of treatments $W_1$, $W_2$, and $W_3$ were 93.6%, 96.48%, and 96.64%, respectively. Survival rates increased with the irrigation amount. Drip irrigation led to significantly higher survival rates than hole irrigation ($p < 0.05$). However, there was no significant difference between optimized and conventional irrigation methods ($p > 0.05$). These results demonstrate that drip irrigation was more effective than hole irrigation but that drip irrigation did not significantly increase the seedling survival rate due to over-irrigation (Figure 2).

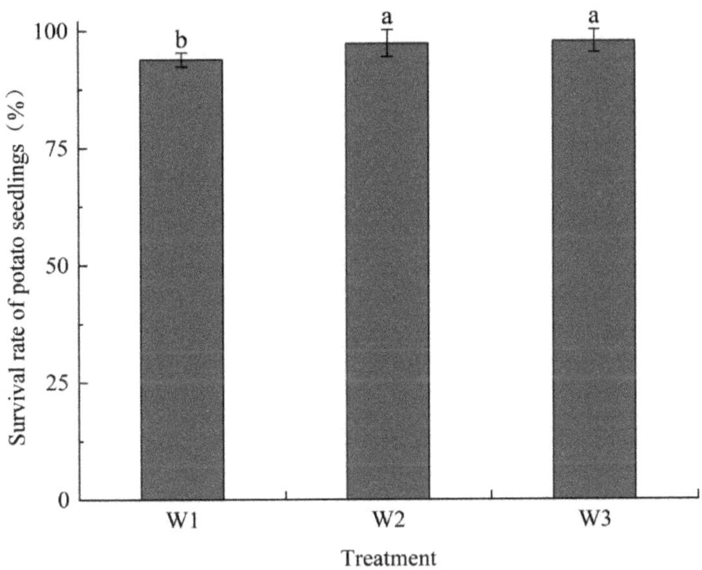


**Figure 2.** Survival rate of sweet potato seedlings after transplanting under different treatments. Different lowercase letters indicate significant differences ($p < 0.05$).

### 2.2.2. Biomass

The above- and belowground dry weight biomass of seedlings increased as the irrigation amount increased in all three sampling periods. Both the above- and belowground dry weight biomass were significantly lower in $W_1$ than in either drip irrigation treatment ($W_2$ and $W_3$), but there was no significant difference between the $W_2$ and $W_3$ treatments ($p > 0.05$). The results indicate that drip irrigation can significantly enhance the above- and belowground biomass of sweet potato in comparison to burrow irrigation. However, there was no discernible impact of irrigation volume on biomass under the drip irrigation method (Table 1).

**Table 1.** Changes in aboveground and belowground biomass under different treatments.

| Part | Treatment | Sampling Day | | |
|---|---|---|---|---|
| | | 27 d | 39 d | 50 d |
| Aboveground dry weight (g plant$^{-1}$) | $W_1$ | 6.22 ± 0.58 b | 15.82 ± 1.85 b | 45.11 ± 3.82 b |
| | $W_2$ | 11.33 ± 1.60 a | 23.57 ± 1.64 a | 58.09 ± 3.48 a |
| | $W_3$ | 11.58 ± 1.25 a | 26.53 ± 2.51 a | 62.12 ± 4.61 a |
| Belowground dry weight (g plant$^{-1}$) | $W_1$ | 0.31 ± 0.02 b | 2.39 ± 0.22 b | 13.55 ± 1.64 b |
| | $W_2$ | 0.40 ± 0.03 ab | 9.12 ± 0.67 a | 22.53 ± 2.05 a |
| | $W_3$ | 0.54 ± 0.15 a | 10.06 ± 0.85 a | 24.56 ± 2.20 a |

Note: The data in the table are the mean ± standard deviation. Different lowercase letters in the same column indicate significant differences ($p < 0.05$).

### 2.2.3. LAI

LAI increased linearly with seedling fertility in all treatments, with a gradual increase in the difference between the $W_2$ and $W_3$ treatments. The LAI was highest in the $W_3$ treatment at all fertility stages, generally decreasing in the order $W_3 > W_2 > W_1$. At 27 days;

the LAI was significantly higher under drip irrigation than under hole irrigation ($p < 0.05$); whereas there was no significant difference between $W_2$ and $W_3$ ($p > 0.05$). At 50 days, the LAI was significantly higher in $W_3$ than in $W_1$ and $W_2$ ($p < 0.05$) but with no significant difference between optimized drip irrigation and hole irrigation ($p > 0.05$; Table 2).

**Table 2.** Variations in the leaf area index under different treatments.

| Treatment | Sampling Day | | |
|---|---|---|---|
| | 27 d | 39 d | 50 d |
| $W_1$ | 0.73 ± 0.18 b | 1.27 ± 0.29 b | 2.41 ± 0.25 b |
| $W_2$ | 1.08 ± 0.22 a | 1.53 ± 0.23 ab | 2.70 ± 0.20 b |
| $W_3$ | 1.40 ± 0.21 a | 2.06 ± 0.47 a | 4.53 ± 0.76 a |

Note: The data in the table are the mean ± standard deviation. Different lowercase letters in the same column indicate significant differences ($p < 0.05$).

### 2.2.4. Photosynthetic Parameters

At 27 days after transplantation, the seedling net photosynthetic rate (Pn), intercellular $CO_2$ content ($C_i$), and transpiration rate (Tr) varied significantly among treatments ($p < 0.05$). However, as the fertility period progressed, the $W_1$ treatment exhibited significantly lower values than both the $W_2$ and $W_3$ treatments ($p < 0.05$). Differences between the $W_2$ and $W_3$ treatments gradually decreased throughout the study period until there was no significant difference ($p > 0.05$). There were no significant differences in stomatal conductance (Gs) between the $W_2$ and $W_3$ treatments ($p > 0.05$), and the differences gradually decreased between the drip irrigation and hole irrigation treatments. At all fertility stages, water utilization efficiency (WUE) was significantly higher in $W_1$ than in the other treatments ($p < 0.05$), with no significant difference between $W_2$ and $W_3$ ($p > 0.05$; Table 3).

**Table 3.** Variations in the photosynthetic parameters under different treatments.

| Sampling Day | Treatment | Pn ($\mu mol \cdot m^{-2} \cdot s^{-1}$) | Ci ($\mu mol \cdot mol^{-1}$) | Gs ($mmol \cdot m^{-2} \cdot s^{-1}$) | Tr ($\mu mol \cdot m^{-2} \cdot s^{-1}$) | WUE ($\mu mol \cdot mmol^{-1}$) |
|---|---|---|---|---|---|---|
| 27 d | $W_1$ | 11.14 ± 1.08 c | 59.36 ± 4.25 c | 143 ± 12.54 b | 2.06 ± 0.14 c | 4.95 ± 0.19 a |
| | $W_2$ | 13.48 ± 1.13 b | 76.84 ± 5.18 b | 162.85 ± 11.54 a | 2.15 ± 0.18 b | 4.15 ± 0.27 b |
| | $W_3$ | 14.26 ± 1.24 a | 84.26 ± 6.24 a | 176.48 ± 12.42 a | 2.34 ± 0.12 a | 3.86 ± 0.24 b |
| 39 d | $W_1$ | 13.78 ± 1.05 b | 72.57 ± 11.43 c | 150.50 ± 19.90 b | 2.45 ± 0.17 b | 5.15 ± 0.29 a |
| | $W_2$ | 14.96 ± 1.44 ab | 92.80 ± 3.29 b | 172.29 ± 11.96 a | 2.76 ± 0.16 a | 4.96 ± 0.38 b |
| | $W_3$ | 15.45 ± 1.90 a | 117.00 ± 5.82 a | 184.23 ± 13.62 a | 2.98 ± 0.12 a | 4.42 ± 0.32 b |
| 50 d | $W_1$ | 13.00 ± 1.16 b | 138.60 ± 12.78 b | 170.64 ± 15.99 a | 3.70 ± 0.46 b | 5.40 ± 0.47 a |
| | $W_2$ | 15.22 ± 1.29 a | 154.43 ± 11.09 a | 187.52 ± 10.85 a | 3.78 ± 0.19 a | 5.03 ± 0.48 b |
| | $W_3$ | 16.88 ± 1.35 a | 163.64 ± 9.67 a | 194.83 ± 9.90 a | 3.94 ± 0.23 a | 4.85 ± 0.51 b |

Note: The data in the table are the mean ± standard deviation. Different lowercase letters in the same column indicate significant differences ($p < 0.05$).

## *2.3. Treatment Effects on Seedling Roots Morphology*

At 14 days after transplantation, root morphology indices were significantly higher under hole irrigation than under drip irrigation ($p < 0.05$), with no significant difference between the optimized and conventional drip irrigation treatments ($p > 0.05$). The total root length, root surface area, and root volume of sweet potato seedlings increased with the amount of irrigation water supplied, and the highest number of root tips was observed in the $W_2$ treatment. At 27 days, root morphology indices were higher in the $W_2$ treatment than in the other treatments, and total root length and root volume differed significantly from those measured at 14 days ($p < 0.05$). At 39 days, all treatments exhibited a trend of increasing and then decreasing with increasing irrigation. Root surface area, root volume, and root tip number differed significantly among treatments ($p < 0.05$). Total root length was significantly lower at 39 days than at 14 days, and the $W_1$ and $W_2$ treatments differed significantly between both time points ($p < 0.05$; Figure 3).

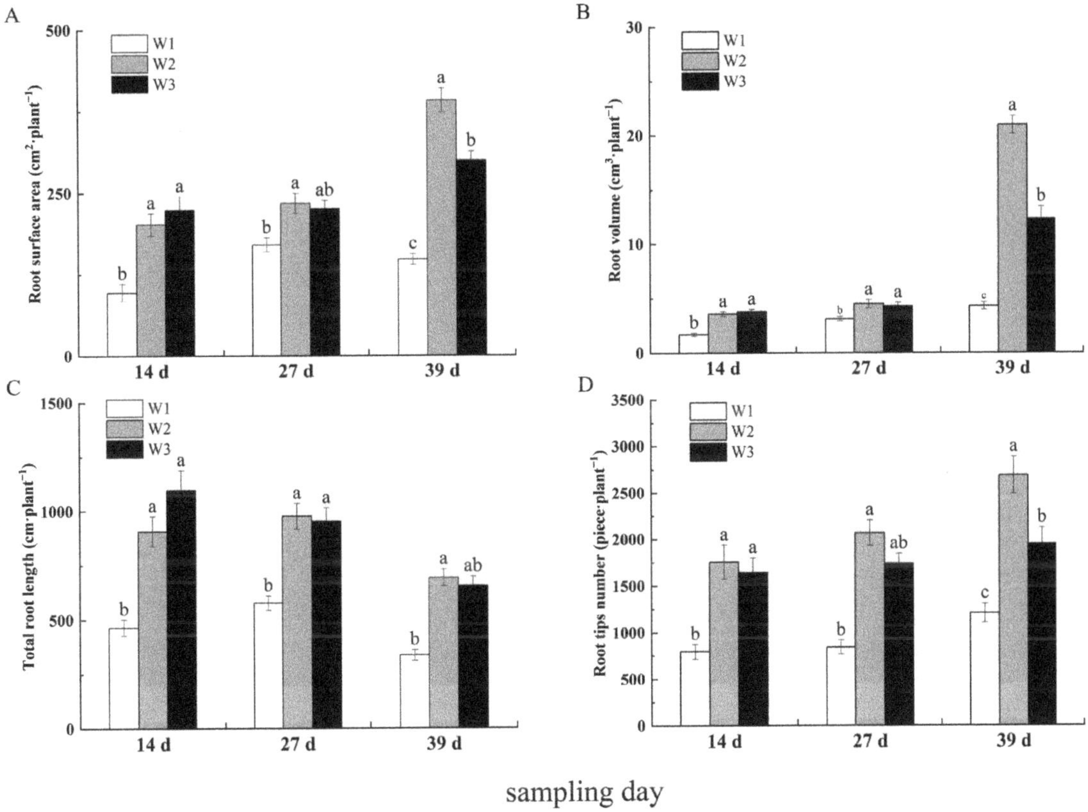


**Figure 3.** Variation in the root morphological parameters under different treatments. Note: (**A**) root surface area (cm$^2$ plant$^{-1}$); (**B**) root volume (cm$^3$ plant$^{-1}$); (**C**) total root length (cm plant$^{-1}$); (**D**) toot tips number (piece plant$^{-1}$). $W_1$, $W_2$, and $W_3$ represent different treatments. Different lowercase letters in each treatment showed significant difference ($p < 0.05$).

### 2.4. Treatment Effects on Seedling $^{13}C$ Accumulation and Partition Rates

At all sampling points, the $^{13}C$ accumulation and partitioning rates, numbers of tuberous roots, total $^{13}C$ accumulation, and root–crown $^{13}C$ accumulation ratio were significantly higher in the $W_2$ treatment than in the other treatments. Among seedling organs, $^{13}C$ accumulation was lowest in fibrous roots, and among treatments, it was highest in $W_2$. Differences between treatments gradually decreased with plant growth, eventually becoming non-significant ($p > 0.05$). $^{13}C$ accumulation in tubers and aboveground parts was significantly higher in $W_2$ than in the other treatments. $^{13}C$ accumulation rates were significantly higher in $W_1$ than in the other treatments ($p < 0.05$), whereas there was no significant difference between treatments $W_2$ and $W_3$ ($p > 0.05$). The root–crown $^{13}C$ accumulation ratio was significantly higher under optimized drip irrigation than under either hole irrigation or traditional drip irrigation ($p < 0.05$). Total $^{13}C$ accumulation was highest in $W_2$, followed by $W_3$ and $W_1$, with significant differences among treatments ($p < 0.05$; Table 4). These results indicated that moderate irrigation facilitated the transfer of photosynthetic products into sweet potato tubers. However, both excessive and insufficient irrigation impeded $^{13}C$ accumulation and distribution.

**Table 4.** Effects of different treatments on the $^{13}C$ accumulation and distribution rate of sweet potato.

| Sampling Day | Treatment | Accumulation $\mu g \cdot Plant^{-1}$ (Allocation %) | | | Accumulation Root–Shoot Ratio R/S | Total Accumulation ($\mu g \cdot plant^{-1}$) |
|---|---|---|---|---|---|---|
| | | Fibrous Root | Storage Root | Aboveground Tissues | | |
| 50 d | $W_1$ | 1.54 ± 0.12 a (10.13) | 3.26 ± 0.61 c (21.44) | 10.40 ± 1.18 b (68.43) | 0.46 | 15.20 ± 1.91 c |
| | $W_2$ | 1.13 ± 0.19 b (4.61) | 8.97 ± 0.93 a (36.00) | 14.41 ± 0.52 a (59.39) | 0.70 | 24.51 ± 1.64 a |
| | $W_3$ | 1.42 ± 0.20 ab (6.97) | 5.29 ± 0.49 b (25.98) | 13.65 ± 1.23 a (67.05) | 0.49 | 20.36 ± 1.28 b |
| 150 d | $W_1$ | 0.22 ± 0.03 a (1.02) | 14.53 ± 1.65 c (67.52) | 6.77 ± 1.24 b (37.46) | 2.17 | 21.52 ± 1.53 c |
| | $W_2$ | 0.17 ± 0.02 a (0.47) | 26.08 ± 2.25 a (72.71) | 9.62 ± 0.56 a (26.82) | 2.72 | 35.87 ± 2.65 a |
| | $W_3$ | 0.19 ± 0.02 a (0.65) | 19.73 ± 1.83 b (67.25) | 9.42 ± 0.87 a (32.10) | 2.11 | 29.34 ± 2.21 b |

Note: The data in the table are the mean ± standard deviation. Different lowercase letters in the same column indicate significant differences ($p < 0.05$).

### 2.5. Treatment Effects on Sweet Potato Yield Components and Deformation

#### 2.5.1. Sweet Potato Yield

Sweet potato final yield and yield component indices were significantly higher under optimized drip irrigation than either conventional drip irrigation or hole irrigation. Optimized drip irrigation increased the yield by 42.98% and 4.76% compared with hole irrigation and traditional drip irrigation, respectively, although the difference between the drip irrigation treatments was not significant. Sweet potato weight per plant and potato yield per plant differed significantly between hole irrigation and optimized drip irrigation ($p < 0.05$), with no significant difference between hole irrigation and conventional drip irrigation ($p > 0.05$; Table 5).

**Table 5.** Differences in the yield and yield components of sweet potato under different treatments.

| Treatment | Yield ($t \cdot ha^{-1}$) | Root Weight ($g \cdot Plant^{-1}$) | Number of Storage Roots Plant |
|---|---|---|---|
| $W_1$ | 25.57 ± 2.41 b | 610.28 ± 17.31 b | 2.51 ± 0.30 b |
| $W_2$ | 36.56 ± 2.79 a | 781.68 ± 27.38 a | 4.00 ± 0.12 a |
| $W_3$ | 34.90 ± 5.80 a | 740.40 ± 24.70 ab | 2.96 ± 0.76 b |

Note: The data in the table are the mean ± standard deviation. Different lowercase letters in the same column indicate significant differences ($p < 0.05$).

#### 2.5.2. Deformation

Sweet potato deformation and specific gravity were significantly higher in $W_2$ than in $W_1$ and $W_3$ ($p < 0.05$), with no significant difference between $W_1$ and $W_3$ ($p > 0.05$). However, in both drip irrigation treatments, the deformity rate increased significantly as the irrigation amount increased ($p < 0.05$) until that under optimized drip irrigation was significantly lower than that under the other treatments ($p < 0.05$), with no significant difference between hole irrigation and traditional drip irrigation ($p > 0.05$; Table 6). This result indicated that over-irrigation adversely affected sweet potato morphology under both subsurface and drip irrigation.

**Table 6.** Sweet potato commodity rate, proportion of commodity potato, and deformity rate.

| Treatment | Commodity Rate (%) | Proportion of Commodity Potato (%) | Deformity Rate (%) |
|---|---|---|---|
| $W_1$ | 64.39 ± 2.01 b | 68.80 ± 5.74 b | 11.37 ± 3.18 a |
| $W_2$ | 86.02 ± 1.59 a | 85.52 ± 9.44 a | 3.35 ± 2.78 b |
| $W_3$ | 70.32 ±11.31 b | 67.93 ± 6.07 b | 10.31 ± 0.55 a |

Note: The data in the table are the mean ± standard deviation. Different lowercase letters in the same column indicate significant differences ($p$ < 0.05).

### *2.6. Redundancy Analysis (RDA) of Soil Environment, Root Morphology, and Various Physiological Indicators in the Seedling Root Zone*

RDA was conducted to examine correlations between root zone soil environment indicators and root morphology indicators among treatments. The results showed that the root zone soil environment and root morphology were significantly positively correlated with the drip irrigation treatment and negatively correlated with the hole irrigation treatment (Figure 4). These results indicated that the irrigation method was an important factor influencing the soil environment and seedling root growth.

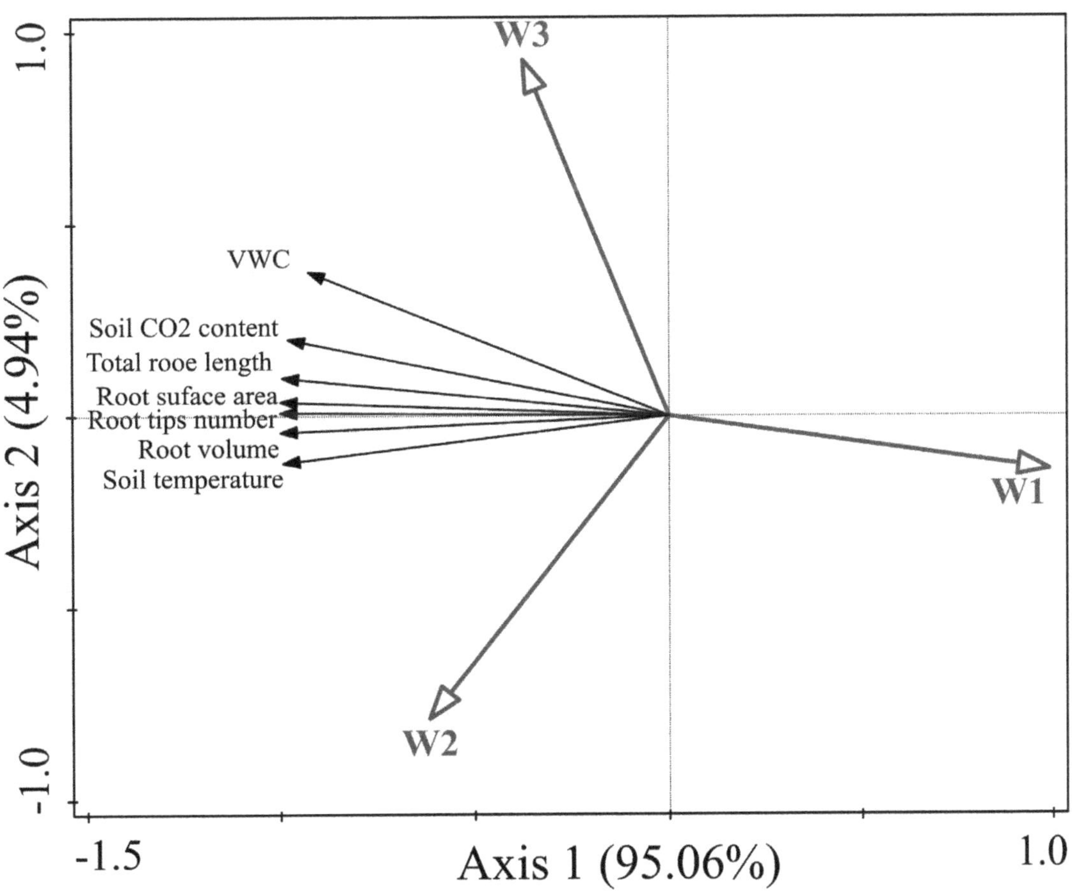


**Figure 4.** RDA analysis of soil internal environment and root morphological indices under different treatments. The coordinates of the first axis in the figure explain 95.06% of the variance, respectively, and the significances (according to Monte Carlo permutation tests) of all canonical axes were PA = 0.012, PB = 0.018, and PC = 0.026.

Correlation analysis also showed that the seedling leaf Pn, Ci, Gs, above- and belowground dry weight, and sweet potato yield were significantly positively correlated with the $W_2$ treatment ($p < 0.05$), whereas the deformity rate, LAI, Tr, and WUE were significantly positively correlated with the $W_3$ treatment ($p < 0.05$). All indicators were negatively correlated with the $W_1$ treatment, except for the deformity rate. The deformity rate was significantly positively correlated with the $W_1$ treatment ($p < 0.05$) and negatively correlated with all other indicators (Figure 5). Thus, irrigation treatment was an important factor influencing photosynthesis, carbon transfer, and yield in sweet potato seedlings.

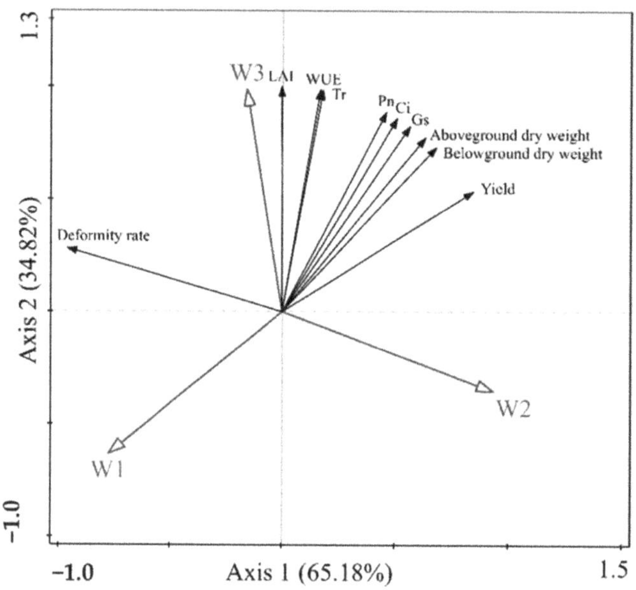


**Figure 5.** RDA analysis of physiological indices and agronomic indices of sweet potato under different treatments. The coordinates of the first axis in the figure explain 65.18% of the variance, respectively, and the significances (according to Monte Carlo permutation tests) of all canonical axes were PA = 0.012, PB = 0.018, and PC = 0.026.

## 3. Discussion

### *3.1. Treatment Effects on Soil Moisture-Oxygen-Temperature in the Root Zone*

This is the first study to monitor the VWC, temperature, $CO_2$ content, and root morphology of sweet potato seedlings in the field during the post-transplantation rooting period. Penman's formula is employed to calculate the quantity of irrigation required based on the actual crop evapotranspiration (ETc). This necessitates the consideration of meteorological factors, including light, temperature, and water vapor saturation. However, in this experiment, only a single irrigation was conducted, and ETc was not employed to calculate the irrigation volume. The results showed that the soil moisture content fluctuated, with a decreasing trend throughout the reproductive period, except during extreme weather events that caused significant increases or decreases in moisture content. Soil VWC did not differ significantly among the two drip irrigation treatments in the first month; however, all drip irrigation treatments had significantly higher water contents than hole irrigation. From 30 to 50 days after transplantation, VWC differed significantly among treatments and increased continuously with the amount of water supplied. This finding is consistent with those of previous studies [16–18] and suggests that increasing irrigation volume leads to higher soil VWC. However, in both drip irrigation treatments, VWC did not differ significantly between 0 and 30 days after transplantation. This phenomenon is attributable to the sandy loam soil at the study site. Single-application hole irrigation provides a large volume of water within a short time span, which can lead to rapid water infiltration and loss. In contrast, drip irrigation allows water to slowly penetrate the soil, maintaining the VWC within a certain range [19,20].

Good soil aeration is crucial for root differentiation and growth. Roots differentiate easily in well-aerated soil; whereas in poorly aerated soils, excessive $CO_2$ can be toxic to the root system, hindering growth [21]. Our results indicate that the soil $CO_2$ content was significantly higher in the drip irrigation treatments than under hole irrigation within the first month after transplanting. However, differences between treatments were significant at 30–50 days after transplanting, and $CO_2$ concentrations were significantly higher in $W_3$ than in $W_2$ and $W_1$. An increase in drip irrigation volume can hinder soil aeration, leading to a reduction in $O_2$ content and an increase in $CO_2$ content, which can have an adverse effect on root system growth, ultimately impacting tuber yield [22]. A previous study demonstrated that proper aeration promotes the transport of photosynthesis assimilation products from the leaves to tuberous roots, increasing tuber yield [23]. The amount of irrigation water supplied directly affects soil aeration and indirectly affects the degree of soil compactness, which, in turn, affects the expansion of the sweet potato root system. A larger irrigation volume leads to a greater soil firmness, which can cause sweet potato deformation [24,25]. In this study, the length and volume of sweet potato roots were significantly lower under hole irrigation than under drip irrigation because hole irrigation is performed in a single application, which causes a sudden increase in the soil VWC in the root zone, increasing the localized mechanical resistance that inhibits sweet potato root growth [26].

Sweet potatoes are a thermophilic crop, such that temperature plays a crucial role in their growth. We detected no significant difference in soil temperature among treatments, as soil temperature is heavily influenced by air temperature, and at a certain air temperature, soil moisture affects soil temperature [27]. The main reason for the effect of soil moisture on soil temperature is that the specific heat capacity of water is greater than that of soil; as the irrigation volume increases, so do the soil water content and specific heat capacity of the soil, resulting in a gradual decrease in soil temperature [28]. Li Siping analyzed the relevant indicators of sweet potato in the middle growth stage through RDA, indicating that RDA can characterize the relationship between soil water, air heat, and treatment [29]. In this experiment, a single irrigation event was employed, and meteorological conditions did not influence soil water content or $CO_2$ concentration. However, soil temperature was found to be more susceptible to climatic influences [30]. Overall, drip irrigation maintains the granular structure of the soil and contributes to preserving the quality of water, fertilizer, gases, and heat within the soil. Drip irrigation is also the most effective method for achieving soil water and gas phase coordination [31]. Because all soil factors are interdependent, it is necessary to consider the comprehensive influence of all factors on sweet potato growth and yield to select a suitable irrigation method that will create a favorable growth environment, which will ultimately result in high tuber yield.

### 3.2. Treatment Effects on Sweet Potato Seedling Growth

Insufficient soil moisture can impede the elongation and expansion of the sweet potato root system, which can affect root differentiation. Conversely, excessive soil moisture will not further increase the values of root system morphological indicators and can lead to root rot and seedling death [32]. Our results showed that sweet potato root system morphological indices were highest in the $W_2$ treatment at 50 days. Increases in LAI and water supply directly affect plant photosynthesis rates [33]. We found that LAI did not significantly differ between the $W_2$ and $W_3$ treatments from 0 to 10 days after transplanting. However, after 20 days, the $W_2$ treatment exhibited a rapid growth trend, with a significantly higher growth rate than the $W_3$ treatment. The amount of irrigation water had a significant impact on photosynthesis in the sweet potato seedlings, with significantly higher Pn and Ci values in the $W_3$ treatment, which received a larger amount of drip irrigation than the other treatments. A previous study also reported that over-irrigation led to a linearly increasing (decreasing) trend in the transfer and distribution of aboveground (belowground) sweet potato photosynthesis products [34]. Another study reported that WUE peaked at 50% of the field water-holding capacity and decreased as the soil water content increased, with no significant difference at 75% and 100% of the field water-holding capacity, suggesting that WUE does not continue to increase with soil water content [35]. Similarly, we observed the highest WUE in the $W_1$ treatment and no significant difference between $W_2$ and $W_3$ ($p > 0.05$).

Increasing the amount of water used for seedling irrigation led to an increase in both above- and belowground seedling dry weight at 50 days after transplanting. However, the highest final yield at 150 days was observed under the $W_2$ treatment. This result was attributed to the susceptibility of aboveground sweet potato stems and leaves to wilting under over-irrigation, which decreases tuber yield [36]. At 50 and 150 days after transplanting, $^{13}$C accumulation and partitioning rates were significantly higher in belowground seedling parts under the $W_2$ treatment than in the other treatments. At 50 days, aboveground plant parts showed fluctuating growth rates during overflooding, as confirmed by the $^{13}$C accumulation and partitioning rates at 150 days after transplanting. Aboveground growth was significantly greater in the $W_3$ treatment than in the other treatments. In contrast, the under-irrigated treatment restricted sweet potato growth, negatively impacting yield. A previous study reported that sweet potato yield was significantly higher in well-aerated soils and could be significantly improved in poorly aerated soils through $^{13}$C-labeling of assimilates [37,38]. Thus, moderate irrigation facilitates the establishment of the sweet potato root system and improves nutrient transport from aboveground to belowground plant parts.

Excessive drip irrigation can adversely affect plant morphology and crop yield. Determining the optimal drip irrigation supply supports high plant quality and yields, while conserving water resources [39]. Although we detected no significant difference in seedling survival rates between traditional and optimized drip irrigation treatments in this study, the final yield and the rate of commercially marketable sweet potatoes were significantly higher in the $W_2$ treatment than in the $W_3$ treatment, with a significantly higher rate of tuber deformation in the $W_3$ treatment. As we did not monitor indices for mid- and late-stage sweet potato growth in this study, future research should explore the effects of different irrigation quantities on sweet potato agronomy and quality from molecular and genetic perspectives.

## 4. Materials and Methods

### 4.1. Experimental Samples and Sample Plots

The experiment was planted on 18 May 2021 and harvested on 18 October 2021. The main fresh-eating sweet potato variety in northern China, Yanshu 25, was selected. A field experiment was conducted at the Nancun Experimental Base of Qingdao Agricultural University (36°26′ N, 120°04′ E). The area has a warm, temperate, semi-humid monsoon climate, with precipitation concentrated in July to August, rain and heat at the same time, an average annual temperature of 12–14 °C, and an average annual rainfall of 630.98 mm. The soil type of the experimental site is brown loam, and the basic physical and chemical properties of the soil are shown in Table 7. The meteorological station was located 1 km from the experimental site. Average monthly rainfall and temperature in Namchon Town for the past three years is shown in Table 8.

**Table 7.** Basic physical and chemical properties of the soil.

| Soil Bulk Density | Proportion of Clay Soil Sand | Electrical Conductivity | pH | Organic Matter ($g \cdot kg^{-1}$) | Alkaline Hydrolysis Nitrogen ($mg \cdot kg^{-1}$) | *p* Content ($mg \cdot kg^{-1}$) | K Content ($mg \cdot kg^{-1}$) |
|---|---|---|---|---|---|---|---|
| 1.27 | 7:40:53 | 0.80 | 8.02 | 8.23 | 57.91 | 10.43 | 82.01 |

**Table 8.** Average monthly rainfall and temperature in Namchon Town for the past three years.

| | Jan. | Feb. | Mar. | Apr. | May. | Jun. | Jul. | Aug. | Sept. | Oct. | Nov. | Dec. |
|---|---|---|---|---|---|---|---|---|---|---|---|---|
| Rainfall/mm | 7.64 | 12.33 | 14.56 | 30.67 | 50.7 | 72.68 | 153.26 | 174.2 | 56.58 | 24.95 | 23.99 | 9.41 |
| temperatures/°C | −1.91 | 0.78 | 6.28 | 12.83 | 18.74 | 23.00 | 26.03 | 25.58 | 21.44 | 14.82 | 7.13 | 0.32 |

### 4.2. Experimental Design

Three treatments were set up in the experiment: hole irrigation (i.e., "watering hole water"). The amount of water for the local farmers was about 800 mL per plant, $5 \times 10^4$ seedlings per hectare, and the total amount of water for the fixed seedlings was calculated to be about 39.6 m$^3$ ha$^{-1}$, which was recorded as $W_1$. Optimized drip irrigation treatment was recorded as W2, with a volume of 180 m$^3$ ha$^{-1}$. This was calculated according to the water holding capacity of the field, which was 65–70%. Traditional drip irrigation treatment was recorded as W3, with a volume of 450 m$^3$ ha$^{-1}$. This was calculated according to the water holding capacity of the field, which was 85–90%. The three treatments represent the past, future, and present water management methods of sweet potato after transplanting. A randomized block design, each treatment set up three replicates, and a total of nine plots were set up, with a single plot area of 30 m$^2$ (6 m × 5 m), each plot having 5 ridges, plant spacing of 0.2 m, and ridge spacing of 0.85 m. Before planting, nitrogen fertilizer ($CO(NH_2)_2$), phosphate fertilizer ($K_2HPO_4$), and potassium fertilizer ($K_2SO_4$) were applied, and the dosages were 120 kg hm$^{-2}$, 75 kg hm$^{-2}$, and 150 kg hm$^{-2}$, respectively, as base fertilizer.

### 4.3. Measuring Indicators and Methods

#### 4.3.1. Dynamic Monitoring of Soil Volumetric Water Content, Soil Temperature, and Soil $CO_2$ Concentration

We used soil temperature and humidity sensors (HJH-WSD-02, Weihai, China) to adopt a stratified monitoring method, setting a soil moisture and temperature observation point every 15 cm from the top of the ridge to dynamically monitor soil moisture and temperature within the corresponding range. Prior to transplanting sweet potatoes, the sensor probe was inserted at the appropriate position on the ridge of each plot according to the set observation points, and volumetric soil moisture and soil temperature were measured at 0–30 cm. The soil temperature and soil moisture sensors were set to collect data every 30 min. We used a soil $CO_2$ sensor (HJH-$CO_2$-01, Weihai, China) to monitor soil $CO_2$ levels in a stratified manner, with a soil $CO_2$ monitoring point set every 15 cm from the top of the ridge to dynamically monitor soil $CO_2$ levels in the corresponding area. The device was inserted between the plants on the ridge of each plot prior to sweet potato transplanting to measure soil $CO_2$ at 0–30 cm. The sensor was set to collect data every 30 min. All sensors were connected to the data collection internet of things (IOT) (HJH-WG-01, Weihai, China) gateway and set to upload data every 30 min.

#### 4.3.2. Determination of Physiological Indices and Photosynthetic Parameters of Sweet Potato

Sweet potato biomass determination: Five sweet potato plants were randomly selected from each sample plot at 27, 39, and 50 days after transplanting. The fresh weight was obtained by separating the fresh stems and leaves from the stored roots, and then drying them in an oven at 105 °C for 30 min and at 80 °C for 48 h until a constant weight was achieved. The dry weight was then determined [40,41].

Leaf area index (LAI): For each treatment, we selected uniform and representative plants, removed all leaves, and weighed the petioles. The perforation method was used to determine [42].

Determination of photosynthetic parameters of sweet potato: at 9:00–11:00 in the morning, using CIRAS 3 type photosynthetic apparatus, each treatment measured 5 strains of sweet potato main stem at the top of the 4th to 5th fully expanded leaves. Net photosynthetic rate (Pn), transpiration rate (Tr), stomatal conductance (Gs), and intercellular $CO_2$ concentration (Ci) index were measured [43].

The $^{13}C$ labeling method of sweet potato leaves and the determination of $^{13}C$ accumulation and distribution rate: The labeling test was conducted on sweet potato leaves 42 days after transplanting. $Ba^{13}CO_3$ reagent was reacted with phosphoric acid in a reaction device to produce $^{13}CO_2$. The labeled $^{13}CO_2$ gas was then collected in a bag. Representative plants with uniform growth were selected from each plot. The $^{13}CO_2$ was labeled on the 4th to 5th fully expand leaves of the longest main stem of the plants between 9:00 and 11:00 on a sunny day. The measurement was taken using the gas bag method [44,45]. The $\delta^{13}C$ and total carbon content (C %) were determined via IRMS in the stable isotope laboratory of the Department of Plant Science, University of California, Davis, CA, USA.

#### 4.3.3. Determination of Root Morphological Indices of Sweet Potato

Samples were collected on the 14th, 27th, 39th, and 50th days after transplanting. Immediately after field sampling, sweet potato root systems were washed with distilled water. All root systems were scanned using an Epson v850 Pro root scanner(v850 Pro, Epson Inc., Nagano Prefecture, Japan) with a resolution of 400 bpi. Root morphological parameters were analyzed using Win RHIZO software (WinRHIZO 2024a) [46].

#### 4.3.4. Determination of Survival Rate of Potato Seedlings, Yield and Deformation

Seedling survival rate: The number of surviving potato seedlings in each plot was counted, and the ratio of surviving seedlings to the total number of transplants one week after transplantation was taken as the survival rate of potato seedlings.

Sweet potato commercialization rates: Sweet potato are classified into three grades based on the weight of their storage roots: Grade A, weighing 150–600 g; Grade B, weighing more than 600 g; and Grade C, weighing less than 150 g (Big Data Platform for Sweet Potato Industry; http://www.isweetpoto.jaas.ac.cn, 2 April 2024). Marketable sweet potato storage roots are classified into grades A and B, with Grade A having the highest economic value and being of the highest quality. Grade C storage roots are generally considered to have low market value. The commercial potato rate is determined by calculating the percentage of Grade A and Grade B storage roots.

Deformity rate: 150 plants were randomly selected from each treatment at the time of harvest to count the proportion of pieces with deformities (split, irregular shape, etc.), and the deformity rate of sweet potato was obtained statistically.

Final yield: At the time of harvest, the number of individual plants and the fresh weight of tubers were recorded for each sample in each treatment. The total weight was then weighed in order to calculate the yield.

#### 4.3.5. Calculation Formula

$$\text{Seedling survival rate} = \text{number of seedlings surviving after transplanting}/\text{total number of transplants}$$

$$\text{Sweet potato commercialization rate} = \text{number of commercial potatoes}/\text{total number of sweet potato}$$

$$\text{Sweet potato deformity rate} = \text{number of deformed potatoes}/\text{total number of sweet potato}$$

$$^{13}\text{C isotope ratio within the sample } R_{sample} = (\delta^{13}C/1000 + 1) \times R_{scale}\ (R_{scale} \text{ is the carbon isotope ratio of the standard material, } R_{scale} = 1.0783)$$

$$^{13}\text{C accumulation in each organ} = R_{sample}/(R_{sample} + 1) \times C\% \times \text{biomass dry weight (g)}$$

$$^{13}\text{C allocation rate (\%)} = {}^{13}\text{C accumulation in the organ}/\text{total } {}^{13}\text{C accumulation in the plant} \times 100\%$$

### 4.4. Data Analysis

The data were organized using Excel 2019 software. To clarify the differences between treatments, mathematical and statistical analyses of the various indicators and related data were performed using SPSS 26.0. The LSR method was used to compare the significance of the differences between the means of the indicators of each treatment. To clarify the correlation between treatments on soil water, air, and heat in the root zone and sweet potato growth, we performed RDA (redundancy) analysis using Canoco 5.0.

## 5. Conclusions

The results of the experiment confirmed the hypothesis that soil moisture directly affects sweet potato growth and yield, while temperature plays an indirect role. Compared to hole irrigation, drip irrigation improved the root zone soil hydrothermal and root morphology properties, LAI, and photosynthetic capacity of sweet potato seedlings in this study. These improvements enhanced sweet potato seedling growth, as well as tuber quality and yield.

Moderate drip irrigation effectively decreased the rate of sweet potato deformation, promoting commercial yield and the transport of photosynthetic products from aboveground to belowground plant parts. However, excessive drip irrigation is not conducive to sweet potato growth but rather causes an imbalance of reservoir sources and wastes water by promoting vigorous aboveground growth. Therefore, we recommend a spring drip irrigation rate of 180 $m^3$ $ha^{-1}$ for sweet potato cultivation in northern China.

**Author Contributions:** Conceptualization, H.L.; methodology, S.H. and H.L.; investigation, S.H., L.Z., M.Q. and T.Z.; resources, Q.L.; writing—original draft preparation, S.H.; writing—review and editing, S.H.; editing and project administration, H.L. and T.Y. All authors have read and agreed to the published version of the manuscript.

**Funding:** This research was funded by the National Key R&D Program of China (2023 YFD 1902702), the China Agriculture Research System (No. CARS-10-GW11), and the Technology Benefit Program of Qingdao (No. 24-1-8-XDNY-23-nsh).

**Data Availability Statement:** Data derived from public domain resources.

**Conflicts of Interest:** No conflicts of interest exist in the submission of this manuscript. The work described herein was original research that has not been previously published, and is not under consideration for publication elsewhere, in whole or in part. All authors listed have approved the enclosed manuscript for publication.

## References

1. He, H.; Yin, M.; Chen, A.; Liu, J.; Xie, X.; Yang, Z. optimal allocation of water resources from the wide-mild water shortage perspective. *Water* **2019**, *10*, 1289. [CrossRef]
2. Li, Y.; Du, X.; Li, D.; Zhang, X. Research on the safe utilization of agricultural water resources in the Yellow River Basin. *Water Sci. Technol. Water Supply* **2022**, *22*, 5281. [CrossRef]
3. Robert, H.P.L. Effects of prolonged restriction in water supply on photosynthesis, shoot development and storage root yield in sweet potato. *Physiol. Plant.* **2008**, *134*, 99–109. [CrossRef] [PubMed]
4. Hou, M.; Wang, X.; Zhang, Y.; Tang, W.; Ma, D.; Li, Q. Study on Correlation Analysis of Main Economic Characters of Sweetpotato and Meteorological Factors. *Agric. Food Sci.* **2013**, *28*, 243–247. [CrossRef]
5. Zhang, Y.; Tang, Y.; Hu, Y.; Feng, S.; Wang, F.; Wang, Z. Effects of film mulching and soil wetted percentage of drip irrigation on soil hydrothermal conditions and sweet potato growth. *Eur. J. Agron.* **2023**, *151*, 126979. [CrossRef]
6. Zhao, L.; Zhang, L.L.; Cui, N.B.; Liang, C.; Feng, Y. The Evaluation of Regional Water-Saving Irrigation Development Level in Humid Regions of Southern China. *Water* **2019**, *11*, 172. [CrossRef]
7. Chen, R. Research on Drip Irrigation System Based on Self-Controlled System. *J. Phys. Conf. Ser.* **2022**, *2386*, 012046. [CrossRef]
8. Tan, M.; Li, W.; Zong, R.; Li, X.; Han, Y.; Luo, P.; Dhital, Y.P.; Lin, H.; Li, H.; Wang, Z. Long-term mulched drip irrigation enhances the stability of soil aggregates by increasing organic carbon stock and reducing salinity. *Soil Tillage Res.* **2024**, *240*, 106069. [CrossRef]
9. Can, K.M.; Ahmet, K.; Cuneyt, D. Effects of water deficit on growth and performance of drip irrigated sweet potato varieties. *J. Sci. Food Agric.* **2020**, *101*, 2961–2973. [CrossRef]
10. Zhu, B.; Cheng, W. Impacts of drying-wetting cycles on rhizosphere respiration and soil organic matter decomposition. *Soil Biol. Biochem.* **2013**, *2013*, 89–96. [CrossRef]

11. Han, H.; Pan, R.; Buitrago, S.; Abou-Elwafa, S.F.; Peng, Y.; Liu, Y.; Zhang, W.Y.; Yang, X.S. The Physiological Basis of Genotypic Variations in Low-Oxygen Stress Tolerance in the Vegetable Sweet Potato. *Russ. J. Plant Physiol.* **2021**, *68*, 1236–1246. [CrossRef]
12. Pan, R.; Liu, Y.; Buitrago, S.; Jiang, W.; Gao, H.; Han, H.; Wu, C.; Wang, Y.; Zhang, W.; Yang, X. Adventitious root formation is dynamically regulated by various hormones in leaf-vegetable sweetpotato cuttings. *J. Plant Physiol.* **2020**, *253*, 153267. [CrossRef]
13. Villordon, A.; Labonte, D.; Solis, J.; Firon, N. Characterization of Lateral Root Development at the Onset of Storage Root Initiation in 'Beauregard' Sweetpotato Adventitious Roots. *HortScience Publ. Am. Soc. Hortic. Sci.* **2012**, *47*, 961–968. [CrossRef]
14. Moyano, F.E.; Manzoni, S.; Chenu, C. Responses of soil heterotrophic respiration to moisture availability: An exploration of processes and models. *Soil Biol. Biochem.* **2013**, *59*, 72–85. [CrossRef]
15. Jiang, Y.; Shi, C.; Wang, Z.; Wang, C.; Liu, H. Effects of plastic film mulching on arable layer soil temperature, moisture and yield of sweet potato. *Chin. J. Eco-Agric.* **2014**, *22*, 627–634. [CrossRef]
16. Barker, J.B.; Franz, T.E.; Heeren, D.M.; Neale, C.M.U.; Luck, J.D. Soil water content monitoring for irrigation management: A geostatistical analysis. *Agric. Water Manag.* **2017**, *188*, 36–49. [CrossRef]
17. Xiang, L.; Zhang, J.; Li, J. Effects of irrigation amount and irrigation frequency on root growth, yield and nutrient uptake of tomato. *J. Northwest A F Univ. (Nat. Sci. Ed.)* **2024**, *8*, 897–907. [CrossRef]
18. Chen, L.; Wang, L.; Song, N.; Zhang, Q.; Pei, X. Effects of Irrigation Amount and Times on Water Consumption Characteristics and Biomass of Alfalfa. *J. Soil Water Conserv.* **2009**, *23*, 36. [CrossRef]
19. Song, R.; Chu, G.; Zhang, R.; Bai, L.; Yang, J. Effects of sand mulching on soil infiltration, evaporation, and salt distribution. *Acta Pedol. Sin.* **2012**, *49*, 282–288. [CrossRef]
20. Zhao, Y.; Wang, Y.; Ma, L. Study on effects of different dripper discharge on wetting soil characteristic of drip irrigation. *Agric. Res. Arid Areas* **2010**, *28*, 30–35. [CrossRef]
21. Jin, C.; Lei, H.; Chen, J.; Xiao, Z.; Leghari, S.J.; Yuan, T.; Pan, H. Effect of Soil Aeration and Root Morphology on Yield under Aerated Irrigation. *Agronomy* **2023**, *13*, 369. [CrossRef]
22. Siqinbatu; Kitaya, Y.; Hirai, H.; Endo, R.; Shibuya, T. Effects of water contents and $CO_2$ concentrations in soil on growth of sweet potato. *Field Crops Res.* **2013**, *152*, 36–43. [CrossRef]
23. Shi, C.; Wang, Z.; Guo, F.; Yu, S. Effects of soil aeration on nutrient uptake, $^{14}C$ assimilate distribution and yield of sweetpotato. *Acta Agric. Nucl. Sin.* **2002**, *16*, 232–236. [CrossRef]
24. Shan, Z.; Luo, X.; Fan, W.; Mo, F. Effects of different irrigation amount and irrigation methods on soil physical and chemical properties and cassava growth were studied. *Guangxi Water Conserv. Hydropower* **2015**, 25–29. (In Chinese) [CrossRef]
25. Passioura, J.B. Soil conditions and plant growth. *Plant Cell Environ.* **2002**, *25*, 311–318. [CrossRef]
26. Watanabe, K.; Ozaki, K.; Yashiki, T. Studies on the Effects of Soil Physical Conditions on the Growth and Yield of Crop Plants: VII. Effects of soil air composition and soil bulk density and their interaction on the growth of sweet potato. *Jpn. J. Crop Sci.* **1968**, *37*, 65–69. [CrossRef]
27. Chen, R.; Meng, M.; Liang, H.; Zhang, J.; Wang, Y.; Wang, Z. Effects of Different Treatments of Irrigation and Fertilization on the Yield and Nitrogen Utilization Characteristic of Potato. *Chin. Agric. Sci. Bull.* **2012**, *28*, 196–201. [CrossRef]
28. Jarrah, M.; Mayel, S.; Franko, U.; Kuka, K. Effects of Agricultural Management Practices on the Temporal Variability of Soil Temperature under Different Crop Rotations in Bad Lauchstaedt-Germany. *Agronomy* **2022**, *12*, 1199. [CrossRef]
29. Li, S.; Zhao, L.; Sun, N.; Liu, Q.; Li, H. Photosynthesis product allocation and yield in sweetpotato with different irrigation levels at mid-season. *Agric. Water Manag.* **2021**, *246*, 106708. [CrossRef]
30. Zhang, H.; Yuan, N.; Ma, Z.; Huang, Y. Understanding the Soil Temperature Variability at Different Depths: Effects of Surface Air Temperature, Snow Cover, and the Soil Memory. *Adv. Atmos. Sci.* **2021**, *38*, 493–503. [CrossRef]
31. Zappa, L.; Dari, J.; Modanesi, S.; Quast, R.; Brocca, L.; Lannoy, G.D.; Massari, C.; Seguí, P.Q.; Ortiz, A.B.; Dorigo, W. Benefits and pitfalls of irrigation timing and water amounts derived from satellite soil moisture. *Agric. Water Manag.* **2024**, *295*, 108773. [CrossRef]
32. Taduri, S.; Bheemanahalli, R.; Wijewardana, C.; Lone, A.A.; Meyers, S.L.; Shankle, M.; Gao, W.; Reddy, K.R. Sweetpotato cultivars responses to interactive effects of warming, drought, and elevated carbon dioxide. *Front. Genet.* **2023**, *13*, 1080125. [CrossRef] [PubMed]
33. Huan, L.; Jinqiang, W.; Qing, L. Photosynthesis product allocation and yield in sweet potato with spraying exogenous hormones under drought stress. *J. Plant Physiol.* **2020**, *253*, 153265. [CrossRef] [PubMed]
34. Bandara, G.; Raja, R.; Shankle, M. Quantifying Growth and Developmental Responses of Sweetpotato to Mid- and Late-Season Temperature. *Agron. J.* **2015**, *107*, 1854. [CrossRef]
35. Hui, Z.; Yongchun, Z.; Yunwang, N.; Congzhi, Z. Research progress on the regulation of soil and fertilizer on the growth of sweet potato. *Chin. J. Soil Sci.* **2012**, *43*, 995–1000. [CrossRef]
36. Shi, C.; Wang, Z.; Zhao, B.Q.; Guo, F.; Yu, S. Effect of potassium nutrition on some physiological characteristics and yield formation of sweet potato. *Plant Natrition Fertil. Sci.* **2002**, *8*, 81–85. [CrossRef]
37. Liu, H.; Zhang, B.-B.; Shi, C.; Liu, Y.; Si, C. Reason exploration for soil aeration promoting photosynthate transportation between sink and source in sweet potato. *Acta Agron. Sin.* **2019**, *46*, 462–471. [CrossRef]
38. Wang, S.; Liu, Q.; Shi, Y.; Li, H. Interactive Effects of Nitrogen and Potassium on Photosynthesis Product Distribution and Accumulation of Sweet Potato. *Sci. Agric. Sin.* **2017**, *50*, 2706–2716. [CrossRef]

39. Akkamis, M.; Caliskan, S. Effects of Different Irrigation Levels and Nitrogen Fertilization on Some Physiological Indicators of Potato. *Potato Res.* **2023**. [CrossRef]
40. Zhou, M.; Hu, S.; Wang, S.; Yin, T.; Liu, Q.; Li, H. Sweet potato yield and quality characteristics affected by different late-season irrigation levels. *J. Sci. Food Agric.* **2024**. [CrossRef]
41. Picha, D. Crude Protein, Minerals, and Total Carotenoids in Sweet Potatoes. *J. Food Sci.* **1985**, *50*, 1768. [CrossRef]
42. Haghshenas, A.; Emam, Y. Accelerating leaf area measurement using a volumetric approach. *Plant Methods* **2022**, *18*, 61. [CrossRef] [PubMed]
43. Zhou, M.; Sun, Y.; Wang, S.; Liu, Q.; Li, H. Photosynthesis Product Allocation and Yield in Sweet Potato in Response to Different Late-Season Irrigation Levels. *Plants* **2023**, *12*, 1780. [CrossRef] [PubMed]
44. De Lima, J.; Balestrini, V.; Giannecchini, I.; De Oliveira, G.; Williams, T. $^{13}$C labelling reveals details of the soybean (*Glycine max* (L.) Merrill) seedling metabolic network. *Theor. Exp. Plant Physiol.* **2024**, *36*, 325–337. [CrossRef]
45. Saranga, Y.; Flash, I.; Paterson, A.H.; Yakir, D. Carbon isotope ratio in cotton varies with growth stage and plant organ. *Plant Sci.* **1999**, *142*, 47–56. [CrossRef]
46. Wang, J.; Li, H.; Liu, Q.; Xiang, D. Effects of drought stress on root development and physiological characteristics of sweet potato at seedling stage. *Chin. J. Appl. Ecol.* **2019**, *30*, 3155–3163. [CrossRef]

*plants*

*Article*

# Optimization of a Lower Irrigation Limit for Lettuce Based on Comprehensive Evaluation: A Field Experiment

**Maomao Hou [1,†], Houdong Zhang [1,†], Hiba Shaghaleh [2], Jingnan Chen [2], Fenglin Zhong [1], Yousef Alhaj Hamoud [2,*] and Lin Zhu [3,*]**

1 College of Horticulture, Fujian Agriculture and Forest University, Fuzhou 350000, China
2 College of Agricultural Science and Engineering, Hohai University, Nanjing 210098, China
3 Nanjing Institute of Environmental Sciences, Nanjing 210000, China
* Correspondence: yousef-hamoud11@hotmail.com (Y.A.H.); zhulin@nies.org (L.Z.)
† These authors contributed equally to this work.

**Abstract:** When optimizing irrigation methods, much consideration is given to crop growth indicators while less attention has been paid to soil's gaseous carbon (C) and nitrogen (N) emission indicators. Therefore, adopting an irrigation practice that can reduce emissions while maintaining crop yield and quality is of great interest. Thus, open-field experiments were conducted from September 2020 to January 2022 using a single-factor randomized block design with three replications. The lettuce plants ("*Feiqiao Lettuce No.1*") were grown using four different irrigation methods established by setting the lower limit of drip irrigation to 75%, 65%, and 55% of soil water content at field capacity corresponding to DR1, DR2, and DR3, respectively. Furrow irrigation (FI) was used as a control. Crop growth indicators and soil gas emissions were observed. Results showed that the mean lettuce yield under DR1 (64,500 kg/ha) was the highest, and it was lower under DR3 and FI. The lettuces under DR3 showed greater concentrations of crude fiber, vitamin C, and soluble sugar, and a greater nitrate concentration. Compared with FI, the DR treatments were more conducive to improving the comprehensive quality of lettuce, including the measured appearance and nutritional quality. Among all the irrigation methods, FI had the maximum cracking rate of lettuce, reaching 25.3%, 24.6%, and 22.7%, respectively, for the three continuous seasons. The stem cracking rates under DR2 were the lowest—only 10.1%, 14.4%, and 8.2%, respectively, which were decreased to nearly half compared with FI. The entropy model detected that the weight coefficient evaluation value of DR2 was the greatest, reaching 0.93, indicating that the DR2 method has the optimal benefits under comprehensive consideration of water saving, yield increase, quality improvement, and emission reduction.

**Keywords:** lettuce; drip irrigation; emission reduction; entropy weight coefficient

## 1. Introduction

Irrigation is the leading water resource consumer and a key factor affecting soil greenhouse gas emissions [1]. Irrigation can change the conditions of soil moisture, temperature, oxygen, and microorganisms, thus influencing the production and release of greenhouse gases such as carbon dioxide ($CO_2$), methane ($CH_4$), and nitrous oxide ($N_2O$) [2,3]. Different irrigation approaches affect the soil's carbon © and nitrogen (N) emissions differently [4]. Earlier studies have shown that water-saving irrigation techniques such as drip irrigation (DI) and micro sprinkler irrigation can reduce $CH_4$ emissions but may increase $N_2O$ emissions; however, conventional irrigation techniques such as flood irrigation have shown opposite impacts [5,6]. In addition, the irrigation amount also markedly influences the C/N discharges [7,8]. However, lower C and N discharges are rarely considered when designing irrigation tactics.

Lettuce is an important vegetable crop. It is rich in soluble sugar, vitamin C (Vc), and mineral nutrients. Irrigation impacts the crop's development indicators, such as leaf area index, plant height, and stem diameter, and quality indicators, such as fiber, nitrate, and soluble solids contents of lettuce [9]. Therefore, crop quality, yield, and water use efficiency should be considered when adopting an irrigation method for lettuce, similar to other vegetable crops. However, there is limited information on the stem cracking rate of lettuce. Stem cracking of lettuce is caused by excessive irrigation after high temperatures and long-term drought, resulting in the rapid growth of fleshy stems and thus bursting of the external xylem. Recently, stem cracking has been the main problem restricting the development of the lettuce industry in southern China [10]. Therefore, the cracking rate under different irrigation methods is important and should be considered when optimizing water-saving irrigation approaches. Thus, when adopting drip irrigation strategies, yield and quality indicators and environmental indicators such as greenhouse gas emissions should be considered [11].

Water-efficient irrigation strategies should achieve environmental sustainability while enhancing water use efficiency and conservation. The optimization of irrigation practice is a decision that involves multiple objectives and multiple indicators. Multiple objectives refer to multiple irrigation systems designed by the experiment, and multiple indicators refer to the indicators that need to be considered, such as crop yield, quality, and water use efficiency [12]. The optimization of irrigation systems often relies on the designer's expertise, which leads to the reasonable and scientific adoption of an irrigation strategy. Therefore, several mathematical models, including the projection pursuit classification, the analytic hierarchy process, and the fuzzy comprehensive evaluation method, have been introduced to the decision-makers of agricultural plans. Thus, earlier research has used the entropy evaluation method to optimize agricultural water conservancy tactics. However, the application of the entropy method in the optimization of irrigation systems for lettuce has not been investigated yet. Therefore, the current study hypothesizes that the different irrigation strategies would impact lettuce yield, quality, and carbon and nitrogen emissions. To test the assumption mentioned above, the present study evaluated the effects of different designed irrigation systems on lettuce yield, quality, water use efficiency, and soil C and N emissions. This study used the entropy weight coefficient evaluation method for selecting the optimal irrigation system for water saving, yield and quality improvements, and greenhouse gas emission reduction as a theoretical and practical base.

## 2. Materials and Methods

### 2.1. Experimental Site

The experiment was conducted in the Fruit Science and Technology Field of the Modern Area (23°57′38″ N, 117°20′5″ E) of Yun Xiao County in Fujian Province, China, from September 2020 to January 2022. The altitude of Yunxiao is 320 m. The annual average temperature of the experimental field is 21.3 °C. The absolute maximum temperature is 38.1 °C, and the absolute minimum temperature is −0.2 °C. The annual precipitation in Fujian Province is 1730.6 mm. However, during the cultivation period of lettuce, an uneven distribution of rainfall often occurs. The frost-free period is 347 days. The soil type is ferrallitic soil. Soil was sampled from the top layer (0–20 cm) in the experimental site and analyzed. Soil properties are a pH of 5.9, available N content of 90.2 mg/kg, available P content of 12.2 mg/kg, available K content of 152.3 mg/kg, bulk density of 1.26 g/cm$^3$, and field capacity of 0.28 g/g. The monthly rainfall is displayed in Figure 1.

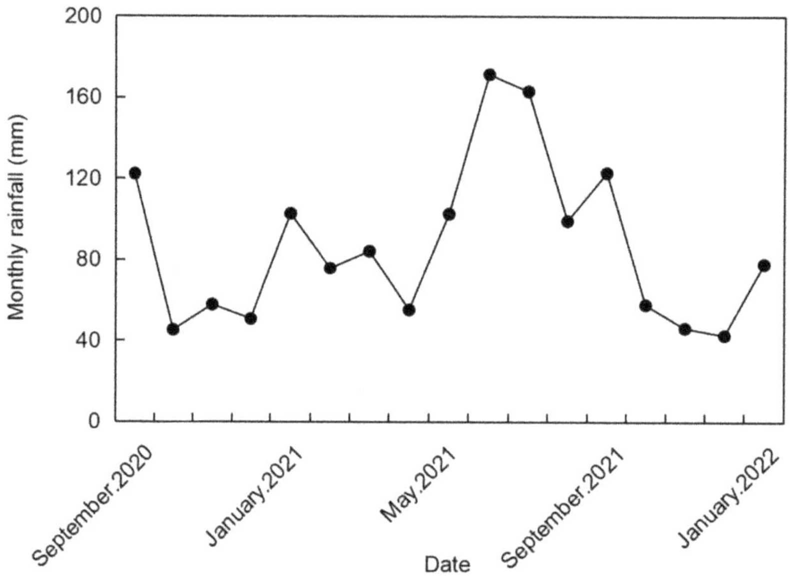


**Figure 1.** The monthly rainfall.

*2.2. Cropping Practices, Experimental Design, and Treatments*

The lettuce variety used for this study was "*Feiqiao Lettuce No.1*". The seeds were placed in pure water for 8 h, then encased and saved at a temperature of 5 °C for 10 h to break dormancy. During this period, the seeds were flipped once to facilitate germination. Then, the seeds were removed and set in cold conditions for about 20 h. When 2/3 of the seeds were well exposed, they were moved to the seedling tray. Seedlings with four to five leaves were transplanted in the field. The lettuce was irrigated after transplanting and when the plant reached the rosette stage.

The plot area was 32 $m^2$ (4 m × 8 m). The experimental area was separated into 12 experimental blocks, and treatment used three blocks. Also, each block used the ridge planting system (height of 20 cm and width of 60 cm). The seedlings were transplanted at three ridges in each plot, with a plant-to-plant distance of 35 cm and a row-to-row interval of 30 cm. To avoid water seepage between treatments, soil ridges without any lettuce planting were placed between the FI block and the other irrigation blocks.

This experiment was conducted on the open field and employed a single-factor randomized block design with four treatments and three repetitions. The main plot was drip irrigation (DI) with three different lower irrigation limits corresponding to soil water contents at 55%, 65%, and 75% of the field capacity, named DR3, DR2, and DR1, respectively. The upper irrigation limitation of all DI was set at a soil water content of 95% of the field capacity. Furrow irrigation (FI) was employed as a control treatment, where the water was supplied to the soil ridge at 1/3 furrow height. The soil water content was measured daily using the drying method after collecting the soil samples from the top layer (8–10 cm), and irrigation was initiated when the soil moisture content reached the low limit of irrigation.

For each DI system, the water amount for DR1, DR2, or DR3 was calculated using the following equation [13]:

$$M = S \times \mathrm{r} \times h \times Q \times \left(q^1 - q^2\right)/0.95$$

where $M$ represents the irrigation amount ($m^3$); $S$ represents the irrigated area ($m^2$); r represents the bulk density ($kg/m^3$); $h$ represents the planned wetting layer depth (0.2 m);

$Q$ is the field capacity (%); $q^1$ represents the upper irrigation limit, $q^2$ represents the irrigation lower limit (75%, 65% or 55%), and 0.95 is the irrigation coefficient.

In local practice, farmers employ FI, and their method is to irrigate until reaching a water level that is 1/3 of the height of the furrow. In their practice, the amount of water supply for FI is similar to the calculated amount under DR2 treatment in this study.

The experiment was repeated for three seasons. The considered growth stage divisions are shown in Table 1.

**Table 1.** Detailed information on irrigation practices implemented during different growth stages.

| Season Order | Date | Growth Stages | Watering Periods | Mean Temperate (°C) |
|---|---|---|---|---|
| First season | 15 September 2020–17 September 2020 | Seed germination | | 29.87 |
| | 18 September 2020–12 October 2020 | Seedling stage | | 26.56 |
| | 13 October 2020–16 November 2020 | Rosette stage | ☆ | 23.82 |
| | 17 November 2020–26 December 2020 | Fleshy stem expansion stage | ☆ | 19.55 |
| | 27 December 2020–30 December 2020 | Harvest stage | | 16.75 |
| Second season | 16 January 2021–18 January 2021 | Seed germination | | 14.01 |
| | 19 January 2021–17 February 2021 | Seedling stage | | 18.16 |
| | 18 February 2021–28 March 2021 | Rosette stage | ☆ | 19.43 |
| | 29 March 2021–8 May 2021 | Fleshy stem expansion stage | ☆ | 23.04 |
| | 9 May 2021–13 May 2021 | Harvest stage | | 28.78 |
| Third season | 20 September 2021–22 September 2021 | Seed germination | | 30.66 |
| | 23 September 2021–15 October 2021 | Seedling stage | | 29.43 |
| | 16 October 2021–21 November 2021 | Rosette stage | ☆ | 22.59 |
| | 22 November 2021–2 January 2022 | Fleshy stem expansion stage | ☆ | 19.11 |
| | 3 January 2022–6 January 2022 | Harvest stage | | 16.25 |

Note: ☆ denotes that plants were grown under different irrigation treatments during a specific growth stage.

Detailed information on the irrigation events, interval, quota (irrigation input at each irrigation event), and total water use implemented in this study is shown in Table 2.

**Table 2.** The irrigation times, interval, quota, and total water used during different growth stages.

| Season Order | Treatments | The Stage from Rosette to Fleshy Stem Expansion | | | | Total Water Use (mm) |
|---|---|---|---|---|---|---|
| | | Irrigation Times | Irrigation Interval (d) | Irrigation Quota (mm) | Irrigation Amount (mm) | |
| First season | DR1 | 8 | 8.7 | 15.8 | 126.5 | 299.1 |
| | DR2 | 5 | 14.0 | 23.7 | 118.6 | 291.2 |
| | DR3 | 3 | 23.5 | 31.6 | 94.9 | 267.5 |
| | FI | 5 | 14.0 | 23.7 | 118.6 | 291.2 |
| Second season | DR1 | 7 | 11.1 | 15.8 | 110.6 | 275.7 |
| | DR2 | 4 | 19.5 | 23.7 | 94.8 | 259.9 |
| | DR3 | 2 | 39.0 | 31.6 | 63.2 | 228.3 |
| | FI | 4 | 19.5 | 23.7 | 94.8 | 259.9 |
| Third season | DR1 | 8 | 9.5 | 15.8 | 126.5 | 284.1 |
| | DR2 | 5 | 15.2 | 23.7 | 118.6 | 276.2 |
| | DR3 | 3 | 25.3 | 31.6 | 94.9 | 252.5 |
| | FI | 5 | 15.2 | 23.7 | 118.6 | 276.2 |

Note: DR1, DR2, and DR3 were drip irrigation treatments, and FI was furrow irrigation.

Urea, calcium superphosphate, and potassium sulfate as chemical fertilizers were supplied to the experimental plots as basal and topdressing fertilizers. The chemical fertilizers used in this study had total doses of 675 kg/ha, 600 kg/ha, and 375 kg/ha for urea, calcium superphosphate, and potassium sulfate, respectively. The entire superphosphate fertilizer was supplied as the base fertilizer. However, potassium sulfate and urea were submitted as a basal fertilizer at the rate of (20%), a first topdressing at the rate of (40%) during the rosette stage, and a second topdressing at the rate of 40% in the fleshy stem expansion stage. The fertilizers were applied to a soil depth of 6 cm using a fertilization machine. Plant disease protection was applied during the whole development phase based on the local farmers' practice, and the same practice was used for all treatments.

### 2.3. Plant Sampling and Measurements

For all three seasons, six lettuce plants were randomly selected from each block to measure the leaf area index (LAI), yield, and stem cracking rate. The plant canopy analyzer (LAI-2200, LI-COR, 4647 SUPERIOR ST. LINCOLN, NE 68504 LOUISIANA USA, Lincoln, NE, USA) was used to measure the LAI. The LAI was calculated based on the attenuation of sky scattered radiation after passing through the canopy at five zenith angle intervals (0–13°, 16–28°, 32–43°, 47–58°, 61–74°; the central zenith angles are 7°, 23°, 38°, 53°, and 68°, respectively) measured using a "fish eye" optical sensor [14]. The lettuce quality indicators, including crude fiber, vitamin, soluble sugar, and nitrate content, were measured. The crude fiber was determined using the paradigm washing method; the soluble sugar content was determined using the anthrone sulfate method; the nitrate was determined using the phenol disulfonic acid colorimetric method; and the Vc was determined using the method of 2,6-dichloro indophenol titration [15].

### 2.4. Measurements and Determination of C and N Emission Fluxes

In this study, a self-made gas collection apparatus was used (Figure 2). The device had a diameter of 30 cm and a height of 80 cm and consisted of a body and a base with a cover constructed from PVC material and wrapped with a silver reflective film. The device base was inserted in the soil, and the upper surface of the base contained a water tank to provide a seal. Prior to gas collection, the water tank was filled with purified water, and the cover was placed on top of the base. A thermometer and a port for collecting gas were fixed on the side of the cover. Also, a small fan powered by an external battery was positioned inside the cover to distribute the emitted gas evenly. The C and N emission flux and the accumulated C and N emission amounts were calculated according to a standard method [16]. The gas samples were collected during the third season.

**Figure 2.** The self-made device.

### 2.5. Application of the Entropy Weight Coefficient Model

The entropy weight coefficient model was applied according to the evalution indicators and the steps are as follows [17]:

Assuming that there are $m$ kinds of lettuce irrigation treatment and $n$ evaluation indexes in this experiment, the matrix for evaluation can be formed by using $m$ kinds of lettuce irrigation treatment corresponding to $n$ different indicators (according to the design of the experiment in this study, $m = 4$, $n = 10$):

$$R = (r_{ij})_{m \times n} \tag{1}$$

In the formula, $r_{ij}$ is the $j$th indicator value under the $i$th irrigation scheme. For a certain indicator $r_j$, there is information entropy:

$$E_j = -\sum_{i=1}^{m} p_{ij} \ln p_{ij}, (j = 1, 2, 3, \ldots n) \tag{2}$$

where, $p_{ij} = r_{ij} / \sum_{i-1}^{m} r_{ij}$.

The $j$th indicator's entropy value can be calculated using:

$$e_j = \frac{1}{\ln m} E_j,\ (j = 1, 2, 3, \ldots, n) \tag{3}$$

The $j$th indicator's objective weight can be calculated using:

$$\theta_j = (1 - e_j) / \sum_{i=1}^{n} (1 - e_j),\ (j = 1, 2, 3, \ldots, n) \tag{4}$$

According to the calculations, it can be found that $0 \leq \theta_j \leq 1$, $\sum_{j=1}^{n} \theta_j = 1$.

In irrigation method decisions, the decision-makers' experience is also important when formulating an irrigation system. Therefore, the subjective weight could be combined with the objective weight $\theta_j$ ($j$ = 1 to $n$) of the evaluation indicators to form a new indicator weight:

$$\alpha_j = \theta_j \overline{\omega_j} / \sum_{j=1}^{n} \theta_j \overline{\omega_j},\ (j = 1, 2, 3, \ldots, n) \tag{5}$$

If the optimum value in each list of indicators in the matrix R is $r_j{}^*$, the evaluation data can be normalized in terms of the $r_j{}^*$ value, and the $r_j{}^*$ value is commonly influenced by the nature of the indexes.

In this experiment, the lettuce yield, nitrogen fertilizer use rate, irrigation water utilization efficiency, crude fiber, soluble sugar content, and Vc content are profitability indicators, which comply with the criterion of "the greater, the better".

The stem cracking rate, nitrate content, and carbon and nitrogen emissions are the loss indexes, which comply with the criterion of "the smaller, the better". The standard index value $d_{ij}$ is formed by:

$$d_{ij} = \begin{cases} \frac{r_{ij}}{r_j{}^*}, & r_j{}^* = \max\{r_{ij}\} \\ \frac{r_j{}^*}{r_{ij}}, & r_j{}^* = \min\{r_{ij}\} \end{cases} \tag{6}$$

The evaluation value using the entropy weight method for the irrigation schemes can be calculated as:

$$\lambda_i = \sum_{j=1}^{n} \alpha d_{ij},\ i = 1, 2, 3, \ldots, m. \tag{7}$$

### *2.6. Statistical Analysis*

Data were statistically analyzed using the SPSS (version 17.0) software. Significant differences were calculated in terms of Duncan's multiple range test. The significant differences among indicator values of the different treatments in the same season were compared. Each value used for optimizing irrigation systems is the average of the three seasons except the C and N emissions.

## 3. Results and Analysis

### *3.1. Effect of the Different Irrigation Treatments on the Lettuce Yield*

The lettuces' responses to different irrigation systems were similar between the first and second seasons, where the lettuce yield under DR1 showed a high level, reaching

67,666.5 and 63,528 kg/ha, respectively (Figure 3). Also, the yield level in the first season under DR1 was significantly higher ($p < 0.05$) than under all other treatments. Moreover, there was no significant ($p > 0.05$) difference in lettuce yield between DR3 (58,836 kg/ha) and FI (57,264 kg/ha) treatments. In the third season, the lettuce yields' responses to different irrigation methods were slightly different from the first and second seasons, showing that the yield levels under DR1 and DR2 were higher than those in DR3 and FI. Also, there was no significant ($p > 0.05$) difference in lettuce yield between DR1 and DR2; also no noticeable difference was found between DR3 treatment and FI. From the average yield of the three seasons, DR1 resulted in the greatest average yield of 64,500 kg/ha, followed by DR2, with an average yield of 62,700 kg/ha. However, the yield of DR3 and FI were relatively low, with an average yield of 57,360 kg/ha and 57,060 kg/ha compared to the other treatments.

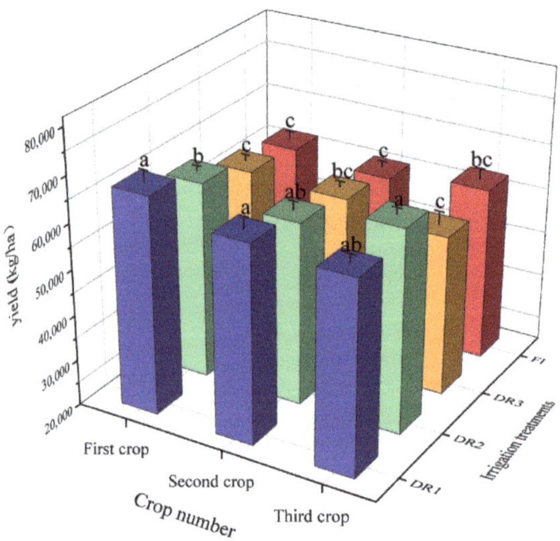


**Figure 3.** Effects of different irrigation methods on the lettuce yield. DR1, DR2, and DR3 represent lower irrigation limits at soil moisture contents of 75%, 65%, and 55% of the field capacity. FI is a furrow irrigation method using the same irrigation quota and regime as the DR2 treatment. The data are mean ± standard deviation. The significant differences among yield values of the different treatments in the same season were compared. The different letters (a, b, c) suggest significant differences at 0.05 level in terms of Duncan's multiple range test.

### 3.2. *Effect of the Different Irrigation Treatments on Lettuce LAI*

The LAI presented three stages, including a slow growth stage from 2 to 29 days after transplant (DAT), a rapid growth stage from 29 to 57 DAT, and after 57 DAT (Figure 4). The differences in LAI of lettuce among different treatments began to be obvious at 43 DAT. For the first and third seasons, the highest LAI was detected under the DR1 treatment, followed by the DR2 treatment, while the lowest was in the DR3 treatment. The LAI of lettuce under FI was lower than that under DR2 treatment with the same irrigation quota. However, in the second season, the LAI under DR2 was slightly higher than under DR1, and other regularities were consistent with the first and third seasons.

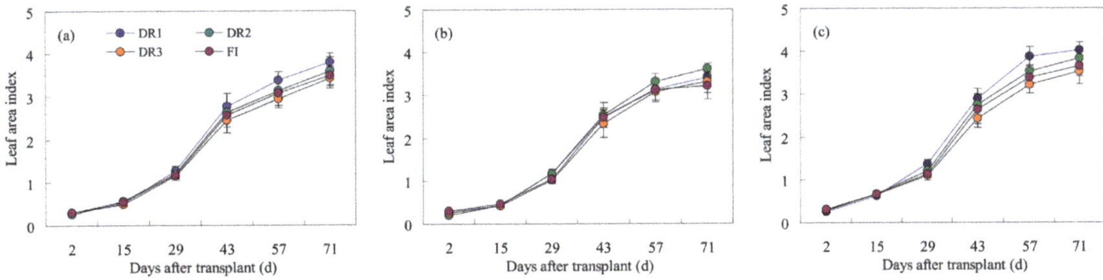


**Figure 4.** Impacts of different irrigation schemes on the LAI of lettuce. ((**a**) the first season, (**b**) the second season, and (**c**) the third season) represent the first, second, or third cultivation season of lettuce. DR1, DR2, and DR3 represent lower irrigation limits at soil moisture contents of 75%, 65%, and 55% of the field capacity. FI is a furrow irrigation method using the same irrigation quota and regime as the DR2 treatment.

Although there were clear differences in the LAI among different irrigation treatments and seasons, LAI indices under DR1 and DR2 were higher than those under DR3 and FI (Figure 5). The average LAI at 71 DAT confirmed these results. Also, it was noted that the average LAI of three seasons under DR1 treatment was at 71 DAT, reaching the highest value of 3.75, followed by DR2 (3.67). Whereas there was no significant difference in the LAI of lettuce among DR1, DR2, and FI, the LAI of lettuce under DR3 was the lowest (3.42) and was significantly ($p < 0.05$) lower in comparison to that under DR1 or DR2 treatment.

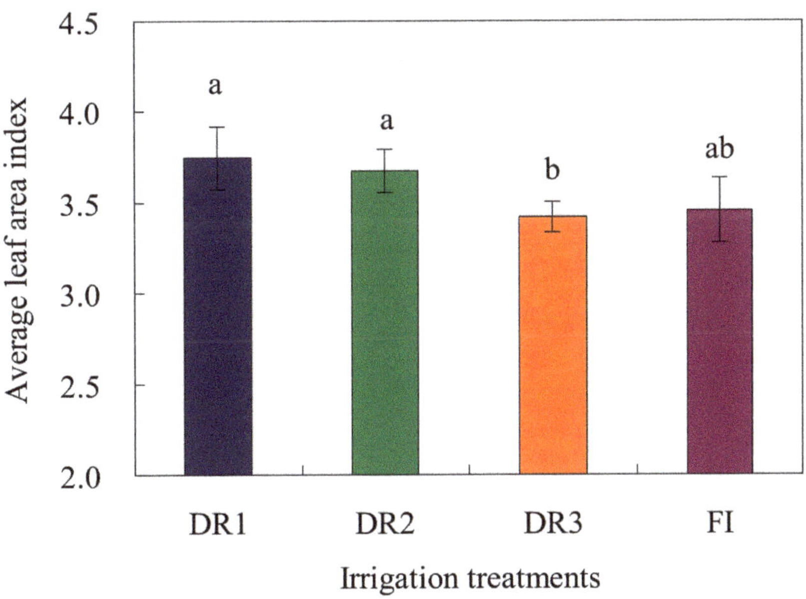


**Figure 5.** The average leaf area index of three seasons of lettuce under different irrigation methods (71 days after transplant). DR1, DR2, and DR3 represent lower irrigation limits at soil moisture contents of 75%, 65%, and 55% of the field capacity. FI is a furrow irrigation method using the same irrigation quota and regime as the DR2 treatment. The data are mean ± standard deviation. The letters, such as 'a' and 'b', suggest that there is a significant difference at the 0.05 level in terms of Duncan's multiple range test.

### 3.3. *Effect of the Different Irrigation Treatments on the Quality of Lettuce*

Among the different drip irrigation treatments, DR3 obtained higher contents of crude fiber, soluble sugar, and Vc (Table 3). This effect was evident in the third season, as the three indicators (crude fiber, soluble sugar, and Vc content) under DR3 were significantly ($p < 0.05$) greater than those under DR1 or DR2 treatments. However, while improving quality indicators, DR3 increased the nitrate content, with nitrate content reaching 412.8, 481.8, and 482.7 mg/kg for three seasons, respectively. The quality indicator values showed a downward trend as the lower irrigation limit increased.

**Table 3.** Effect of the different irrigation methods on the quality of lettuce for the three seasons.

| Season Order | Treatment Applied | Crude Fiber (%) | Soluble Sugar (%) | Nitrate (mg/kg) | Vitamin c (mg/kg) |
|---|---|---|---|---|---|
| First season | DR1 | 2.67 ± 0.21 b | 4.39 ± 0.12 c | 334.8 ± 9.5 c | 65.2 ± 5.6 b |
| | DR2 | 3.22 ± 0.25 a | 4.92 ± 0.08 a | 362.8 ± 13.2 b | 83.6 ± 6.7 a |
| | DR3 | 3.56 ± 0.32 a | 5.08 ± 0.15 a | 412.8 ± 21.8 a | 84.8 ± 6.8 a |
| | FI | 3.11 ± 0.13 a | 4.66 ± 0.09 b | 388.4 ± 24.2 ab | 64.1 ± 5.4 b |
| Second season | DR1 | 2.78 ± 0.25 c | 4.60 ± 0.15 b | 419.1 ± 21.6 b | 75.1 ± 4.2 b |
| | DR2 | 3.81 ± 0.32 ab | 5.35 ± 0.35 a | 422.7 ± 19.1 b | 93.7 ± 7.1 a |
| | DR3 | 3.92 ± 0.11 a | 5.02 ± 0.18 a | 481.8 ± 29.7 a | 87.6 ± 4.2 a |
| | FI | 3.71 ± 0.06 b | 4.45 ± 0.19 b | 440.7 ± 31.5 ab | 77.1 ± 2.9 b |
| Third season | DR1 | 3.22 ± 0.15 c | 4.63 ± 0.15 b | 392.6 ± 19.2 c | 73.3 ± 3.9 c |
| | DR2 | 3.62 ± 0.08 b | 4.61 ± 0.19 b | 412.3 ± 25.6 bc | 85.7 ± 2.5 b |
| | DR3 | 3.86 ± 0.11 a | 5.26 ± 0.36 a | 482.7 ± 18.3 a | 92.3 ± 1.5 a |
| | FI | 3.77 ± 0.15 ab | 4.72 ± 0.29 ab | 434.6 ± 15.8 b | 67.5 ± 4.8 c |

Note: DR1, DR2, and DR3 represent lower irrigation limits at soil moisture contents of 75%, 65%, and 55% of the field capacity. FI is a furrow irrigation method using the same irrigation quota and regime as the DR2 treatment. The different letters (a, b, c) suggest significant differences at 0.05 level in terms of Duncan's multiple range test. The significant differences among quality indicator values of the different treatments in the same season were compared.

The different Irrigation methods also have a significant impact on the quality indicators of lettuce. For the first and second seasons, drip irrigation under the same irrigation quota was beneficial to increase the contents of crude fiber, soluble sugar, and Vc in lettuce while reducing the nitrate content. There was no significant difference ($p < 0.05$) in crude fiber or soluble sugar content between DR2 and FI in the third season. However, the DR2 significantly ($p < 0.05$) increased the Vc content and decreased the nitrate content. From the comprehensive performance of lettuce quality, drip irrigation was more efficient in promoting quality improvement compared to furrow irrigation.

### 3.4. *Effect of the Different Irrigation Treatments on the Cracking Rate of Lettuce*

There was an obvious relationship between the cracking rate of lettuce and irrigation methods. The cracking rate under FI was the highest, with the cracking rates for the three seasons reaching 25.3%, 24.6%, and 22.7%, respectively, and were significantly ($p < 0.05$) greater compared to that under the other treatments (Figure 6). Among drip irrigation treatments, DR2 showed the lowest rate of stem cracking compared to other treatments, especially in the first and third seasons. The stem cracking rates under DR2 were 10.1%, 14.4%, and 8.2%, respectively, for the three seasons, which were decreased by nearly half compared to FI. There was no significant difference in the lettuce stem cracking rate between DR1 and DR3. Drip irrigation had a significant effect on reducing the lettuce stem cracking rate, and DR2 was optimal in controlling the cracking rate.

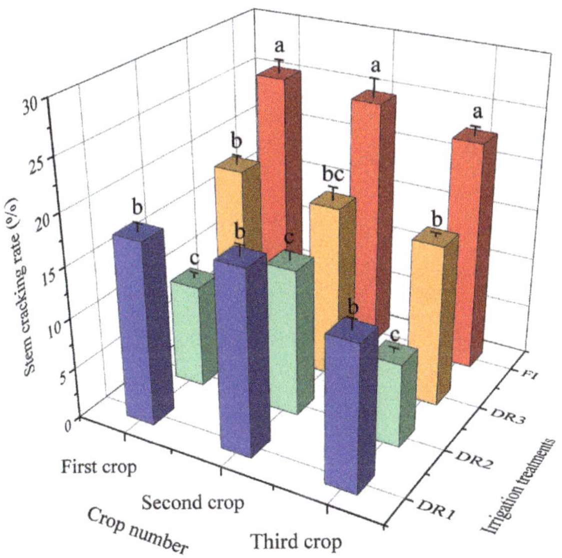


**Figure 6.** Effects of the different irrigation systems on the cracking rate of lettuce stems for the three seasons. DR1, DR2, and DR3 represent lower irrigation limits at soil moisture contents of 75%, 65%, and 55% of the field capacity. FI is a furrow irrigation method using the same irrigation quota and regime as the DR2 treatment. The data used in the figure are mean ± standard deviation. The different letters like a, b, and c suggest that there were significant differences at the 0.05 level in terms of Duncan's multiple range test. The significant differences among the stem cracking rate values of the different treatments in the same season were compared.

### *3.5. Optimization of Irrigation Systems Based on Water Conservation and Emission Reduction*

The evaluation indicators for lettuce irrigation systems are shown in Table 4. The yield, stem cracking rate, irrigation water efficiency, and lettuce quality indicators such as crude fiber, soluble sugar, nitrate, and Vc in the table are averaged for the three seasons. The C emissions are the sum of accumulated C emissions converted by $CO_2$ and $CH_4$. The N emissions are the sum of accumulated N emissions converted by $N_2O$ and NO.

**Table 4.** The evaluation indicators for lettuce grown under different irrigation systems.

| Treatment | Yield (kg/ha) | Stem Cracking Rate (%) | Nitrogen Use Efficiency (%) | Irrigation Water Use Efficiency (kg/m$^3$) | Crude Fiber (%) | Soluble Sugar (%) | Nitrate (mg/kg) | Vc (mg/kg) | C Emission (g C/m$^2$) | N Emission (kg/ha) |
|---|---|---|---|---|---|---|---|---|---|---|
| DR1 | 64,500 | 16.7 | 41.6 | 22.52 | 2.89 | 4.54 | 382.16 | 71.21 | 344.7 | 2.10 |
| DR2 | 62,700 | 10.9 | 41.5 | 22.73 | 3.55 | 4.96 | 399.26 | 87.68 | 459.6 | 2.41 |
| DR3 | 57,360 | 17.2 | 37.3 | 22.98 | 3.78 | 5.12 | 459.12 | 88.23 | 483.4 | 2.99 |
| FI | 57,060 | 24.2 | 34.7 | 20.68 | 3.53 | 4.61 | 421.23 | 69.58 | 382.7 | 3.95 |

Note: DR1, DR2, and DR3 represent lower irrigation limits at soil moisture contents of 75%, 65%, and 55% of the field capacity. FI is a furrow irrigation method using the same irrigation quota and regime as the DR2 treatment.

Among the different drip irrigation treatments, DR2 showed the highest evaluation value, reaching 0.93 (Figure 7), indicating that considering the four dimensions of water saving, yield increase, quality improvement, and emission reduction, the comprehensive benefits of DR2 treatment were optimal. From the perspective of indicator composition, the yield, nitrogen fertilizer utilization efficiency, and irrigation water utilization efficiency were all suboptimal, and the C and N emission reduction effect was also at the suboptimal level under DR2. However, the stem cracking rate indicator under DR2 was optimal among the four treatments. These factors led to the best evaluation result for DR2. The evaluation value under FI was lower compared to DR2, which was only 0.58. Based on the entropy weight coefficient evaluation model, DR2 can be proposed as a water-saving and emission-reduction irrigation method for lettuce fields, with an optimal low limit of drip irrigation at a soil water content of 65% of the field capacity.

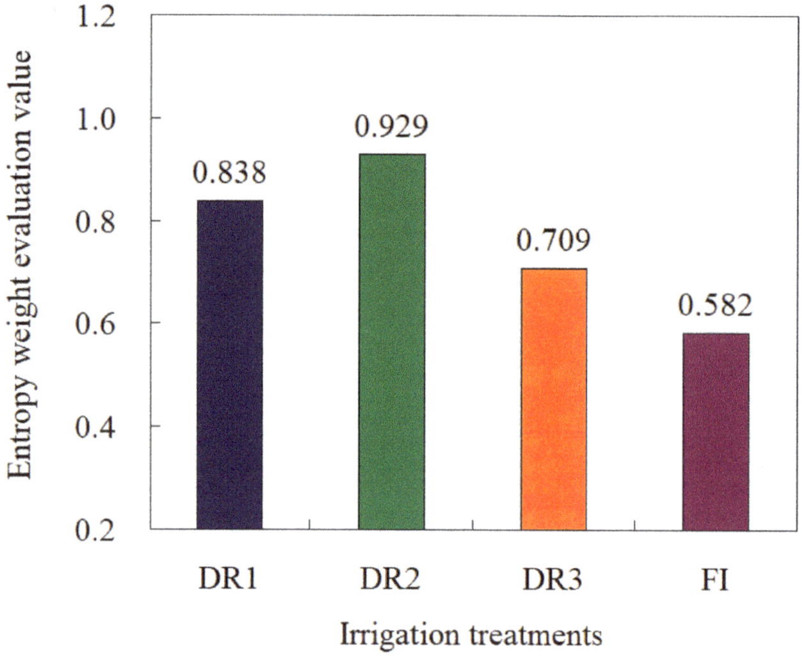


**Figure 7.** Evaluation value of entropy weight method for comprehensive benefits of water saving, yield increasing, quality improving, and emission reduction with different irrigation schemes. DR1, DR2, and DR3 represent lower irrigation limits at soil moisture contents of 75%, 65%, and 55% of the field capacity. FI is a furrow irrigation method using the same irrigation quota and regime as the DR2 treatment.

## 4. Discussion

The leaves largely reflect the physiological and biochemical conditions of plants. Therefore, the LAI is an important indicator for measuring the growth characteristics of plant populations. An earlier study has shown that with sufficient water supply, the crop leaves are relatively thin, and the leaf area increases significantly [18]. On the contrary, under conditions of soil water stress, the leaves are thicker, and the leaf growth rate is slower [19]. In this study, the total water use was the highest under DR1, and the LAI under DR1 was also the greatest among the four treatments. This result was in line with the regularity summarized by previous studies [20,21].

Crop yield is formed after the solar energy is effectively transformed into chemical energy and accumulated in the crop [22]. Unreasonable soil water supply will directly affect the transformation efficiency of solar energy on crop plants, and will ultimately be reflected in the yield [23]. In this study, the average yield of DR1 was at a high level, indicating that within a certain range, irrigation frequency and irrigation amount are positively correlated with crop yield, which is consistent with previous research conclusions [22,24]. The average yields of DR3 and FI lettuce were at the middle levels. The reason may be that the DR3 can save water; however, the lettuce, due to insufficient soil moisture, suffered damage in physiological processes and, eventually, the yield formation. Conversely, under FI, soil developed saturation zones, which hinders the process of obtaining oxygen by the root system and limits the growth and development of the fleshy stem, thus reducing yield.

It is believed that the quality indicators of vegetables include taste indicators, nutritional indicators, and constraint indicators [25,26]. Due to the difficulty in quantifying taste indicators, this study selected soluble solids as the quality indicator that reflects taste. Meanwhile, the nutritional indicators, including crude fiber and Vc, the constraint indicator nitrate, were selected for the analysis of the comprehensive quality of lettuce. Herein, the average soluble sugar content of three seasons of lettuce was higher under DR3 than under other irrigation treatments, which may be due to the fact that the amount of water supply throughout the entire growth period under DR3 treatment was the lowest, which resulted in a relatively high concentration of soluble sugar in the fleshy stems of the lettuce. A previous study has shown that an increase in crude fiber is often related to low water supply, high temperature, and low fertilizer [21]. Consistently, in this study, the treatment (DR3) with a lower irrigation amount obtained the higher crude fiber content. The correlation between the average Vc content and irrigation systems in this study was similar to that between other quality indicators and irrigation methods. Our study found that the contents of crude fiber, soluble sugar, nitrate, and Vc were at the highest levels under DR3, indicating that no treatment in this study could increase the lettuce quality while reducing nitrate content. In addition, DR3 has the lowest lettuce yield, indicating that changing the irrigation method from "small quota but high frequency" to "high quota but low frequency" will decrease the lettuce yield, but improve the values of quality indicators.

The rate of stem cracking under DR3 was relatively high, which may be due to the intermittent drought accelerating the lignification of the outer epidermis of lettuce. Also, a greater irrigation quota under DR3 caused the rapid growth of the fleshy stem, and the difference in growth rate between the epidermis and fleshy stem led to the formation of stem cracking, which is consistent with the result of earlier studies [27,28]. In addition, the rate of stem cracking under DR1 was also at a high level, which may be related to local production habits.

Earlier studies on crop quality often focused on the nutritional quality or the harmful indicators such as hormones, heavy metals, and pesticide residues [29,30]. Limited attention has been paid to the marketable indicators such as the stem cracking rate. In addition, when designing the irrigation system, little attention has been paid to C and N emission indicators. Therefore, the scientific evaluation of constructing indicators of designed irrigation systems to achieve a comprehensive evaluation of this system is a major issue that needs to be addressed at present. Previous studies have found that nitrogen fertilizer input in crops may be an effective strategy to reduce carbon and nitrogen emissions, and crop yield and quality are also important indicators of crop harvest [31]. In this study, crop yield, quality, water, and nitrogen fertilizer utilization efficiency, as well as the C and N emissions as evaluation indicators, were selected. The entropy evaluation model was used to select the most suitable lower limit of irrigation for water conservation and emission reduction in lettuce fields. This study emphasizes the importance of considering C and N emissions when designing an irrigation scheme under the current environmental background of the greenhouse effect. In addition, tuberous crops commonly cause significant disturbance to soils, which greatly affects the emission of gaseous C and N from the soil, and irrigation strategies are closely related to crop root growth. Therefore, investigating suitable emission-reduced irrigation strategies for tuberous crops is more representative. Previous studies have shown that the entropy evaluation method is more accurate when there are more indicators and schemes [32]. In this experiment, four irrigation systems were implemented, which is generally at a low level compared to previous studies. One shortcoming of this study is that we only had monthly rainfall data from local meteorological stations, while the daily rainfall was not measured, which made the total rainfall at each reproductive stage unclear. The results revealed that the DR2 treatment had the optimal comprehensive benefits in consideration of water saving, yield increase, quality improvement, and emission reduction. However, the soil's mechanical composition is crucial for soil water availability.

Therefore, future research should pay attention to the soil texture and its relation to the irrigation schedule, the crop yield, and the quality of lettuce.

## 5. Conclusions

The present study revealed that drip irrigation treatments were more conducive to improving the comprehensive quality of lettuce than furrow irrigation. Among the drip irrigation treatments, DR3 obtained the higher contents of crude fiber, Vc, and soluble sugar. For all irrigation treatments, DR1 increased the crop yield significantly, and FI was found to largely increase the cracking rate of lettuce. The results of the entropy method indicated that the DR2 treatment has the optimal comprehensive benefits in consideration of water conservation, yield and quality improvement, and emission reduction. We highlighted the significance of considering soil C and N emissions when formulating an irrigation scheme, especially for tuberous crops. Further research can focus on the process and mechanism of the impact of root enlargement on soil C and N emissions under the influence of irrigation.

**Author Contributions:** Investigation, H.Z. and F.Z.; Writing—original draft, M.H., J.C. and H.S.; Writing—review & editing, L.Z. and Y.A.H. All authors have read and agreed to the published version of the manuscript.

**Funding:** The Seed Industry Innovation and Industrialization Project in Fujian Province (zycxny2021009); the Rural Revitalization Vegetable Industry Service Project of Fujian Agriculture and Forestry University (11899170118); the Fujian Modern Agricultural Vegetable Industry System Construction Project (2019-897); facilities suitable for mechanized harvesting green leafy vegetables introduction, screening, and demonstration promotion (2022-P-001).

**Institutional Review Board Statement:** Not applicable.

**Informed Consent Statement:** Not applicable.

**Data Availability Statement:** The data presented in this study are available on request from the corresponding author. The data are not publicly available due to the safety of the project. The relevant data are confidential and cannot be published in this article.

**Conflicts of Interest:** The authors declare no conflicts of interest.

## References

1. Hamad, A.A.A.; Ni, L.; Shaghaleh, H.; Elsadek, E.; Hamoud, Y.A. Effect of Carbon Content in Wheat Straw Biochar on $N_2O$ and $CO_2$ Emissions and Pakchoi Productivity Under Different Soil Moisture Conditions. *Sustainability* **2023**, *15*, 5100. [CrossRef]
2. Abalos, D.; Sanchez-Martin, L.; Garcia-Torres, L.; van Groenigen, J.W.; Vallejo, A. Management of irrigation frequency and nitrogen fertilization to mitigate GHG and NO emissions from drip-fertigated crops. *Sci. Total Environ.* **2014**, *490*, 880–888. [CrossRef] [PubMed]
3. Guo, S.; Qi, Y.; Luo, X. Effects of drip irrigation on soil $CO_2$ and $N_2O$ emissions and their global warming potentials of spring wheat field in arid region. *J. Agro Environ. Sci.* **2016**, *35*, 792–800.
4. Hamad, A.A.A.; Wei, Q.; Wan, L.; Xu, J.; Hamoud, Y.A.; Li, Y.; Shaghaleh, H. Subsurface Drip Irrigation with Emitters Placed at Suitable Depth Can Mitigate $N_2O$ Emissions and Enhance Chinese Cabbage Yield under Greenhouse Cultivation. *Agronomy* **2022**, *12*, 745. [CrossRef]
5. Ye, X.H.; Han, B.; Li, W.; Zhang, X.C.; Zhang, Y.L.; Lin, X.G.; Zou, H.T. Effects of different irrigation methods on nitrous oxide emissions and ammonia oxidizers microorganisms in greenhouse tomato fields. *Agric. Water Manag.* **2018**, *203*, 115–123. [CrossRef]
6. Yu, Y.; Jiao, Y.; Yang, W.; Song, C.; Zhang, J.; Liu, Y. Mechanisms underlying nitrous oxide emissions and nitrogen leaching from potato fields under drip irrigation and furrow irrigation. *Agric. Water Manag.* **2022**, *260*, 107270. [CrossRef]
7. Gultekin, R.; Avağ, K.; Görgişen, C.; Öztürk, Ö.; Yeter, T.; Alsan, P.B. Effect of deficit irrigation practices on greenhouse gas emissions in drip irrigation. *Sci. Hortic.* **2023**, *310*, 111757. [CrossRef]
8. Chen, H.; Shang, Z. Effects of irrigation amounts on soil $CO_2$, $N_2O$ and $CH_4$ emissions in greenhouse tomato field. *Chin. J. Appl. Ecol.* **2019**, *30*, 102–111.
9. Yavuz, D.; Seymen, M.; Kal, Ü.; Atakul, Z.; Tanrıverdi, Ö.B.; Türkmen, Ö.; Yavuz, N. Agronomic and physio-biochemical responses of lettuce to exogenous sodium nitroprusside (SNP) applied under different irrigation regimes. *Agric. Water Manag.* **2023**, *277*, 108127. [CrossRef]
10. Zhong, F.; Lin, Y. Research and Innovation on Optimal Utilization of Agricultural Water and Land Resources. *China Resour. Compr. Util.* **2022**, *40*, 108–110.

11. Hamad, A.A.A.; Wei, Q.; Xu, J.; Hamoud, Y.A.; He, M.; Shaghaleh, H.; Wei, Q.; Li, X.; Qi, Z. Managing Fertigation Frequency and Level to Mitigate $N_2O$ and $CO_2$ Emissions and $NH_3$ Volatilization from Subsurface Drip-Fertigated Field in a Greenhouse. *Agronomy* **2022**, *12*, 1414. [CrossRef]
12. Kolkiewicz, A.; Rice, G.; Xie, Y. Projection pursuit based tests of normality with functional data. *J. Stat. Plan. Inference* **2021**, *211*, 326–339. [CrossRef]
13. Jin, Q.; You, J.; Xie, M.; Qiu, Y.; Lei, S.; Ding, Q.; Chen, J. Drip Irrigation Reduced Fertilizer Nitrogen Loss from Lettuce Field—A Case Study Based on N-15 Tracing Technique. *Water* **2022**, *14*, 675. [CrossRef]
14. Monteith, J.L. Light distribution and photosynthesis in field crops. *Ann. Bot.* **1965**, *29*, 17–37. [CrossRef]
15. Chen, Z. Effects of Different Phosphorus Fertilizers on Growth and Quality of Chinese Cabbage and Lettuce. *Mod. Agric. Technol.* **2023**, *9*, 49–57.
16. Shao, Y.; Chen, J.; Wang, L.; Hou, M.; Chen, D. Effects of fermented organic fertilizer application on soil N2O emission under the vegetable rotation in polyhouse. *Environ. Res.* **2021**, *200*, 111491. [CrossRef] [PubMed]
17. Shao, G. Entropy weight coefficient evaluation of irrigation and drainage scheme based on analysis of quality and rain ingredients of tomat. *J. Drain. Irrig. Mach. Eng.* **2012**, *30*, 733–736.
18. Talpur, M.M.A.; Shaghaleh, H.; Ali Adam Hamad, A.; Chang, T.; Zia-ur-Rehman, M.; Usman, M.; Alhaj Hamoud, Y. Effect of Planting Geometry on Growth, Water Productivity, and Fruit Quality of Tomatoes under Different Soil Moisture Regimes. *Sustainability* **2023**, *15*, 9526. [CrossRef]
19. Hamoud, Y.A.; Guo, X.; Wang, Z.; Shaghaleh, H.; Chen, S.; Hassan, A.; Bakour, A. Effects of irrigation regime and soil clay content and their interaction on the biological yield, nitrogen uptake and nitrogen-use efficiency of rice grown in southern China. *Agric. Water Manag.* **2019**, *213*, 934–946. [CrossRef]
20. Hansona, B.R.; Schwankl, L.J.; Schulbach, K.F.; Pettygrove, G.S. A comparison of furrow, surface drip, and subsurface drip irrigation on lettuce yield and applied water. *Agric. Water Manag.* **1997**, *33*, 139–157. [CrossRef]
21. Xiao, B. Analysis of the Effects of Different Irrigation Techniques on Soil Moisture and Cotton Leaf Area Index. *Agric. Technol.* **2020**, *40*, 33–35.
22. Zhou, D. *Effects of Different Water and Fertilizers Treatment on Water Use Efficiency and Nutrient Absorption Characteristics of Asparagus Lettuce*; Gansu Agricultural University: Lanzhou, China, 2014.
23. Jian, Z.; Huanjie, C. Effect of water-controlled partial wetting irrigation on the growth and water use efficiency of leaf lettuce. *Agric. Res. Arid. Areas* **2008**, *6*, 70–75.
24. Wang, S.; Jia, X. Response Mechanism of Crops to Drought Stress and Measures for Improving Drought Resistance of Crops: Research Progress. *Chin. Agric. Sci. Bull.* **2022**, *38*, 31–44.
25. Sharabiani, V.R.; Sabzi, S.; Pourdarbani, R.; Solis-Carmona, E.; Hernández-Hernández, M.; Hernández-Hernández, J.L. Non-Destructive Prediction of Titratable Acidity and Taste Index Properties of Gala Apple Using Combination of Different Hybrids ANN and PLSR-Model Based Spectral Data. *Plants* **2020**, *9*, 1718. [CrossRef]
26. Nektarios, P.A.; Ischyropoulos, D.; Kalozoumis, P.; Savvas, D.; Yfantopoulos, D.; Ntoulas, N.; Tsaniklidis, G.; Goumenaki, E. Impact of substrate depth and fertilizer type on growth, production, quality characteristics and heavy metal contamination of tomato and lettuce grown on urban green roofs. *Sci. Hortic.* **2022**, *305*, 111318. [CrossRef]
27. Lan, S. Discussion on pollution-free cultivation techniques for winter and spring lettuce. *Agric. Technol.* **2019**, *39*, 115–116.
28. Sun, S.; Gu, C. Causes and preventive measures of lettuce stem cracking. *Hebei Agric.* **2015**, *2*, 27–28.
29. Hu, Y.M.; Yang, S.Q.; Qian, X.; Li, Z.X.; Fan, Y.C.; Manevski, K.; Chen, Y.; Gao, W. Bibliometric Network Analysis of Crop Yield Gap Research over the Past Three Decades. *Agriculture* **2023**, *13*, 2105. [CrossRef]
30. Bossolani, J.W.; Meneghette, H.H.A.; Sanches, I.R.; Dos Santos, F.L.; Parra, L.F.; Lazarini, E. 5-year Effects of Potassium Fertilization Levels on Nutrition Balance and Yield in the Maize/Soybean Succession. *Commun. Soil Sci. Plant Anal.* **2022**, *53*, 2017–2028. [CrossRef]
31. Jiang, Z.H.; Zhong, Y.M.; Yang, J.P.; Wu, Y.X.Y.; Li, H.; Zheng, L. Effect of nitrogen fertilizer rates on carbon footprint and ecosystem service of carbon sequestration in rice production. *Sci. Total Environ.* **2019**, *670*, 210–217. [CrossRef] [PubMed]
32. Duan, Y.; Mu, H.; Li, N.; Li, L.; Xue, Z. Research on Comprehensive Evaluation of Low Carbon Economy Development Level Based on AHP-Entropy Method: A Case Study of Dalian. *Energy Procedia* **2016**, *104*, 468–474. [CrossRef]

plants

MDPI

*Article*

# Physiological Responses of a Grapefruit Orchard to Irrigation with Desalinated Seawater

**Josefa M. Navarro [1], Alberto Imbernón-Mulero [2,*], Juan M. Robles [1], Francisco M. Hernández-Ballester [1], Vera Antolinos [1], Belén Gallego-Elvira [2] and José F. Maestre-Valero [2]**

[1] Irrigation and Stress Physiology Group, Department of Bioeconomy, Water and Environment, Murcia Institute of Agri-Food Research and Development (IMIDA), c/Mayor s/n, 30150 Murcia, Spain; josefam.navarro2@carm.es (J.M.N.); juanm.robles@carm.es (J.M.R.); franciscom.hernandez4@carm.es (F.M.H.-B.); verantolinos@gmail.com (V.A.)

[2] Agricultural Engineering Center, Technical University of Cartagena, Paseo Alfonso XIII 48, 30203 Cartagena, Spain; belen.gallego@upct.es (B.G.-E.); josef.maestre@upct.es (J.F.M.-V.)

* Correspondence: alberto.imbernon@edu.upct.es; Tel.: +34-968-177-731

**Abstract:** Desalinated seawater (DSW) has emerged as a promising solution for irrigation in regions facing water scarcity. However, adopting DSW may impact the existing cultivation model, given the presence of potentially harmful elements, among other factors. A three-year experiment was carried out to assess the short-term effects of four irrigation waters—freshwater (FW), DSW, a mix 1:1 of FW and DSW (MW), and DSW with low boron (B) concentration (DSW–B)—on a 'Rio Red' grapefruit orchard. These irrigation waters exhibited varying levels of phytotoxic elements, some potentially harmful to citrus trees. Sodium ($Na^+$) and chloride ($Cl^-$) concentrations exceeded citrus thresholds in all treatments, except in DSW−B, whilst B exceeded toxicity levels in DSW and MW treatments. Leaf concentrations of $Cl^-$ and $Na^+$ remained low in all treatments, whereas B approached toxic levels only in DSW and MW–irrigated trees. The rapid growth of the trees, preventing excessive accumulation through a dilution effect, protected the plants from significant impacts on nutrition and physiology, such as gas exchange and chlorophyll levels, due to phytotoxic elements accumulation. Minor reductions in photosynthesis in DSW–irrigated trees were attributed to high B in leaves, since $Cl^-$ and $Na^+$ remained below toxic levels. The accelerated tree growth effectively prevented the substantial accumulation of phytotoxic elements, thereby limiting adverse effects on tree development and yield. When the maturation of trees reaches maximal growth, the potential accumulation of phytotoxic elements is expected to increase, potentially influencing tree behavior differently. Further study until the trees reach maturity is imperative for comprehensive understanding of the long-term effects of desalinated seawater irrigation.

**Keywords:** boron; chloride; citrus; leaf toxicity; sodium; water management

## 1. Introduction

In recent years, water scarcity has become a major concern for farmers in arid and semi-arid regions [1]. Owing to widespread climate change, the reduction in freshwater resources is threatening the sustainability of irrigated agriculture, and hence food security [2,3]. Furthermore, the upcoming scenario of exorbitant food demand, added to the intensification of water deficits, requires a new agricultural management model to be found, based on alternative water resources to encourage current agri-food production in water-starved areas [2,4]. The Mediterranean area suffers from a severe water scarcity and it is one of the most water-stressed areas in Europe [4–6], yet it is one of the regions with the highest fruit and vegetables production and exportation rates worldwide. This has been achievable due to its excellent edaphoclimatic conditions and very competitive agriculture strategies, which allow for extremely efficient use of the available water supplies [2]. Given this scenario, the use of unconventional water resources such as reclaimed water and desalinated seawater (DSW) is today considered as a promising stimulus for irrigation in this area [7,8]. However, its use for irrigation must be undertaken with caution, since it may condition the current agriculture framework [1,2]. One of the main drawbacks of its use lies in its chemical composition, as DSW is characterized as having a very low concentration of essential nutrients ($Ca^{2+}$, $Mg^{2+}$, or $SO_4^{2-}$), albeit with a high concentration of B, $Na^+$, and $Cl^-$ [7,9]. Proper management is therefore required since most woody crops are rather sensitive to B, $Cl^-$, and $Na^+$ leaf toxicity [10].

The southeast of Spain is a major citrus producer, due to its warm Mediterranean climate and fertile soil that create an ideal environment for citrus cultivation [11,12]. Because of the high structural deficit of water in southeastern Spain, the desalination of seawater has been promoted politically to sustain socioeconomic progress and food production in this important citrus-producing area [9]. However, using DSW for citrus irrigation requires caution due to its high salt content, since particularly citrus have been shown to be vulnerable to soil salinity and extremely sensitive to the concentration of B [10]; waters with a B concentration above 0.5 $mgL^{-1}$ can induce toxicity problems [9,13]. Common toxic effects of a high B concentration in citrus trees are leaf burn, yellowing, and browning, fruit with sunken areas, damaged root systems impacting water and nutrient uptake, and premature flower and fruit drop, leading to reduced yield [14]. Moreover, many studies have shown that citrus trees are sensitive to high levels of $Cl^-$ and/or $Na^+$ in soil and water [15,16], so the excessive salt amounts in DSW could damage plant roots, thereby hindering nutrient absorption and tree growth. In principle, when citrus are irrigated with high salts concentrations, the soil and leaf accumulations may lead to damage in the tree by reducing the net assimilation of $CO_2$; they can also damage leaves and roots, disrupt nutrient absorption, and cause water stress, ultimately influencing tree growth, development, and above all, productivity [16–18]. Overall, their ability to control these environmental stresses seems to depend on the rootstock [18,19]. Recent research conducted under controlled conditions has shown that the rootstock can modulate the citrus plant response to DSW irrigation [20,21].

When referring specifically to physiological and nutritional effects, the studies available in the literature to date have mainly evaluated the detrimental effects of using DSW for citrus irrigation under controlled conditions. The research showed that high soil salinity and the accumulation of B, $Cl^-$, and $Na^+$ in citrus trees can impede the uptake of certain nutrients due to chemical competition and produce specific toxicity problems in the plants [20,21]. When DSW was used to irrigate a mandarin orchard, non-significant toxic responses (leaf toxic ions accumulation, low plant water relations, or low crop gas exchange) were observed; this was likely due to the young age of the trees [3]. On the contrary, other studies in lemon trees irrigated with DSW for three consecutive years showed reductions in stomatal conductance and net photosynthesis, along with a noticeable accumulation of B in leaves [22]. In any case, the variety of factors that seem to affect the crops' response to DSW irrigation can hinder the establishment of sound conclusions; further information regarding its primary effects is therefore still needed.

Sustainable water management and balancing plant needs are clearly crucial for agricultural and environmental health. Therefore, the present study aims to initiate a long-term experiment to provide guidance on the effects of DSW utilization in semi-arid areas where low-quality waters are commonly used for irrigation, particularly for crops potentially sensitive to DSW, such as citrus. Our hypothesis is that the irrigation of grapefruits with DSW could alter the physiological response of the tree compared to conventional irrigation due to the accumulation of phytotoxic elements, particularly boron, in the soil and plant. The main objective of this experiment was to assess, for the first time, the main physiological implications of irrigating a 'Rio Red' grapefruit orchard with DSW. Therefore, the effects of the accumulation of phytotoxic elements on the soil and the plant due to the irrigation with DSW for three consecutive seasons were evaluated, as well as their implications on the gas exchange parameters, tree water relations, vegetative growth, and fruit yield. The significance of this study lies in its contribution to sustainable water management practices, offering crucial insights into the utilization of DSW in semi-arid regions for irrigation, particularly for sensitive crops like citrus. Grapefruit was selected due to its relatively limited research attention, although it holds great importance due to its relevant production in Europe and the Mediterranean area. In addition, it is known for its susceptibility to excessive B levels. Two previously described agronomical practices [9] were assessed in this study in order to determine the potential risk of crop damage associated with the high concentration of B in DSW. These practices aim to diminish the B concentration in DSW, either by blending DSW and water with a low B content, or by reducing its concentration with an on-farm reverse osmosis (RO) system [23].

## 2. Results and Discussion

### 2.1. Irrigation Water Quality

The water quality of the different water sources used during the experimental trial showed notable differences (Table 1 and Figure 1). After a one-stage RO process at the coastal desalination plant to produce the DSW, a large removal of salts occurred and the EC of this water fell to 0.89 dS $m^{-1}$ (Table 1). This value was significantly lower than that registered by the water provided by the Campo de Cartagena Irrigators Community (FW), which presented the highest EC of the experiment. On the contrary, the DSW–B treatment had the lowest value since a second on-farm RO stage was conducted.

**Table 1.** The electrical conductivity (EC, dS $m^{-1}$) and average ionic composition (mg $L^{-1}$) of the water sources used during the experimental work (FW: fresh water; MW: mixed water; DSW: desalinated seawater; and DSW–B: DSW with reduced boron). Data were averaged from 40 samples taken between June 2019 and December 2022.

| Water Source | EC | $Ca^{2+}$ | $Mg^{2+}$ | $K^+$ | $NO_3^-$ | $PO_4^{3-}$ | $SO_4^{2-}$ |
|---|---|---|---|---|---|---|---|
| FW | 1.23 d | 71.4 d | 42.7 d | 5.88 b | 4.92 b | 0.88 bc | 209.2 c |
| MW | 1.00 c | 45.8 c | 20.4 c | 6.22 b | 2.57 c | 1.15 c | 90.6 b |
| DSW | 0.89 b | 24.5 b | 5.3 b | 6.89 b | 1.31 d | 0.36 ab | 7.4 a |
| DSW–B | 0.17 a | 2.7 a | 1.0 a | 0.41 a | 0.78 a | 0.16 a | 1.3 a |
| ANOVA | *** | *** | *** | *** | *** | ** | *** |

** $p < 0.01$; *** $p < 0.001$; ns: not significant. Different letters indicate significant differences according to Duncan's multiple range test at the 95% confidence level.

One of the main issues about the quality of DSW concerns the lack of essential nutrients for the plants, such as $Ca^{2+}$, $Mg^{2+}$, and $SO_4^{2-}$ (Table 1). Desalination through RO processes not only separates the undesirable salts from the water, but also removes minerals that are essential nutrients for plant growth [7]. Therefore, DSW significantly reduced the concentration of nutrients with regard to FW (Table 1), and MW presented intermediate concentrations, as would be expected after mixing both sources of water at 50%. Moreover, DSW–B received two RO processes (coastal and on-farm), and the final concentrations of nutrients were much lower than those of the rest of the treatments. However, to guarantee agricultural needs, the missing essential nutrients ($Ca^{2+}$, $Mg^{2+}$, and $SO_4^{2-}$) were implemented in all treatments through a fertilization program similar to that applied by local farmers to citrus orchards.

The main objective of the RO process is the elimination of undesirable salts in seawater, mainly $Na^+$ and $Cl^-$, both ions represent the main source of salt in seawater (10.8 g $L^{-1}$ and 19.5 g $L^{-1}$, respectively) [24]. However, in addition to the high percentage of $Na^+$ and $Cl^-$ retained by the RO membranes (99.8 and 98.9%), respectively [25], after the desalination process, the final concentrations of these ions still remained high (Figure 1). In this experiment, DSW contained the highest $Na^+$ and $Cl^-$ concentrations of all the treatments in the first half of the experiment, whilst MW had values between those found in DSW and in FW. Additionally, DSW–B presented the lowest $Na^+$ and $Cl^-$ concentrations after the removal of salts produced by the second RO process (Figure 1). The average concentration of $Na^+$ measured during the experimental period in all treatments, except in DSW–B, exceeded the maximum thresholds established for citrus irrigation (115 mg $L^{-1}$) [26] and hence, might cause detrimental effects on crops. With regard to the $Cl^-$ concentrations, except for the DSW–B treatment, all the treatments presented concentrations above the 152 mg $L^{-1}$ established as the threshold for citrus by some authors [27], but below the 238 mg $L^{-1}$ [28] and 350 mg $L^{-1}$ thresholds proposed by others [29].

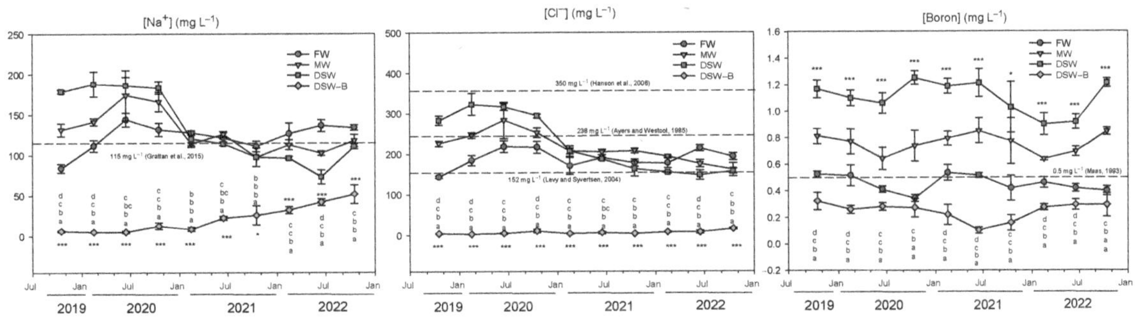


**Figure 1.** Concentrations of the phytotoxic elements, $Na^+$, $Cl^-$, and B in the water resources used during the experimental work (FW: fresh water; MW: mixed water; DSW: desalinated seawater; and DSW–B: DSW with reduced boron). The dashed line in the figure determines the phytotoxicity threshold suggested for $Na^+$ [26], $Cl^-$ [27–29], and B [16]. Data were taken between July 2019 and December 2022. Each point represents the average of four samples. * $p < 0.05$; *** $p < 0.001$; ns: not significant. For each date, different letters indicate significant differences according to Duncan's multiple range test at the 95% confidence level.

Conversely, as RO membranes are poorly selective for B (50–80% at normal feedwater pH values) [30], B concentrations were high after the desalination process (Figure 1). In this experiment, DSW presented the highest B concentrations throughout the experiment (1.1 mg $L^{-1}$ on average), whilst the concentration in FW was significantly lower (0.45 mg $L^{-1}$). This could lead to phytotoxicity and detrimental effects on crop productivity if DSW were used as the sole source of irrigation [31]. The B concentration in MW (0.76 mg $L^{-1}$) was between those recorded for both DSW and FW, whereas the value registered for DSW–B was the lowest one (0.24 mg $L^{-1}$). Thus, the average B concentration measured during the experimental period in DSW and MW exceeded the maximum threshold of 0.5 mg $L^{-1}$ referenced for citrus irrigation [16], whereas in FW it was very close to that threshold and in DSW–B it was far below it (Figure 1). These concentrations were found to be within the range reported in previous studies [3,22,32]. In fact, recent studies [20–22] have detected toxic B levels in the leaf and a decline in physiological parameters in lemon trees with similar DSW water quality properties. In this sense, the second RO stage performed on the farm was able to decrease the overall DSW concentration of B below the phytotoxic threshold (Figure 1), thereby avoiding possible detrimental effects on the tree.

### 2.2. *Evolution of Toxic Elements in the Soil*

The sodium concentration in the soil ranged from 100 to 200 mg $L^{-1}$ at the beginning of the experiment (Figure 2) and fluctuated throughout the experiment. After three years of irrigation, its final concentration in soil was 2–3 times higher (280–460 mg $L^{-1}$). Soils irrigated with DSW had the highest $Na^+$ concentration overall (average of 365 ± 36 mg $L^{-1}$), whereas those irrigated with DSW–B had the lowest (average of 128 ± 20 mg $L^{-1}$). The soils irrigated with FW presented an intermediate $Na^+$ concentration (213 ± 20 mg $L^{-1}$), whilst those irrigated with MW had slightly higher $Na^+$ levels (245 ± 26 mg $L^{-1}$). These $Na^+$ values in the soil were significantly correlated with the different $Na^+$ concentrations of the irrigation waters used in the four treatments (Table 2).

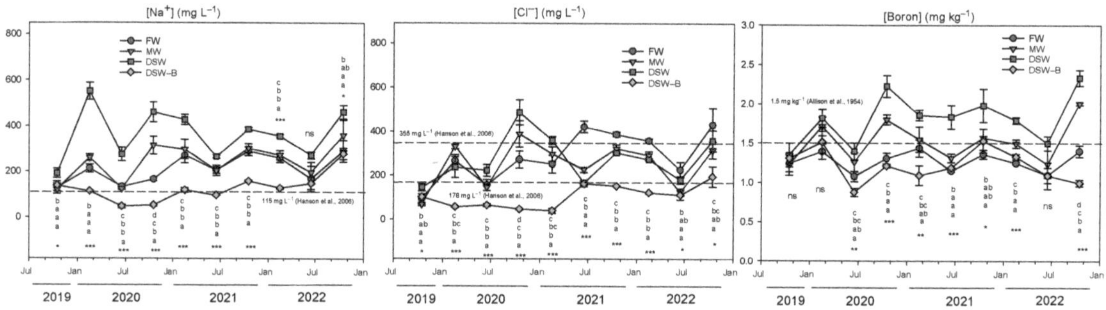


**Figure 2.** Evolution of the concentration of water-soluble $Na^+$ and $Cl^-$ and the extractable B in soil samples collected at the 0–0.50 m depth and at 0.30 m from the emitter. The dashed line in the figure determines the phytotoxicity threshold suggested for $Na^+$ [29], $Cl^-$ [29], and B [33]. Initials represent the four irrigation treatments (FW: fresh water; MW: mixed water; DSW: desalinated seawater; and DSW–B: DSW with reduced boron). Data were recorded between October 2019 and December 2022. Each point represents the average of six samples. * $p < 0.05$; ** $p < 0.01$; *** $p < 0.001$; ns: not significant. For each date, different letters indicate significant differences according to Duncan's multiple range test at the 95% confidence level.

**Table 2.** Matrix of Pearson's correlation coefficients obtained between the concentrations of $Na^+$, $Cl^-$, and B found in the spring bud leaves. Data were collected three times per year (February, June, and October) from October 2019 to October 2022.

| | $[Na^+]_{water}$ | $[Na^+]_{soil}$ | $[Na^+]_{leaf}$ | $[Cl^-]_{water}$ | $[Cl^-]_{soil}$ | $[Cl^-]_{leaf}$ | $[B]_{water}$ | $[B]_{soil}$ | $[B]_{leaf}$ |
|---|---|---|---|---|---|---|---|---|---|
| $[Na^+]_{water}$ | - | **0.525 ***** | 0.180 | **0.982 ***** | **0.547 ***** | 0.187 | **0.661 ***** | **0.365 *** | 0.129 |
| $[Na^+]_{soil}$ | **0.525 ***** | - | 0.107 | **0.517 **** | **0.725 ***** | 0.021 | **0.720 ***** | **0.812 **** | **0.521 **** |
| $[Na^+]_{leaf}$ | 0.180 | 0.107 | - | 0.234 | 0.0756 | 0.203 | 0.236 | 0.245 | **0.529 ***** |
| $[Cl^-]_{water}$ | **0.982 ***** | **0.517 **** | 0.234 | - | **0.526 ***** | 0.204 | **0.691 ***** | **0.380 *** | 0.113 |
| $[Cl^-]_{soil}$ | **0.547 ***** | **0.725 ***** | 0.076 | **0.526 ***** | - | −0.244 | **0.425 **** | **0.588 ***** | **0.439 **** |
| $[Cl^-]_{leaf}$ | 0.187 | 0.021 | 0.203 | 0.204 | −0.244 | - | 0.293 | 0.024 | 0.135 |
| $[B]_{water}$ | **0.661 ***** | **0.720 ***** | 0.236 | **0.691 ***** | **0.420 **** | 0.293 | - | **0.725 ***** | 0.302 |
| $[B]_{soil}$ | **0.365 *** | **0.812 **** | 0.245 | **0.380 *** | **0.588 ***** | 0.024 | **0.725 ***** | - | **0.570 ***** |
| $[B]_{leaf}$ | 0.129 | **0.521 **** | **0.529 ***** | 0.113 | **0.439 **** | 0.135 | 0.302 | **0.570 ***** | - |

The numbers in bold indicate significant differences. *, **, and *** indicate significant differences at the 0.05, 0.01, and 0.001 levels of probability, respectively.

Similarly, the $Cl^-$ concentrations at the beginning of the experiment were between 50 and 150 mg $L^{-1}$, rising to 200–450 mg $L^{-1}$ at the end of the experiment, i.e., increasing by 2–5 times (Figure 2). However, unlike with the $Na^+$ accumulation in the soil, the highest $Cl^-$ accumulation was not found in soils irrigated with DSW, since they presented similar $Cl^-$ concentrations to those of the FW and MW treatments. Only soils irrigated with DSW–B had soil $Cl^-$ concentrations below the lower limit of the range 178–355 mg $L^{-1}$ beyond which many sensitive plants are injured [29], whereas in the rest of treatments the concentrations were almost always below the higher limit. Although only the soils of the DSW–B treatment had significantly lower $Cl^-$ concentrations than the rest of treatments throughout the experiment, a significant correlation was found between the $Cl^-$ levels of the water and the $Cl^-$ accumulation in the soil (Table 2).

With regard to B, its soil concentration fluctuated throughout the experiment, but remained almost constant at the final measurements (Figure 2). In fact, the soil B concentration increased slightly, from 1.2–1.3 mg $kg^{-1}$ soil at the beginning of the experimental period to 1.0–2.3 mg $kg^{-1}$ soil at the end, increasing by only 1.7 times in the soils from the DSW treatment. Those soils irrigated with DSW had the highest B concentration due to its higher concentration in the water source. Recent studies using DSW for citrus irrigation also found important accumulations of B in soils irrigated with DSW [21], albeit without the B content exceeding medium–high levels for agricultural soils (0.21–1.10 mg $kg^{-1}$ soil) [34]. Consequently, the lower the proportion of DSW in the irrigation blend, the lower the concentration of B in the soil (DSW > MW > FW > DSW–B; Figure 2). In this sense, a strong positive correlation was found between the B concentrations of the water used in the treatments of this experiment and the soil B concentration (Table 2). Some of the B levels found in the experiment could prove toxic for citrus. The range between B deficiency and toxicity is extremely narrow compared to other elements [35] and the B concentrations found in all of the treatments of this experiment were between 0.9 and 2.3 mg $kg^{-1}$, the vast majority of these values being between 1.2 and 3.0 mg $kg^{-1}$ soil; such a range of concentration in soil has been deemed to be very high [34]. Citrus trees are considered to be very sensitive to B toxicity [26] and some authors have proposed 1.5 mg $kg^{-1}$ soil of hot-water-extractable B as the threshold for sensitive plant species [33]. In fact, only soils irrigated with DSW–B (average of 1.2 mg $kg^{-1}$ soil) or FW (average of 1.3 mg $kg^{-1}$ soil) had B concentrations below the threshold of 1.5 mg $kg^{-1}$ soil, whilst the remaining treatments (1.5 mg $kg^{-1}$ soil and 1.8 mg $kg^{-1}$ soil for MW and DSW, respectively) had B concentrations above this level, and could thus present B toxicity problems.

### 2.3. *The Accumulation and Partitioning of Phytotoxic Elements in the Trees*

The trees were irrigated with water of different characteristics throughout the experiment (Table 1 and Figure 1). Since the amounts of the phytotoxic elements ($Na^+$, $Cl^-$, and B) in the four treatments were different, their concentrations in the trees were studied. The evolution of these elements in spring bud leaves showed similar behavior throughout the three years of the experiment, so the average data of the three years were evaluated.

Despite being the oldest leaves on the tree, during the first months of the experiment the old leaves were still very young, and they did not accumulate $Na^+$ in any treatment until June 2020 (almost constant concentration, 0.10–0.15%, Figure 3). Subsequently, the $Na^+$ presence progressively increased up to 0.16–0.22% in October 2020. From that moment on, the $Na^+$ concentrations were higher in DSW-irrigated trees than in the rest of the treatments, although significant differences were only occasionally found. The spring bud leaves accumulated slightly lower amounts of $Na^+$ than older leaves and, although there were no differences with the rest of treatments, DSW-irrigated trees had the highest $Na^+$ levels throughout the year (Figure 3). In any case, in both old and young leaves, concentrations remained below the maximum $Na^+$ threshold of 0.25% proposed for citrus [26], even though the $Na^+$ concentration in the water for all the treatments (except for DSW–B, Figure 1) exceeded the toxic threshold of 115 mg $L^{-1}$ proposed for citrus [26]. Moreover, although the irrigation with different $Na^+$ concentrations produced different $Na^+$ accumulations in the soil (Figure 2 and Table 2), no significant correlations were found between the water or soil $Na^+$ concentrations and the $Na^+$ concentration in the leaves. Additionally, higher $Na^+$ levels due to DSW irrigation were not found in the root examined in May 2021 (Figure 3). Therefore, the higher $Na^+$ concentration of soils irrigated with DSW during the experiment was not enough to produce an increase in $Na^+$ in the roots nor in the spring bud leaves, and only the oldest leaves occasionally had higher $Na^+$ levels than in the rest of the treatments.

The higher concentration of $Na^+$ in DSW could produce an imbalance of essential nutrients since direct competition between ions can reduce the nutrient uptake by plants [36]. Previous studies with DSW in citrus irrigation have shown high $Na^+/Ca^{2+}$, $Na^+/K^+$, and $Na^+/Mg^{2+}$ ratios in plant tissues when DSW was used for citrus irrigation [20]. In our experiment, since the DSW irrigation did not cause a large accumulation of $Na^+$ in the leaves, non-relevant nutritional disorders were found. In this sense, the irrigation with DSW only affected the mineral nutrition of old leaves, which had the highest $Na^+/Ca^{2+}$ ratio (Figure 4), mainly due to the higher amount of $Na^+$ taken up by the roots to the leaves. However, the $Na^+/Mg^{2+}$ ratio was the highest in old leaves and roots of DSW-irrigated trees (Figure 4) due to the increase in $Na^+$ concentrations in leaves -. Similar findings were found in spring bud leaves but without significant differences. On the other hand, $K^+$ does not always decline in citrus leaves under salinity [37]; in our experiment, no increase in the $Na^+/K^+$, ratio was found in roots or leaves due to the high $Na^+$ concentrations of the DSW treatment (Figure 4).

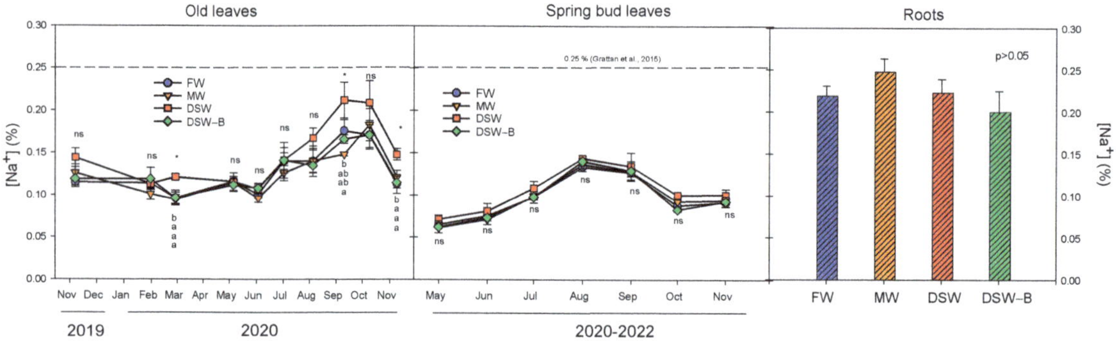


**Figure 3.** Effect of the irrigation water on the evolution of the concentration of $Na^+$ in old leaves (2019–2020), spring bud leaves (average of 2020–2022), and roots (May 2021). The dashed line in the figure determines the phytotoxicity threshold suggested for $Na^+$ [26]. Initials represent the four irrigation treatments (FW: fresh water; MW: mixed water; DSW: desalinated seawater; and DSW–B: DSW with reduced boron). * $p < 0.05$; ns: not significant. For each date, different letters indicate significant differences according to Duncan's multiple range test at the 95% confidence level.

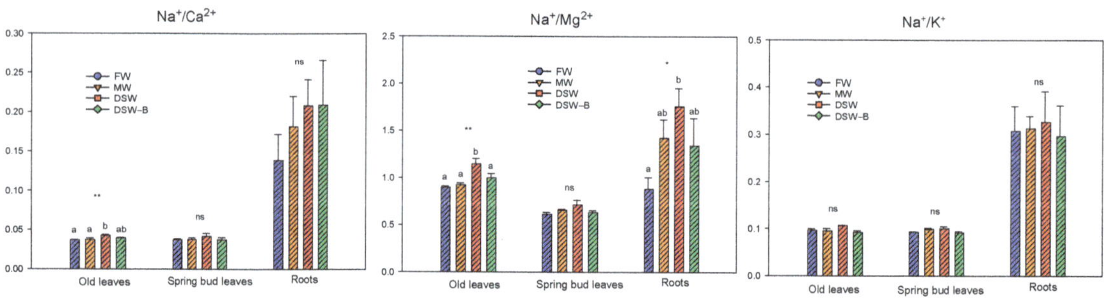


**Figure 4.** Effects of the irrigation water on the $Na^+/Ca^{2+}$, $Na^+/Mg^{2+}$, and $Na^+/K^+$ ratios in old and spring bud leaves, and in roots. Initials represent the four irrigation treatments (FW: fresh water; MW: mixed water; DSW: desalinated seawater; and DSW–B: DSW with reduced boron). Data were averaged from samples taken between November 2019 and November 2020 for old leaves, and samples taken between June 2020 and December 2022 for spring bud leaves. Roots were sampled in May 2021. * $p < 0.05$; ** $p < 0.01$; ns: not significant. In each tissue, different letters indicate significant differences according to Duncan's multiple range test at the 95% confidence level.

The $Cl^-$ concentration in the treatments (except in DSW–B, Figure 1) fell within the range of 152–238 mg $L^{-1}$, proposed as a range in which $Cl^-$ toxicity can occur in citrus [26]. However, these concentrations were below the threshold of 350 mg $L^{-1}$ proposed by other authors [29]. Despite these high $Cl^-$ concentrations in the irrigation water, the leaf $Cl^-$ concentrations found in the trees of all treatments were well below the leaf toxic threshold of 0.6% proposed for citrus [38] (Figure 5). Although the oldest leaves accumulated the highest $Cl^-$ concentrations, no clear buildups were observed in the leaves of the DSW treatment, whilst the DSW-irrigated trees had the highest $Cl^-$ concentrations in the spring bud leaves. Nevertheless, a low correlation was found between the $Cl^-$ of the irrigation water and the $Cl^-$ found in the leaves (Table 2), since leaf $Cl^-$ levels were similar in all the treatments (except the young leaves of the DSW), even in those trees irrigated with very low

$Cl^-$ (trees irrigated with DSW–B, Figure 1). Therefore, although the $Cl^-$ accumulation in soil was significantly correlated with the $Cl^-$ of the irrigation water (Table 2 and Figure 2), the leaf $Cl^-$ levels were not correlated with those in the irrigation water or soil. Unlike what happened with the $Na^+$ in roots, the DSW-irrigated trees had the highest $Cl^-$ concentrations in roots, whilst the lowest were found in the DSW–B treatment, in accordance with the concentrations of $Cl^-$ in the irrigation water of the treatments prior to root sampling in May 2021 (Figure 1). Thus, the $Cl^-$ accumulated in the soil due to the irrigation with high $Cl^-$ concentrations significantly increased its root concentration but without being enough to achieve a high presence in the leaves.

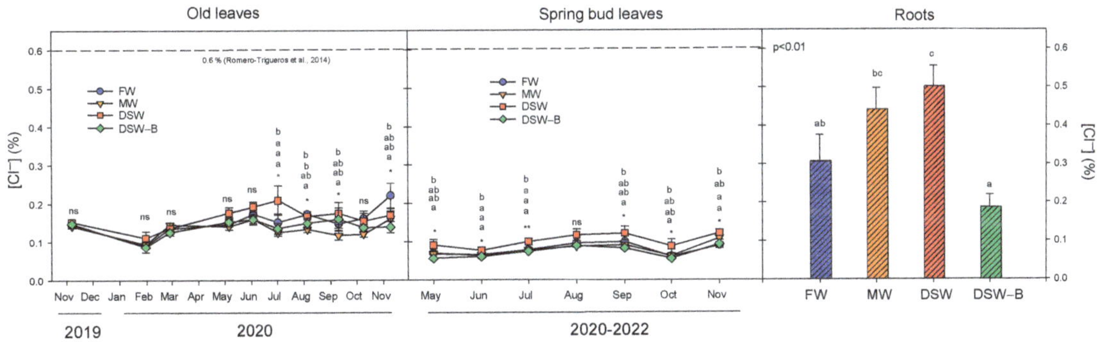


**Figure 5.** Effect of the irrigation water on the evolution of the concentration of $Cl^-$ in old leaves (2019–2020), spring bud leaves (average of 2020–2022), and roots (May 2021). The dashed line in the figure determines the phytotoxicity threshold suggested for $Cl^-$ [38]. Initials represent the four irrigation treatments (FW: fresh water; MW: mixed water; DSW: desalinated seawater; and DSW–B: DSW with reduced boron). * $p < 0.05$; ** $p < 0.01$; ns: not significant. For each date, different letters indicate significant differences according to Duncan's multiple range test at the 95% confidence level.

Although the B concentration in DSW and MW exceeded the maximum threshold of 0.5 mg $L^{-1}$ (Figure 1) proposed for citrus [14,39], the oldest leaves were still too young to accumulate large amounts during the first months of the experiment. Therefore, until June 2020, B levels remained approximately constant between 90 and 140 mg $kg^{-1}$ DW (Figure 6); these values were slightly higher than the threshold of 100 mg $kg^{-1}$ DW above which B can become toxic, but still in the low end of the 100–300 mg $kg^{-1}$ DW range, where slight to moderate damage can occur [26]. Boron concentrations in old leaves significantly increased from June to August 2020 up to concentrations close to 250 mg $kg^{-1}$ DW, with the range of 250–260 mg $kg^{-1}$ DW being the critical leaf B level in citrus when toxicity occurs [40]. The B concentration in spring bud leaves increased progressively throughout the year: after two months it was almost three times the spring concentration and four times their B concentration in late autumn (Figure 6). This reveals how B was quickly accumulated in the young tissues of the plant. The spring bud leaves of DSW- and MW-irrigated trees reached their highest B concentration in autumn (over 200 mg $kg^{-1}$ DW), accumulating significantly more B than in the FW or DSW–B treatments (Figure 6). These high concentrations found in spring leaves 6–7 months after sprouting were very close to the critical range of 250–260 mg $kg^{-1}$ DW [41]. The lowest B values in spring bud leaves were found in the trees irrigated with DSW–B, since the B concentration in that type of water was the lowest (Figure 1). Despite this, Supplementary Figure S1 shows images of buds, young leaves, and old leaves from the DSW treatment, with no phytotoxic effects observed in any of the cases. No direct correlation was found between B in the irrigation water and the leaves (Table 2). However, its concentration in the irrigation water was significantly correlated with its presence in the soil, and in turn with the leaf B concentration, as has been previously found in other studies using DSW for citrus irrigation [22]. This also indicates a higher B mobility in soil solution under the DSW treatment than under the other treatments. On the other hand, in our experiment, roots had much lower B than leaves, and the roots with the highest concentration were those of DSW-irrigated trees (Figure 6). Other authors have also found relatively low B levels in roots compared to those in leaves, and non-visible symptoms of toxicity developed in roots even at very high supply concentrations [14,42,43]. In our experiment, B leaf concentration increased with its presence in the soil as well as with the exposure time. In addition, and since the major concentration was observed in leaves as compared to roots, the data revealed that B was distributed in the plant predominantly by the transpiration stream, accumulating in leaf tissues, as has been established in previous studies [41–44].

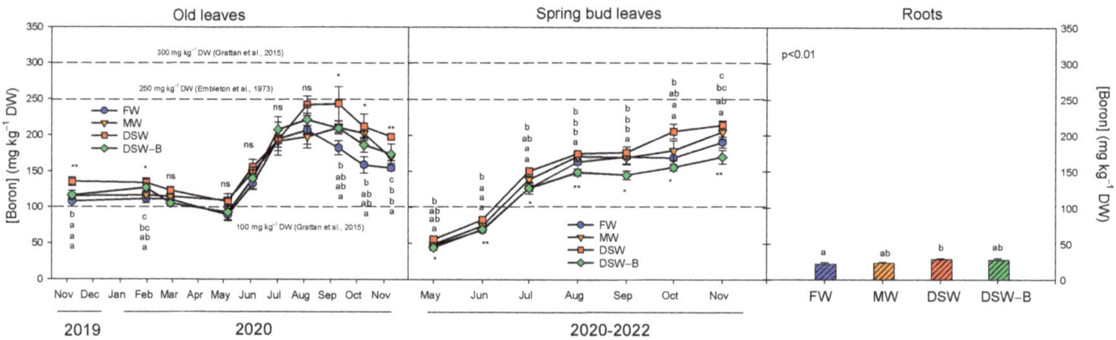


**Figure 6.** Effect of the irrigation water on the evolution of the concentration of boron in old leaves (2019–2020), spring bud leaves (average of 2020–2022), and roots (May 2021). The dashed line in the figure determines the phytotoxicity threshold suggested for B [26,40]. Initials represent the four irrigation treatments (FW: fresh water; MW: mixed water; DSW: desalinated seawater; and DSW–B: DSW with reduced boron). * $p < 0.05$; ** $p < 0.01$; ns: not significant. For each date, different letters indicate significant differences according to Duncan's multiple range test at the 95% confidence level.

### 2.4. Plant Water Status and Physiological Responses

Plant water status was periodically measured in mature leaves from May to November (2020–2022 period), and the average of the data registered during the three years of the experiment is shown in Figure 7. In terms of the threshold value in citrus trees for the midday stem water potential ($\Psi_{stem}$, which directly reflects the plant's water status), all treatments in this study registered values ranging between −0.6 and −1.4 MPa during the May–November period (Figure 7), which is close to data previously reported [45]. As has been described for citrus [46], the $\Psi_{stem}$ values suffered a seasonal behavior: a progressive decrease during the spring–summer months as a response to the elevated vapor pressure deficit (VPD), reaching their lowest values in August, and a later recovery due to the decrease in ETc and to rainfall in the subsequent months. Similar responses were found in the leaf water potential ($\Psi_{leaf}$), with slightly lower values than for $\Psi_{stem}$ (between −0.8 and −1.9 MPa), with the lowest values being achieved in August, and a slower recovery after the summer, as was observed for leaf water potential. In general, no clear differences were found between treatments in $\Psi_{stem}$ or $\Psi_{leaf}$ throughout the experiment (Figure 7). The seasonal decrease in $\Psi_{leaf}$ did not produce a similar decrease in osmotic potential ($\Pi$), and only a slight decrease was detected in $\Pi$ during the months of elevated VPD. This decrease in $\Pi$ was not enough to avoid changes in the turgor values, and leaf turgor suffered a similar seasonal decrease to that described for $\Psi_{leaf}$ (Figure 7).

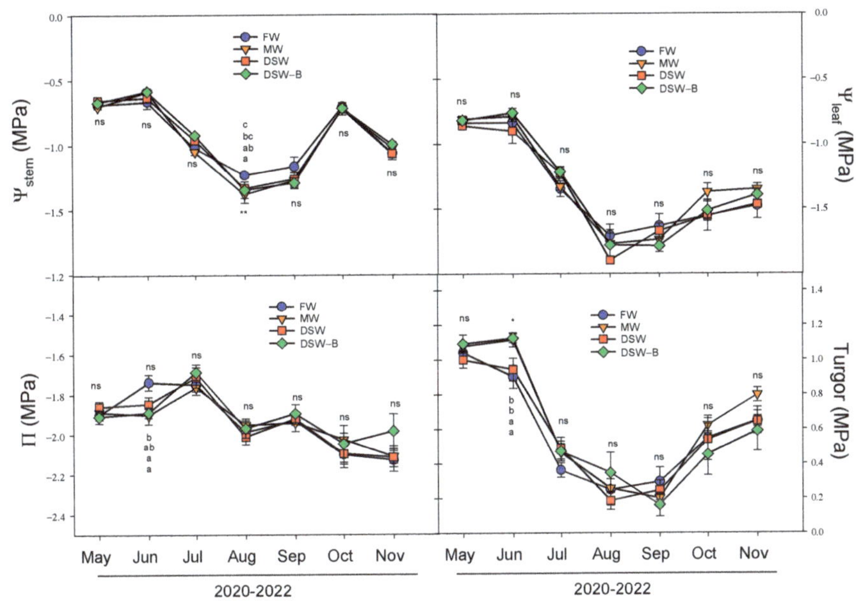


**Figure 7.** Effect of the irrigation water on the evolution of the midday stem water potential ($\Psi_{stem}$), leaf water potential ($\Psi_{leaf}$), osmotic potential (Π), and leaf turgor in spring bud leaves (average of 2020, 2021, and 2022) throughout the experiment. Initials represent the four irrigation treatments (FW: fresh water; MW: mixed water; DSW: desalinated seawater; and DSW–B: DSW with reduced boron). * $p < 0.05$; ** $p < 0.01$; ns: not significant. For each date, different letters indicate significant differences according to Duncan's multiple range test at the 95% confidence level.

The slight reduction in Π throughout the year was only due to the leaf accumulation of the phytotoxic elements $Na^+$, $Cl^-$, and B (Figures 4, 6 and 7), since no contribution of organic solutes as proline or quaternary ammonium compounds (QACs) was detected in response to the seasonal osmotic potential decrease (Figure 8). In general, no differences were found in the concentration of proline or QACs throughout the experiment with the irrigation using different types of water. Despite the results reported by other authors, which determined that QACs are involved in the adaptation of citrus to B toxicity [41], we found no changes in QACs in the leaves of DSW-irrigated trees and only those trees irrigated with DSW or MW had significantly higher proline concentrations in June than those of FW or DSW–B (Figure 8).

In addition to the plant water status analysis, a similar study of the photosynthetic machinery (gas exchange, chlorophyll fluorescence, and chlorophyll concentration) was also carried out for the same period (May–November, 2020–2022). During the summer months, the strong increase in VPD due to the low air humidity and high temperature produces a typical midday stomatal closure, even in well-watered citrus trees [47]. These high VPD conditions reduce stomatal conductance and photosynthesis while simultaneously increasing plant water losses through transpiration. In this experiment, the gas exchange parameters were measured before the midday stomatal closure (between 08:00 and 10:00 GMT), and a seasonal behavior was observed (Figure 9). Stomatal conductance ($g_s$) increased from 0.06 mol $H_2O$ $m^{-2}s^{-1}$ (registered in May) to 0.17 mol $H_2O$ $m^{-2}s^{-1}$ (August), and slowly decreased after that maximum value. The water loss due to transpiration ($E$) presented a similar behavior throughout the same period, with values ranging from 1.3 (May) to 4.1 mmol $H_2O$ $m^{-2}s^{-1}$ (August). However, the photosynthetic rate ($A$) did not follow the same seasonal pattern and it increased progressively, reaching values from 7.4 to 14.1 μmol $CO_2$ $m^{-2}s^{-1}$ registered in May and November, respectively (Figure 9).

As a consequence, the intrinsic water use efficiency ($A/g_s$) also increased from August to November, showing the intrinsic capacity of the leaves to exacerbate the photosynthesis rate even with a reduction in the stomatal conductance (Figure 9).

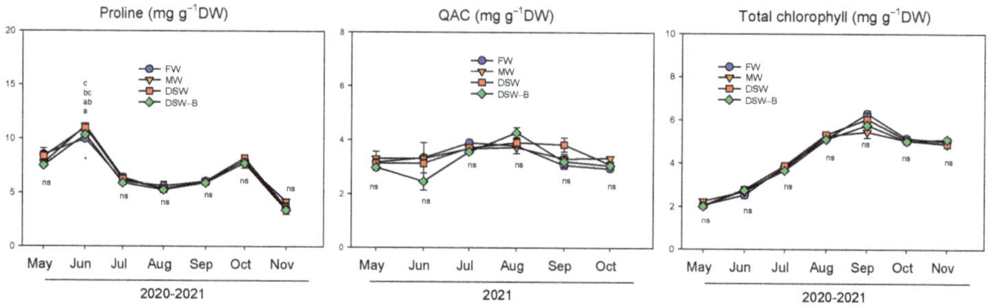


**Figure 8.** Effect of the irrigation water on the evolution of the proline, quaternary ammonium compounds (QACs), and total chlorophyll concentration in mature leaves (average of 2020, 2021 for proline and chlorophyll, and 2021 for QACs). Initials represent the four irrigation treatments (FW: fresh water; MW: mixed water; DSW: desalinated seawater; and DSW–B: DSW with reduced boron). * $p < 0.05$; ns: not significant. For each date, different letters indicate significant differences according to Duncan's multiple range test at the 95% confidence level.

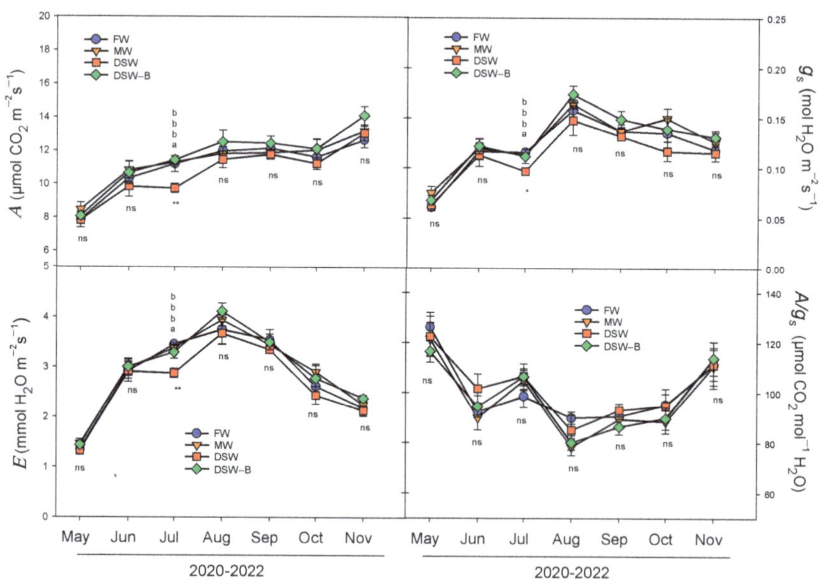


**Figure 9.** Effect of the irrigation water on the evolution of net photosynthesis (A), stomatal conductance ($g_s$), transpiration rate (E), and intrinsic water use efficiency ($A/g_s$) in mature leaves (average of 2020, 2021, and 2022) throughout the experiment. Initials represent the four irrigation treatments (FW: fresh water; MW: mixed water; DSW: desalinated seawater; and DSW–B: DSW with reduced boron). * $p < 0.05$; ** $p < 0.01$; ns: not significant. For each date, different letters indicate significant differences according to Duncan's multiple range test at the 95% confidence level.

On the other hand, photosynthesis is one of the main metabolic processes impaired by excessive B [41–44] due to both stomata and non-stomata limitations. In our experiment, slight differences were found in July in photosynthetic parameters due to the type of water used for irrigation (Figure 9). Since B accumulation is influenced by transpiration, the reduction in the water transport and transpiration under high external B is a mechanism of B tolerance [48]. In this experiment, there were no large reductions in transpiration, so this tolerance mechanism did not work. In general, lower $g_s$ on DSW-irrigated trees with respect to the rest of the trees was observed throughout the experiment, which produced lower $E$ and $A$, although they were only significantly different in July. Previous experiments with young lemon trees under controlled conditions showed a decrease in $A$ but not in $g_s$ or $E$ due to the irrigation with DSW, which produced a decrease in the intrinsic water-use efficiency [20,21]. However, no decrease in $A/g_s$ was observed due to the irrigation with DSW in this experiment. Other authors have described a similar decrease in $A$ when DSW was used for citrus irrigation [22], although they also found a higher decrease in $g_s$ by the accumulation of B in the leaf, which produced an increase in the intrinsic water-use efficiency.

The decline in $A$ produced by irrigation with the DSW could be attributed to other internal non-stomatal factors related to $Cl^-$, $Na^+$, or B toxicity, such as the reduction in the chlorophyll concentration, alterations in the carboxylation efficiency, reduced activity of photosynthetic enzymes, or alterations in the photochemical efficiency of photosystem II [20,21]. Although an excess of B in citrus trees can result in alterations of photosynthetic pigments [49], in our experiment no reduction was detected in the chlorophyll concentration by the irrigation with different types of water (Figure 8). In general, no changes were observed in the chlorophyll fluorescence parameters studied, so the photochemical machinery was apparently not damaged by the phytotoxic elements in leaves of DSW-irrigated trees either. Therefore, the proportion of the absorbed energy being used in photochemistry ($\Phi_{PSII}$) was not modified by the DSW treatment (Figure 10). However, due to the reduction in the photosynthetic rate in July, the $A/\Phi_{PSII}$ ratio decreased in DSW-irrigated trees, indicating that reactive oxygen species (ROS) can be generated, since $O_2$ can operate as an alternative acceptor, depleting the electron excess that is not utilized in metabolic processes [50]. Although oxidative stress was not evaluated in this study, previous results have shown that citrus trees have different capacities to overcome this oxidative stress generated by irrigation with DSW depending on the rootstock [20,21].

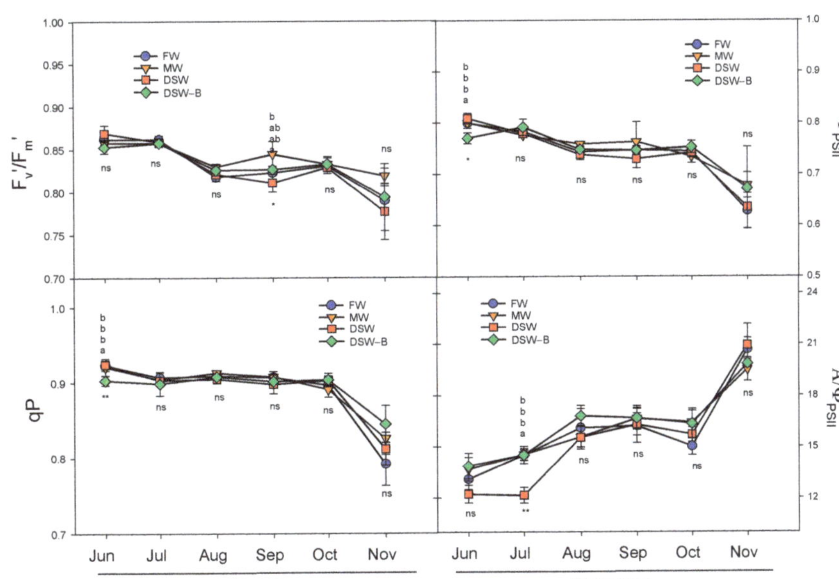


**Figure 10.** Effect of the irrigation water on the evolution of the efficiency of the antennas from PSII ($F'_v/F'_m$), photochemical efficiency of PSII ($\Phi_{PSII}$), photochemical quenching (qP), and $A/\Phi_{PSII}$ ratio

in spring bud leaves (average of 2020 and 2021) throughout the experiment. Initials represent the four irrigation treatments (FW: fresh water; MW: mixed water; DSW: desalinated seawater; and DSW–B: DSW with reduced boron). * $p < 0.05$; ** $p < 0.01$; ns: not significant. For each date, different letters indicate significant differences according to Duncan's multiple range test at the 95% confidence level.

### 2.5. Plant Growth and Fruit Yield

The effect of irrigation with the different water sources on tree growth was evaluated throughout the experiment by conducting periodic measurements of the canopy (Figure 11). Our 'Rio Red' grapefruit was grafted onto a five-year-old *Citrus macrophylla* rootstock in June 2019, so the tree growth was very rapid. Supplementary Figure S2 displays the evolution of the overall size of the trees throughout the experiment. Trees grew exponentially during the first part of the experiment, producing an increase in the tree canopy from 0.1 $m^3$ measured in November 2019 to almost 4 $m^3$ one year later (Figure 11); after that, linear growth in the second half of the experiment increased the canopy volume from 4 to 8 $m^3$. However, no significant changes in tree growth were observed with the treatments throughout the experiment, although a slightly higher growth of DSW–B-irrigated trees was observed from the beginning of 2021 onwards (Figure 11). Unfortunately, only limited information is available regarding the effect of irrigation with DSW in citrus trees on its agronomic behavior in the field. Under controlled conditions, young lemon trees irrigated with DSW had decreased shoot growth (mainly the leaf size) [20,21], whilst no effects were found on plant growth when similar water to that in our DSW–B treatment was tested [20].

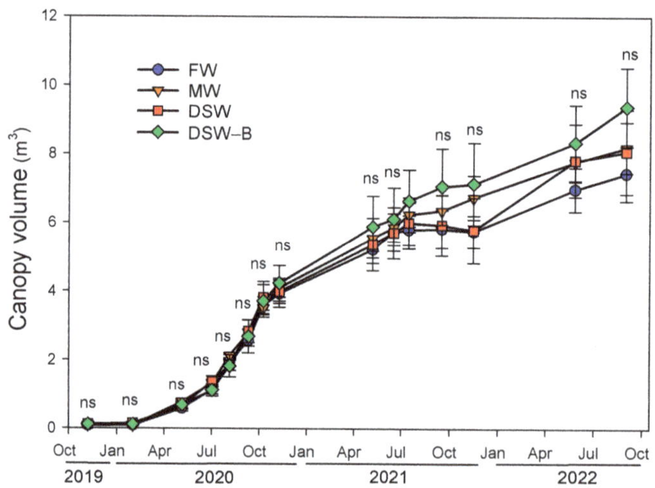


**Figure 11.** Effect of the irrigation water on the evolution of the canopy volume of the trees throughout the experiment (from November 2019 to December 2022). Initials represent the four irrigation treatments (FW: fresh water; MW: mixed water; DSW: desalinated seawater; and DSW–B: DSW with reduced boron). ns: not significant at the 95% confidence level.

Table 3 shows the effect of the irrigation with the different water sources on the fruit yield parameters obtained as the average yield achieved in 2021 and 2022. Due to the youth of the trees, no yield was obtained in 2020, and the yields of 2021 and 2022 were still low compared to commercial grapefruit yields. No significant effect of the irrigation with different water sources was observed in any of the fruit yield parameters studied. Previous results with similar treatments in a young orchard of mandarin trees showed no differences in the total yield or in the crop load during the first two years of the experiment [3]. It is therefore possible that differences may be observed as the tree production increases and stabilizes with maturity.

**Table 3.** Effect of the irrigation water on the fruit yield parameters during the experimental period for each irrigation treatment (FW: fresh water; MW: mixed water; DSW: desalinated seawater; and DSW–B: DSW with reduced boron). The data are averages of 2021 and 2022.

| Water Source | Yield (kg tree$^{-1}$) | Fruit Number | Fruit Weight (g) |
|---|---|---|---|
| FW | 85.1 ± 4.2 | 229.5 ± 16.1 | 380.5 ± 13.7 |
| MW | 81.0 ± 6.2 | 222.0 ± 15.4 | 367.6 ± 5.9 |
| DSW | 91.5 ± 4.7 | 246.5 ± 16.7 | 382.8 ± 22.8 |
| DSW–B | 90.4 ± 6.6 | 248.4 ± 18.4 | 368.8 ± 22.8 |
| ANOVA | ns | ns | ns |

The values after the ± sign represent the standard deviation of the data. ns: not significant at 95% confidence level.

The high tree growth rate during the three-year study prevented a major accumulation of phytotoxic elements, by means of a dilution effect, avoiding toxic concentrations. It is likely that under a more stationary scenario where the trees have reached their maximum growth, leaves could accumulate phytotoxic elements in higher concentrations, which may alter the different nutritional, physiological, and metabolic processes, and ultimately affect tree growth and yield.

## 3. Materials and Methods

### 3.1. Experimental Plot, Vegetal Material, and Crop Management

The study was performed over three consecutive years (from June 2019 to December 2022) in a commercial farm located in Torre Pacheco, Murcia, Spain (37°47′30″ N; 1°03′85″ W; 30 m above sea level). The climate is semi-arid Mediterranean, featuring predominantly warm, arid summers and mild winters, with a mean annual rainfall of about 400 mm during the three years of the experiment, and a total annual reference evapotranspiration ($ET_0$), calculated via the Penman–Monteith method, of about 1300 mm (three-year average).

The experiment was performed in a 0.28 ha 'Rio Red' grapefruit orchard (*Citrus* × *paradisi* Mac.) newly grafted on *Citrus macrophylla* Wester rootstock aged five years at the beginning of the experiment (June 2019), with a tree-spacing of 5.5 m × 3.5 m.

The irrigation system consisted of two polyethylene drip lines laid on the soil surface, one on either side of the tree. Four self-pressure compensating emitters per tree provided a discharge of 4 L h$^{-1}$ each. The irrigation doses were scheduled based on the daily crop evapotranspiration accumulated during the previous week. No leaching fraction was added to the irrigation doses, regardless of water salinity. All the treatments received the same amounts of fertilizer (N-$P_2O_5$-$K_2O$-CaO, MgO), supplied through the drip irrigation system. This enabled the possible effects of the incorporation of DSW to be analyzed; such effects could have been hidden if the fertilizer supply had been adjusted based on the nutrient content of the irrigation water. Pest control and pruning practices were those commonly used by growers in the area, and no weeds were allowed to develop within the orchard.

### 3.2. Treatments and Experimental Design

The treatments were established by supplying four water sources to the irrigation system: (i) fresh water (FW), provided by the Campo de Cartagena irrigators community; (ii) DSW provided by the coastal desalination plant of Escombreras (30 km from the farm); (iii) mixed water (MW), a 50% blend of FW and DSW (1:1); and (iv) DSW with a low B concentration (DSW–B). DSW from the Escombreras plant presented a B concentration of around 1.05 mg L$^{-1}$, and an on-farm reverse osmosis system treated the DSW to reduce the B concentration (between 0.2 and 0.4 mg L$^{-1}$) in the DSW–B treatment [23].

A total of four treatments were applied, corresponding to the four water resources previously described: FW, DSW, MW, and DSW–B. The lay-out of the experiment took the form of three completely randomized blocks with four experimental plots in each one (one treatment per plot). In each experimental plot (of 4 × 3 trees), border trees were excluded from the study to eliminate potential edge effects, and only the two central trees were studied.

### 3.3. Water Quality and Soil Analysis

Water samples were collected in glass bottles for each water source monthly throughout the experiment, transported in an icebox to the laboratory and stored at 5 °C until the physical and chemical analyses. The electrical conductivity (EC) was measured with a conductivity instrument GLP-31 (Crison Instruments S.A., Barcelona, Spain). Inductively coupled plasma (ICP-MS Agilent 7900, Agilent Technologies, Santa Clara, CA, USA) was used to determine the $Na^+$, $K^+$, and B concentrations. Chloride was quantified by ion chromatography with a liquid chromatograph (Thermo Scientific Dionex, Model ICS-2100, Thermo Scientific, Basel, Switzerland).

Three soil samples (one per block) for each irrigation treatment were collected every four months (in February, June, and October) in the area surrounding one central tree, at 0–0.50 m depths, and 0.30 m away from the emitter. Soluble salt contents were determined in the saturated paste extract [51]. Inductively coupled plasma (ICP-MS Agilent 7900) was used to determine the concentration of water soluble $Na^+$. Extractable B was determined in soil samples by refluxing 20 g soil with 40 mL hot water (boiling) for a period of 5 min. One aliquot from the filtered extract was then used for measuring B using inductively coupled plasma (ICP-MS Agilent 7900).

### 3.4. Plant Mineral Analysis

Leaf samples were periodically taken for phytotoxic elements analysis. Old leaves were sampled during the first year of the experiment, with the oldest leaves on the tree being considered as the old leaves. On the other hand, the leaves that sprouted in spring (spring bud leaves) were also studied in 2020, 2021, and 2022. For each sampling, the two central trees of each experimental plot were sampled, giving a total of six samples per treatment (six trees per irrigation system). The leaf samples were rinsed in deionized water, freeze-dried, and ground for analytical determinations. In May 2021, fine roots of the 0–0.25 m layer were sampled in the same trees described above, they were carefully separated from the soil, rinsed with deionized water, oven-dried at 60 °C, and ground. The ground tissues (leaves and roots) were ashed, dissolved in 0.7 N $HNO_3$, and $Na^+$, $K^+$, $Ca^{2+}$, $Mg^{2+}$, and B were determined by inductively coupled plasma optical emission spectrometry (Varian ICP-OES Vista MPX, Varian Inc., Palo Alto, USA). Chloride was extracted from 50 mg of ground plant material with 25 mL of deionized water and measured by ion chromatography with a liquid chromatograph (Model ICS-3000, Thermo Fisher Scientific Inc., Logan, UT, USA).

### 3.5. Plant Water Relations

Midday stem water potential ($\Psi_{stem}$) was measured periodically in one mature, fully-expanded leaf in the middle third of the tree, in the same trees as the leaf mineral study (six trees per irrigation system). The leaves were enclosed within foil-covered plastic and aluminum envelopes at least two hours prior to the midday measurement [52]. In the same trees and on the same days as the $\Psi_{stem}$ determination, the leaf water potential ($\Psi_{leaf}$) was also measured in similar unwrapped leaves. Both $\Psi_{stem}$ and $\Psi_{leaf}$ were measured at noon (12:00–14:00), using a pressure chamber (model 3000; Soil Moisture Equipment Corp., Santa Barbara, CA, USA) and following the recommendations of Turner [53]. Following the $\Psi_{leaf}$ measurement, the leaves were frozen and stored at −20 °C for determination of the leaf osmotic potential (Π) by a Wescor 5520 vapor pressure osmometer (Wescor, Logan, UT, USA). The leaf turgor potential was calculated as the difference between $\Psi_{leaf}$ and Π.

### 3.6. Gas Exchange Parameters and Chlorophyll Fluorescence

Gas exchange measurements were taken periodically, coinciding with the plant water study, between 08:00 and 10:00 GMT, in daylight hours (to avoid high afternoon temperatures and air vapor pressure deficit). Measurements were made on healthy, fully expanded mature leaves (one leaf per tree, for the same six trees described above) exposed to the sun in exterior mid-canopy positions. The leaf net photosynthesis rate ($A$), stomatal conductance ($g_s$), and leaf transpiration ($E$) were measured with a portable photosynthesis system (Li-6400, Li-Cor, Lincoln, NE, USA) equipped with a broad leaf chamber (6.0 $cm^2$). The air flow rate inside the leaf chamber was 500 μmol $s^{-1}$. The reference $CO_2$ concentration was fixed at 400 μmol $mol^{-1}$ through a $CO_2$ injector system (6400-01 LiCOR, Lincoln, NE, USA). All measurements were made using a red–blue light source (6400-02B light emitting diode; Li-COR, Lincoln, NE, USA) and the PPFD was fixed at 1200 μmol $m^{-2}$ $s^{-1}$, which exceeds the saturating light intensity for photosynthesis of citrus leaves [54].

Chlorophyll fluorescence measurements were performed in parallel to those of the gas exchange parameters, on the same days and in the same leaves, using a pulse-modulated field-fluorescence monitoring system (FMS-2, Hansatech Instruments, Norfolk, UK). The chlorophyll fluorescence kinetics of leaves adapted to light were studied and the following parameters were measured: the antenna efficiency of PSII [$F_v'/F_m' = (F_m' - F_0')/F_m'$], the quantum efficiency of PSII [$\Phi_{PSII} = (F_m' - F_s)/F_m'$], and the photochemical quenching coefficient [$qP = (F_m' - F_s)/(F_m' - F_0')$], where $F_v'$ is the variable fluorescence, $F_m'$ is the maximum value when all reaction centers are closed after a pulse of saturating light (12,000 μmol $m^{-2}$ $s^{-1}$ for 0.8 s), $F_0'$ is the minimal fluorescence in the light-adapted state that is obtained by turning off the actinic light temporarily and applying a pulse of far-red light (735 nm) to drain the electrons from PSII, and $F_s$ is the steady-state fluorescence yield.

### 3.7. Osmolytes and Chlorophyll Determinations

The leaf samples for mineral analysis were used for osmolytes and chlorophyll analysis. Proline was extracted from 50 mg of leaf tissue with sulfosalicylic acid (3%) and quantified according to the protocol described by Bates et al. [55]. Quaternary ammonium compounds (QAC) were extracted from dry tissue with 1 M $H_2SO_4$ and quantified using a glycine betaine standard curve, according to the method described by Grieve and Grattan [56].

Chlorophyll contents were estimated using the procedure described by Inskeep and Blomm [57], extracting 20 mg of ground material with N,N-dimethylformamide and measuring the absorbance at 664.5 and 647 nm in a Shimadzu UV-1800 spectrophotometer (Shimadzu Corporation, Kyoto, Japan).

### 3.8. Vegetative Growth and Yield Fruit

Measurements were made in the two central trees of each experimental plot. The trunk circumference of the rootstock, tree canopy height, and perimeter were periodically measured during the experimental period in the same six trees per treatment. Canopy volume (V) was calculated considering the canopy diameter (D) and height (H), according to the equation of Turrell [58]: $V = 0.5238\ H\ D^2$, where V, D, and H are in $m^3$, m, and m, respectively.

In early February 2022 and late December 2022 (2021 and 2022 yields, respectively), the individual tree yield was measured for six trees per treatment (two trees per experimental plot. The number of fruits and the total fruit weight of each tree were measured.

### 3.9. Statistical Analysis

All data were analyzed using the analysis of variance (ANOVA) procedures in Statgraphics Plus 5.1 software (Statistical Graphics Corporation, Warrenton, VA, USA). When there was a significant effect (value of $p < 0.05$), means were separated using Duncan's multiple range test. The Pearson correlation coefficients between the concentrations of phytotoxic elements in water, soil, and leaves were calculated using the same statistical software.

## 4. Conclusions

The data obtained in this study highlight the interaction between irrigation water quality and plant behavior. While the investigation revealed varying concentrations of phytotoxic elements in the irrigation waters used for citrus, with notably elevated levels of boron, the rapid growth of the trees seemed to mitigate potential adverse effects due to a dilution effect. Despite boron concentrations nearing toxicity thresholds in leaf tissues, the overall impact on tree growth and fruit yield was not significant.

This research has practical implications for agricultural management, particularly in semi-arid regions reliant on alternative water sources like desalinated seawater (DSW). The findings suggest that while DSW irrigation may introduce elevated levels of certain phytotoxic elements, the rapid growth rates can buffer against harmful effects, at least in the short to medium term.

However, the study also highlights the need for continued investigation, especially over extended periods, to fully elucidate the potential long-term impacts of DSW irrigation on citrus cultivation. Future research should delve deeper into understanding how prolonged exposure to elevated phytotoxic element concentrations might affect nutritional, physiological, and metabolic processes within the citrus trees. Moreover, examining the implications of DSW irrigation under more stable growth conditions, where trees have reached maturity, could provide valuable insights into potential risks.

In summary, while the initial hypothesis regarding the alteration of physiological responses due to DSW irrigation was not fully supported by this study, the applied nature of the research underscores the importance of ongoing research. Longer-term investigations will be essential to gain more knowledge of the relationship between water quality, plant behavior, and agricultural productivity, ultimately guiding sustainable irrigation practices in semi-arid environments.

**Supplementary Materials:** The following supporting information can be downloaded at: https://www.mdpi.com/article/10.3390/plants13060781/s1, Figure S1: Shape of bud (A), new leaves (B) and old leaves (C) of a grapefruit tree from the orchard. Figure S2: Grapefruit trees in the orchard in November 2019 (A), July 2020 (B), July 2021 (C), May 2022 (D), and December 2022 before (E) and after yield (F), indicating the current average height of the trees.

**Author Contributions:** Conceptualization, J.M.N., J.M.R., A.I.-M. and J.F.M.-V.; methodology, J.M.N., J.M.R., F.M.H.-B., A.I.-M. and J.F.M.-V.; data curation, J.M.N., J.M.R., F.M.H.-B. and A.I.-M.; formal analysis, J.M.N., V.A. and A.I.-M.; funding acquisition, J.M.N., B.G.-E. and J.F.M.-V.; investigation, all authors; validation, J.M.N., J.M.R., F.M.H.-B., V.A. and A.I.-M.; visualization, J.M.N., J.M.R., F.M.H.-B. and A.I.-M.; project administration, J.M.N. and J.F.M.-V.; writing—original draft, J.M.N.; writing—review and editing: all authors. All authors have read and agreed to the published version of the manuscript.

**Funding:** This research and the APC were funded by the projects SEARRISOST (RTC-2017-6192-2) and SEA4CROP (PID2020-118492RA-C22), the Ministry for Science and Innovation and the State Research Agency (MCIN/AEI/10.13039/501100011033, Spain), and the European Regional Development Fund (ERDF, EU). B. Gallego-Elvira acknowledges the support from the Spanish Ministry of Universities ("Beatriz Galindo" Fellowship BEAGAL18/00081).

**Data Availability Statement:** Data are contained within the article and Supplementary Materials.

**Acknowledgments:** A. Imbernón-Mulero acknowledges the support for his PhD work from the project SEA4CROP and the predoctoral program of the Technical University of Cartagena (RV-484/21, UPCT, Spain). The collaboration of Catedra Trasvase y sostenibilidad—Jose Manuel Claver is acknowledged. We would like to thank J.A. Palazón for his assistance in the field, and E.M. Arques for her support in laboratory analyses.

**Conflicts of Interest:** The authors declare no conflicts of interest.

## References

1. Martínez-Alvarez, V.; Imbernón-Mulero, A.; Gallego-Elvira, B.; Soto-García, M.; Maestre-Valero, J.F. Multidisciplinary assessment of the agricultural supply of desalinated seawater in south-eastern Spain. *Desalination* **2023**, *548*, 116252. [CrossRef]
2. Martinez-Mate, M.A.; Martin-Gorriz, B.; Martínez-Alvarez, V.; Soto-García, M.; Maestre-Valero, J.F. Hydroponic system and desalinated seawater as an alternative farm-productive proposal in water scarcity areas: Energy and greenhouse gas emissions analysis of lettuce production in southeast Spain. *J. Clean. Prod.* **2018**, *172*, 1298–1310. [CrossRef]

3. Maestre-Valero, J.F.; Martínez-Álvarez, V.; Jódar-Conesa, F.J.; Acosta, J.A.; Martin-Gorriz, B.; Robles, J.M.; Pérez-Pérez, J.G.; Navarro, J.M. Short-term response of young mandarin trees to desalinated seawater irrigation. *Water* **2020**, *12*, 159. [CrossRef]
4. Pellicer-Martinez, F.; Martínez-Paz, J.M. Climate change effects on the hydrology of the headwaters of the Tagus River: Implications for the management of the Tagus-Segura transfer. *Hydrol. Earth Syst. Sci.* **2018**, *22*, 6473–6491. [CrossRef]
5. European Environmental Agency (EEA). Use of Freshwater Resources in Europe. 2022. Available online: www.eea.europa.eu/airs/2018/resource-efficiency-and-low-carbon-economy/freshwater-use (accessed on 19 January 2023).
6. Ministerio de Agricultura, Pesca y Alimentación (MAPA). Informe Anual de Indicadores, Agricultura Pesca y Alimentación. 2021. Available online: https://www.mapa.gob.es/es/ministerio/servicios/analisis-y-prospectiva/iai2021_version_final_web_tcm30-626537.pdf (accessed on 2 February 2023).
7. Martínez-Álvarez, V.; Martin-Gorriz, B.; Soto-García, M. Seawater desalination for crop irrigation—A review of current experiences and revealed key issues. *Desalination* **2016**, *381*, 58–70. [CrossRef]
8. Redondo-Orts, J.A.; López-Ortiz, M.I. The economic impact of drought on the irrigated crops in the Segura River Basin. *Water* **2020**, *12*, 2955. [CrossRef]
9. Martínez-Álvarez, V.; González-Ortega, M.J.; Martin-Gorriz, B.; Soto-García, M.; Maestre-Valero, J.F. The use of desalinated seawater for crop irrigation in the Segura River Basin (south-eastern Spain). *Desalination* **2017**, *422*, 153–164. [CrossRef]
10. Hilal, N.; Kim, G.J.; Somerfield, C. Boron removal from saline water: A comprehensive review. *Desalination* **2011**, *273*, 23–35. [CrossRef]
11. Navarro, L. The Spanish citrus industry. *Acta Hortic.* **2015**, *1065*, 41–48. [CrossRef]
12. Ben Amor, R.; de Miguel Gomez, M.D. Competitiveness of Spanish orange sector in the Mediterranean area. *Acta Hortic.* **2020**, *1292*, 23–30. [CrossRef]
13. Voutchkov, N.; Semiat, R. Seawater desalination. In *Advanced Membrane Technology and Applications*, 1st ed.; Li, N.N., Fane, A.G., Ho, W.W., Matsuura, T., Eds.; John Wiley &Sons, Inc.: Hoboken, NJ, USA, 2008; pp. 47–86.
14. Nable, R.O.; Banuelos, G.S.; Paull, J.G. Boron toxicity. *Plant Soil* **1997**, *193*, 181–198. [CrossRef]
15. Storey, R.; Walker, R.R. Citrus and salinity. *Sci. Hort.* **1999**, *78*, 39–81. [CrossRef]
16. Maas, E.V. Salinity and citriculture. *Tree Physiol.* **1993**, *12*, 195–216. [CrossRef]
17. García-Sánchez, F.; Jifon, J.; Carvajal, M.; Syvertsen, J.P. Gas exchange, chlorophyll and nutrient contents in relation to $Na^+$ and $Cl^-$ accumulation in 'Sunburst' mandarin grafted on different rootstock. *Plant Sci.* **2002**, *162*, 705–712. [CrossRef]
18. Garcia-Sanchez, F.; Syvertsen, J.P. Salinity tolerance of Cleopatra mandarin and Carrizo citrange citrus rootstock seedling is affected by $CO_2$ enrichment during growth. *J. Am. Soc. Hortic. Sci.* **2006**, *131*, 24–31. [CrossRef]
19. Mesquita, G.L.; Zambrosi, F.C.B.; Tanaka, F.A.O.; Boaretto, R.M.; Quaggio, J.A.; Ribeiro, R.V.; Mattos, D., Jr. Anatomical and physiological responses of citrus trees to varying boron availability are dependent on rootstock. *Front. Plant Sci.* **2016**, *7*, 224. [CrossRef] [PubMed]
20. Navarro, J.M.; Antolinos, V.; Robles, J.M.; Botía, P. Citrus irrigation with desalinated seawater under a climate change scenario. *Front. Plant Sci.* **2022**, *13*, 909083. [CrossRef] [PubMed]
21. Navarro, J.M.; Antolinos, V.; Botía, P.; Robles, J.M. Deficit irrigation applied to lemon trees grafted on two rootstocks and irrigated with desalinated seawater. *Plants* **2023**, *12*, 2300. [CrossRef] [PubMed]
22. Vera, A.; Moreno, J.L.; García, C.; Nicolás, E.; Bastida, F. Agro-physiological and soil microbial responses to desalinated seawater irrigation in two crops. *Ecotoxicol. Environ. Saf.* **2023**, *250*, 114507. [CrossRef] [PubMed]
23. Imbernón-Mulero, A.; Gallego-Elvira, B.; Martínez-Alvarez, V.; Martin-Gorriz, B.; Molina-del-Toro, R.; Jodar-Conesa, F.J.; Maestre-Valero, J.F. Boron removal from desalinated seawater for irrigation with an on-farm reverse osmosis system in southeastern Spain. *Agronomy* **2022**, *12*, 611. [CrossRef]
24. Murugaiyan, K.; Sivakumar, K. Seasonal variation in elemental composition of *Stoechospermum marginatum* (Ag.) Kutz and *Sargassum wightii* (Greville Mscr.) J.G. Agardh in relation to chemical composition of seawater. *Colloids Surf. B* **2008**, *64*, 140–144. [CrossRef]
25. Jones, E.; Qadir, M.; van Vliet, M.T.; Smakhtin, V.; Kang, S.M. The state of desalination and brine production: A global outlook. *Sci. Total Environ.* **2019**, *657*, 1343–1356. [CrossRef]
26. Grattan, S.R.; Diaz, F.J.; Pedrero, F.; Vivaldi, G.A. Assessing the suitability of saline wastewaters for irrigation of *Citrus* spp.: Emphasis on boron and specific-ion interactions. *Agric. Water Manag.* **2015**, *157*, 48–58. [CrossRef]
27. Levy, Y.; Syvertsen, J. Irrigation water quality and salinity effects in citrus trees. *Hortic. Rev.* **2004**, *30*, 37–82. [CrossRef]
28. Ayers, R.S.; Westcot, D.W. *Water Quality for Agriculture*; FAO Irrigation and Drainage Paper 29 Rev. 1; Food and Agriculture Organization of the United Nations: Rome, Italy, 1985; 174p.
29. Hanson, B.; Grattan, S.; Fulton, A. *Agricultural Salinity and Drainage*; Water Management Series Publication No. 3375; University of California: Davis, CA, USA, 2006; p. 157.
30. Dydo, P.; Turek, M.; Milewski, A. Removal of boric acid, monoborate and boron complexes with polyols by reverse osmosis membrane. *Desalination* **2013**, *334*, 39–45. [CrossRef]
31. Martínez-Álvarez, V.; Maestre-Valero, J.F.; González-Ortega, M.J.; Gallego-Elvira, B.; Martin-Gorriz, B. Characterization of the agricultural supply of desalinated seawater in southeastern Spain. *Water* **2019**, *11*, 1233. [CrossRef]
32. Díaz, F.J.; Tejedor, M.; Jiménez, C.; Grattan, S.R.; Dorta, M.; Hernández, J.M. The imprint of desalinated seawater on recycled wastewater: Consequences for irrigation in Lanzarote Island, Spain. *Agric. Water Manag.* **2013**, *116*, 62–72. [CrossRef]

33. Allison, L.E.; Brown, J.W.; Hayward, H.E.; Richards, L.A.; Bernstein, L.; Fireman, M.; Pearson, G.A.; Wilcox, L.V.; Bower, C.A.; Hatcher, J.T.; et al. *Diagnosis and Improvement of Saline and Alkali Soils*; Agriculture Handbook 60; United States Department of Agriculture: Washington, DC, USA, 1954; p. 160.
34. de Abreu, C.A.; van Raij, B.; de Abreu, M.F.; González, A.P. Routine soil testing to monitor heavy metals and boron. *Sci. Agric.* **2005**, *62*, 564–571. [CrossRef]
35. Eaton, F.M. Deficiency, toxicity and accumulation of boron in plants. *J. Agric. Res.* **1944**, *69*, 237–277.
36. Grattan, S.R.; Grieve, C.M. Mineral element acquisition and growth response of plants grown in saline environments. *Agric. Ecosyst. Environ.* **1992**, *38*, 275–300. [CrossRef]
37. Behboudian, M.H.; Törökfalvy, E.; Walker, R.R. Effects of salinity on ionic content, water relations, and gas exchange parameters in some citrus scion-rootstock combinations. *Sci. Hort.* **1986**, *28*, 105–116. [CrossRef]
38. Romero-Trigueros, C.; Nortes, P.A.; Pedrero, F.; Mounzer, O.; Alarcón, J.J.; Bayona, J.M.; Nicolás, E. Assessment of the viability of using saline reclaimed water in grapefruit in medium to long term. *Span. J. Agric. Res.* **2014**, *12*, 1137–1148. [CrossRef]
39. Maas, E. Crop salt tolerance. In *Salinity Assessment and Management*; Tanji, K.K., Ed.; American Society of Civil Engineers: New York, NY, USA, 1990.
40. Embleton, T.W.; Jones, W.W.; Labanauskas, C.K.; Reuther, W. Leaf analysis as a diagnostic tool and guide to fertilization. In *The Citrus Industry*; Reuther, V., Ed.; University of California Press: Berkeley, CA, USA, 1973; Volume III, pp. 183–210.
41. Simón-Grao, S.; Nieves, M.; Martínez-Nicolás, J.J.; Cámara-Zapata, J.M.; Alfosea-Simón, M.; García-Sánchez, F. Response of three citrus genotypes used as rootstocks grown under boron excess conditions. *Ecotoxicol. Environ. Saf.* **2018**, *159*, 10–19. [CrossRef]
42. Simón-Grao, S.; Nieves, M.; Martínez-Nicolás, J.J.; Alfosea-Simón, M.; Cámara-Zapata, J.M.; Fernández-Zapata, J.C.; García-Sánchez, F. Arbuscular mycorrhizal symbiosis improves tolerance of Carrizo citrange to excess boron supply by reducing leaf B concentration and toxicity in the leaves and roots. *Ecotoxicol. Environ. Saf.* **2019**, *173*, 322–330. [CrossRef]
43. Papadakis, I.E.; Dimassi, N.; Bosabalidis, A.M.; Therios, I.N.; Patakas, A.; Giannakoula, A. Boron toxicity in Clementine mandarin plants on two rootstocks. *Plant Sci.* **2004**, *166*, 539–547. [CrossRef]
44. Gimeno, V.; Simon, I.; Nieves, M.; Martinez, V.; Camara-Zapata, J.M.; Garcia, A.L.; García-Sánchez, F. The physiological and nutritional responses to an excess of boron by Verna lemon trees that were grafted on four contrasting rootstocks. *Trees* **2012**, *26*, 1513–1526. [CrossRef]
45. Ortuño, M.F.; García-Orellana, Y.; Conejero, W.; Ruiz, M.C.; Alarcón, J.J.; Torrecillas, A. Stem and leaf water potentials, gas exchange, sap flow, and trunk diameter fluctuations for detecting water stress in lemon trees. *Trees* **2006**, *20*, 1–8. [CrossRef]
46. Ribeiro, R.V.; Machado, E.C.; Santos, M.G.; Oliveira, R.F. Seasonal and diurnal changes in photosynthetic limitation of young sweet orange trees. *Environ. Exp. Bot.* **2009**, *66*, 203–211. [CrossRef]
47. Veste, M.; Ben-Gal, A.; Shani, U. Impact of thermal tress and high VPD on gas exchange and chlorophyll fluorescence of Citrus grandis under desert conditions. *Acta Hortic.* **2000**, *531*, 143–149. [CrossRef]
48. Macho-Rivero, M.A.; Herrera-Rodríguez, M.B.; Brejcha, R.; Schäffner, A.R.; Tanaka, N.; Fujiwara, T.; González-Fontes, A.; Camacho-Cristóbal, J.J. Boron toxicity reduces water transport from root to shoot in *Arabidopsis* plants. Evidence for a reduced transpiration rate and expression of major PIP aquaporin genes. *Plant Cell Physiol.* **2018**, *59*, 841–849. [CrossRef] [PubMed]
49. Huang, J.-H.; Cai, Z.-J.; Wen, S.-X.; Guo, P.; Ye, X.; Lin, G.-Z.; Chen, L.-S. Effects of boron toxicity on root and leaf anatomy in two *Citrus* species differing in boron tolerance. *Trees* **2014**, *28*, 1653–1666. [CrossRef]
50. Cakmak, I.; Römheld, V. Boron deficiency-induced impairments of cellular functions in plants. *Plant Soil.* **1997**, *193*, 71–83. [CrossRef]
51. Rhoades, J.D. Soluble salts. In *Methods of Soil Analysis Part 2, Chemical and Microbiological Properties*; Page, A.L., Miller, R.H., Kenney, D.R., Eds.; Agronomy Monograph; Soil Science Society of America: Madison, WI, USA, 1982; Volume 9, pp. 167–178.
52. McCutchan, H.; Shackel, K.A. Stem-water potential as a sensitive indicator of water stress in prune trees (*Prunus domestica* L. cv French). *J. Am. Soc. Hortic. Sci.* **1992**, *117*, 607–611. [CrossRef]
53. Turner, N.C. Measurements of plant water status by pressure chamber technique. *Irrig. Sci.* **1988**, *9*, 289–308. [CrossRef]
54. Sinclair, T.T.; Allen, L.H. Carbon dioxide and water vapour exchange of leaves on field-grown citrus trees. *J. Exp. Bot.* **1982**, *33*, 1166–1175. Available online: https://www.jstor.org/stable/23690488 (accessed on 15 January 2023).
55. Bates, L.; Waldren, R.; Teare, I.D. 1973. Rapid determination of free proline for water stress studies. *Plant Soil* **1973**, *39*, 205–207. [CrossRef]
56. Grieve, C.; Grattan, S. Rapid assay for determination of water soluble quaternary ammonium compounds. *Plant Soil* **1983**, *70*, 303–307. [CrossRef]
57. Inskeep, W.P.; Bloom, P.R. Extinction coefficient of chlorophyll a and b in N,N-Dimethylformamide and 80% acetone. *Plant Physiol.* **1985**, *77*, 483–485. [CrossRef]
58. Turrell, F.M. Growth and photosynthesis area of citrus. *Bot. Gaz.* **1961**, *122*, 284–298. [CrossRef]

plants

MDPI

Article

# Evaluating the Effect of Deficit Irrigation on Yield and Water Use Efficiency of Drip Irrigation Cotton under Film in Xinjiang Based on Meta-Analysis

Qi Xu [1,2,†], Xiaomei Dong [1,†], Weixiong Huang [1,3], Zhaoyang Li [1,2], Tongtong Huang [1], Zaijin Song [4], Yuhui Yang [1,2,*] and Jinsai Chen [5,*]

1 College of Water Conservancy and Architectural Engineering, Tarim University, Alar 843300, China; 10757222200@stumail.taru.edu.cn (Q.X.); 120130023@taru.edu.cn (X.D.); huangwx@cug.edu.cn (W.H.); 120130021@taru.edu.cn (Z.L.); 10757223118@stumail.taru.edu.cn (T.H.)
2 Key Laboratory of Modern Agricultural Engineering, Tarim University, Alar 843300, China
3 MOE Key Laboratory of Groundwater Quality and Health, School of Environmental Studies, China University of Geosciences, Wuhan 430078, China
4 Research Office, Beijing City University, Beijing 100193, China; songzaijin@bcu.edu.cn
5 College of Agronomy and Biotechnology, China Agricultural University, Beijing 100193, China
* Correspondence: 120130022@taru.edu.cn (Y.Y.); jinsaichen@cau.edu.cn (J.C.)
† These authors contributed equally to this work.

**Abstract:** Water scarcity constrains the sustainable development of Chinese agriculture, and deficit irrigation as a new irrigation technology can effectively alleviate the problems of water scarcity and water use inefficiency in agriculture. In this study, the drip irrigation cotton field under film in Xinjiang was taken as the research object. Meta-analysis and machine learning were used to quantitatively analyze the effects of different farm management practices, climate, and soil conditions on cotton yield and water use efficiency under deficit irrigation, to investigate the importance of the effects of different factors on cotton yield and water use efficiency, and to formulate appropriate optimization strategies. The results showed that deficit irrigation significantly increased cotton water use efficiency (7.39%) but decreased cotton yield (−15.00%) compared with full irrigation. All three deficit irrigation levels (80~100% FI, 60~80% FI, and 40~60% FI; FI: full irrigation) showed a significant decrease in cotton yield and a significant increase in water use efficiency. Under deficit irrigation, cotton yield reduction was the smallest and cotton water use efficiency increased the most when planted with one film, two tubes, a six-row cropping pattern, an irrigation frequency ≥10 times, a nitrogen application of 300~400 kg·ha$^{-1}$, and a crop density ≥240,000 per hectare, and planted with the Xinluzhong series of cotton varieties; deficit irrigation in areas with average annual temperature >10 °C, annual evapotranspiration >2000 mm, annual precipitation <60 mm, and with loam, sandy soil had the least inhibition of cotton yield and the greatest increase in cotton water use efficiency. The results of the random forest showed that the irrigation amount and nitrogen application had the greatest influence on cotton yield and water use efficiency. Rational irrigation based on optimal management practices under conditions of irrigation not less than 90% FI is expected to achieve a win–win situation for both cotton yield and water use efficiency. The above results can provide the best strategy for deficit irrigation and efficient water use in drip irrigation cotton under film in arid areas.

**Keywords:** drip irrigation under film; meta-analysis; machine learning; cotton yield; water use efficiency; trade-off analysis

## 1. Introduction

Cotton is an important cash crop in the world and occupies an important position in world agricultural development [1]. Cotton cultivation in China is mainly distributed in the Yangtze River Basin, the Yellow River Basin, and the northwestern interior. In recent years, with the reduction in cotton planting areas in the Yangtze River and Yellow River basins, Xinjiang has become the largest cotton area in China [2]. The cotton production in Xinjiang accounts for about 90.2% of the total national production [3]. Drip irrigation cotton under film is a combination of drip irrigation technology and film covering technology, which ensures a suitable water and heat environment for cotton growth, reduces the negative impact of soil salinity on plants, promotes seedling growth and development, and significantly improves cotton yield [4,5].

With the successive emergence of problems such as rapid population growth and over-exploitation of groundwater in recent years, China is facing the increasingly severe challenge of water scarcity [6,7]. Water scarcity largely restricts the development of agriculture. According to statistics, China's agricultural water consumption accounts for about 70% of the total water consumption, with agricultural irrigation accounting for about 90% of the agricultural water use [8,9]. Climate change also constrains the development of water resources, and climate change, mainly global warming, has become one of the most important environmental issues in the world today. Climate change has a direct impact on precipitation, evaporation, soil moisture, etc., making regional water scarcity problems more pronounced [10]. Deficit irrigation is a new type of irrigation technology proposed in response to water scarcity and inefficiency in water use. Deficit irrigation does not pursue the highest yield per unit area and allows for a certain limit of yield reduction. Relevant studies on crops such as alfalfa and wheat have shown that deficit irrigation significantly increases water productivity but reduces crop yields compared to full irrigation [11,12]. Zhang et al. (2016) explored the effects of plant density on cotton growth and yield under deficit irrigation through field experiments, and the results showed that appropriately increasing plant density under deficit irrigation in arid regions is a promising solution to save water without reducing yield [13]; Hou et al. (2021) investigated the response of cotton yield to different irrigation amounts based on a two-year field trial [14]; He et al. (2022) investigated the effects of different irrigation depths on cotton yield and water use efficiency through a field trial [15]; Wang et al. (2018) investigated the effects of different irrigation depths and cotton varieties on cotton yield and water utilization efficiency [16]. In addition, some studies have shown that both soil texture and climatic conditions significantly affect crop yield and water utilization efficiency [17]. Numerous scholars have already conducted multifaceted studies on deficit irrigation, but due to the limitations of experimental conditions and sites, most of the studies have only explored the effects of a small number of factors on cotton yield. Moreover, there are not yet many integrated studies exploring the changing patterns of the yield and water use efficiency of drip irrigation cotton under film. Therefore, a comprehensive analytical method is needed to deeply explore and systematically analyze the existing studies and summarize the effects of deficit irrigation under different conditions, so as to better apply deficit irrigation technology in future practice.

Meta-analysis is the application of statistical concepts and methods to synthesize and quantify the results of a study and to dissect clear patterns of relationships among the variables of interest. Cheng et al. (2021) analyzed the response of cotton yield and water use efficiency to insufficient irrigation under different conditions based on the integration of a global scale but did not systematically study factors such as cotton varieties, planting densities, and cropping patterns [18]. Therefore, this study takes under-film drip irrigation cotton fields as the research object, evaluates the comprehensive effects of deficit irrigation on cotton yield and water use efficiency based on the meta-analysis method, analyzes the importance of the effects of different factors on cotton yield and water use efficiency based on the random forest model, explores the possibility of realizing the mutual balance of the two indexes, and develops suitable optimization schemes through the existing data. The objective of this study is to provide the optimal deficit irrigation strategy and to realize the efficient use of water resources in drip irrigation cotton under film in arid areas.

## 2. Results

### 2.1. Overview of the Database

The database was derived from 50 literature examples including 280 pairs of observations on cotton yield and 168 pairs of water use efficiency under deficit irrigation conditions, and the frequency distributions of the effect values of cotton yield and water use efficiency conformed to a normal distribution. Compared with full irrigation, deficit irrigation significantly increased cotton water use efficiency by 7.39% (95% *CI*: 5.59~9.18%) but significantly reduced cotton yield (−15.00%, 95% *CI*: −16.36~−13.65%) (Figure 1). The subgroup data can be further analyzed for the specific effects of different farm management practices, soil conditions, and climatic factors on cotton yield and water use efficiency.

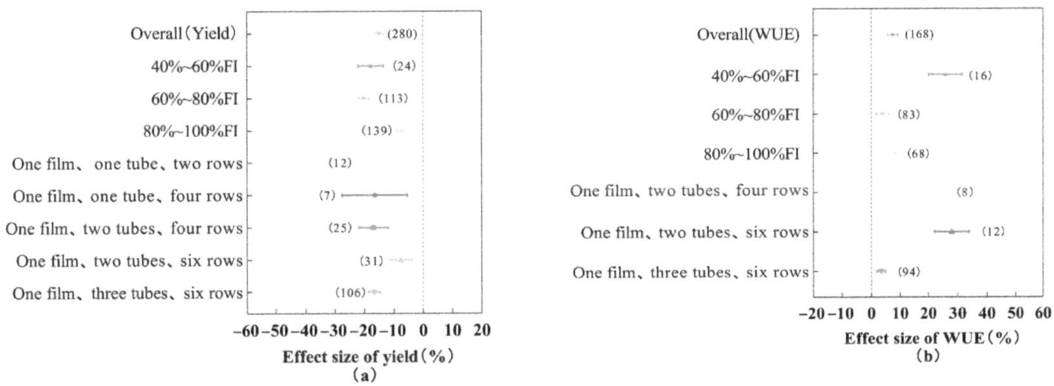


**Figure 1.** Effects of different irrigation amounts and drip tape modes on cotton yield (**a**) and water use efficiency (**b**). Note: FI is the fully irrigated level. Dots and error lines represent response ratios and their 95% confidence intervals, respectively; non-overlapping of confidence intervals between different subgroups means that the results are significant, and the opposite is not significant. Values in parentheses represent sample sizes, same as below.

### 2.2. Effects of Different Irrigation Amounts and Drip Tape Modes on Cotton Yield and Water Use Efficiency

When the irrigation amount was in the range of 80~100% FI, deficit irrigation reduced cotton yield the least (−7.46%, 95% *CI*: −9.15~−5.77%), which was significantly better than 40~60% FI (−17.74%) and 60~80% FI (−20.21%). As the irrigation amount decreased, the cotton water use efficiency showed a tendency of decreasing and then increasing, as shown in the following: at the irrigation amount of 80~100% FI, the improvement of deficit irrigation on cotton water use efficiency was 8.41% (95% *CI*: 5.86~10.96%); at the irrigation amount of 60%~80% FI, the improvement of deficit irrigation on cotton water use efficiency was 3.83% (95% *CI*: 1.60~6.07%); at the irrigation amount of 40~60% FI, the improvement of deficit irrigation on cotton water use efficiency was 25.95% (95% *CI*: 20.18~31.72%). When drip irrigation tapes were laid in the form of one film and two tubes in six rows, deficit irrigation had the smallest reduction in cotton yield (−7.65%, 95% *CI*: −11.57~−3.73%) and the largest improvement in cotton water use efficiency (28.10%, 95% *CI*: 22.24~33.96%), which was significantly better than one film and two tubes in four rows (16.54%) and one film and three tubes in six rows (3.47%). It can be seen that, under deficit irrigation, both irrigation amount and drip tape mode significantly affected cotton yield and water use efficiency; with the increase in deficit level, both cotton yield and water use efficiency showed the trend of decreasing and then increasing; the drip tape mode of one film, two tubes, and six rows could significantly inhibit the reduction in cotton yield and significantly improve the water use efficiency of cotton.

### 2.3. Effect of Different Irrigation Frequencies and Nitrogen Applications on Cotton Yield and Water Use Efficiency

As shown in Figure 2, the reduction in cotton yield by deficit irrigation at irrigation frequencies of ≥10 times (−12.08%, 95% *CI*: −16.21~−7.95%) was lower than the magnitude of reduction in cotton yield at irrigation frequencies of <10 times (−15.29%); and the magnitude of improvement in cotton water use efficiency by deficit irrigation at irrigation frequencies of ≥10 times (13.90%, 95% *CI*: 7.69~20.12%) was higher than the enhancement of cotton water use efficiency at irrigation frequencies of <10 times (5.83%). The reduction in cotton yield with deficit irrigation was lower at nitrogen applications of 300~400 kg·ha$^{-1}$ (−13.91%, 95% *CI*: −17.23~−10.59%) than at ≤200 kg·ha$^{-1}$ (−14.95%), 200~300 kg·ha$^{-1}$ (−15.46%), and >400 kg·ha$^{-1}$ (−17.60%). With the increase in nitrogen applications, the improvement of cotton water use efficiency by deficit irrigation showed a trend of increasing and then decreasing, as shown in the following: under deficit irrigation, cotton water use efficiency was improved by 5.31%, 7.56%, and 10.55% when nitrogen applications were ≤200 kg·ha$^{-1}$, 200~300 kg·ha$^{-1}$, and 300~400 kg·ha$^{-1}$, respectively, but cotton water use efficiency showed an inhibitory effect (−4.19%) when the nitrogen application was >400 kg·ha$^{-1}$. Due to its small sample size ($n$ = 4), a larger-scale study was needed to obtain more accurate results. In conclusion, under deficit irrigation, both irrigation frequency and nitrogen application had significant effects on cotton yield and water use efficiency; an irrigation frequency of ≥10 times and a nitrogen application of 300~400 kg·ha$^{-1}$ could effectively inhibit cotton yield reduction and significantly improve cotton water use efficiency.

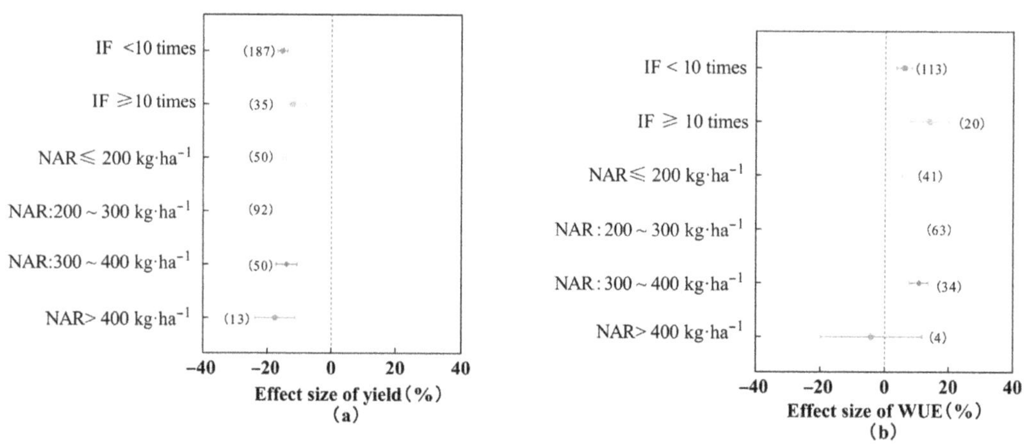


**Figure 2.** Effects of different irrigation frequencies and nitrogen application rates on cotton yield (**a**) and water use efficiency (**b**). Note: dots and error lines represent response ratios and their 95% confidence intervals, respectively; non-overlapping of confidence intervals between different subgroups means that the results are significant, and the opposite is not significant. Values in parentheses represent sample sizes. IF: irrigation frequency, NAR: nitrogen application rate.

### 2.4. Effects of Different Varieties and Planting Densities on Cotton Yield and Water Use Efficiency

As can be seen in Figure 3. Tahe No. 2 cotton seed showed the advantage of increased yield (3.97%) and efficiency (49.36%) under deficit irrigation; Zhaofeng No. 1 cotton seed showed the disadvantage of reduced yield (−41.47%) and efficiency (−15.23%) under deficit irrigation, and the overall effect was not significant due to the small sample sizes of the two types of cotton seed. Under deficit irrigation, sowing Xinluzhong-series cotton seeds significantly suppressed cotton yield reduction (−12.68%, 95% *CI*: −15.95~−9.41%), which was more effective than Xinluzao-series cotton seeds (−17.52%) and Xinken cotton seeds (−14.00%). Sowing Xinluzhong-series cotton seeds significantly improved cotton water use efficiency (16.55%, 95% *CI*: 13.72~19.39%), which was better than that of Xinluzao-series cotton seeds (3.60%) and Xinken cotton seeds (4.96%). With the increase in cotton planting density, the reduction in cotton yield by deficit irrigation tended to decrease significantly. This was shown as follows: the reduction in cotton yield was 19.04% (95% *CI*: −22.69~−15.38%) at a cotton planting density <240,000 plants per hectare and 6.36% (95% *CI*: −11.09~−1.63%) at a cotton planting density ≥240,000 plants per hectare, whereas the cotton water use efficiency did not differ significantly among the planting densities of different levels. In conclusion, the Xinluzhong series of cotton varieties could significantly inhibit the cotton yield reduction and significantly increase the cotton water utilization efficiency under deficit irrigation conditions, and the crop density of ≥240,000 plants per hectare could significantly inhibit the cotton yield reduction.

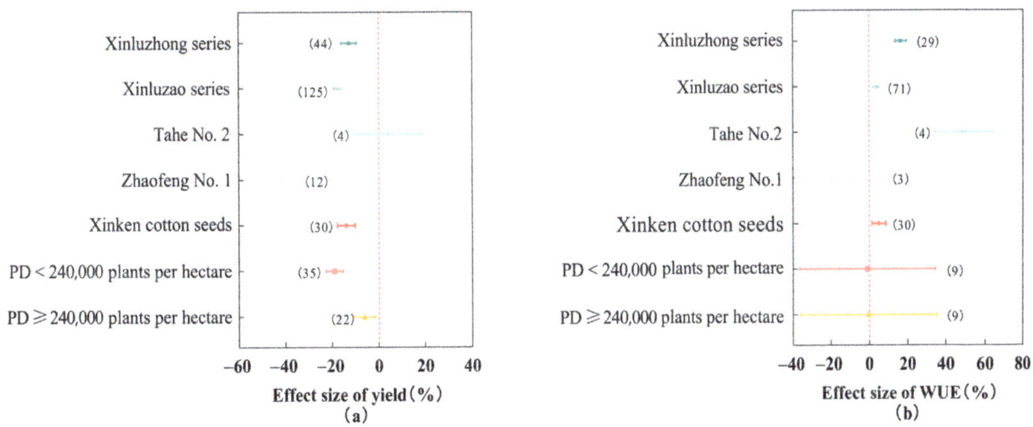


**Figure 3.** Effects of different varieties and planting density on cotton yield (**a**) and water use efficiency (**b**). Note: dots and error lines represent response ratios and their 95% confidence intervals, respectively; non-overlapping of confidence intervals between different subgroups means that the results are significant, and the opposite is not significant. Values in parentheses represent sample sizes. PD: planting density.

### *2.5. Effects of Different Climatic and Soil Conditions on Cotton Yield and Water Use Efficiency*

As shown in Figure 4, the reduction in cotton yield due to deficit irrigation in areas with a mean annual temperature >10 °C (−8.15%, 95% *CI*: −10.45~−5.85%) was significantly lower than that in areas with a mean annual temperature ≤10 °C (−15.31%), and the enhancement of cotton water use efficiency due to deficit irrigation (17.80%, 95% *CI*: 15.24~20.36%) was significantly higher (1.99%) than in areas with a mean annual temperature ≤10 °C. With the increase in annual evaporation, the reduction in cotton yield by deficit irrigation gradually decreased and the improvement of cotton water use efficiency gradually increased, as shown in the following: under deficit irrigation, the reduction in cotton yield at an annual evaporation of 1500~2000 mm was 14.90% (95% *CI*: −17.90~−11.89%), and the reduction in cotton yield at an annual evaporation >2000 mm was 10.88% (95% *CI*: −15.35~−6.42%). The increases in cotton water use efficiency were −3.03% for an annual evaporation of 1500~2000 mm (which showed an inhibitory effect) and 14.84% for an annual evaporation of >2000 mm. With the increase in annual precipitation, the decrease in deficit irrigation on cotton yield gradually increased, and the increase in cotton water use efficiency gradually decreased, as shown in the following: under deficit irrigation conditions and annual precipitation amounts of <60 mm, 60~200 mm, and >200 mm, cotton yield decreased by 8.32%, 15.42%, and 18.72%, respectively; cotton water use efficiency increased by 18.43%, 2.27%, and −1.14%, respectively (at an annual precipitation >200 mm, the cotton water use efficiency showed an inhibitory effect). Therefore, deficit irrigation in areas with average annual temperature >10 °C, annual evaporation >2000 mm, and annual precipitation <60 mm had the least inhibitory effect on cotton yield and the greatest increase in cotton water use efficiency.

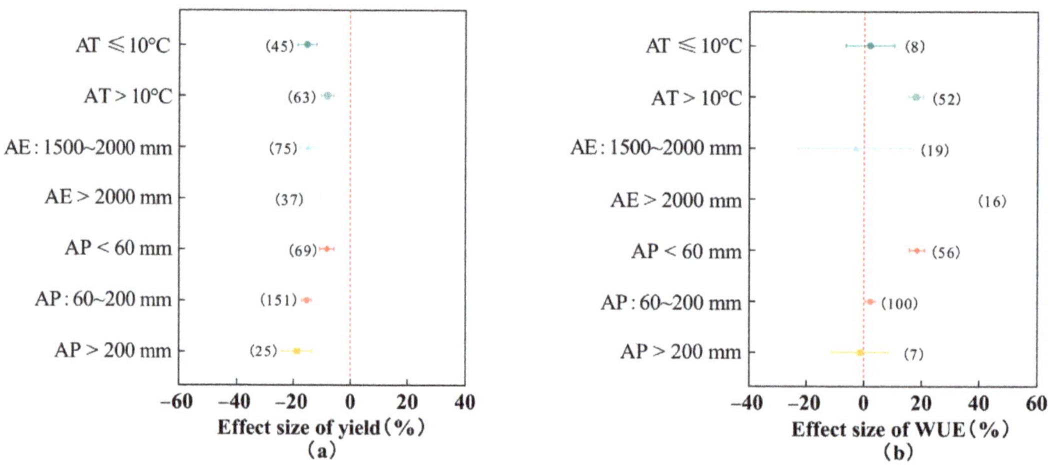


**Figure 4.** Effects of different climatic conditions on cotton yield (**a**) and water use efficiency (**b**). Note: dots and error lines represent response ratios and their 95% confidence intervals, respectively; non-overlapping of confidence intervals between different subgroups means that the results are significant, and the opposite is not significant. Values in parentheses represent sample sizes. AT: annual temperature, AE: annual evaporation, AP: annual precipitation.

As shown in Figure 5, deficit irrigation had the least reduction in cotton yield when the soil was of a loam texture (−8.73%, 95% *CI*: −14.02~−3.44%), followed by a sandy soil texture (−13.98%), while the greatest reduction in cotton yield was observed when the soil was of a sandy loam texture (−17.63%). The greatest improvement in cotton water use efficiency was observed in soil with a medium loam texture (13.51%, 95% *CI*: −67.26~94.28%), but the overall boost was not significant due to the small sample size ($n$ = 2) and the 95% confidence interval containing zero. Deficit irrigation in soil with a sandy soil texture significantly improved cotton water use efficiency (8.83%, 95% *CI*: 7.03~10.63%), and the enhancement was better than that in a sandy loam soil texture (7.93%) and clay loam soil texture (3.94%). Cotton yield did not differ significantly among the different levels of initial soil organic carbon, but the cotton field soil initial organic carbon content of 5.8~11.6 $g \cdot kg^{-1}$ significantly improved cotton water use efficiency (3.41%, 95% *CI*: 1.13~5.70%). Deficit irrigation at a soil available nitrogen content $\leq$60 $mg \cdot kg^{-1}$ in a cotton field could effectively inhibit cotton yield reduction (−14%), and the effect was better than that of a soil available nitrogen content of 60~120 $mg \cdot kg^{-1}$ (−17.37%); the increase in cotton water use efficiency at a soil available nitrogen content $\leq$60 $mg \cdot kg^{-1}$ in a cotton field (3.11%) was greater than that of a soil available nitrogen content of 60~120 $mg \cdot kg^{-1}$ (0.69%), but there was no significant difference between the two levels. In conclusion, deficit irrigation in loam and sandy soils showed the least reduction in cotton yield, and deficit irrigation in sandy soils showed the greatest increase in cotton water use efficiency.

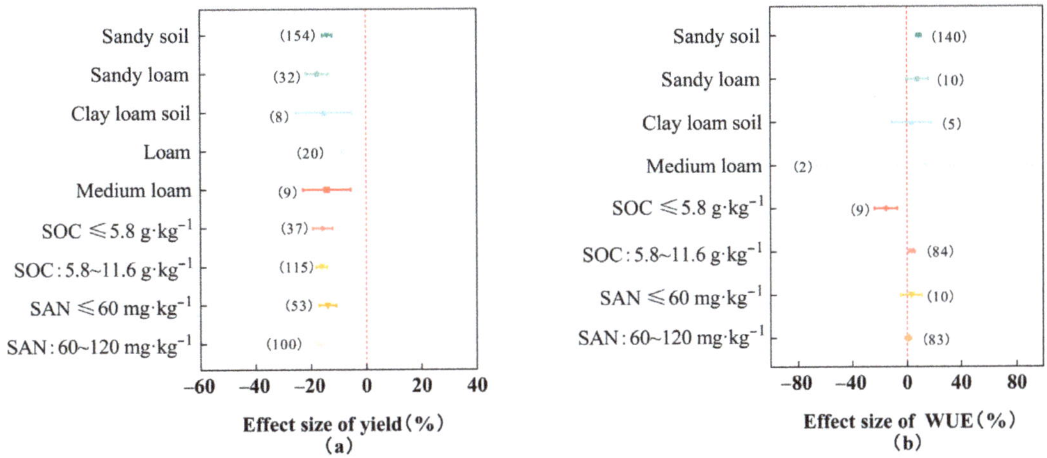


**Figure 5.** Effects of different soil conditions on cotton yield (**a**) and water use efficiency (**b**). Note: dots and error lines represent response ratios and their 95% confidence intervals, respectively; non-overlapping of confidence intervals between different subgroups means that the results are significant, and the opposite is not significant. Values in parentheses represent sample sizes. SOC: soil organic carbon, SAN: soil available nitrogen.

### *2.6. Trade-Offs between Cotton Yield and Water Use Efficiency under Deficient Irrigation and Their Key Influencing Factors*

In order to better understand the equilibrium relationship between deficit irrigation on cotton yield and water use efficiency, in this study, 168 data points of cotton yield and water use efficiency were plotted in the plane of Figure 6. All the paired effect values of yield and water use efficiency were classified into four regions, where region I represents the increase in both cotton yield and water use efficiency, region II represents the increase in cotton yield and water use efficiency decrease, region III represents the decrease in both cotton yield and water use efficiency, region IV represents the increase in water use efficiency with a decrease in cotton yield, and region V represents the realization of an increase in cotton water use efficiency under the condition that the cotton yield is reduced by less than 15.00%. Among which, region I represents a win–win situation, although region I is the expected goal of practice, but with only 14 pairs of data, accounting for the total number of strongholds of 8.44%. The low percentage of data volume also indicates that it is difficult to realize the mutual improvement of cotton yield and water use efficiency by deficit irrigation under the existing management measures, and appropriate irrigation strategies should be adopted to realize the trade-off between cotton yield and water use efficiency. The absence of data points in region II also showed that there was no increase in cotton yield but a decrease in water use efficiency under deficit irrigation. Data points in regions III, IV, and V accounted for 19.27%, 72.29%, and 45.78% of the total, respectively. Although it is difficult to achieve a win–win situation for both cotton yield and water use efficiency, the 45.78% data share within region V indicates that there was a slight decrease in cotton yield and an increase in water use efficiency in most of the studies.

Importance analysis of the factors affecting cotton yield and water use efficiency was carried out using random forest, and the results are shown in Figure 7. Relative importance represents the degree of influence of each factor on cotton yield and water use efficiency, and the larger the value, the greater the importance of the influence of the variable [19]. The * sign in the figure represents the significance level of each variable. As shown in the figure, the factors that had a greater effect on cotton yield and reached a highly significant level were irrigation amount, nitrogen application, and average annual temperature. The factors that had a significant effect on cotton water use efficiency at the highly significant level were nitrogen application, initial soil organic carbon content, and irrigation amount.

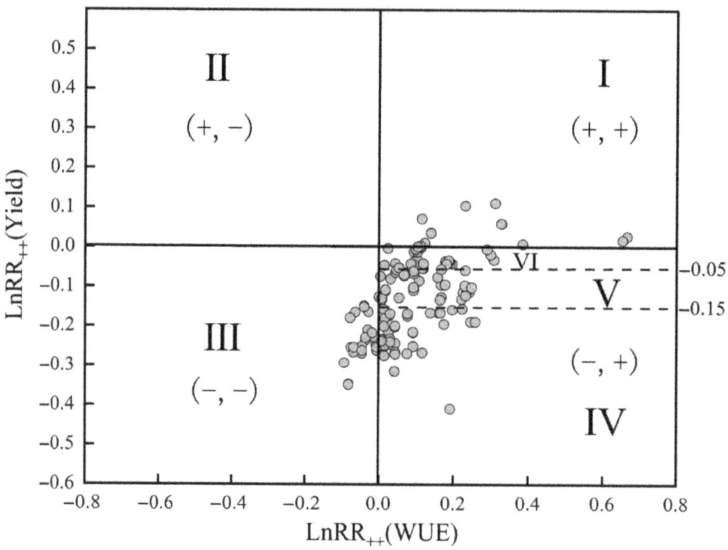


**Figure 6.** Effect distribution of cotton yield and water use efficiency on deficit irrigation response based on all paired data in regions I, II, III, IV, and V. Note: the area with 5% cotton yield reduction and water use efficiency greater than 0 is divided into VI. The data points in the figure are the matching data points of cotton yield and water use efficiency, and the horizontal line in the figure represents the critical line of 5% cotton yield reduction and 15% cotton yield reduction.

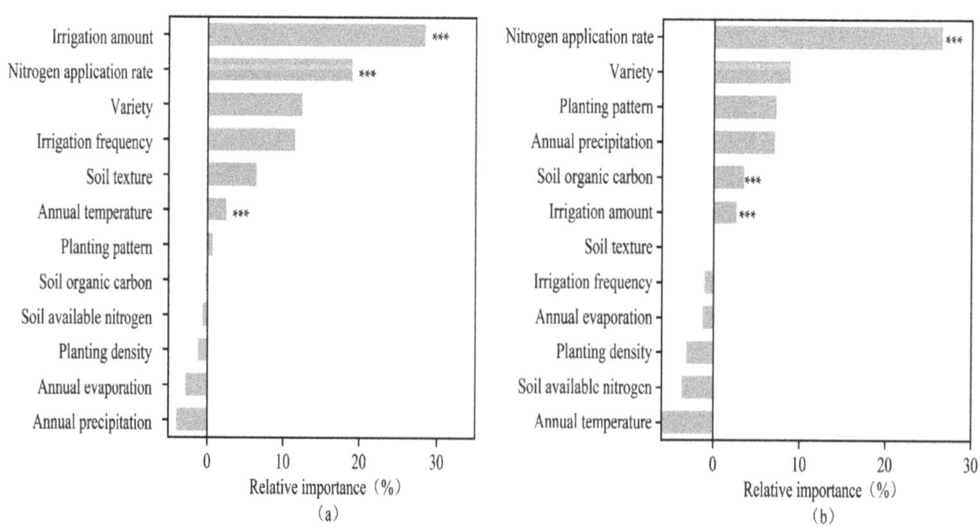


**Figure 7.** The relative importance of each factor to cotton yield (**a**) and water use efficiency (**b**) under the random forest model, with the *** sign in the figure representing the significance level.

In this study, the area with a 5% cotton yield reduction and water use efficiency >0 was divided into region VI, which represents the lowest cotton yield reduction acceptable to this study. Regions I and VI were the target regions, and this study analyzed the data from different groups of irrigation amount, nitrogen application, and average annual temperature, which had significant effects on cotton yield, in the target and other regions (Table 1). Among them, the proportion of data points in the target region was much larger than that in other regions for a nitrogen application of 300~400 kg·ha$^{-1}$ and a nitrogen application of >400 kg·ha$^{-1}$, while the other groups showed that the proportion of data in the target region was smaller than that in other regions, or the proportion of data in the target region was similar to that in other regions.

**Table 1.** Proportions of different subgroups of irrigation amount, nitrogen application rate, and annual mean temperature in the target area and other regions.

| Impact Factor | Category | Target Zone | Other Zone |
|---|---|---|---|
| Irrigation amount, % | 40~60% FI | 13.33% | 86.67% |
| | 60~80% FI | 23.73% | 76.27% |
| | 80~100% FI | 55.74% | 44.26% |
| Nitrogen application rate, kg·ha$^{-1}$ | ≤200 | 21.21% | 78.79% |
| | 200~300 | 31.25% | 68.75% |
| | 300~400 | 62.07% | 37.93% |
| | >400 | 100% | 0% |
| Average annual temperature, °C | ≤10 | 50.00% | 50.00% |
| | >10 | 58.00% | 42.00% |

In order to construct a suitable deficit irrigation strategy to realize the rational use of agricultural water resources and promote the win–win situation of cotton yield and water use efficiency in the target area, this study conducted subgroup analysis of irrigation water quantity in the target area. The results, as shown in Table 2, showed that the mean effect size of deficit irrigation on cotton yield in the target area was −0.99%, and the negative effect of deficit irrigation on cotton yield gradually decreased with the increase in irrigation amount, and the decrease in deficit irrigation on cotton yield was the smallest (−0.16%) and was significantly larger than that of −0.99% when the irrigation amount was 90%~100% FI. In other words, rational irrigation according to optimal management measures under an irrigation amount of not less than 90% FI is expected to achieve a win–win situation of cotton yield and water use efficiency.

**Table 2.** Analysis of optimized irrigation volumes within the target region.

| Impact Factor | Category | Effect Value of Cotton Yield | 95% Confidence Interval | Data Sample Size | Mean Effect Size |
|---|---|---|---|---|---|
| Irrigation amount, % | 80~90% FI | −1.23% | −3.53~1.07% | 25 | −0.99% |
| | 90~100% FI | −0.16% | −4.86~4.53% | 11 | |

## 3. Discussion

### *3.1. The Combined Effect of Deficient Irrigation on Cotton Yield and Water Use Efficiency*

In this study, the combined effect of deficit irrigation on cotton yield and water use efficiency was quantified based on the meta-analysis method by using the valid data of 280 pairs of cotton yield and 168 pairs of cotton water use efficiency from 50 literature examples. The results of the study showed that deficit irrigation increased cotton water use efficiency (7.39%) but significantly reduced cotton yield (−15.00%) compared to full irrigation. Different management practices (irrigation amount, drip tape mode, etc.), climate, and soil conditions significantly affected cotton yield and water use efficiency. Therefore, it is necessary to analyze the effects of different subgroups on cotton yield and water use efficiency and develop suitable optimization strategies under deficit irrigation to minimize the reduction in cotton yield and significantly improve cotton water use efficiency.

### 3.2. Effects of Different Farm Management Practices on Cotton Yield and Water Use Efficiency

The research content of this experiment is about drip irrigation cotton under film, which is a combination of drip irrigation technology and film covering technology [20], to ensure the appropriate water and heat environment for cotton growth, so that the cotton fertility period can be advanced and cotton yield can be significantly increased [4]. The results of this study showed that the drip tape mode of one film, two tubes, and six rows could significantly inhibit cotton yield reduction and significantly improve cotton water utilization efficiency. The planting pattern of one film, two tubes, and six rows has a good thermal insulation effect, and the appropriate soil temperature is favorable to the early growth and development of cotton [21]. Some studies have shown that the drip tape mode of double tubes is better than single tubes in terms of the salt control effect [22,23]. Too much soil salt will inhibit the absorption of nutrients by the cotton root system, thus affecting the normal growth and development of cotton; too much salt leads to a decrease in the infiltration type of soil water, which significantly reduces the soil water content, leading to the shortage of water caused by the drought stress of cotton. Therefore, the planting pattern of one film, two tubes, and six rows under deficit irrigation can create a desalinated soil environment for the cotton root zone, significantly improve the water utilization efficiency of cotton, and inhibit the reduction in cotton yield. This study showed that deficit irrigation reduced cotton yield by 15% and increased water use efficiency by 7.39%. An irrigation ration of 80~100% FI had the smallest reduction in cotton yield (7.46%), and the appropriate range of moisture changes did not significantly reduce cotton yield [24]; this study found that a 20% to 25% water deficit had less effect on cotton yield [25,26], which may be due to the fact that drip irrigation under film reduces the ground evaporation between plants and allows the soil to remain loose and water-laden [27]. When the irrigation amount of 80~100% FI decreased to 60~80% FI, the cotton yield decreased significantly, but the water use efficiency did not increase significantly. It has been suggested that an increase in water use efficiency under limited water conditions does not equate to high yields, even if yields are significantly reduced, which is mainly dependent on the transpiration and photosynthetic capacity of leaves in response to water stress [11,28].

Under deficit irrigation, with the increase in irrigation frequency, the yield reduction of cotton gradually decreases, and the improvement of water use efficiency gradually increases. Appropriately increasing the frequency of irrigation can effectively reduce the risk of cotton yield reduction. Studies by some scholars have shown that an increase in the frequency of irrigation can increase crop yield [29,30]. High-frequency irrigation concentrates water in a small area of the root zone, which promotes the growth of the cotton root system and the uptake and utilization of nutrients. Previous studies have shown that the appropriate amount of nitrogen application can increase cotton yield and water utilization efficiency [18]. A study of the coupling of water and nitrogen conducted by Zhong Deng et al. (2015) in south Xinjiang showed that a nitrogen application of 300 kg·ha$^{-1}$ promoted cotton yield at different irrigation water levels [31]. Appropriate fertilizer application contributes to crop growth and development and photosynthesis [32]; it helps to improve crop water use efficiency [33]. Excessive nitrogen application leads to a plant nutrient imbalance detrimental to cotton growth and yield formation and aggravates nitrogen loss and environmental burdens [34]. In this study, we showed that under deficit irrigation, an irrigation frequency $\geq$10 times and a nitrogen application of 300~400 kg·ha$^{-1}$ increased cotton water use efficiency the most and significantly suppressed cotton yield reduction.

The overall effect of the Tahe No. 2 and Zhaofeng No. 1 cotton varieties was not significant due to the small sample size, and a larger study is needed to obtain more accurate results. Therefore, this study focuses on the effects of Xinluzhong-series cotton varieties, Xinluzao-series cotton varieties, and Xinken cotton 1 on cotton yield and water use efficiency. As can be seen from Figure 4, Under deficit irrigation conditions, Xinluzhong-series cotton varieties had the smallest reduction in cotton yield and the largest increase in water use efficiency compared with the other two types of cotton varieties. Xinluzhong-series and Xinluzao-series cotton varieties belong to the same Xinjiang cotton varieties.

Xinluzhong-series cotton varieties are medium-maturity land cotton varieties, which are generally stronger in disease resistance and yield than the Xinluzao-series cotton varieties. Xinken cotton 1 and Xinluzao-series cotton varieties belong to the same category of early maturity land cotton, but they are more suitable to be planted in the Hexi Corridor cotton area [35]. As far as crop planting density is concerned, it has been shown that under deficit irrigation, high planting density can conserve water without decreasing seed cotton yield [13,18,36]. Chen's study showed that limited irrigation conditions and high planting density can obtain larger crop yields [37], which are all consistent with the results of this study. The results of this study showed that a crop planting density of $\geq$240,000 plants per hectare could significantly inhibit cotton yield reduction.

### *3.3. Effect of Different Climatic and Soil Conditions on Cotton Yield and Water Use Efficiency*

The response of cotton yield and water use efficiency to deficit irrigation was significantly influenced by climate and soil conditions. The results of this study showed that deficit irrigation in areas with an average annual temperature >10 °C, an annual evaporation >2000 mm, and an annual precipitation <60 mm had the least inhibition of cotton yield and the greatest increase in cotton water use efficiency. Cotton is a thermophilic crop, and a higher-temperature environment is favorable for cotton growth and water use; low temperature will prolong the cotton fertility period and, thus, lead to a decrease in cotton yield [38]. Some studies have shown that a high evaporation rate reduces plant moisture and significantly reduces boll yield [39], which is inconsistent with the results of the present study, probably because the study area is an arid zone with low precipitation and high evaporation, and the cotton species is highly drought-tolerant. The reduction in precipitation leads to a decrease in evapotranspiration of the cotton species, which in turn increases the water use efficiency. As far as soil texture is concerned, loam and sandy soils under deficit irrigation conditions showed the least reduction in cotton yield, and sandy soils showed the greatest increase in cotton water use efficiency. Sandy soil has the advantages of rough particles and good permeability. Some scholars' studies have concluded that rough-textured soil has higher hydraulic conductivity, which is more favorable to plant root growth [12]; loam soil combines the advantages of both clay and sandy soil [40,41], has strong resistance to stress and drought and flooding, and can inhibit the reduction in cotton yields under deficit irrigation conditions. In terms of soil initial organic carbon and soil available nitrogen content, there is no significant effect on cotton yield and water use efficiency among different subgroups under deficit irrigation conditions. A too-high initial soil fertility will lead to vigorous microbial activity in the soil, which will consume soil oxygen and reduce cotton yield; a too-low initial soil fertility will also inhibit cotton growth. In addition, the drip irrigation cotton under film was supplied with fertilizer by drip application with water through the fertilizer tank, and the drip application of fertilizers with water ensured that the water and fertilizer solutions were uniformly distributed in the soil, which allowed the nutrients to be fully utilized. This may also be the reason why the initial soil fertility did not have a significant effect on cotton yield and water use efficiency.

## 4. Materials and Methods

### *4.1. Data Collection*

The literature data adopted in this study were obtained from the China National Knowledge Infrastructure (www.cnki.net) and Web of Science (www.webofscience.com). The keywords of the literature search included cotton, drip irrigation under film, drip irrigation, cotton yield, and yield. Excluding the water quality of irrigation, such as brackish water, magnetized water, etc., we screened scholars' published journal papers and master's doctoral dissertations from 1 January 1960 to 31 December 2022. According to the needs of the research topic, the literature screening for inclusion needed to meet the following criteria: (1) the test site should be a field trial within the scope of Xinjiang, potting and indoor trials should be excluded, and the test object should be cotton; (2) the number of trial replicates should be ≥3; (3) cotton yields recorded in the study cases were seed cotton yields; (4) multiple replicated observations under the same treatment were averaged; (5) when there were studies with additional factor effects in the trials, they were considered as independent trials separately included in the database; (6) for multi-year experimental studies in the same region in the literature, they were considered to be mutually independent trials separately included in the database. In addition, based on scholars' studies on deficit irrigation in Xinjiang [37,42,43], this study defined the range of complete irrigation (full irrigation) as between 368 and 425 mm, and the irrigation amount below this range was considered as deficit irrigation and classified into three levels (40~60% FI, 60~80% FI, and 80~100% FI, FI being the fully irrigated level). The irrigation amount described in this study is the sum of successive irrigation volumes over the entire cotton reproductive period, which consists of four periods: seedling, bud, boll blossom, and maturity. The data in the literature tables were extracted directly, and for the data in the bar charts and line graphs, the software Get Data 2.20 was used to digitize the images before extraction [44]. According to the above criteria, 50 documents and 448 sets of valid data were finally screened and obtained in this study. The names of the literature included in this study and the distribution of the test sample sites were documented in the additional material.

### 4.2. Meta-Analysis

In order to ensure the reliability of the study, the weighted response ratio (*RR*) was used as a statistical indicator in this study. The natural logarithm of the weighted response ratio (*LnRR*) was used to express the degree of influence of a particular driving factor, and its 95% confidence interval (95% *CI*) was calculated. Its calculation formula is [45]:

$$RR = Xe/Xc \tag{1}$$

Taking cotton yield as an example, *Xe* is the average yield of cotton under deficit irrigation (kg·ha$^{-1}$); *Xc* is the average yield of cotton under full irrigation (kg·ha$^{-1}$).

$$LnRR = Ln(Xe/Xc) \tag{2}$$

If $LnRR > 0$, it means that deficit irrigation has a positive effect on cotton yield, which shows the promotion effect; if $LnRR < 0$, it means that deficit irrigation has a negative effect on cotton yield, which shows the inhibition effect, and the larger the value of *LnRR* is, the more obvious the effect is [46].

The response ratios of each study were weighted together to produce an average weighted response ration (*RR*++). In addition, variance (*V*), weighted factor (*W*), *RR*++ standard deviation, and 95% confidence intervals were obtained by calculating Equations (3)–(7).

$$RR_{++} = \frac{\sum_{i=1}^{m}\sum_{j=1}^{ki} W_{ij}RR_{ij}}{\sum_{i=1}^{m}\sum_{j=1}^{ki} W_{ij}} \tag{3}$$

where *m* is the number of subgroups, e.g., different varieties or soil texture subgroups, and *ki* represents the number of pairs of data in the *i*th subgroup.

$$V = \frac{SDe^2}{NeXe^2} + \frac{SDc^2}{NcXc^2} \tag{4}$$

$SDe^2$ and $SDc^2$ represent the standard deviation of the test and control groups, respectively. For missing *SD* values in the literature, 1/10 of the mean of the corresponding test or control group was used as a substitute; *Ne* and *Nc* represent the number of samples in the test and control groups [47].

$$W = \frac{1}{V} \quad (5)$$

$$S(RR_{++}) = \sqrt{\frac{1}{\sum_{i=1}^{m}\sum_{j=1}^{ki} W_{ij}}} \quad (6)$$

$$95\%CI = RR_{++} \pm 1.96S(RR_{++}) \quad (7)$$

To better understand the results of the meta-analysis, 95% confidence intervals were transformed into percent change by ($e^{(LnRR++)-1)} \times 100\%$). When the 95% *CI* was far from the zero line, it indicated that the treatment group was significantly different from the control group ($p < 0.05$); when the 95% *CI* intersected with the zero line, it indicated that there was no significant difference between the treatment group and the control group ($p > 0.05$).

### 4.3. Publication Bias Test

The chi-square test (*CST*) was performed on the data to clarify whether there was clear heterogeneity among treatments. According to the results, the corresponding fixed-effect model or random-effect model was adopted; if $p > 0.05$, it indicated that there was no significant difference in the results of the studies, and the fixed-effect model was adopted for the calculations. On the contrary, the random-effect model was adopted for the calculations. The bias test was conducted by the fail-safe number (*Nfs*) method; if *Nfs* > 5*n* + 10 (*n* is the logarithm of data), it means that there is no publication bias, and the conclusion is credible [11]. The significance tests of cotton yield and water use efficiency in this study have *p* values < 0.001, which indicates that a random-effects model should be adopted to carry out subgroup analyses to explain the heterogeneity in depth; the *Nfs* values of cotton yield and water use efficiency are 48,172.8 and 3945.7, respectively. The numbers of pairs of data are 280 and 168 pairs, the *Nfs* is much larger than 5*n* + 10, and the cotton yield and water use efficiency effect values frequencies conformed to normal distribution (Figure 8), which indicated that there was no publication bias in this study and that the conclusions were reliable.

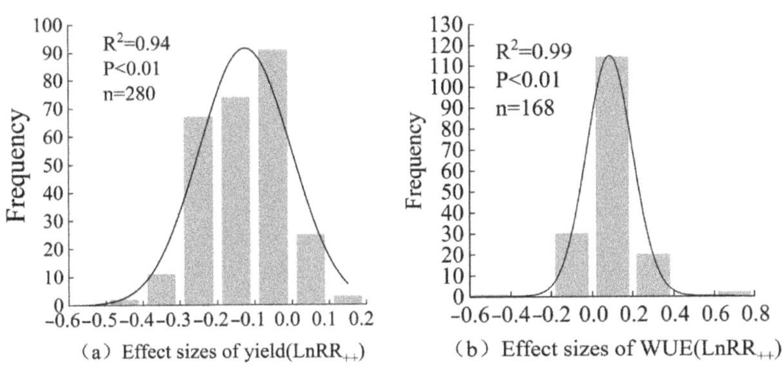


**Figure 8.** Frequency distribution of the effects of deficit irrigation on cotton yield (**a**) and water use efficiency (**b**). *n* is the number of sample sizes, and *p* is the significance level.

### 4.4. Subgroup Analysis

Different farm management practices (irrigation amount, planting mode, variety, etc.) have a large impact on cotton yield and water use efficiency, while many factors such as climate and soil conditions also have an effect on them. Therefore, in this study, the literature data were grouped in multiple ways for irrigation amount [11], drip tape mode, irrigation frequency [38], nitrogen application [18], variety, planting density [37], average annual temperature [12], annual evapotranspiration [48], annual precipitation [49,50], soil texture, and initial soil fertility [51,52] (Table 3).

**Table 3.** Data grouping.

| Categorical Variables | Groups |
|---|---|
| Irrigation amount, % | 40~60% FI; 60~80% FI; 80~100% FI |
| Drip tape mode | one film, one tube, two rows; one film, one tube, four rows; one film, two tubes, four rows; one film, two tubes, six rows; one film, three tubes, six rows |
| Irrigation frequency, times | <10; ≥10 |
| Nitrogen application rate, kg·ha$^{-1}$ | ≤200; 200~300; 300~400; >400 |
| Variety | Xinluzhong series; Xinluzao series; Tahe No. 1; Zhao feng No. 2; Xinken cotton seeds |
| Planting density, plants·ha$^{-1}$ | <240,000; ≥240,000 |
| Annual temperature, °C | ≤10; >10 |
| Annual evaporation, mm | 1500~2000; >2000 |
| Annual precipitation, mm | <60; 60~200; >200 |
| Soil texture | Sandy soil; Sandy loam; Loam; Clay loam soil; Medium loam |
| Soil organic carbon, g·kg$^{-1}$ | ≤5.8; 5.8~11.6 |
| Soil available nitrogen, mg·kg$^{-1}$ | ≤60; 60~120 |

### *4.5. Statistical Analysis*

Microsoft Office 2016 software (Microsoft, Redmond, WA, USA) was used to complete the database and some of the data calculations; MetaWin 2.1 (State University of New York at Stony Brook, New York, NY, USA) was used to complete the meta-analysis of the random-effects model; GraphPad Prism 8 (GraphPad Software, Boston, MA, USA) and Origin 2021 (OriginLab, Northampton, MA, USA) were used to complete the graphs; and R 4.3.2 statistical software (The University of Auckland, Auckland, New Zealand) was used to perform the random forest importance analysis.

## 5. Conclusions

In this study, the combined effects of deficit irrigation on cotton yield and water use efficiency were analyzed in depth. The results showed that deficit irrigation reduced cotton yield by 15.00% and increased water use efficiency by 7.39% compared with full irrigation. The specific effects of different farm management practices and climatic conditions on cotton yield and water use efficiency were further analyzed by subgroup data. The results showed that all three deficit irrigation levels showed significant decreases in cotton yield and significant increases in water use efficiency. Under deficit irrigation, the conditions of a drip tape mode with one film, two tubes, and six rows, an irrigation frequency of ≥10 times, and a nitrogen application of 300~400 kg·ha$^{-1}$ could effectively suppress the cotton yield reduction and significantly improve the cotton water use efficiency; the sowing of the Xinluzhong series of cotton seeds and a crop density of ≥240,000 plants per hectare effectively inhibited the reduction in cotton yield and significantly improved the cotton water use efficiency. In the area with an average annual temperature >10 °C, an annual evaporation >2000 mm, an annual precipitation <60 mm, and loam and sandy soil, deficit irrigation had the smallest inhibition of cotton yield and the largest increase in cotton water use efficiency. Among them, irrigation amount and nitrogen application had the greatest effect on cotton yield and water use efficiency. Rational irrigation based on optimal management practices under conditions of irrigation not less than 90% FI is expected to achieve a win–win situation for both cotton yield and water use efficiency.

**Supplementary Materials:** The following supporting information can be downloaded at https://www.mdpi.com/article/10.3390/plants13050640/s1. Section S1. Classification of variables; Figure S1. Literature Screening Flowchart; Table S1. Description of categorical variables; Table S2. Sample size and significance test for categorical variables; Table S3. Distribution of titles and locations of the literature included in the database.

**Author Contributions:** Writing the initial draft, Q.X. and X.D.; critical review of the work, W.H. and Z.L.; critical review and revision of the work, T.H. and Z.S.; critical revision of the work, Y.Y. and J.C. All authors have read and agreed to the published version of the manuscript.

**Funding:** Bingtuan Science and Technology Program (2022BC009, Z.L.), China Agricultural University-Tarim University Joint Research Fund (1122214, Y.Y.).

**Data Availability Statement:** The data already exist in the manuscript and Supplementary Materials.

**Conflicts of Interest:** The authors declare no conflict of interest.

## References

1. Tan, H.; Ma, Q. A tentative discussion on the constraints on production and development of cotton and countermeasures. In Proceedings of the International Conference on Frontiers in Educational Technologies and Management Sciences, London, UK, 18–20 October 2017. [CrossRef]
2. Yang, Z.N.; Tang, J.J.; Yu, M.; Zhang, Y.; Abbas, A.; Wang, S.D.; Bagadeem, S. Sustainable cotton production through increased competitiveness: Analysis of comparative advantage and influencing factors of cotton production in Xinjiang, China. *Agronomy* **2022**, *12*, 2239. [CrossRef]
3. Announcement of national bureau of statistics on cotton production in 2022. *China Inf. News* 2022. [CrossRef]
4. Xu, F.P.; Li, Y.K.; Ren, S.M. Investigation and discussion of drip irrigation under mulch in Xinjiang uygur autonomous region. *Trans. Chin. Soc. Agric. Eng.* **2003**, *19*, 25–27. [CrossRef]
5. Feng, L.; Dai, J.L.; Tian, L.W.; Zhang, H.J.; Li, W.J.; Dong, H.Z. Review of the technology for high-yielding and efficient cotton cultivation in the northwest inland cotton-growing region of China. *Field Crops Res.* **2017**, *208*, 18–26. [CrossRef]
6. Jiang, Y. China's water scarcity. *J. Environ. Manag.* **2009**, *90*, 3185–3196. [CrossRef]
7. Liu, J.G.; Yang, H.; Gosling, S.N.; Kummu, M.; Florke, M.; Pfister, S.; Hanasaki, N.; Wada, Y.; Zhang, X.X.; Zheng, C.M.; et al. Water scarcity assessments in the past, present, and future. *Earth's Future* **2017**, *5*, 545–559. [CrossRef] [PubMed]
8. Guo, J.K.; Shi, Z.L.; He, M.Q.; Zhang, X.Q.; Zhang, A.M.; Jia, X. Development of water-saving wheat cultivars to limit water shortage in North China—A case study of Hebei Province. *Chin. J. Eco-Agric.* **2010**, *18*, 876–879. [CrossRef]
9. Wang, S.C. Living with nature in harmony: Water resources problem and countermeasures in China. *J. Beijing Norm. Univ. (Nat. Sci.)* **2009**, *35*, 441–445.
10. Li, F.P.; Zhang, G.X.; Dong, L.Q. Studies for impact of climate change on hydrology and water resources. *Sci. Geogr. Sin.* **2013**, *33*, 457–464. [CrossRef]
11. Yu, L.Y.; Zhao, X.N.; Gao, X.D.; Siddique, K.H.M. Improving/maintaining water-use efficiency and yield of wheat by deficit irrigation: A global meta-analysis. *Agric. Water Manag.* **2020**, *228*, 105906. [CrossRef]
12. Li, M.N.; Zhang, Y.L.; Ma, C.Z.; Sun, H.R.; Ren, W.; Wang, X.G. Maximizing the water productivity and economic returns of alfalfa by deficit irrigation in China: A meta-analysis. *Agric. Water Manag.* **2023**, *287*, 108454. [CrossRef]
13. Zhang, D.M.; Luo, Z.; Liu, S.H.; Li, W.J.; Tang, W.; Dong, H.Z. Effects of deficit irrigation and plant density on the growth, yield and fiber quality of irrigated cotton. *Field Crops Res.* **2016**, *197*, 1–9. [CrossRef]
14. Hou, X.H.; Fan, J.L.; Hu, W.H.; Zhang, F.C.; Yan, F.L.; Xiao, C.; Li, Y.P.; Cheng, H.L. Optimal irrigation amount and nitrogen rate improved seed cotton yield while maintaining fiber quality of drip-fertigated cotton in northwest China. *Ind. Crops Prod.* **2021**, *170*, 113710. [CrossRef]
15. He, P.R.; Yu, S.E.; Zhang, F.C.; Ma, T.; Ding, J.H.; Chen, K.W.; Chen, X.; Dai, Y. Effects of soil water regulation on the cotton yield, fiber quality and soil salt accumulation under mulched drip irrigation in southern Xinjiang, China. *Agronomy* **2022**, *12*, 1246. [CrossRef]
16. Wang, H.D.; Wu, L.F.; Cheng, M.H.; Fan, J.L.; Zhang, F.C.; Zou, Y.F.; Chau, H.W.; Gao, Z.J.; Wang, X.K. Coupling effects of water and fertilizer on yield, water and fertilizer use efficiency of drip-fertigated cotton in northern Xinjiang, China. *Field Crops Res.* **2018**, *219*, 169–179. [CrossRef]
17. Xie, H.L.; Yang, Y.P.; Dong, Y.; Wang, T. Analysis on international development trends of alfalfa. *Chin. Bull. Bot.* **2021**, *56*, 740. [CrossRef]
18. Cheng, M.H.; Wang, H.D.; Fan, J.L.; Zhang, S.H.; Wang, Y.L.; Li, Y.P.; Sun, X.; Yang, L.; Zhang, F.C. Water productivity and seed cotton yield in response to deficit irrigation: A global meta-analysis. *Agric. Water Manag.* **2021**, *255*, 107027. [CrossRef]
19. Rodriguez-Galiano, V.F.; Ghimire, B.; Rogan, J.; Chico-Olmo, M.; Rigol-Sanchez, J.P. An assessment of the effectiveness of a random forest classifier for land-cover classification. *ISPRS J. Photogramm. Remote Sens.* **2012**, *67*, 93–104. [CrossRef]
20. Wang, Z.; Fan, B.; Guo, L. Soil salinization after long-term mulched drip irrigation poses a potential risk to agricultural sustainability. *Eur. J. Soil Sci.* **2019**, *70*, 20–24. [CrossRef]

21. Wang, D.W.; Wang, Z.H.; Lv, T.B.; Zong, R.; Zhu, Y.; Zhang, J.Z.; Wang, T.Y. Effects of drip tape modes on soil hydrothermal conditions and cotton yield (*Gossypium hirsutum* L.) under machine-harvest patterns. *PeerJ.* **2021**, *9*, 12004. [CrossRef]
22. Zhao, X.N.; Othmanli, H.; Schiller, T.; Zhao, C.Y.; Sheng, Y.; Zia, S.; Muller, J.; Stahr, K. Water use efficiency in saline soils under cotton cultivation in the tarim river basin. *Water* **2015**, *7*, 3103–3122. [CrossRef]
23. Wang, C.S.; Yang, P.N.; Yu, Y.M.; Di, F.Y.; Huang, F.C. Effect of different pipe arrangements on soil water-salt transport and yield of cotton under mulched drip-irrigation. *Agric. Res. Arid. Areas* **2016**, *34*, 38–45. [CrossRef]
24. Yang, B.F.; Yang, G.Z.; Feng, L.; Han, Y.C.; Lei, Y.P.; Fan, Z.Y.; Wang, Z.B.; Li, Y.B. Effects of deficit irrigation on cotton growth and water use efficiency: A review. *Chin. J. Appl. Ecol.* **2021**, *32*, 1112–1118. [CrossRef]
25. Dağdelen, N.; Başal, H.; Yılmaz, E.; Gvrbvz, T.; Akcay, S. Different drip irrigation regimes affect cotton yield, water use efficiency and fiber quality in western Turkey. *Agric. Water Manag.* **2009**, *96*, 111–120. [CrossRef]
26. Chai, Q.; Gan, Y.T.; Turner, N.C.; Zhang, R.Z.; Yang, C.; Niu, Y.N.; Siddique, K.H.M. Water-saving innovations in Chinese agriculture. *Adv. Agron.* **2014**, *126*, 149–201. [CrossRef]
27. Wang, Q.J.; Wang, W.Y.; Wang, Z.R.; Zhang, J.F.; Li, Y. Determination of technique parameters for saline-alkali soil through drip irrigation under film. *Trans. Chin. Soc. Agric. Eng.* **2001**, *17*, 47–50. [CrossRef]
28. Flexas, J.; Díaz-Espejo, A.; Conesa, M.A.; Coopman, R.E.; Douthe, C.; Gago, J.; Galle, A.; Galmes, J.; Medrano, H.; Ribas-Carbo, M.; et al. Mesophyll conductance to $CO_2$ and rubisco as targets for improving intrinsic water use efficiency in $C_3$ plants. *Plant Cell Environ.* **2016**, *39*, 965–982. [CrossRef]
29. Wang, Z.H.; Yang, B.L.; Xie, X.W.; Wang, Z.Y.; Xie, H.Z.; Dong, X.J. Effects of different irrigation regimes of drip irrigation under plastic film on sugar beet yield and water use efficiency. *Trans. Chin. Soc. Agric. Eng.* **2019**, *35*, 158–166. [CrossRef]
30. Hunsaker, D.J.; Clemmens, A.J.; Fangmeier, D.D. Cotton response to high frequency surface irrigation. *Agric. Water Manag.* **1998**, *37*, 55–74. [CrossRef]
31. Deng, Z.; Zhai, G.L.; Zong, J.; Lv, M.C.; Li, Y.; Feng, J.J.; Cai, J.M.; Zhang, W.Z. Effects of water and nitrogen regulation on root and shoot growth characteristics and yield of cotton in arid area. *Soil Fertil. Sci. China* **2015**, 57–64. [CrossRef]
32. Albrizio, R.; Todorovic, M.; Matic, T.; Stellacci, A.M. Comparing the interactive effects of water and nitrogen on durum wheat and barley grown in a Mediterranean environment. *Field Crops Res.* **2010**, *115*, 179–190. [CrossRef]
33. Wang, X.K.; Fan, J.L.; Xing, Y.Y.; Xu, G.C.; Wang, H.D.; Deng, J.; Wang, Y.F.; Zhang, F.C.; Li, P.; Li, Z.B. The effects of mulch and nitrogen fertilizer on the soil environment of crop plants. *Adv. Agron.* **2019**, *153*, 121–173. [CrossRef]
34. Sun, Z.M.; Wu, Z.J.; Chen, L.J.; Liu, Y.G. Research advances in nitrogen fertilization and its environmental effects. *Chin. J. Soil Sci.* **2006**, *37*, 782–786. [CrossRef]
35. Wang, Z.J.; Dong, Y.M.; Li, Y.Z.; Zhao, Z.Q.; Sun, G.Q.; Xie, Z.M. Breeding and cultivation techniques of cotton variety xinkenmian 1 with early-maturing and high-yield. *China Cotton* **2018**, *45*, 30–31. [CrossRef]
36. Liu, S.H.; Peng, Y.; Peng, X.F.; Luo, Z.; Dong, H.Z. Effects of regulated deficit irrigation and plant density on plant growth and yield and fiber quality of cotton in dry land area. *Cotton Sci.* **2016**, *28*, 184–188. [CrossRef]
37. Chen, Z.K.; Niu, Y.P.; Zhao, R.H.; Han, C.L.; Han, H.Y. The combination of limited irrigation and high plant density optimizes canopy structure and improves the water use efficiency of cotton. *Agric. Water Manag.* **2019**, *218*, 139–148. [CrossRef]
38. Jia, Y.Y.; Huang, W.B.; Yang, B.F.; Li, X.F.; Wang, G.P.; Han, Y.C.; Wang, Z.B.; Li, Y.B.; Feng, L. A meta-analysis of yield and irrigation water productivity of cotton under deficit irrigation in China. *Cotton Sci.* **2023**, *35*, 195–210. [CrossRef]
39. Sawan, Z.M. Climatic variables: Evaporation, sunshine, relative humidity, soil and air temperature and its adverse effects on cotton production. *Inf. Process. Agric.* **2018**, *5*, 134–148. [CrossRef]
40. Zhang, Z.; Ma, G.X.; Hai, X.Y.; Zhang, D.M.; Zhang, G.L.; Wang, F.; Zhang, Q.; Lv, X. Effect of nitrogen fertilization and soil texture on cotton root traits and yields under conventional drip irrigation. *Soils* **2018**, *50*, 622–627. [CrossRef]
41. Han, Z.L.; Zhang, D.J.; Shao, Y.; Wang, W.F.; Hao, Z.Z.; Chen, J.W.; Wang, J.R. Effects of different soil textures on growth and physiological characteristics of wheat. *J. Henan Agric. Sci.* **2015**, *44*, 23–26. [CrossRef]
42. Wu, L.F.; Zhang, F.C.; Zhou, H.M.; Suo, Y.S.; Xue, F.D.; Zhou, J.W.; Liang, F. Effect of drip irrigation and fertilizer application on water use efficiency and cotton yield in north of Xinjiang. *Trans. Chin. Soc. Agric. Eng.* **2014**, *30*, 137–146. [CrossRef]
43. Koudahe, K.; Sheshukov, A.Y.; Aguilar, J.; Djaman, K. Irrigation-water management and productivity of cotton: A review. *Sustainability* **2021**, *13*, 10070. [CrossRef]
44. Ainsworth, E.A. Rice production in a changing climate: A meta-analysis of responses to elevated carbon dioxide and elevated ozone concentration. *Glob. Change Biol.* **2008**, *14*, 1642–1650. [CrossRef]
45. Zhao, X. Study on the Effect of Soil Carbon Sequestration and Emission Reduction and Its Potential of Conservation Tillage Farmland in China Based on Meta-Analysis. Maters' Thesis, China Agricultural University, Beijing, China, 2017.
46. Dieleman, W.I.J.; Luyssaert, S.; Rey, A.; Deangelis, P.; Barton, C.V.M.; Broadmeadow, M.S.J.; Broadmeadow, S.B.; Chigwerewe, K.S.; Crookshanks, M.; Dufrene, E. Soil N modulates soil C cycling in $CO_2$-fumigated tree stands: A meta-analysis. *Plant Cell Environ.* **2010**, *33*, 2001–2011. [CrossRef]
47. Chen, H.H.; Li, X.C.; Hu, F.; Shi, W. Soil nitrous oxide emissions following crop residue addition: A meta-analysis. *Glob. Change Biol.* **2013**, *19*, 2956–2964. [CrossRef]
48. Gu, X.B.; Cai, H.J.; Fang, H.; Li, Y.P.; Chen, P.P.; Li, Y.N. Effects of degradable film mulching on crop yield and water use efficiency in China: A meta-analysis. *Soil Tillage Res.* **2020**, *202*, 104676. [CrossRef]

49. Wang, Q.; Zhai, P.M.; Qin, D.H. New perspectives on 'warming–wetting' trend in Xinjiang, China. *Adv. Clim. Change Res.* **2020**, *11*, 252–260. [CrossRef]
50. Committee on Physical Geography of China, Chinese Academy of Sciences. *Physical Geography of China: Climate*; Science Press: Beijing, China, 1984.
51. Bian, Q.Q.; Wang, Y.N.; Chen, J.J.; Qiao, S.C.; Hu, L.L.; Yin, Y.M.; Yang, X.P.; Yang, Y.F. Effects of potassium application on yield and potassium balance of sweet potato field in China: A meta-analysis. *J. Plant Nutr. Fertil.* **2022**, *28*, 1509–1519. [CrossRef]
52. Lv, C.L.; Chen, Y.H.; He, W.T.; Zhang, S.Y.; Jiang, N.; Fan, D.J.; Yang, H.W.; Yang, D.L.; Zou, G.Y. Response of soil phosphorus availability to long-term application of organic fertilizer under maize cropping system: A meta-analysis. *J. Agro-Environ. Sci.* **2022**, *41*, 2011–2022. [CrossRef]

plants

MDPI

Article

# Safe Farming: Ultrafine Bubble Water Reduces Insect Infestation and Improves Melon Yield and Quality

**Jo-Chi Hung [1,2], Ning-Juan Li [3], Ching-Yen Peng [3], Ching-Chieh Yang [3,*] and Swee-Suak Ko [1,2,*]**

1 Agricultural Biotechnology Research Center, Academia Sinica, Taipei 115, Taiwan; rochehorng@gmail.com
2 Biotechnology Center in Southern Taiwan, Academia Sinica, Tainan 711, Taiwan
3 Season Agricultural Technology Co., Ltd., Tainan 711, Taiwan; ningjuan.farm@gmail.com (N.-J.L.); pchingyen@gmail.com (C.-Y.P.)
* Correspondence: ccyang.farm@gmail.com (C.-C.Y.); sweesuak@gate.sinica.edu.tw (S.-S.K.); Tel.: +886-6-2393639 (C.-C.Y.); +886-6-3032280 (ext. 414) (S.-S.K.)

**Abstract:** Melon pest management relies on the excessive application of pesticides. Reducing pesticide spraying has become a global issue for environmental sustainability and human health. Therefore, developing a new cropping system that is sustainable and eco-friendly is important. This study found that melon seedlings irrigated with ultrafine water containing $H_2$ and $O_2$ (UFW) produced more root hairs, increased shoot height, and produced more flowers than the control irrigated with reverse osmosis (RO) water. Surprisingly, we also discovered that UFW irrigation significantly reduced aphid infestation in melons. Based on cryo-scanning electron microscope (cryo-SEM) observations, UFW treatment enhanced trichome development and prevented aphid infestation. To investigate whether it was $H_2$ or $O_2$ that helped to deter insect infestation, we prepared UF water enrichment of $H_2$ (UF+$H_2$) and $O_2$ (UF+$O_2$) separately and irrigated melons. Cryo-SEM results indicated that both UF+$H_2$ and UF+$O_2$ can increase the density of trichomes in melon leaves and petioles. RT-qPCR showed that UF+$H_2$ significantly increased the gene expression level of the trichome-related gene *GLABRA2* (*GL2*). We planted melons in a plastic greenhouse and irrigated them with ultrafine water enrichment of hydrogen (UF+$H_2$) and oxygen (UF+$O_2$). The SPAD value, photosynthetic parameters, root weight, fruit weight, and fruit sweetness were all better than the control without ultrafine water irrigation. UFW significantly increased trichome development, enhanced insect resistance, and improved fruit traits. This system thus provides useful water management for pest control and sustainable agricultural production.

**Keywords:** jasmonate; muskmelon; nanobubbles water; pest damage; trichome; yield

## 1. Introduction

Ultrafine bubble water (UFW), also known as nanobubble or microbubble water, contains small molecules less than 100 nm in diameter which can carry gases on their surfaces [1]. UFW water penetrates the soil quickly and can be taken up by roots more effectively, enhancing plant growth and development. It has been widely used in crop production [2,3]. It has been reported that plants irrigated with UFW have increased seed germination [4–6], show significantly enhanced rooting and adventitious root development [7], and enhanced root nutrient uptake and increased nutrient use efficiency [8]. Many reports have shown UFW irrigation increased crops yield and quality, such as in rice [9], lettuce [10], tomatoes [11,12], cucumbers [13], muskmelon [14], and strawberries [15].

Ultrafine water enriched with hydrogen prolongs the vase life and the quality of cut flowers [16,17]. Also, it extends the shelf life of kiwifruit [18] and strawberries [19]. It also plays an essential role in plant tolerance to abiotic stresses. Hydrogen pretreatment can improve the salt stress resistance of rice and *Arabidopsis* [20]. It has been reported that molecular hydrogen ($H_2$) has antioxidant activity, removing reactive oxygen species (ROS) and reactive nitrogen species (RNS) and reducing free radical toxicity [21,22].

Melon or muskmelon (*Cucumis melo* L.) is a popular fruit consumed worldwide, and it has a significant economic value in the global market. Melon crops are susceptible to infestation by a variety of insects, such as aphids, thrips, whiteflies, cucumber beetles, and spider mites [23]. Aphids are tiny insects that suck the sap from the plant and can cause stunted growth, curled leaves, transmit viruses, and decrease crop yield. Pathogens such as aphid-transmitted melon cucumber mosaic virus (CMV) and watermelon mosaic virus-2 (WMV-2) [24] cause severe damage to melon plants and lead to reduced yields and decreased fruit quality. Consequently, farmers spray pesticides frequently, which causes a food safety issue. Therefore, the development of a new cropping system for the Sustainability Assessment of Farming and the Environment (SAFE) [25] in melon production is very important.

Trichomes are hair-like outgrowth on the surface of plant organs such as leaves, stems, and flowers. Trichomes act as a physical barrier against herbivorous insects by deterring their ability to feed on the plant and reducing insect movement. Plants with higher trichome density are known to be more resistant to insects [26] and have a strong positive correlation between trichome density and insect resistance [27]. Moreover, glandular trichomes can also produce volatile compounds that are toxic or repellent to insects [28]. Jasmonic acid (JA) is a herbivory-induced hormone that participates in terpene biosynthesis [29]. Methyl jasmonate (Me-JA) treatment significantly enhanced the expression of several monoterpene and sesquiterpene synthases. Research showed that the knockout of an HD-ZIP IV transcription factor (TF), woolly (wo), led to a significant defect in trichomes and a reduction of terpene levels and is associated with insect resistance in tomatoes [30]. Me-JA induced type VI glandular trichome formation on the newly expanding tomato leaves, thus decreasing herbivore insect populations [31]. The gene regulation network controlling trichome development is complex [32]. It is regulated by GLABRA1 (GL1), GLABRA2 (GL2), GLABRA3 (GL3), and TRANSPARENT TESTA GLABRA1 (TTG). The loss of function of these TFs showed glabrous phenotypes [33–39]. GLABRA3 (GL3) is a wound-induced trichome formation acting downstream of the JA signaling pathway [35]. TRIPTYCHON (TRY) is a negative regulator of trichome and root hair development [40]. It has been reported that JAZ is required for jasmonate-meditated glandular trichome development in *Nicotiana benthamiana* [41] and rice [42].

The application of hydrogen in agriculture has attracted much attention over the last decade and has several prospects [43]. However, to the best of our knowledge, there is no report on UFW-induced JA and increasing trichome development. The aims of the present work were as follows: (i) to observe whether UFW improves melon seedling growth and fruit production; (ii) to understand whether UFW enhances pest resistance; and (iii) to know whether UFW affects the JA-pathway and induces trichome development in melons.

## 2. Results

### 2.1. UFW Treatment Improved the Growth of Melon Seedlings

To understand the effects of ultrafine water (UFW) on seed germination and seedling growth, we tested four lines of melon seeds: M1, M2, M3, and Camilla. Forty seeds per line were imbibed in UFW and RO water overnight, and then placed in square Petri dishes containing UFW and reverse osmosis (RO) water as a control (CK), respectively. The dishes were then placed in a growth chamber in the dark and set to a constant temperature of 28 °C. After germination for one day, the melon seeds in UFW produced longer and more root hairs than CK (Figure 1A). The germination rates of UFW-treated seeds of M2, M3, and Camilla were higher than those of CK (Figure 1B). We transplanted the germinated melon seeds into a #104 plug tray filled with peat moss and raised the seedlings in the greenhouse. Treatment with UFW produced more vigorous roots and seedlings than CK at 7 days after transplantation (Figure 1C).

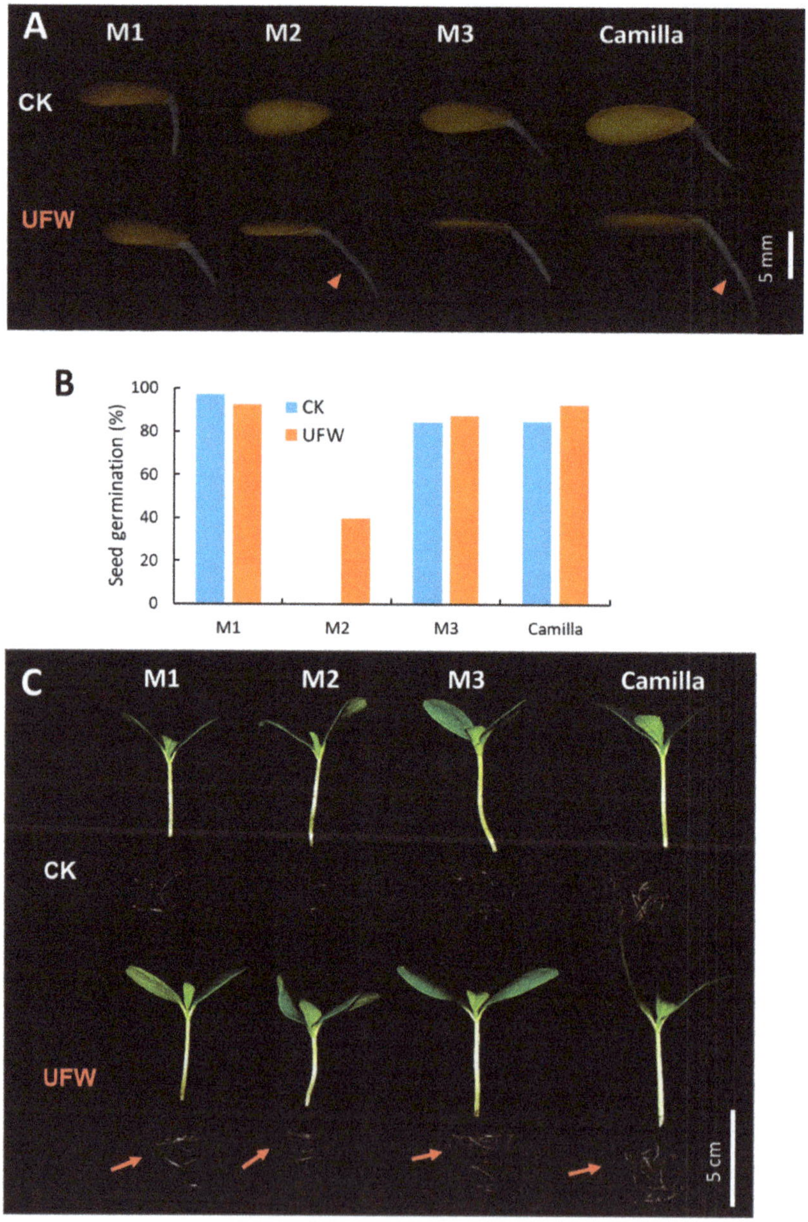


**Figure 1.** Ultrafine water affected seed germination and rooting of melons. (**A**) Effects of ultrafine water (UFW) on melon seed germination. Four melon varieties, each with 40 seeds, were germinated in Petri dishes containing RO water and UFW. Arrows show the presence of root hairs on the root at 1 day after seed germination. (**B**) Germination rate of melon seeds at 7 days after germination. (**C**) Melon seedlings grown in plug trays containing peat moss at 7 days after sowing (DAS). Arrows show vigorous root development.

*2.2. UFW Reduced Aphid Infestation of Seedlings*

In order to understand the effect of UFW on the growth of melon seedlings, we transplanted melon seedlings from plugs into pots and placed them in the same growth chamber to grow, but irrigated them with RO water and UFW. At 14 days after transplanting, the plant heights of the UFW-irrigated melon lines M2 and M3 were higher and produced more flowers than those of CK (Figure 2).

We found that aphids attacked melon seedlings in the growth chamber 14 days after transplantation (DAT). Surprisingly, it was found that the aphid density in the leaves and flower buds of melons irrigated with UFW was lower than that of the control group (Figure 3A–C). We performed cryo-SEM and observed that the trichomes of CK were fewer and drooping, and the mouthparts of aphids could easily reach the leaf surface (Figure 3D). However, the UFW-irrigated melon plants have upright and dense trichomes that interfered with aphid movement and feeding (Figure 3E).

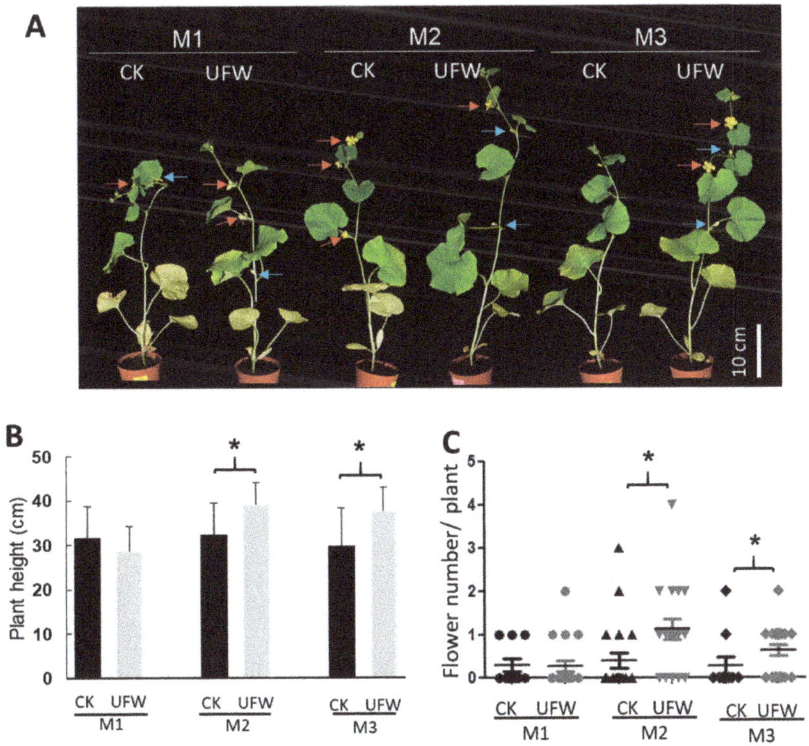


**Figure 2.** UFW irrigation affected the growth of melon seedlings. (**A**) The phenotype of melon potted plants at 14 days after transplantation (DAT). Red arrows indicate the fresh flowers, blue arrows indicate the wilting flowers. Bars, 10 cm. (**B**) Plant height of melons. Error bars represent the standard error of the mean ($n$ = 10–21 per treatment). (**C**) Scatter plot of flower number per plant at 14 DAT. Horizontal lines indicate mean values ($n$ = 10–21). *, significant differences between CK and UFW treatment were determined using Student's $t$-test at $p < 0.05$ (**B**,**C**).

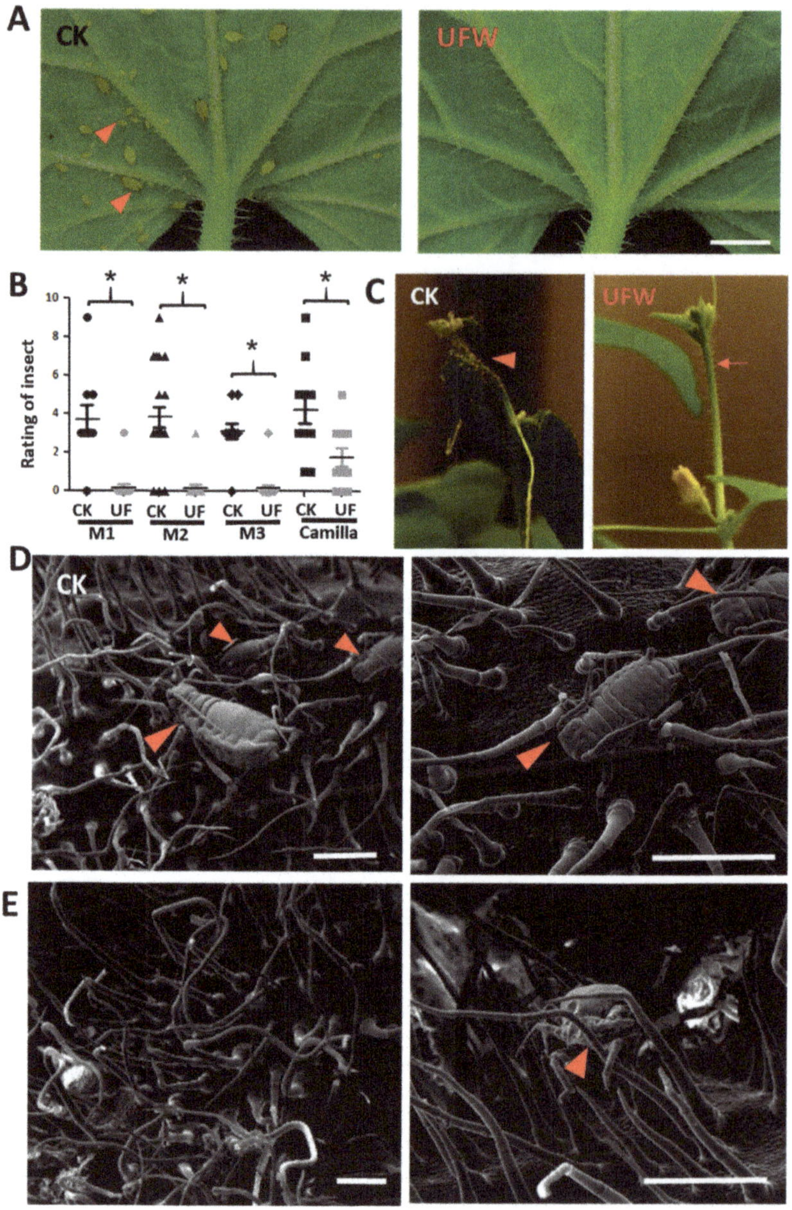


**Figure 3.** UFW irrigation affected aphid infestation on melon seedlings. (**A**) Phenotype of melon leaves attacked by aphids 14 days after transplantation. Scale bars, 2 mm. (**B**) Scatter plot of aphid infestation rating. A rating of 0 indicates no aphids were observed, and 9 indicates a high aphid density. Horizontal lines indicate mean values ($n$ = 10–21). *, significant differences between CK and UFW treatment were determined using Student's $t$-test at $p < 0.05$. (**C**) Aphids attacked the young flower buds of melon (arrowhead). Trichomes development after UFW treatment (arrow). (**D**) Cryo-SEM showed aphid infestation on flower buds of CK (**D**) and UFW (**E**). The arrowheads point to the aphids. Scale bars, 500 μm (**D**,**E**).

### 2.3. *Effect of Hydrogen-Rich (UF+$H_2$) or Oxygen-Rich (UF+$O_2$) Ultrafine Water on Trichome Development*

Our previous experiments showed that UFW containing both $H_2$ and $O_2$ significantly increased trichome density and deterred aphid infestation (Figure 3). We were, thus, interested to know whether this phenomenon was due to the effect of $H_2$ or $O_2$ molecules. Hence, we prepared the UF water enrichment of pure hydrogen (UF+$H_2$), pure oxygen (UF+$O_2$), and RO water (CK), and irrigated melon cultivar "Camilla", respectively. Melon plants irrigated with UF+$H_2$ produced taller and denser trichomes on the petioles, leaf veins, leaves, and leaf tip compared with CK. Meanwhile, UF+$O_2$ irrigated plants had longer trichomes than CK (Figure S1). Under a dissecting microscope, we could observe that melons irrigated with UF+$H_2$ produced longer and denser trichomes on the mid-rib than those irrigated with UF+$O_2$ or CK. We took pictures and measured trichome density in petioles and found that UF+$H_2$ and UF+$O_2$ irrigation significantly increased trichome density (Figure 4A,D).

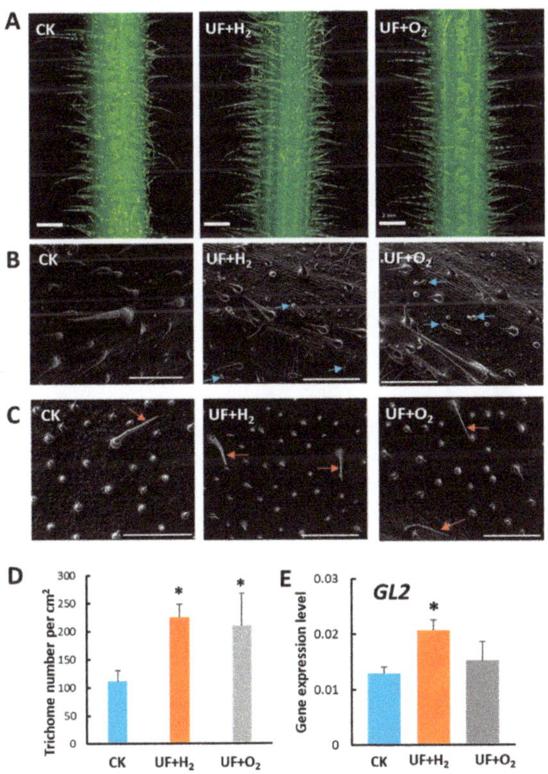


**Figure 4.** Hydrogen-rich or oxygen-rich ultrafine water irrigation affected the development of trichomes in melon cv. "Camilla". (**A**) Dissecting microscope observation of the development of trichomes in melon petioles after irrigation with ultrafine water enrichment of hydrogen (UF+$H_2$), oxygen (UF+$O_2$), and RO water (Ck), respectively. Bars, 2 mm. (**B**) Cryo-scanning electron microscope (cryo-SEM) showing trichomes on the midribs of the melons. Blue arrows indicate the presence of granular trichomes. Bars, 500 µm. (**C**) Cryo-SEM showed the development of trichomes on the abaxial of newly established young leaves of melon (red arrows). Bars, 500 µm. (**D**) Trichome density in melon petioles irrigated with RO water, UF+$H_2$, and UF+$O_2$, $n$ = 3 to 6. (**E**) RT-qPCR showed *GLABRA2* (*GL2*) gene expression patterns in young melon leaves irrigated with UF+$H_2$, UF+$O_2$, and RO water control (CK). *, significant differences between CK and UFW treatment were determined using Student's $t$-test at $p < 0.05$ (**D**,**E**).

### 2.4. Enrichment of Hydrogen-Induced Jasmonic Acid Accumulation

We detected the JA and MeJA contents in melon leaves and found that JA was significantly increased (6.9-fold) under UF+$H_2$ treatment compared to the CK (Figure 5A). Although UF+$H_2$ and UF+$O_2$ slightly increased the MeJA content, there was no statistically significant difference at $p < 0.05$ (Figure 5B). Our RT-qPCR showed that *JASMONATE ZIM DOMAIN PROTEIN* (*JAZ*) and *JA carboxyl methyltransferase* (*JMT*) were upregulated after UF+$H_2$ treatment but there was no statistically significant difference at $p < 0.05$ (Figure 5C).

### 2.5. Effect of UF Water on Photosynthesis Parameters, Fruit Yield and Quality

We grew melons in a greenhouse to understand the effect of UF+$H_2$ and UF+$O_2$ irrigation on melon fruit production. All crop management practices were similar except for irrigation water use. We measured chlorophyll content and photosynthesis parameters using a SPAD meter and Li-600 Porometer/Fluorometers meter, respectively. The results showed that $H_2$ and $O_2$ enrichment significantly increased SPAD values (representing chlorophyll content) and stomatal conductance (gsw). UF+$O_2$ increased the quantum yield of PSII calculated from fluorescence (ΦPSII) and electron transport rate (ETR) (Figure 6).

During the growth process of melons in the late harvest stage, melon plants irrigated with UF+$H_2$ or UF+$O_2$ retained more green leaves (Figure 7A, arrows) than CK irrigated with tap water. After irrigation with UFW, the root system of the melon plants developed more vigorously, and the fresh weight and dry weight of the roots were significantly higher than the control (Figure 7B–D). UF+$H_2$ irrigation increased the melon fruit size and weight (Figure 7E,F). Furthermore, UF+$H_2$ and UF+$O_2$ irrigation were both able to increase melon fruit sweetness (Figure 7G).

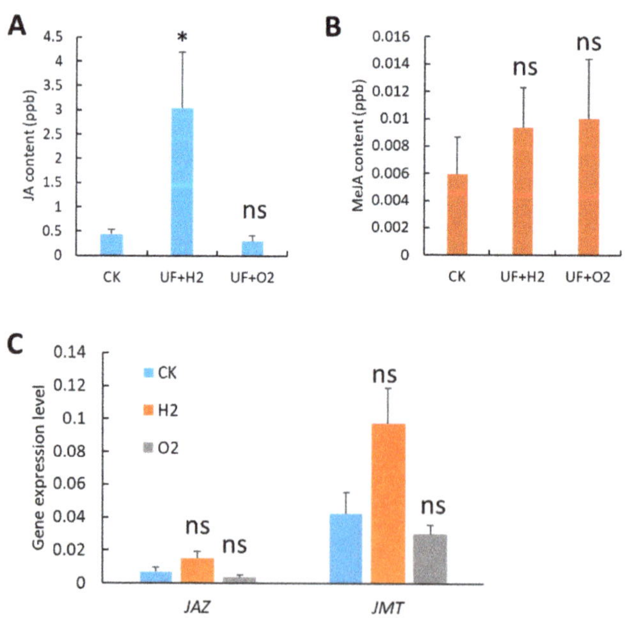


**Figure 5.** Melon irrigated with hydrogen- and oxygen-rich ultrafine water altered jasmonic acid (JA) and methyl-JA (MeJA) contents, and gene expression patterns. (**A**) JA content. (**B**) MeJA content. (**C**) The gene expression level of *JASMONATE ZIM DOMAIN PROTEIN* (*JAZ*) and *JA carboxyl methyltransferase* (*JMT*). The gene expression level was normalized to two housekeeping genes: *Actin* (MELO3C023264) and *ADP ribosylation factor 1* (*ADP*, MELO3C023630). Error bars represent the standard error of the mean ($n = 3$). Student's $t$-test was used to find the significant difference between CK and UF+$H_2$ or UF+$O_2$ treatment. *, $p < 0.05$; ns, not significant.

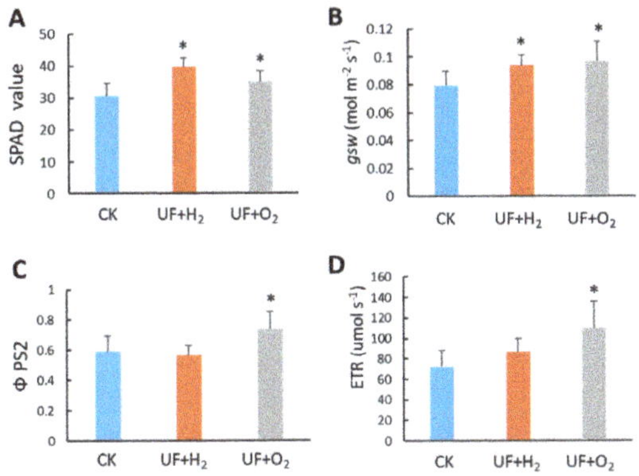


**Figure 6.** Ultrafine water irrigation affected the photosynthesis capacity of melons. (**A**) Chlorophyll content in melons. The SPAD value was measured on the 4th leaf at the late stage of fruit maturity. $n = 4$ plants. The Li600 Porometer/Fluorometers meter detected the photosynthesis parameters of (**B**) stomatal conductance (gsw); (**C**) ΦPSII, the quantum yield of PSII calculated from fluorescence; and (**D**) the electron transport rate (ETR) of L1 melon leaves. Student's *t*-test was used to find the significant difference between UFW and the regular tap water (CK). *, $p < 0.05$. Error bars represent the standard error of the mean ($n = 4$).

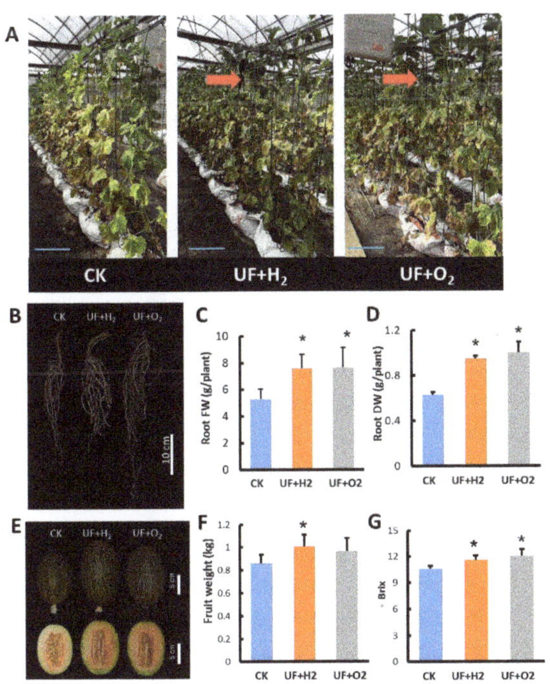


**Figure 7.** UFW irrigation affected fruit weight and sweetness of melon cv. "Camilla". (**A**) Melons were planted in a greenhouse. Photo taken 42 days after pollination. Bars, 20 cm. (**B**) Root morphology at harvest stage. (**C**) Root fresh weight of each plant. (**D**) Root dry weight per plant. (**E**) Melon fruits at 5 days after harvest. (**F**) Average fruit weight of melon. (**G**) The sweetness of melon fruits. UF+$H_2$, hydrogen-rich ultrafine water irrigation. UF+$O_2$, oxygen-rich ultrafine water irrigation. CK, irrigated with tap water. Bars, standard deviation of 22 plants. Student's *t*-test was used to find significant difference between CK and UF+$H_2$ or UF+$O_2$ treatment. *, $p < 0.05$.

## 3. Discussion

This study showed that UFW irrigation significantly improved melon seed germination, seedling growth, and enhanced root development. Our greenhouse experiment also showed that the UFW enrichment of hydrogen (UF+$H_2$) or oxygen (UF+$O_2$) produced higher root biomass than the control without UFW treatment. The robust root system contributed to plant growth and development. Previous studies have demonstrated that hydrogen-rich water increased auxin and GA3 biosynthesis and enhanced root development [44]. It regulated heme oxygenase-1/carbon monoxide pathways and increased root development [13]. Some researchers have suggested that hydrogen has antioxidant properties, which can help to reduce oxidative stress in plants, improve nutrient uptake by plants, and improve overall plant growth and development [6,8,10,13,22]. Hydrogen molecules are not easy to apply. Nonetheless, water electrolysis produces hydrogen gas, which is easily fused into ultrafine water and provides a good solution for agricultural applications.

Our data indicated that UFW positively affects crop production compared to the previous reports on cucumber [45] and maize [46]. This study showed that hydrogen enrichment water is better than oxygen. Compared to $H_2$-rich water, studies of $O_2$-rich water on plant growth are rare. Recently, a report highlighted that the nanobubble water enrichment of $O_2$ improves soil structure and microbial diversity, thereby increasing tomato yield [47]. The UFW enrichment of $O_2$ could enhance oxygen delivery to soil and promote aerobic respiration [48]. Some reports indicated that high $O_2$ content in UFW does not necessarily lead to better crop performance. In a previous study on maize treated with dissolved oxygen (DO) concentrations of 10, 20, and 30 mg/L, a moderate DO concentration of 20 mg/L had the highest root growth and yield [46]. In this study, we grew melons in a well-ventilated soilless medium of peat moss, which may reduce the positive effects of UFW. A more significant impact would be expected if the UFW irrigated a high-density clay field with poor aeration. It is hypothesized that UF+$O_2$ may benefit plant survival under flooding-induced hypoxic conditions.

We observed that melons irrigated with UF+$H_2$ or UF+$O_2$ retained green leaves, and in the later stages of melon development, the leaves contained higher chlorophyll. This is a beneficial trait that can increase the rate of photosynthesis and produce more assimilates for fruit development, thereby increasing the fruit weight and sweetness of the melon (Figure 7). As reported previously, hydrogen enrichment water increased strawberry fruit flavor and quality [15].

Non-glandular trichomes have been reported to play a role in the mechanical defense against insects, while glandular trichomes secrete metabolites such as terpene [49]. In this study, glandular trichomes were found in the midrib of melon leaves after irrigation with UFW (UF+$H_2$ and UF+$O_2$) but not in the control (Figure 4). To our knowledge, this is the first report showing that ultrafine water can increase trichome density and induce glandular trichome development. We discovered that UF+$H_2$ can induce JA-biosynthesis genes and enhance root and trichome development. Trichomes deter herbivores and reduce insect damage. In the future, it will be worth investigating what secondary metabolites were induced after UFW treatment.

JA is known to be involved in trichome development [30,31]. In this study, we found that a trichome initiation marker gene *GL2* was significantly upregulated in young leaves after UF+$H_2$ treatment (Figure 4E), further supporting the notion that hydrogen may induce JA and enhance trichome initiation to prevent herbivory infestation and improve plant growth. Our data show that irrigating melon with UFW ($H_2$ and $O_2$) improved the resistance of melons to aphid infestation (Figure 3). Furthermore, we found that the UFW enrichment of $H_2$ plays an important role in trichome development due to the upregulation of JA biosynthesis genes and increased JA accumulation in the plants irrigated with the UFW enrichment of $H_2$ (Figure 5). This enhanced trichome formation and deterred insects or interfered with their feeding and growth, making the plants less susceptible to damage. Furthermore, reducing pest infestation will reduce systemic viral infection problems. Overall, these data indicate that hydrogen-rich water upregulates JA-pathway marker genes and increases melon trichome development, supported by the upregulation of *GL2* transcripts. A high density of globular trichomes may help to resist aphids. In the future, it will be worth conducting more extensive research on the underlying mechanisms through which UFW induces JA, enhances trichome development, and confers insect resistance.

We demonstrated that UFW increased trichome density and prevented aphid feeding (Figures 3 and 4), induced more flower development, increased fruit weight, and increased the sweetness of melon fruits (Figure 7). All these beneficial effects contribute to melon crop production. These results demonstrate that hydrogen-rich water management has excellent potential as a natural, non-toxic pest and disease control treatment in crop plants. This might reduce pesticide spraying and improve food security. The UFW-induced JA response facilitates the establishment of a natural defense system in crop plants against insect attacks. Therefore, agriculture is safe without relying on pesticides. It is an environmentally friendly agricultural practice that increases crop yields and fruit quality and reduces pest damage. In addition, UFW has hydrophobic and surface charge properties that enhance the release and absorption of soil nutrients, thereby reducing fertilizer demand [9]. This will reduce the carbon footprint in crop production and enable sustainable agricultural production. Some studies emphasize enhancing insect resistance through genetic engineering trichome genes, but due to consumer concern about the biosafety of the genetically modified organisms (GMO), here we proposed a hydrogen-rich UFW irrigation method which would be more acceptable to consumers as it does not include GMO. Furthermore, hydrogen-rich water is safe and easy to use [50].

## 4. Materials and Methods

### 4.1. Ultrafine Water Preparation

We used a Hydrogen–Oxygen Ultrafine bubble system model HOU-3 (Season Agricultural Technology Co., Ltd., Tainan, Taiwan) to prepare ultrafine bubble water (UFW) enriched with hydrogen or oxygen. Hydrogen-rich water (UF+$H_2$) was prepared using reverse osmosis (RO) water to obtain 1000 ppb $H_2$ molecules, and oxygen-rich water (UF+$O_2$) water was prepared using RO water to contain 10 mg/L $O_2$. Hydrogen concentration was determined with a portable Dissolved Hydrogen Meter (Trustlex Co., Ltd., ENH-1000, Tokyo, Japan). Dissolved oxygen (DO) content was determined using a Dissolved Oxygen Meter (Lutron Co., Ltd., PDO-519, Taipei, Taiwan).

### 4.2. Plant Materials and Growth Conditions

Melon (*Cucumis melo* L.) seeds were provided by Known-you Seed Co., Ltd. (Kaohsiung, Taiwan). Melon lines 6792T-744 (M1), 6792T-LD (M2), 6792T-LQ (M3), and a popular and high-quality melon cultivar "Camilla" were used in this study. To test the effect of UFW (containing $H_2$ and $O_2$) on seed germination, a total of 40 melon seeds per line were imbibed in RO water and UFW overnight and sown on a wetted tissue paper in a $125 \times 125 \times 20$ mm square Petri dish (SPL, Gyeonggi-do, Republic of Korea). Seed germination rates were recorded one day after sowing. Germination was defined as when the root length was over half the seed length. The germinated seeds were transferred into a #104 plug tray containing peat moss (Known-you Seed). Then, the plug seedling was transplanted into a plastic pot (7.5 cm width $\times$ 7.5 cm height) containing 140 mL peat moss. The melon seedlings varied from 10 to 21 per line and were raised in a greenhouse at a temperature of $24 \pm 4$ °C.

The commercial melon cultivar "Camilla" was used to evaluate the effect of UF+H2, UF+$O_2$, and a tap water control on the growth and fruit production in a plastic greenhouse of the Biotechnology Center in Southern Taiwan (AS-BCST) (23°06′14.4″ N 120°17′31.2″ E). We transplanted two melon seedlings at the four-leaf stage into a package of 80 L peat moss (Known-you Seed). A total of 22 seedlings were planted for each treatment. In the plastic greenhouse, we used tap water to prepare hydrogen-rich water (UF+$H_2$) and oxygen-rich water (UF+$O_2$), and the control was ordinary tap water. Water was supplied using a drip irrigation system. In this study, melons were grown vertically to keep the fruit cleaner and healthier. Flowers were pollinated at 13 to 16 nodes, retaining one fruit per plant. When the number of nodes on the mother vine reached 26, we removed the top growth point. According to the weather and soil moisture conditions, we supplied the appropriate water amount (500 mL~1000 mL per day per plant) during planting to ensure good plant growth and avoid fruit cracking in the later stages of fruit maturity, which may reduce fruit quality.

### 4.3. Insect Materials

Cotton aphids were collected from melon plants in a field of Tainan, Taiwan (23°14′56.0″ N 120°19′32.5″ E). The insects were reared in Camilla melon pot plants and placed in an insect cage. The cage was placed in a greenhouse at a temperature of 24 ± 4 °C. M1, M2, M3, and Camilla melon pot seedlings at four leave stages, of which each genotype has 10 to 21 seedlings, were placed in a walk-in growth chamber (25 ± 2 °C, 12 h photoperiod). Four aphids were released onto each melon plant. Each treatment has more than ten plants. The pot seedlings were irrigated with RO water and UFW, respectively. We assessed aphid population development at 14 days after infestation, and the number of aphids was rated per plant. A rating of 0 indicates no aphids were observed, and 9 indicates a high aphid density.

### 4.4. Observation of Trichome Density

To quantify trichome number, we collected the newly established leaves (L1) from 3 to 6 plants of the melon 'Camilla', excised the middle part of the petioles, and obtained images of trichomes under a dissecting microscope (Leica S9D, Hamburg, Germany) at 10× magnification. We counted the number of trichomes and calculated the average number of trichomes per square centimeter.

### 4.5. Cryo-Scanning Electron Microscopy

The first newly developed melon leaves (L1) were used to observe trichome development and aphid infestation on melon seedlings. The abaxial sides of the leaves were observed via cryogenic scanning electron microscopy (cryo-SEM), using a FEI Quanta 200/Quorum PP2000TR FEI 2007 high-resolution SEM at the Plant Cell Biology Core Laboratory in the Institute of Plant and Microbial Biology, Academia Sinica, Taipei, Taiwan. Briefly, leaf section samples containing insect tarsals were loaded onto frozen specimen holders and cryo-fixed in slush nitrogen (−210 °C), then rapidly transferred to a cryo-unit in the frozen state. Samples were imaged by cryo-SEM at an accelerating voltage of 20 kV.

### 4.6. Total RNA Isolation and Real-Time PCR

Total RNAs of melon leaf tissues were isolated using the TRIzol Plus RNA purification kit (Thermo Fisher Scientific, San Jose, CA, USA), treated with DNase (Promega, Madison, WI, USA), and the first-strand cDNA was synthesized using an M-MLV Reverse Transcriptase cDNA synthesis kit (Promega). Quantitative real-time PCR (RT–qPCR) reactions were performed using 2× KAPA SYBR FAST master mix (KAPA Biosystems, Wilmington, MA, USA), and performed for 35 cycles in a volume of 15 µL using complementary DNA reverse transcribed from 2 ng of total RNA. The RT-qPCR was performed using a CFX96 Real-Time PCR detection system (Bio-Rad, Hercules, CA, USA) and quantification analysis was performed using the CFX Manager Software version 3.1 (Bio-Rad). *β-Actin* (MELO3C023264) and *ADP ribosylation factor 1* (*ADP*, MELO3C023630) were used as reference genes for normalization. The primers used in this study are listed in Supplementary Table S1. Each sample had three biological replicates.

### 4.7. Detected JA and Methyl-JA Content

Melon leaf samples were snap-frozen in liquid nitrogen and ground to a fine powder with a pestle and mortar. The powder (500 mg) was suspended in 2.5 mL of ice-cold 50%

MeOH (−20 °C). The extracts were vortexed for 5 min and then centrifuged at 3 °C at 4000 rpm for 15 min. The supernatants were collected and the pellets were re-extracted using 500 µL ice-cold 50% MeOH (−20 °C). The supernatants were combined and applied to Sep-Pak Vac 3 mL C18 200 mg cartridges (Waters, Milford, CT, USA) for sample clean-up and concentration. The cartridges were conditioned with 2 mL of MeOH and equilibrated with 2 mL of water, then 3 mL of sample was applied to the C18 cartridge. The Solid Phase Extraction (SPE) cartridges were eluted with 1 mL of 100% acetonitrile to release the MeJA, followed by a 1 mL clean-up with MeOH. The eluates from the cartridges were filtered through 0.22 µm filters, transferred into chromatography vials, and detected using an ultra-performance liquid chromatography high-resolution tandem mass spectrometry (UPLC-HRMS/MS) (Thermo Fisher Scientific). UPLC separation was carried out on a BEH C18 column (2.5 × 100 mm, 1.7 µm, Waters) at a 0.3 mL/min flow rate. The column oven temperature was 40 °C. The gradient program was applied using 0.1% formic acid (FA) in water (phase A), and 0.1% FA in ACN (phase B). The sample injection volume was 20 µL.

### 4.8. SPAD Value and Photosynthesis Rate of Melon

A non-destructive portable chlorophyll (Chl) meter SPAD-502 Plus (Konica Minolta Optics, Osaka, Japan) was used to measure the chlorophyll content of the fourth newly established melon leaf (L4) at 45 days after pollination. To determine the photosynthesis rate, we used a Li-600 Porometer/Fluorometers meter (Li-COR, Lincoln, NE, USA) to measure stomatal conductance (gsw), electron transport rate (ETR), and the quantum yield of PSII calculated from fluorescence parameters (ΦPSII) of the L1 leaves at 23 days after pollination.

### 4.9. Statistical Analysis

Student's *t*-test was used to compare the difference between CK and UF+$H_2$ or UF+$O_2$ treatment among different genotypes. *p* values of less than 5% were considered statistically significant.

## 5. Conclusions

This study shows that ultrafine hydrogen-rich water can significantly promote root development and increase the fruit yield and the sweetness of melons. As shown in the paper, we found that ultrafine water (UFW) significantly increased JA accumulation, increased *GL2* gene expression, and may induce trichome development to reduce insect infestation. UFW is easy to use. Farmers can incorporate UFW into their irrigation systems. This could be a promising non-pesticide crop protection method. It is beneficial to high-economic crops and organic farming. Farmers can work towards safer and more sustainable melon production systems. For basic research, the underlying molecular mechanisms of how ultrafine water promotes insect resistance deserve detailed investigation.

**Supplementary Materials:** The following supporting information can be downloaded at: https://www.mdpi.com/article/10.3390/plants13040537/s1, Supplementary Figure S1. Morphology of trichomes in different tissues of melon after ultrafine water irrigation; Supplementary Table S1. List of primers used in this study.

**Author Contributions:** S.-S.K. and C.-C.Y. designed and supervised the study; J.-C.H., N.-J.L. and C.-Y.P. performed the experiments; J.-C.H. analyzed and prepared the data. S.-S.K. and J.-C.H. wrote the article. All authors have read and agreed to the published version of the manuscript.

**Funding:** This study was supported by a cooperative research program between Academia Sinica Agricultural Biotechnology Research Center and Season Agricultural Technology Co., Ltd., funding by Season Agricultural Technology Co., Ltd. (14T-1090311-1C).

**Data Availability Statement:** Data supporting the findings of this study are available within this paper and its Supplementary Data published online.

**Acknowledgments:** We thank Known-you Seed Co., Ltd. for providing melon seeds for this study; the genetically modified greenhouse of the Biotechnology Center of Southern Taiwan (AS-BCST) for providing greenhouse and technical support; the Mass Analysis Core facility at AS-BCST for the determination of JA and Me-JA contents; Chyi-Chuann Chen for taking cryo-SEM pictures; and Miranda Loney for language editing.

**Conflicts of Interest:** Authors Ning Juan Li, Ching Yen Peng, and Ching Chieh Yang, were employed by the company Season Agricultural Technology Co., Ltd., Tainan, Taiwan. The remaining authors declare that the research was conducted in the absence of any commercial or financial relationships that could be construed as a potential conflict of interest.

## References

1. Ushikubo, F.Y.; Furukawa, T.; Nakagawa, R.; Enari, M.; Makino, Y.; Kawagoe, Y.; Shiina, T.; Oshita, S. Evidence of the existence and the stability of nano-bubbles in water. *Colloids Surf. A* **2010**, *361*, 31–37. [CrossRef]
2. Li, L.N.; Zeng, Y.; Cheng, X.; Shen, W.B. The applications of molecular hydrogen in horticulture. *Horticulturae* **2021**, *7*, 513. [CrossRef]
3. Takahata, J.; Takaki, K.; Satta, N.; Takahashi, K.; Fujio, T.; Sasaki, Y. Improvement of growth rate of plants by bubble discharge in water. *Jpn. J. Appl. Phys.* **2014**, *54*, 01AG07. [CrossRef]
4. Lee, S.Y.; Jeong, S.; Cho, A.R.; Shim, M.S.; Chung, Y.K.; Kim, Y.J. Germination and seedling growth response of sprouts and leafy vegetables after applying oxygen nanobubble water. *J. People Plants Environ.* **2021**, *24*, 607–617. [CrossRef]
5. Oshita, S.; Boerzhijin, S.; Kameya, H.; Yoshimura, M.; Sotome, I. Promotion effects of ultrafine bubbles/ nanobubbles on seed germination. *Nanomaterials* **2023**, *13*, 1677. [CrossRef] [PubMed]
6. Liu, S.; Oshita, S.; Makino, Y.; Wang, Q.; Kawagoe, Y.; Uchida, T. Oxidative Capacity of Nanobubbles and its effect on seed germination. *ACS Sustain. Chem. Eng.* **2015**, *4*, 1347–1353. [CrossRef]
7. Zhu, Y.; Liao, W. A positive role for hydrogen gas in adventitious root development. *Plant Signal Behav.* **2016**, *11*, e1187359. [CrossRef]
8. Xue, S.; Marhaba, T.; Zhang, W. Nanobubble Watering Affects Nutrient Release and Soil Characteristics. *ACS Agric. Sci. Technol.* **2022**, *2*, 453–461. [CrossRef]
9. Wang, Y.; Wang, S.; Sun, J.; Dai, H.; Zhang, B.; Xiang, W.; Hu, Z.; Li, P.; Yang, J.; Zhang, W. Nanobubbles promote nutrient utilization and plant growth in rice by upregulating nutrient uptake genes and stimulating growth hormone production. *Sci. Total Environ.* **2021**, *800*, 149627. [CrossRef] [PubMed]
10. Park, J.S.; Kurata, K. Application of microbubbles to hydroponics solution promotes lettuce growth. *Horttechnology* **2009**, *19*, 212–215. [CrossRef]
11. Xue, S.; Gao, J.; Liu, C.; Marhaba, T.; Zhang, W. Unveiling the potential of nanobubbles in water: Impacts on tomato's early growth and soil properties. *Sci. Total Environ.* **2023**, *903*, 166499. [CrossRef]
12. Marcelino, K.R.; Ling, L.; Wongkiew, S.; Nhan, H.T.; Surendra, K.; Shitanaka, T.; Lu, H.; Khanal, S.K. Nanobubble technology applications in environmental and agricultural systems: Opportunities and challenges. *Crit. Rev. Environ. Sci. Technol.* **2023**, *53*, 1378–1403. [CrossRef]
13. Lin, Y.; Zhang, W.; Qi, F.; Cui, W.; Xie, Y.; Shen, W. Hydrogen-rich water regulates cucumber adventitious root development in a heme oxygenase-1/carbon monoxide-dependent manner. *J. Plant Physiol.* **2014**, *171*, 1–8. [CrossRef]
14. He, J.; Liu, Y.Z.; Wang, T.Z.; Chen, W.J.; Liu, B.; Zhou, Y.P.; Li, Y.K. Effects of nanobubble in subsurface drip irrigation on the yield, quality, irrigation water use efficiency and nitrogen partial productivity of watermelon and muskmelon. *Int. Agrophysics* **2022**, *36*, 163–171. [CrossRef] [PubMed]
15. Li, L.; Wang, J.; Jiang, K.; Kuang, Y.; Zeng, Y.; Cheng, X.; Liu, Y.; Wang, S.; Shen, W. Preharvest application of hydrogen nanobubble water enhances strawberry flavor and consumer preferences. *Food Chem.* **2022**, *377*, 131953. [CrossRef] [PubMed]
16. Li, L.; Yin, Q.; Zhang, T.; Cheng, P.; Xu, S.; Shen, W. Hydrogen nanobubble water delays petal senescence and prolongs the vase life of cut carnation (*Dianthus caryophyllus* L.) flowers. *Plants* **2021**, *10*, 1662. [CrossRef] [PubMed]
17. Ren, P.J.; Jin, X.; Liao, W.B.; Wang, M.; Niu, L.J.; Li, X.P.; Xu, X.T.; Zhu, Y.C. Effect of hydrogen-rich water on vase life and quality in cut lily and rose flowers. *Hortic. Environ. Biotechnol.* **2017**, *58*, 576–584. [CrossRef]
18. Hu, H.; Li, P.; Wang, Y.; Gu, R. Hydrogen-rich water delays postharvest ripening and senescence of kiwifruit. *Food Chem.* **2014**, *156*, 100–109. [CrossRef] [PubMed]
19. Jin, Z.W.; Liu, Z.Y.; Chen, G.M.; Li, L.N.; Zeng, Y.; Cheng, X.; Pathier, D.; Xu, G.Y.; Shen, W.B. Molecular hydrogen-based irrigation extends strawberry shelf life by improving the synthesis of cell wall components in fruit. *Postharvest Biol. Technol.* **2023**, *206*, 112551. [CrossRef]
20. Xie, Y.; Mao, Y.; Lai, D.; Zhang, W.; Shen, W. $H_2$ enhances arabidopsis salt tolerance by manipulating ZAT10/12-mediated antioxidant defence and controlling sodium exclusion. *PLoS ONE* **2012**, *7*, e49800. [CrossRef] [PubMed]
21. Russell, G.; Zulfiqar, F.; Hancock, J.T. Hydrogenases and the role of molecular hydrogen in plants. *Plants* **2020**, *9*, 1136. [CrossRef]
22. Iijima, M.; Yamashita, K.; Hirooka, Y.; Ueda, Y.; Yamane, K.; Kamimura, C. Ultrafine bubbles alleviated osmotic stress in soybean seedlings. *Plant Prod. Sci.* **2022**, *25*, 218–223. [CrossRef]

23. Webb, S.E. *Insect Management for Cucurbits (Cucumber, Squash, Cantaloupe, and Watermelon): ENY-460/IN168, Rev. 9/2005*; EDIS: Graz, Austria, 2005.
24. Luis Alonso-Prados, J.; Luis-Arteaga, M.; Alvarez, J.M.; Moriones, E.; Batlle, A.; Laviña, A.; García-Arenal, F.; Fraile, A. Epidemics of aphid-transmitted viruses in melon crops in Spain. *Eur. J. Plant Pathol.* **2003**, *109*, 129–138. [CrossRef]
25. Van Cauwenbergh, N.; Biala, K.; Bielders, C.; Brouckaert, V.; Franchois, L.; Cidad, V.G.; Hermy, M.; Mathijs, E.; Muys, B.; Reijnders, J. SAFE—A hierarchical framework for assessing the sustainability of agricultural systems. *Agric. Ecosyst. Environ.* **2007**, *120*, 229–242. [CrossRef]
26. Dalin, P.; Ågren, J.; Björkman, C.; Huttunen, P.; Kärkkäinen, K. Leaf trichome formation and plant resistance to herbivory. In *Induced Plant Resistance to Herbivory*; Springer: Berlin/Heidelberg, Germany, 2008; pp. 89–105.
27. Handley, R.; Ekbom, B.; Ågren, J. Variation in trichome density and resistance against a specialist insect herbivore in natural populations of *Arabidopsis thaliana*. *Ecol. Entomol.* **2005**, *30*, 284–292. [CrossRef]
28. Lin, S.Y.; Trumble, J.T.; Kumamoto, J. Activity of volatile compounds in glandular trichomes of Lycopersicon species against two insect herbivores. *J. Chem. Ecol.* **1987**, *13*, 837–850. [CrossRef] [PubMed]
29. Hong, G.J.; Xue, X.Y.; Mao, Y.B.; Wang, L.J.; Chen, X.Y. Arabidopsis MYC2 interacts with DELLA proteins in regulating sesquiterpene synthase gene expression. *Plant Cell* **2012**, *24*, 2635–2648. [CrossRef] [PubMed]
30. Hua, B.; Chang, J.; Wu, M.; Xu, Z.; Zhang, F.; Yang, M.; Xu, H.; Wang, L.J.; Chen, X.Y.; Wu, S. Mediation of JA signalling in glandular trichomes by the woolly/SlMYC1 regulatory module improves pest resistance in tomato. *Plant Biotechnol. J.* **2021**, *19*, 375–393. [CrossRef]
31. Boughton, A.J.; Hoover, K.; Felton, G.W. Methyl jasmonate application induces increased densities of glandular trichomes on tomato, Lycopersicon esculentum. *J. Chem. Ecol.* **2005**, *31*, 2211–2216. [CrossRef]
32. Pattanaik, S.; Patra, B.; Singh, S.K.; Yuan, L. An overview of the gene regulatory network controlling trichome development in the model plant, Arabidopsis. *Front. Plant Sci.* **2014**, *5*, 259. [CrossRef]
33. Ohashi, Y.; Oka, A.; Rodrigues-Pousada, R.; Possenti, M.; Ruberti, I.; Morelli, G.; Aoyama, T. Modulation of phospholipid signaling by GLABRA2 in root-hair pattern formation. *Science* **2003**, *300*, 1427–1430. [CrossRef]
34. Oppenheimer, D.G.; Herman, P.L.; Sivakumaran, S.; Esch, J.; Marks, M.D. A Myb Gene Required for Leaf Trichome Differentiation in Arabidopsis Is Expressed in Stipules. *Cell* **1991**, *67*, 483–493. [CrossRef]
35. Yoshida, Y.; Sano, R.; Wada, T.; Takabayashi, J.; Okada, K. Jasmonic acid control of GLABRA3 links inducible defense and trichome patterning in Arabidopsis. *Development* **2009**, *136*, 1039–1048. [CrossRef]
36. Zhao, M.; Morohashi, K.; Hatlestad, G.; Grotewold, E.; Lloyd, A. The TTG1-bHLH-MYB complex controls trichome cell fate and patterning through direct targeting of regulatory loci. *Development* **2008**, *135*, 1991–1999. [CrossRef]
37. Rerie, W.G.; Feldmann, K.A.; Marks, M.D. The GLABRA2 gene encodes a homeo domain protein required for normal trichome development in Arabidopsis. *Genes Dev.* **1994**, *8*, 1388–1399. [CrossRef]
38. Morohashi, K.; Zhao, M.; Yang, M.; Read, B.; Lloyd, A.; Lamb, R.; Grotewold, E. Participation of the Arabidopsis bHLH factor GL3 in trichome initiation regulatory events. *Plant Physiol.* **2007**, *145*, 736–746. [CrossRef] [PubMed]
39. Bernhardt, C.; Lee, M.M.; Gonzalez, A.; Zhang, F.; Lloyd, A.; Schiefelbein, J. The bHLH genes GLABRA3 (GL3) and ENHANCER OF GLABRA3 (EGL3) specify epidermal cell fate in the Arabidopsis root. *Development* **2003**, *130*, 6431–6439. [CrossRef]
40. Schellmann, S.; Schnittger, A.; Kirik, V.; Wada, T.; Okada, K.; Beermann, A.; Thumfahrt, J.; Jurgens, G.; Hulskamp, M. TRIPTYCHON and CAPRICE mediate lateral inhibition during trichome and root hair patterning in Arabidopsis. *EMBO J.* **2002**, *21*, 5036–5046. [CrossRef] [PubMed]
41. Yan, X.; Cui, L.; Liu, X.; Cui, Y.; Wang, Z.; Zhang, H.; Chen, L.; Cui, H. NbJAZ3 is required for jasmonate-meditated glandular trichome development in Nicotiana benthamiana. *Physiol. Plant* **2022**, *174*, e13666. [CrossRef] [PubMed]
42. Sun, B.; Shang, L.; Li, Y.; Zhang, Q.; Chu, Z.; He, S.; Yang, W.; Ding, X. Ectopic Expression of OsJAZs Alters Plant Defense and Development. *Int. J. Mol. Sci.* **2022**, *23*, 4581. [CrossRef] [PubMed]
43. English, N.J. Environmental Exploration of Ultra-Dense Nanobubbles: Rethinking Sustainability. *Environments* **2022**, *9*, 33. [CrossRef]
44. Wu, Q.; Su, N.N.; Huang, X.; Ling, X.P.; Yu, M.; Cui, J.; Shabala, S. Hydrogen-rich water promotes elongation of hypocotyls and roots in plants through mediating the level of endogenous gibberellin and auxin. *Funct. Plant Biol.* **2020**, *47*, 771–778. [CrossRef]
45. Zhu, Y.; Liao, W.; Wang, M.; Niu, L.; Xu, Q.; Jin, X. Nitric oxide is required for hydrogen gas-induced adventitious root formation in cucumber. *J. Plant Physiol.* **2016**, *195*, 50–58. [CrossRef]
46. Zhou, Y.; Li, Y.; Liu, X.; Wang, K.; Muhammad, T. Synergistic improvement in spring maize yield and quality with micro/nanobubbles water oxygation. *Sci. Rep.* **2019**, *9*, 5226. [CrossRef] [PubMed]
47. Chen, W.J.; Bastida, F.; Liu, Y.Z.; Zhou, Y.P.; He, J.; Song, P.; Kuang, N.K.; Li, Y.K. Nanobubble oxygenated increases crop production via soil structure improvement: The perspective of microbially mediated effects. *Agric. Water Manag.* **2023**, *282*, 108263. [CrossRef]
48. Wang, S.; Liu, Y.; Lyu, T.; Pan, G.; Li, P. Aquatic macrophytes in morphological and physiological responses to the nanobubble technology application for water restoration. *ACS EST Water* **2020**, *1*, 376–387. [CrossRef]

49. Wang, F.; Park, Y.L.; Gutensohn, M. Glandular trichome-derived sesquiterpenes of wild tomato accessions (*Solanum habrochaites*) affect aphid performance and feeding behavior. *Phytochemistry* **2020**, *180*, 112532. [CrossRef] [PubMed]
50. Hancock, J.T.; LeBaron, T.W.; May, J.; Thomas, A.; Russell, G. Molecular Hydrogen: Is This a Viable New Treatment for Plants in the UK? *Plants* **2021**, *10*, 2270. [CrossRef] [PubMed]

MDPI AG
Grosspeteranlage 5
4052 Basel
Switzerland
Tel.: +41 61 683 77 34

*Plants* Editorial Office
E-mail: plants@mdpi.com
www.mdpi.com/journal/plants

www.ingramcontent.com/pod-product-compliance
Lightning Source LLC
Chambersburg PA
CBHW061548220926
51850CB00005B/284
* 9 7 8 3 7 2 5 8 6 9 2 8 2 *